Lecture Notes in Computer Science 8617

Commenced Publication in 1973
Founding and Former Series Editors:
Gerhard Goos, Juris Hartmanis, and Jan van Leeuwen

Editorial Board

David Hutchison
  Lancaster University, UK

Takeo Kanade
  Carnegie Mellon University, Pittsburgh, PA, USA

Josef Kittler
  University of Surrey, Guildford, UK

Jon M. Kleinberg
  Cornell University, Ithaca, NY, USA

Alfred Kobsa
  University of California, Irvine, CA, USA

Friedemann Mattern
  ETH Zurich, Switzerland

John C. Mitchell
  Stanford University, CA, USA

Moni Naor
  Weizmann Institute of Science, Rehovot, Israel

Oscar Nierstrasz
  University of Bern, Switzerland

C. Pandu Rangan
  Indian Institute of Technology, Madras, India

Bernhard Steffen
  TU Dortmund University, Germany

Demetri Terzopoulos
  University of California, Los Angeles, CA, USA

Doug Tygar
  University of California, Berkeley, CA, USA

Gerhard Weikum
  Max Planck Institute for Informatics, Saarbruecken, Germany

T0224228

Juan A. Garay   Rosario Gennaro (Eds.)

Advances in Cryptology – CRYPTO 2014

34th Annual Cryptology Conference
Santa Barbara, CA, USA, August 17-21, 2014
Proceedings, Part II

 Springer

Volume Editors

Juan A. Garay
Yahoo Labs
701 First Avenue
Sunnyvale, CA 94089, USA
E-mail: garay@yahoo-inc.com

Rosario Gennaro
The City College of New York
160 Convent Avenue
New York, NY 10031, USA
E-mail: rosario@cs.ccny.cuny.edu

ISSN 0302-9743   e-ISSN 1611-3349
ISBN 978-3-662-44380-4   e-ISBN 978-3-662-44381-1
DOI 10.1007/978-3-662-44381-1
Springer Heidelberg New York Dordrecht London

Library of Congress Control Number: 2014944726

LNCS Sublibrary: SL 4 – Security and Cryptology

© International Association for Cryptologic Research 2014

This work is subject to copyright. All rights are reserved by the Publisher, whether the whole or part of
the material is concerned, specifically the rights of translation, reprinting, reuse of illustrations, recitation,
broadcasting, reproduction on microfilms or in any other physical way, and transmission or information
storage and retrieval, electronic adaptation, computer software, or by similar or dissimilar methodology
now known or hereafter developed. Exempted from this legal reservation are brief excerpts in connection
with reviews or scholarly analysis or material supplied specifically for the purpose of being entered and
executed on a computer system, for exclusive use by the purchaser of the work. Duplication of this publication
or parts thereof is permitted only under the provisions of the Copyright Law of the Publisher's location,
in ist current version, and permission for use must always be obtained from Springer. Permissions for use
may be obtained through RightsLink at the Copyright Clearance Center. Violations are liable to prosecution
under the respective Copyright Law.
The use of general descriptive names, registered names, trademarks, service marks, etc. in this publication
does not imply, even in the absence of a specific statement, that such names are exempt from the relevant
protective laws and regulations and therefore free for general use.
While the advice and information in this book are believed to be true and accurate at the date of publication,
neither the authors nor the editors nor the publisher can accept any legal responsibility for any errors or
omissions that may be made. The publisher makes no warranty, express or implied, with respect to the
material contained herein.

Typesetting: Camera-ready by author, data conversion by Scientific Publishing Services, Chennai, India

Printed on acid-free paper

Springer is part of Springer Science+Business Media (www.springer.com)

Preface

CRYPTO 2014, the 34rd Annual International Cryptology Conference, was held August 17–21, 2014, on the campus of the University of California, Santa Barbara. The event was sponsored by the International Association for Cryptologic Research (IACR) in cooperation with the UCSB Computer Science Department.

The program represents the recent significant advances and trends in all areas of cryptology. Out of 227 submissions, 60 were included in the program; these two-volume proceedings contains the revised versions of all the papers. Two of the papers shared a single presentation slot in the program. The program also included two invited talks. On Monday, Mihir Bellare from UCSD delivered the IACR Distinguished Lecture, entitled "Caught in Between Theory and Practice." On Wednesday, Yael Tauman Kalai from Microsoft Research New England spoke about "How to Delegate Computations: The Power of No-Signalling Proofs." As usual, the rump session took place on Tuesday evening, and was chaired by Dan Bernstein and Tanja Lange.

This year's program continued the trend started last year of trying to accommodate as many high-quality submissions as possible, yielding a high number of accepted papers. As a result, sessions were also held on Tuesday and Thursday afternoons, and presentations were kept short (20 minutes per paper, including questions and answers). The option of having parallel sessions, which would allow for longer presentations and an early adjournment on Thursday, was also discussed and decided against, since we assessed that our research field is still sufficiently homogeneous and the community would benefit from the option of attending all the talks. However, we believe that future Program Committees should continue to explore possible options to implement some form of parallel sessions.

The submissions were reviewed by a Program Committee (PC) consisting of 38 leading researchers in the field, in addition to the two co-chairs. Each PC member was allowed to submit one paper, plus an additional one if co-authored with a junior researcher (a student or a postdoc). PC-authored submissions were held to higher standards during the review process. Papers were reviewed in a double-blind fashion. Initially, each paper was assigned to three reviewers (four for PC-authored papers); during the discussion phase, when necessary, extra reviews were solicited. The process also included a rebuttal phase after preliminary reviews were finalized, where authors received them and were given the option to comment on the reviews within a window of several days. The authors' comments were then taken into account in the discussions within the PC and the final reviews. Despite being labor-intensive, we feel the rebuttal phase was a worthwhile process as it resulted in the significantly better understanding of many submissions. As part of the discussion phase, the PC held a 1.5-day in-person meeting on May 15 and 16 in Copenhagen, Denmark, right after Eurocrypt.

We would like to sincerely thank the authors of all submissions—those whose papers made it into the program and those whose papers did not. Our deep appreciation also goes out to the PC members, who invested an extraordinaty amount of time in reviewing papers, interacting with the authors via the rebuttal mechanism, and participating in so many discussions on papers, their contribution, and the state of the art in their areas of expertise. We also sympathize with the occasional frustration from seeing decisions go against personal recommendations and preferences, in spite of all the hard work.

We are also indebted to the many external reviewers who significantly contributed to the comprehensive evaluation of the submissions. A list of PC members and external reviewers appears after this note. Despite all our efforts, the list of external reviewers may contain errors or omissions; we apologize for that in advance.

We would like to thank Sasha Boldyreva, the general chair, for working closely with us throughout the whole process and providing the much needed support at every step, including artfully creating and maintaining the website and taking care of all aspects of the conference's logistics—especially the in-person PC meeting arrangements.

As always, special thanks are due to Shai Halevi for his tireless support regarding the *websubrev* software, which we used for the whole conference planning and operation, including paper submission and evaluation and interaction among PC members and with the authors. Alfred Hofmann and his colleagues at Springer provided a meticulous service for the timely production of these proceedings.

Finally, we would like to thank Google, Microsoft Research, and the National Science Foundation for their generous support.

August 2014   Juan A. Garay
                                                    Rosario Gennaro

CRYPTO 2014

The 34rd International Cryptology Conference

Sponsored by the *International Association for Cryptologic Research*

General Chair

Alexandra Boldyreva   Georgia Institute of Technology, USA

Program Co-Chairs

Juan A. Garay   Yahoo Labs, USA
Rosario Gennaro   The City College of New York – CUNY, USA

Program Committee

Yevgeniy Dodis	New York University, USA
Orr Dunkelman	University of Haifa, Israel
Serge Fehr	CWI, The Netherlands
Pierre-Alain Fouque	Université Rennes I, France
Craig Gentry	IBM Research, USA
Vipul Goyal	MSR India
Nadia Heninger	University of Pennsylvania, USA
Thomas Holenstein	ETH, Switzerland
Yuval Ishai	Technion, Israel
Dimitar Jetchev	EPFL, Switzerland
Aggelos Kiayias	University of Athens, Greece
Kaoru Kurosawa	Ibaraki University, Japan
Alexander May	Ruhr-Universität Bochum, Germany
Ilya Mironov	MSR, USA
Payman Mohassel	University of Calgary, Canada
Jörn Müller-Quade	Karlruhe Institute of Technology, Germany
María Naya-Plasencia	Inria Paris-Rocquencourt, France
Claudio Orlandi	Aarhus University, Denmark
Rafael Pass	Cornell University, USA
Christopher Peikert	Georgia Institute of Technology, USA
Krzysztof Pietrzak	Institute of Science and Technology, Austria
Leonid Reyzin	Boston University, USA
Ron Rivest	MIT, USA

Amit Sahai   UCLA, USA
Gil Segev   Hebrew University, USA
Elaine Shi   University of Maryland, USA
Tom Shrimpton   Portland State University, USA
Alice Silverberg   UC Irvine, USA
Marc Stevens   CWI, The Netherlands
Katsuyuki Takashima   Mitsubishi Electric, Japan
Stefano Tessaro   UC Santa Barbara, USA
Vinod Vaikuntanathan   MIT, USA
Gilles Van Assche   STMicroelectronics, Belgium
Muthu Venkitasubramanian   University of Rochester, USA
Ivan Visconti   University of Salerno, Italy
Bogdan Warinschi   University of Bristol, UK
Brent Waters   UT Austin, USA
Vassilis Zikas   ETH, Switzerland

External Reviewers

Michel Abdalla	Anne Canteaut	Leo Ducas
Masayuki Abe	Ignacio Cascudo	Alina Dudeanu
Arash Afshar	David Cash	Markus Duermuth
Divesh Aggarwal	Dario Catalano	Frédéric Dupuis
Martin Albrecht	Andr Chailloux	Aner Ben Efraim
Joel Alwen	Nishanth Chandran	Xiong Fan
Scott Ames	Jie Chen	Antonio Faonio
Prabhanjan Ananth	Cheng Chen	Sebastian Faust
Daniel Apon	Céline Chevalier	Dario Fiore
George Argyros	Kai-Min Chung	Marc Fischlin
Gilad Asharov	Aloni Cohen	Georg Fuchsbauer
Nuttapong Attrapadung	Henry Cohn	Benjamin Fuller
Christian Badertscher	Sandro Coretti	Jun Furukawa
Abhishek Banerjee	Jean-Sebastien Coron	Steven Galbraith
Carsten Baum	Craig Costello	Nicolas Gama
Amos Beimel	Dana Dachman-Soled	Chaya Ganesh
Mihir Bellare	Joan Daemen	Peter Gaži
David Bernhard	Ivan Damgård	Ran Gelles
Dan Bernstein	Bernardo David	Essam Ghadafi
Guido Bertoni	Gregory Demay	Sasha Golovnev
Raghav Bhaskar	Yi Deng	Sergey Gorbunov
Joppe Bos	Itai Dinur	Dov Gordon
Elette Boyle	Nico Doettling	Robert Granger
Brandon Broadnax	Rafael Dowsley	Jens Groth
Christina Brzuska	Chandan Dubey	Divya Gupta
Ran Canetti	Alexandre Duc	Tim Gneysu

Shai Halevi
Sean Hallgren
Moritz Hardt
Brett Hemenway
Yan Huang
Jan Hazla
William Skeith III
Vincenzo Iovino
Takashi Ito
Ioana Ivan
Tibor Jager
Abhishek Jain
David Jao
Stanislaw Jarecki
Mahavir Jhawar
Antoine Joux
Marc Joye
Yael Kalai
Seny Kamara
Jean-Gabriel Kammerer
Pierre Karpman
Jonathan Katz
Yutaka Kawai
Nathan Keller
Dakshita Khurana
Eike Kiltz
Thorsten Kleinjung
Vlad Kolesnikov
Venkata Koppula
Daniel Kraschewski
Hugo Krawczyk
Sara Krehbiel
Abishek
   Kumarasubramaniam
Ranjit Kumaresan
Robin Künzler
Tanja Lange
Gregor Leander
Nikos Leonardos
Anthony Leverrier
Kevin Lewi
Allison Bishop Lewko
Benoit Libert
Huijia (Rachel) Lin
Yehuda Lindell

Feng-Hao Liu
Adriana Lopez-Alt
Steve Lu
Stefan Lucks
Atul Luykx
Vadim Lyubashevsky
Mohammad Mahmoody
Hemanta Maji
Alex Malozemoff
Mohammad Mammody
Christian Matt
Daniele Micciancio
Andrea Miele
Eric Miles
Andrew Miller
Brice Minaud
Toru Nakanishi
Jesper Buus Nielsen
Valeria Nikolaenko
Tobias Nilges
Ryo Nishimaki
Adam O'Neill
Wakaha Ogata
Cristina Onete
Pascal Paillier
Omkant Pandey
Omer Paneth
Dimitris Papadopoulos
Charalampos
   Papamanthou
Sunoo Park
Anat
   Paskin-Cherniavsky
Valerio Pastro
Kenny Paterson
Michal Peeters
Ludovic Perret
Christophe Petit
Le Trieu Phong
Stefano Pironio
Manoj Prabhakaran
Ananth Raghunathan
Kim Ramchen
Vanishree Rao
Pavel Raykov

Mariana Raykova
Christian Rechberger
Oded Regev
Thomas Ristenpart
Ben Riva
Mike Rosulek
Aaron Roth
Yannis Rouselakis
saeed Sadeghian
Yusuke Sakai
Katerina Samari
Alessandra Scafuro
Christian Schaffner
Thomas Schneider
Lior Seeman
Nicolas Sendrier
Karn Seth
Yannick Seurin
Barak Shani
Nigel Smart
Ben Smith
Florian Speelman
François-Xavier
   Standaert
Damien Stehlé
John Steinberger
Noah
   Stephens-Davidowitz
Mario Strefler
Takeshi Sugawara
Koutarou Suzuki
Björn Tackmann
Qiang Tang
Sidharth Telang
Aris Tentes
Isamu Teranishi
R. Seth Terashima
Abhradeep Guha
   Thakurta
Justin Thaler
Emmanuel Thom
Mehdi Tibouchi
Jean-Pierre Tillich
Joana Treger
Roberto Trifiletti

Eran Tromer
Yiannis Tselekounis
Hoang Viet Tung
Dominique Unruh
Berkant Ustaoglu
Prashant Vasudevan
Thomas Vidick

Dhinakaran
  Vinayagamurthy
Akshay Wadia
Gaven Watson
Hoeteck Wee
Daniel Wichs
Shota Yamada

Kazuki Yoneyama
Thomas Zacharias
Hila Zarosim
Mark Zhandry
Bingsheng Zhang
Hong-Sheng Zhou
Jens Zumbrägel

Table of Contents – Part II

Side Channels and Leakage Resilience II

Information-Theoretic Security

Key Exchange and Secure Communication

Zero Knowledge

Composable Security

Secure Computation – Foundations

Secure Computation – Implementations

Table of Contents – Part I

Side Channels and Leakage Resilience I

Obfuscation I

FHE

Quantum Position Verification
in the Random Oracle Model

Dominique Unruh

University of Tartu, Tartu, Estonia

Abstract. We present a quantum position verification scheme in the random oracle model. In contrast to prior work, our scheme does not require bounded storage/retrieval/entanglement assumptions. We also give an efficient position-based authentication protocol. This enables secret and authenticated communication with an entity that is only identified by its position in space.

1 Introduction

What Is Position Verification? Consider the following setting: A device P wishes to access a location-based service. This service should only be available to devices in a certain spacial region **P**, e.g., within a sports stadium. The service provider wants to be sure no malicious device outside **P** accesses the service. In other words, we need a protocol such that a prover P can prove to a verifier V that P is at certain location. Such a protocol is called a *position verification* (PV) scheme. A special case of position verification is *distance bounding*: P proves that he is within a distance δ of V. In its simplest form, this is done by V sending a random message r to P, and P has to send it back immediately. If r comes back to V in time t, P must be within distance $tc/2$ where c is the speed of light. In general, however, it may not be practical to require a device V in the middle of a spherical region **P**. (E.g., **P** might be a rectangular room.) In general PV, thus, we assume several verifier devices $V_1, \ldots, V_n$, and a prover P somewhere in the convex hull of $V_1, \ldots, V_n$. The verifiers should then interact with P in such a way that based on the response times of P, they can make sure that P is at the claimed location (a kind of triangulation). Unfortunately, [5] showed that position verification based on *classical* cryptography cannot be secure, even when using computational assumptions, if the prover has several devices at different locations (collusion). [4] showed impossibility in the quantum setting, but only for information-theoretically secure protocols. Whether a protocol in the computational setting exists was left open.[1] In this work, we close this gap and give a simple protocol in the random oracle model.

Applications. The simplest application of PV is just for a device to prove that it is at a particular location to access a service. In a more advanced setting, location can be used for authentication: a prover can send a message which is guaranteed to have originated within a particular region (position-based

---

[1] But both [5,4] give positive results assuming bounded retrieval/entanglement, see "related work" below.

J.A. Garay and R. Gennaro (Eds.): CRYPTO 2014, Part II, LNCS 8617, pp. 1–18, 2014.
© International Association for Cryptologic Research 2014

authentication, PBA). Finally, when combining PBA with quantum key distribution (QKD), an encrypted message can be sent in such a way that only a recipient at a certain location can decrypt it. (E.g., think of sending a message to an embassy – you can make sure that it will be received only in the embassy, even if you do not know the embassy's public key.) More applications are position-based multi-party computation and position-based PKIs, see [5].

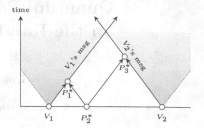

Fig. 1. Message flow in [4,11]. Security is only guaranteed if no entanglement is created before the shaded region. The scheme can be attacked if P_2^* sends EPR pairs to P_1^*, P_3^* who then can execute the attack from [8, Section 1].

Our Contribution. We present the first PV and PBA schemes secure against colluding provers that do not need bounded storage/retrieval/entanglement assumptions. (Cf. "related work" below.) Our protocols use quantum cryptography and are proven secure in the (quantum) random oracle model, and they work in the 3D setting. (Actually, in any number of dimensions, as well as in curved spacetime.[2]) Using [4], this also immediately implies position-based QKD. (And we even get *everlasting security*, i.e., if the adversary breaks the hash function *after* the protocol run, he cannot break the secrecy of the protocol.)

We also introduce a methodology for analyzing quantum circuits in spacetime which we believe simplifies the rigorous analysis of protocols that are based on the speed of light (such as, e.g., PV or relativistic commitments [7,6]). And for the first time (to our knowledge), a security analysis uses adaptive programming of the quantum random oracle (in our PBA security proof).[3]

Related Work. [5] showed a general impossibility of computationally secure PV in the classical setting; [4] showed the impossibility of information-theoretically secure PV in the quantum setting. [5] proposed computationally secure protocols for PV and position-based key exchange *in the bounded retrieval model*. Their model assumes that a party can only retrieve part of a large message reaching it. In particular, a party cannot forward a message ("reflection attacks" in the language of [5]); this may be difficult to ensure in practice because a mirror might be such a forwarding device. [4,11] provide a quantum protocol that is secure if the adversary can have no/limited entanglement before receiving the verifiers' messages. (I.e., in the message flow diagram Figure 1, only in the shaded areas.) In particular, using the message flow drawn in Figure 1, the attack from

---

[2] At the first glance, taking curvature of spacetime into account might seem like overkill. But for example GPS needs to take general relativity into account to ensure precise positioning (see, e.g., [1]). There is no reason to assume that this would not be the case for long-distance PV.

[3] The semi-constant distribution technique from [13] programs the random oracle *before* the first adversary invocation, i.e., only *non-adaptive* programming is possible.

[8, Section 1] can be applied, even though no entanglement is created before the protocol start ($t = 0$) and no entanglement needs to be stored. This makes the assumption difficult to justify. Our protocol is an extension of theirs, essentially adding one hash function application. [4] also gives a generic transformation from PV to PBA; however, their construction is considerably less efficient than our specialized one and does not achieve concurrent security (see the discussion after Definition 7 below). Furthermore, the protocols from [4,11] only work in the one-dimensional setting. ([4] has a construction for the 3D case, but their proof seems incorrect, see the full version [12] for a discussion.)

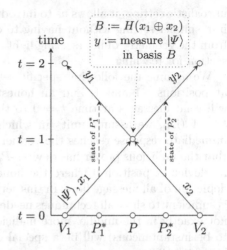

Fig. 2. One-dimensional PV protocol. Dotted lines indicate additional message flows of the adversary P_1^*, P_2^*.

Organization. In Section 2 we first explain our scheme in the 1D case. In Section 3.1 we explain the difficulties occurring in the 3D case which we solve in Sections 3.2 and 3.3. In Section 4 we present our PBA scheme. Full proofs and further discussion are deferred to the full version [12].

1.1 Preliminaries

$\omega(x)$ denotes the Hamming weight of x. $h(p) = -p \log p - (1 - p) \log(1 - p)$ denotes the binary entropy. $|x|$ denotes the absolute value or cardinality of x. $\|x\|$ denotes the Euclidean norm. $x \xleftarrow{\$} M$ means x is uniformly random from M, and $x \leftarrow A()$ means x is chosen by algorithm A.

For a background in quantum mechanics, see [9]. But large parts of this paper should be comprehensible without detailed knowledge on quantum mechanics. For $x \in \{0,1\}^n$, $|x\rangle$ denotes the quantum state x encoded in the computational basis, and $|\Psi\rangle$ denotes arbitrary quantum states (not necessarily in the computational basis). $\langle\Psi|$ is the conjugate transpose of $|\Psi\rangle$. For $B \in \{0,1\}^n$, $|x\rangle_B$ denotes x encoded in the bases specified by B, more precisely $|x\rangle_B = H^{B_1}|x_1\rangle \otimes \cdots \otimes H^{B_n}|x_n\rangle$ where H is the Hadamard matrix. An *EPR pair* has state $\frac{1}{\sqrt{2}}|00\rangle + \frac{1}{\sqrt{2}}|11\rangle$. $\mathrm{TD}(\rho, \rho')$ denotes the trace distance between states ρ, ρ'. Given a (quantum) oracle algorithm A and a function H, $A^H()$ means that A has oracle access to H and can query H on different inputs in superposition. This is important for modeling the quantum random oracle correctly [3].

2 1D Position Verification

In this section, we consider the case of one-dimensional PV only. That is, all verifiers and the honest and malicious provers live on a line. Although this is an unrealistic setting, it allows us to introduce our construction and proof technique in a simpler setting without having to consider the additional subtleties arising from the geometry of intersecting light cones. We also suggest the content of this section for teaching.

We assume the following specific setting: There are two verifiers V_1 and V_2 at positions -1 and 1, and an honest prover P at position 0. The verifiers will send messages at time $t = 0$ to the prover P, who receives them at time $t = 1$ (i.e., we assume units in which the speed of light is $c = 1$), and his immediate response reaches the verifiers at time $t = 2$. In an attack, we assume that the malicious prover has devices P_1^* and P_2^* left and right of position 0, but no device at position 0 where the honest prover is located. See Figure 2 for a depiction of all message flows in this setting. This setting simplifies notation and is sufficient to show all techniques needed in the 1D case. The general 1D case (P not exactly in the middle, more malicious provers, not requiring P's responses to be instantaneous) will be a special case of the higher dimensional theorems in Section 3.3.

In this setting, we use the following PV scheme:

Definition 1 (1D position verification). *Let n (number of qubits) and ℓ (bit length of classical challenges) be integers, $0 \leq \gamma < 1/2$ (fraction of allowed errors). Let $H : \{0,1\}^\ell \to \{0,1\}^n$ be a hash function (modeled as a quantum random oracle).*
- *Before time $t = 0$, verifier V_1 picks uniform $x_1, x_2 \in \{0,1\}^\ell$, $\hat{y} \in \{0,1\}^n$ and forwards x_2 to V_2 over a secure channel.*
- *At time $t = 0$, V_1 sends $|\Psi\rangle$ and x_1 to P. Here $B := H(x_1 \oplus x_2)$, $|\Psi\rangle := |\hat{y}\rangle_B$. And V_2 sends x_2 to P.*
- *At time $t = 1$, P receives $|\Psi\rangle, x_1, x_2$, computes $B := H(x_1 \oplus x_2)$, measures $|\Psi\rangle$ in basis B to obtain outcome y_1, and sends y_1 to V_1 and $y_2 := y_1$ to V_2. (We assume all these actions are instantaneous, so P sends y_1, y_2 at time $t = 1$.)*
- *At time $t = 2$, V_1 and V_2 receive y_1, y_2. Using secure channels, they check whether $y_1 = y_2$ and $\omega(y_1 - \hat{y}) \leq \gamma n$. If so (and y_1, y_2 arrived in time), they accept.*

We can now prove security in our simplified setting.

Theorem 2 (1D position verification). *Assume P_1^* and P_2^* perform at most q queries to H. Then in an execution of V_1, V_2, P_1^*, P_2^* with V_1, V_2 following the protocol from Definition 1, the probability that V_1, V_2 accept is at most*[4]

$$2q2^{-\ell/2} + \left(2^{h(\gamma)} \frac{1 + \sqrt{1/2}}{2}\right)^n.$$

Proof. To prove this theorem, we proceed using a sequence of games. The first game is the original protocol execution, and in the last game, we will be able to show that $\Pr[\mathsf{Accept}]$ is small. Here we abbreviate the event "$y_1 = y_2$ and $\omega(y_1 - \hat{y}) \leq \gamma n$" as "Accept".

---

[4] This probability is negligible if $\gamma \leq 0.037$ and n, ℓ are superlogarithmic.

Game 1. *An execution as described in Theorem 2.*

As a first step, we use EPR pairs to delay the choice of the basis B. This is a standard trick that has been used in QKD proofs and other settings. By choosing B sufficiently late, we will be able to argue below that B is independent of the state of P_1^* and P_2^*.

Game 2. *As in Game 1, except that V_1 prepares n EPR pairs, with their first qubits in register X and their second qubits in Y. Then V_1 sends X at time $t = 0$ instead of sending $|\Psi\rangle$. At time $t = 2$, V_1 measures Y in basis $B := H(x_1 \oplus x_2)$, the outcome is $\hat{y}$.*

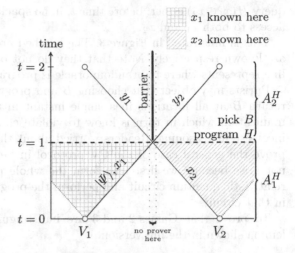

Fig. 3. Spacetime diagram depicting various steps of the proof of Theorem 2

Note in particular that V_1, V_2 never query H before time $t = 2$. (But P_1^*, P_2^* might, of course.) It is easy to verify (and well-known) that for any $B \in \{0,1\}$, preparing a qubit $X := |y\rangle_B$ for random $y \in \{0,1\}$ is perfectly indistinguishable (when given X, y, B) from producing an EPR pair XY, and then measuring Y in bases B to get outcome y. Thus $\Pr[\text{Accept} : \text{Game 1}] = \Pr[\text{Accept} : \text{Game 2}]$.

The problem now is that, although we have delayed the time when the basis B is used, the basis is still chosen early: At time $t = 0$, the values x_1, x_2 are chosen, and those determine B via $B = H(x_1 \oplus x_2)$. We have that neither P_1^* nor P_2^* individually knows B, but that does not necessarily exclude an attack. (For example, [8, Section 1] gives an efficient attack for the case that H is the identity, even though in this case B would still not be known to P_1^* nor P_2^* individually before time $t = 1$.) We can only hope that H is a sufficiently complex function such that computationally, B is "as good as unknown" before time $t = 1$ (where x_1 and x_2 become known to both P_1^*, P_2^*). The next game transformation formalizes this:

Game 3. *As in Game 1, except that at time $t = 1$, the value $B \xleftarrow{\$} \{0,1\}^n$ is chosen, and the random oracle is reprogrammed to return $H(x_1 \oplus x_2) = B$ after $t = 1$.*

To clarify this, if $H_0 : \{0,1\}^\ell \to \{0,1\}^n$ denotes a random function chosen at the very beginning of the execution, then at time $t \leq 1$, $H(x) = H_0(x)$ for all $x \in \{0,1\}^\ell$, while at time $t > 1$, $H(x_0 \oplus x_1) = B$ and $H(x) = H_0(x)$ for all $x \neq x_0 \oplus x_1$.

Intuitively, the change between Games 2 and 3 cannot be noticed because before time $t = 1$, the verifiers never query $H(x_1 \oplus x_2)$, and the provers cannot query $H(x_1 \oplus x_2)$ either: before time t, in no spacial location the prover will have access to both x_1 and x_2.

This is illustrated in Figure 3: The hatched areas represent where x_1 and x_2 are known respectively. Note that they do not overlap. The dashed horizontal line represents where the random oracle is programmed ($t = 1$).

Purists may object that choosing B and programming the random oracle to return B at all locations in a single instant in time needs superluminal communication which in turn is know to violate causality and might thus lead to inconsistent reasoning. Readers worried about this aspect should wait until we prove the general case of the PV protocol in Section 3.3, there this issue will not arise because we first transform the whole protocol execution into a non-relativistic quantum circuit and perform the programming of the random oracle in that circuit.

To prove that Games 2 and 3 are indistinguishable, we use the following lemma shown in the full version.

Lemma 3. *Let $H : \{0,1\}^\ell \to \{0,1\}^n$ be a random oracle. Let (A_1, A_2) be oracle algorithms sharing state between invocations that perform at most q queries to H. Let C_1 be an oracle algorithm that on input (j, x) does the following: Run $A_1^H(x)$ till the j-th query to H, then measure the argument of that query in the computational basis, and output the measurement outcome. (Or $\perp$ if no j-th query occurs.) Let*

$$P_A^1 := \Pr[b' = 1 : H \xleftarrow{\$} (\{0,1\}^\ell \to \{0,1\}^n), x \leftarrow \{0,1\}^\ell, A_1^H(x), b' \leftarrow A_2^H(x, H(x))]$$

$$P_A^2 := \Pr[b' = 1 : H \xleftarrow{\$} (\{0,1\}^\ell \to \{0,1\}^n), x \leftarrow \{0,1\}^\ell, B \xleftarrow{\$} \{0,1\}^n,$$
$$A_1^H(x), H(x) := B, b' \leftarrow A_2^H(x, B)]$$

$$P_C := \Pr[x = x' : H \xleftarrow{\$} (\{0,1\}^\ell \to \{0,1\}^n), x \leftarrow \{0,1\}^\ell, j \xleftarrow{\$} \{1, \ldots, q\}, x' \leftarrow C_1^H(j, x)]$$

Then $|P_A^1 - P_A^2| \leq 2q\sqrt{P_C}$.

In other words, an adversary can only notice that the random oracle is reprogrammed at position x if he can guess x before the reprogramming takes place.

To apply Lemma 3 to Games 2 and 3, let $A_1^H(x)$ be the machine that executes verifiers and provers from Game 2 until time $t = 1$ (inclusive). When V_1 chooses x_1, x_2, $A_1^H(x)$ chooses $x_1 \xleftarrow{\$} \{0,1\}^\ell$ and $x_2 := x \oplus x_1$. And let $A_2^H(x, B)$ be the machine that executes verifiers and provers after time $t = 1$. When V_1 queries $H(x_1 \oplus x_2)$, A_2^H uses the value B instead. In the end, A_2^H returns 1 iff $y_1 = y_2$ and $\omega(\hat{y} - y_1) \leq \gamma n$. (See Figure 3 for the time intervals handled by A_1^H, A_2^H.) Since V_1, V_2 make no oracle queries except for $H(x_1 \oplus x_2)$, and since P_1^*, P_2^* make at most q oracle queries, we have that A_1^H, A_2^H perform at most q queries.

By construction, $P_A^1 = \Pr[\mathsf{Accept} : \text{Game 2}]$. And $P_A^2 = \Pr[\mathsf{Accept} : \text{Game 3}]$. And $P_C = \Pr[x' = x_1 \oplus x_2 : \text{Game 4}]$ for the following game:

Game 4. *Pick $j \xleftarrow{\$} \{1, \ldots, q\}$. Then execute Game 2 till time $t = 1$ (inclusive), but stop at the j-th query and measure the query register. Call the outcome x'.*

Since Game 4 executes only till time $t = 1$, and since till time $t = 1$, no gate can be reached by both x_1, x_2 (note: at time $t = 1$, at position 0 both x_1, x_2 could be known, but no malicious prover may be at that location), the probability that $x_1 \oplus x_2$ will be guessed is bounded by $2^{-\ell}$. Hence $\Pr[x' = x_1 \oplus x_2 : \text{Game 3}] \leq 2^{-\ell}$. (This argument was a bit nonrigorous; we will be more precise in the proof of the generic case, in the proof of Theorem 6.)

Thus by Lemma 3, we have

$$\left| \Pr[\text{Accept} : \text{Game 2}] - \Pr[\text{Accept} : \text{Game 3}] \right| = |P_A^1 - P_A^2| \leq 2q\sqrt{P_C}$$
$$= 2q\sqrt{\Pr[x' = x_1 \oplus x_2 : \text{Game 4}]} \leq 2q 2^{-\ell/2}. \quad (1)$$

We continue to modify Game 3.

Game 5. *Like Game 3, except that for time $t > 1$, we install a barrier at position 0 (i.e., where the honest prover P would be) that lets no information through.*

The barrier is illustrated in Figure 3 with a thick vertical line.

Time $t = 1$ is latest time at which information from position 0 could reach the verifiers V_1, V_2 at time $t \leq 2$. Since we install the barrier only for time $t > 1$, whether the barrier is there or not cannot influence the measurements of V_1, V_2 at time $t = 2$. And Accept only depends on these measurements. Thus $\Pr[\text{Accept} : \text{Game 3}] = \Pr[\text{Accept} : \text{Game 5}]$.

Let ρ be the state of the execution of Game 5 directly after time $t = 1$ (i.e., after the gates at times $t \leq 1$ have been executed). Then ρ is a threepartite state consisting of registers Y, L, R where Y is the register containing the EPR qubits which will be measured to give $\hat{y}$ (cf. Game 2), and L and R are the quantum state left and right of the barrier respectively. Then $\hat{y}$ is the result of measuring Y in basis B, and y_1 is the result of applying some measurement M_1 to L (consisting of all the gates left of the barrier), and y_2 is the result of applying some measurement M_2 to R. Notice that due to the barrier, M_1 and M_2 operate only on L and R, respectively, without interaction between those two. We have thus:

$$\Pr[\text{Accept} : \text{Game 5}] = \Pr[y_1 = y_2 \text{ and } \omega(\hat{y} - y_1) \leq \gamma n : B \xleftarrow{\$} \{0,1\}^n, YLR \leftarrow \rho,$$
$$\hat{y} \leftarrow M^B(Y), y_1 \leftarrow M_1(L), y_2 \leftarrow M_2(R)]$$

where $YLR \leftarrow \rho$ means initializing YLR with state ρ. And M^B is a measurement in bases B. And $\hat{y} \leftarrow M^B(Y)$ means measuring register Y using measurement M^B and assigning the result to $\hat{y}$. And $y_1 \leftarrow M_1(L), y_2 \leftarrow M_2(R)$ analogously.

The rhs of this equation is a so-called monogamy of entanglement game, and [11] shows that the rhs is bounded by $\left(2^{h(\gamma)} \frac{1+\sqrt{1/2}}{2}\right)^n$. Thus $\Pr[\text{Accept} :$ Game 5$] \leq \left(2^{h(\gamma)} \frac{1+\sqrt{1/2}}{2}\right)^n$. And from (1) and the equalities between games, we have $\left| \Pr[\text{Accept} : \text{Game 1}] - \Pr[\text{Accept} : \text{Game 5}] \right| \leq 2q 2^{-\ell/2}$.

Thus altogether $\Pr[\text{Accept} : \text{Game 1}] \leq 2q 2^{-\ell/2} + \left(2^{h(\gamma)} \frac{1+\sqrt{1/2}}{2}\right)^n$. $\qquad \square$

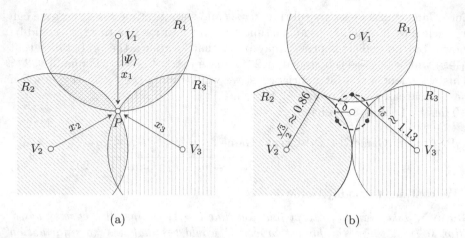

(a)   (b)

Fig. 4. The geometry of space at time t_δ (i.e., when B first becomes known). Left for $\delta = 0$, right for $\delta = \sqrt{4 - \sqrt{12}} - \frac{1}{2} \approx 0.23$.

3 Position Verification in Higher Dimensions

3.1 Difficulties

Excepting special cases where the honest prover happens to lie on a line between two verifiers, one-dimensional PV with two verifiers is not very useful. We therefore need to generalize the approach to three dimensions. It turns out that some non-trivialities occur here. For n-dimensional PV we need at least $n+1$ verifiers.[5] To illustrate the problems occurring in the higher dimensional case, we sketch what happens if we try to generalize the protocol and proof from Section 2 to the 2D case.

In the 2D case we need at least three verifiers V_1, V_2, V_3. Let's assume that they are arranged in a equilateral triangle, each at distance 1 from an honest prover P in the center. (Cf. Figure 4 (a).) V_1 sends a quantum state $|\Psi\rangle$, and all V_i send a random x_i. At time $t = 1$, all x_i are received by P who computes $B := H(x_1 \oplus x_2 \oplus x_3)$ and measures $|\Psi\rangle$ in basis B, yielding the value y to be sent to V_1, V_2, V_3.

Now as in Section 2 we can argue that before time $t = 1$, there is no point in space where all x_1, x_2, x_3 are known. Hence $B := H(x_1 \oplus x_2 \oplus x_3)$ will not be queried before $t = 1$. Hence by programming the random oracle (using Lemma 3) we can assume that the basis B is chosen randomly only at time $t = 1$. In Section 2 we then observed that space is partitioned into two disjoint regions:

---

[5] PV (in Euclidean space) can only work if the prover P is in the convex hull C of the verifiers. Otherwise, if we project P onto the hypersurface H separating C from P, we get a point P' that is closer to any point of C than P. Since the convex hull of n provers can at most be $n - 1$ dimensional, we need at least $n + 1$ provers to get an n dimensional convex hull.

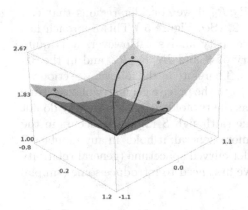
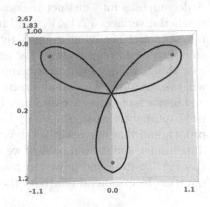

Fig. 5. The surface S in spacetime at which B is sampled. The dots floating over S denote when the verifiers need to receive y (i.e., the dots are at time 2 and space V_1, V_2, V_3). The thick black lines enclose the areas R_1, R_2, R_3 on S from which the verifiers can be reached in time. (Right: top view. In PDF: click figures for interaction.)

Region L from which light can reach V_1 by time $t = 2$, and region R from which light can reach V_2 by time $t = 2$. The results from [11] then imply that the correct y cannot be obtained from two independent (but possibly entangled) quantum registers L and R simultaneously. What happens if we apply this reasoning in the 2D case? Figure 4 (a) depicts the three regions R_1, R_2, R_3 of points that can reach V_1, V_2, V_3 until time $t = 2$. These regions are not disjoint! We cannot argue that measuring y in each of these regions violates the monogamy of entanglement, y does not result from measuring separate quantum registers.

Can we fix this? The most obvious consequence would be to weaken the security claim: "A malicious prover which has devices anywhere except at point P *or distance δ from P* cannot make the verifiers accept." Then the time t_δ when the random oracle is programmed is the earliest time at which some point at distance δ from P has access to all x_1, x_2, x_3. Then R_1, R_2, R_3 are the regions from which light can travel to V_1, V_2, V_3 within time $2 - t_\delta$. We can compute that they are disjoint iff $\delta > \sqrt{4 - \sqrt{12}} - \frac{1}{2} \approx 0.23$. (Cf. Figure 4 (b).) This means that the malicious prover is only guaranteed to be within a circle of diameter 2δ, which is about 46% of the distance between prover and verifier. In the 3D case, using a numerical calculation, we even get $\delta \approx 0.38$.

Can we improve on this bound? Indeed, when we said that the B is sampled at time $t = 1$, this was not a tight analysis. At time $t = 1$, the query $B = H(x_1 \oplus x_2 \oplus x_3)$ can only occur at point P. The farther away from P we get, the later we get all of x_1, x_2, x_3. Thus, if we plot the earliest time of querying B as a function of space, we get a surface S in 3D spacetime (Figure 5) which is not a plane. Now, instead of considering the state of the provers at time $t = 1$, we consider the state of the prover on S. (I.e., the state of all devices of the prover at points in spacetime in S.) We ask the reader to take it on trust for the moment this is actually a well-defined state. And now we can again ask whether

S decomposes into distinct regions R_1, R_2, R_3 if we consider regions that can reach the verifiers V_1, V_2, V_3 by time $t = 2$. (See Figure 5.) This approach has the potential of giving a much tighter security analysis. However, it is quite complicated to reason about the geometry of S and R_1, R_2, R_3, and in the 3D case things will get even more complicated. Therefore in the following section we will take an approach that abstracts away from the precise geometry of spacetime and uses a more generic reasoning. This has the twofold advantage that we do not need to analyze what S actually looks like (although S implicitly occurs in the proof), and that our result will be much more general: it holds in any number of dimensions, and it even holds if we consider curved spacetime (general relativity theory). To state and prove our results, we first need to introduce some (simple) notation from general relativity theory.

3.2   Circuits in Spacetime

Spacetime is the set of all locations in space and time. That is, intuitively spacetime consists of all tuples $(t, x_1, \ldots, x_n)$ where t is the time and $x_1, \ldots, x_n$ is the position in space. Such a location in spacetime is called an *event*. Relativity theory predicts that there is no natural distinction between the time coordinate t and the space coordinates $x_1, \ldots, x_n$. (In a similar way as in "normal" space there is no reason why three particular directions in space are coordinates.) As it turns out, for analyzing our PV protocol, we do not need to know the structure of spacetime, so in the following spacetime will just be some set of events, with no particular structure.[6] However, the reader may of course assume throughout the paper that spacetime consists of events $(t, x_1, \ldots, x_n)$ with $t, x_1, \ldots, x_n \in \mathbb{R}$. This is called *flat spacetime*.

The geometry of spacetime (to the extent needed here) is described by a partial order on the events: We say x *causally precedes* y ($x \prec y$) iff information originating from event x can reach event y. Or in other words, if you can get from x to y traveling at most the speed of light. In flat spacetime, this relation is familiar: $(t_x, x_1, \ldots, x_n) \prec (t_y, y_1, \ldots, y_n)$ iff $t_x \leq t_y$ and $\|(x_1, \ldots, x_n) - (y_1, \ldots, y_n)\| \leq t_y - t_x$.

Given this relation, we can define the *causal future* $C^+(x)$ of an event x as the set of all events reachable from x, $C^+(x) := \{y : x \prec y\}$. Similarly, we define the *causal past* $C^-(x) := \{y : y \prec x\}$.

In the case of flat spacetime, the causal future of $x = (t, x_1, \ldots, x_n)$ is an infinite cone with its point at x and extending towards the future. Thus it is also called a future *light cone*. Similarly the causal past of x is an infinite cone with its point at x extending into the past.

This language allows us to express quantum computations in space that do not transfer information faster than light. A *spacetime circuit* is a quantum circuit

---

[6] For readers knowledgeable in general relativity: We do assume that spacetime is a Lorentzian manifold which is time-orientable (otherwise the notions of causal future/-past would not make send) without closed causal curves (at least in the spacetime region where the protocol is executed; otherwise quantum circuits may end up having loops).

where every gate is at a particular event. There can only be a wire from a gate at event x to a gate at event y if x causally precedes y ($x \prec y$). Note that since $\prec$ is a partial order and thus antisymmetric, this ensures that a circuit cannot be cyclic. Note further that there is no limit to how much computation can be performed in an instant since $\prec$ is reflexive. We can model malicious provers that are not at the location of an honest prover by considering circuits with no gates in $\mathbf{P}$, where $\mathbf{P}$ is a region in spacetime. (This allows for more finegrained specifications than, e.g., just saying that the malicious prover is not within δ distance of the honest prover. For example, $\mathbf{P}$ might only consist of events within a certain time interval; this means that the malicious prover is allowed to be at any space location outside that time interval.) Notice that a spacetime circuit is also just a normal quantum circuit if we forget where in spacetime gates are located. Thus transformations on quantum circuits (such as changing the execution order of commuting gates) can also be applied to spacetime circuits, the result will be a valid circuit, though possibly not a spacetime circuit any more.

3.3 Achieving Higher-Dimensional Position Verification

We can now formulate the definition of secure PV in higher dimensions using the language from the previous section.

Definition 4 (Sound position verification). *Let $\mathbf{P}$ be a region in spacetime. A position verification protocol is* sound *for $\mathbf{P}$ iff for any non-uniform polynomial-time[7] spacetime circuit P^* that has no gates in $\mathbf{P}$, the following holds: In an interaction between the verifiers and P^*, the probability that the verifiers accept (the* soundness error*) is negligible.*

The smaller the region $\mathbf{P}$ is, the better the protocol localizes the prover. Informally, we say the protocol has higher *precision* if $\mathbf{P}$ is smaller.

Next, we describe the generalization of the protocol in Section 2. In this generalization, only two of the verifiers check whether the answers of the prover are correct. Although we believe that we get higher precision if more verifiers check the answers, it is an open problem to prove that.

Definition 5 (Position verification protocol). *Let P be a prover, and P° an event in spacetime (P° specifies where and when the honest prover performs its computation). Let $V_1, \ldots, V_r$ be verifiers. Let $V_1^+, \ldots, V_r^+$ be events in spacetime that causally precede P°. (V_i^+ specifies where and when the verifier V_i sends its challenge.) Let V_1^-, V_2^- be events in spacetime such that P° causally precedes V_1^-, V_2^-. (V_i^- specifies where and when V_i expects the prover's response.)*

Let n (number of qubits) and ℓ (bit length of classical challenges) be integers, and $0 \leq \gamma < \frac{1}{2}$ (fraction of allowed errors). Let $H : \{0,1\}^\ell \to \{0,1\}^n$ be a hash function (modeled as a quantum random oracle).

---

[7] Non-uniform polynomial-time means that we are actually considering a family of circuits of polynomial size in the security parameter, consisting only of standard gates (from some fixed universal set) and oracle query gates. In addition, we assume that the circuit is given an (arbitrary) initial quantum state that does not need to be efficiently computable.

- *The verifiers choose uniform $x_1, \ldots, x_r \in \{0,1\}^\ell$, $\hat{y} \in \{0,1\}^n$. (By communicating over secure channels.)*
- *At some event that causally precedes P°, V_0 sends $|\Psi\rangle$ to P. Here $B :=$ $H(x_1 \oplus \cdots \oplus x_r)$, $|\Psi\rangle := |\hat{y}\rangle_B$.*
- *For $i = 1, \ldots, r$: V_r sends x_r to P at event V_r^+.*
- *At event P°, P will have $|\Psi\rangle, x_1, \ldots, x_r$. Then P computes $B := H(x_1 \oplus \cdots \oplus x_r)$, measures $|\Psi\rangle$ in basis B to obtain outcome y_1, and sends y_1 to V_1 and $y_2 := y_1$ to V_2.*
- *At events V_1^-, V_2^-, V_1 and V_2 receive y_1, y_2. Using secure channels, the verifiers check whether $y_1 = y_2$ and $\omega(y_1 - \hat{y}) \leq \gamma n$. If so (and y_1, y_2 indeed arrived at V_1^-, V_2^-), the verifiers accept.*

In the protocol description, for simplicity we assume that V_1, V_2 are the receiving verifiers. However, there is no reason not to choose other two verifiers, or even additional verifiers not used for sending. Similarly, $|\Psi\rangle$ could be sent by any verifier, or by an additional verifier. In the analysis, we only use the events at which different messages are sent/received, not which verifier device sends which message.

Note that this protocol also allows for realistic provers that cannot perform instantaneous computations: In this case, one chooses the events V_1^-, V_2^- such that the prover's messages can still reach them even if the prover sends y_1, y_2 with some delay.

We can now state the main security result:

Theorem 6. *Assume that $\gamma \leq 0.037$ and n, ℓ are superlogarithmic.*

Then the PV protocol from Definition 5 is sound for $\mathbf{P} := \bigcap_{i=1}^r C^+(V_i^+) \cap C^-(V_1^-) \cap C^-(V_2^-)$. (In words: There is no event in spacetime outside of $\mathbf{P}$ at which one can receive the messages x_i from all V_i, and send messages that will be received in time by V_1, V_2.)

Concretely, if the malicious prover performs at most q oracle queries, then the soundness error is at most $\nu := \left(2^{h(\gamma)} \frac{1+\sqrt{1/2}}{2} \right)^n + 2q2^{-\ell/2}$.

Notice that the condition on the locations of the provers is tight: If $E \in \bigcap_{i=1}^r C^+(V_i^+) \cap C^-(V_1^-) \cap C^-(V_2^-) \setminus \mathbf{P} \neq \varnothing$, then the protocol could even be broken by a malicious prover with a single device: P^* could be at event E, receive $x_1, \ldots, x_r$, compute y_1, y_2 honestly, and send them to V_1, V_2 in time. The same reasoning applies to any protocol where only two verifiers receive. Our protocol is thus optimal in terms of precision under all such protocols.

Proof of Theorem 6. In the following, we write short C^+_i for $C^+(V_i^+)$ and C^-_i for $C^-(V_i^-)$. We also write $\bigcap$ instead of $\bigcap_{i=1}^r$. The precondition of the theorem then becomes: $\bigcap C^+_i \cap C^-_1 \cap C^-_2 \subseteq \mathbf{P}$. Let Ω denote all of spacetime.

We now partition the gates in the spacetime circuit P^* into several disjoint sets of gates (subcircuits), depending on where they are located in spacetime. For each subcircuit, we also give an rough intuitive meaning; those meanings are not precisely what the subcircuits do but help to guide the intuition in the proof.

Subcircuit	Region in spacetime	Intuition
P^*_{pre}	$(C^-{}_1 \cup C^-{}_2) \setminus \bigcap C^+{}_i$	Precomputation
$P^*_{\mathbf{P}}$	$\bigcap C^+{}_i \cap C^-{}_1 \cap C^-{}_2$	Gates in $\mathbf{P}$ (empty)
P^*_1	$\bigcap C^+{}_i \cap C^-{}_1 \setminus C^-{}_2$	Computing y_1
P^*_2	$\bigcap C^+{}_i \cap C^-{}_2 \setminus C^-{}_1$	Computing y_2
P^*_{post}	$\Omega \setminus C^-{}_1 \setminus C^-{}_2$	After protocol end

Note that all those subcircuits are disjoint, and their union is all of Ω. The subcircuits have analogues in the proof in the one-dimensional case. P^*_{pre} corresponds to the gates below the dashed line in Figure 3; P^*_1 to the gates above the dashed line and left of the barrier; P^*_2 above the dashed line and right of the barrier; P^*_{post} to everything that is above the picture. This correspondance is not exact, because as discussed in Section 3.1, the dashed line needs to be replaced by a surface S (Figure 5) which is not flat. In our present notation, S is the border between P^*_{pre} and the other subcircuits.

In addition, in some abuse of notation, by V_1 we denote the circuit at V^-_1 that receives y_1. Similar for V_2.

By definition of spacetime circuits, there can only be a wire from gate G_1 to gate G_2 if G_1, G_2 are at events E_1, E_2 with $E_1 \prec E_2$ (E_1 causally precedes E_2). Thus, by definition of causal futures and the transitivity of $\prec$, there can be no wire leaving $C^+{}_i$. Similarly, there can be no wire entering $C^-{}_i$. These two facts are sufficient to check the following facts:

$$P^*_1, P^*_2, P^*_{\mathrm{post}} \nrightarrow P^*_{\mathrm{pre}}, \quad P^*_1 \nrightarrow P^*_2, \quad P^*_2 \nrightarrow P^*_1,$$
$$P^*_1 \nrightarrow V_2, \quad P^*_2 \nrightarrow V_1, \quad P^*_{\mathrm{post}} \nrightarrow P^*_1, P^*_2, V_1, V_2. \tag{2}$$

Here $A \nrightarrow B$ means that there is no wire from subcircuit A to subcircuit B.

Given these subcircuits, we can write the execution of the protocol as the following quantum circuit:

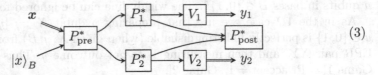

$$\tag{3}$$

Here x is short for $x_1, \ldots, x_r$. And we have omitted wires between subcircuits that are in the transitive hull of the wires drawn. (E.g., there can be a wire from P^*_{pre} to V_1, but we did not draw it because we drew wires from P^*_{pre} to P^*_1 to V_1.) Note that $P^*_{\mathbf{P}}$ does not occur in this circuit, because it contains no gates (it consists of gates in $\bigcap C^+{}_i \cap C^-{}_1 \cap C^-{}_2 = \mathbf{P}$ which by assumption contains no gates).

From (2) it follows that no wires are missing in (3). In particular, (2) implies that the quantum circuit is well-defined. If we did not have, e.g., $P^*_1 \nrightarrow P^*_{\mathrm{pre}}$, there might be wires between P^*_1 and P^*_{pre} in both directions; the result would not be a quantum circuit. We added arrow heads in (2), these are only to stress that the wires indeed go in the right directions, below we will follow the usual left-to-right convention in quantum circuits and omit the arrow heads.

The circuit (3) now encodes all information dependencies that we will need, we can forget that (3) is a spacetime circuit and treat it as a normal quantum circuit.

We now proceed to analyze the protocol execution using a sequence of games. The original execution can be written as follows:

Game 1 (Protocol execution). *Pick* $x_1, \ldots, x_r \xleftarrow{\$} \{0,1\}^\ell$, $\hat{y} \xleftarrow{\$} \{0,1\}^n$, $H \xleftarrow{\$} Fun$ *where Fun is the set of functions* $\{0,1\}^\ell \rightarrow \{0,1\}^n$. *Let* $B := H(x_1 \oplus \cdots \oplus x_r)$. *Execute circuit (3) resulting in* y_1, y_2. *Let* accept $:= 1$ *iff* $y_1 = y_2$ *and* $\omega(y_1 - \hat{y}) \le \gamma n$.

To prove the theorem, we need to show that $\Pr[\text{accept} = 1 : \text{Game 1}] \le \nu$.

As in the proof of the 1D case, we now delay the choice of x by using EPR pairs. And we remove the subcircuit P^*_{post} which clearly has no effect on the outputs y_1, y_2.

Game 2 (Using EPR pairs). *Pick* $x_1, \ldots, x_r \xleftarrow{\$} \{0,1\}^\ell$, $H \xleftarrow{\$} Fun$. *Let* $B := H(x_1 \oplus \cdots \oplus x_r)$. *Execute circuit (4) resulting in* y_1, y_2.
Let accept $:= 1$ *iff* $y_1 = y_2$ *and* $\omega(y_1 - \hat{y}) \le \gamma n$.

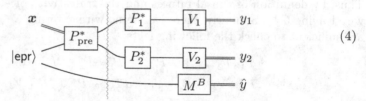

$$(4)$$

Here $|\text{epr}\rangle$ is the state consisting of n EPR pairs, i.e., $|\text{epr}\rangle = 2^{-n/2} \sum_{x \in \{0,1\}^n} |x\rangle \otimes |x\rangle$. The top and bottom wire originating from $|\text{epr}\rangle$ represent the first and last n qubits, respectively. And M^B is the gate that measures n qubits in bases $B \in \{0,1\}^n$. The wiggly line can be ignored for now.

As in the 1D case, we use that preparing a qubit $X := |y\rangle_B$ for random $y \in \{0,1\}$ is perfectly indistinguishable (when given X, y, B) from producing an EPR pair XY, and then measuring Y to get outcome y. Thus $\Pr[\text{accept} = 1 : \text{Game 1}] = \Pr[\text{accept} = 1 : \text{Game 2}]$.

Again like in the 1D case, we will now reprogram the random oracle. That is, instead of computing $B := H(x_1 \oplus \cdots \oplus x_r)$, we pick $B \xleftarrow{\$} \{0,1\}^n$ at some point in the execution and then program the random oracle via $H(x_1 \oplus \cdots \oplus x_r) := B$. The question is: at which point shall we program the random oracle? In the 1D case, we used the fact that before time $t = 1$ (dashed line in Figure 3), there is no event at which both x_1 and x_2 are known. An analogous reasoning can be done in the present setting: since P^*_{pre} consists only of gates outside $\bigcap C^+_i$, it means that any gate in P^*_{pre} is outside some C^+_i and thus does not have access to x_i. (We will formally prove this later.) So we expect that left of the wiggly line in (4), $H(x_1 \oplus \cdots \oplus x_r)$ occurs with negligible probability only. In other words, the wiggly line corresponds to the surface S discussed in Section 3.1. In fact, if we draw the border between P^*_{pre} and the remaining gates, we get exactly Figure 5 (in the 2D case at least). However, the approach of decomposing spacetime into subcircuits removes the necessity of dealing with the exact geometry of S.

Formally, we will need to apply Lemma 3. Given a function H and values x, B, let $H_{x \mapsto B}$ denote the function identical to H, except that $H_{x \mapsto B}(x) = B$. Let $A_1^H(x)$ denote the oracle machine that picks $x_1, \ldots, x_{r-1} \xleftarrow{\$} \{0,1\}^\ell$ and sets $x_r := x \oplus x_1 \oplus \cdots \oplus x_{r-1}$ and prepares the state $|\mathrm{epr}\rangle$ and then executes P_{pre}^*. Let $A_2^H(x, B)$ denote the oracle machine that, given the state from A_1^H, executes $P_1^*, P_2^*, V_1, V_2, M^B$ with oracle access to $H_{x \mapsto B}$ instead of H, sets accept $:= 1$ iff $y_1 = y_2$ and $\omega(y_1 - \hat{y}) \leq \gamma n$, and returns accept. Let C_1, P_A^1, P_A^2, P_C be defined as in Lemma 3. Then by construction, $P_A^1 = \Pr[\mathrm{accept} = 1 : \text{Game 2}]$ (using the fact that $H = H_{x \mapsto H(x)}$). And $P_A^2 = \Pr[\mathrm{accept} = 1 : \text{Game 3}]$ for the following game:

Game 3 (Reprogramming H). *Pick* $x_1, \ldots, x_r \xleftarrow{\$} \{0,1\}^\ell$, $H \xleftarrow{\$} Fun$. *Execute circuit (4) until the wiggly line (with oracle access to H). Pick $B \xleftarrow{\$} \{0,1\}^n$. Execute circuit (4) after the wiggly line (with oracle access to $H_{x \mapsto B}$) resulting in $y_1, y_2, \hat{y}$. Let* accept $:= 1$ *iff $y_1 = y_2$ and $\omega(y_1 - \hat{y}) \leq \gamma n$.*

And finally $P_C = \Pr[x' = x_1 \oplus \cdots \oplus x_r : \text{Game 4}]$ for the following game:

Game 4 (Guessing $x_1 \oplus \cdots \oplus x_r$). *Pick* $x_1, \ldots, x_r \xleftarrow{\$} \{0,1\}^\ell$, $H \xleftarrow{\$} Fun$, *and $j \xleftarrow{\$} \{1, \ldots, q\}$. Prepare $|\mathrm{epr}\rangle$ and execute circuit P_{pre}^* until the j-th query to H. Measure the argument x' of that query.*

By Lemma 3, we have $|P_A^1 - P_A^2| \leq 2q\sqrt{P_C}$. Thus, abbreviating $x = x_1 \oplus \cdots \oplus x_r$ as guessX, we have

$$\left| \Pr[\mathrm{accept} = 1 : \text{Game 2}] - \Pr[\mathrm{accept} = 1 : \text{Game 3}] \right|$$
$$\leq 2q\sqrt{\Pr[\mathrm{guessX} : \text{Game 4}]}. \quad (5)$$

We now focus on Game 3. Let ρ_{YLR} denote the state in circuit (4) at the wiggly line (for random $x_1, \ldots, x_r, H$). Let L refer to the part of ρ_{YLR} that is on the wires entering P_1^*, and R refer to the part of ρ_{LR} on the wires entering P_2^*. Let Y refer to the lowest wire (containing EPR qubits). Notice that we have now reproduced the situation from the 1D case where space is split into two separate registers R and L, and the computation of y_1, y_2 is performed solely on R, L, respectively. In fact, we have now also identified the regions R_1, R_2 from the discussion in Section 3.1 (Figure 5): R_1 is the boundary between P_{pre}^* and P_1^*; analogously R_2. (R_3 from Figure 5 has no analogue here because V_3 does not receive here.) For given B, let $M_L(B)$ be the POVM operating on L consisting of P_1^* and V_1. (M_L can be modeled as a POVM because P_1^* and V_1 together return only a classical value and thus constitute a measurement.) Let $M_R(B)$ be the POVM operating on R consisting of P_2^* and V_2. Then we can rewrite Game 3 as:

Game 5 (Monogamy game). *Prepare ρ_{YLR}. Pick $B \xleftarrow{\$} \{0,1\}^n$. Apply measurement $M_L(B)$ to L, resulting in y_1. Apply measurement $M_R(B)$ to R, resulting in y_2. Measure Y in basis B, resulting in $\hat{y}$. Let* accept $:= 1$ *iff $y_1 = y_2$ and $\omega(y_1 - \hat{y}) \leq \gamma n$.*

Then $\Pr[\mathsf{accept} = 1 : \text{Game } 3] = \Pr[\mathsf{accept} = 1 : \text{Game } 5]$. Furthermore, Game 5 is again a monogamy of entanglement game, and [11] shows that $\Pr[\mathsf{accept} = 1 : \text{Game } 5] \leq \left(2^{h(\gamma)}\frac{1+\sqrt{1/2}}{2}\right)^n$. We can furthermore show (see the full version [12]) that $\Pr[\mathsf{guessX} : \text{Game } 4] \leq 2^{-\ell}$. With (5) we get

$$\Pr[\mathsf{accept} = 1 : \text{Game } 1] \leq \left(2^{h(\gamma)}\frac{1+\sqrt{1/2}}{2}\right)^n + 2q2^{-\ell/2} = \nu.$$

Numerically, we can verify that for $\gamma \leq 0.037$, we have $2^{h(\gamma)}\frac{1+\sqrt{1/2}}{2} < 1$ and thus ν is negligible (for superlogarithmic n, ℓ and polynomially bounded q). $\square$

In Flat Spacetime. Theorem 6 tells us where in spacetime a prover can be that passes verification. (Region **P**.) However, the theorem is quite general; it is not immediate what this means in the concrete setting of flat spacetime. In the full version [12] we derive specialized criteria for flat spacetime and show that Theorem 6 implies that a prover can be precisely localized by verifiers arranged as a tetrahedron.

4   Position-Based Authentication

Position verification is, in itself, a primitive of somewhat limited use. It guarantees that no prover outside the region **P** can pass the verification. Yet nothing forbids a prover to just wait until some other honest party has successfully passed position verification, and then to impersonate that honest party. To realize the applications described in the introduction, we need a stronger primitive that not only proves that a prover is at a specific location, but also allows him to bind this proof to specific data. (The difference is a bit like that between identification schemes and message authentication schemes.) Such a primitive is be *position-based authentication*. This guarantees that the malicious prover cannot authenticate a message m unless he is in region **P** (or some honest party at location m wishes to authenticate that message).

Definition 7 (Secure position-based authentication). *A position-based authentication (PBA) scheme is a PV scheme where provers and verifiers get an additional argument m, a message to be authenticated.*

*Let **P** be a region in spacetime. A position-based authentication (PBA) protocol is sound for **P** iff for any non-uniform polynomial-time spacetime circuit P^* that has no gates in **P**, the probability that the challenge verifiers (soundness error) accept is negligible in the following execution:*

*P^* picks a message m^* and then interacts with honest verifiers (called the challenge verifiers) on input m^*. Before, during, and after that interaction, P^* may spawn instances of the honest prover and honest verifiers, running on inputs $m \neq m^*$. These instances run concurrently with P^* and the challenge verifiers and P^* may arbitrarily interact with them. Note that the honest prover/honest verifier instances may have gates in **P**.*

PBA was already studied in [4]. They give a generic transformation to convert a PV protocol into a PBA. The generic solution has two drawbacks, though:

- It needs $\Omega(\ell\mu)$ invocations of the PV protocol for
  ell-bit messages and $2^{-\mu}$ security level. (Our protocol below will need only one invocation.)
- It is only secure if a single instance of the honest prover runs concurrently. If the malicious prover can suitably interleave several instances of the honest prover, he can authenticate arbitrary messages.

(We do not know whether their solution gives adaptive security, i.e., whether the adversary can choose m^* and the honest provers' inputs m depending on communication he has seen before.) Although we do not have a generic transformation from PV to PBA that solves these issues, a small modification of our PV protocol leads to an efficient PBA secure against concurrent executions of the honest prover:

Definition 8 (Position-based authentication protocol). *The protocol is the same as in Definition 5, with the following modification only: Whenever in Definition 5, the verifier or prover queries $B := H(x_1 \oplus \cdots \oplus x_r)$, here he queries $B := H(x_1 \oplus \cdots \oplus x_r \| m)$ instead. (Where m is the message to be authenticated.) We also require that the verifiers do not start sending the messages x_i or expect y_1, y_2 before all V_i got m, and that $V_1^+ \neq V_2^+$ (i.e., V_1, V_2 do not send x_1, x_2 from the same location in space at the same time, a natural assumption).*

Theorem 9. *Assume that $\gamma \leq 0.037$ and n, ℓ are superlogarithmic.*

Then the PBA protocol from Definition 8 is sound for $\mathbf{P} := \bigcap_{i=1}^{r} C^+(V_i^+) \cap C^-(V_1^-) \cap C^-(V_2^-)$. (In words: There is no event in spacetime outside of $\mathbf{P}$ at which one can receive the messages x_i from all V_i, and send messages that will be received in time by V_1, V_2.)

Concretely, if the malicious prover performs at most q oracle queries, then the soundness error is at most $\left(2^{h(\gamma)}\frac{1+\sqrt{1/2}}{2}\right)^n + 6q2^{-\ell/2}$.

The main difference to Theorem 6 is that now oracle queries are performed even within $\mathbf{P}$ (by the honest provers). We thus need to show that these queries do not help the adversary. The main technical challenge is that the message m^* is chosen adaptively by the adversary. The proof is given in the full version [12].

Position-Based Quantum Key Distribution. Once we have PBA, we immediately get position-based quantum key distribution, and thus we can send messages that can only be decrypted by someone within region $\mathbf{P}$. We refer to [4] who describe how to do this, their construction applies to arbitrary PBA schemes. (As long as it has adaptive security, since in the QKD protocol, the adversary can influence the messages to be authenticated.)

Acknowledgements. We thank Serge Fehr and Andris Ambainis for valuable discussions. Dominique Unruh was supported by the Estonian ICT program 2011-2015 (3.2.1201.13-0022), the European Union through the European Regional Development Fund through the sub-measure "Supporting the development of R&D

of info and communication technology", by the European Social Fund's Doctoral Studies and Internationalisation Programme DoRa, by the Estonian Centre of Excellence in Computer Science, EXCS. We also used Sage [10] and PPL [2] for calculations and experiments, and the Sage Cluster funded by National Science Foundation Grant No. DMS-0821725.

References

1. Ashby, N.: General relativity in the global positioning system. Matters of Gravity (newsletter of the Topical Group in Gravitation of the APS), 9 (1997), http://www.phys.lsu.edu/mog/mog9/node9.html (accessed: February 07, 2014) (Archived by WebCite at http://www.webcitation.org/6ND19QXJ3)
2. Bagnara, R., Hill, P.M., Zaffanella, E.: The Parma Polyhedra Library: Toward a complete set of numerical abstractions for the analysis and verification of hardware and software systems. Science of Computer Programming 72(1-2), 3–21 (2008), http://bugseng.com/products/ppl/
3. Boneh, D., Dagdelen, Ö., Fischlin, M., Lehmann, A., Schaffner, C., Zhandry, M.: Random oracles in a quantum world. In: Lee, D.H., Wang, X. (eds.) ASIACRYPT 2011. LNCS, vol. 7073, pp. 41–69. Springer, Heidelberg (2011)
4. Buhrman, H., Chandran, N., Fehr, S., Gelles, R., Goyal, V., Ostrovsky, R., Schaffner, C.: Position-based quantum cryptography: Impossibility and constructions. In: Rogaway, P. (ed.) CRYPTO 2011. LNCS, vol. 6841, pp. 429–446. Springer, Heidelberg (2011)
5. Chandran, N., Goyal, V., Moriarty, R., Ostrovsky, R.: Position based cryptography. In: Halevi, S. (ed.) CRYPTO 2009. LNCS, vol. 5677, pp. 391–407. Springer, Heidelberg (2009)
6. Kaniewski, J., Tomamichel, M., Hänggi, E., Wehner, S.: Secure bit commitment from relativistic constraints. IEEE Trans. on Inf. Theory 59(7), 4687–4699 (2013)
7. Kent, A.: Unconditionally secure bit commitment by transmitting measurement outcomes. Phys. Rev. Lett. 109(13), 130501 (2012)
8. Kent, A., Munro, W.J., Spiller, T.P.: Quantum tagging: Authenticating location via quantum information and relativistic signaling constraints. Phys. Rev. A 84, 012326 (2011)
9. Nielsen, M., Chuang, I.: Quantum Computation and Quantum Information, 10th anniversary edn. Cambridge University Press, Cambridge (2010)
10. Stein, W., et al.: Sage Mathematics Software (Version 5.12). The Sage Development Team (2014), http://www.sagemath.org
11. Tomamichel, M., Fehr, S., Kaniewski, J., Wehner, S.: One-sided device-independent QKD and position-based cryptography from monogamy games. In: Johansson, T., Nguyen, P.Q. (eds.) EUROCRYPT 2013. LNCS, vol. 7881, pp. 609–625. Springer, Heidelberg (2013)
12. Unruh, D.: Quantum position verification in the random oracle model. IACR ePrint 2014/118, Full version of this paper (February 2014), http://eprint.iacr.org/2014/118
13. Zhandry, M.: Secure identity-based encryption in the quantum random oracle model. In: Safavi-Naini, R., Canetti, R. (eds.) CRYPTO 2012. LNCS, vol. 7417, pp. 758–775. Springer, Heidelberg (2012)

Single-Shot Security for One-Time Memories in the Isolated Qubits Model

Yi-Kai Liu

Applied and Computational Mathematics Division,
National Institute of Standards and Technology (NIST),
Gaithersburg, MD, USA
yi-kai.liu@nist.gov

Abstract. One-time memories (OTM's) are simple, tamper-resistant cryptographic devices, which can be used to implement sophisticated functionalities such as one-time programs. Can one construct OTM's whose security follows from some physical principle? This is not possible in a fully-classical world, or in a fully-quantum world, but there is evidence that OTM's can be built using "isolated qubits" — qubits that cannot be entangled, but can be accessed using adaptive sequences of single-qubit measurements.

Here we present new constructions for OTM's using isolated qubits, which improve on previous work in several respects: they achieve a stronger "single-shot" security guarantee, which is stated in terms of the (smoothed) min-entropy; they are proven secure against adversaries who can perform arbitrary local operations and classical communication (LOCC); and they are efficiently implementable.

These results use Wiesner's idea of conjugate coding, combined with error-correcting codes that approach the capacity of the q-ary symmetric channel, and a high-order entropic uncertainty relation, which was originally developed for cryptography in the bounded quantum storage model.

Keywords: Quantum cryptography, information theory, local operations and classical communication (LOCC), oblivious transfer, one-time programs.

1 Introduction

One-time memories (OTM's) are a simple type of tamper-resistant cryptographic hardware. An OTM has the following behavior: a user Alice can write two messages s and t into the OTM, and then give the OTM to another user Bob; Bob can then choose to read either s or t from the OTM, but he can only learn one of the two messages, not both. A single OTM is not especially exciting by itself, but when many OTM's are combined in an appropriate way, they can be used to implement *one-time programs*, which are a powerful form of secure computation [3,4,5,6]. (Roughly speaking, a one-time program is a program that can be run exactly once, on an input chosen by the user. After running once,

J.A. Garay and R. Gennaro (Eds.): CRYPTO 2014, Part II, LNCS 8617, pp. 19–36, 2014.
© International Association for Cryptologic Research 2014

the program "self-destructs," and it never reveals any information other than the output of the computation.)

Can one construct OTM's whose security follows from some physical principle? At first glance, the answer seems to be "no." OTM's cannot exist in a fully classical world, because information can always be copied without destroying it. One might hope to build OTM's in a quantum world, where the no-cloning principle limits an adversary's ability to copy an unknown quantum state. However, this is also impossible, because an OTM can be used to perform oblivious transfer with information-theoretic security, which is ruled out by various "no-go" theorems [7,8,9,10].

One way around these no-go theorems is to try to construct protocols that are secure against restricted classes of quantum adversaries, e.g., adversaries who can only perform k-local measurements [11], or adversaries who only have bounded or noisy quantum storage [12,13,14,15,16,17]. More recently, Liu has proposed a construction for OTM's in the *isolated qubits model* [1], where the adversary is only allowed to perform local operations and classical communication (LOCC). That is, the adversary can perform single-qubit quantum operations, including single-qubit measurements, and can make adaptive choices based on the classical information returned by these measurements; but the adversary cannot perform entangling operations on sets of two or more qubits. (Honest parties are also restricted to LOCC operations.) The isolated qubits model is motivated by recent experimental work using solid-state qubits, such as nitrogen vacancy (NV) centers; see [1] for a more complete discussion of this model, and [18] for earlier work on implementing quantum money using NV centers.[1]

In this paper we show a new construction and security analysis for OTM's in the isolated qubits model, which improves on the results of [1] in several respects. First, we show a stronger "single-shot" security guarantee, which is stated in terms of the (smoothed) min-entropy [19,20]. This shows that a constant fraction of the message bits remain hidden from the adversary. This stronger statement is necessary for most cryptographic applications; note that the previous results of [1] were not sufficient, as they used the Shannon entropy.

Second, we prove security against general LOCC adversaries, who can perform arbitrary measurements (including weak measurements), and can measure each qubit multiple times. This improves on the results of [1], which only showed security against 1-pass LOCC adversaries that use 2-outcome measurements. Our new security proof is based solely on the definition of the isolated qubits model, without any additional assumptions.

Third, we show a construction of OTM's that is efficiently implementable, i.e., programming and reading out the OTM can be done in polynomial time. This improves on the construction in [1], which was primarily an information-theoretic

---

[1] Note that the devices constructed in [1], and in this paper, are more precisely described as *leaky* OTM's, because they can leak additional information to the adversary. It is not known whether such leaky OTM's are sufficient to construct one-time programs as defined in [3]. We will discuss this issue in Section 1.2; for now, we will simply refer to our devices as OTM's.

result, using random error-correcting codes that did not allow efficient decoding. (In fact, our new construction is quite flexible, and does not depend heavily on the choice of a particular error-correcting code. Our OTM's can be constructed using any code that satisfies two simple requirements: the code must be linear over $GF(2)$, and it must approach the capacity of the q-ary symmetric channel. We show one such code in this paper; several more sophisticated constructions are known [22,23,24].)

We will describe our OTM construction in the following section. Here, we briefly comment on some related work. Note that OTM's cannot make use of standard techniques such as privacy amplification. This is because OTM's are non-interactive and asynchronous: all of the communication between Alice and Bob occurs at the beginning, while the adversary can wait until later to attack the OTM. (To do privacy amplification, Alice would have to first force the adversary to take some action, and then send one more message to Bob. This trick is very natural in protocols for quantum key distribution and oblivious transfer, but it is clearly impossible in the case of an OTM.) As we will see below, the security of our OTM's follows from rather different arguments. (A similar issue was studied recently in [17], albeit with a weaker, non-adaptive adversary.)

In addition, it is a long-standing open problem to prove strong upper-bounds on the power of LOCC operations. Previous results in this area include demonstrations of "nonlocality without entanglement" [25] (see [26] for a recent survey), and constructions of data-hiding states [27,28,29,30]. Our OTM's are not directly comparable to these earlier results, as the security requirements for our OTM's are quite different.

1.1 Our Construction

We now describe our OTM construction, which is based on Wiesner's idea of conjugate coding [21]. Our OTM will store two messages $s, t \in \{0,1\}^{\ell}$, and will use $n \lg q$ qubits, where q is a (large) power of 2. Let $C : \{0,1\}^{\ell} \rightarrow \{0,1\}^{n \lg q}$ be any error-correcting code that satisfies the following two requirements: C is linear over $GF(2)$, and C approaches the capacity of the q-ary symmetric channel $\mathcal{E}_q$ with error probability $p_e := \frac{1}{2} - \frac{1}{2q}$ (where the channel treats each block of $\lg q$ bits as a single q-ary symbol). Note that, when q is large, the capacity of the channel $\mathcal{E}_q$ is roughly $1 - p_e$, which is roughly $\frac{1}{2}$, so we have $n \lg q \approx 2\ell$.

Given two messages s and t, let $C(s)$ and $C(t)$ be the corresponding codewords, and view each codeword as n blocks consisting of $\lg q$ bits. We prepare the qubits in the OTM as follows. For each $i = 1, 2, \ldots, n$,

- Let $\gamma_i \in \{0,1\}$ be the outcome of a fair and independent coin toss.
- If $\gamma_i = 0$, prepare the i'th block of qubits in the standard basis state corresponding to the i'th block of $C(s)$.
- If $\gamma_i = 1$, prepare the i'th block of qubits in the Hadamard basis state corresponding to the i'th block of $C(t)$.

To recover the first message s, we measure every qubit in the standard basis, which yields a string of measurement outcomes $z \in \{0,1\}^{n \lg q}$, and then we run the decoding algorithm for C. To recover the second message t, we measure every qubit in the Hadamard basis, then follow the same procedure. It is easy to see that all of these procedures require only single-qubit state preparations and single-qubit measurements, which are allowed in the isolated qubits model.[2]

(We remark that this OTM construction uses blocks of qubits, rather than individual qubits as in [21] and [1]. That is, we set q large, instead of using $q = 2$. This difference seems to help our security proof, although it is not clear whether it affects the actual security of the scheme.)

We now sketch the proofs of correctness and security for this OTM. With regard to correctness, note that an honest player who wanted to learn s will obtain measurement outcomes that have the same distribution as the output of the q-ary symmetric channel $\mathcal{E}_q$ acting on $C(s)$; hence the decoding algorithm will return s. A similar argument holds for t.

To prove security, we consider adversaries that make *separable* measurements (which include LOCC measurements as a special case). The basic idea is to consider the distribution of the messages s and t, conditioned on one particular measurement outcome z obtained by the adversary. Since the adversary is separable, the corresponding POVM element M_z will be a tensor product of single-qubit operators $\bigotimes_{a=1}^{n \lg q} R_a$ (up to normalization). Now, one can imagine a fictional adversary that measures the qubits one at a time, and happens to observe this same string of single-qubit measurement outcomes $R_1, R_2, \ldots, R_{n \lg q}$. This event leads to the same conditional distribution of s and t. But the fictional adversary is easier to analyze, because it is non-adaptive, it measures each qubit only once, and the measurements can be done in arbitrary order.

Now, our proof will be based on the following intuition. In order to learn both messages s and t, the adversary will want to determine the basis choices $\gamma = (\gamma_1, \gamma_2, \ldots, \gamma_n)$, so that he will know which blocks of qubits should be measured in the standard basis, and which blocks of qubits should be measured in the Hadamard basis. The choice of the code C is crucial to prevent the adversary from doing this; for instance, if the adversary could predict some of the bits in the codewords $C(s)$ and $C(t)$, he could then measure the corresponding qubits, and gain some information about which bases were used to prepare them. (Note moreover that the adversary has full knowledge of C, before he measures any of the qubits.) We will argue that certain properties of the code C prevent the adversary from learning these basis choices γ perfectly, and that this in turn limits the adversary's knowledge of the messages s and t.

Since C is a linear code over $GF(2)$, it has a generator matrix G, which has rank ℓ. Thus there must exist a subset of ℓ bits of the codeword $C(s)$ that look uniformly random, assuming the message s was chosen uniformly at random; and a similar statement holds for $C(t)$. Now, let A be the subset of ℓ qubits that encode these bits of $C(s)$ and $C(t)$. We can imagine that the fictional adversary happens to measure these qubits *first*. Therefore, during these first ℓ steps, the fictional adversary learns nothing about which bases had been used to prepare

---

[2] We note in passing that Winter's "gentle measurement lemma" [31] does not imply an attack on this OTM using LOCC operations; see the full paper [2] for details.

the state, i.e., the basis choices γ are independent of the fictional adversary's measurement outcomes.

One can then show that the conditional distribution of s and t after these first ℓ steps of the fictional adversary is related to the distribution of measurement outcomes when the state $\bigotimes_{a \in A} R_a$ is measured in a random basis. This kind of situation has been studied previously, in connection with cryptography in the bounded quantum storage model. In particular, we can use a high-order entropic uncertainty relation from [16] to show a lower-bound on the smoothed min-entropy of this distribution. We then use trivial bounds to analyze the remaining $n \lg q - \ell$ steps of the fictional adversary. Roughly speaking, we get a bound of the form:

$$H_\infty^\varepsilon(S, T | Z) \gtrsim \tfrac{1}{2} \ell, \tag{1}$$

for any separable adversary (where Z denotes the adversary's measurement outcome). Thus, while the OTM may leak some information, it still hides a constant fraction of the bits of the messages s and t. For more details, see Section 3.

Finally, we show one construction of a code C that satisfies the above requirements and is efficiently decodable. The basic idea is to fix some $q_0 < q$, first encode the messages s and t using a random linear code $C_0 : \{0,1\}^\ell \to \{0,1\}^{n \lg q_0}$, then encode each block of $\lg q_0$ bits using a fixed linear code $C_1 : \{0,1\}^{\lg q_0} \to \{0,1\}^{\lg q}$. The code C_1 is used to detect the errors made by the q-ary symmetric channel; these corrupted blocks of bits are then treated as erasures, and we can decode C_0 by solving a linear system of equations, which can be done efficiently. Moreover, choosing C_0 to be a random linear encode ensures that, with high probability, C approaches the capacity of the q-ary symmetric channel. For more details, see Section 4.

1.2   Outlook

The results of this paper can be summarized as follows: we construct OTM's based on conjugate coding, which achieve a fairly strong ("single-shot") notion of security, are secure against general LOCC adversaries, and can be implemented efficiently. These results are a substantial improvement on previous work [1].

We view these results as a first step in a broader research program that aims to develop practical implementations of isolated qubits, one-time memories, and ultimately one-time programs. We now comment briefly on some different aspects of this program.

Experimental realization of isolated qubits is quite challenging, though there has been recent progress in this direction [39,40]. Broadly speaking, isolated qubits seem to be at an intermediate level of difficulty, somewhere between photonic quantum key distribution (which already exists as a commercial product), and large-scale quantum computers (which are still many years in the future).

Working with quantum devices in the lab also raises the question of fault-tolerance: can our OTM's be made robust against minor imperfections in the qubits? We believe this can be done, by slightly modifying our OTM construction: we would use a slightly noisier channel to describe the imperfect measurements made by an honest user, and we would choose the error-correcting code C accordingly. The proof of security would still hold against LOCC adversaries who can make perfect measurements. There is plenty of "slack" in the security bounds, to allow this modification to the OTM's.

In addition, one may wonder whether our OTM's are secure against so-called "k-local" adversaries [11], which can perform entangled measurements on small numbers of qubits (thus going outside the isolated qubits model). There is some reason to be optimistic about this: while we have mainly discussed separable adversaries in this paper, our security proof actually works for a larger set of adversaries, who can generate entanglement among some of the qubits, but are still separable across the partition defined by the subset A (as described in the proof). Also, from a physical point of view, k-local adversaries are quite natural. In particular, even when one can perform entangling operations on pairs of qubits, it may be hard to entangle large numbers of qubits, due to error accumulation.

Finally, let us turn to the construction of one-time programs. Because our OTM's leak some information, it is not clear whether they are sufficient to construct one-time programs. There are a couple of approaches to this problem. On one hand, one can try to strengthen the security proof, perhaps by proving constraints on the *types* of information that an LOCC adversary can extract from the OTM. We conjecture that, when our OTM's are used to build one-time programs as in [3], the specific information that is relevant to the security of the one-time program does in fact remain hidden from an LOCC adversary.

On the other hand, one can try to strengthen the OTM constructions, in order to eliminate the leakage. As noted previously, standard privacy amplification (e.g., postprocessing using a randomness extractor) does not work in this setting, because the adversary also knows the seed for the extractor. However, there are other ways of solving this problem, for instance by assuming the availability of a random oracle, or by using something similar to leakage-resilient encryption [32,33] (but with a different notion of leakage, where the "leakage function" is restricted to use only LOCC operations, but is allowed access to side-information).

2   Preliminaries

2.1   Notation

For any natural number n, let $[n]$ denote the set $\{1, 2, \ldots, n\}$. Let $\lg(x) = \log_2(x)$ denote the logarithm with base 2.

For any random variable X, let P_X be the probability density function of X, that is, $P_X(x) = \Pr[X = x]$. Likewise, define $P_{X|Y}(x|y) = \Pr[X = x|Y = y]$, etc. For any event $\mathcal{E}$, define $P_{\mathcal{E}X}$ to be the probability density function of X smoothed by $\mathcal{E}$, that is $P_{\mathcal{E}X}(x) = \Pr[X = x$ and $\mathcal{E}$ occurs].

We say that C is a binary code with codeword length n and message length k if C is a subset of $\{0, 1\}^n$ with cardinality 2^k. We say that C has minimum distance $d = \min_{x,y \in C} d_H(x, y)$, where $d_H(\cdot, \cdot)$ denotes the Hamming distance.

We say that C is a binary linear code if C is a linear subspace of $GF(2)^n$. (Note, $GF(2)$ and $\{0,1\}$ denote the same set, but we will write $GF(2)$ in situations where we use arithmetic operations.) In this case, there exists a matrix $G \in GF(2)^{k \times n}$, such that the map $x \mapsto x^T G$ is a bijection from $GF(2)^k$ to the code subspace C. We will overload the notation and use C to denote the map $x \mapsto x^T G$; then the codewords consist of the strings $C(x)$ for all $x \in GF(2)^k$.

2.2 The q-ary Symmetric Channel

The q-ary symmetric channel with error probability p_e acts as follows: given an input $x \in GF(q)$, it returns an output $y \in GF(q)$, with conditional probabilities $\Pr(y|x) = 1 - p_e$ (if $y = x$) and $\Pr(y|x) = p_e/(q-1)$ (if $y \neq x$). The capacity of this channel, measured in q-ary symbols per channel use, is given by [23]:

$$L(p_e) = 1 + (1 - p_e) \log_q(1 - p_e) + p_e \log_q(p_e) - p_e \log_q(q-1)$$
$$= 1 - \frac{h_2(p_e)}{\lg q} - p_e \frac{\lg(q-1)}{\lg q} \geq 1 - \frac{1}{\lg q} - p_e, \tag{2}$$

where $h_2(\cdot)$ is the binary entropy function.

2.3 LOCC Adversaries and Separable Measurements

An LOCC adversary is an adversary that uses only local operations and classical communication (LOCC). Here, "local operations" consist of quantum operations on single qubits, and "classical communication" refers to the adversary's ability to choose each single-qubit operation adaptively, depending on classical information, such as measurement outcomes, that were obtained from previous single-qubit operations. However, the adversary is not allowed to make adaptive choices that depend on quantum information, or perform entangling operations on multiple qubits.

Formally, an LOCC adversary can be described as follows. Consider a system of n qubits. The adversary makes a sequence of steps, labelled by $i = 1, 2, 3, \ldots$. At step i, the adversary chooses one of the qubits $q_i \in [n]$, and performs a general quantum measurement $\mathcal{M}_i$ on that qubit; this returns a measurement outcome, which is described by a classical random variable Z_i. The adversary's choices of q_i and $\mathcal{M}_i$ can depend on $Z_1, Z_2, \ldots, Z_{i-1}$. Also, note that the adversary can perform weak measurements, and can measure the same qubit multiple times. Finally the adversary discards the qubits, and outputs the sequence of measurement outcomes $Z_1, Z_2, Z_3, \ldots$.

A POVM measurement $\mathcal{M} = \{M_z \mid z = 1, 2, 3, \ldots\}$ is called *separable* if every POVM element M_z can be written as a tensor product of single-qubit operators. It is easy to see that any LOCC adversary can be simulated by a separable measurement, i.e., for any LOCC adversary $\mathcal{A}$, there exists a separable POVM measurement $\mathcal{M}$, such that for every quantum state ρ, the output of $\mathcal{M}$ acting on ρ has the same distribution as the output of $\mathcal{A}$ acting on ρ [38].

2.4 Leaky OTM's

We will use the following definition of a leaky OTM [1].

Definition 1. *Fix some class of adversary strategies* $\mathbb{M}$, *some leakage parameter* $\delta \in [0,1]$, *and some failure probability* $\varepsilon \in [0,1]$. *A leaky one-time memory (leaky OTM) with parameters* $(\mathbb{M}, \delta, \varepsilon)$ *is a device that has the following behavior. Suppose that the device is programmed with two messages* s *and* t *chosen uniformly at random in* $\{0,1\}^\ell$; *and let* S *and* T *be the random variables containing these messages. Then:*

1. *Correctness: There exists an honest strategy* $\mathcal{M}^{(1)} \in \mathbb{M}$ *that interacts with the device and recovers the message* s *with probability* $\geq 1 - \varepsilon$. *Likewise, there exists an honest strategy* $\mathcal{M}^{(2)} \in \mathbb{M}$ *that recovers the message* t *with probability* $\geq 1 - \varepsilon$.
2. *Leaky security: For every strategy* $\mathcal{M} \in \mathbb{M}$, *if* Z *is the random variable containing the classical information output by* $\mathcal{M}$, *then* $H_\infty^\varepsilon(S, T|Z) \geq (1 - \delta)\ell$.

Here H_∞^ε is the smoothed conditional min-entropy, which is defined as follows [19,20]:

$$H_\infty^\varepsilon(X|Y) = \max_{\mathcal{E}:\, \Pr(\mathcal{E}) \geq 1 - \varepsilon} \min_{x,y} \left[-\lg\big[P_{\mathcal{E}X|Y}(x|y)\big] \right], \tag{3}$$

where the maximization is over all events $\mathcal{E}$ (defined by the conditional probabilities $P_{\mathcal{E}|XY}$) such that $\Pr(\mathcal{E}) \geq 1 - \varepsilon$. Observe that a lower-bound of the form $H_\infty^\varepsilon(X|Y) \geq h$ implies that there exists an event $\mathcal{E}$ with $\Pr(\mathcal{E}) \geq 1 - \varepsilon$ such that, for all x and y, $\Pr[\mathcal{E}, X = x|Y = y] \leq 2^{-h}$.

The definition of a leaky OTM is weaker than that of an ideal OTM in two important respects: it assumes that the messages s and t are chosen uniformly at random, independent of all other variables; and it allows the adversary to obtain partial information about both s and t, so long as the adversary still has $(1 - \delta)k$ bits of uncertainty (as measured by the smoothed min-entropy). We suspect that this *definition* of a leaky OTM is not strong enough to construct one-time programs (although we conjecture that our actual *constructions* of OTM's in Sections 3 and 4 are, in fact, strong enough for this purpose).

2.5 Uncertainty Relations for the Min-entropy

We will use an uncertainty relation from [16], with a slight modification to describe quantum systems that consist of many non-identical subsystems:

Theorem 1. *Consider a quantum system with Hilbert space* $\bigotimes_{i=1}^{\ell_0} \mathbb{C}^{d_i}$, *i.e., the system can be viewed as a collection of* ℓ_0 *subsystems, where the* i*'th subsystem has Hilbert space dimension* d_i.

For each $i \in [\ell_0]$, *let* B_i *be a finite collection of orthonormal bases for* $\mathbb{C}^{d_i}$, *and suppose that these bases satisfy the following uncertainty relation: for every quantum state* ρ *on* $\mathbb{C}^{d_i}$, $|B_i|^{-1} \sum_{\omega \in B_i} H(P_\omega) \geq h_i$, *where* P_ω *is the distribution of measurement outcomes when* ρ *is measured in basis* ω.

Now let ρ be any quantum state over $\bigotimes_{i=1}^{\ell_0} \mathbb{C}^{d_i}$, let $\Theta = (\Theta_1, \ldots, \Theta_{\ell_0})$ be chosen uniformly at random from $B_1 \times \cdots \times B_{\ell_0}$, and let $X = (X_1, \ldots, X_{\ell_0})$ be the measurement outcome when ρ is measured in basis Θ (i.e., each X_i is the outcome of measuring subsystem i in basis Θ_i).

Then, for any $\tau > 0$, and any $\lambda_1, \ldots, \lambda_{\ell_0} \in (0, \frac{1}{2})$, we have:

$$H_\infty^\varepsilon(X|\Theta) \geq -\tau + \sum_{i=1}^{\ell_0} (h_i - \lambda_i), \qquad (4)$$

where $\varepsilon \leq \exp(-2\tau^2/c)$, and $c = \sum_{i=1}^{\ell_0} 16\left(\lg \frac{|B_i|d_i}{\lambda_i}\right)^2$.

The proof is essentially the same as in [16]; it uses a martingale argument and Azuma's inequality, but it allows the martingale to have different increments at each step.

In addition, we will use the following chain rule for the smoothed min-entropy [20]:

$$H_\infty^{\varepsilon+\varepsilon'}(X|Y) > H_\infty^\varepsilon(X,Y) - H_0(Y) - \lg(\tfrac{1}{\varepsilon'}). \qquad (5)$$

3   One-Time Memories

We now show the correctness and security of the OTM construction described in Section 1.1. Recall that this OTM uses $n \lg q$ qubits, stores two messages of length ℓ, and uses an error-correcting code C. We will show how to set n and q, and how to choose the code C.

Let us introduce some notation. We view the code C as a function $C : \{0,1\}^\ell \to \{0,1\}^{n \lg q}$. We view each codeword $x \in \{0,1\}^{n \lg q}$ as a sequence of n blocks, where each block is a binary string of length $\lg q$. We write the codeword as $x = (x_{ij})_{i \in [n], j \in [\lg q]}$, and we write the i'th block as $x_i = (x_{ij})_{j \in [\lg q]}$. Finally, let H be the Hadamard gate acting on a single qubit.

We now prepare the qubits in the OTM as follows. For each $i = 1, 2, \ldots, n$,

- Let $\gamma_i \in \{0,1\}$ be the outcome of a fair and independent coin toss.
- If $\gamma_i = 0$, prepare the i'th block of qubits in the state $|C(s)_i\rangle$.
- If $\gamma_i = 1$, prepare the i'th block of qubits in the state $H^{\otimes(\lg q)}|C(t)_i\rangle$.

To recover the first message s, we measure every qubit in the standard basis, which yields a string of measurement outcomes $z \in \{0,1\}^{n \lg q}$, and then we run the decoding algorithm for C. To recover the second message t, we measure every qubit in the Hadamard basis, obtain a string of measurement outcomes z, and again run the decoding algorithm for C.

We will prove the following general theorem, which works for any code C that satisfies certain properties:

Theorem 2. Let $q \geq 2$ be any power of 2. Let $\mathcal{E}_q$ be the q-ary symmetric channel with error probability $p_e = (1/2) - (1/2q)$. Let $\ell \geq 1$ and $n \geq 1$, and let $C : \{0,1\}^\ell \to \{0,1\}^{n \lg q}$ be any error-correcting code that satisfies the following two requirements:

1. C can transmit information reliably over the channel $\mathcal{E}_q$ (where the channel treats each block of $\lg q$ bits as a single q-ary symbol).
2. C is a linear code over $GF(2)$.

Then the above OTM stores two messages $s, t \in \{0, 1\}^\ell$, and has the following properties:

1. The OTM behaves correctly for honest parties.
2. For any small constants $0 < \lambda \ll \frac{1}{2}$, $0 < \tau_0 \ll 1$, and $0 < \delta \ll 1$, the following statement holds. Suppose the messages s and t are chosen independently and uniformly at random in $\{0, 1\}^\ell$. For any separable adversary,[3] we have the following security bound:

$$H_\infty^{\delta+\varepsilon}(S, T|Z)$$
$$\geq \left((\tfrac{1}{2} - \lambda) - 4\tau_0 \left(1 + \tfrac{1}{\sqrt{\lg q}}(1 + \lg \tfrac{1}{\lambda}) \right) + (2 - \tfrac{1}{\alpha}) \right) \cdot \ell - \lg \tfrac{1}{\delta} \quad (6)$$
$$\gtrsim \left(\tfrac{1}{2} + (2 - \tfrac{1}{\alpha}) \right) \cdot \ell.$$

Here S and T are the random variables describing the two messages, Z is the random variable representing the adversary's measurement outcome, we have $\varepsilon \leq \exp(-2\tau_0^2 \ell / \lg q)$, and $\alpha = \ell/(n \lg q)$ is the rate of the code C.

Note that, to get a strong security bound, one must use a code C whose rate α is large. It is useful to ask, then, how large α can be. Let L_q denote the capacity of the channel $\mathcal{E}_q$, measured in q-ary symbols per channel use. Using a good code C, we can hope to have rate $\alpha \approx L_q$. Moreover, L_q is lower-bounded by:

$$L_q \geq 1 - \tfrac{1}{\lg q} - p_e = \tfrac{1}{2} - \tfrac{1}{\lg q} + \tfrac{1}{2q} \approx \tfrac{1}{2}, \quad (7)$$

which is nearly tight when q is large. So we can hope to have $\alpha \approx \frac{1}{2}$, in which case our security bound becomes:

$$H_\infty^{\delta+\varepsilon}(S, T|Z) \gtrsim \tfrac{1}{2}\ell. \quad (8)$$

3.1 Correctness for Honest Parties

We first show the "correctness" part of Theorem 2. Without loss of generality, suppose we want to recover the first message s. (A similar argument applies if we want to recover the second message t.) Let $z \in \{0, 1\}^{n \lg q}$ be the string of measurement outcomes obtained by measuring each qubit in the standard basis. Observe that z is the output of a q-ary symmetric channel $\mathcal{E}_q$ with error probability $p_e = (1/2) - (1/2q)$, acting on the string $C(s) \in \{0, 1\}^{n \lg q}$ (viewed as a sequence of n symbols in $GF(q)$). Since the code C can transmit information reliably over this channel, it follows that we can recover s.

---

[3] Note that this includes LOCC adversaries as a special case.

3.2   Security against Separable Adversaries

We now show the "security" part of Theorem 2. Let us first introduce some notation (see Figure 1). Suppose the OTM is programmed with two messages s and t that are chosen independently and uniformly at random in $\{0,1\}^{\ell}$. Let S and T be the random variables representing these messages. Let Γ be the random variable representing the coin flips $\gamma = (\gamma_1, \ldots, \gamma_n)$ used in programming the OTM. C denotes the error-correcting code, which maps $\{0,1\}^{\ell}$ to $\{0,1\}^{n \lg q}$. "Select" is an operation that maps $\{0,1\}^{n \lg q} \times \{0,1\}^{n \lg q}$ to $\{0,1\}^{n \lg q}$, depending on the value of Γ, as follows:

$$\text{Select}(x, y)_{i,j} = \begin{cases} x_{i,j} & \text{if } \Gamma_i = 0, \\ y_{i,j} & \text{if } \Gamma_i = 1, \end{cases} \qquad \text{for all } i \in [n], \ j \in [\lg q]. \qquad (9)$$

"Select" outputs a string of $n \lg q$ classical bits, which are converted into $n \lg q$ qubits (in the standard basis states $|0\rangle$ and $|1\rangle$). H denotes a Hadamard gate controlled by the value of Γ; that is, for each $i \in [n]$ and $j \in [\lg q]$, if $\Gamma_i = 1$, then H is applied to the (i,j)'th qubit.

Fix any separable adversary $\mathcal{A}$, let L be the number of possible outcomes that can be observed by the adversary, and let $\mathcal{M} = \{M_z \mid z \in [L]\}$ be the separable POVM measurement performed by the adversary. Let Z be the random variable representing the adversary's output; so Z takes values in $[L]$.

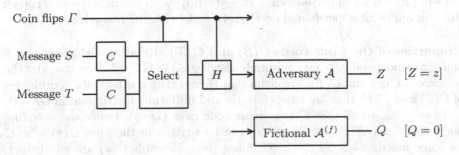

Fig. 1. OTM with separable adversary $\mathcal{A}$, and "fictional" adversary $\mathcal{A}^{(f)}$. In the proof, we will analyze the distributions of S and T conditioned on the events $Z = z$ and $Q = 0$.

Fix some small constant $\delta > 0$. We say that a measurement outcome $z \in [L]$ is "negligible" if $\Pr[Z = z] \leq (\delta/2^{n \lg q}) \operatorname{tr}(M_z)$. Note that the probability of observing any of these "negligible" measurement outcomes is small:

$$\Pr[Z \text{ is "negligible"}] = \sum_{z \text{ "negl."}} \Pr[Z = z] \leq (\delta/2^{n \lg q}) \sum_{z \text{ "negl."}} \operatorname{tr}(M_z) \leq \delta. \quad (10)$$

The proof will proceed as follows: for all messages $s, t \in \{0,1\}^{\ell}$, and for all measurement outcomes $z \in [L]$ that are not "negligible," we will upper-bound $\Pr[S = s, T = t | Z = z]$. This will imply a lower-bound on $H_{\infty}^{\delta}(S, T|Z)$, which is what we desire.

A Fictional Adversary. We begin by fixing some measurement outcome $z \in [L]$ that is not "negligible." Since the adversary performed a separable measurement, we can write the corresponding POVM element M_z as a tensor product of single-qubit operators. In particular, we can write $M_z = \text{tr}(M_z) \bigotimes_{i=1}^{n} \bigotimes_{j=1}^{\lg q} R_{ij}$, where each R_{ij} is a single-qubit operator, positive semidefinite, with trace 1.

We now construct a fictional adversary $\mathcal{A}^{(f)}$, which we will use in the proof. The fictional adversary acts in the following way: for each qubit $(i,j) \in [n] \times [\lg q]$, it performs the POVM measurement $\{R_{ij}, I - R_{ij}\}$ on qubit (i,j), which yields a binary measurement outcome Q_{ij} (where $Q_{ij} = 0$ corresponds to the POVM element R_{ij}, and $Q_{ij} = 1$ corresponds to $I - R_{ij}$). Let us write the vector of measurement outcomes as $Q = (Q_{ij})_{i \in [n], j \in [\lg q]}$, which takes values in $\{0,1\}^{n \lg q}$. Let 0 denote the vector $(0, 0, \ldots, 0) \in \{0,1\}^{n \lg q}$.

Intuitively, the event $Q = 0$ (in an experiment using the fictional adversary) corresponds to the event $Z = z$ (in an experiment using the real adversary). More precisely, for any $s, t \in \{0,1\}^{\ell}$, we have

$$
\begin{aligned}
P_{ST|Z}(s,t|z) &= \frac{P_{Z|ST}(z|s,t) P_{ST}(s,t)}{P_Z(z)} \\
&= \frac{P_{Q|ST}(0|s,t) \, \text{tr}(M_z) P_{ST}(s,t)}{P_Q(0) \, \text{tr}(M_z)} = P_{ST|Q}(s,t|0).
\end{aligned}
\tag{11}
$$

We will proceed by upper-bounding $P_{ST|Q}(s,t|0)$ (with the fictional adversary); this will imply an upper-bound on $P_{ST|Z}(s,t|z)$ (with the real adversary).

Properties of the Codewords $C(S)$ and $C(T)$. Recall that the messages S and T are independently and uniformly distributed in $GF(2)^{\ell}$. Now consider the codewords $C(S)$ and $C(T)$. We claim that there exists a subset of ℓ coordinates of $C(S)$ and $C(T)$ that are independently and uniformly distributed in $GF(2)^{\ell}$.

To see this, recall that C is a linear code over $GF(2)$. Hence the encoding operation $C : GF(2)^{\ell} \to GF(2)^{n \lg q}$ can be written in the form $C(x) = x^T G$ for some matrix $G \in GF(2)^{\ell \times n \lg q}$. Since the codewords $C(x)$ are all distinct, the matrix G must have row-rank ℓ. Hence the column-rank of G must also be ℓ, so there exists a subset of ℓ columns of G that are linearly independent over $GF(2)$. Let us denote this subset by $A \subset [n] \times [\lg q]$, $|A| = \ell$.

Now look at those coordinates of $C(S)$ and $C(T)$ that correspond to the subset A; we write these as $C(S)_A = (C(S)_{ij})_{(i,j) \in A}$ and $(C(T)_{ij})_{(i,j) \in A}$. It follows that $C(S)_A$ and $C(T)_A$ are independently and uniformly distributed in $GF(2)^{\ell}$.

Behavior of the Fictional Adversary on the Subset of Qubits A. We now analyze the behavior of the fictional adversary on those qubits belonging to the subset A. Without loss of generality, we can assume that the fictional adversary measures the qubits in the subset A first, and then measures the remaining qubits in the subset $([n] \times [\lg q]) \setminus A$. (This follows because the fictional adversary is *non-adaptive*, in that it makes all its decisions about what measurements to perform, before seeing any of the results of the measurements; and because all of the measurements commute with one another, since each measurement only involves a single qubit.)

For convenience, let $B = ([n] \times [\lg q]) \setminus A$. Let $Q_A = (Q_{ij})_{(i,j) \in A}$ denote the measurement outcomes of the qubits in the subset A, and let $Q_B = (Q_{ij})_{(i,j) \in B}$ denote the measurement outcomes of the qubits in the subset B.

We claim that the OTM's coin tosses Γ, conditioned on the event $Q_A = 0$, are still uniformly distributed in $\{0,1\}^n$. This is a fairly straightforward calculation; see the full paper [2] for details.

Using the Uncertainty Relation. We will upper-bound these probabilities $P_{ST|\Gamma Q_A}(s,t|\gamma,0)$, using an entropic uncertainty relation. The basic idea is to consider another experiment, where one runs the OTM and the fictional adversary "backwards" in time. This experiment can be analyzed using the uncertainty relation in Theorem 1 (originally due to [16]).

We now describe this new experiment (see Figure 2). One prepares the quantum state $\bigotimes_{(i,j) \in A} R_{ij}$, one chooses a uniformly random sequence of measurement bases $\Theta = (\Theta_1, \ldots, \Theta_n)$ (where $\Theta_i = 0$ denotes the standard basis and $\Theta_i = 1$ denotes the Hadamard basis), and then one measures each qubit $(i,j) \in A$ in the basis Θ_i to get a measurement outcome X_{ij} (which can be either 0 or 1).

Intuitively, the state $\bigotimes_{(i,j) \in A} R_{ij}$ corresponds to the fictional adversary's measurement outcome $Q_A = 0$, the random bases Θ correspond to the OTM's coin flips Γ, and the measurement outcomes X correspond to those bits $C(S)_A$ and $C(T)_A$ used in the OTM. (Note that the OTM's coin flips Γ are uniformly distributed, even when one conditions on the event $Q_A = 0$, as shown in the previous section.)

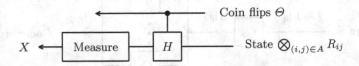

Fig. 2. In order to understand the behavior of the fictional adversary, conditioned on the event $Q_A = 0$, we consider an analogous experiment, where the state $\bigotimes_{(i,j) \in A} R_{ij}$ is measured in a random basis. We will analyze this using an entropic uncertainty relation.

To make this intuition precise, we will first show that:

$$H_\infty^\varepsilon(S,T|\Gamma, Q_A = 0) = H_\infty^\varepsilon(X|\Theta) + \ell. \tag{12}$$

(See the full version [2] for details.) Then note that conditioning on Γ can only reduce the entropy, hence we have:

$$H_\infty^\varepsilon(S,T|Q_A = 0) \geq H_\infty^\varepsilon(X|\Theta) + \ell. \tag{13}$$

We then use Theorem 1 to show a lower-bound on $H_\infty^\varepsilon(X|\Theta)$; see the full paper [2] for details.

Combining All the Pieces. The fictional adversary's complete sequence of measurement outcomes is denoted by $Q = (Q_A, Q_B)$. So far we have analyzed the adversary's actions on those qubits belonging to the subset A, and we have shown a lower-bound on $H^\varepsilon_\infty(S, T|Q_A = 0)$. Now, we will show a lower-bound on $H^\varepsilon_\infty(S, T|Q = 0)$. To do this, we bound the adversary's actions on the subset B in a more-or-less trivial way, using the fact that $\Pr[Q = 0] = \Pr[Z = z]/\operatorname{tr}(M_z) \geq \delta/2^{n \lg q}$, since z was assumed to be "non-negligible."

We will then consider the real adversary, and show a lower-bound on $H^{\delta+\varepsilon}_\infty(S, T|Z)$. Here we use the following identity that relates the real adversary and the fictional adversary (see equation (11)):

$$H^\varepsilon_\infty(S, T|Z = z) = H^\varepsilon_\infty(S, T|Q = 0). \tag{14}$$

Finally we combine these results to prove the theorem; see the full paper [2] for details.

4   Efficient Implementations of One-Time Memories

In the previous section, we showed that one-time memories can be constructed from any code that approaches the capacity of the q-ary symmetric channel, and is linear over $GF(2)$. In this section, we will construct codes that have these properties, and moreover can be encoded and decoded efficiently. Using these codes, we will get efficient implementations of one-time memories.

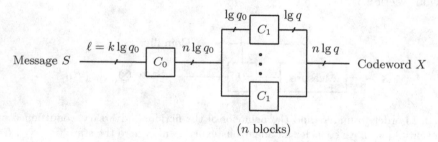

(n blocks)

Fig. 3. Efficient codes for the q-ary symmetric channel, based on erasure coding and error detection

There are several known constructions for codes that approach the capacity of the q-ary symmetric channel, and are efficiently decodable [22,23,24]. To illustrate how these techniques can be applied in our setting, we will describe one simple approach, which is based on erasure coding and error detection [23]. (See Figure 3.)

The basic idea is to take the message s, encode it using a code C_0 that outputs a string of q_0-ary symbols (where $q_0 < q$), and then encode each q_0-ary symbol using a code C_1 that outputs a q-ary symbol. The code C_1 is used to detect errors made by the q-ary symmetric channel; once detected, these errors can be treated as erasures. The code C_0 is then used to correct these erasures, which is relatively straightforward. For instance, we can choose C_0 to be a random linear code; then we can decode in the presence of erasure errors by solving a linear system of equations, which we can do efficiently.

We now describe the construction in detail. Let $k \geq 2$ be an integer, let $p_e \in (0, 1)$, and choose any small constants $0 < \varepsilon \ll 1, 0 < \delta \ll 1$ and $0 < \theta \ll 1$. Define:

$$n = \left\lfloor \frac{k}{1 - p_e - \theta} \right\rfloor, \tag{15}$$

$$q = 2^c, \quad c = \lg q = \left\lfloor \tfrac{2}{\delta} \right\rfloor \left\lceil \varepsilon n + \lg(n p_e) \right\rceil, \tag{16}$$

$$q_0 = 2^{c_0}, \quad c_0 = \lg q_0 = \left\lceil \tfrac{2}{\delta} - 2 \right\rceil \left\lceil \varepsilon n + \lg(n p_e) \right\rceil. \tag{17}$$

Note that our setting is slightly unusual, in that we will be constructing codes for the q-ary symmetric channel where q is not fixed. In particular, $\lg q$ (the number of bits used to describe each q-ary symbol) grows polynomially with the codeword length n, which is proportional to the message length k.

We will construct a code $C : \{0, 1\}^{k \lg q_0} \to \{0, 1\}^{n \lg q}$ as follows:

1. Choose a uniformly random matrix $G_0 \in GF(2)^{k \lg q_0 \times n \lg q_0}$, and define a code $C_0 : \{0, 1\}^{k \lg q_0} \to \{0, 1\}^{n \lg q_0}$ by setting $C_0(s) = s^T G_0$.
2. Fix any full-rank matrix $G_1 \in GF(2)^{\lg q_0 \times \lg q}$, and define a code $C_1 : \{0, 1\}^{\lg q_0} \to \{0, 1\}^{\lg q}$ by setting $C_1(v) = v^T G_1$.
3. Define $C(s) = C_1 \circ C_0(s)$, where we view $C_0(s) \in \{0, 1\}^{n \lg q_0}$ as a sequence of n blocks of $\lg q_0$ bits, and C_1 acts separately on each of these blocks. Equivalently, we can write $C(s) = s^T G_0(\bigoplus_{i=1}^{n} G_1)$, where $\bigoplus_{i=1}^{n} G_1$ denotes a direct sum of n copies of the matrix G_1.

We use the following decoding algorithm:

1. Given a string $z \in \{0, 1\}^{n \lg q}$, write it as a sequence of n blocks of $\lg q$ bits: $z = (z_{ij})_{i \in [n], j \in [\lg q]}$.
2. For each $i \in [n]$, try to decode the q-ary symbol $z_i \in \{0, 1\}^{\lg q}$, i.e., try to find some $v \in \{0, 1\}^{\lg q_0}$ such that $C_1(v) = z_i$. Let b_i be the result (or set $b_i = *$ if z_i lies outside the image of C_1). Thus we get a string $b = (b_1, b_2, \ldots, b_n) \in (\{0, 1\}^{\lg q_0} \cup \{*\})^n$.
3. Try to decode the string b, treating the $*$ symbols as erasures, i.e., try to find some $a \in \{0, 1\}^{k \lg q_0}$ such that, for all $i \in [n]$ such that $b_i \neq *$, and for all $j \in [\lg q]$, $C_0(a)_{ij} = b_{ij}$. If a solution exists, output it; if there are multiple solutions, choose any one of them and output it; otherwise, abort.

Finally, we introduce some more notation. Let us choose a message (represented by a random variable S) uniformly at random in $\{0, 1\}^{k \lg q_0}$. Let $\mathcal{E}_q$ be the q-ary symmetric channel with error probability p_e. We take the message S, encode it using the code C, transmit it through the channel $\mathcal{E}_q$, then run the decoding algorithm, and get an estimate of the original message; call this $\hat{S}$.

We prove the following statement (see the full paper [2] for details):

Theorem 3. *Let $k \geq 2$ be an integer, let $p_e \in (0,1)$, and choose any small constants $0 < \varepsilon \ll 1$, $0 < \delta \ll 1$ and $0 < \theta \ll 1$. Let us construct the code $C : \{0,1\}^{k \lg q_0} \to \{0,1\}^{n \lg q}$ as described above. Then C has the following properties:*

1. *With high probability (over the choice of the random matrix G_0), C can transmit information reliably over the q-ary symmetric channel $\mathcal{E}_q$ with error probability p_e.*
   More precisely, choose any small constant τ such that $0 < \tau < \theta$, and choose any large constant $\lambda \gg 1$. Then, with probability $\geq 1 - \frac{1}{\lambda}$ (over the choice of G_0), the code C can transmit information over the channel $\mathcal{E}_q$, and the probability of decoding failure is bounded by:

$$\Pr[\hat{S} \neq S] \leq \lambda\big(e^{-2\tau^2 n} + 2^{-\varepsilon n} + 2^{(-n\theta + n\tau + 1)\lg q_0}\big) \leq e^{-\Omega(n)}. \qquad (18)$$

2. *C is a linear code over $GF(2)$.*
3. *C has rate $\alpha := \frac{k \lg q_0}{n \lg q} \geq (1 - p_e - \theta)(1 - \delta)$. (Note that this approaches the capacity of the channel $\mathcal{E}_q$, as shown in equation (2), when q is large.)*
4. *The encoding and decoding algorithms for C run in time polynomial in $n \lg q$. (Also note that $\lg q$ grows at most linearly with n, and n is proportional to k.)*

Finally, we can take the code C constructed above (for $p_e = \frac{1}{2}$), and combine it with the OTM construction of Theorem 2, to get the following result:

Corollary 1. *For any $k \geq 2$, and for any small constant $0 < \mu \ll 1$, there exists an OTM construction that stores two messages $s, t \in \{0,1\}^\ell$, where $\ell = \Theta(k^2)$, and has the following properties:*

1. *The OTM behaves correctly for honest parties.*
2. *The OTM can be implemented in time polynomial in k.*
3. *Let $0 < \delta \ll 1$ be any small constant. Suppose the messages s and t are chosen independently and uniformly at random in $\{0,1\}^\ell$. For any separable adversary,[4] we have the following security bound:*

$$H_\infty^{\delta+\varepsilon}(S,T|Z) \geq (\tfrac{1}{2} - \mu)\ell - \lg \tfrac{1}{\delta}. \qquad (19)$$

Here S and T are the random variables describing the two messages, Z is the random variable representing the adversary's measurement outcome, and we have $\varepsilon \leq \exp(-\Omega(k))$.

Acknowledgements. It is a pleasure to thank Serge Fehr, Stephen Jordan, Maris Ozols, Rene Peralta, Eren Sasoglu, Christian Schaffner, Barbara Terhal, Alexander Vardy, and several anonymous reviewers, for helpful suggestions about this work. Some of these discussions took place at the Schloss Dagstuhl – Leibniz Center for Informatics. This paper is a contribution of NIST, an agency of the US government, and is not subject to US copyright.

---

[4] Note that this includes LOCC adversaries as a special case.

References

1. Liu, Y.-K.: Building one-time memories from isolated qubits. In: 5th Conference on Innovations in Theoretical Computer Science (ITCS 2014), pp. 269–286 (2014)
2. Liu, Y.-K.: Single-shot security for one-time memories in the isolated qubits model. ArXiv:1402.0049
3. Goldwasser, S., Kalai, Y.T., Rothblum, G.N.: One-Time Programs. In: Wagner, D. (ed.) CRYPTO 2008. LNCS, vol. 5157, pp. 39–56. Springer, Heidelberg (2008)
4. Goyal, V., Ishai, Y., Sahai, A., Venkatesan, R., Wadia, A.: Founding Cryptography on Tamper-Proof Hardware Tokens. In: Micciancio, D. (ed.) TCC 2010. LNCS, vol. 5978, pp. 308–326. Springer, Heidelberg (2010)
5. Bellare, M., Hoang, V.T., Rogaway, P.: Adaptively Secure Garbling with Applications to One-Time Programs and Secure Outsourcing. In: Wang, X., Sako, K. (eds.) ASIACRYPT 2012. LNCS, vol. 7658, pp. 134–153. Springer, Heidelberg (2012)
6. Broadbent, A., Gutoski, G., Stebila, D.: Quantum one-time programs. In: Canetti, R., Garay, J.A. (eds.) CRYPTO 2013, Part II. LNCS, vol. 8043, pp. 344–360. Springer, Heidelberg (2013)
7. Lo, H.-K., Chau, H.F.: Is quantum bit commitment really possible? Phys. Rev. Lett. 78, 3410 (1997)
8. Lo, H.-K.: Insecurity of quantum secure computations. Phys. Rev. A 56(2), 1154–1162 (1997)
9. Mayers, D.: Unconditionally secure quantum bit commitment is impossible. Phys. Rev. Lett. 78, 3414–3417 (1997)
10. Buhrman, H., Christandl, M., Schaffner, C.: Complete Insecurity of Quantum Protocols for Classical Two-Party Computation. Phys. Rev. Lett. 109, 160501 (2012)
11. Salvail, L.: Quantum Bit Commitment from a Physical Assumption. In: Krawczyk, H. (ed.) CRYPTO 1998. LNCS, vol. 1462, pp. 338–353. Springer, Heidelberg (1998)
12. Damgaard, I., Fehr, S., Salvail, L., Schaffner, C.: Cryptography In the Bounded Quantum-Storage Model. In: FOCS 2005, pp. 449–458 (2005)
13. Koenig, R., Terhal, B.M.: The Bounded Storage Model in the Presence of a Quantum Adversary. IEEE Trans. Inf. Th. 54(2) (2008)
14. Damgård, I.B., Fehr, S., Salvail, L., Schaffner, C.: Secure Identification and QKD in the Bounded-Quantum-Storage Model. In: Menezes, A. (ed.) CRYPTO 2007. LNCS, vol. 4622, pp. 342–359. Springer, Heidelberg (2007)
15. Wehner, S., Schaffner, C., Terhal, B.: Cryptography from Noisy Storage. Phys. Rev. Lett. 100, 220502 (2008)
16. Damgård, I.B., Fehr, S., Renner, R., Salvail, L., Schaffner, C.: A Tight High-Order Entropic Quantum Uncertainty Relation with Applications. In: Menezes, A. (ed.) CRYPTO 2007. LNCS, vol. 4622, pp. 360–378. Springer, Heidelberg (2007)
17. Bouman, N.J., Fehr, S., González-Guillén, C., Schaffner, C.: An all-but-one entropic uncertainty relation, and application to password-based identification. In: Kawano, Y. (ed.) TQC 2012. LNCS, vol. 7582, pp. 29–44. Springer, Heidelberg (2012)
18. Pastawski, F., Yao, N.Y., Jiang, L., Lukin, M.D., Cirac, J.I.: Unforgeable Noise-Tolerant Quantum Tokens. Proc. Nat. Acad. Sci. 109, 16079–16082 (2012)
19. Renner, R.: Security of Quantum Key Distribution. PhD thesis, ETH Zurich (2005)
20. Renner, R., Wolf, S.: Simple and Tight Bounds for Information Reconciliation and Privacy Amplification. In: Roy, B. (ed.) ASIACRYPT 2005. LNCS, vol. 3788, pp. 199–216. Springer, Heidelberg (2005)
21. Wiesner, S.: Conjugate coding. ACM SIGACT News 15(1), 78–88 (1983); original manuscript written circa 1970

22. Bleichenbacher, D., Kiayias, A., Yung, M.: Decoding of Interleaved Reed Solomon Codes over Noisy Data. In: Baeten, J.C.M., Lenstra, J.K., Parrow, J., Woeginger, G.J. (eds.) ICALP 2003. LNCS, vol. 2719, pp. 97–108. Springer, Heidelberg (2003)

23. Shokrollahi, A.: Capacity-approaching codes on the q-ary symmetric channel for large q. In: ITW 2004, pp. 204–208 (2004)

24. Brown, A., Minder, L., Shokrollahi, M.A.: Improved Decoding of Interleaved AG Codes. In: Smart, N.P. (ed.) Cryptography and Coding 2005. LNCS, vol. 3796, pp. 37–46. Springer, Heidelberg (2005)

25. Bennett, C.H., DiVincenzo, D.P., Fuchs, C.A., Mor, T., Rains, E., Shor, P.W., Smolin, J.A., Wootters, W.K.: Quantum nonlocality without entanglement. Phys. Rev. A 59, 1070–1091 (1999)

26. Childs, A.M., Leung, D., Mancinska, L., Ozols, M.: A framework for bounding nonlocality of state discrimination. Comm. Math. Phys. 323, 1121–1153 (2013)

27. DiVincenzo, D.P., Leung, D.W., Terhal, B.M.: Quantum Data Hiding. IEEE Trans. Inf. Theory 48(3), 580–599 (2002)

28. Eggeling, T., Werner, R.F.: Hiding Classical Data in Multipartite Quantum States. Phys. Rev. Lett. 89, 097905 (2002)

29. DiVincenzo, D.P., Hayden, P., Terhal, B.M.: Hiding Quantum Data. Found. Phys. 33(11), 1629–1647 (2003)

30. Hayden, P., Leung, D., Smith, G.: Multiparty data hiding of quantum information. Phys. Rev. A 71, 062339 (2005)

31. Winter, A.: Coding theorem and strong converse for quantum channels. IEEE Trans. Inform. Theory 45(7), 2481–2485 (1999)

32. Akavia, A., Goldwasser, S., Vaikuntanathan, V.: Simultaneous Hardcore Bits and Cryptography against Memory Attacks. In: Reingold, O. (ed.) TCC 2009. LNCS, vol. 5444, pp. 474–495. Springer, Heidelberg (2009)

33. Naor, M., Segev, G.: Public-Key Cryptosystems Resilient to Key Leakage. In: Halevi, S. (ed.) CRYPTO 2009. LNCS, vol. 5677, pp. 18–35. Springer, Heidelberg (2009)

34. Nielsen, M.A., Chuang, I.L.: Quantum Computation and Quantum Information. Cambridge University Press (2000)

35. Maassen, H., Uffink, J.: Generalized Entropic Uncertainty Relations. Phys. Rev. Lett. 60, 1103 (1988)

36. Wehner, S., Winter, A.: Entropic uncertainty relations - A survey. New J. Phys. 12, 025009 (2010)

37. Dubhashi, D.P., Panconesi, A.: Concentration of Measure for the Analysis of Randomized Algorithms. Cambridge University Press (2009)

38. Horodecki, R., Horodecki, P., Horodecki, M., Horodecki, K.: Quantum Entanglement. Rev. Mod. Phys. 81, 865–942 (2009)

39. Saeedi, K., et al.: Room-Temperature Quantum Bit Storage Exceeding 39 Minutes Using Ionized Donors in Silicon-28. Science 342(6160), 830–833 (2013)

40. Dreau, A., et al.: Single-Shot Readout of Multiple Nuclear Spin Qubits in Diamond under Ambient Conditions. Phys. Rev. Lett. 110, 060502 (2013)

How to Eat Your Entropy and Have It Too – Optimal Recovery Strategies for Compromised RNGs

Yevgeniy Dodis[1,*], Adi Shamir[2], Noah Stephens-Davidowitz[1],
and Daniel Wichs[3,**]

[1] Dept. of Computer Science, New York University, New York, NY, USA
dodis@cs.nyu.edu, noahsd@gmail.com
[2] Dept. of Computer Science and Applied Mathematics, Weizmann Institute,
Rehovot, Israel
adi.shamir@weizmann.ac.il
[3] Dept. of Computer Science, Northeastern University, Boston, MA, USA
wichs@ccs.neu.edu

Abstract. We study random number generators (RNGs) with input, RNGs that regularly update their internal state according to some auxiliary input with additional randomness harvested from the environment. We formalize the problem of designing an efficient recovery mechanism from complete state compromise in the presence of an active attacker. If we knew the timing of the last compromise and the amount of entropy gathered since then, we could stop producing any outputs until the state becomes truly random again. However, our challenge is to recover within a time proportional to this optimal solution even in the hardest (and most realistic) case in which (a) we know nothing about the timing of the last state compromise, and the amount of new entropy injected since then into the state, and (b) any premature production of outputs leads to the total loss of all the added entropy *used by the RNG*. In other words, the challenge is to develop recovery mechanisms which are guaranteed to save the day as quickly as possible after a compromise we are not even aware of. The dilemma is that any entropy used prematurely will be lost, and any entropy which is kept unused will delay the recovery.

After formally modeling RNGs with input, we show a nearly optimal construction that is secure in our very strong model. Our technique is inspired by the design of the Fortuna RNG (which is a heuristic RNG construction that is currently used by Windows and comes without any formal analysis), but we non-trivially adapt it to our much stronger adversarial setting. Along the way, our formal treatment of Fortuna enables us to improve its entropy efficiency by almost a factor of two, and to show that our improved construction is essentially tight, by proving a rigorous lower bound on the possible efficiency of any recovery mechanism in our very general model of the problem.

Keywords: Random number generators, RNGs with input.

* Research partially supported by gifts from VMware Labs and Google, and NSF grants 1319051, 1314568, 1065288, 1017471.
** Research partially supported by gift from Google and NSF grants 1347350, 1314722.

J.A. Garay and R. Gennaro (Eds.): CRYPTO 2014, Part II, LNCS 8617, pp. 37–54, 2014.
© International Association for Cryptologic Research 2014

1   Introduction

Randomness is essential in many facets of cryptography, from the generation of long-term cryptographic keys, to sampling local randomness for encryption, zero-knowledge proofs, and many other randomized cryptographic primitives. As a useful abstraction, designers of such cryptographic schemes assume a source of (nearly) uniform, unbiased, and independent random bits of arbitrary length. In practice, however, this theoretical abstraction is realized by means of a *Random Number Generator* (RNG), whose goal is to quickly accumulate entropy from various physical sources in the environment (such as keyboard presses or mouse movement) and then convert it into the required source of (pseudo) random bits. We notice that a highly desired (but, alas, rarely achieved) property of such RNGs is their ability to quickly recover from various forms of *state compromise*, in which the current state S of the RNG becomes known to the attacker, either due to a successful penetration attack, or via side channel leakage, or simply due to insufficient randomness in the initial state. This means that the state S of practical RNGs should be periodically refreshed using the above-mentioned physical sources of randomness I. In contrast, the simpler and much better-understood theoretical model of pseudorandom generators (PRGs) does not allow the state to be refreshed after its initialization. To emphasize this distinction, we will sometimes call our notion an "RNG with input", and notice that virtually all modern operating systems come equipped with such an RNG with input; e.g., /dev/random [21] for Linux, Yarrow [14] for MacOs/iOS/FreeBSD and Fortuna [10] for Windows [9].

Unfortunately, despite the fact that they are widely used and often referred to in various standards [2,8,13,16], RNGs with input have received comparatively little attention from theoreticians. The two notable exceptions are the works of Barak and Halevi [1] and Dodis et al. [5]. The pioneering work of [1] emphasized the importance of rigorous analysis of RNGs with input and laid their first theoretical foundations. However, as pointed out by [5], the extremely clean and elegant security model of [1] ignores the "heart and soul" issue of most real-world RNGs with input, namely, their ability to gradually "accumulate" many low-entropy inputs I into the state S at the same time that they lose entropy due to premature use. In particular, [5] showed that the construction of [1] (proven secure in their model) may always fail to recover from state compromise when the entropy of each input $I_1, \ldots, I_q$ is sufficiently small, *even for arbitrarily large q*.

Motivated by these considerations, Dodis et al. [5] defined an improved security model for RNGs with input, which explicitly guaranteed eventual recovery from any state compromise, provided that the *collective* fresh entropy of inputs $I_1, \ldots, I_q$ crosses some security threshold γ^*, *irrespective of the entropies of individual inputs I_j*. In particular, they demonstrated that Linux's /dev/random does not satisfy their stronger notion of *robustness* (for similar reasons as the construction of [1]), and then constructed a simple scheme which is provably robust in this model. However, as we explain below, their robustness model did not address the issue of efficiency of the recovery mechanism when the RNG is being *continuously used* after the compromise.

The Premature Next Problem. In this paper, we extend the model of [5] to address some additional desirable security properties of RNGs with input not captured by this model. The main such property is resilience to the "*premature next* attack". This general attack, first explicitly mentioned by Kelsey, Schneier, Wagner, and Hall [15], is applicable in situations in which the RNG state S has accumulated an insufficient amount of entropy e (which is very common in bootup situations) and then must produce some outputs R via legitimate "next" calls in order to generate various system keys. Not only is this R not fully random (which is expected), but now the attacker can potentially use R to recover the current state S by brute force, effectively "emptying" the e bits of entropy that S accumulated so far. Applied iteratively, this simple attack, when feasible, can prevent the system from ever recovering from compromise, irrespective of the total amount of fresh entropy injected into the system since the last compromise.

At first, it might appear that the only way to prevent this attack is by discovering a sound way to estimate the current entropy in the state and to use this estimate to block the premature next calls. This is essentially the approach taken by Linux's /dev/random and many other RNGs with input. Unfortunately, sound entropy estimation is hard or even infeasible [10,20] (e.g., [5] showed simple ways to completely fool Linux's entropy estimator). This seems to suggest that the modeling of RNGs with input should consider each premature next call as a full state compromise, and this is the highly conservative approach taken by [5] (which we will fix in this work).

Fortuna. Fortunately, the conclusion above is overly pessimistic. In fact, the solution idea already comes from two very popular RNGs mentioned above, whose designs were heavily affected by the desire to overcome the premature next problem: Yarrow (designed by Schneier, Kelsey and Ferguson [14] and used by MacOS/iOS/FreeBSD), and its refinement Fortuna (subsequently designed by Ferguson and Schneier [10] and used by Windows [9]). The simple but brilliant idea of these works is to partition the incoming entropy into multiple entropy "pools" and then to cleverly use these pools at vastly different rates when producing outputs, in order to guarantee that at least one pool will eventually accumulate enough entropy to guarantee security before it is "prematurely emptied" by a next call. (See Section 4 for more details.)

Ferguson and Schneier provide good security intuition for their Fortuna "pool scheduler" construction, assuming that all the RNG inputs $I_1, \ldots, I_q$ have the same (unknown) entropy and that each of the pools can losslessly accumulate all the entropy that it gets. (They suggest using iterated hashing with a cryptographic hash function as a heuristic way to achieve this.) In particular, if q is the upper bound on the number of inputs, they suggest that one can make the number of pools $P = \log_2 q$, and recover from state compromise (with premature next!) at the loss of a factor $O(\log q)$ in the amount of fresh entropy needed.

Our Main Result. Inspired by the idea of Fortuna, we formally extend the prior RNG robustness notion of [5] to *robustness against premature next*. Unlike Ferguson and Schneier, we do so without making any restrictive assumptions such as requiring that the entropy of all the inputs I_j be constant. (Indeed, these entropies can be adversarily chosen, as in the model of [5], and can be *unknown* to the RNG.) Also, in our formal and general security model, we do not assume ideal entropy accumulation or inherently rely on cryptographic hash functions. In fact, our model is syntactically very similar to the prior RNG model of [5], except: (1) a premature next call is not considered an unrecoverable state corruption, but (2) in addition to the (old) "entropy penalty" parameter γ^*, there is a (new) "time penalty" parameter $\beta \geq 1$, measuring how long it will take to recover from state compromise relative to the optimal recovery time needed to receive γ^* bits of fresh entropy. (See Figures 2 and 3.)

To summarize, our model formalizes the problem of designing an efficient recovery mechanism from state compromise as an online optimization problem. If we knew the timing of the last compromise and the amount of entropy gathered since then, we could stop producing any outputs until the state becomes truly random again. However, our challenge is to recover within a time proportional to this optimal solution even in the hardest (and most realistic) case in which (a) we know nothing about the timing of the last state compromise, and the amount of new entropy injected since then into the state, and (b) any premature production of outputs leads to the total loss of all the added entropy *used by the RNG*, since the attacker can use brute force to enumerate all the possible low-entropy states. In other words, the challenge is to develop recovery mechanisms which are guaranteed to save the day as quickly as possible after a compromise we are not even aware of. The dilemma that we face is that *any entropy used prematurely will be lost, and any entropy which is kept unused will delay the recovery*.

After extending our model to handle premature next calls, we define the generalized Fortuna construction, which is provably robust against premature next. Although heavily inspired by actual Fortuna, the syntax of our construction is noticeably different (See Figure 5), since we prove it secure in a stronger model and without any idealized assumptions (like perfect entropy accumulation, which, as demonstrated by the attacks in [5], is not a trivial thing to sweep under the rug). In fact, to obtain our construction, we: (a) abstract out a rigorous security notion of a (pool) *scheduler*; (b) show a formal composition theorem (Theorem 2) stating that a secure scheduler can be composed with any robust RNG in the prior model of [5] to achieve security against premature next; (c) obtain our final RNG by using the provably secure RNG of [5] and a Fortuna-like scheduler (proven secure in our significantly stronger model). In particular, the resulting RNG is secure in the standard model, and only uses the existence of standard PRGs as its sole computational assumption.

Constant-Rate RNGs. In Section 5.3, we consider the actual constants involved in our construction, and show that under a reasonable setting or parameters, our RNG will recover from compromise in $\beta = 4$ times the number of steps it takes to get 20 to 30 kB of fresh entropy. While these numbers are a bit high, they are also obtained in an extremely strong adversarial model. In contrast, remember that Ferguson and Schneier informally analyzed the security of Fortuna in a much simpler case in which entropy drips in at a constant rate. While restrictive, in Section 6 we also look at the security of generalized Fortuna (with a better specialized scheduler) in this model, as it could be useful in some practical scenarios and allow for a more direct comparison with the original Fortuna. In this simpler constant entropy dripping rate, we estimate that our RNG (with standard security parameters) will recover from a complete compromise immediately after it gets about 2 to 3 kB of entropy (see the full version for details [6]), which is comparable to [10]'s (corrected) claim, but without assuming ideal entropy accumulation into the state. In fact, our optimized constant-rate scheduler beats the original Fortuna's scheduler by almost a factor of 2 in terms of entropy efficiency.

Rate Lower Bound. We also show that any "Fortuna-like construction" (which tries to collect entropy in multiple pools and cleverly utilize them with an arbitrary scheduler) must lose at least a factor $\Omega(\log q)$ in its "entropy efficiency", even in the case where all inputs I_j have an (unknown) *constant-rate* entropy. This suggests that the original scheduler of Fortuna (which used $\log q$ pools which evenly divide the entropy among them) is asymptotically optimal in the constant-rate case (as is our improved version).

Semi-Adaptive Set-Refresh. As a final result, we make progress in addressing another important limitation of the model of Dodis et al. [5] (and our direct extension of the current model that handles premature nexts). Deferring technical details to the full version [6], in that model the attacker $\mathcal{A}$ had very limited opportunities to adaptively influence the samples produced by another adversarial quantity, called the *distribution sampler* $\mathcal{D}$. As explained there and in [5], *some* assumption of this kind is necessary to avoid impossibility results, but it does limit the applicability of the model to some real-world situations. As the initial step to removing this limitation, in the full version we introduce the "semi-adaptive set-refresh" model and show that both the original RNG of [5] and our new RNG are provably secure in this more realistic adversarial model [6].

Other Related Work. As we mentioned, there is very little literature focusing on the design and analysis of RNGs with inputs in the standard model. In addition to [1,5], some analysis of the Linux RNG was done by Lacharme, Röck, Strubel and Videau [17]. On the other hand, many works showed devastating attacks on various cryptographic schemes when using weak randomness; some notable examples include [4,7,11,12,15,18,19].

2   Preliminaries

Entropy. For a discrete distribution X, we denote its *min-entropy* by $\mathbf{H}_\infty(X) = \min_x\{-\log \Pr[X = x]\}$. We also define worst-case min-entropy of X conditioned on another random variable Z by in the following conservative way: $\mathbf{H}_\infty(X|Z) =$

$-\log([\max_{x,z} \Pr[X = x | Z = z]])$. We use this definition instead of the usual one so that it satisfies the following relation, which is called the "chain rule": $\mathbf{H}_\infty(X, Z) - \mathbf{H}_\infty(Z) \geq \mathbf{H}_\infty(X|Z)$.

Pseudorandom Functions and Generators. We say that a function $\mathbf{F}$: $\{0,1\}^\ell \times \{0,1\}^m \to \{0,1\}^m$ is a (deterministic) $(t, q_{\mathbf{F}}, \varepsilon)$-*pseudorandom function* (PRF) if no adversary running in time t and making $q_{\mathbf{F}}$ oracle queries to $\mathbf{F}(\text{key}, \cdot)$ can distinguish between $\mathbf{F}(\text{key}, \cdot)$ and a random function with probability greater than ε when key $\overset{\$}{\leftarrow} \{0,1\}^\ell$. We say that a function $\mathbf{G} : \{0,1\}^m \to \{0,1\}^n$ is a (deterministic) (t, ε)-*pseudorandom generator* (PRG) if no adversary running in time t can distinguish between $\mathbf{G}(\text{seed})$ and uniformly random bits with probability greater than ε when seed $\overset{\$}{\leftarrow} \{0,1\}^m$.

Game Playing Framework. For our security definitions and proofs we use the code-based game-playing framework of [3]. A game GAME has an initialize procedure, procedures to respond to adversary oracle queries, and a finalize procedure. A game GAME is executed with an adversary $\mathcal{A}$ as follows: First, initialize executes, and its outputs are the inputs to $\mathcal{A}$. Then $\mathcal{A}$ executes, its oracle queries being answered by the corresponding procedures of GAME. When $\mathcal{A}$ terminates, its output becomes the input to the finalize procedure. The output of the latter is called the output of the game, and we let $\text{GAME}^{\mathcal{A}} \Rightarrow y$ denote the event that this game output takes value y. $\mathcal{A}^{\text{GAME}}$ denotes the output of the adversary and $\text{Adv}_{\mathcal{A}}^{\text{GAME}} = 2 \times \Pr[\text{GAME}^{\mathcal{A}} \Rightarrow 1] - 1$. Our convention is that Boolean flags are assumed initialized to false and that the running time of the adversary $\mathcal{A}$ is defined as the total running time of the game with the adversary in expectation, including the procedures of the game.

3   RNG with Input: Modeling and Security

In this section we present formal modeling and security definitions for RNGs with input, largely following [5].

Definition 1 (RNG with input). *An RNG with input is a triple of algorithms* $\mathcal{G} = (\text{setup}, \text{refresh}, \text{next})$ *and a triple* $(n, \ell, p) \in \mathbb{N}^3$ *where* n *is the state length,* ℓ *is the output length and* p *is the input length of* $\mathcal{G}$:

- setup: *a probabilistic algorithm that outputs some public parameters* seed *for the generator.*
- refresh: *a deterministic algorithm that, given* seed, *a state* $S \in \{0,1\}^n$ *and an input* $I \in \{0,1\}^p$, *outputs a new state* $S' = \text{refresh}(\text{seed}, S, I) \in \{0,1\}^n$.
- next: *a deterministic algorithm that, given* seed *and a state* $S \in \{0,1\}^n$, *outputs a pair* $(S', R) = \text{next}(\text{seed}, S)$ *where* $S' \in \{0,1\}^n$ *is the new state and* $R \in \{0,1\}^\ell$ *is the output.*

Before moving to defining our security notions, we notice that there are two adversarial entities we need to worry about: the *adversary* $\mathcal{A}$ whose task is (intuitively) to distinguish the outputs of the RNG from random, and the *distribution sampler* $\mathcal{D}$ whose task is to produce inputs $I_1, I_2, \ldots$, which have high entropy *collectively*, but somehow help $\mathcal{A}$ in breaking the security of the RNG. In other words, the distribution sampler models potentially adversarial environment (or "nature") where our RNG is forced to operate.

3.1 Distribution Sampler

The distribution sampler $\mathcal{D}$ is a *stateful and probabilistic* algorithm which, given the current state σ, outputs a tuple (σ', I, γ, z) where: (a) σ' is the new state for $\mathcal{D}$; (b) $I \in \{0,1\}^p$ is the next input for the refresh algorithm; (c) γ is some *fresh entropy estimation* of I, as discussed below; (d) z is the *leakage* about I given to the attacker $\mathcal{A}$. We denote by $q_{\mathcal{D}}$ the upper bound on number of executions of $\mathcal{D}$ in our security games, and say that $\mathcal{D}$ is *legitimate* if

$$\mathbf{H}_\infty(I_j \mid I_1, \ldots, I_{j-1}, I_{j+1}, \ldots, I_{q_{\mathcal{D}}}, z_1, \ldots, z_{q_{\mathcal{D}}}, \gamma_0, \ldots, \gamma_{q_{\mathcal{D}}}) \geq \gamma_j \qquad (1)$$

for all $j \in \{1, \ldots, q_{\mathcal{D}}\}$ where $(\sigma_i, I_i, \gamma_i, z_i) = \mathcal{D}(\sigma_{i-1})$ for $i \in \{1, \ldots, q_{\mathcal{D}}\}$ and $\sigma_0 = 0$.[1]

Dodis et al. provide a detailed discussion of the distribution sampler in [5], which we also include in the full version of this paper for completeness [6]. In particular, note that the distribution sampler $\mathcal{D}$ is required to output a lower bound γ on the min-entropy of I. These entropy estimates will be used in the security game. In particular, we of course cannot guarantee security unless the distribution sampler has provided the challenger with some minimum amount of entropy. Many implemented RNGs try to get around this problem by attempting to estimate the entropy of a given distribution directly in some ad-hoc manner. However, entropy estimation is impossible in general and computationally hard even in very special cases [20]. Note that these entropy estimates will be used only in the security game, and are *not* given to the refresh and next procedures. By separating entropy estimation from security, [5] provides a meaningful definition of security without requiring the RNG to know anything about the entropy of the sampled distributions.

3.2 Security Notions

We define the game $\mathsf{ROB}(\gamma^*)$ in our game framework. We show the initialize and finalize procedures for $\mathsf{ROB}(\gamma^*)$ in Figure 1. The attacker's goal is to guess the correct value b picked in the initialize procedure with access to several oracles, shown in Figure 2. Dodis et al. define the notion of *robustness* for an RNG with input [5]. In particular, they define the parametrized security game $\mathsf{ROB}(\gamma^*)$ where γ^* is a measure of the "fresh" entropy in the system when security should be expected. Intuitively, in this game $\mathcal{A}$ is able to view or change the state of the RNG (get-state and set-state), to see output from it (get-next), and to update

---

[1] Since conditional min-entropy is defined in the worst-case manner, the value γ_j in the bound below should not be viewed as a random variable, but rather as an arbitrary fixing of this random variable.

it with a sample I_j from $\mathcal{D}$ ($\mathcal{D}$-refresh). In particular, notice that the $\mathcal{D}$-refresh oracle keeps track of the fresh entropy in the system and declares the RNG to no longer be corrupted only when the fresh entropy c is greater than γ^*. (We stress again that the entropy estimates γ_i and the counter c are not available to the RNG.) Intuitively, $\mathcal{A}$ wins if the RNG is not corrupted and he correctly distinguishes the output of the RNG from uniformly random bits.

---

proc. initialize
seed $\xleftarrow{\$}$ setup; $\sigma \leftarrow 0$; $S \xleftarrow{\$} \{0,1\}^n$
$c \leftarrow n$; corrupt $\leftarrow$ false; $b \xleftarrow{\$} \{0,1\}$
OUTPUT seed

proc. finalize(b^*)
IF $b = b^*$ RETURN 1
ELSE RETURN 0

Fig. 1. Procedures initialize and finalize for $\mathcal{G} = (\text{setup}, \text{refresh}, \text{next})$

---

proc. $\mathcal{D}$-refresh
$(\sigma, I, \gamma, z) \xleftarrow{\$} \mathcal{D}(\sigma)$
$S \leftarrow \text{refresh}(S, I)$
$c \leftarrow c + \gamma$
IF $c \geq \gamma^*$,
    corrupt $\leftarrow$ false
OUTPUT (γ, z)

proc. get-state
$c \leftarrow 0$; corrupt $\leftarrow$ true
OUTPUT S

proc. set-state(S^*)
$c \leftarrow 0$; corrupt $\leftarrow$ true
$S \leftarrow S^*$

proc. next-ror
$(S, R_0) \leftarrow \text{next}(S)$
$R_1 \xleftarrow{\$} \{0,1\}^\ell$
IF corrupt = true,
    $c \leftarrow 0$
    RETURN R_0
ELSE OUTPUT R_b

proc. get-next
$(S, R) \leftarrow \text{next}(S)$
IF corrupt = true,
    $c \leftarrow 0$
OUTPUT R

Fig. 2. Procedures in $\text{ROB}(\gamma^*)$ for $\mathcal{G} = (\text{setup}, \text{refresh}, \text{next})$

Definition 2 (Security of RNG with input). *A pseudorandom number generator with input $\mathcal{G} = (\text{setup}, \text{refresh}, \text{next})$ is called $((t, q_\mathcal{D}, q_R, q_S), \gamma^*, \varepsilon)$-robust if for any attacker $\mathcal{A}$ running in time at most t, making at most $q_\mathcal{D}$ calls to $\mathcal{D}$-refresh, q_R calls to next-ror or get-next and q_S calls to get-state or set-state, and any legitimate distribution sampler $\mathcal{D}$ inside the $\mathcal{D}$-refresh procedure, the advantage of $\mathcal{A}$ in game $\text{ROB}(\gamma^*)$ is at most ε.*

Notice that in $\text{ROB}(\gamma^*)$, if $\mathcal{A}$ calls get-next when the RNG is still corrupted, this is a "premature" get-next and the entropy counter c is reset to 0. Intuitively, [5] treats information "leaked" from an insecure RNG as a total compromise. We modify their security definition and define the notion of *robustness against premature next* with the corresponding security game $\text{NROB}(\gamma^*, \gamma_{\max}, \beta)$.

Our modified game $\mathsf{NROB}(\gamma^*, \gamma_{\max}, \beta)$ has identical initialize and finalize procedures to [5]'s $\mathsf{ROB}(\gamma^*)$ (Figure 1). Figure 3 shows the new oracle queries. The differences with $\mathsf{ROB}(\gamma^*)$ are highlighted for clarity.

In our modified game, "premature" get-next calls do not reset the entropy counter. We pay a price for this that is represented by the parameter $\beta \geq 1$. In particular, in our modified game, the game does not immediately declare the state to be uncorrupted when the entropy counter c passes the threshold γ^*. Instead, the game keeps a counter T that records the number of calls to $\mathcal{D}$-refresh since the last set-state or get-state (or the start of the game). When c passes γ^*, it sets $T^* \leftarrow T$ and the state becomes uncorrupted only after $T \geq \beta T^*$ (of course, provided $\mathcal{A}$ made no additional calls to get-state or set-state). In particular, while we allow extra time for recovery, notice that we do *not* require any additional entropy from the distribution sampler $\mathcal{D}$.

Intuitively, we allow $\mathcal{A}$ to receive output from a (possibly corrupted) RNG and, therefore, to potentially learn information about the state of the RNG without any "penalty". However, we allow the RNG additional time to "mix the fresh entropy" received from $\mathcal{D}$, proportional to the amount of time T^* that it took to get the required fresh entropy γ^* since the last compromise.

As a final subtlety, we set a maximum $\gamma_{\max}$ on the amount that the entropy counter can be increased from one $\mathcal{D}$-refresh call. This might seem strange, since it is not obvious how receiving too much entropy at once could be a problem. However, $\gamma_{\max}$ will prove quite useful in the analysis of our construction. Intuitively, this is because it is harder to "mix" entropy if it comes too quickly. Of course $\gamma_{\max}$ is bounded by the length of the input p, but in practice we often expect it to be substantially lower. In such cases, we are able to prove much better performance for our RNG construction, *even if $\gamma_{\max}$ is unknown to the RNG*. In addition, we get very slightly better results if some upper bound on $\gamma_{\max}$ is incorporated into the construction.

Definition 3 (Security of RNG with input against premature next).
A pseudorandom number generator with input $\mathcal{G} = (\mathsf{setup}, \mathsf{refresh}, \mathsf{next})$ is called $((t, q_{\mathcal{D}}, q_R, q_S), \gamma^*, \gamma_{\max}, \varepsilon, \beta)$-premature-next robust if for any attacker $\mathcal{A}$ running in time at most t, making at most $q_{\mathcal{D}}$ calls to $\mathcal{D}$-refresh, q_R calls to next-ror or get-next and q_S calls to get-state or set-state, and any legitimate distribution sampler $\mathcal{D}$ inside the $\mathcal{D}$-refresh procedure, the advantage of $\mathcal{A}$ in game $\mathsf{NROB}(\gamma^*, \gamma_{\max}, \beta)$ is at most ε.

Relaxed Security Notions. We note that the above security definition is quite strong. In particular, the attacker has the ability to arbitrarily set the state of $\mathcal{G}$ many times. Motivated by this, we present several relaxed security definitions that may better capture real-world security. These definitions will be useful for our proofs, and we show in Section 4.2 that we can achieve better results for these weaker notions of security:

- $\mathsf{NROB}_{\mathsf{reset}}(\gamma^*, \gamma_{\max}, \beta)$ is $\mathsf{NROB}(\gamma^*, \gamma_{\max}, \beta)$ in which oracle calls to set-state are replaced by calls to reset-state. reset-state takes no input and simply sets

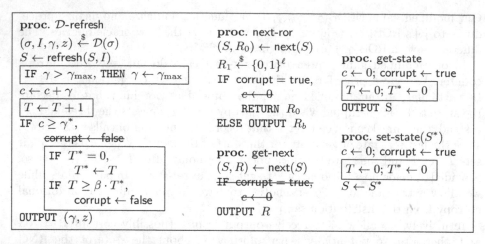

Fig. 3. Procedures in $\mathsf{NROB}(\gamma^*, \gamma_{\max}, \beta)$ for $\mathcal{G} = (\mathsf{setup}, \mathsf{refresh}, \mathsf{next})$ with differences from $\mathsf{ROB}(\gamma^*)$ highlighted

the state of $\mathcal{G}$ to some fixed state S_0, determined by the scheme and sets the entropy counter to zero.[2]

- $\mathsf{NROB}_{1\mathsf{set}}(\gamma^*, \gamma_{\max}, \beta)$ is $\mathsf{NROB}(\gamma^*, \gamma_{\max}, \beta)$ in which the attacker may only make one set-state call at the beginning of the game.
- $\mathsf{NROB}_{0\mathsf{set}}(\gamma^*, \gamma_{\max}, \beta)$ is $\mathsf{NROB}(\gamma^*, \gamma_{\max}, \beta)$ in which the attacker may not make any set-state calls.

We define the corresponding security notions in the natural way (See Definition 3), and we call them respectively *robustness against resets*, *robustness against one* set-state, and *robust without* set-state.

4 The Generalized Fortuna Construction

At first, it might seem hopeless to build an RNG with input that can recover from compromise in the presence of premature next calls, since output from a compromised RNG can of course reveal information about the (low-entropy) state. Surprisingly, Ferguson and Schneier presented an elegant away to get around this issue in their Fortuna construction [10]. Their idea is to have several "pools of entropy" and a special "register" to handle next calls. As input that potentially has some entropy comes into the RNG, any entropy "gets accumulated" into one pool at a time in some predetermined sequence. Additionally, some of the pools may be used to update the register. Intuitively, by keeping some of the entropy away from the register for prolonged periods of time, we hope to allow one pool to accumulate enough entropy to guarantee security, even if the adversary makes arbitrarily many premature next calls (and therefore potentially learns the entire

---

[2] Intuitively, this game captures security against an attacker that can cause a machine to reboot.

state of the register). The hope is to schedule the various updates in a clever way such that this is guaranteed to happen, and in particular Ferguson and Schneier present an informal analysis to suggest that Fortuna realizes this hope in their "constant rate" model (in which the entropy γ_i of each input is an unknown constant).

In this section, we present a generalized version of Fortuna in our model and terminology. In particular, while Fortuna simply uses a cryptographic hash function to accumulate entropy and implicitly assumes perfect entropy accumulation, we will explicitly define each pool as an RNG with input, using the robust construction from [5] (and simply a standard PRG as the register). And, of course, we do not make the constant-rate assumption. We also explicitly model the choice of input and output pools with a new object that we call a scheduler, and we define the corresponding notion of scheduler security. In addition to providing a formal model, we achieve strong improvements over Fortuna's implicit scheduler.

As a result, we prove formally in the standard model that the generalized Fortuna construction is premature-next robust when instantiated with a number of robust RNGs with input, a secure scheduler, and a secure PRG.

4.1   Schedulers

Definition 4. *A scheduler is a deterministic algorithm SC that takes as input a key* skey *and a state* $\tau \in \{0,1\}^m$ *and outputs a new state* $\tau' \in \{0,1\}^m$ *and two pool indices,* in, out $\in \mathbb{N} \cup \{\bot\}$.

We call a scheduler keyless if there is no key. In this case, we simply omit the key and write $SC(\tau)$. We say that SC has P pools if for any skey and any state τ, if $(\tau', \text{in}, \text{out}) = SC(\text{skey}, \tau)$, then in, out $\in [0, P-1] \cup \{\bot\}$.

Given a scheduler SC with skey and state τ, we write

$$SC^q(\text{skey}, \tau) = (\text{in}_j(SC, \text{skey}, \tau), \text{out}_j(SC, \text{skey}, \tau))_{j=1}^q \tag{2}$$

to represent the sequence obtained by computing $(\text{in}, \text{out}, \tau) \leftarrow SC(\text{skey}, \tau)$ repeatedly, a total of q times. When SC, skey, and τ are clear or implicit, we will simply write in_j and out_j. We think of in_j as a pool that is to be "filled" at time j and out_j as a pool to be "emptied" immediately afterwards. When $\text{out}_j = \bot$, no pool is emptied.

For a scheduler with P pools, we define security game $\mathsf{SGAME}(P, q, w_{\max}, \alpha, \beta)$ as in Figure 4. In the security game, there are two adversaries, a sequence sampler $\mathcal{E}$ and an attacker $\mathcal{A}$. We think of the sequence sampler $\mathcal{E}$ as a simplified version of the distribution sampler $\mathcal{D}$ that is only concerned with the entropy estimates $(\gamma_i)_{i=1}^q$. $\mathcal{E}$ simply outputs a sequence of weights $(w_i)_{i=1}^q$ with $0 \leq w_i \leq w_{\max}$, where we think of the weights as normalized entropies $w_i = \gamma_i/\gamma^*$ and $w_{\max} = \gamma_{\max}/\gamma^*$.

The challenger chooses a key skey at random. Given skey and $(w_i)_{i=1}^q$, $\mathcal{A}$ chooses a start state for the scheduler τ_0, resulting in the sequence $(\text{in}_i, \text{out}_i)_{i=1}^q$.

```
proc. SGAME
w_1, ..., w_q ← E
skey ←$ {0,1}^|skey|
τ_0 ← A(skey, (w_i)_{i=1}^q)
(in_i, out_i)_{i=1}^q ← SC^q(skey, τ_0)
c ← 0; c_0 ← 0, ..., c_{P-1} ← 0; T* ← 0
FOR T in 1, ..., q,
    IF w_T > w_max, THEN OUTPUT 0
    c ← c + w_T; c_{in_T} ← c_{in_T} + w_T
    IF out ≠ ⊥,
        IF c_{out_T} ≥ 1, THEN OUTPUT 0
        ELSE c_{out_T} ← 0
    IF c ≥ α
        IF T* = 0, THEN T* ← T
        IF T ≥ β·T*, THEN OUTPUT 1
OUTPUT 0
```

Fig. 4. SGAME$(P, q, w_{\max}, \alpha, \beta)$, the security game for a scheduler $\mathcal{SC}$

Each pool has an accumulated weight c_j, initially 0, and the pools are filled and emptied in sequence; on the T-th step, the weight of pool in_T is increased by w_T and pool out_T is emptied (its weight set to 0), or no pool is emptied if $\mathsf{out} = \perp$. If at some point in the game a pool whose weight is at least 1 is emptied, the adversary loses. (Remember, 1 here corresponds to γ^*, so this corresponds to the case when the underlying RNG recovers.) We say that such a pool is a *winning pool at time T against* $(\tau_0, (w_i)_{i=1}^q)$. On the other hand, the adversary wins if $\sum_{i=1}^{T^*} w_i \geq \alpha$ and the game reaches the $(\beta \cdot T^*)$-th step (without the challenger winning). Finally, if neither event happens, the adversary loses.

Definition 5 (Scheduler security). *A scheduler $\mathcal{SC}$ is $(t, q, w_{\max}, \alpha, \beta, \varepsilon)$-*secure *if it has P pools and for any pair of adversaries $\mathcal{E}, \mathcal{A}$ with cumulative run-time t, the probability that $\mathcal{E}, \mathcal{A}$ win game* SGAME$(P, q, w_{\max}, \alpha, \beta)$ *is at most ε. We call $r = \alpha \cdot \beta$ the* competitive ratio *of $\mathcal{SC}$.*[3]

We note that schedulers are non-trivial objects. Indeed, in the full version of the paper [6], we prove the following lower bounds, which in particular imply that schedulers can only achieve superconstant competitive ratios $r = \alpha \cdot \beta$.

Theorem 1. *Suppose that $\mathcal{SC}$ is a $(t, q, w_{\max}, \alpha, \beta, \varepsilon)$-secure scheduler running in time $t_{\mathcal{SC}}$. Let $r = \alpha \cdot \beta$ be the competitive ratio. Then, if $q \geq 3$, $\varepsilon < 1/q$, $t = \Omega(q \cdot (t_{\mathcal{SC}} + \log q))$, and $r \leq w_{\max}\sqrt{q}$, we have that*

[3] The intuition for the competitive ratio $r = \alpha \cdot \beta$ (which will be explicit in Section 6) comes from the case when the sequence sampler $\mathcal{E}$ is restricted to constant sequences $w_i = w$. In that case, r bounds the ratio between the time taken by $\mathcal{SC}$ to win and the time taken to receive a total weight of one.

$$r > \log_e q - \log_e(1/w_{\max}) - \log_e \log_e q - 1 \,, \qquad (3)$$

$$\alpha > \frac{w_{\max}}{w_{\max}+1} \cdot \frac{\log_e(1/\varepsilon) - 1}{\log_e \log_e(1/\varepsilon) + 1} \,. \qquad (4)$$

4.2 The Composition Theorem

Our generalized Fortuna construction consists of a scheduler SC with P pools, P entropy pools $\mathcal{G}_i$, and register ρ. The $\mathcal{G}_i$ are themselves RNGs with input and ρ can be thought of as a much simpler RNG with input which just gets uniformly random samples. On a refresh call, Fortuna uses SC to select one pool $\mathcal{G}_{\text{in}}$ to update and one pool $\mathcal{G}_{\text{out}}$ from which to update ρ. next calls use only ρ.

Formally, we define a generalized Fortuna construction as follows: Let SC be a scheduler with P pools and let pools $\mathcal{G}_i = (\text{setup}_i, \text{refresh}_i, \text{next}_i)$ for $i = 0, \ldots, P-1$ be RNGs with input. For simplicity, we assume all the RNGs have input length p and output length ℓ, and the same setup procedure, $\text{setup}_i = \text{setup}_{\mathcal{G}}$. We also assume $\mathbf{G} : \{0,1\}^\ell \to \{0,1\}^{2\ell}$ is a pseudorandom generator (without input). We construct a new RNG with input $\mathcal{G}(SC, (\mathcal{G}_i)_{i=0}^{P-1}, \mathbf{G}) = (\text{setup}, \text{refresh}, \text{next})$ as in Figure 5.

---

proc. setup :
$\text{seed}_{\mathcal{G}} \overset{\$}{\leftarrow} \text{setup}_{\mathcal{G}}()$
$\text{skey} \overset{\$}{\leftarrow} \{0,1\}^{|\text{skey}|}$
OUTPUT $\text{seed} = (\text{skey}, \text{seed}_{\mathcal{G}})$

proc.next(seed, S) :
PARSE $\left(\tau, S_\rho, (S_i)_{i=0}^{P-1}\right) \leftarrow S$
$(S_\rho, R) \leftarrow \mathbf{G}(S_\rho)$
OUTPUT $\left(S = \left(\tau, S_\rho, (S_i)_{i=0}^{P-1}\right), R\right)$

proc. refresh(seed, S, I) :
PARSE $(\text{skey}, \text{seed}_{\mathcal{G}}) \leftarrow \text{seed}$
PARSE $\left(\tau, S_\rho, (S_i)_{i=0}^{P-1}\right) \leftarrow S$
$(\tau, \text{in}, \text{out}) \leftarrow SC(\text{skey}, \tau)$
$S_{\text{in}} \leftarrow \text{refresh}_{\text{in}}(\text{seed}_{\mathcal{G}}, S_{\text{in}}, I)$
$(S_{\text{out}}, R) \leftarrow \text{next}_{\text{out}}(\text{seed}_{\mathcal{G}}, S_{\text{out}})$
$S_\rho \leftarrow S_\rho \oplus R$
OUTPUT $S = \left(\tau, S_\rho, (S_i)_{i=0}^{P-1}\right)$

Fig. 5. The generalized Fortuna construction

We prove the following composition theorem in the full version of this paper [6].

Theorem 2. *Let $\mathcal{G}$ be an RNG with input constructed as above. If the scheduler SC is a $(t_{SC}, q_D, w_{\max}, \alpha, \beta, \varepsilon_{SC})$-secure scheduler with P pools and state length m, the pools $\mathcal{G}_i$ are $((t, q_D, q_R = q_D, q_S), \gamma^*, \varepsilon)$-robust RNGs with input and the register $\mathbf{G}$ is $(t, \varepsilon_{\text{prg}})$-pseudorandom generator, then $\mathcal{G}$ is $((t', q_D, q_R', q_S), \alpha \cdot \gamma^*, w_{\max} \cdot \gamma^*, \varepsilon', \beta)$-premature-next robust where $t' \approx \min(t, t_{SC})$ and $\varepsilon' = q_D^2 \cdot q_S \cdot (q_D \cdot \varepsilon_{SC} + P \cdot 2^m \cdot \varepsilon + q_R' \varepsilon_{\text{prg}})$.*

For our weaker security notions, we achieve better ε':

- *For $\text{NROB}_{\text{reset}}$, $\varepsilon' = q_D^2 \cdot q_S \cdot (q_D \cdot \varepsilon_{SC} + P \cdot \varepsilon + q_R' \varepsilon_{\text{prg}})$.*
- *For $\text{NROB}_{\text{1set}}$, $\varepsilon' = q_D \cdot \varepsilon_{SC} + P \cdot 2^m \cdot \varepsilon + q_R' \varepsilon_{\text{prg}}$.*
- *For $\text{NROB}_{\text{0set}}$, $\varepsilon' = q_D \cdot \varepsilon_{SC} + P \cdot \varepsilon + q_R' \varepsilon_{\text{prg}}$.*

5   Instantiating the Construction

5.1   A Robust RNG with Input

Recall that our construction of a premature-next robust RNG with input still requires a robust RNG with input. We therefore present [5]'s construction of such an RNG.

Let $\mathbf{G} : \{0,1\}^m \to \{0,1\}^{n+\ell}$ be a (deterministic) pseudorandom generator where $m < n$. Let $[y]_1^m$ denote the first m bits of $y \in \{0,1\}^n$. The [5] construction of an RNG with input has parameters n (state length), ℓ (output length), and $p = n$ (sample length), and is defined as follows:

- setup(): Output seed $= (X, X') \leftarrow \{0,1\}^{2n}$.
- $S' = \mathsf{refresh}(S, I)$: Given seed $= (X, X')$, current state $S \in \{0,1\}^n$, and a sample $I \in \{0,1\}^n$, output: $S' := S \cdot X + I$, where all operations are over $\mathbb{F}_{2^n}$.
- $(S', R) = \mathsf{next}(S)$: Given seed $= (X, X')$ and a state $S \in \{0,1\}^n$, first compute $U = [X' \cdot S]_1^m$. Then output $(S', R) = \mathbf{G}(U)$.

Theorem 3 ([5, Theorem 2]). *Let $n > m, \ell, \gamma^*$ be integers and $\varepsilon_{\mathsf{ext}} \in (0,1)$ such that $\gamma^* \geq m + 2\log(1/\varepsilon_{\mathsf{ext}}) + 1$ and $n \geq m + 2\log(1/\varepsilon_{\mathsf{ext}}) + \log(q_\mathcal{D}) + 1$. Assume that $\mathbf{G} : \{0,1\}^m \to \{0,1\}^{n+\ell}$ is a deterministic $(t, \varepsilon_{\mathsf{prg}})$-pseudorandom generator. Let $\mathcal{G} = (\mathsf{setup}, \mathsf{refresh}, \mathsf{next})$ be defined as above. Then $\mathcal{G}$ is a $((t', q_\mathcal{D}, q_R, q_S), \gamma^*, \varepsilon)$-robust RNG with input where $t' \approx t$, $\varepsilon = q_R(2\varepsilon_{\mathsf{prg}} + q_\mathcal{D}^2 \varepsilon_{\mathsf{ext}} + 2^{-n+1})$.*

Dodis et al. recommend using AES in counter mode to instantiate their PRG, and they provide a detailed analysis of its security with this instantiation. (See [5, Section 6.1].) We notice that our construction only makes next calls to their RNG during our refresh calls, and Ferguson and Schneier recommend limiting the number of refresh calls by simply allowing a maximum of ten per second [10]. They therefore argue that it is reasonable to set $q_\mathcal{D} = 2^{32}$ for most security cases (effectively setting a time limit of over thirteen years). So, we can plug in $q_\mathcal{D} = q_R = q_S = 2^{32}$.

In this setting, guidelines of [5, Section 6.1] show that their construction can provide a pseudorandom 128-bit string after receiving γ_0^* bits of entropy with γ_0^* in the range of 350 to 500, depending on the desired level of security.

5.2   Scheduler Construction

To apply Theorem 2, we still need a secure scheduler (as defined in Section 4.1). Our scheduler will be largely derived from Ferguson and Schneier's Fortuna construction [10], but improved and adapted to our model and syntax. In our terminology, Fortuna's scheduler $\mathcal{SC}_\mathcal{F}$ is keyless with $\log_2 q$ pools, and its state is a counter τ. The pools are filled in a "round-robin" fashion. Every $\log_2 q$ steps, Fortuna empties the maximal pool i such that 2^i divides $\tau / \log_2 q$.

$\mathcal{SC}_\mathcal{F}$ is designed to be secure against some unknown but *constant* sequence of weights $w_i = w.$[4] We modify Fortuna's scheduler so that it is secure against

---

[4] We analyze their construction against constant sequences more carefully in Section 6.

```
proc. SC(skey, τ) :
IF τ ≠ 0 mod P/w_max, THEN out ← ⊥
ELSE out ← max{out : τ = 0 mod 2^out · P/w_max}
in ← F(skey, τ)
τ' ← τ + 1 mod q
OUTPUT (τ', in, out)
```

Fig. 6. Our scheduler construction

arbitrary (e.g., not constant) sequence samplers by replacing the round-robin method of filling pools with a pseudorandom sequence.

Assume for simplicity that $\log_2 \log_2 q$ and $\log_2(1/w_{max})$ are integers. We let $P = \log_2 q - \log_2 \log_2 q - \log_2(1/w_{max})$. We denote by skey the key for some pseudorandom function $\mathbf{F}$ whose range is $\{0, \ldots, P - 1\}$. Given a state $\tau \in \{0, \ldots, q - 1\}$ and a key skey, we define $SC(\text{skey}, \tau)$ formally in Figure 6. In particular, the input pool is chosen pseudorandomly such that $\text{in} = \mathbf{F}(\text{skey}, \tau)$. When $\tau = 0 \mod P/w_{max}$, the output pool is chosen such that out is maximal with $2^{\text{out}} \cdot P/w_{max}$ divides τ. (Otherwise, there is no output pool.)

The following theorem is proven in the full version of this paper [6].

Theorem 4. *If the pseudorandom function $\mathbf{F}$ is $(t, q, \varepsilon_{\mathbf{F}})$-secure, then for any $\varepsilon \in (0, 1)$, the scheduler SC defined above is $(t', q, w_{max}, \alpha, \beta, \varepsilon_{SC})$-secure with $t' \approx t$, $\varepsilon_{SC} = q \cdot (\varepsilon_{\mathbf{F}} + \varepsilon)$, $\alpha = 2 \cdot (w_{max} \cdot \log_e(1/\varepsilon) + 1) \cdot (\log_2 q - \log_2 \log_2 q - \log_2(1/w_{max}))$, and $\beta = 4$.*

Remark. *Note that we set $P = \log_2 q - \log_2 \log_2 q - \log_2(1/w_{max})$ for the sake of optimization. In practice, $w_{max} = \gamma_{max}/\gamma^*$ may be unknown, in which case we can safely use $\log_2 q - \log_2 \log_2 q$ pools at a very small cost. So, one can safely instantiate our scheduler (and the corresponding RNG) without a known bound on w_{max}, and still benefit if w_{max} happens to be low in practice.*

Instantiation and Concrete Numbers. To instantiate the scheduler in practice, we suggest using AES as the PRF $\mathbf{F}$. As in [5], we ignore the computational error term $\varepsilon_{\mathbf{F}}$ and set $\varepsilon_{SC} \approx q\varepsilon$.[5] In our application, our scheduler will be called only on refresh calls to our generalized Fortuna RNG construction, so we again set $q = 2^{32}$. It seems reasonable for most realistic scenarios to set $w_{max} = \gamma_{max}/\gamma^* \approx 1/16$ and $\varepsilon_{SC} \approx 2^{-192}$, but we provide values for other w_{max} and ε as well:

ε_{SC}	q	w_{max}	α	P	ε_{SC}	q	w_{max}	α	P	ε_{SC}	q	w_{max}	α	P
2^{-128}	2^{32}	1/64	115	21	2^{-192}	2^{32}	1/64	144	21	2^{-256}	2^{32}	1/64	174	21
2^{-128}	2^{32}	1/16	367	23	2^{-192}	2^{32}	1/16	494	23	2^{-256}	2^{32}	1/16	622	23
2^{-128}	2^{32}	1/4	1445	25	2^{-192}	2^{32}	1/4	2000	25	2^{-256}	2^{32}	1/4	2554	25
2^{-128}	2^{32}	1	6080	27	2^{-192}	2^{32}	1	8476	27	2^{-256}	2^{32}	1	10,871	27

---

[5] See [5] for justification for such an assumption.

5.3   Putting It All Together

Now, we have all the pieces to build an RNG with input that is premature-next robust (by Theorem 2). Again setting $q = 2^{32}$ and assuming $w_{\max} = \gamma_{\max}/\gamma^* \approx 32/500 \approx 1/16$, our final scheme can output a secure 128-bit key in four times the amount of time that it takes to receive roughly 20 to 30 kilobytes of entropy.

6   Constant-Rate Adversaries

We note that the numbers that we achieve in Section 5.3 are not ideal. But, our security model is also very strong. So, we follow Ferguson and Schneier [10] and consider the weaker model in which the distribution sampler $\mathcal{D}$ is restricted to a constant entropy rate. Analysis in this model suggests that our construction may perform much in practice. Indeed, if we think of the distribution sampler $\mathcal{D}$ as essentially representing nature, this model may not be too unreasonable.

Constant-Rate Model. We simply modify our definitions in the natural way. We say that a distribution (resp., sequence) sampler is *constant* if, for all i, $\gamma_i = \gamma$ (resp., $w_i = w$) for all i for some fixed γ (resp., w). We say that a scheduler is $(t, q, w_{\max}, r, \varepsilon)$-*secure against constant sequences* if, for some[6] α, β such that $\alpha \cdot \beta = r$ it is $(t, q, w_{\max}, \alpha, \beta, \varepsilon)$-secure when the sequence sampler $\mathcal{E}$ is required to be constant. When $\varepsilon = 0$ and the adversaries are allowed unbounded computation (as is the case in our construction), we simply leave out the parameters t and ε. We similarly define premature-next robustness for RNGs with input.

In the full version [6], we the note that our composition theorem, Theorem 2, applies equally well in the constant-rate case. This allows us to construct an RNG with input that is premature-next robust against constant adversaries with much better parameters.

Optimizing Fortuna's Scheduler. Ferguson and Schneier essentially analyze the security of a scheduler that is a deterministic version of our scheduler from Section 5.2, with pseudorandom choices replaced by round-robin choices [10]. (This is, of course, where we got the idea for our scheduler.) They conclude that it achieves a competitive ratio of $2 \log_2 q$. However, the correct value is $3 \log_2 q$.[7] Ferguson and Schneier's model differs from ours in that they do not consider adversarial starting times τ_0 between the emptying of pools. Taking this (important) consideration into account, it turns out that $\mathcal{SC}_\mathcal{F}$ achieves a competitive ratio of $r_\mathcal{F} = 3.5 \log_2 q$ in our model.[8]

Interestingly, the pseudocode in [10] actually describes a potentially stronger scheduler than the one that they analyzed. Instead of emptying just

---

[6] We note that when the sequence sampler $\mathcal{E}$ must be constant, $(t, q, w_{\max}, \alpha, \beta, \varepsilon)$-security is equivalent to $(t, q, w_{\max}, \alpha', \beta', \varepsilon)$-security if $\alpha \cdot \beta = \alpha' \cdot \beta'$.

[7] There is an attack: Let $w = 1/(2^i + 1)$ and start Fortuna's counter so that pool $i+1$ is emptied after $2^i \cdot \log_2 q$ steps. Clearly, $\mathcal{SC}_\mathcal{F}$ takes $(2^i + 2^{i+1}) \cdot \log_2 q = 3 \cdot 2^i \cdot \log_2 q$ total steps to finish, achieving a competitive ratio arbitrarily close to $3 \log_2 q$.

[8] This follows from the analysis of our own scheduler in the full version [6].

pool i, this new scheduler empties *each* pool j with $j \leq i$. Although Ferguson and Schneier did not make use of this in their analysis, we observe that this would lead to significantly improved results provided that the scheduler could "get credit" for all the entropy from *multiple* pools. While our model cannot syntactically capture the notion of multiple pools being emptied at once, we notice that it can simulate a multiple-pool scheduler by simply treating any set of pools that is emptied together at a given time as one new pool.

In the full version of this paper, we make this observation concrete and further optimize the scheduler of Fortuna to obtain the following result [6].

Theorem 5. *For any integer $b \geq 2$, there exists a keyless scheduler SC_b that is $(q, w_{\max}, r_b)$-secure against constant sequences where*

$$r_b = \left(b + \frac{w_{\max}}{b} + \frac{1 - w_{\max}}{b^2} \right) \cdot \left(\log_b q - \log_b \log_b q - \log_b(1/w_{\max}) \right) . \qquad (5)$$

In particular, with $w_{\max} = 1$ and $q \to \infty$, $b = 3$ is optimal with

$$r_3 \approx 2.1 \log_2 q \approx \frac{r_F}{1.66} \approx \frac{r_2}{1.19} \approx \frac{r_4}{1.01} . \qquad (6)$$

We note that SC_b performs even better in the non-asymptotic case. For example, in the case that Ferguson and Schneier analyzed, $q = 2^{32}$ and $w_{\max} = 1$, we have $r_3 \approx 58.2 \approx \frac{r_F}{1.9}$, saving almost half the entropy compared to Fortuna.

References

1. Barak, B., Halevi, S.: A model and architecture for pseudo-random generation with applications to /dev/random. In: Proceedings of the 12th ACM Conference on Computer and Communications Security, CCS 2005, pp. 203–212. ACM, New York (2005)
2. Barker, E., Kelsey, J.: Recommendation for random number generation using deterministic random bit generators. NIST Special Publication 800-90A (2012)
3. Bellare, M., Rogaway, P.: The security of triple encryption and a framework for code-based game-playing proofs. In: Vaudenay, S. (ed.) EUROCRYPT 2006. LNCS, vol. 4004, pp. 409–426. Springer, Heidelberg (2006)
4. CVE-2008-0166. Common Vulnerabilities and Exposures (2008)
5. Dodis, Y., Pointcheval, D., Ruhault, S., Vergniaud, D., Wichs, D.: Security analysis of pseudo-random number generators with input: /dev/random is not robust. In: Proceedings of the 2013 ACM SIGSAC Conference on Computer Communications Security, CCS 2013, pp. 647–658. ACM, New York (2013)
6. Dodis, Y., Shamir, A., Stephens-Davidowitz, N., Wichs, D.: How to eat your entropy and have it too – optimal recovery strategies for compromised rngs. Cryptology ePrint Archive, Report 2014/167 (2014), http://eprint.iacr.org/
7. Dorrendorf, L., Gutterman, Z., Pinkas, B.: Cryptanalysis of the random number generator of the windows operating system. IACR Cryptology ePrint Archive 2007, 419 (2007)
8. Eastlake, D., Schiller, J., Crocker, S.: RFC 4086 - Randomness Requirements for Security (June 2005)

9. Ferguson, N.: Private communication (2013)
10. Ferguson, N., Schneier, B.: Practical Cryptography, 1st edn. John Wiley & Sons, Inc., New York (2003)
11. Gutterman, Z., Pinkas, B., Reinman, T.: Analysis of the linux random number generator. In: Proceedings of the 2006 IEEE Symposium on Security and Privacy, SP 2006, pp. 371–385. IEEE Computer Society, Washington, DC (2006)
12. Heninger, N., Durumeric, Z., Wustrow, E., Halderman, J.A.: Mining your Ps and Qs: Detection of widespread weak keys in network devices. In: Proceedings of the 21st USENIX Security Symposium (August 2012)
13. Information technology - Security techniques - Random bit generation. ISO/IEC18031:2011 (2011)
14. Kelsey, J., Schneier, B., Ferguson, N.: Yarrow-160: Notes on the design and analysis of the yarrow cryptographic pseudorandom number generator. In: Heys, H.M., Adams, C.M. (eds.) SAC 1999. LNCS, vol. 1758, pp. 13–33. Springer, Heidelberg (2000)
15. Kelsey, J., Schneier, B., Wagner, D., Hall, C.: Cryptanalytic attacks on pseudorandom number generators. In: Vaudenay, S. (ed.) Fast Software Encryption, FSE 1998. LNCS, vol. 1372, pp. 168–188. Springer, Heidelberg (1998)
16. Killmann, W., Schindler, W.: A proposal for: Functionality classes for random number generators. AIS 20 / AIS31 (2011)
17. Lacharme, P., Röck, A., Strubel, V., Videau, M.: The linux pseudorandom number generator revisited. IACR Cryptology ePrint Archive 2012, 251 (2012)
18. Lenstra, A.K., Hughes, J.P., Augier, M., Bos, J.W., Kleinjung, T., Wachter, C.: Public keys. In: Safavi-Naini, R., Canetti, R. (eds.) CRYPTO 2012. LNCS, vol. 7417, pp. 626–642. Springer, Heidelberg (2012)
19. Nguyen, Shparlinski: The insecurity of the digital signature algorithm with partially known nonces. Journal of Cryptology 15(3), 151–176 (2002)
20. Sahai, A., Vadhan, S.P.: A complete problem for statistical zero knowledge. J. ACM 50(2), 196–249 (2003)
21. Wikipedia. /dev/random (2004), http://en.wikipedia.org/wiki//dev/random (accessed February 09, 2014)

Cryptography with Streaming Algorithms*

Periklis A. Papakonstantinou and Guang Yang

Institute for Theoretical Computer Science,
Tsinghua University,
Beijing 100084, China

Abstract. We put forth the question of whether cryptography is feasible using streaming devices. We give constructions and prove lower
bounds. In streaming cryptography (not to be confused with stream-
ciphers) everything—the keys, the messages, and the seeds—are huge
compared to the internal memory of the device. These streaming algorithms have small internal memory size and make a constant number
of passes over big data maintained in a constant number of read/write
external tapes. Typically, the internal memory size is $O(\log n)$ and we
use 2 external tapes; whereas 1 tape is provably insufficient. In this setting we cannot compute instances of popular intractability assumptions.
Nevertheless, we base cryptography on these assumptions by employing
non-black-box techniques, and study its limitations.

We introduce new techniques to obtain unconditional lower bounds
showing that no super-linear stretch pseudorandom generator exists, and
no Public Key Encryption (PKE) exists with private-keys of size sublinear in the plaintext length.

For possibility results, assuming the existence of one-way functions
computable in NC^1—e.g. factoring, lattice assumptions—we obtain
streaming algorithms computing one-way functions and pseudorandom
generators. Given the Learning With Errors (LWE) assumption we construct PKE where both the encryption and decryption are streaming
algorithms. The starting point of our work is the groundbreaking work
of Applebaum-Ishai-Kushilevitz on Cryptography in NC^0. In the end,
our developments are technically orthogonal to their work; e.g. there is
a PKE where the decryption is a streaming algorithm, whereas no PKE
decryption can be in NC^0.

Keywords: streaming, lower bound, big data, randomized encoding,
non-black-box, PRG, PKE.

1 Introduction

In most cryptosystems the keys can be assumed to reside in a local memory
provided with unlimited access. What if access to the keys is not for free? Suppose

---

* This work was supported in part by the National Basic Research Program of China
Grant 2011CBA00300, 2011CBA00301, the National Natural Science Foundation of
China Grant 61033001, 61350110536, 61361136003.

J.A. Garay and R. Gennaro (Eds.): CRYPTO 2014, Part II, LNCS 8617, pp. 55–70, 2014.
© International Association for Cryptologic Research 2014

that the key is very long and, together with everything else in the input, is stored as a stream that can be sequentially scanned only a few times. Is it possible to compute cryptographically secure functions in this way?

More formally, we consider the possibility of cryptography against arbitrary polynomial time adversaries, who are as powerful as usual, using a (less powerful) streaming algorithm which has access to a bounded internal memory, and to external read/write tapes (*RW streams*) where we quantify on the number of passes over them. These RW streams are commonly thought (see e.g. [19,17] and references within) to correspond to hard disk drives or other sequentially accessed buffers. The question of cryptography using streaming devices is motivated in practice in settings where the keys and messages are huge (e.g. authenticating big data), whereas theoretically it falls in the fundamental study of cryptography with rudimentary resources.

Below, we give an overview of our results, then we discuss related work and several subtleties of streaming cryptography, and finally we compare to previous work in randomized encodings.

Our results. For the rest of this paper and unless mentioned otherwise, a *streaming algorithm* has 2 RW streams (external tapes), over which it makes $O(1)$ many passes, and it uses $O(\log n)$ internal memory size[1] for input length n. These are optimal parameters, within constants, under which streaming cryptography may exist (we show that 1 read/write stream is not sufficient). We devise streaming constructions of private- and public-key primitives by synthesizing various previous works along with new techniques necessary for streaming. This possibility is quite unexpected for distinct reasons. We also introduce technically novel machinery and study the limitations of private- and public-key cryptography in this setting.

Impossibility results. We show the impossibility of super-linear stretch streaming pseudorandom generator, and we also obtain a linear (in the security parameter) lower bound on the key-size of streaming PKE. The proof technique is inspired by [19,17]. However, a cryptography lower bound is *not for a specific function* as per usual streaming lower bounds. It must hold for all cryptographically secure functions realizing the same primitive.

Possibility results. Given that one-way functions exist in NC^1, e.g. based on factoring or lattice assumptions, we construct one-way functions and pseudorandom generators by streaming algorithms that use: internal memory of size $O(\log n)$, 2 RW streams (one contains the input and the other one is auxiliary), 5 passes in total for one-way functions and 7 for pseudorandom generators.

Starting from the Learning With Errors (LWE) assumption [8] and based on the constructions in [8,7] we construct an Indistinguishable under Chosen-Plaintext Attack (IND-CPA) secure (or semantically secure [21]) Public-Key

---

[1] Logarithmic memory size precludes uninteresting trivialities that can happen for size $\omega(\log n)$, when one assumes the existence of very hard functions. In this case, in principle the question is not about streaming.

Encryption (PKE), where both the encryption and the decryption are streaming algorithms. This is mainly a feasibility result. Improved key-lengths, and a CCA-secure PKE are very interesting open questions.

For the existence of streaming one-way functions and pseudorandom generators, the assumption can be relaxed to existence of one-way functions in Logspace. For PKE we rely on LWE, a concrete intractability assumption.

Relation to previous work. The construction of streaming one-way functions from an arbitrary one-way function in NC^1 relies on Barrington's characterization [20]. In particular, computing a boolean NC^1 function reduces in some very local sense to computing the product $\sigma_1\sigma_2\cdots\sigma_m$ of permutations from $Sym(5)$, the symmetric group over 5 letters. Kilian [23] encoded $\sigma = \sigma_1\cdots\sigma_m$ as $\hat\sigma = \langle\sigma_1 r_1, r_1^{-1}\sigma_2 r_2, \ldots, r_{m-1}^{-1}\sigma_m\rangle$ for uniformly random chosen r_i's. Then, given $\hat\sigma$ we can efficiently "decode" σ, whereas no additional information about $\sigma_1, \cdots, \sigma_m$ can be extracted. The seminal work of Applebaum-Ishai-Kushilevitz [10,9,11,3] provides a paradigm for dealing with general forms of similar *randomized encodings*.

Constructing a pseudorandom generator is more complex. One could have tried to implement as a streaming algorithm the steps of the celebrated [14] construction. Indeed, such a streaming pseudorandom generator is *non-trivially* achievable given certain entropy parameters of hashed values. But, there is no obvious way to streaming-compute these values, neither can it be circumvented by creating many copies as in [14]. We instead adapt [5,16] that bypass this obstacle and also buys us efficiency over [14]. Both [5] and [16] can be non-trivially modified into streaming algorithms. We use [5] because it is simpler and gives better parameters.

Regarding Public Key Encryption systems, we base our construction on [8,7], where the original constructions are not streaming computable.

Why streaming cryptography is not immediate from concrete hardness assumptions, such as lattice assumptions? By modifying [17] we see that multiplying a matrix by a vector requires $\Omega(\log n)$ many passes if the number of streams is constant and the internal memory logarithmic. This limitation is circumvented by taking randomized encodings of NC^1 computations of such functions (these are non-black-box constructions since the computation itself is encoded). We note that [12] ruled out families of black-box streaming cryptography constructions and it conjectured impossibility of streaming cryptography with a constant number of passes, which we refute in this paper. Thus, the possibility of streaming cryptography is unexpected.

One more reason that makes streaming cryptography counter-intuitive is that no single-stream algorithm with internal memory size $O(\log n)$ and $O(1)$-many passes computes a one-way function. However, by adding a second stream we can bring the number of passes down to a constant, and this strongly contrasts folk wisdom[2] in streaming computation.

---

[2] It was believed that for common types of functions, if when adding a second tape helps then permuting the input in the single-stream model will help as well. But a permuted one-way function is also one-way.

The multiple read/write (RW) stream model we consider here is closely related to the reversals-parameterized Turing Machines [2,18], except that we only make forward scans. To the best of our knowledge, before our work for $o(\log n)$ many passes in the literature there were only lower bounds, e.g. [19,17]. This multiple stream model generalizes the single-stream model, aka "online model", which has been scrutinized for quite a while. In the study of randomness [15] gives lower bounds in the single-stream model as well as constructions for online universal hash functions, extractors, and condensers. Also, a restricted form of the single-stream model with read-only (RO) input was studied in [13,6].

Table 1. Cryptography with Streaming Algorithms vs Cryptography in NC^0

	Streaming Model	NC^0
one-way function pseudorandom generator (PRG)	✓	✓
PKE (**Enc & Dec**)	✓	×
linear-stretch PRG	?	from Alekhnovitch's assumption [22]
super-linear-stretch PRG	×	?

Streaming Cryptography vs NC^0 and Locality. Streaming cryptography and [10,9,1,11,3] rely on randomized encodings, but they are incomparable in a number of places. There are obvious things streaming algorithms can do (e.g. sample almost uniformly from $\mathsf{Sym}(5)$) but NC^0 cannot, whereas, generally, NC^0 functions with underlying dependency graphs of $\mathsf{poly}(n)$ treewidth cannot be computed by streaming algorithms. This holds in particular for circuits associated with cellular automata (CA) as they appear in [3], where the treewidth is $\Omega(\sqrt{n})$. Furthermore, there are concrete technical separations between streaming and highly parallel cryptography. For example, IND-CPA secure PKE with streaming encryption and decryption exists whereas no NC^0 decryption is possible (and for AC^0 is still open). The CAs constructions are based on the concrete DRLC assumption (see Section 7), whereas even for [1] (these CAs make a single step that makes them a special case of constant input/output locality circuits [1]) it is impossible to start from general encodings. Our streaming private-key primitives are from generic assumptions.

2   Preliminaries

We use capital bold letters, e.g. $\mathbf{A}$, to denote matrices, and use lower case bold letters, e.g. $\mathbf{x}$, for column vectors, and correspondingly $\mathbf{x}^T$ for row vectors. Let $\mathbb{Z}_q := \mathbb{Z}/q\mathbb{Z} = \{0, 1, 2, \cdots, q-1\}$ be the ring of integers with addition and multiplication modulo q.

Probability distributions are denoted by calligraphic letters, e.g. $\mathcal{D}$. We use $x \leftarrow \mathcal{D}$ to denote that x is sampled from $\mathcal{D}$, and $x \in_R S$ when x is sampled uniformly from the set S. $\mathcal{U}_n$ denotes the uniform distribution over $\{0,1\}^n$ and $\mathcal{U}_S$ is the uniform distribution over the set S.

In this work all complexity classes are function classes (but we prefer to write e.g. NC^0 instead of FNC^0). Logspace denotes the set of all functions computable by a Turing Machine (transducer) with a read-only input, $O(\log n)$ large working tape, and a write-only output tape.

For $i \in \mathbb{Z}^{\geq 0}$, NC^i denotes the set of functions computable by families of poly-size boolean circuits with constant fan-in gates and $O(\log^i n)$ depth for input length n. A family of circuits is (log-space) uniform if there is a (log-space) Turing machine such that on input 1^n, $n \in \mathbb{Z}^{\geq 0}$, it generates the description of the corresponding circuit in the family.

An (s, p, t) *streaming algorithm* is a Turing machine that has internal memory $s(n)$, t-many unbounded external RW streams which we can scan from left to right for $p(n)$ passes. Unless mentioned otherwise for a *streaming algorithm* $s(n) = O(\log n)$, $t = 2$, and $p = O(1)$. A function is *streaming computable* if it can be computed by a streaming algorithm. An *oblivious* streaming algorithm is one where the head movement and the internal memory depend only on the time step.

Private and public key primitives. $f : \{0,1\}^* \to \{0,1\}^*$ is a (T, ϵ)-*secure one-way function* for $T = T(n), \epsilon = \epsilon(n)$ if f is polynomial time computable and for all sufficiently large n, for every time T randomized algorithm $\mathcal{A}$, we have $\Pr_{y \leftarrow f(\mathcal{U}_n)}[f(\mathcal{A}(y)) = y] < \epsilon$.

A polynomial time computable function $G : \{0,1\}^* \to \{0,1\}^*$ is a (T, ϵ)-*pseudorandom generator* if $\forall x, |G(x)| > |x|$ and $G(\mathcal{U}_n)$ is (T, ϵ)-*pseudorandom*, i.e. for sufficiently large n and for every time T randomized algorithm $\mathcal{D}$, there is $\left| \Pr[\mathcal{D}(G(\mathcal{U}_n)) = 1] - \Pr[\mathcal{D}(\mathcal{U}_{|G(1^n)|}) = 1] \right| < \epsilon$. For simplicity, we omit (T, ϵ) for computational security, i.e. when $T = n^{\omega(1)}, \epsilon = n^{-\omega(1)}$.

A *public-key encryption (PKE) system* consists of three polynomial time algorithms **KeyGen**, **Enc**, **Dec**, for key-generation, encryption and decryption respectively, where the key-generation and the encryption are probabilistic. (i) **KeyGen** takes the security parameter 1^n as input and it generates a public encryption key $\mathcal{PK}$ and a private decryption key $\mathcal{SK}$; (ii) **Enc** outputs a ciphertext c on input $(\mathcal{PK}, m)$, for every m drawn from the message space; (iii) **Dec** takes $(\mathcal{SK}, c)$ as input to decrypt m from c with overwhelming probability over the random choices of **Enc**. We say that a PKE system is streaming computable if both **Enc** and **Dec** are streaming algorithms.

In a PKE system, IND-CPA security is defined in the following security experiment as a game between a Challenger and an Adversary:

- The challenger runs **KeyGen** and uses its random choices to generate a public $\mathcal{PK}$ and a private $\mathcal{SK}$ key, and reveals the $\mathcal{PK}$ to the adversary.
- The adversary chooses two equal-length messages $\mathbf{x}_0$ and $\mathbf{x}_1$, and sends them to the challenger.
- The challenger flips an unbiased coin $b \in_R \{0,1\}$, computes $\mathbf{c} = \mathbf{Enc}(\mathcal{PK}, \mathbf{x}_b)$ and gives the ciphertext $\mathbf{c}$ to the adversary.
- The adversary outputs $b' \in \{0,1\}$ based on $\mathcal{PK}$ and $\mathbf{c}$, and it wins if and only if $b' = b$.

Here the adversary is a probabilistic polynomial time algorithm. The PKE is *IND-CPA secure* if for every $k \in \mathbb{R}$, we have $\Pr[b' = b] < \frac{1}{2} + \frac{1}{N^k}$, when $N = |\mathbf{x}_i|$ is sufficiently large.

Randomized Encoding. For a function $f : \{0,1\}^n \to \{0,1\}^m$, the function $\hat{f} : \{0,1\}^n \times \{0,1\}^\rho \to \{0,1\}^{m'}$ is a *randomized encoding* of f if the following conditions hold

1. For every $x \in \{0,1\}^n$, the output distribution $\hat{f}(x, \mathcal{U}_\rho)$ uniquely determines a $f(x)$, i.e. $\hat{f}(x,r) \neq \hat{f}(x',r')$ for any r,r', as long as $f(x) \neq f(x')$.
2. The output distribution is fully determined by the encoded value $f(x)$, i.e. if $f(x) = f(x')$ then $\hat{f}(x,\mathcal{U}_\rho)$ and $\hat{f}(x',\mathcal{U}_\rho)$ are identically.
3. $|\rho| = \mathsf{poly}(n)$ and there are $\mathsf{poly}(n)$-time algorithms to decode $f(x)$ from any sample in $\hat{f}(x,\mathcal{U}_\rho)$, and to sample from $\hat{f}(x,\mathcal{U}_\rho)$ when given $f(x)$.

Intuitively, (1) means that $\hat{f}(x,r)$ contains all information about $f(x)$, and (2) asserts that $\hat{f}(x,\mathcal{U}_\rho)$ reveals no extra information about x other than the value of $f(x)$. Putting these two together we have that if f is a one-way function then $\hat{f}$ is also one-way [10].

3 Warm-Up: How to Construct Streaming One-Way Functions?

We present a generic compiler (Section 3.2) that maps every $f \in \mathsf{NC}^1$ to its streaming randomized encoding $\hat{f}$. Due to a very useful coincidence regarding the specific encoding we use, and after a little "massaging" we get $\hat{f}$ computable with 2 streams (the reader is encouraged to think ahead to see where the issue is). Corollary 1 immediately follows by [10].

Theorem 1. *Every function $f \in \mathsf{NC}^1$ has a randomized encoding function $\hat{f}$ which is oblivious streaming computable with 5 passes.*

Corollary 1. *A streaming one-way function exists if a one-way function exists in* Logspace.

Here is an advanced remark. The construction in the proof of Theorem 1 relies on a specific randomized encoding that also causes a polynomial blow-up compared to the regular output of Barrington (see below). unavoidable (for this technique) and why the AIK encoding [10] cannot be used.

3.1 Background: NC^1 to Width-5 Branching Programs

Let us now recall the definition of a bounded-width permutation branching program.

Definition 1. *A width-w permutation branching program is a sequence of $m = m(n)$ instructions $B_n = (s_1, \langle j_1, \sigma_1, \tau_1 \rangle) \cdots (s_m, \langle j_m, \sigma_m, \tau_m \rangle)$, where for every $1 \leq i \leq m$, $j_i \in \{1, \cdots, n\}$, $\sigma_i, \tau_i \in \mathsf{Sym}(w)$. Here $\mathsf{Sym}(w)$ refers to the group of permutations over $[w] = \{1, 2, \cdots, w\}$. On input $x = (x_1, \cdots, x_n) \in \{0,1\}^n$, B_n is evaluated as $B_n(x) = s_1 \cdot s_2 \cdot \cdots \cdot s_m$, where $s_i = \sigma_i$ if $x_{j_i} = 1$ and $s_i = \tau_i$ if $x_{j_i} = 0$.*

A function $f : \{0,1\}^n \to \{0,1\}$ is recognized by B_n if there exists a cycle $\theta \in \mathsf{Sym}(w)$, such that $\forall x \in \{0,1\}^n$, $B_n(x) = \theta$ when $f(x) = 1$, and $B_n(x) = e$ is the identity permutation when $f(x) = 0$.

Everything in the following theorem holds as well for log-space uniform branching programs.

Theorem 2 (Barrington's Theorem [20]). *Any boolean function f computable by a family of depth d and fan-in 2 circuits can be recognized by a family of width-5 permutation branching programs for $m \leq 4^d$. In particular, $m = \mathsf{poly}(n)$ for $f \in \mathsf{NC}^1$ and input length n.*

Thus, evaluating $f : \{0,1\}^n \to \{0,1\}$ on input x reduces to deciding whether $B_n(x) = s_1 s_2 \cdots s_m$ is the identity. Define $\hat{f} : \{0,1\}^n \times \mathsf{Sym}(5)^{m-1} \to \mathsf{Sym}(5)^m$ as $\hat{f}(x; r) = \langle \pi_1, \cdots, \pi_m \rangle = \langle s_1 r_1, r_1^{-1} s_2 r_2, \cdots, r_{m-2}^{-1} s_{m-1} r_{m-1}, r_{m-1}^{-1} s_m \rangle$, where $r_i \in_R \mathsf{Sym}(5)$, s_i follows the i-th instruction in B_n and m is the length of B_n. Then, $\hat{f}$ is a randomized encoding of f, since $\langle \pi_1, \cdots, \pi_m \rangle$ uniformly distributes over $\mathsf{Sym}(5)^m$ conditioned on $\pi_1 \pi_2 \cdots \pi_m = s_1 s_2 \cdots s_m$.

We define **Sample** : $\{0,1\}^q \to \mathsf{Sym}(5)$ to be the algorithm that samples $r_i \in_R \mathsf{Sym}(5)$ within statistical distance $2^{-\Omega(q)}$ using $q = q(n)$ (read-once) random bits. Then, every permutation in $\mathsf{Sym}(5)$ is identified by its unique binary ID from $\{0,1\}^7$. Thus, $\hat{f}$ is represented in binary as $\widehat{f} : \{0,1\}^n \times (\{0,1\}^q)^{m-1} \to (\{0,1\}^7)^m$ that induces a loss of at most $2^{-\Omega(q(n))}$ in the output distribution. It remains to make $\widehat{f}$ streaming computable. The issue is that non-consecutive s_i's may be arbitrarily associated with the same input bit, so we must do something about this.

3.2   Streaming Computable Randomized Encoding

Our streaming algorithm is based on the following observations:

- fixing any poly-time invertible permutation ψ over $\{1, \ldots, m\}$, $g(x; r) = \langle \pi_{\psi(1)}, \cdots, \pi_{\psi(m)} \rangle$ is a one-way function as long as $\hat{f}$ is a one-way function, recalling that $\hat{f}(x; r) = \langle \pi_1, \cdots, \pi_m \rangle$;
- a permutation branching program (e.g. B_n) recognizes exactly the same function after inserting *dummy instructions* like $(s, < j, e, e >)$; that is, $(s_1, \langle j_1, \sigma_1, \tau_1 \rangle) \cdots (s_m, \langle j_m, \sigma_m, \tau_m \rangle)$ recognizes exactly the same function as $(s_1, \langle j_1, \sigma_1, \tau_1 \rangle) \cdots (s, < j, e, e >) \cdots (s_m, \langle j_m, \sigma_m, \tau_m \rangle)$.

Due to space limitations we omit the (not hard) full proof of Theorem 1. Here is a sufficiently detailed outline. By the second observation, we may replace in B_n, the length m branching program that recognizes f, the first instruction $(s_1, \langle j_1, \sigma_1, \tau_1 \rangle)$ with $(s_1', \langle 1, e, e \rangle) \cdots (s_{j_1}', \langle j_1, \sigma_1, \tau_1 \rangle) \cdots (s_n', \langle n, e, e \rangle)$, i.e. $s_{j_1}' = s_1$, whereas $s_i' = e$ for $i \neq j_1$, so that $s_1' \cdots s_n' = s_1$ for every input. The advantage of the new instructions is that $\forall i \in \{1, 2, \ldots, n\}$, s_i' depends on exactly x_i. With similar tricks for $s_2, \cdots, s_m$, we get a length $mn = \mathsf{poly}(n)$ new branching program B_n' and $s_1', \cdots, s_{mn}'$ with oblivious input dependency. In what follows, we use B_n and s_i instead of B_n' and s_i' for simplicity.

In a single pass, we compute the s_i's in the order: $s_1, s_{n+1}, \cdots, s_{mn-n+1}$, $s_2, s_{n+2}, \cdots, s_{mn-n+2}, \cdots \cdots, s_n, s_{2n}, \cdots, s_{mn}$ (sorted by their dependency on x, which coincide the subscripts modular n). Then, for $f : \{0,1\}^n \to \{0,1\}$, we apply the first observation to construct the oblivious streaming computable randomized encoding $\widehat{f} : \{0,1\}^n \times (\{0,1\}^q)^{mn-1} \to (\{0,1\}^7)^{mn}$ as follows

$$\widehat{f}(x; y_1, \ldots, y_{mn-1}) =$$
$$\left\langle s_1 r_1,\ r_n^{-1} s_{n+1} r_{n+1}, \cdots, r_{mn-n}^{-1} s_{mn-n+1} r_{mn-n+1} \cdots, \cdots, r_{mn-1}^{-1} s_{mn} \right\rangle$$

where $r_i = \mathbf{Sample}(y_i)$, r_i^{-1} is the inverse of r_i, and s_i is a function of $x_{(i \bmod n)}$ for $i = 1, 2, \cdots, mn$.

When $f(x) = \langle f_1(x), f_2(x), \cdots, f_{\ell(n)}(x) \rangle$ has $\ell(n)$ output bits, we design $\widehat{f} = \langle \widehat{f}_1, \cdots, \widehat{f}_{\ell(n)} \rangle$, which consists of an individual randomized encoding $\widehat{f}_i$ one for each f_i. It is not too hard to globally rearrange the output bits and obtain the final streaming computable function $\widehat{f}$.

4   Streaming Pseudorandom Generators

The encoding in Section 3.2 does not preserve pseudorandomness, simply because $2^7 \nmid |\mathsf{Sym}(5)|$. In fact, Barrington's theorem holds also for the non-solvable $\mathsf{Sym}(w)$, $w \geq 5$ but there is no $k \in \mathbb{Z}$ such that $2^k | (w!)$. Yet, we provide a rather technical adaptation of [5] to build streaming pseudorandom generators from any streaming one-way function f.

Theorem 3. *Let $f : \{0,1\}^n \to \{0,1\}^m$ be a streaming one-way function. Then, there is a streaming computable pseudorandom generator G requiring 2 additional passes to the streaming algorithm for $\langle f^{(1)}, \cdots, f^{(\ell t)} \rangle$, for ℓ, t defined as below*
Moreover, if $m = O(n)$, then $\ell = \frac{n}{\log n}, t = O(n^2 \log^2 n)$, and the seed length of G is $O(n^6 \log^3 n)$.

In fact, the construction in Section 3.2 gives an oblivious streaming one-way function, which implies that evaluating polynomial many copies of f does not need more passes than f.

There are four steps in the [5] construction: 1) next-block pseudoentropy generation; 2) entropy equalization; 3) converting Shannon entropy to min-entropy and amplifying the gap; 4) randomness extraction. The first three steps remain intact for our streaming algorithm, whereas the fourth has to be modified.

In a nutshell, the first step constructs the generator $G_{nb}(\mathbf{s}) = \langle f(\mathbf{s}), \mathbf{s} \rangle$ for a random seed $s \leftarrow \mathcal{U}_n$. For notation convenience, let $f : \{0,1\}^n \to \{0,1\}^{m-n}$, so that $G_{nb} : \{0,1\}^n \to \{0,1\}^m$. The second step concatenates the outputs of G_{nb} on ℓ independent seeds to get $z^{(1)}, \ldots, z^{(\ell)}$, and randomly shifts (by discarding from the head and the tail) blocks to convert total entropy into the entropy in individual blocks, via $EQ : \{0,1\}^{\log m} \times (\{0,1\}^m)^\ell \to \{0,1\}^{m_\ell}$ for $m_\ell = m(\ell-1)$ and

$$EQ\left(j, z^{(1)}, \cdots, z^{(\ell)}\right) := \left\langle z_j^{(1)}, \cdots, z_m^{(1)}, \cdots, z_1^{(\ell)}, \cdots, z_{j-1}^{(\ell)} \right\rangle$$

Let $\mathcal{X} := EQ\left(J, G_{nb}(\mathcal{U}_n^{(1)}), \cdots, G_{nb}(\mathcal{U}_n^{(\ell)})\right)$ for $J = \mathcal{U}_{\log m}$, the third step concatenates t independent copies of $\mathcal{X}$ within each block (we do this step virtually by allowing non-consecutive cells in one block), i.e. $\mathcal{Y} = \left\langle \left(\mathcal{X}_1^{(1)}, \cdots, \mathcal{X}_1^{(t)}\right), \cdots, \left(\mathcal{X}_{m_\ell}^{(1)}, \cdots, \mathcal{X}_{m_\ell}^{(t)}\right)\right\rangle$, thus every block has high pseudo-min-entropy conditioned on previous blocks. The details for these steps can be found in [5,16].

Now, let us give an informal but accurate description of the part that has to be non-trivially modified for the streaming construction. The fourth step requires evaluating a single random universal hash function $\mathbf{H} : \{0,1\}^t \to \{0,1\}^{\alpha_t - \log^2 n}$ on every block of $\mathcal{Y}$ to extract randomness, where α_t is the next-block pseudo-min-entropy of each block. This step is difficult since streaming algorithms cannot re-read the code of $\mathbf{H}$. To that end, we prove that each block can be associated with a different linear hash function $\mathbf{H}^{(i)} \in \{0,1\}^{t \times (\alpha_t - \log^2 n)}$. Then, in the streaming algorithm we use $\widehat{\mathbf{H}^{(i)}}$, the randomized encoding of $\mathbf{H}^{(i)}$, to extract randomness. Let $\mathbf{H}_j^{(i)}$ be the j-th row of $\mathbf{H}^{(i)}$, and $\mathbf{R}_j^{(i)}$ the j-th row of the random input $\mathbf{R}^{(i)} \in \{0,1\}^{(t-1) \times (\alpha_t - \log^2 n)}$, then $\widehat{\mathbf{H}^{(i)}}\left(\left(\mathcal{X}_i^{(1)}, \cdots, \mathcal{X}_i^{(t)}\right), \mathbf{R}^{(i)}\right) = \left\langle \mathcal{X}_i^{(1)} \mathbf{H}_1^{(i)} + \mathbf{R}_1^{(i)}, \ldots, \mathcal{X}_i^{(j)} \mathbf{H}_j^{(i)} + \mathbf{R}_{j-1}^{(i)} + \mathbf{R}_j^{(i)}, \ldots, \mathcal{X}_i^{(t)} \mathbf{H}_t^{(i)} + \mathbf{R}_{t-1}^{(i)} \right\rangle$ where all the additions are modular 2. Combining the four steps, G appears as follows

$$G\left(\mathbf{J}^{(1)}, \cdots, \mathbf{J}^{(t)}, \mathcal{U}_n^{(1)}, \cdots, \mathcal{U}_n^{(t\ell)}; \mathbf{H}; \mathbf{R}\right)$$
$$= \Big\langle \mathbf{H}, \; \mathcal{X}_1^{(1)} \mathbf{H}_1^{(1)} + \mathbf{R}_1^{(1)}, \qquad\qquad \cdots, \mathcal{X}_{m_\ell}^{(1)} \mathbf{H}_1^{(m_\ell)} + \mathbf{R}_1^{(m_\ell)},$$
$$\cdots, \mathcal{X}_1^{(j)} \mathbf{H}_j^{(1)} + \mathbf{R}_{j-1}^{(1)} + \mathbf{R}_j^{(1)}, \qquad \cdots, \mathcal{X}_{m_\ell}^{(j)} \mathbf{H}_j^{(m_\ell)} + \mathbf{R}_{j-1}^{(m_\ell)} + \mathbf{R}_j^{(m_\ell)},$$
$$\cdots, \mathcal{X}_1^{(t)} \mathbf{H}_t^{(1)} + \mathbf{R}_{t-1}^{(1)}, \qquad\qquad \cdots, \mathcal{X}_{m_\ell}^{(t)} \mathbf{H}_t^{(m_\ell)} + \mathbf{R}_{t-1}^{(m_\ell)} \Big\rangle$$

Note that we use a family of linear hash functions because it is not clear how to implement with streaming algorithms the description-succinct hash family in [5,16]. This causes loss in efficiency (which contrasts the purpose of [5,16]), but here we strive for a streaming feasibility result which is not at all obvious how to get. Theorems 1 and 3, and [10] yield:

Corollary 2. *If there is a one-way function in* Logspace, *then there exists a pseudorandom generator which is streaming computable with 7 passes.*

5   Limitations for Super-Linear Stretch Pseudorandom Generators

We devise a new lower bounding methodology, which a central technical contribution of this work. We first use this to show that streaming computable super-linear stretch pseudorandom generators do not exist. Note that $n^{1-\varepsilon}$ stretch is easy to achieve by running in parallel $n^{1-\varepsilon}$ copies of a single-bit stretch pseudorandom generator on independent seeds.

Theorem 4. *Suppose $\ell(n) = \omega(n)$ and $G : \{0,1\}^n \to \{0,1\}^{\ell(n)}$ is a pseudorandom generator. Then, no streaming algorithm can compute G.*

We prove Theorem 4 by analyzing the information flow in the computation, and by partitioning appropriately the output into blocks, we upper bound the entropy transferred to each output block from the input. Intuitively, a single block cannot collect much entropy and therefore it cannot induce large stretch.

Our proof makes use of the following observation and the concept of a dependency graph originally introduced in [19,17] (in fact, [24]). We tailor them for cryptographic applications to partition the computation into $p + 1$ phases corresponding to p passes.

Observation 5 ([24]). *When a tape cell is written, its content only depends on the internal memory and the t cells currently being scanned by the heads of the external streams. Moreover, those t cells are written before this pass, since no cell can be visited twice before making a new pass.*

5.1   Dependency Graphs and Dependency Trees

First, we provide the definition of dependency graph. Due to space limitations we omit concrete examples and diagrams of dependency graphs and trees.

Definition 2. *Fix a streaming algorithm G which on input x it makes $\leq p$ passes over t external tapes. The dependency graph, denoted by $\Gamma(x)$, is a directed graph with $p + 1$ levels. Each beginning of a new pass (at any tape) is associated with a distinct level. The i-th level in $\Gamma(x)$ contains all nodes labeled (v, i) if the tape cell v has ever been visited before the i-th pass begins. We assume that all input cells are written at the beginning. (e.g. $\{(v, 1) \mid v$ is an input cell$\}$ contains exactly all the nodes in level 1, and $\{(v, p+1) \mid v$ is written in the computation of G on input x $\}$ for level $p + 1$.) $\Gamma(x)$ has edges $(u, i) \to (v, i+1)$ iff there is a head reading (u, i) when $(v, i+1)$ is being written. Furthermore, there is always an edge $(u, i) \to (u, i+1)$ as long as (u, i) is in $\Gamma(x)$.*

In the dependency graph, each level represents a single phase in the computation. Therefore the nodes (except for those at level 1) have in-degree at most t, while all edges are heading to the next level. Intuitively, those directed edges depict the information flow excluding the internal memory.

We also remark it possible that old passes are not yet finished when a new pass begins. In this case, old passes will be processed in the new level.

5.2 Overview of the Lower-Bound

We first introduce the definition of *blocks*. Intuitively, blocks are used to package the entropy from the input, and the dependency of blocks describes the information flow during the computation, except for the information carried in the bounded internal memory. By analyzing the dependency of blocks we have devised an elegant information-theoretic way of upper bounding the amount of entropy in part of the output.

Definition 3. *A block is an equivalence class consisting of all nodes corresponding to tape cells at the same level on the same tape such that they depend on exactly the same set of blocks at the previous level. Specifically, an input block refers to a set of nodes at the first level corresponding to consecutive tape cells on the input tape.*

Note that if two cells, from the same tape and the same level, have the same dependency on blocks at the previous level, then any cell in between would have exactly the same dependency, because the dependency changes "monotonically". Therefore, our partition of blocks is well-defined such that every block consists of only consecutive tape cells.

Then, we partition the input x into b input blocks as $x = (x_1, x_2, \cdots, x_b)$, and use a corollary of Proposition 3.1 in [19] to bound the number of blocks.

Proposition 1. *Partition x into b input blocks and let $\Gamma(x)$ be the dependency graph. Then, the number of blocks at level i in $\Gamma(x)$ is bounded $\leq (b+1)t^{i-1}$, where t denotes the number of tapes.*

Due to space limitation, here we give only a proof sketch of Theorem 4.

Proof (Proof Sketch of Theorem 4). By Proposition 1, there are at most $(b+1)t^p = O(b)$ blocks at level $p+1$ because both t, p are constants.

We consider $G : \{0,1\}^n \to \{0,1\}^{\ell(n)}$ where $\ell(n) = \omega(n)$. Partition the input equally into $b = \lceil n/\log n \rceil$ blocks, so that each input block has length $\leq \log n$. Recalling that there are $O(b)$ output blocks, there is an output block v with $\ell(n)/O(b) = \omega(\log n)$ many bits in expectation.

However, every output block depends on $O(1)$ input blocks when t, p are both constants. That is, the block v has expected length $\omega(\log n)$, while it only receives $O(\log n)$ bits of entropy from $O(1)$ input blocks plus another $O(\log n)$ bits from the internal memory. This immediately suggests an advised distinguisher $\mathcal{D}_A$:

$\mathcal{D}_A$: DISTINGUISHING $G(\mathcal{U}_n)$ FROM $\mathcal{U}_{\ell(n)}$ (with advice A):

1 For all $z \in A$, check whether z is a sub-string of the input;
2 If find any $z \in A$ in the input, output 1;
3 Otherwise output 0.

The advice A is a poly(n) long list containing all strings that are sufficiently long (i.e. $\Omega(\ell(n)/b) = \omega(\log n)$) and could appear in the block v. $\mathcal{D}_A(G(\mathcal{U}_n)) = 1$

if the block v is sufficiently long to be captured by A, which happens with probability $\Omega(1/b)$ by Markov's inequality. On the other hand, $\Pr[\mathcal{D}_A(\mathcal{U}_{\ell(n)}) = 1] < \mathsf{poly}(n) \cdot \ell(n) \cdot 2^{-\Omega(\ell(n)/b)} = 2^{-\omega(n)}$. Therefore

$$\Pr\left[\mathcal{D}_A\big(G(\mathcal{U}_n)\big) = 1\right] - \Pr\left[\mathcal{D}_A(\mathcal{U}_{\ell(n)}) = 1\right] \geq \Omega(\frac{1}{b}) - 2^{-\omega(n)} = \Omega(\log n/n)$$

For a uniform distinguisher, the advice A is efficiently generated as follows: enumerate every input block and internal memory on every possible dependency tree of v, then simulate the computation of v, and add only sufficiently long output substrings to the list. Although suffering from a polynomial blow-up than optimal, such advice A suffices for the above argument of $\mathcal{D}_A$.

6   Public-Key Encryption in the Streaming Model

We construct an IND-CPA secure PKE system based on Regev's LWE assumption together with the PKE construction in [8], where the encryption and the decryption algorithms are streaming algorithms, henceforth called *streaming PKE*. The private keys contain a good deal of redundancy. We also show that large private-keys are necessary. A lower bound on the length of the private key is given in Theorem 7 (using the technique introduced in Section 5), though there is still a gap with our construction.

Theorem 6. *Given the decision-LWE assumption (Assumption 1), the construction in Section 6.1 is an IND-CPA secure PKE. Moreover, both the encryption and the decryption algorithms are streaming computable.*

Theorem 7. *For every IND-CPA secure PKE whose decryption scheme is a streaming algorithm, the private-key has length $\Omega(N)$, where N is the length of the plaintext.*

The main challenge of a streaming PKE is the decryption algorithm. The techniques we developed so far do not apply, because the decryption algorithm should output exactly the plaintext rather than any code.

We construct our streaming PKE based on the decision-LWE assumption. The intuition of such assumption is exposited in [8], which also gives reductions from worst-case lattice problems (by now these lattice assumptions and reductions are common place).

Definition 4 (LWE problem). *Let $q = q(n) \leq \mathsf{poly}(n)$, consider a list of equations $b_i = \langle \mathbf{s}, \mathbf{a}_i \rangle + e_i \pmod{q}$ for $i = 1, 2, \cdots, \mathsf{poly}(n)$, where $\mathbf{s} \in \mathbb{Z}_q^n$, $\mathbf{a}_i \in_R \mathbb{Z}_q^n$ and $b_i \in \mathbb{Z}_q$. If furthermore $e_i \in \mathbb{Z}_q$ follows a discrete Gaussian distribution[3] with parameter α, we denote by $\mathsf{search\text{-}LWE}_{q,\alpha}$ the problem of recovering $\mathbf{s}$ from such equations. In $\mathsf{decision\text{-}LWE}_{q,\alpha}$ the goal is to distinguish $(\mathbf{a}, \langle \mathbf{s}, \mathbf{a} \rangle + e \mod q)$ from $\mathcal{U}_{\mathbb{Z}_q^{n+1}}$ with non-negligible advantage, when both $\mathbf{s}, \mathbf{a} \in_R \mathbb{Z}_q^n$.*

---

[3] A discrete Gaussian distribution over $\mathbb{Z}_q$ is defined by $D_{\mathbb{Z}_q,\alpha}(x) = \rho_\alpha(x/q)/\rho_\alpha(\mathbb{Z}_q)$, where $\rho_\alpha(x) = \sum_{k=-\infty}^{\infty} \alpha^{-1} \exp(-\pi(\frac{x+k}{\alpha})^2)$ follows a continuous Gaussian distribution, and $\rho_\alpha(\mathbb{Z}_q) = \sum_{x \in \mathbb{Z}_q} \rho_\alpha(x/q)$.

Assumption 1 (cf. [8,7]). *When $\alpha \geq 2\sqrt{n}$, search-$\mathsf{LWE}_{q,\alpha}$ cannot be solved in probabilistic polynomial time with non-negligible probability. If $\alpha \geq \omega(\sqrt{n}\log n)$ then decision-$\mathsf{LWE}_{q,\alpha}$ cannot be solved in probabilistic polynomial time with non-negligible advantage.*

6.1   The Construction

In our construction the public and private keys are "streaming useable" forms of the following two matrices: $\mathbf{A}$ and a random matrix $\mathbf{D}$. Matrix $\mathbf{A}$ is statistically close to uniform, and at the same time orthogonal to $\begin{bmatrix} \mathbf{I} \\ \mathbf{D} \end{bmatrix}$. The latter consists of short vectors which cannot be retrieved from a uniformly random matrix (this is the lattice hardness assumption).

KeyGen: Pick a matrix $\mathbf{D} \in \mathbb{Z}_p^{(m-w)\times w}$ uniformly at random from $\{0, \pm 1\}^{(m-w)\times w}$. Uniformly at random pick $\overline{\mathbf{A}} \in \mathbb{Z}_q^{n\times(m-w)}$, and compute $\mathbf{A} \in \mathbb{Z}_q^{n\times m}$ as $\mathbf{A} = [-\overline{\mathbf{A}}\mathbf{D} \mid \overline{\mathbf{A}}] \bmod q$. Let $\begin{bmatrix} \mathbf{I} \\ \mathbf{D} \end{bmatrix} = [\mathbf{d}_1, \cdots, \mathbf{d}_w]$. Here $k = \lceil 2\log n\rceil, q = 2^k, m = 3nk, w = nk$, for the security parameter n.

Output N copies of $\mathbf{A}$ as the public key, and nN copies of $\mathbf{d}_1, \cdots, \mathbf{d}_w$ as the private key. Each copy of $\mathbf{A}$ is written in row-first order, i.e. $(a_{11}, a_{12}, \cdots, a_{1m}, a_{21}, \cdots, a_{2m}, \cdots, a_{n1}, \cdots, a_{nm})$.

Enc: On input $\mathbf{x} = (x^{(1)}, \cdots, x^{(N)}) \in \{0,1\}^N$, for $i = 1, 2, \cdots, N$, uniformly choose $\mathbf{s}_i \in_R \mathbb{Z}_q^n$ and $\mathbf{x}_i \in_R \{qx^{(i)}/2\} \times \mathbb{Z}_q^{w-1}$.

Sample $\mathbf{e}_i \in \mathbb{Z}_q^m$ for $i = 1, 2, \cdots, N$, where each entry $e_{ij} \sim \mathcal{D}_\alpha$ follows the discrete Gauss distribution with mean 0 and standard deviation α, for $j = 1, 2, \cdots, m$.

For every $i = 1, 2, \cdots, N$, sequentially output y_i, where y_i is a randomized encoding of $\mathbf{s}_i^T\mathbf{A} + \mathbf{e}_i^T + (\mathbf{x}_i^T, 0) \bmod q$. That is, for $\mathbf{R} \in_R \mathbb{Z}_q^{(n-1)\times m}$, realizing $\mathbf{e}_i^T, (\mathbf{x}_i^T, 0)$ as $1 \times m$ row vectors, and recalling that $\mathbf{A}$ is an $n \times m$ matrix, we define y_i is the row-first order of $\mathbf{Y}_i$ as follows

$$\mathbf{Y}_i = \begin{bmatrix} s_{i1} & & \\ & \ddots & \\ & & s_{in} \end{bmatrix} \cdot \mathbf{A} + \begin{bmatrix} \mathbf{R} \\ (\mathbf{x}^T, 0) \end{bmatrix} + \begin{bmatrix} \mathbf{e}_i^T \\ -\mathbf{R} \end{bmatrix}$$

Dec: Given the ciphertext $\{y_i\}_{i=1,2,\cdots,N}$ and the decryption key nN copies of $\mathbf{d}_1$. We compute $b = [1\ 1\ \cdots\ 1]_{1\times n}\mathbf{Y}_i\mathbf{d}_1 \bmod q$ and output $x^{(i)} = \lfloor 2b/q + 1/2\rfloor \bmod 2$ for every $i = 1, \cdots, N$.

Comparison with [8]. The above construction is similar to the PKE construction in [8]. We borrow from [7] the key generation and encryption algorithms which enable us to turn them into streaming computable encryption/decryption. Note that [7], unlike us, achieves a CCA-secure PKE. Currently, we do not know how to perform ciphertext validity checks (as in e.g. [7]) in a streaming fashion.

This Public-Key Encryption scheme is statistically correct and IND-CPA secure, and it has both encryption and decryption in a streaming fashion.

7 Conclusions and Some Remarks on Practical Constructions

Our work leaves open the possibility of streaming cryptography for a number of popular private and public-key primitives. As a next step we propose to study the streaming possibility for the following cryptographic primitives: (i) linear-stretch pseudorandom generators, (ii) CCA-secure PKE, (iii) signature schemes, and (iv) message authentication.

It is also open whether the number of passes we achieve (see Table 2 below) are optimal, and also simultaneously improve the seed-efficiency of streaming pseudorandom generators from NC^1 one-way functions. For example, our generic streaming one-way function is done with 5 passes, whereas when starting from a concrete assumption (see below) we can do it with 4, which is optimal.

Some remarks on practicality. Randomized encodings generally use huge amounts of randomness (typically $\Omega(n^4)$) for input length n, and thus our generic compilers can be understood as feasibility results. In practice, starting from concrete intractability assumptions we can do much better. Here is a practical example which in fact resembles a lot the one in [3] (but a few model-specific differences – our model is not two dimensional but things are arranged similarly).

Assumption 2 (Decoding Random Linear Codes (DRLC)). *A random linear code f_{code} is defined as $f_{code} : (\mathbf{A}, \mathbf{x}, \mathbf{e}) \mapsto (\mathbf{A}, \mathbf{A}\mathbf{x} + \mathbf{e})$, where $\mathbf{A} \in \mathsf{GF}(2)^{m \times n}$, $\mathbf{x} \in \mathsf{GF}(2)^n$, $\mathbf{e} \in \mathsf{GF}(2)^m$. Choose positive constants κ, ϵ, δ such that $\kappa = \frac{n}{m} < 1 - H_2\big((1 + \epsilon)\delta\big)$, where $H_2(p) = -p \log_2 p - (1 - p) \log_2(1 - p)$ for $p < 1/2$ and $H_2(p) = 1$ otherwise. If $\mathbf{A}, \mathbf{x}$ are chosen uniformly at random, while $\mathbf{e}$ has at most $\frac{\delta m}{2}$ one-entries, then f_{code} is a one-way function.*

Theorem 8. *Suppose that the DRLC assumption holds true. Then, there exists a one-way function F computable by a streaming algorithm with 2 streams, 4 passes and $O(\log n)$ internal memory. Furthermore, if the DRLC input is of size N the corresponding input size for F is $n \leq 2N$.*

Proof (Construction outline). Suppose the random bits $(r_{11}, r_{21}, \cdots, r_{mn})$ are given on the extra stream (this is without loss of generality/not necessary), and parse the input stream as $(x_1, a_{11}, a_{21}, \cdots, a_{m1}, \cdots, x_n, a_{1n}, \cdots, a_{mn}, e_1, \cdots, e_m)$.

In the first pass (over two streams) we compute $(a_{11}x_1 + r_{11}, \cdots, a_{m1}x_1 + r_{m1}, a_{12}x_2 + r_{12}, \cdots, a_{m2}x_2 + r_{m2}, \cdots, a_{1n}x_n + r_{1n}, \cdots, a_{mn}x_n + r_{mn}, e_1, \cdots, e_m)$.

In the next pass we compute $(a_{11}x_1 + r_{11}, \cdots, a_{m1}x_1 + r_{m1}, a_{12}x_2 + r_{12} - r_{11}, \cdots, a_{m2}x_2 + r_{m2} - r_{m1}, \cdots, a_{1n}x_n + r_{1n} - r_{1(n-1)}, \cdots, a_{mn}x_n + r_{mn} - r_{m(n-1)}, e_1 - r_{1n}, \cdots, e_m - r_{mn})$. Thus, a randomized encoding of $\mathbf{A}\mathbf{x} + \mathbf{e}$ is computed with 4 passes over 2 streams.

Table 2. OWF & PRG from any OWF in Logspace; PKE from LWE

	# of passes	external tapes
one-way function	5	1 RO & 1 RW
pseudorandom generator	7	2 RW
	15	1 RO & 1 RW
PKE **Enc**	5	1 RO & 1 RW
PKE **Dec**	2	key & cipher in different tapes

We conclude with a note on the practicality of the multi-stream model. One physical analog of a stream is a hard-disk or a disk-array. Although it makes sense to think of physical disks to be of size $2n$ or $3n$, for an input of length n, under no stretch of imagination n^3 is reasonable size. For more than one stream we believe that this *stream-size* parameter should be added to the other parameters: number of streams, number of passes, internal memory size. In this paper all constructions make ≤ 9 passes and the stream size never exceeds $2 \times$ input length. In practice, though the stream size is even more important and in the sense that perhaps we might be able to tolerate slightly super-constant many passes given that the stream size stays linear throughout the computation.

References

1. Applebaum, B., Ishai, Y., Kushilevitz, E.: Cryptography with constant input locality. Journal of Cryptology, 429–469; In: Menezes, A. (ed.) CRYPTO 2007. LNCS, vol. 4622, pp. 92–110. Springer, Heidelberg (2007)
2. Chen, J., Yap, C.-K.: Reversal complexity. SIAM Journal on Computing 20(4), 622–638 (1991)
3. Applebaum, B., Ishai, Y., Kushilevitz, E.: Cryptography by Cellular Automata or How Fast Can Complexity Emerge in Nature? In: ICS, pp. 1–19 (2010)
4. Impagliazzo, R., Levin, L.A., Luby, M.: In: Symposium on Theory of Computing (STOC), pp. 12–24 (1989)
5. Vadhan, S.P., Zheng, C.J.: Characterizing pseudoentropy and simplifying pseudorandom generator constructions. In: Symposium on Theory of Computing (STOC), pp. 817–836 (2012)
6. Yu, X., Yung, M.: Space Lower-Bounds for Pseudorandom-Generators. In: Structure in Complexity Theory Conference, pp. 186–197 (1994)
7. Micciancio, D., Peikert, C.: Trapdoors for lattices: Simpler, tighter, faster, smaller. In: Pointcheval, D., Johansson, T. (eds.) EUROCRYPT 2012. LNCS, vol. 7237, pp. 700–718. Springer, Heidelberg (2012)
8. Regev, O.: On lattices, learning with errors, random linear codes, and cryptography. In: Symposium on Theory of Computing (STOC), pp. 84–93 (2005)
9. Applebaum, B., Ishai, Y., Kushilevitz, E.: Computationally Private Randomizing Polynomials and Their Applications. Computational Complexity 15(2), 115–162 (2006)
10. Applebaum, B., Ishai, Y., Kushilevitz, E.: Cryptography in NC^0. SIAM Journal of Computing (SICOMP) 36(4), 845–888 (2006)

11. Applebaum, B., Ishai, Y., Kushilevitz, E.: On pseudorandom generators with linear stretch in NC^0. Computational Complexity 17(1), 38–69 (2008)
12. Bronson, J., Juma, A., Papakonstantinou, P.A.: Limits on the stretch of non-adaptive constructions of pseudo-random generators. In: Ishai, Y. (ed.) TCC 2011. LNCS, vol. 6597, pp. 504–521. Springer, Heidelberg (2011)
13. Kharitonov, M., Goldberg, A.V., Yung, M.: Lower Bounds for Pseudorandom Number Generators. In: Foundations of Computer Science (FOCS), pp. 242–247 (1989)
14. Håstad, J., Impagliazzo, R., Levin, L.A., Luby, M.: A Pseudorandom Generator from any One-way Function. SIAM Journal of Computing (SICOMP) 28(4), 1364–1396 (1999)
15. Bar-Yossef, Z., Reingold, O., Shaltiel, R., Trevisan, L.: Streaming Computation of Combinatorial Objects. In: Annual IEEE Conference on Computational Complexity (CCC), vol. 17 (2002)
16. Haitner, I., Reingold, O., Vadhan, S.: Efficiency improvements in constructing pseudorandom generators from one-way functions. In: Symposium on Theory of Computing (STOC), pp. 437–446 (2010)
17. Grohe, M., Hernich, A., Schweikardt, N.: Lower bounds for processing data with few random accesses to external memory. Journal of the ACM 56(3): Art. 12, 58 (2009)
18. Hernich, A., Schweikardt, N.: Reversal complexity revisited. Theoretical Computer Science 401(1-3), 191–205 (2008)
19. Beame, P., Huynh, T.: The Value of Multiple Read/Write Streams for Approximating Frequency Moments. ACM Transactions on Computation Theory 3(2), 6 (2012)
20. Barrington, D.A.: Bounded-width polynomial-size branching programs recognize exactly those languages in NC^1. Journal of Computer and System Sciences 38(1), 150–164 (1989)
21. Goldwasser, S., Micali, S.: Probabilistic Encryption and How to Play Mental Poker Keeping Secret All Partial Information. In: Symposium on Theory of Computing (STOC), pp. 365–377 (1982)
22. Alekhnovich, M.: More on average case vs approximation complexity. In: Foundations of Computer Science (FOCS), pp. 298–307 (2003)
23. Kilian, J.: Founding cryptography on oblivious transfer. In: Symposium on Theory of Computing (STOC), pp. 20–31 (1988)
24. Grohe, M., Schweikardt, N.: Lower bounds for sorting with few random accesses to external memory. In: Symposium on Principles of Database Systems (PODS), pp. 238–249 (2005)

The Impossibility of Obfuscation
with Auxiliary Input or a Universal Simulator

Nir Bitansky[1,*], Ran Canetti[1,2,**], Henry Cohn[3], Shafi Goldwasser[4,5],
Yael Tauman Kalai[3], Omer Paneth[2,***], and Alon Rosen[6,†]

[1] Tel Aviv University, Tel Aviv, Israel
[2] Boston University, Boston, USA
[3] Microsoft Research New England, Cambridge, USA
[4] Massachusetts Institute of Technology, Cambridge, USA
[5] Weizmann Institute of Science, Rehovot, Israel
[6] IDC Herzliya, Herzliya, Israel

Abstract. In this paper we show that indistinguishability obfuscation for general circuits implies, somewhat counterintuitively, strong impossibility results for virtual black box obfuscation. In particular, it implies:

- The impossibility of average-case virtual black box obfuscation with auxiliary input for any circuit family with super-polynomial pseudo-entropy (for example, many cryptographic primitives). Impossibility holds even when the auxiliary input depends only on the public circuit family, and not which circuit in the family is being obfuscated.
- The impossibility of average-case virtual black box obfuscation with a universal simulator (with or without any auxiliary input) for any circuit family with super-polynomial pseudo-entropy.

These bounds significantly strengthen the impossibility results of Goldwasser and Kalai (FOCS 2005).

1 Introduction

The study of *program obfuscation*—a method that transforms a program (say, described as a Boolean circuit) into a form that is executable, but otherwise completely unintelligible—has been a longstanding research direction in cryptography. Barak et al. [BGI+01] formalized a number of security notions for this task. The strongest and most applicable of these notions is *virtual black box*

* Supported by an IBM Ph.D. Fellowship, the Check Point Institute for Information Security, and the Israeli Ministry of Science and Technology.
** Supported by the Check Point Institute for Information Security, an NSF EAGER grant, and NSF Algorithmic Foundations grant no. 1218461.
*** Supported by the Simons award for graduate students in theoretical computer science and NSF Algorithmic Foundations grant no. 1218461.
† Supported by ISF grant no. 1255/12 and by the ERC under the EU's Seventh Framework Programme (FP/2007-2013) ERC Grant Agreement no. 307952.

J.A. Garay and R. Gennaro (Eds.): CRYPTO 2014, Part II, LNCS 8617, pp. 71–89, 2014.
© International Association for Cryptologic Research 2014

(VBB) obfuscation, which requires that any adversary trying to learn information from an obfuscated program cannot do better than a simulator that is given only black-box access to the program. Barak et al. constructed contrived function families that cannot be VBB obfuscated, thus ruling out a universal obfuscator, but they left open the possibility that large classes of programs might still be obfuscated. Subsequently, VBB obfuscators were produced only for a number of restricted (and mostly simple) classes of programs [Can97, CD08, CRV10, BR13a, BBC+14].

In contrast, recent progress for more relaxed notions of obfuscation suggests a much more positive picture: Garg et al. [GGH+13] proposed a candidate construction for *indistinguishability obfuscation* for *all* circuits. This notion requires only that it is hard to distinguish an obfuscation of C_0 from an obfuscation of C_1, where C_0 and C_1 are circuits of the same size that compute the same function [BGI+01]. Indeed, unlike the case of VBB obfuscation, there are no known impossibility theorems for indistinguishability obfuscation. Furthermore, the Garg et al. construction and variants thereof were shown to satisfy the VBB guarantee in ideal algebraic oracle models [CV13, BR13b, BGK+13].

Although indistinguishability obfuscation might initially sound arcane, it is surprisingly powerful. For example, it amounts to *best possible* obfuscation [GR07], in the sense that anything that can be hidden by some obfuscator will be hidden by every indistinguishability obfuscator. Subsequent to [GGH+13], a flood of results have appeared showing that indistinguishability obfuscation suffices for many applications [SW13, GGH+13, HSW13, GGJS13].

Still, for many program classes the meaningfulness and applicability of indistinguishability obfuscation is unclear. Thus, understanding which classes of programs are VBB obfuscatable remains of central importance. Aiming towards such a characterization, Goldwasser and Kalai [GK05] proved strong limitations on VBB obfuscation for a broad class of *pseudo-entropic programs*, including many cryptographic functions, such as pseudo-random functions and certain natural instances of encryption and signatures. They showed the impossibility of a form of VBB security with respect to adversaries that have some a priori *auxiliary information*. When the auxiliary information depends on the actual obfuscated program, they showed that no class of pseudo-entropic functions can be obfuscated, assuming VBB obfuscation for a simple class of *point-filter functions*. For auxiliary information that depends only on the class of programs to be obfuscated, they gave an unconditional result, but only for a restricted class of programs (those that evaluate *NP-filter functions*).

This Work in a Nutshell. We strengthen the impossibility results for VBB obfuscation with auxiliary input, and we suggest another interpretation of auxiliary-input obfuscation. In a somewhat strange twist, our negative results are based on indistinguishability obfuscation, which is typically viewed positively. Specifically:

- We weaken the conditions for the impossibility of *dependent* auxiliary-input VBB obfuscation to *witness encryption*, which in turn follows from *indistinguishability obfuscation*.

- We extend the impossibility of *independent* auxiliary-input VBB obfuscation to *all* pseudo-entropic functions, assuming *indistinguishability obfuscation*.
- We observe that auxiliary-input VBB obfuscation is equivalent to a natural formulation of VBB obfuscation with universal simulation. This equivalence provides conceptual support for the significance of our impossibility results.

In the rest of the introduction, we introduce the notion of universal simulation and further discuss the notion of auxiliary-input VBB obfuscation. Then, we provide an overview of the results and sketch the proof techniques involved.

Universal Simulators. The definition of VBB obfuscation requires that for each PPT adversary A, there exists a PPT simulator S that succeeds in simulating the output of A when A is given the obfuscation $\mathcal{O}(f)$ but S is given only black-box access to f. This definition does not say how hard (or easy) it is to find the corresponding simulator S for a given adversary A. When security with black-box access to the function depends on computational hardness assumptions, this definition leaves open the possibility that the obfuscation could be broken in practice without providing an algorithm that breaks these assumptions.

A stronger and arguably more meaningful definition requires that there exist an efficient transformation from an adversary to its corresponding simulator, or equivalently a *universal* PPT simulator capable of simulating any PPT adversary A given the code of A. We will refer to such a definition as VBB obfuscation with a *universal simulator*.

As we said above, we will show that VBB obfuscation with a universal simulator is impossible for function families with super-polynomial pseudo-entropy if general indistinguishability obfuscation is possible.

Auxiliary Input. The definition of VBB security with auxiliary inputs, originally considered in [GK05], is a strengthening of VBB security, corresponding to a setting in which the adversary may have some additional a priori information.

Allowing auxiliary input is crucial when obfuscation is used together with other components in a larger scheme or protocol. Consider, for example, a zero-knowledge protocol in which one of the prover's messages to the verifier contains an obfuscated program $\mathcal{O}(f)$. To prove that the protocol is zero-knowledge, we would like to show that every verifier V has a zero-knowledge simulator S_{zk} that can simulate V's view of the protocol. Intuitively, S_{zk} would rely on the security of $\mathcal{O}$ by thinking of V as an "obfuscation adversary" that is trying to learn information from $\mathcal{O}(f)$. Such an adversary has an "obfuscation simulator" $S_{\mathcal{O}}$ that can learn the same information given only black-box access to f, and S_{zk} can try to use $S_{\mathcal{O}}$. The problem is that the view of V does not depend only on the code of V, but also on auxiliary input to V, such as other prover messages and the statement being proven. An obfuscation definition that does not allow auxiliary input is insufficient to handle this case.

The problem can be avoided by using a definition that guarantees the existence of an obfuscation simulator that can simulate the view of V given any auxiliary input. If the obfuscated program f depends on other prover messages or on the statement, then we require security with respect to *dependent* auxiliary input. Otherwise *independent* auxiliary input suffices. The paper [GK05] considered both of these notions. In the case of dependent auxiliary input, the virtual black box property is required to hold even when the auxiliary input given to the adversary and simulator depends on the actual, secret circuit being obfuscated. In the case of independent auxiliary input, this requirement is weakened: the auxiliary input may depend only on the family of circuits, which is public. The actual circuit to be obfuscated is chosen randomly from the family, independently of the auxiliary input given to the adversary and simulator.

More precisely, an obfuscator $\mathcal{O}$ for a function family $\mathcal{F}$ is (worst-case) VBB secure with *dependent* auxiliary inputs if for every probabilistic polynomial-time (PPT) adversary A, there exists a PPT simulator S such that for every $f \in \mathcal{F}$ and every auxiliary input aux (which may depend on the function f), the output of $\mathsf{A}(\mathcal{O}(f), \mathsf{aux}(f))$ is computationally indistinguishable from $\mathsf{S}^f(\mathsf{aux}(f))$. The average-case analogue of this definition requires that the output of $\mathsf{A}(\mathcal{O}(f), \mathsf{aux}(f))$ be computationally indistinguishable from $\mathsf{S}^f(\mathsf{aux}(f))$ for a *random* function $f \leftarrow \mathcal{F}$.

VBB security with *independent* auxiliary inputs is defined only with respect to an average-case definition.[1] An obfuscator $\mathcal{O}$ for a function family $\mathcal{F}$ is average-case VBB secure with *independent* auxiliary inputs if for every PPT adversary A, there exists a PPT simulator S such that for every auxiliary input aux and for a random $f \leftarrow \mathcal{F}$, the output of $\mathsf{A}(\mathcal{O}(f), \mathsf{aux})$ is computationally indistinguishable from $\mathsf{S}^f(\mathsf{aux})$.

For the case of dependent auxiliary input, Goldwasser and Kalai [GK05] showed that functions with super-polynomial pseudo-entropy cannot be VBB obfuscated, assuming that a different class of *point filter functions* can be VBB obfuscated. For the weaker notion of VBB obfuscation with independent auxiliary input, they showed a more restricted impossibility result for a subclass of functions called *filter functions*. Our results extend these theorems, assuming indistinguishability obfuscators exist.

1.1   Overview of Results and Techniques

First we prove that VBB security with a universal simulator is equivalent to VBB security with auxiliary inputs, which is the obfuscation version of the known equivalence for zero-knowledge proofs [Ore87]. More specifically, we prove that worst-case VBB security with a universal simulator is equivalent to worst-case VBB security with *dependent* auxiliary inputs, and that average-case VBB security with a universal simulator is equivalent to average-case VBB security with *independent* auxiliary inputs. To be consistent with the literature, when we refer to VBB security we always consider the worst-case version. When we would like to consider the average-case version we refer to it as average-case VBB.

---

[1] It is not clear how to enforce that the auxiliary input is independent of the function in a worst-case definition.

Informal Lemma 1. *A candidate obfuscator is a (worst-case) VBB obfuscator with a universal simulator for a class of functions F if and only if it is a (worst-case) VBB obfuscator for F with dependent auxiliary inputs.*

Informal Lemma 2. *A candidate obfuscator is an average-case VBB obfuscator with a universal simulator for a class of functions F if and only if it is an average-case VBB obfuscator for F with independent auxiliary inputs.*

We state and prove these results as Lemmas 1 and 2 in Section 3.

The above two lemmas imply that in order to obtain negative results for VBB obfuscation with a universal simulator, it suffices to obtain negative results for VBB obfuscation with auxiliary inputs.

New Impossibility Results. We show that indistinguishability obfuscation implies that any function family with super-polynomial pseudo-entropy *cannot* be VBB obfuscated with auxiliary input. Loosely speaking, a function family F has super-polynomial pseudo-entropy if it is difficult to distinguish a genuine function in F from one that has been randomly modified in some locations: for every polynomial p there exists a polynomial-size set I of inputs such that no efficient adversary can distinguish between a random function $f \leftarrow F$ and such a function with its values on I replaced with another random variable with min-entropy p. We refer the reader to Definition 7 for the precise definition, but note that such families include all pseudo-random function families. They also include all semantically secure secret-key or public-key encryption schemes or secure digital signature schemes, provided that the randomness is generated by using a (secret) pseudo-random function. (See Claim 4.0.1 in [GK05].)

Recently, the notion of witness encryption was put forth by Garg et al. [GGSW13]. It was observed by Goldwasser et al. [GKP+13] that an extractable version of witness encryption can be used to obfuscate the class of point-filter functions with respect to dependent auxiliary inputs. Thus, together with [GK05], this shows that the existence of an extractable witness encryption scheme implies that *any* function with super-polynomial pseudo-entropy cannot be obfuscated with respect to dependent auxiliary inputs.

Here we show that the proof of [GK05] actually implies that witness encryption, *without* the extractability property, suffices to prove that all functions with super-polynomial pseudo-entropy are not obfuscatable with respect to dependent auxiliary inputs.

Informal Theorem 3. *Assume the existence of a witness encryption scheme. Then no function family with super-polynomial pseudo-entropy has an average-case VBB obfuscator with respect to dependent auxiliary input.*

The idea behind the proof is that functions with high pseudo-entropy cannot be efficiently compressed; i.e., given oracle access to such a function, one cannot produce a small circuit for it. The reason is that functions with genuinely high entropy cannot be compressed at all (let alone efficiently), and no efficient algorithm can distinguish them from those with high pseudo-entropy.

Using this observation, the proof works as follows. Suppose we wish to construct an obfuscation $\mathcal{O}(f)$ of a function f that has high pseudo-entropy on a polynomial-size set I of inputs. We use witness encryption to encrypt a random bit b so that it can be read only by someone who knows a circuit of size at most $|\mathcal{O}(f)|$ for the values of f on I. Given this encryption of b as auxiliary input, knowledge of the circuit $\mathcal{O}(f)$ suffices to decrypt b. However, black-box access to f is not enough to produce any small circuit, and so VBB security is violated.

We note that this theorem is true in a strong sense: for *any* secret predicate $\pi(f)$ that is not learnable from black-box access to f, there exists an adversary and auxiliary input $\mathsf{aux}(f)$ such that given $\mathcal{O}(f)$ and $\mathsf{aux}(f)$, the adversary efficiently recovers $\pi(f)$, whereas given $\mathsf{aux}(f)$ and oracle access to f, it is computationally hard to recover $\pi(f)$. Moreover, the theorem holds even if we restrict $\mathsf{aux}(f)$ to be an efficiently computable function of f.

It was shown by Garg et al. [GGSW13] (using different terminology) that indistinguishability obfuscation for point-filter functions implies the existence of witness encryption. Thus, the informal theorem above can be restated as follows: assuming the existence of indistinguishability obfuscation for point-filter functions, functions with super-polynomial pseudo-entropy are not average-case VBB obfuscatable with respect to dependent auxiliary inputs.

For independent auxiliary input, we make use of a different hypothesis, namely indistinguishability obfuscation for *puncturable pseudo-random functions* (see Definition 6). Roughly speaking, these are pseudo-random functions for which we can produce alternate keys that effectively randomize the output for a specified input while leaving the rest of the function unchanged.

Informal Theorem 4. *Assume the existence of indistinguishability obfuscation for a class of puncturable pseudo-random functions. Then no function family with super-polynomial pseudo-entropy has an average-case VBB obfuscator with respect to independent auxiliary input.*

The proof of this theorem is a little more subtle than the previous proof. Suppose we are trying to obfuscate a circuit family with high pseudo-entropy on a set I of inputs. The auxiliary input will be $i\mathcal{O}(K_s)$, where $i\mathcal{O}$ denotes indistinguishability obfuscation and K_s is a circuit that takes another circuit $\tilde{C}$ as input and applies a puncturable pseudo-random function G_s to the values $\tilde{C}(I)$ of $\tilde{C}$ on I. Here, s is a random key.

Now, let $\mathcal{O}(C)$ be a candidate obfuscation of a circuit C. By definition, applying the auxiliary circuit $i\mathcal{O}(K_s)$ to $\mathcal{O}(C)$ yields $K_s(C)$ (i.e., $\mathsf{G}_s(C(I))$), but we will show that $K_s(C)$ cannot be computed using only black-box access to C. If it could, then we could replace the C oracle with suitable random values Y on I and still get the answer $\mathsf{G}_s(Y)$, by the definition of pseudo-entropy. Then we could modify the auxiliary input to be $i\mathcal{O}(K_s^*)$, where the pseudo-random function in K_s^* has been punctured to randomize its value at Y. The reason this modification is allowable is that with high probability, K_s and K_s^* define the same function (Y has entropy too high to be compressible to any small circuit, so no input $\tilde{C}$ to K_s^* will ever satisfy $\tilde{C}(I) = Y$). Thus, $i\mathcal{O}(K_s)$ and $i\mathcal{O}(K_s^*)$ are indistinguishable. However, by construction K_s^* does not determine the value $\mathsf{G}_s(Y)$, which is a contradiction.

We state and prove these results more formally as Theorems 1 and 2. Together with Lemmas 1 and 2, they immediately yield impossibility results for VBB obfuscation with a universal simulator. In particular, Theorem 1 and Lemma 1 imply the following corollary.

Corollary 1. *Assume the existence of a witness encryption scheme. Then no function family with super-polynomial pseudo-entropy has a VBB obfuscator with a universal simulator.*

As was the case for Theorem 1, this corollary is true in a strong sense: for *any* secret predicate $\pi(f)$ that is not learnable from black-box access to f, there exists an adversary that efficiently recovers $\pi(f)$ given $\mathcal{O}(f)$, whereas given the code of the adversary and given oracle access to f, it is computationally hard to recover $\pi(f)$.

Theorem 2 and Lemma 2 imply the following corollary.

Corollary 2. *Assume the existence of indistinguishability obfuscation for a class of puncturable pseudo-random functions. Then no function family with super-polynomial pseudo-entropy has an average-case VBB obfuscator with a universal simulator.*

2 Preliminaries

Let $\mathcal{F} = \{f_s\}$ be a family of polynomial-size circuits. In what follows, we write $\mathcal{F} = \bigcup_{k \in \mathbb{N}} \mathcal{F}_k$ with $\mathcal{F}_k = \{f_s\}_{s \in \{0,1\}^k}$. Each circuit f_s will have size $\mathsf{poly}(|s|)$, where poly denotes an unspecified, polynomially-bounded function.

Definition 1 (VBB obfuscation with universal simulator). *Let $\mathcal{F} = \{f_s\}$ be a family of polynomial-size circuits. We say that a probabilistic algorithm $\mathcal{O}$ (mapping circuits to circuits) is an obfuscation of $\mathcal{F}$ with a universal simulator if the following conditions hold:*

– *Correctness: For every function $f_s \in \mathcal{F}$ and every possible input x,*

$$\mathcal{O}(f_s)(x) = f_s(x).$$

– *Polynomial slowdown: There exists a polynomial p such that for every $f_s \in \mathcal{F}$,*

$$|\mathcal{O}(f_s)| \leq p(|f_s|).$$

– *Security with a universal simulator: There exists a (possibly non-uniform) PPT S such that for every (possibly non-uniform) PPT A, every predicate π, every $k \in \mathbb{N}$, and every $s \in \{0,1\}^k$,*

$$\left| \Pr[\mathsf{A}(\mathcal{O}(f_s)) = \pi(s)] - \Pr[\mathsf{S}^{f_s}(\mathsf{A}) = \pi(s)] \right| = \mathsf{negl}(k), \tag{1}$$

where the probabilities are over the random coin tosses of A and S. Here $\mathsf{negl}(k)$ denotes an unspecified, negligible function (i.e., $|\mathsf{negl}(k)| = O(1/k^c)$ for each constant $c > 0$).

We say that $\mathcal{O}$ is an **average-case** obfuscation of $\mathcal{F}$ with a universal simulator if Equation (1) holds for **random** $s \leftarrow \{0,1\}^k$; in other words, it means there exists a (possibly non-uniform) PPT S such that for every (possibly non-uniform) PPT A, every predicate π, and every $k \in \mathbb{N}$,

$$\left| \Pr[\mathsf{A}(\mathcal{O}(f_s)) = \pi(s)] - \Pr[\mathsf{S}^{f_s}(\mathsf{A}) = \pi(s)] \right| = \mathsf{negl}(k),$$

where the probabilities are over $s \leftarrow \{0,1\}^k$ and over the coin tosses of A and S.

When A is non-uniform, the notation $\mathsf{S}^{f_s}(\mathsf{A})$ of course means that S is given a circuit for A for inputs of the appropriate size. When A is uniform, it means the same thing as in the non-uniform case; equivalently, S is given the code for A together with $1^{\mathsf{time}(\mathsf{A}(\mathcal{O}(f_s)))}$ to ensure that it is allowed enough time.

In Definition 1, we have conflated the circuit size parameter k and the security parameter of the obfuscation method. One could distinguish between them at the cost of more notation, but this conflation is harmless for impossibility theorems.

Definition 2 (VBB obfuscation with auxiliary inputs). Let $\mathcal{F} = \{f_s\}$ be a family of polynomial-size circuits. We say that a probabilistic algorithm $\mathcal{O}$ is an obfuscation of $\mathcal{F}$ with (dependent) auxiliary inputs if it satisfies the correctness and polynomial slowdown conditions of Definition 1, and in addition it satisfies the following security requirement:

– **Security with auxiliary inputs:** For every (possibly non-uniform) PPT A, there exists a (possibly non-uniform) PPT S such that for every predicate π, every $k \in \mathbb{N}$, every $s \in \{0,1\}^k$, and every auxiliary input $\mathsf{aux}(s)$ of size $\mathsf{poly}(k)$,

$$\left| \Pr[\mathsf{A}(\mathcal{O}(f_s), \mathsf{aux}(s)) = \pi(s, \mathsf{aux}(s))] - \Pr[\mathsf{S}^{f_s}(\mathsf{aux}(s)) = \pi(s, \mathsf{aux}(s))] \right| = \mathsf{negl}(k),$$

where the probabilities are over the random coin tosses of A and S. We write $\mathsf{aux}(s)$ as a function of s for clarity, but this is implicit in the quantification.

We say that $\mathcal{O}$ is an **average-case** obfuscation of $\mathcal{F}$ with (dependent) auxiliary inputs if the above equation holds for **random** $s \leftarrow \{0,1\}^k$; namely, if for every (possibly non-uniform) PPT A there exists a (possibly non-uniform) PPT S such that for every predicate π, every $k \in \mathbb{N}$, and every auxiliary input $\mathsf{aux}(s)$ of size $\mathsf{poly}(s)$ (and allowed to depend on s),

$$\left| \Pr[\mathsf{A}(\mathcal{O}(f_s), \mathsf{aux}(s)) = \pi(s, \mathsf{aux}(s))] - \Pr[\mathsf{S}^{f_s}(\mathsf{aux}(s)) = \pi(s, \mathsf{aux}(s))] \right| = \mathsf{negl}(k),$$

where the probabilities are over $s \leftarrow \{0,1\}^k$ and over the random coin tosses of A and S.

In the definition above we allowed the auxiliary input to depend on the function being obfuscated. In what follows we define VBB obfuscation with *independent* auxiliary inputs, where we restrict the auxiliary input to be *independent* of the function being obfuscated. For this definition, only the average-case version makes sense.

Definition 3 (Average-case VBB obfuscation with independent auxiliary inputs). *Let* $\mathcal{F} = \{f_s\}$ *be a family of polynomial-size circuits. We say that* $\mathcal{O}$ *is an obfuscation of* $\mathcal{F}$ *with independent auxiliary inputs if it satisfies the correctness and polynomial slowdown conditions of Definition 1, and in addition it satisfies the following security requirement:*

- *Average-case security with independent auxiliary input: For every (possibly non-uniform) PPT* A*, there exists a (possibly non-uniform) PPT* S *such that for every predicate* π*, every* $k \in \mathbb{N}$*, and every auxiliary input* $\mathsf{aux} \in \{0,1\}^{\mathsf{poly}(k)}$*,*

$$\left| \Pr[\mathsf{A}(\mathcal{O}(f_s), \mathsf{aux}) = \pi(s, \mathsf{aux})] - \Pr[\mathsf{S}^{f_s}(\mathsf{aux}) = \pi(s, \mathsf{aux})] \right| = \mathsf{negl}(k),$$

where the probabilities are over $s \leftarrow \{0,1\}^k$ *and the coin tosses of* A *and* S*.*

Definition 4 (Witness encryption). *A witness encryption scheme for an* NP *language* $\mathcal{L}$ *with corresponding witness relation* $\mathcal{R}_{\mathcal{L}}$ *is a pair of PPT algorithms* (Enc, Dec) *such that the following conditions hold:*

- *Correctness: For all* $(x, w) \in \mathcal{R}_{\mathcal{L}}$ *and every* $b \in \{0,1\}$*,*

$$\Pr[\mathsf{Dec}(\mathsf{Enc}_x(1^k, b), w) = b] = 1 - \mathsf{negl}(k).$$

- *Semantic Security: For every* $x \notin \mathcal{L}$ *and every (possibly non-uniform) PPT adversary* A*,*

$$\left| \Pr[\mathsf{A}(\mathsf{Enc}_x(1^k, 0)) = 1] - \Pr[\mathsf{A}(\mathsf{Enc}_x(1^k, 1)) = 1] \right| = \mathsf{negl}(k),$$

where the probability is over the random coin tosses of Enc *and* A*.*

Definition 5 (Indistinguishability obfuscation). *Let* $\mathcal{C}$ *be a family of polynomial-size circuits. A PPT algorithm* $i\mathcal{O}$ *is said to be an indistinguishability obfuscator for* $\mathcal{C}$ *if it satisfies the correctness and polynomial slowdown conditions of Definition 1, and in addition it satisfies the following security requirement:*

- *Indistinguishability: For all* $C, C' \in \mathcal{C}$ *that are of the same size and define the same function,* $i\mathcal{O}(C)$ *and* $i\mathcal{O}(C')$ *are computationally indistinguishable. More formally, for every (possibly non-uniform) PPT distinguisher* D*,*

$$\left| \Pr[\mathsf{D}(i\mathcal{O}(C)) = 1] - \Pr[\mathsf{D}(i\mathcal{O}(C')) = 1] \right| = \mathsf{negl}(k),$$

where the probability is over the random coin tosses of $i\mathcal{O}$ *and* D*.*

We next define puncturable pseudo-random functions, following [SW13]. We consider a simple case in which any PRF might be punctured at a single point.

Definition 6 (Puncturable PRFs). *Let* ℓ, m *be polynomially bounded length functions. An efficiently computable family of functions*

$$\mathcal{G} = \left\{ \mathsf{G}_s \colon \{0,1\}^{m(k)} \to \{0,1\}^{\ell(k)} \;\middle|\; s \in \{0,1\}^k, k \in \mathbb{N} \right\},$$

associated with an efficient (probabilistic) key sampler $\mathsf{Gen}_\mathcal{G}$, *is a puncturable PRF if there exists a puncturing algorithm* Punc *that takes as input a key* $s \in \{0,1\}^k$ *and a point* $x^* \in \{0,1\}^{m(k)}$ *and outputs a punctured key* s_{x^*} *so that the following conditions are satisfied:*

- **Functionality is preserved under puncturing:** *For every* $x^* \in \{0,1\}^{m(k)}$, *if we sample* s *from* $\mathsf{Gen}_\mathcal{G}(1^k)$ *and let* $s_{x^*} = \mathsf{Punc}(s, x^*)$, *then* G_s *and* $\mathsf{G}_{s_{x^*}}$ *have the same values at every point other than* x^* *with probability 1.*

- **Indistinguishability at punctured points:** *The two ensembles*

$$\left\{ \left(x^*, s_{x^*}, \mathsf{G}_s(x^*)\right) \,\middle|\, s \leftarrow \mathsf{Gen}_\mathcal{G}(1^k), s_{x^*} = \mathsf{Punc}(s, x^*) \right\}_{x^* \in \{0,1\}^{m(k)}, k \in \mathbb{N}},$$

$$\left\{ \left(x^*, s_{x^*}, u\right) \,\middle|\, s \leftarrow \mathsf{Gen}_\mathcal{G}(1^k), s_{x^*} = \mathsf{Punc}(s, x^*), u \leftarrow \{0,1\}^{\ell(k)} \right\}_{x^* \in \{0,1\}^{m(k)}, k \in \mathbb{N}}$$

are computationally indistinguishable by (possibly non-uniform) PPT distinguishers.

To be explicit, we include x^* in the distribution; throughout, we shall assume for simplicity that a punctured key s_{x^*} includes x^* in the clear. As shown in [BGI13, BW13, KPTZ13], the pseudo-random functions from [GGM86] yield puncturable PRFs as defined above.

Definition 7 (Pseudo-entropy of a circuit class). *Let* $p = p(k)$ *be a polynomial. We say that a class of circuits* $\mathcal{C} = \bigcup_{k \in \mathbb{N}} \mathcal{C}_k$ *has pseudo-entropy at least* $p = p(k)$, *if there exists a polynomial* $t = t(k)$ *and a subset* $I_k \subseteq \{0,1\}^k$ *of size* $t(k)$, *and for every* $C \in \mathcal{C}_k$ *there exists a random variable* $Y^C = (Y_i)_{i \in I_k} \in \{0,1\}^{I_k}$, *such that the following conditions hold:*

1. *The random variable* Y^C *has statistical min-entropy at least* $p(k)$. *In other words, each of its values occurs with probability at most* $2^{-p(k)}$.
2. *For every (possibly non-uniform) PPT distinguisher* D,

$$\left| \Pr[\mathsf{D}^C(1^k) = 1] - \Pr[\mathsf{D}^{C \circ Y^C}(1^k) = 1] \right| = \mathsf{negl}(k),$$

where $C \circ Y^C$ *denotes an oracle that agrees with* C *except that* Y^C *replaces the values of* C *for inputs in* I_k. *Here the probabilities are over* $C \leftarrow \mathcal{C}_k$, *the random variable* Y^C, *and the random coin tosses of* D.

We say that $\mathcal{C}$ has super-polynomial pseudo-entropy if it has pseudo-entropy at least p for every polynomial p, and we then call the circuits in $\mathcal{C}$ pseudo-entropic.

3   Equivalence between a Universal Simulator and Auxiliary Inputs

In this section we show that VBB obfuscation with a universal simulator is equivalent to VBB obfuscation with auxiliary inputs. Specifically, we prove the following two lemmas.

Lemma 1. *Let $\mathcal{F} = \{f_s\}$ be a family of polynomial-size circuits. Then $\mathcal{O}$ is a VBB obfuscator for $\mathcal{F}$ with a universal simulator if and only if it is a VBB obfuscator for $\mathcal{F}$ with dependent auxiliary inputs.*

Lemma 2. *Let $\mathcal{F} = \{f_s\}$ be a family of polynomial-size circuits. Then $\mathcal{O}$ is an average-case VBB obfuscator for $\mathcal{F}$ with a universal simulator if and only if it is an average-case VBB obfuscator for $\mathcal{F}$ with independent auxiliary inputs.*

Proof of Lemma 1

($\Rightarrow$): Suppose that $\mathcal{O}$ is a VBB obfuscator for $\mathcal{F}$ with a universal simulator. Namely, there exists a (possibly non-uniform) PPT S such that for every (possibly non-uniform) PPT A, every predicate π, every $k \in \mathbb{N}$ and every $s \in \{0,1\}^k$,

$$\left| \Pr[A(\mathcal{O}(f_s)) = \pi(s)] - \Pr[S^{f_s}(A) = \pi(s)] \right| = \mathsf{negl}(k),$$

where the probabilities are over the random coin tosses of A and S.

We will prove that $\mathcal{O}$ is a VBB obfuscator for $\mathcal{F}$ with dependent auxiliary inputs. To this end, fix any (possibly non-uniform) PPT adversary A. Let S_A be the PPT simulator defined as follows: for every auxiliary input $\mathsf{aux}(s)$, $S_A^{f_s}(\mathsf{aux}(s))$ runs the universal simulator S^{f_s} on input $A_{\mathsf{aux}(s)}$, where $A_{\mathsf{aux}(s)}$ is the (non-uniform) adversary that simulates A with auxiliary input $\mathsf{aux}(s)$. We need to prove that for every predicate π, every $k \in \mathbb{N}$, and every $s \in \{0,1\}^k$,

$$\left| \Pr[A(\mathcal{O}(f_s), \mathsf{aux}(s)) = \pi(s, \mathsf{aux}(s))] - \Pr[S_A^{f_s}(\mathsf{aux}(s)) = \pi(s, \mathsf{aux}(s))] \right| = \mathsf{negl}(k),$$

where the probabilities are over the random coin tosses of A and S.

To do so, we check that

$$
\begin{aligned}
&\left| \Pr[A(\mathcal{O}(f_s), \mathsf{aux}(s)) = \pi(s, \mathsf{aux}(s))] - \Pr[S_A^{f_s}(\mathsf{aux}(s)) = \pi(s, \mathsf{aux}(s))] \right| \\
&= \left| \Pr[A(\mathcal{O}(f_s), \mathsf{aux}(s)) = \pi(s, \mathsf{aux}(s))] - \Pr[S^{f_s}(A_{\mathsf{aux}(s)}) = \pi(s, \mathsf{aux}(s))] \right| \\
&\leq \left| \Pr[A(\mathcal{O}(f_s), \mathsf{aux}(s)) = \pi(s, \mathsf{aux}(s))] - \Pr[A_{\mathsf{aux}(s)}(\mathcal{O}(f_s)) = \pi(s, \mathsf{aux}(s))] \right| \\
&\quad + \left| \Pr[A_{\mathsf{aux}(s)}(\mathcal{O}(f_s)) = \pi(s, \mathsf{aux}(s))] - \Pr[S^{f_s}(A_{\mathsf{aux}(s)}) = \pi(s, \mathsf{aux}(s))] \right| \\
&= \mathsf{negl}(k),
\end{aligned}
$$

where the first equation follows by the definition of S_A, the inequality follows from the triangle inequality, and the last equation follows from the definition of $A_{\mathsf{aux}(s)}$ and from the fact that $\mathcal{O}$ is VBB secure with the universal simulator S.

($\Leftarrow$): Suppose that $\mathcal{O}$ is a VBB obfuscator for $\mathcal{F}$ with dependent auxiliary inputs. Namely, for every (possibly non-uniform) PPT A there exists a (possibly non-uniform) PPT S such that for every predicate π, every $k \in \mathbb{N}$, every $s \in \{0,1\}^k$, and every auxiliary input $\mathsf{aux}(s)$ of size $\mathsf{poly}(k)$,

$$\left| \Pr[A(\mathcal{O}(f_s), \mathsf{aux}(s)) = \pi(s, \mathsf{aux}(s))] - \Pr[S^{f_s}(\mathsf{aux}(s)) = \pi(s, \mathsf{aux}(s))] \right| = \mathsf{negl}(k),$$

where the probabilities are over the random coin tosses of A and S. We prove that $\mathcal{O}$ is a VBB obfuscator for $\mathcal{F}$ with a universal simulator. To this end, let A^*

be a universal PPT adversary that interprets its auxiliary input $\mathsf{aux} = \mathsf{aux}(s)$ as a (possibly non-uniform) PPT adversary and runs this adversary. (As pointed out after Definition 1, we must interpret this carefully regarding running times in the uniform case.) The fact that $\mathcal{O}$ is a VBB obfuscator with dependent auxiliary inputs implies that there is a PPT simulator S such that for every predicate π, every $k \in \mathbb{N}$, every $s \in \{0,1\}^k$, and every auxiliary input $\mathsf{aux}(s)$ of size $\mathsf{poly}(k)$,

$$\left| \Pr[\mathsf{A}^*(\mathcal{O}(f_s), \mathsf{aux}(s)) = \pi(s, \mathsf{aux}(s))] - \Pr[\mathsf{S}^{f_s}(\mathsf{aux}(s)) = \pi(s, \mathsf{aux}(s))] \right| = \mathsf{negl}(k),$$
(2)

where the probabilities are over the random coin tosses of A^* and S. We claim that S is a universal simulator for $\mathcal{O}$. Namely, we claim that for every (possibly non-uniform) PPT adversary A, every predicate π, every $k \in \mathbb{N}$, and every $s \in \{0,1\}^k$,

$$\left| \Pr[\mathsf{A}(\mathcal{O}(f_s)) = \pi(s)] - \Pr[\mathsf{S}^{f_s}(\mathsf{A}) = \pi(s)] \right| = \mathsf{negl}(k).$$

To see why, note that

$$
\begin{aligned}
& \left| \Pr[\mathsf{A}(\mathcal{O}(f_s)) = \pi(s)] - \Pr[\mathsf{S}^{f_s}(\mathsf{A}) = \pi(s)] \right| \\
& \leq \left| \Pr[\mathsf{A}(\mathcal{O}(f_s)) = \pi(s)] - \Pr[\mathsf{A}^*(\mathcal{O}(f_s), \mathsf{A}) = \pi(s)] \right| \\
& \quad + \left| \Pr[\mathsf{A}^*(\mathcal{O}(f_s), \mathsf{A}) = \pi(s)] - \Pr[\mathsf{S}^{f_s}(\mathsf{A}) = \pi(s)] \right| \\
& = \left| \Pr[\mathsf{A}^*(\mathcal{O}(f_s), \mathsf{A}) = \pi(s)] - \Pr[\mathsf{S}^{f_s}(\mathsf{A}) = \pi(s)] \right| \\
& = \mathsf{negl}(k),
\end{aligned}
$$

where the inequality follows from the triangle inequality, the next equation follows from the definition of A^*, and the last equation follows from Equation (2).   □

The proof of Lemma 2 is almost identical to that of Lemma 1 (see arXiv paper 1401.0348v3 at http://arXiv.org/abs/1401.0348v3 for the details).

4   Impossibility for Obfuscation with Auxiliary Inputs

As mentioned in the introduction, Goldwasser and Kalai [GK05] proved that either point-filter functions are not obfuscatable with dependent auxiliary inputs or *all* function families with sufficient pseudo-entropy are not obfuscatable with dependent auxiliary inputs. It was recently observed by Goldwasser et al. [GKP+13] that extractable witness encryption implies that point-filter functions are obfuscatable with dependent auxiliary inputs, and thus that any function family with sufficient pseudo-entropy is not obfuscatable with dependent auxiliary inputs. We now show that the same impossibility result (with essentially the same proof as in [GK05]) can be obtained assuming the existence of witness encryption, without any extractability property.

Theorem 1. *Assume the existence of a witness encryption scheme for an NP-complete language. Then no function family with super-polynomial pseudo-entropy has an average-case VBB obfuscator with respect to dependent auxiliary input.*

In fact, the proof rules out average-case obfuscation if we restrict the auxiliary input to be efficiently computable given the function (or even oracle access to the function).

Theorem 2. *Assume the existence of indistinguishability obfuscation for a class of puncturable pseudo-random functions. Then no function family with super-polynomial pseudo-entropy has an average-case VBB obfuscator with respect to independent auxiliary input.*

We describe the specific class for which we need indistinguishability obfuscation in the proof of the theorem.

Theorems 1 and 2, together with Lemmas 1 and 2, immediately yield impossibility results for VBB obfuscation with a universal simulator. In particular, Theorem 1 and Lemma 1 imply the following corollary.

Corollary 1. *Assume the existence of a witness encryption scheme for an NP-complete language. Then no function family with super-polynomial pseudo-entropy has a VBB obfuscator with a universal simulator.*

Theorem 2 and Lemma 2 imply the following corollary.

Corollary 2. *Assume the existence of indistinguishability obfuscation for a class of puncturable pseudo-random functions. Then no function family with super-polynomial pseudo-entropy has an average-case VBB obfuscator with a universal simulator.*

All that remains is to prove Theorems 1 and 2. For notation in both proofs, let $C = \bigcup_{k \in \mathbb{N}} C_k$ be a class of circuits with super-polynomial pseudo-entropy such that each $C \in C_k$ maps $\{0,1\}^{\ell(k)}$ to $\{0,1\}^{\ell'(k)}$. Let $\mathcal{O}$ be any candidate obfuscator for C, and let $m(k)$ be a polynomial such that $|\mathcal{O}(C)| \leq m(k)$ for every $C \in C_k$.

4.1   Proof of Theorem 1

The fact that C has super-polynomial pseudo-entropy implies that it has pseudo-entropy at least $m(k) + k$. In particular, recalling Definition 7, this implies that there exists a polynomial $t = t(k)$ and a subset $I_k \subseteq \{0,1\}^k$ of size $t(k)$ such that for every C there exists a random variable $Y^C = (Y_1, \ldots, Y_t)$ such that the following conditions hold:

1. The random variable Y^C has statistical min-entropy at least $m(k) + k$.
2. For every (possibly non-uniform) PPT distinguisher D,

$$\left| \Pr[\mathsf{D}^C(1^k) = 1] - \Pr[\mathsf{D}^{C \circ Y^C}(1^k) = 1] \right| = \mathsf{negl}(k),$$

where $C \circ Y^C$ denotes an oracle that agrees with C except that Y^C replaces the values of C for inputs in I_k. Here the probabilities are over $C \leftarrow C_k$, the random variable Y^C, and the random coin tosses of D.

We define an NP language $\mathcal{L}$ by

$$\mathcal{L} = \{(x_i)_{i \in I_k} \mid k \in \mathbb{N} \text{ and } \exists \text{ circuit } C \text{ with } |C| \le p(k) \text{ and } C(i) = x_i \text{ for } i \in I_k\}.$$

Set $x = (C(i))_{i \in I_k}$ and let $\mathsf{aux}(C) = \mathsf{Enc}_x(1^k, b)$, where $b \leftarrow \{0, 1\}$ is a random bit and Enc is a witness encryption for the language $\mathcal{L}$. Note that the fact that there is a witness encryption for an NP-complete language implies that there is a witness encryption for every NP language, and in particular for $\mathcal{L}$.

Given $\mathcal{O}(C)$ and $\mathsf{aux}(C) = \mathsf{Enc}_x(1^k, b)$, one can efficiently decrypt b with probability $1 - \mathsf{negl}(k)$, since $\mathcal{O}(C)$ is a valid witness of x. It remains to prove the following claim.

Claim. For any (possibly non-uniform) PPT adversary S which takes as input $\mathsf{aux}(s) = \mathsf{Enc}_x(1^k, b)$ and has black-box access to C,

$$\Pr[\mathsf{S}^C(\mathsf{Enc}_x(1^k, b)) = b] \le \frac{1}{2} + \mathsf{negl}(k).$$

Proof. Suppose for the sake of contradiction that there exists a PPT adversary S such that

$$\Pr[\mathsf{S}^C(\mathsf{Enc}_x(1^k, b)) = b] \ge \frac{1}{2} + \epsilon(k)$$

for some non-negligible function ϵ, where the probability is over random $C \leftarrow \mathcal{C}_k$, the choice of b, and the randomness of Enc.

Let D be the distinguisher that, given oracle access to C, does the following. First, it computes $x = (C(i))_{i \in I_k}$ by querying the oracle $t(k)$ times. Then it computes $\mathsf{Enc}_x(1^k, b)$ and simulates $\mathsf{S}^C(\mathsf{Enc}_x(1^k, b))$ to arrive at its output.

By assumption,

$$\Pr[\mathsf{D}^C(1^k) = b] \ge \frac{1}{2} + \epsilon(k).$$

Thus, because $\mathcal{C}$ has super-polynomial pseudo-entropy,

$$\Pr[\mathsf{D}^{C \circ Y^C}(1^k) = b] \ge \frac{1}{2} + \epsilon(k) + \mathsf{negl}(k). \tag{3}$$

When it is given oracle access to $C \circ Y^C$, D replaces x with $x^* = Y^C$, and at the end it is trying to recover b from $\mathsf{Enc}_{x^*}(1^k, b)$.

Note however that x^* has min-entropy $m(k) + k$, and so the probability that it is in $\mathcal{L}$ is at most 2^{-k}. (For each of the at most $2^{m(k)}$ circuits of size $m(k)$ in the definition of $\mathcal{L}$, the probability of obtaining x^* is at most $2^{-m(k)-k}$.) Thus, Equation (3) contradicts the semantic security of the underlying witness-encryption scheme.

Remark 1. Note that for any secret predicate π that is not learnable from black-box access to the circuit, we could have taken the auxiliary input to be $\mathsf{aux}(C) = \mathsf{Enc}_x(1^k, b)$ where $b = \pi(C)$ (as opposed to being truly random). In this case, there exists a PPT adversary A that given the obfuscated circuit $\mathcal{O}(C)$ and the auxiliary input $\mathsf{aux}(C)$ outputs $\pi(C)$ with probability 1, whereas any PPT simulator cannot learn $\pi(C)$ from $\mathsf{aux}(C)$ and black-box access to C.

Using Lemma 1, we conclude that for any secret predicate π that is not learnable from black-box access to the circuit and for any circuit C there exists an adversary $\mathsf{A}_{\mathsf{aux}(C)}$ that outputs $\pi(C)$ with probability 1, whereas any universal simulator S, which is given black box access to C and takes as input the code of $\mathsf{A}_{\mathsf{aux}(C)}$, cannot learn the predicate $\pi(C)$.

Thus our negative result is a strong one: VBB obfuscation with a universal simulator cannot conceal *any* secret predicate that is not learnable from black-box access to the circuit.

4.2 Proof of Theorem 2

We first describe an auxiliary-input distribution ensemble $\mathcal{Z}$ and a PPT adversary A such that given $z \leftarrow \mathcal{Z}$ and an obfuscation of $C \leftarrow \mathcal{C}$, A always learns some predicate $\pi(C, z)$. Then, we show that any PPT simulator that is only given oracle access to C fails to learn the predicate.

The Auxiliary Input Distribution $\mathcal{Z}$. By assumption, $\mathcal{C}$ has pseudo-entropy at least $m(k) + k$. Let $\{I_k\}_{k \in \mathbb{N}}$ be the sets guaranteed by Definition 7, where I_k is of polynomial size $t(k)$, and let $\mathcal{G}$ be a puncturable one-bit PRF family

$$\mathcal{G} = \left\{ \mathsf{G}_s \colon \{0,1\}^{\ell'(k) \cdot t(k)} \to \{0,1\} \mid s \in \{0,1\}^k, k \in \mathbb{N} \right\}.$$

We define two circuit families

$$\mathcal{K} = \left\{ K_s \colon \{0,1\}^{m(k)} \to \{0,1\} \mid s \in \{0,1\}^k, k \in \mathbb{N} \right\},$$

$$\mathcal{K}^* = \left\{ K^*_{s_{x^*}} \colon \{0,1\}^{m(k)} \to \{0,1\} \mid s \in \{0,1\}^k, x^* \in \{0,1\}^{\ell'(k) \cdot t(k)}, k \in \mathbb{N} \right\}.$$

Given a circuit $\tilde{C} \colon \{0,1\}^\ell \to \{0,1\}^{\ell'}$ of size m as input, the circuit K_s computes $x := \tilde{C}(I_k) := (\tilde{C}(i))_{i \in I_k}$ and outputs $\mathsf{G}_s(x)$. See Figure 1.

Hardwired: a PRF key $s \in \{0,1\}^k$ and the set I_k.
Input: a circuit $\tilde{C} \colon \{0,1\}^\ell \to \{0,1\}^{\ell'}$, where $|\tilde{C}| = m(k)$.
  1. Compute $x = \tilde{C}(I_k)$.
  2. Return $\mathsf{G}_s(x)$.

Fig. 1. The circuit K_s

The circuit $K^*_{s_{x^*}}$, has a hardwired PRF key s_{x^*} that was derived from s by puncturing it at the point x^*. It operates the same as K_s, except that when $x = x^*$, it outputs an arbitrary bit, say, 0. See Figure 2. In particular, if $x^* \neq \tilde{C}(I_k)$ for all circuits $\tilde{C} \in \{0,1\}^{m(k)}$, then $K^*_{s_{x^*}}$ and K_s compute the exact same function.

Hardwired: a punctured PRF key $s_{x^*} = \mathsf{Punc}(s, x^*)$ and the set I_k.
Input: a circuit $\tilde{C} \colon \{0,1\}^\ell \to \{0,1\}^{\ell'}$, where $|\tilde{C}| = m(k)$.
1. Compute $x = \tilde{C}(I_k)$.
2. If $x \neq x^*$, return $\mathsf{G}_{s_{x^*}}(x)$.
3. If $x = x^*$, return 0.

Fig. 2. The circuit $K^*_{s_{x^*}}$

We are now ready to define our auxiliary-input distribution $\mathcal{Z} = \{Z_k\}_{k \in \mathbb{N}}$. Let $d = d(k)$ be the maximal size of circuits in either $\mathcal{K}$ or $\mathcal{K}^*$, corresponding to security parameter k. Denote by $[K]_d$ a circuit K padded with zeros to size d, and by $[\mathcal{K}]_d$ the class of circuits where every circuit $K \in \mathcal{K}$ is replaced with $[K]_d$. Let $i\mathcal{O}$ be an indistinguishability obfuscator for the class $[\mathcal{K} \cup \mathcal{K}^*]_d$.

The distribution Z_k simply consists of an obfuscated (padded) circuit K_s for a randomly generated s. See Figure 3.

1. Sample $s \leftarrow \mathsf{Gen}_{\mathcal{G}}(1^k)$.
2. Sample an obfuscation $z \leftarrow i\mathcal{O}([K_s]_{d(k)})$.
3. Output z.

Fig. 3. The auxiliary input distribution Z_k

The Adversary A and Predicate π. The adversary A, given auxiliary input $z = [i\mathcal{O}(K_s)]_{d(k)}$ and an obfuscation $\mathcal{O}(C)$ with $C \in \mathcal{C}_k$, outputs

$$z(\mathcal{O}(C)) = K_s(\mathcal{O}(C)) = \mathsf{G}_s(\mathcal{O}(C)(I_k)) = \mathsf{G}_s(C(I_k)),$$

where the above follows by the definition of K_s and the functionality of $i\mathcal{O}$ and $\mathcal{O}$.

Thus, A always successfully outputs the predicate

$$\pi(C, K_s) = K_s(C) = \mathsf{G}_s(C(I_k)).$$

Adversary A Cannot Be Simulated. We prove the following claim implying that the candidate obfuscator $\mathcal{O}$ for the class $\mathcal{C}$ fails to meet the virtual black box requirement:

Claim. For any PPT simulator S,

$$\Pr_{\substack{C \leftarrow \mathcal{C}_k \\ z \leftarrow Z_k}} [S^C(z) = \pi(C, z)] \leq \frac{1}{2} + \mathsf{negl}(k).$$

Proof. Assume towards contradiction that there exists a PPT simulator S that learns $\pi(C, z)$ with probability $\frac{1}{2} + \epsilon(k)$, for some non-negligible ϵ. We show how to use S to break either the pseudo-entropy of C or the pseudo-randomness at punctured points of $\mathcal{G}$.

According to the definition of Z_k,

$$\Pr\left[S^C(i\mathcal{O}([K_s]_d) = G_s(C(I_k))\right] \geq \frac{1}{2} + \epsilon(k),$$

where the probability is over $C \leftarrow \mathcal{C}_k$, $s \leftarrow \mathsf{Gen}_\mathcal{G}(1^k)$, and the random coin tosses of S.

Now, for every $C \in \mathcal{C}_k$, let $Y^C = (Y_1, \ldots, Y_t)$ be the random variable guaranteed by the pseudo-entropy of values in I_k (Definition 7). We first consider an alternative experiment in which the oracle C is replaced with an oracle $C \circ Y^C$ that behaves like C on all points outside I_k, and on points in I_k answers according to Y^C. We claim that

$$\Pr\left[S^{C \circ Y^C}(i\mathcal{O}([K_s]_d) = G_s(Y^C)\right] \geq \frac{1}{2} + \epsilon(k) - \mathsf{negl}(k),$$

where the probability is over $C \leftarrow \mathcal{C}_k$, the random variable Y^C, $s \leftarrow \mathsf{Gen}_\mathcal{G}(1^k)$, and the coin tosses of S. Indeed, this follows directly from the pseudo-entropy guarantee (Definition 7), together with the fact that a distinguisher can sample s and compute $i\mathcal{O}([K_s]_d)$ on its own.

Next, we change the above experiment so that instead of an indistinguishability obfuscation of K_s, the simulator gets an indistinguishability obfuscation of the circuit $K^*_{s^*_x}$, where s is punctured at the point $x^* = Y^C$. We claim that

$$\Pr\left[S^{C \circ Y^C}(i\mathcal{O}([K^*_{s_{x^*}}]_d) = G_s(Y^C)\right] \geq \frac{1}{2} + \epsilon(k) - \mathsf{negl}(k),$$

where the probability is over $C \leftarrow \mathcal{C}_k$, the random variable Y^C, $s \leftarrow \mathsf{Gen}_\mathcal{G}(1^k)$, and the coin tosses of S, $x^* = Y^C$, and $s_{x^*} = \mathsf{Punc}(s, x^*)$. Indeed, recalling that Y^C has min-entropy $m(k) + k$ for every $C \in \mathcal{C}_k$, there does not exist a circuit $\tilde{C}$ such that $x^* := Y^C = \tilde{C}(I_k)$, except with negligible probability 2^{-k}. However, recall that in this case K_s and $K^*_{s_{x^*}}$ have the exact same functionality, and thus the above follows by the indistinguishability obfuscation guarantee.

It is now left to note that S predicts with noticeable advantage the value of G_s at the punctured point x^*, and thus violates the pseudo-randomness at punctured points requirement (Definition 6).

References

[BBC+14] Barak, B., Bitansky, N., Canetti, R., Kalai, Y.T., Paneth, O., Sahai, A.: Obfuscation for evasive functions. In: Lindell, Y. (ed.) TCC 2014. LNCS, vol. 8349, pp. 26–51. Springer, Heidelberg (2014)

[BGI+01]   Barak, B., Goldreich, O., Impagliazzo, R., Rudich, S., Sahai, A., Vadhan, S., Yang, K.: On the (im)possibility of obfuscating programs. In: Kilian, J. (ed.) CRYPTO 2001. LNCS, vol. 2139, pp. 1–18. Springer, Heidelberg (2001)

[BGI13]   Boyle, E., Goldwasser, S., Ivan, I.: Functional signatures and pseudorandom functions. Cryptology ePrint Archive, Report 2013/401

[BGK+13]   Barak, B., Garg, S., Kalai, Y.T., Paneth, O., Sahai, A.: Protecting obfuscation against algebraic attacks. Cryptology ePrint Archive, Report 2013/631

[BR13a]   Brakerski, Z., Rothblum, G.N.: Black-box obfuscation for d-cnfs. Cryptology ePrint Archive, Report 2013/557

[BR13b]   Brakerski, Z., Rothblum, G.N.: Virtual black-box obfuscation for all circuits via generic graded encoding. Cryptology ePrint Archive, Report 2013/563

[BW13]   Boneh, D., Waters, B.: Constrained pseudorandom functions and their applications. Cryptology ePrint Archive, Report 2013/352

[Can97]   Canetti, R.: Towards realizing random oracles: Hash functions that hide all partial information. In: Kaliski Jr., B.S. (ed.) CRYPTO 1997. LNCS, vol. 1294, pp. 455–469. Springer, Heidelberg (1997)

[CD08]   Canetti, R., Dakdouk, R.R.: Extractable perfectly one-way functions. In: Aceto, L., Damgård, I., Goldberg, L.A., Halldórsson, M.M., Ingólfsdóttir, A., Walukiewicz, I. (eds.) ICALP 2008, Part II. LNCS, vol. 5126, pp. 449–460. Springer, Heidelberg (2008)

[CRV10]   Canetti, R., Rothblum, G.N., Varia, M.: Obfuscation of hyperplane membership. In: Micciancio, D. (ed.) TCC 2010. LNCS, vol. 5978, pp. 72–89. Springer, Heidelberg (2010)

[CV13]   Canetti, R., Vaikuntanathan, V.: Obfuscating branching programs using black-box pseudo-free groups. Cryptology ePrint Archive, Report 2013/500

[GGH+13]   Garg, S., Gentry, C., Halevi, S., Raykova, M., Sahai, A., Waters, B.: Candidate indistinguishability obfuscation and functional encryption for all circuits. In: 54th Annual IEEE Symposium on Foundations of Computer Science (FOCS 2013), pp. 40–49. IEEE Computer Society (2013)

[GGJS13]   Goldwasser, S., Goyal, V., Jain, A., Sahai, A.: Multi-input functional encryption. Cryptology ePrint Archive, Report 2013/727

[GGM86]   Goldreich, O., Goldwasser, S., Micali, S.: How to construct random functions. J. ACM 33(4), 792–807 (1986)

[GGSW13]   Garg, S., Gentry, C., Sahai, A., Waters, B.: Witness encryption and its applications. In: 45th Annual ACM Symposium on Theory of Computing (STOC 2013), pp. 467–476. ACM (2013)

[GK05]   Goldwasser, S., Kalai, Y.T.: On the impossibility of obfuscation with auxiliary input. In: 46th Annual IEEE Symposium on Foundations of Computer Science (FOCS 2005), pp. 553–562. IEEE Computer Society (2005)

[GKP+13]   Goldwasser, S., Kalai, Y.T., Popa, R.A., Vaikuntanathan, V., Zeldovich, N.: Reusable garbled circuits and succinct functional encryption. In: 45th Annual ACM Symposium on Theory of Computing (STOC 2013), pp. 555–564. ACM (2013)

[GR07]   Goldwasser, S., Rothblum, G.N.: On best-possible obfuscation. In: Vadhan, S.P. (ed.) TCC 2007. LNCS, vol. 4392, pp. 194–213. Springer, Heidelberg (2007)

[HSW13]   Hohenberger, S., Sahai, A., Waters, B.: Replacing a random oracle: Full
           domain hash from indistinguishability obfuscation. Cryptology ePrint
           Archive, Report 2013/509
[KPTZ13]   Kiayias, A., Papadopoulos, S., Triandopoulos, N., Zacharias, T.: Del-
           egatable pseudorandom functions and applications. Cryptology ePrint
           Archive, Report 2013/379
[Ore87]   Oren, Y.: On the cunning power of cheating verifiers: Some observations
           about zero knowledge proofs. In: 28th Annual IEEE Symposium on Foun-
           dations of Computer Science (FOCS 1987), pp. 462–471. IEEE Computer
           Society (1987)
[SW13]   Sahai, A., Waters, B.: How to use indistinguishability obfuscation: Deni-
           able encryption, and more. Cryptology ePrint Archive, Report 2013/454

Self-bilinear Map on Unknown Order Groups from Indistinguishability Obfuscation and Its Applications[*]

Takashi Yamakawa[1,2], Shota Yamada[2],
Goichiro Hanaoka[2], and Noboru Kunihiro[1]

[1] The University of Tokyo, Japan
yamakawa@it.k.u-tokyo.ac.jp, kunihiro@k.u-tokyo.ac.jp
[2] National Institute of Advanced Industrial Science and Technology (AIST), Japan
{yamada-shota,hanaoka-goichiro}@aist.go.jp

Abstract. A self-bilinear map is a bilinear map where the domain and target groups are identical. In this paper, we introduce a *self-bilinear map with auxiliary information* which is a weaker variant of a self-bilinear map, construct it based on indistinguishability obfuscation and prove that a useful hardness assumption holds with respect to our construction under the factoring assumption. From our construction, we obtain a multilinear map with interesting properties: the level of multilinearity is not bounded in the setup phase, and representations of group elements are compact, i.e., their size is independent of the level of multilinearity. This is the first construction of a multilinear map with these properties. Note, however, that to evaluate the multilinear map, auxiliary information is required. As applications of our multilinear map, we construct multiparty non-interactive key-exchange and distributed broadcast encryption schemes where the maximum number of users is not fixed in the setup phase. Besides direct applications of our self-bilinear map, we show that our technique can also be used for constructing somewhat homomorphic encryption based on indistinguishability obfuscation and the Φ-hiding assumption.

Keywords: self-bilinear map, indistinguishability obfuscation, multilinear map.

1 Introduction

1.1 Background

Bilinear maps are an important tool in the construction of many cryptographic primitives, such as identity-based encryption (IBE) [2], attribute-based encryption (ABE) [22], non-interactive zero-knowledge (NIZK) proof systems [16] etc. The bilinear maps which are mainly used in cryptography, are constructed on

---

[*] The first and second authors are supported by a JSPS Fellowship for Young Scientists.

J.A. Garay and R. Gennaro (Eds.): CRYPTO 2014, Part II, LNCS 8617, pp. 90–107, 2014.
© International Association for Cryptologic Research 2014

elliptic curve groups. In these constructions, the target group is different from the domain groups.

This leads to the natural question: is it possible to construct a bilinear map where the domain and target groups are identical? Such a bilinear map is called a *self-bilinear map*, and has previously been studied by Cheon and Lee [5]. They showed that a self-bilinear map is useful to construct cryptographic primitives by highlighting that it can be used for constructing a multilinear map [3]. However, in contrast to this useful property, they also proved an impossibility result: the computational Diffie-Hellman (CDH) assumption cannot hold in a group G of *known prime order* if there exists an efficiently computable self-bilinear map on G. This is undesirable for cryptographic applications. The overview of the proof is as follows. Let $e : G \times G \to G$ be a self-bilinear map and g be a generator of G, then we have $e(g^x, g^y) = e(g, g)^{xy} = g^{cxy}$ where c is an integer such that $e(g, g) = g^c$. Then we can compute g^{xy} by computing c-th root of $e(g^x, g^y)$ since G is a prime and known order group.[1] However, their impossibility result cannot be applied for the case that G is a *composite and unknown* order group. This is the setting we focus on in this paper.

1.2 Our Contribution

In this paper, we consider a group of composite and unknown order and construct a *self-bilinear map with auxiliary information* which is a weaker variant of a self-bilinear map, by using indistinguishability obfuscation [10]. Though our self-bilinear map with auxiliary information has a limited functionality compared with a self-bilinear map, we show that it is still useful to construct various cryptographic primitives. Especially, it is sufficient to instantiate some multilinear map-based cryptographic primitives such as multiparty non-interactive key exchange (NIKE), broadcast encryption and attribute-based encryption for circuits. Our multiparty NIKE and distributed broadcast encryption schemes are the first schemes where all algorithms can be run independent of the number of users. We also show that our technique can be used for constructing a somewhat homomorphic encryption scheme for NC^1 circuits.

Applications of Our Self-bilinear Map with Auxiliary Information. As applications of our self-bilinear map with auxiliary information, we construct a multilinear map. From our construction, we obtain multiparty NIKE, distributed broadcast encryption and ABE for circuits schemes. The details follow.

– **Multilinear map.** We can construct a multilinear map by iterated usage of a self-bilinear map. Since our variant of a self-bilinear map in this paper requires auxiliary information to compute the map, the resulting multilinear map also inherits this property. However, we show that it is sufficient to replace existing multilinear maps in some applications which are given

---

[1] Here, we consider only the case in which c is known. However, [5] proved that the CDH assumption does not hold even if c is unknown as long as G is a group of known prime order.

below. Moreover, our multilinear map has an interesting property that existing multilinear maps do not have: the level of multilinearity is *not bounded at the instance generation phase* and representations of group elements are *compact*, i.e., their sizes are independent of the level of multilinearity.

- **Multiparty NIKE.** We construct a multiparty NIKE scheme where the maximum number of users is not fixed in the setup phase. In particular, the size of both the public parameters and a public key generated by a user are independent of the number of users. The construction is a natural extension of the Diffie-Hellman key exchange by using our multilinear map [7,3]. We note that [4] also constructed multiparty NIKE schemes based on indistinguishability obfuscation. However, in their schemes, the setup algorithm or key generation algorithm have to take the number of users as input unlike ours.

- **Distributed broadcast encryption.** Distributed broadcast encryption is broadcast encryption where a user can join the system by himself without the assistance of a (semi) trusted third party holding a master key. We construct a distributed broadcast encryption scheme where the maximum number of users is not fixed in the setup phase based on our multiparty NIKE scheme. In particular, the size of both the public parameters and a ciphertext overhead are independent of the number of users. We note that [4] also constructed a distributed broadcast encryption scheme based on indistinguishability obfuscation. However, in their scheme, the setup algorithm have to take the number of users as input unlike ours.

- **ABE for circuits.** We construct an ABE scheme for general circuits by using our multilinear map. The construction is an analogue of the scheme in [11]. Note that this is *not* the first ABE scheme for general circuits based on indistinguishability obfuscation since a indistinguishability obfuscation implies witness encryption [10], and [12] constructed ABE for circuits based on witness encryption. We also note that Gorbunov et al. [15] constructed attribute based encryption for circuit based on the standard learning with errors (LWE) assumption.

The above results can be interpreted as evidence that our multilinear map can replace existing multilinear maps in some applications based on the multilinear CDH assumption since all of the above constructions are simple analogue of multilinear map-based constructions.

Besides direct applications of our self-bilinear map with auxiliary information, we construct a somewhat homomorphic encryption scheme by using a similar technique. Our somewhat homomorphic encryption scheme is chosen plaintext (CPA) secure, NC^1 homomorphic and compact.

Note that all known candidate constructions of indistinguishability obfuscation are far from practical, and hence, the above constructions are mostly of theoretical interest.

Technical Overview. Here, we give a technical overview of our result. Our basic idea is to avoid the impossibility result of self-bilinear maps which is explained above by considering a group of *composite and unknown order*. Note that even if we consider such a group, many decisional assumptions such as the decisional Diffie-Hellman (DDH) assumption cannot hold if there exists an efficiently computable self-bilinear map on the group. Therefore we consider only computational assumptions such as the CDH assumption. For a Blum integer N, we consider the group $\mathbb{QR}_N^+$ of signed quadratic residues [17]. On this group, we consider a self-bilinear map $e : \mathbb{QR}_N^+ \times \mathbb{QR}_N^+ \to \mathbb{QR}_N^+$ which is defined as $e(g^x, g^y) := g^{2xy}$. The reason why we define it in this manner is that we want to ensure that the CDH assumption holds in $\mathbb{QR}_N^+$, even if e is efficiently computable. That is, even if we can compute $e(g^x, g^y) = g^{2xy}$, it is difficult to compute g^{xy} from it since the Rabin function is hard to invert under the factoring assumption. However, given only the group elements g^x and g^y, we do not know how to compute $e(g^x, g^y)$ efficiently. To address this, we introduce *auxiliary information* τ_y for each element $g^y \in \mathbb{QR}_N^+$ which enables us to compute a map $e(\cdot, g^y)$ efficiently. This leads to the notion of *self-bilinear map with auxiliary information* which we introduce in this paper.

The problem is how to define auxiliary information τ_y which enables us to compute $e(\cdot, g^y)$ efficiently. The most direct approach is to define τ_y as a circuit that computes the $2y$-th power. However, if we define τ_y as a "natural" circuit that computes the $2y$-th power, then we can extract $2y$ from τ_y, and thus we can compute y. This clearly enables us to compute g^{xy}, which breaks the CDH assumption.

A more clever way is to define τ_y as a circuit that computes the t_y-th power where $t_y = 2y \pm \mathrm{ord}(\mathbb{QR}_N^+)$.[2] In this way, it seems that τ_y does not reveal y since t_y is a "masked" value of $2y$ by $\mathrm{ord}(\mathbb{QR}_N^+)$ which is an unknown odd number. This idea is already used by Seurin [25] to construct a trapdoor DDH group. Actually, he proved that even if t_y is given in addition to g^x and g^y, it is still difficult to compute g^{xy}. In this way, it seems that we can construct a self-bilinear map with auxiliary information. However, this creates a problem: we do not have an efficient algorithm to compute t_y from y without knowing the factorization of N. If such an algorithm does not exist, then we cannot instantiate many bilinear map-based primitives using the resulting map such as the 3-party Diffie-Hellman key exchange [19].

To overcome the above difficulty, we use *indistinguishability obfuscation*. An indistinguishability obfuscator $(i\mathcal{O})$ is an efficient randomized algorithm that makes circuits C_0 and C_1 computationally indistinguishable if they have exactly the same functionality.

We observe that a circuit that computes the $2y$-th power and a circuit that computes the t_y-th power for an element of $\mathbb{QR}_N^+$ have exactly the same functionality since we have $t_y = 2y \pm \mathrm{ord}(\mathbb{QR}_N^+)$. Therefore if we obfuscate these circuits by $i\mathcal{O}$, then the resulting circuits are computationally indistinguishable. Then we define auxiliary information τ_y as an obfuscation of a circuit that computes the $2y$-th power. With this definition, it is clear that τ_y can be computed from y efficiently, and the above mentioned problem is solved. Moreover, τ_y is

---

[2] In the definition of t_y, whether $+$ or $-$ is used depends on y. See [25] for more details.

computationally indistinguishable from an obfuscation of a circuit that computes the t_y-th power. Therefore it must still be difficult to compute g^{xy} even if τ_y is given in addition to g^x and g^y.

Thus we obtain a self-bilinear map with auxiliary information on $\mathbb{QR}_N^+$ while ensuring that the auxiliary information does not allow the CDH assumption to be broken. Moreover, by extending this, we can prove that an analogue of multilinear CDH assumption holds with respect to a multilinear map which is constructed from our self-bilinear map with auxiliary information based on $i\mathcal{O}$ and the factoring assumption.

1.3 Related Work

In cryptography, bilinear maps on elliptic curves were first used for breaking the discrete logarithm problem on certain curves [21]. The first constructive cryptographic applications of a bilinear map are given in [19,24,2]. Since then, many constructions of cryptographic primitives based on a bilinear map have been proposed.

Boneh and Silverberg [3] considered a multilinear map which is an extension of a bilinear map, and showed its usefulness for constructing cryptographic primitives though they did not give a concrete construction of multilinear maps. Garg et al. [8] proposed a candidate construction of multilinear maps based on ideal lattices. Coron et al. [6] proposed another construction over the integers.

The notion of indistinguishability obfuscation was first proposed by Barak et al. [1]. The first candidate construction of indistinguishability obfuscation was proposed by Garg et al. [10]. Since then, many applications of indistinguishability obfuscation have been proposed [23,9,18,14,4].

2 Preliminaries

2.1 Notations

We use $\mathbb{N}$ to denote the set of all natural numbers, and $[n]$ to denote the set $\{1, \ldots n\}$ for $n \in \mathbb{N}$. If S is a finite set, then we use $x \xleftarrow{\$} S$ to denote that x is chosen uniformly at random from S. If $\mathcal{A}$ is an algorithm, we use $x \leftarrow \mathcal{A}(y; r)$ to denote that x is output by $\mathcal{A}$ whose input is y and randomness is r. We often omit r. We say that a function $f(\cdot) : \mathbb{N} \to [0, 1]$ is negligible if for all positive polynomials $p(\cdot)$ and all sufficiently large $\lambda \in \mathbb{N}$, we have $f(\lambda) < 1/p(\lambda)$. We say f is overwhelming if $1 - f$ is negligible. We say that an algorithm $\mathcal{A}$ is efficient if there exists a polynomial p such that the running time of $\mathcal{A}$ with input length λ is less than $p(\lambda)$. For two integers $x \neq 0$ and y, we say that x and y are negligibly close if $|x - y|/x$ is negligible. For a set S and a random variable x over S, we say that x is almost random on S if the statistical distance between the distribution of x and the uniform distribution on S is negligible. For a circuit C, we denote the size of C by $|C|$. For a wire w which is an output wire of a gate, we denote the first input incoming wire of the gate by $A(w)$ and the second incoming wire of the gate by $B(w)$. We use λ to denote the security parameter.

2.2   Indistinguishability Obfuscator

Here, we recall the definition of an indistinguishability obfuscator [10,23].

Definition 1. *(Indistinguishability Obfuscator.) Let C_λ be the class of circuits of size at most λ. An efficient randomized algorithm $i\mathcal{O}$ is called an indistinguishability obfuscator for P/poly if the following conditions are satisfied:*

- *For all security parameters $\lambda \in \mathbb{N}$, for all $C \in C_\lambda$, we have that*

$$\Pr[\forall x \ C'(x) = C(x) : C' \leftarrow i\mathcal{O}(\lambda, C)] = 1.$$

- *For any (not necessarily uniform) efficient algorithm $\mathcal{A} = (\mathcal{A}_1, \mathcal{A}_2)$, there exists a negligible function α such that the following holds: if $\mathcal{A}_1(1^\lambda)$ always outputs (C_0, C_1, σ) such that we have $C_0, C_1 \in C_\lambda$ and $\forall x \ C_0(x) = C_1(x)$, then we have*

$$|\Pr[\mathcal{A}_2(\sigma, i\mathcal{O}(\lambda, C_0)) = 1 : (C_0, C_1, \sigma) \leftarrow \mathcal{A}_1(1^\lambda)]$$
$$- \Pr[\mathcal{A}_2(\sigma, i\mathcal{O}(\lambda, C_1)) = 1 : (C_0, C_1, \sigma) \leftarrow \mathcal{A}_1(1^\lambda)]| \leq \alpha(\lambda)$$

Note that a candidate construction of $i\mathcal{O}$ that satisfies the above definition is given in [10].

2.3   Group of Signed Quadratic Residues

Here, we recall the definition and some properties of a group of signed quadratic residues [17] that we mainly work with in this paper. An integer $N = PQ$ is called a Blum integer if P and Q are distinct primes and $P \equiv Q \equiv 3 \bmod 4$ holds. Let $\mathsf{RSAGen}(1^\lambda)$ be an efficient algorithm which outputs a random $\ell_N(\lambda)$-bit Blum integer $N = PQ$ and its factorization (P, Q) so that the length of P and Q are the same and we have $\gcd(P - 1, Q - 1) = 1$. For simplicity, we often omit λ and simply denote $\ell_N(\lambda)$ as ℓ_N. We say that the factoring assumption holds with respect to RSAGen if for any efficient adversary $\mathcal{A}$, $\Pr[x \in \{P, Q\} : (N, P, Q) \leftarrow \mathsf{RSAGen}(1^\lambda), x \leftarrow \mathcal{A}(1^\lambda, N)]$ is negligible. We define the group of quadratic residues as $\mathbb{QR}_N := \{u^2 : u \in \mathbb{Z}_N^*\}$. Note that $\mathbb{QR}_N$ is a cyclic group of order $(P - 1)(Q - 1)/4$ if N is output by $\mathsf{RSAGen}(1^\lambda)$.

For any subgroup $H \in \mathbb{Z}_N^*$, we define its signed group as $H^+ := \{|x| : x \in H\}$ where $|x|$ is the absolute value of x when it is represented as an element of $\{-(N-1)/2, \ldots, (N-1)/2\}$. This is certainly a group by defining a multiplication as $x \circ y := |(xy \bmod N)|$ for $x, y \in H^+$. For simplicity, we often denote multiplications on H^+ as usual multiplication when it is clear that we are considering a signed group. If H is a subgroup of $\mathbb{QR}_N$, then $H \cong H^+$ by the natural projection since $-1 \notin \mathbb{QR}_N$. In particular, $\mathbb{QR}_N^+$ is a cyclic group of order $(P - 1)(Q - 1)/4$. We call $\mathbb{QR}_N^+$ a group of signed quadratic residues. A remarkable property of $\mathbb{QR}_N^+$ is that it is efficiently recognizable. That is, there exists an efficient algorithm that determines whether a given string is an element of $\mathbb{QR}_N^+$ or not [17].

3   Self-bilinear Maps

In this section, we recall the definition of a self-bilinear map [5]. Next, we introduce the notion of *self-bilinear map with auxiliary information* which is a weaker variant of a self-bilinear map. Finally we define hardness assumptions with respect to a multilinear map which is constructed from a self-bilinear map.

3.1   Definition of a Self-bilinear Map

First, we recall the definition of a self-bilinear map. A self-bilinear map is a bilinear map where the domain and target groups are identical. The formal definition is as follows.

Definition 2. *(Self-bilinear Map [5]) For a cyclic group G, a self-bilinear map $e : G \times G \to G$ has the following properties.*

- *For all $g_1, g_2 \in G$ and $\alpha \in \mathbb{Z}$, it holds that*

$$e(g_1^{\alpha}, g_2) = e(g_1, g_2^{\alpha}) = e(g_1, g_2)^{\alpha}.$$

- *The map e is non-degenerate, i.e, if $g_1, g_2 \in G$ are generators of G, then $e(g_1, g_2)$ is a generator of G.*

In addition to the above, we usually require that e is efficiently computable. As shown in [5], we can construct an n-multilinear map for any integer $n \geq 2$ from a self-bilinear map e. This can be seen by easy induction: suppose that an n-multilinear map e_n can be constructed from a self-bilinear map e, then we can construct an $(n + 1)$-multilinear map e_{n+1} by defining

$$e_{n+1}(g_1, \ldots, g_n, g_{n+1}) := e(e_n(g_1, \ldots, g_n), g_{n+1}).$$

3.2   Self-bilinear Map with Auxiliary Information

Instead of constructing a self-bilinear map, we construct a *self-bilinear map with auxiliary information* which is a weaker variant of a self-bilinear map. In a self-bilinear map with auxiliary information, the map is efficiently computable only if "auxiliary information" is given. That is, when we compute $e(g^x, g^y)$, we require auxiliary information τ_x for g^x or τ_y for g^y. This is the difference from an "ideal" self-bilinear map in which $e(g^x, g^y)$ can be computed only from g^x and g^y. We formalize a self-bilinear map with auxiliary information as a set of algorithms $\mathcal{SBP} = (\mathsf{InstGen}, \mathsf{Sample}, \mathsf{AIGen}, \mathsf{Map}, \mathsf{AIMult})$ and a set R of integers.

$\mathsf{InstGen}(1^{\lambda}) \to \mathsf{params}$: $\mathsf{InstGen}$ takes the security parameter 1^{λ} as input and outputs the public parameters params which specifies an efficiently recognizable cyclic group G on which the group operation is efficiently computable. We require that an approximation $\mathsf{Approx}(G)$ of $\mathrm{ord}(G)$ can be computed efficiently from params and that $\mathsf{Approx}(G)$ is negligibly close to $\mathrm{ord}(G)$. Additionally, params specifies sets T_x^{ℓ} of auxiliary information for all integers x and $\ell \in \mathbb{N}$.

Sample(params) $\to g$: Sample takes params as input and outputs an almost random element g of G. The self-bilinear map $e : G \times G \to G$ is defined with respect to the element g.

AIGen(params, ℓ, x) $\to \tau_x$: AIGen takes params, level ℓ and an integer $x \in R$ as input, and outputs corresponding auxiliary information $\tau_x \in T_x^\ell$.

Map(params, g^x, τ_y) $\to e(g^x, g^y)$: Map takes params, $g^x \in G$ and $\tau_y \in \cup_{\ell \in \mathbb{N}} T_y^\ell$ as input and outputs $e(g^x, g^y)$. By using this algorithm iteratively, we can compute $e_n(g_1^{x_1}, \ldots, g_n^{x_n})$ if we are given $g^{x_1}, \ldots, g^{x_n}$ and $\tau_{x_1}, \ldots, \tau_{x_n}$. (Note that not all of these elements are required to evaluate the map.)

AIMult(params, ℓ, τ_x, τ_y) $\to \tau_{x+y}$: AIMult takes params, ℓ, $\tau_x \in T_x^{\ell_1}$, $\tau_y \in T_y^{\ell_2}$ such that $\ell > \max\{\ell_1, \ell_2\}$ as input and outputs $\tau_{x+y} \in T_{x+y}^\ell$.

In addition to the above algorithms, we require for $\mathcal{SBP}$ to satisfy the following property.

Indistinguishability of Auxiliary Information. We require that any efficient algorithm which is given auxiliary information cannot tell whether it is generated by AIGen or AIMult. More formally, for any params $\leftarrow$ InstGen(1^λ), $\ell \in \mathbb{N}$ (which does not depend on λ), natural numbers $\ell_1, \ell_2 < \ell$, integers x, y and z (which are polynomially bounded in λ), such that $z \in R$ and $z \equiv x + y \mod \mathrm{ord}(G)$, and auxiliary information $\tau_x \in T_x^{\ell_1}$ and $\tau_y \in T_y^{\ell_2}$, the following two distributions are computationally indistinguishable:

$$\mathcal{D}_1 = \{\tau_z : \tau_z \leftarrow \mathsf{AIGen}(\mathsf{params}, \ell, z)\}$$

$$\mathcal{D}_2 = \{\tau_{x+y} : \tau_{x+y} \leftarrow \mathsf{AIMult}(\mathsf{params}, \ell, \tau_x, \tau_y)\}.$$

Remark 1. *A level ℓ of auxiliary information grows by at least 1 when AIMult is applied. One can think of it as an analogue of a noise level in the GGH graded encoding [11]. In our construction, the size of auxiliary information grows exponentially in a level ℓ. Therefore an efficient algorithm can only handle auxiliary information of a constant level. Actually, in our applications in this paper, ℓ is set at most 2.*

3.3   Hardness Assumptions

For cryptographic use, we introduce some hardness assumptions. We use $\mathcal{SBP}$ to construct a multilinear map, and thus our hardness assumptions are associated with a multilinear map which is constructed from $\mathcal{SBP}$. In the following, we let $\mathcal{SBP} = ($InstGen, Sample, AIGen, Map, AIMult$)$ be self-bilinear map procedures. First, we define the multilinear computational Diffie-Hellman with auxiliary information (MCDHAI) assumption which is an analogue of the multilinear computational Diffie-Hellman (MCDH) assumption.

Definition 3. *(MCDHAI assumption) We say that the n-MCDHAI assumption holds with respect to $\mathcal{SBP}$ if for any efficient algorithm $\mathcal{A}$,*

$$\Pr[e_n(g, \ldots, g)^{s\Pi_{i=1}^n x_i} \leftarrow \mathcal{A}(\mathsf{params}, g, g^s, g^{x_1}, \ldots, g^{x_n}, \tau_s, \tau_{x_1} \ldots, \tau_{x_n})]$$

is negligible, where params $\leftarrow$ InstGen(1^λ), $g \leftarrow$ Sample(params), $s, x_1, \ldots, x_n \leftarrow$ [Approx(G)], $\tau_s \leftarrow$ AIGen(params, $1, s$), $\tau_{x_i} \leftarrow$ AIGen(params, $1, x_i$) *for all* $i \in [n]$.

We say that the MCDHAI assumption holds with respect to $\mathcal{SBP}$ if the n-MCDHAI assumption holds with respect to $\mathcal{SBP}$ for any integer n which is polynomially bounded in λ.

We also define the multilinear hashed Diffie-Hellman with auxiliary information (MHDHAI) assumption which is an analogue of the multilinear hashed Diffie-Hellman (MHDH) assumption.

Definition 4. *(MHDHAI assumption) We say that the n-MHDHAI assumption holds with respect to $\mathcal{SBP}$ and a family of hash functions $\mathcal{H} = \{H : G \to \{0,1\}^k\}$ if for any efficient algorithm $\mathcal{D}$,*

$$| \Pr[1 \leftarrow \mathcal{D}(\text{params}, g, g^s, g^{x_1}, \ldots, g^{x_n}, \tau_s, \tau_{x_1} \ldots, \tau_{x_n}, H, T)|\beta = 1]$$
$$- \Pr[1 \leftarrow \mathcal{D}(\text{params}, g, g^s, g^{x_1}, \ldots, g^{x_n}, \tau_s, \tau_{x_1} \ldots, \tau_{x_n}, H, T)|\beta = 0]|$$

is negligible, where params $\leftarrow$ InstGen(1^λ), $g \leftarrow$ Sample(params), $s, x_1, \ldots, x_n \leftarrow$ [Approx(G)], $\tau_s \leftarrow$ AIGen(params, $1, s$), $\tau_{x_i} \leftarrow$ AIGen(params, $1, x_i$) *for all* $i \in [n]$, $\beta \xleftarrow{\$} \{0,1\}$ *and* $T \xleftarrow{\$} \{0,1\}^k$ *if* $\beta = 0$, *and otherwise* $T = H(e_n(g, \ldots, g)^{s\Pi_{i=1}^n x_i})$.

We say that the MHDHAI assumption holds with respect to $\mathcal{SBP}$ and $\mathcal{H}$ if the n-MHDHAI assumption holds with respect to $\mathcal{SBP}$ and $\mathcal{H}$ for any integer n which is polynomially bounded in λ.

Note that if the MCDHAI assumption holds with respect to $\mathcal{SBP}$ then the MHD-HAI assumption holds with respect to $\mathcal{SBP}$ and the Goldreich-Levin hardcore bit function [13].

4   Our Construction of a Self-bilinear Map

In this section, we construct a self-bilinear map with auxiliary information by giving a construction of self-bilinear map procedures $\mathcal{SBP}$. We prove that the MCDHAI assumption holds with respect to $\mathcal{SBP}$ if the factoring assumption holds and there exists an indistinguishability obfuscator for $P/poly$.

4.1   Construction

First we prepare some notations for circuits on $\mathbb{QR}_N^+$.

Notation for Circuits on $\mathbb{QR}_N^+$. In the following, for an ℓ_N-bit RSA modulus N and an integer $x \in \mathbb{Z}$, $\mathcal{C}_{N,x}$ denotes a set of circuits $C_{N,x}$ that work as follows. For input $y \in \{0,1\}^{\ell_N}$, $C_{N,x}$ interprets y as an element of $\mathbb{Z}_N$ and returns y^x where the exponentiation is done on $\mathbb{QR}_N^+$ if $y \in \mathbb{QR}_N^+$ and otherwise returns 0^{ℓ_N} (which is interpreted as $\bot$). We define the canonical circuit $\tilde{C}_{N,x}$ in $\mathcal{C}_{N,x}$ in a natural way [3]. For circuits C_1, C_2 whose output can be interpreted as elements

---

[3] There is flexibility to define the canonical circuit. However, any definition works if the size of $\tilde{C}_{N,x}$ is polynomially bounded in λ and $|x|$.

of $\mathbb{QR}_N^+$, $\mathsf{Mult}(C_1, C_2)$ denotes a circuit that computes $C_{\mathsf{mult}}(C_1(x), C_2(y))$ for input (x, y) where C_{mult} is a circuit that computes a multiplication for elements of $\mathbb{QR}_N^+$. If an input of C_{mult} is not a pair of two elements in $\mathbb{QR}_N^+$, then it outputs 0^{ℓ_N}.

Now we are ready to describe our construction. The construction of $\mathcal{SBP}$ is as follows.

$\mathsf{InstGen}(1^\lambda) \to \mathsf{params}$: Run $\mathsf{RSAGen}(1^\lambda)$ to obtain (N, P, Q), and outputs $\mathsf{params} = N$. params defines the underlying group $G = \mathbb{QR}_N^+$ and $\mathsf{Approx}(G) = (N-1)/4$. For an integer x and $\ell \in \mathbb{N}$, the set T_x^ℓ is defined as $T_x^\ell = \{i\mathcal{O}(M_\ell, C_{N,2x}; r) : C_{N,2x} \in \mathcal{C}_{N,2x} \text{ such that } |C_{N,2x}| \leq M_\ell, r \in \{0,1\}^*\}$, where M_ℓ is defined later.

$\mathsf{Sample}(\mathsf{params}) \to g$: Choose a random element $g \in \mathbb{Z}_N^*$, computes g^2 in $\mathbb{Z}_N^*$ and outputs $|g^2|$ where the absolute value is taken when it is represented as an element of $\{-(N-1)/2, \ldots, (N-1)/2\}$. When $\mathsf{params} = N$ and a generator $g \in \mathbb{QR}_N^+$ are fixed, the self-bilinear map e is defined as $e(g^x, g^y) = g^{2xy}$.

$\mathsf{AIGen}(\mathsf{params}, \ell, x) \to \tau_x$: Define the range of x as $R := [(N-1)/2]$. Take the canonical circuit $\tilde{C}_{N,2x} \in \mathcal{C}_{N,2x}$, set $\tau_x \leftarrow i\mathcal{O}(M_\ell, \tilde{C}_{N,2x})$ and output τ_x.

$\mathsf{Map}(\mathsf{params}, g^x, \tau_y) \to e(g^x, g^y)$: Compute $\tau_y(g^x)$ and output it. (Recall that τ_y is a circuit that computes the $2y$-th power for an element of $\mathbb{QR}_N^+$.)

$\mathsf{AIMult}(\mathsf{params}, \ell, \tau_x, \tau_y) \to \tau_{x+y}$: Compute $\tau_{x+y} \leftarrow i\mathcal{O}(M_\ell, \mathsf{Mult}(\tau_x, \tau_y))$ and output it.

Definition of M_ℓ. M_ℓ represents an upper bound of the size of a circuit which is obfuscated by $i\mathcal{O}$ when auxiliary information with level ℓ is generated. To define it, we consider another integer M_ℓ' which represents an upper bound of the size of auxiliary information with level ℓ. We define M_ℓ and M_ℓ' recursively. We define M_0' as an integer which is larger than $\max_{x \in [(N+1)/2]}\{|\tilde{C}_{N,x}|\}$. For $\ell \geq 1$, we define $M_\ell := 2M_{\ell-1}' + |C_{\mathsf{Mult}}|$ and $M_\ell' := poly(M_\ell, \lambda)$ where $poly$ is a polynomial that satisfies $|i\mathcal{O}(M, C)| < poly(M, \lambda)$ for any integer M and circuit C such that $|C| < M$.

Indistinguishability of Auxiliary Information. If $z \equiv x + y \bmod \mathsf{ord}(\mathbb{QR}_N^+)$ holds, then $C_{N,2z}$ and $\mathsf{Mult}(\tau_x, \tau_y)$ have exactly the same functionality. Therefore if we obfuscate these circuits by $i\mathcal{O}$, then the resulting circuits are computationally indistinguishable.

4.2   Hardness Assumptions

We prove that the MCDHAI assumption holds with respect to our construction of a self-bilinear map if $i\mathcal{O}$ is an indistinguishability obfuscator for $P/poly$ and the factoring assumption holds. From that, we can immediately see that the MHDHAI assumption also holds with respect to our construction if we use the Goldreich-Levin hardcore bit function [13] as $\mathcal{H}$.

First, we prove that the MCDHAI assumption holds if $i\mathcal{O}$ is an indistinguishability obfuscator for $P/poly$ and the factoring assumption holds.

Theorem 1. *The MCDHAI assumption holds with respect to $\mathcal{SBP}_{\mathsf{Ours}}$ if the factoring assumption holds with respect to* RSAGen *and* $i\mathcal{O}$ *is an indistinguishability obfuscator for* P/poly.

Proof. For an algorithm $\mathcal{A}$ and an integer n (which is polynomially bounded by the security parameter), we consider the following games.

Game 1. This game is the original n-MCDHAI game. More precisely, it is as follows.

$$(N, P, Q) \leftarrow \mathsf{RSAGen}(1^\lambda)$$
$$g \xleftarrow{\$} \mathbb{QR}_N^+$$
$$s, x_1, \ldots, x_n \xleftarrow{\$} [(N-1)/4]$$
$$\tau_s \leftarrow i\mathcal{O}(M_1, \tilde{C}_{N,2s}), \tau_{x_i} \leftarrow i\mathcal{O}(M_1, \tilde{C}_{N,2x_i}) \text{ for } i \in [n]$$
$$U \leftarrow \mathcal{A}(N, g, g^s, g^{x_1}, \ldots, g^{x_n}, \tau_s, \tau_{x_1} \ldots, \tau_{x_n})$$

Game 1' This game is the same as **Game 1** except that $s, x_1, \ldots, x_n$ are chosen from $[\mathrm{ord}(\mathbb{QR}_N^+)]$.

Game 2' This game is the same as **Game 1'** except that g, s, $x_1, \ldots, x_n$, $\tau_s, \tau_{x_1}, \ldots, \tau_{x_n}$ are set differently. More precisely, it is as follows.

$$(N, P, Q) \leftarrow \mathsf{RSAGen}(1^\lambda)$$
$$h \xleftarrow{\$} \mathbb{QR}_N^+$$
$$g := h^2$$
$$s', x_1', \ldots, x_n' \xleftarrow{\$} [\mathrm{ord}(\mathbb{QR}_N^+)]$$
$$g^s := g^{s'}h, \ g^{x_i} := g^{x_i'}h \text{ for } i \in [n]$$
(This implicitly defines $s \equiv s' + 1/2 \bmod \mathrm{ord}(\mathbb{QR}_N^+)$ and $x_i \equiv x_i' + 1/2 \bmod \mathrm{ord}(\mathbb{QR}_N^+)$).
$$\tau_s \leftarrow i\mathcal{O}(M_1, \tilde{C}_{N,2s'+1}), \tau_{x_i} \leftarrow i\mathcal{O}(M_1, \tilde{C}_{N,2x_i'+1}) \text{ for } i \in [n]$$
$$U \leftarrow \mathcal{A}(N, g, g^s, g^{x_1}, \ldots, g^{x_n}, \tau_s, \tau_{x_1} \ldots, \tau_{x_n})$$

Game 2. This game is the same as **Game 2'** except that $s, x_1, \ldots, x_n$ are chosen from $[(N-1)/4]$.

We say that $\mathcal{A}$ wins if it outputs $U = e_n(g, \ldots, g)^{s\Pi_{i=1}^n x_i}$. For $i = 1, 2$, we let T_i and T_i' be the events that $\mathcal{A}$ wins in **Game** i and **Game** i', respectively. What we want to prove is that $\Pr[T_1]$ is negligible. We prove it by the following lemmas.

Lemma 1. $|\Pr[T_i] - \Pr[T_i']|$ *is negligible for* $i = 1, 2$

Proof. This follows since $(N-1)/4$ is negligibly close to $\mathrm{ord}(\mathbb{QR}_N^+)$.

Lemma 2. $|\Pr[T_1'] - \Pr[T_2']|$ *is negligible if* $i\mathcal{O}$ *is an indistinguishability obfuscator for* P/poly.

Proof. We consider hybrid games $H_0, \ldots H_{n+1}$. A hybrid game H_i is the same as **Game 1'** except that the first i auxiliary information (i.e, $\tau_s, \tau_{x_1}, \ldots, \tau_{x_{i-1}}$) are generated as in **Game 2'**. It is clear that H_0 is identical to **Game 1'** and H_{n+1} is identical to **Game 2'**. Let S_i be the event that $\mathcal{A}$ wins in **Game** H_i.

It suffices to show that $|\Pr[S_i] - \Pr[S_{i-1}]|$ is negligible by the standard hybrid argument. We construct an algorithm $\mathcal{B} = (\mathcal{B}_1, \mathcal{B}_2)$ that breaks the security of $i\mathcal{O}$ for the security parameter M_1 by using $\mathcal{A}$ that distinguishes H_i and H_{i-1}. In the following, we use x_0 to mean s for notational convenience.

$\mathcal{B}_1(1^\lambda)$: $\mathcal{B}_1$ runs $(N, P, Q) \leftarrow \mathsf{RSAGen}(1^\lambda)$, chooses $h \xleftarrow{\$} \mathbb{QR}_N^+$ and $x_0, \ldots,$
$x_n \xleftarrow{\$} [\mathrm{ord}(\mathbb{QR}_N^+)]$ and sets $g := h^2$. $\mathcal{B}_1$ computes $x'_0, \ldots, x'_n \in \mathrm{ord}(\mathbb{QR}_N^+)$ such that $x_j \equiv x'_j + 1/2 \bmod \mathrm{ord}(\mathbb{QR}_N^+)$ for $j = 0, \ldots, n$. (This can be computed since $\mathcal{B}_1$ knows the factorization of N.) Then $\mathcal{B}_1$ sets $C_0 := \tilde{C}_{N, 2x_{i-1}}$,
$C_1 := \tilde{C}_{N, 2x'_{i-1}+1}$ and $\sigma := (N, P, Q, h, g, x_0, \ldots, x_n, x'_0, \ldots, x'_n)$ and outputs (C_0, C_1, σ).

$\mathcal{B}_2(\sigma, C^*)$: $\mathcal{B}_2$ sets

$$\tau_{x_j} \leftarrow \begin{cases} i\mathcal{O}(M_1, \tilde{C}_{N, 2x'_j+1}) & \text{if } j = 0, \ldots, i-2 \\ C^* & \text{if } j = i-1 \\ i\mathcal{O}(M_1, \tilde{C}_{N, 2x_j}) & \text{if } j = i, \ldots, n. \end{cases}$$

Then $\mathcal{B}_2$ runs $\mathcal{A}(N, g, g^{x_0}, \ldots, g^{x_n}, \tau_{x_0}, \ldots, \tau_{x_n})$ to obtain U. If we have $U = e_n(g, \ldots, g)^{\Pi_{i=0}^n x_i}$, then $\mathcal{B}_2$ outputs 1, and otherwise outputs 0.

The above completes the description of $\mathcal{B}$. First, we note that each of g_j ($j = 0, \ldots, n$) is distributed in $\mathbb{QR}_N^+$ independently of each other in all hybrid games H_i for $i = 0, \ldots, n+1$. Therefore $\mathcal{B}$ generates them in exactly the same way as those are generated in the hybrids H_{i-1} and H_i. Then we can see that $\mathcal{B}$ perfectly simulates H_{i-1} if $C^* \leftarrow i\mathcal{O}(M_1, C_0)$ and H_i if $C^* \leftarrow i\mathcal{O}(M_1, C_1)$ from the view of $\mathcal{A}$. If the difference between the probability that $\mathcal{A}$ wins in H_{i-1} and that in H_i is non-negligible, then $\mathcal{B}$ succeeds in distinguish whether C^* is computed as $C^* \leftarrow i\mathcal{O}(M_1, C_0)$ or $C^* \leftarrow i\mathcal{O}(M_1, C_1)$, with non-negligible advantage, and thus breaks the security of $i\mathcal{O}$.

Lemma 3. $\Pr[T_2]$ *is negligible if the factoring assumption holds.*

Proof. Assuming that $\mathcal{A}$ wins in **Game** 2 with non-negligible probability, we construct an algorithm $\mathcal{B}$ that computes $h^{1/2}$ given an RSA modulus N and a random element $h \in \mathbb{QR}_N^+$ with non-negligible probability. This yields the factoring algorithm [17]. The construction of $\mathcal{B}$ is as follows.

$\mathcal{B}(N, h)$: $\mathcal{B}$ sets $g := h^2$ and chooses $s', x'_1, \ldots, x'_n \xleftarrow{\$} [(N-1)/4]$. Then $\mathcal{B}$ sets $g^s := g^{s'}h$, $g^{x_i} := g^{x'_i}h$ for all $i \in [n]$, $\tau_s \leftarrow i\mathcal{O}(M_1, \tilde{C}_{N, 2s'+1})$ and $\tau_{x_i} \leftarrow i\mathcal{O}(M_1, \tilde{C}_{N, 2x'_i+1})$ for all $i \in [n]$. Then $\mathcal{B}$ runs $\mathcal{A}(N, g, g^s, g^{x_1}, \ldots, g^{x_n}, \tau_s, \tau_{x_1} \ldots, \tau_{x_n})$. Let U be the output of $\mathcal{A}$. Then $\mathcal{B}$ computes $X := \Pi_{i=1}^n (2x'_i + 1)$ and outputs $Ug^{-(s'X+(X-1)/2)}$. (Note that X is odd and therefore $(X-1)/2$ is an integer.)

Since $\mathcal{B}$ perfectly simulates **Game** 2 from the view of $\mathcal{A}$, $\mathcal{A}$ outputs $e_n(g, \ldots, g)^{s\Pi_{i=1}^n x_i}$ with non-negligible probability. If it occurs, then we have

$$U = e_n(g, \ldots, g)^{s\Pi_{i=1}^n x_i} = g^{2^{n-1}s\Pi_{i=1}^n x_i} = h^{2^n s\Pi_{i=1}^n x_i} = h^{s\Pi_{i=1}^n 2x_i}$$
$$= h^{(s'+1/2)\Pi_{i=1}^n (2x'_i+1)} = h^{s'X+X/2} = h^{s'X+(X-1)/2+1/2}$$

and therefore we have $Ug^{-(s'X+(X-1)/2)} = h^{1/2}$.

Theorem 1 is proven by the above lemmas.   □

The following is immediate from Theorem 1 and the Goldreich-Levin theorem.

Theorem 2. *The MHDHAI assumption holds with respect to* $\mathcal{SBP}_{\text{Ours}}$ *and the Goldreich-Levin hardcore bit function if the factoring assumption holds with respect to* RSAGen *and* $i\mathcal{O}$ *is an indistinguishability obfuscator for P/poly.*

5   Applications of Our Self-bilinear Map

In Sec. 4, we constructed a self-bilinear map with auxiliary information. In this section, we construct a multilinear map, multiparty NIKE, distributed broadcast encryption and ABE for circuits by using it.

Multilinear Map. Here, we consider a multilinear map which is constructed from a self-bilinear map with auxiliary information. As shown in Sec. 3.1 we can construct a multilinear map by iterated usage of a self-bilinear map. However, if we use a self-bilinear map with auxiliary information as a building block, then the resulting multilinear map has a restricted functionality: we need auxiliary information to compute the map. The concrete formulation is as follows.

Similarly to self-bilinear map procedures in Sec. 3.2, we formalize a multilinear map with auxiliary information as a set of algorithms $\mathcal{SBP} = (\text{InstGen}_{\text{mult}}, \text{Sample}_{\text{mult}}, \text{AIGen}_{\text{mult}}, \text{Map}_{\text{mult}}, \text{AIMult}_{\text{mult}})$ and a set R of integers. $\text{InstGen}_{\text{mult}}$ takes the security parameter as input and outputs the public parameters params which specify an underlying group G and a multilinear map e on it. $\text{Sample}_{\text{mult}}$ takes params as input and outputs an almost random element of G. $\text{AIGen}_{\text{mult}}$ takes params, ℓ, and $x \in R$ as input and outputs auxiliary information τ_x of level ℓ with respect to x. Map_{mult} takes params, $g^{x_1}, \ldots, g^{x_n}, \tau_{x_1}, \ldots, \tau_{x_{n-1}}$ and a level of multilinearity n as input, and outputs $e_n(g^{x_1}, \ldots, g^{x_n})$. $\text{AIMult}_{\text{mult}}$ takes params, an integer ℓ and auxiliary information τ_x and τ_y whose levels are less than ℓ as input and outputs auxiliary information τ_{x+y} of level ℓ with respect to $x + y$. A more precise definition is given in the full version.

In spite of the limitation that it requires auxiliary information to compute the map, a multilinear map with auxiliary information is sufficient to replace existing multilinear maps in some applications. Moreover, our multilinear map has interesting properties that existing multilinear maps do not have: the level of multilinearity is not bounded at the instance generation phase and representations of group elements are compact, i.e., their sizes are independent of the level of multilinearity. By this property, cryptographic primitives which are constructed from our multilinear map inherit these properties too.

Multiparty NIKE. By extending the Diffie-Hellman key exchange [7] to a multilinear setting as in [3], we obtain a multiparty NIKE scheme. By using our multilinear map (with auxiliary information) as a building block, we obtain a multiparty NIKE scheme where the maximum number of users is not fixed in the setup phase. In particular, the size of both the public parameters and a public key generated by a user are independent of the number of users. Note that [4] also constructed multiparty NIKE schemes based on indistinguishability obfuscation. However, in their schemes, the setup algorithm or key generation algorithm have to take the number of users as input unlike ours.

Distributed Broadcast Encryption. It is known that a multiparty NIKE scheme can be converted to a *distributed broadcast encryption* [3,4], where a user can join the system by himself without the assistance of a (semi) trusted third party holding a master key. The conversion is very simple: The setup algorithm runs $\text{Setup}_{\text{NIKE}}(1^\lambda)$ to obtain PP and publishes it. A user who wants to join the system runs $\text{Publish}_{\text{NIKE}}(\text{PP})$ to obtain (pk, sk), publishes pk as his public key and keeps sk as his secret key. A sender who wants to send a message M to a set S of users plays the role of a user of the underlying NIKE, shares a derived key K with users in S and encrypts M to obtain a ciphertext Ψ by a symmetric key encryption scheme using the key K. A ciphertext consists of S, the sender's public key and Ψ. It is easy to prove that the resulting broadcast encryption scheme is CPA secure if the underlying multiparty NIKE scheme is statically secure. In our scheme, as in the multiparty NIKE scheme, all algorithms can be run independently of the number of users. In particular, the size of both the public parameters and a ciphertext overhead are independent of the number of users. This is the first distributed broadcast encryption scheme with this property. Note that [4] also constructed distributed broadcast encryption schemes based on indistinguishability obfuscation. However, in their schemes, the setup algorithm or key generation algorithm have to take the number of users as input unlike ours.

Attribute Based Encryption for Circuits. We can construct ABE for circuits based on our self-bilinear map almost similarly to the scheme in [11]. The concrete construction can be found in the full version.

6   Homomorphic Encryption

In this section, we construct a somewhat homomorphic encryption scheme by using an indistinguishability obfuscator. This is not a direct application of our self-bilinear map. However, the idea behind the construction is similar.

6.1   Definition of Homomorphic Encryption

Here, we recall some definitions for homomorphic encryption. A homomorphic encryption scheme HE consists of the four algorithms (KeyGen, Enc, Eval, Dec). KeyGen takes the security parameter 1^λ as input and outputs a public key pk and a secret key sk. Enc takes a public key pk and a massage $m \in \{0, 1\}$ as input, and outputs a ciphertext c. Eval takes a public key pk, a circuit f with input length ℓ and a set of ℓ ciphertexts $c_1, \ldots, c_\ell$ as input, and outputs a ciphertext c_f. Dec takes a secret key sk and a ciphertext c as input, and outputs a message m. For correctness of the scheme, we require that for all $(pk, sk) \leftarrow \mathsf{KeyGen}(1^\lambda)$ and all $m \in \{0, 1\}$, we have $\mathsf{Dec}(sk, \mathsf{Enc}(pk, m)) = m$ with overwhelming probability.

Next, we define some properties of homomorphic encryption such as the CPA security, $\mathcal{C}$-homomorphism, and compactness.

Definition 5. *(CPA security) We say that a scheme* HE *is CPA secure if for any efficient adversary* $\mathcal{A}$,

$$| \Pr[1 \leftarrow \mathcal{A}(pk, \mathsf{Enc}(pk, 0))] - \Pr[1 \leftarrow \mathcal{A}(pk, \mathsf{Enc}(pk, 1))]|$$

is negligible, where $(pk, sk) \leftarrow \mathsf{KeyGen}(1^\lambda)$.

Definition 6. *($\mathcal{C}$-homomorphism) Let* $\mathcal{C} = \{\mathcal{C}_\lambda\}_{\lambda \in \mathbb{N}}$ *be a class of circuits. A scheme* HE *is $\mathcal{C}$-homomorphic if for any family of circuits* $\{f_\lambda\}_{\lambda \in \mathbb{N}}$ *such that* $f_\lambda \in \mathcal{C}$ *whose input length is* ℓ *and any messages* $m_1, \dots, m_\ell \in \{0, 1\}$,

$$\Pr[\mathsf{Dec}(sk, \mathsf{Eval}(pk, C, c_1, \dots, c_\ell)) \neq C(m_1, \dots, m_\ell)]$$

is negligible, where $(pk, sk) \leftarrow \mathsf{KeyGen}(1^\lambda)$ *and* $c_i \leftarrow \mathsf{Enc}(pk, m_i)$.

Remark 2. *We can also consider the additional property that an output of* Eval *can be used as input of another homomorphic evaluation. This is called "multi-hop" homomorphism, and many fully homomorphic encryption schemes have this property. However, our scheme does not.*

Definition 7. *(Compactness) A homomorphic encryption scheme* HE *is compact if there exists a polynomial poly such that the output length of* Eval *is at most* $poly(\lambda)$-*bit.*

6.2 Φ-Hiding Assumption

Here, we give the definition of the Φ-hiding assumption [20] as follows. Let $\mathsf{RSA}[p \equiv 1 \bmod e]$ be an efficient algorithm which takes the security parameter 1^λ as input and outputs (N, P, Q) where $N = PQ$ is an ℓ_N-bit Blum integer such that $P \equiv 1 \bmod e$ and $\mathbb{QR}_N^+$ is cyclic. Let $\mathcal{P}_\ell$ be the set of all ℓ-bit primes.

Definition 8. *For a constant* c, *we consider the following distributions.*

$$\mathcal{R} = \{(e, N) : e, e' \xleftarrow{R} \mathcal{P}_{c\ell_N}; N \leftarrow \mathsf{RSA}[p \equiv 1 \bmod e'](1^\lambda)\}$$
$$\mathcal{L} = \{(e, N) : e \xleftarrow{R} \mathcal{P}_{c\ell_N}; N \leftarrow \mathsf{RSA}[p \equiv 1 \bmod e](1^\lambda)\}$$

We say that the Φ-hiding assumption holds with respect to RSA *if for any efficient adversary* $\mathcal{A}$, $| \Pr[1 \leftarrow \mathcal{A}(\mathcal{L})] - \Pr[1 \leftarrow \mathcal{A}(\mathcal{R})]|$ *is negligible.*

Parameters. According to [20], N can be factorized in time $O(N^\epsilon)$ where $e \xleftarrow{R} \mathcal{P}_{c\ell_N}$; $N \leftarrow \mathsf{RSA}[p \equiv 1 \bmod e](1^k)$ and $c = 1/4 - \epsilon$. In our scheme, we set c to be the value such that $c\ell_N = \lambda$. This setting avoids the above mentioned attack in a usual parameter setting (e.g., $\ell_N = 1024$ for 80-bit security).

6.3   Our Construction

Here, we construct a somewhat homomorphic encryption scheme by using indistinguishability obfuscation. We use the notation for circuits on $\mathbb{QR}_N^+$ which is given in Sec. 4. In addition to that, here, we use the following notation. For circuits C_1 and C_2 such that an output of C_1 can be interpreted as input for C_2, $C_1 \circ C_2$ denotes the composition of C_1 and C_2, i.e, $C_1 \circ C_2$ is a circuit that computes $C_2(C_1(x))$ for input x. The construction of our homomorphic encryption $\mathsf{HE}_{\mathsf{Ours}} = (\mathsf{KeyGen}, \mathsf{Enc}, \mathsf{Eval}, \mathsf{Dec})$ is as follows.

$\mathsf{KeyGen}(1^\lambda)$: Choose $e \xleftarrow{\$} \mathcal{P}_\lambda$ and $(N, P, Q) \leftarrow \mathsf{RSA}[p \equiv 1 \bmod e](1^\lambda)$. Choose $g \xleftarrow{\$} \mathbb{QR}_N^+$ and compute an integer ρ such that $\rho \equiv 0 \bmod \mathrm{ord}(\mathbb{QR}_N^+)/e$ and $\rho \equiv 1 \bmod e$. It outputs a public key $pk = (N, e, g)$ and a secret key $sk = (\rho, pk)$.

$\mathsf{Enc}(pk, m \in \{0,1\})$: Choose $r \xleftarrow{\$} [(N-1)/4]$, set $c \leftarrow i\mathcal{O}(\mathsf{Max}, \tilde{C}_{N,m+re})$ and output c, where Max is defined as an integer larger than
$$\max_{m \in \{0,1\}, r \in [(N-1)/4]}\{|\tilde{C}_{N,m+re}|\}.$$

$\mathsf{Eval}(pk, f, c_1, \ldots, c_\ell)$: Work only if $c_1, \ldots, c_\ell$ are circuits (i.e., generated by Enc). Convert f into an arithmetic circuit f' on $\mathbb{Z}_e$. (That is, each gate of f' is addition, multiplication or negation on $\mathbb{Z}_e$.)[4] Compute as follows for all wires of f' from wires with lower depth.

- *Input:* Let w be the i-th input wire. Then c_i is assigned to this wire.
- *Addition:* Let w be an output wire of an addition gate. Set $c_w := \mathsf{Mult}(c_{A(w)}, c_{B(w)})$.
- *Multiplication:* Let w be an output wire of a multiplication gate. Set $c_w := c_{A(w)} \circ c_{B(w)}$.
- *Negation:* Let w be an output wire of a negation gate. Set $c_w := C_{N,inv} \circ c_{A(w)}$ where $C_{N,inv}$ is a circuit that computes an inverse on $\mathbb{QR}_N^+$.

Let v be the output wire. Compute $c_{\mathsf{eval}} = c_v(g)$ and output it. Note that it is a group element and not a circuit. Therefore we cannot evaluate it again.

$\mathsf{Dec}(sk, c)$: Work differently depending on whether c is an output of Enc or Eval. If c is an output of Enc, then compute $M = c(g)$. If $M^\rho = 1$, then output 0, and otherwise output 1. If c is an output of Eval, then output 0 if $c^\rho = 1$, and otherwise output 1.

First, we prove the correctness of the scheme. We have $e|\mathrm{ord}(\mathbb{QR}_N^+)$ by the choice of N. Therefore, there exists a subgroup G_e^+ of order e of $\mathbb{QR}_N^+$. We can see that for any element $h \in \mathbb{QR}_N^+$, h^ρ is the G_e^+ component of h. In the decryption, we have $M = i\mathcal{O}(\mathsf{Max}, C_{N,m+re})(g) = g^{m+re}$. Therefore M^ρ is the G_e^+ component of g^m. We can see that G_e^+ component of g is not 1 with overwhelming probability since e is a λ-bit prime. Therefore $M^\rho = 1$ is equivalent to $m = 0$ and $M^\rho \neq 1$ is equivalent to $m = 1$ with overwhelming probability. Thus the correctness follows.

The security of $\mathsf{HE}_{\mathsf{Ours}}$ relies on the Φ-hiding assumption.

---

[4] This can be done since we have $a \wedge b = a \cdot b \bmod e$ and $a \vee b = a + b - a \cdot b \bmod e$ if $a, b \in \{0,1\}$.

Theorem 3. $\mathsf{HE}_{\mathsf{Ours}}$ *is NC^1-homomorphic, compact and CPA secure if the Φ-hiding assumption holds with respect to* RSA *and* $i\mathcal{O}$ *is an indistinguishability obfuscator for* $P/poly$.

Here, we give only an intuitive explanation. The full proof can be found in the full version. The compactness is clear since an output of Eval consists of one group element of $\mathbb{QR}_N^+$. It is easy to see that evaluated ciphertexts are decrypted correctly. The problem is whether Eval works in polynomial time. To see this, we observe that the size of a circuit assigned to a wire of depth i is $O(2^i poly(\lambda))$. Thus if the depth of an evaluated circuit is $O(\log \lambda)$, then the size of the circuit assigned to an output wire is $O(poly(\lambda))$ and thus Eval works in polynomial time. The CPA security is reduced to the ϕ-hiding assumption: by the assumption, if N is replaced with N' such that e does not divide $\mathrm{ord}(\mathbb{QR}_{N'}^+)$, any efficient adversary cannot tell the difference. We can see that $(re \bmod \mathrm{ord}(\mathbb{QR}_{N'}^+))$ is distributed almost uniformly where $r \xleftarrow{\$} [(N'-1)/4]$ since $\gcd(e, \mathrm{ord}(\mathbb{QR}_N^+)) = 1$ holds. Thus $((m+re) \bmod \mathrm{ord}(\mathbb{QR}_{N'}^+))$ is uniformly distributed regardless of the value of m and the ciphertexts of 0 and 1 are distributed almost identically.

Acknowledgment. We would like to thank the anonymous reviewers and members of the study group "Shin-Akarui-Angou-Benkyou-Kai" for their helpful comments. Especially, we would like to thank Satsuya Ohata for his instructive comment on self-bilinear maps, and Takahiro Matsuda and Jacob Schuldt for their detailed proofreading.

References

1. Barak, B., Goldreich, O., Impagliazzo, R., Rudich, S., Sahai, A., Vadhan, S.P., Yang, K.: On the (im)possibility of obfuscating programs. In: Kilian, J. (ed.) CRYPTO 2001. LNCS, vol. 2139, pp. 1–18. Springer, Heidelberg (2001)
2. Boneh, D., Franklin, M.: Identity-based encryption from the weil pairing. In: Kilian, J. (ed.) CRYPTO 2001. LNCS, vol. 2139, pp. 213–229. Springer, Heidelberg (2001)
3. Boneh, D., Silverberg, A.: Applications of multilinear forms to cryptography. Contemporary Mathematics 324, 71–90 (2002)
4. Boneh, D., Zhandry, M.: Multiparty key exchange, efficient traitor tracing, and more from indistinguishability obfuscation. In: Garay, J.A., Gennaro, R. (eds.) CRYPTO 2014, Part I. LNCS, vol. 8616, pp. 480–499. Springer, Heidelberg (2014)
5. Cheon, J.H., Lee, D.H.: A note on self-bilinear maps. Bulletin of the Korean Mathematical Society 46 (2009)
6. Coron, J.-S., Lepoint, T., Tibouchi, M.: Practical multilinear maps over the integers. In: Canetti, R., Garay, J.A. (eds.) CRYPTO 2013, Part I. LNCS, vol. 8042, pp. 476–493. Springer, Heidelberg (2013)
7. Diffie, W., Hellman, M.E.: New directions in cryptography. IEEE Transactions on Information Theory 22(6), 644–654 (1976)
8. Garg, S., Gentry, C., Halevi, S.: Candidate multilinear maps from ideal lattices. In: Johansson, T., Nguyen, P.Q. (eds.) EUROCRYPT 2013. LNCS, vol. 7881, pp. 1–17. Springer, Heidelberg (2013)

9. Garg, S., Gentry, C., Halevi, S., Raykova, M.: Two-round secure MPC from indistinguishability obfuscation. In: Lindell, Y. (ed.) TCC 2014. LNCS, vol. 8349, pp. 74–94. Springer, Heidelberg (2014)
10. Garg, S., Gentry, C., Halevi, S., Raykova, M., Sahai, A., Waters, B.: Candidate indistinguishability obfuscation and functional encryption for all circuits. In: FOCS, pp. 40–49 (2013)
11. Garg, S., Gentry, C., Halevi, S., Sahai, A., Waters, B.: Attribute-based encryption for circuits from multilinear maps. In: Canetti, R., Garay, J.A. (eds.) CRYPTO 2013, Part II. LNCS, vol. 8043, pp. 479–499. Springer, Heidelberg (2013)
12. Garg, S., Gentry, C., Sahai, A., Waters, B.: Witness encryption and its applications. In: STOC, pp. 467–476 (2013)
13. Goldreich, O., Levin, L.A.: A hard-core predicate for all one-way functions. In: STOC, pp. 25–32 (1989)
14. Goldwasser, S., Gordon, S.D., Goyal, V., Jain, A., Katz, J., Liu, F.-H., Sahai, A., Shi, E., Zhou, H.-S.: Multi-input functional encryption. In: Nguyen, P.Q., Oswald, E. (eds.) EUROCRYPT 2014. LNCS, vol. 8441, pp. 578–602. Springer, Heidelberg (2014)
15. Gorbunov, S., Vaikuntanathan, V., Wee, H.: Attribute-based encryption for circuits. In: STOC, pp. 545–554 (2013)
16. Groth, J., Ostrovsky, R., Sahai, A.: Perfect non-interactive zero knowledge for NP. In: Vaudenay, S. (ed.) EUROCRYPT 2006. LNCS, vol. 4004, pp. 339–358. Springer, Heidelberg (2006)
17. Hofheinz, D., Kiltz, E.: The group of signed quadratic residues and applications. In: Halevi, S. (ed.) CRYPTO 2009. LNCS, vol. 5677, pp. 637–653. Springer, Heidelberg (2009)
18. Hohenberger, S., Sahai, A., Waters, B.: Replacing a random oracle: Full domain hash from indistinguishability obfuscation. In: Nguyen, P.Q., Oswald, E. (eds.) EUROCRYPT 2014. LNCS, vol. 8441, pp. 201–220. Springer, Heidelberg (2014)
19. Joux, A.: A one round protocol for tripartite diffie-hellman. In: Bosma, W. (ed.) ANTS 2000. LNCS, vol. 1838, pp. 385–394. Springer, Heidelberg (2000)
20. Kiltz, E., O'Neill, A., Smith, A.: Instantiability of RSA-OAEP under chosen-plaintext attack. In: Rabin, T. (ed.) CRYPTO 2010. LNCS, vol. 6223, pp. 295–313. Springer, Heidelberg (2010)
21. Menezes, A., Okamoto, T., Vanstone, S.A.: Reducing elliptic curve logarithms to logarithms in a finite field. IEEE Transactions on Information Theory 39(5), 1639–1646 (1993)
22. Sahai, A., Waters, B.: Fuzzy identity-based encryption. In: Cramer, R. (ed.) EUROCRYPT 2005. LNCS, vol. 3494, pp. 457–473. Springer, Heidelberg (2005)
23. Sahai, A., Waters, B.: How to use indistinguishability obfuscation: Deniable encryption, and more. In: STOC (2014)
24. Sakai, R., Ohgishi, K., Kasahara, M.: Cryptosystems based on pairing. In: SCIS (2000) (in Japanese)
25. Seurin, Y.: New constructions and applications of trapdoor DDH groups. In: Kurosawa, K., Hanaoka, G. (eds.) PKC 2013. LNCS, vol. 7778, pp. 443–460. Springer, Heidelberg (2013)

On Virtual Grey Box Obfuscation
for General Circuits

Nir Bitansky[1,*], Ran Canetti[1,2,**], Yael Tauman Kalai[3],
and Omer Paneth[2,***]

[1] Tel Aviv University, Tel Aviv, Israel
[2] Boston University, Boston, MA, USA
[3] Microsoft Research New England, Cambridge, MA, USA

Abstract. An obfuscator $\mathcal{O}$ is Virtual Grey Box (VGB) for a class $\mathcal{C}$
of circuits if, for any $C \in \mathcal{C}$ and any predicate π, deducing $\pi(C)$ given
$\mathcal{O}(C)$ is tantamount to deducing $\pi(C)$ given unbounded computational
resources and polynomially many oracle queries to C. VGB obfuscation is
often significantly more meaningful than indistinguishability obfuscation
(IO). In fact, for some circuit families of interest VGB is equivalent to
full-fledged Virtual Black Box obfuscation.

We investigate the feasibility of obtaining VGB obfuscation for general
circuits. We first formulate a natural strengthening of IO, called *strong
IO* (SIO). Essentially, $\mathcal{O}$ is SIO for class $\mathcal{C}$ if $\mathcal{O}(C) \approx \mathcal{O}(C')$ whenever
the pair (C, C') is taken from a distribution over $\mathcal{C}$ where, for all x,
$C(x) \neq C'(x)$ only with negligible probability.

We then show that an obfuscator is VGB for a class $\mathcal{C}$ if and only
if it is SIO for $\mathcal{C}$. This result is unconditional and holds for any $\mathcal{C}$. We
also show that for some circuit collections, SIO implies virtual black-box
obfuscation.

Finally, we formulate a slightly stronger variant of the semantic security property of graded encoding schemes [Pass-Seth-Telang Crypto 14],
and show that existing obfuscators such as the obfuscator of Barak et. al
[Eurocrypt 14] are SIO for all circuits in NC^1, assuming that the underlying graded encoding scheme satisfies our variant of semantic security.

*Put together, we obtain VGB obfuscation for all NC^1 circuits under
assumptions that are almost the same as those used by Pass et. al to
obtain IO for NC^1 circuits. We also show that semantic security is in
essence necessary for showing VGB obfuscation.*

1 Introduction

Program obfuscation, namely the ability to efficiently compile a given program
into a functionally equivalent program that is "unintelligible", is an intriguing

* Supported by an IBM Ph.D. Fellowship, the Check Point Institute for Information
  Security, and the Israeli Ministry of Science and Technology.
** Supported by the Check Point Institute for Information Security, an NSF EAGER
  grant, and NSF Algorithmic Foundations grant no. 1218461.
*** Supported by the Simons award for graduate students in theoretical computer
  science and NSF Algorithmic Foundations grant no. 1218461.

J.A. Garay and R. Gennaro (Eds.): CRYPTO 2014, Part II, LNCS 8617, pp. 108–125, 2014.
© International Association for Cryptologic Research 2014

concept. Indeed, much effort has been devoted to understanding this concept from the definitional aspect, the algorithmic aspect, and the applications aspect. Here let us concentrate on the first two aspects.

Starting with the works of Hada [Had00] and Barak et al. [BGI+01], a number of measures of security for program obfuscation have been proposed. Let us briefly review three notions of interest. The first, *virtual black box (VBB)* obfuscation [BGI+01], requires that having access to the obfuscated program is essentially the same as having access to the program only as black box. Concretely, focusing on programs represented as circuits, an obfuscator $\mathcal{O}$ for a family of circuits is worst-case VBB if for any poly-time adversary $\mathcal{A}$, there exists a poly-time simulator S, such that for any circuit C from the family, and any predicate $\pi(\cdot)$, $\mathcal{A}$ cannot learn $\pi(C)$ from $\mathcal{O}(C)$ with noticeably higher probability than S can, given only oracle access to C. The obfuscator $\mathcal{O}$ is average-case VBB if the above is only required to hold for circuits C that are sampled at random from the family.

While this VBB obfuscation is natural and strong, Barak et al. [BGI+01] showed that this definition, and variants thereof, are unobtainable in general by demonstrating a family of *unobfuscatable functions:* these are functions f where any circuit computing the function inherently leaks secrets that are infeasible to compute given only black box access to f. Moreover it turns out that, under cryptographic assumptions, if the simulator S is universal (or equivalently, works for any adversarial auxiliary input) then VBB obfuscation is unobtainable for *any* circuit family whose functionality has super-polynomial "pseudo entropy" [GK05, BCC+14].

A weaker variant of VBB, called *virtual grey-box (VGB)* [BC10], allows the simulator to be *semi-bounded,* namely it can be computationally unbounded, while still making only a polynomial number of queries to the circuit C. While significantly weaker than VBB in general, VGB is still meaningful for circuits that are unlearnable even by semi-bounded learners. Furthermore, VGB obfuscators for circuits escape the general impossibility results that apply to VBB obfuscators.

A weaker notion yet, called indistinguishability obfuscation (IO) [BGI+01], allows the (now computationally unbounded) simulator to also make an unbounded number of queries to C. Equivalently, $\mathcal{O}$ is an IO for a circuit collection if for any two circuits C_0 and C_1 in the collection, having the same size and functionality, $\mathcal{O}(C_0)$ and $\mathcal{O}(C_1)$ are indistinguishable.

While IO has some attractive properties (e.g., any IO is the "best possible" obfuscation for its class), and some important cryptographic applications [SW13, GGH+13b], the security guarantees provided by IO are significantly weaker than those provided by either VBB or VGB obfuscation.

On the algorithmic level, for many years we had candidate obfuscators only for very simple functions such as point functions and variants. The landscape has changed completely with the recent breakthrough work of [GGH+13b], which proposed a candidate general-purpose obfuscation algorithm for all circuits.

[GGH+13b] show that their scheme resists some simple attacks; but beyond that, they do not provide any analytic evidence for security.

Considerable efforts have been made to analyze the security of the [GGH+13b] obfuscator and variants. The difficulty appears to be in capturing the security properties required from the *graded encodings schemes* [GGH13a, CLT13], which is a central component in the construction. As a first step towards understanding the security of the [GGH+13b] obfuscator, [BR13, BGK+13] consider an ideal algebraic model, where the adversary is given "generic graded encodings" that can only be manipulated via admissible algebraic operations. They show that, in this model, variants of the [GGH+13b] scheme are VBB obfuscators for all poly-size circuits. (We remark that [CV13] construct a VBB general obfuscator with similar properties; however their abstract model is different and does not seem to correspond to any existing cryptographic primitive.)

Still, neither of these idealized constructions or their analyses have, in of themselves, any bearing on the security of obfuscation algorithms in the plain model.

Pass et al. [PTS13] make the first step towards proving the security of a general obfuscation scheme based on some natural hardness assumption in the plain model. Specifically, they define a *semantic security* property for graded encoding schemes, which is aimed at capturing what it means for a graded encoding scheme to "behave essentially as an ideal multi-linear graded encoding oracle". They then show, assuming the existence of such a semantically-secure encoding scheme, that a specially-crafted variant of the [BGK+13] obfuscator, with the graded encoding scheme replaced by a semantically-secure graded encoding scheme, is IO for all circuits.

In this work we address the following question:

What is the strongest form of security for general obfuscation that can be based on natural cryptographic assumptions such as semantically-secure graded encoding?

Our contributions. As our main result we obtain worst-case VGB obfuscation for NC^1, based on almost the same assumptions as those used in [PTS13] to show IO for NC^1. As an intermediate step towards this goal, we put forth a somewhat stronger variant of indistinguishability obfuscation, called *strong IO* (SIO). Informally, an obfuscator $\mathcal{O}$ is SIO for a class of circuits $\mathcal{C}$ if $\mathcal{O}(C) \approx \mathcal{O}(C')$ not only when $C, C' \in \mathcal{C}$ have the same functionality, but also when C and C' come from distributions over circuits in $\mathcal{C}$ that are "close together", in the sense that at any given input x, the probability that $C(x) \neq C'(x)$ is negligible. An alternative view of the definition (which turns out to be equivalent) is that if no adversary (even computationally unbounded) can distinguish oracle access to C from access to C' given only polynomial many queries, then the obfuscated circuits should be indistinguishable as well.

We then show that:

1. Strong IO is in fact *equivalent* to worst-case VGB obfuscation. Furthermore, for certain classes of functions, such as point functions, hyperplanes, or fuzzy

point functions, SIO is equivalent to full-fledged worst-case VBB obfuscation. These equivalences hold unconditionally. We consider this to be the main technical step in this work.

2. Assuming the existence of graded encoding schemes that satisfy a somewhat stronger variant of the semantic security notion of Pass et al. [PTS13], we show that known obfuscation schemes are SIO for all circuits in NC^1. More generally, we show that *any* obfuscator for a class of circuits C that is VBB in the ideal graded encoding model, is SIO in the plain model, when the ideal graded encoding oracle is replaced by a graded encoding scheme that satisfies a variant of the [PTS13] assumption.

We also give evidence for the *necessity* of semantically-secure graded encoding for obtaining VGB. Specifically we show that, assuming the existence of VGB obfuscators for all circuits, there exists *mutlilinear jigsaw puzzles*, a simplified variant of multilinear maps [GGH+13b], that satisfies a form of semantic security. Such mutlilinear jigsaw puzzles are sufficient for obtaining the positive result described in Item 2 above.

Finally, we investigate the plausibility of the semantic security assumption on graded encoding schemes, propose some relaxed variants, and show that our main results can be obtained under all these relaxations. Namely, we first give new evidence for the relative strength of the semantically-secure graded encodings assumption. Specifically, we show that semantically-secure graded encodings are subject to the following limitations:

1. *SAT lower bounds.* We show that semantically-secure graded encodings imply exponential circuit lower bounds for SAT. Such lower bounds are currently not known to follow from IO (even assuming $P \neq NP$).

2. *A generic attack.* We present an attack showing that any graded encoding scheme with certain efficiency properties cannot satisfy semantic security. While the attack does not apply to currently known candidate graded encodings [GGH13a, CLT13], it does point out potential limitations of this notion. We complement this observation by suggesting a natural relaxation of semantic security called *bounded semantic-security* that bypasses this attack. Our main results can be obtained also under this relaxed assumption.

In addition to the above relaxation, we consider several other relaxations, and investigate their relations. We show that our main results can be obtained under all these relaxations.

The rest of the introduction provides a more detailed overview of our results. Section 1.1 presents the implication from SIO to VGB and VBB obfuscation. Section 1.2 provides background on graded encoding schemes and the semantic security assumption. Sections 1.3 presents the construction of strong IO obfuscators from semantically-secure graded encoding schemes, and Section 1.4 describes additional results on the viability of the semantic security assumption of graded encoding schemes, and relations among various variants.

1.1   From Strong IO to VGB and VBB Obfuscation

We first define strong IO a bit more precisely. A distribution $\tilde{C}$ over circuits is said to be ε-*concentrated* around a boolean function f if for any value x in the domain of f we have that $\Pr[\tilde{C}(x) \neq f(x)] \leq \varepsilon$. We say that $\tilde{C}$ is simply *concentrated* if it is ε-concentrated for some negligible function ε. An obfuscator $\mathcal{O}$ is strong IO for a class $\mathcal{C}$ of circuits if $\mathcal{O}(\tilde{C}) \approx \mathcal{O}(\tilde{C}')$ for any two distributions $\tilde{C}, \tilde{C}'$ over circuits in $\mathcal{C}$ that are concentrated around the same function. We stress that these distributions need not be efficiently samplable. We show the following.

Theorem 1 (informal). *An obfuscator is SIO for a class of circuits $\mathcal{C}$ if and only if it is worst-case VGB obfuscator for $\mathcal{C}$.*

Showing that VGB implies SIO is straightforward. In the other, more challenging, direction we construct an (inefficient) simulator $\mathcal{S}$ for any adversary $\mathcal{A}$. Recall that, for *any* circuit $C \in \mathcal{C}$ in the given collection $\mathcal{C}$, the simulator $\mathcal{S}$ should simulate what $\mathcal{A}$ learns from an obfuscation $\mathcal{O}(C)$, given only oracle access to C. The high level idea is as follows: $\mathcal{S}$ will use its oracle to C to gradually reduce the set $\mathcal{K}$ of candidates for the circuit C, starting from $\mathcal{K}_0 = \mathcal{C}$, and ending with a smaller set of candidates

$$\mathcal{K}_i \subsetneq \mathcal{K}_{i-1} \subsetneq \cdots \subsetneq \mathcal{K}_0 = \mathcal{C}.$$

$\mathcal{S}$ will continue this process until it obtains a set $\mathcal{K}^*$ where $\mathcal{A}$ cannot distinguish an obfuscation $\mathcal{O}(C)$ of C from an obfuscation $\mathcal{O}(C')$ of a random circuit C' in $\mathcal{K}^*$.

To carry out this plan, $\mathcal{S}^C$ iteratively performs two main steps: *concentration*, and *majority separation*. In the concentration step $\mathcal{S}$ tries to learn C in a straightforward way: it queries C on a point x_i that splits the current set of candidate circuits $\mathcal{K}_i$ as evenly as possible. Based on the value of $C(x_i)$, $\mathcal{S}$ rules out some of the candidates. This process is repeated until there is no query that necessarily shrinks the set of candidates by a factor of at least $1 - \varepsilon$, where ε is a parameter of the simulation that is chosen such that $1/\varepsilon$ is a polynomial, depending only on $\mathcal{A}$ and on the required simulation accuracy. Note that at the end of the concentration step, $\mathcal{S}$ must reach a set of candidates $\mathcal{K}_j$ that is ε-concentrated. This occurs after at most $\log |\mathcal{C}|/\varepsilon$ queries. The concentration step alone essentially suffices to ensure *average-case* VGB simulation; indeed, we show that for a circuit C chosen at random from a concentrated set $\mathcal{K}_j$, $\mathcal{A}$ cannot compute any predicate $\pi(C)$, given $\mathcal{O}(C)$, better than it can given an obfuscation $\mathcal{O}(C')$ of an independent random $C' \leftarrow \mathcal{K}_j$.

However, the concentration step alone does not guarantee *worst-case* simulation. In particular, $\mathcal{A}$ may have some hardwired information that allows it to distinguish C from a random circuit in $\mathcal{K}_j$. In this case, however, $\mathcal{S}$ can further reduce the set of candidates $\mathcal{K}_j$ by separating any such distinguishable circuit C from the majority $\text{maj}_{\mathcal{K}_j}$. Concretely, we define the set $\mathbb{D}_{\mathcal{A}}(\mathcal{K}_j)$ of *distinguishable circuits* in $\mathcal{K}_j$, as those circuits C in $\mathcal{K}_j$ such that $\mathcal{A}$ can ε-distinguish between $\mathcal{O}(C)$ and $\mathcal{O}(C')$ for a random $C' \leftarrow \mathcal{K}_j$.

In the majority-separation step, the simulator will query its oracle C on a *small* set of roughly $\log |\mathcal{C}|/\varepsilon$ points $L_{\mathcal{K}_j}$ that *separates* all the distinguishable circuits in $\mathbb{D}_\mathcal{A}(\mathcal{K}_j)$ from the majority circuit $\text{maj}_{\mathcal{K}_j}$. This means that, if the oracle C agrees with the majority $\text{maj}_{\mathcal{K}_j}$ on all points $x \in \mathcal{L}_{\mathcal{K}_j}$, then $\mathcal{A}$ cannot tell apart $\mathcal{O}(C)$ form $\mathcal{O}(C')$ for a random $C' \leftarrow \mathcal{K}_j$, in which case, the simulation can be completed. Otherwise, $\mathcal{S}$ manages to separate C from $\text{maj}_{\mathcal{K}_j}$, and obtain a new set of candidates $\mathcal{K}_{j+1} \subsetneq \mathcal{K}_j$ which is necessarily smaller by a $1 - \varepsilon$ factor, since $\mathcal{K}_j$ is ε-concentrated.

By iteratively applying the two steps we either reach some $\mathcal{K}^*$ for which $\mathcal{A}$ cannot distinguish $\mathcal{O}(C)$ from $\mathcal{O}(C')$ for a random $C' \leftarrow \mathcal{K}^*$, or we have completely exhausted the collection $\mathcal{C}$ and found exactly the circuit C. In any case, since we reduce $\mathcal{K}_j$ at each step by a $1 - \varepsilon$ factor, the process must end after at most $\log |\mathcal{C}|/\varepsilon$ steps, and at most $\text{poly}(\log |\mathcal{C}|/\varepsilon)$ queries.

But how do we establish the existence of a small set $L_{\mathcal{K}_j}$ that separates $\mathbb{D}_\mathcal{A}(\mathcal{K}_j)$ from the majority $\text{maj}_{\mathcal{K}_j}$? Here we rely on the fact that $\mathcal{O}$ is a strong IO obfuscator. Specifically, strong IO implies that any subset S of the distinguishable circuits $\mathbb{D}_\mathcal{A}(\mathcal{K}_j)$, cannot be ε-concentrated around $\text{maj}_{\mathcal{K}_j}$, because $\mathcal{A}$ distinguishes $\mathcal{O}(C)$, for $C \leftarrow \mathcal{K}_j$ from $\mathcal{O}(C')$ for $C' \leftarrow S \subseteq \mathbb{D}_\mathcal{A}(\mathcal{K}_j)$.[1] Since no S as above is ε-concentrated around $\text{maj}_{\mathcal{K}_j}$, we can separate all of the circuits in $\mathbb{D}_\mathcal{A}(\mathcal{K}_j)$ from $\text{maj}_{\mathcal{K}_j}$ with at most $\text{poly}(\log |\mathcal{C}|/\varepsilon)$ points, as required.

On the possibility of VBB obfuscation. The simulation strategy described above requires only a polynomial number of queries $\text{poly}(\log |\mathcal{C}|/\varepsilon)$; however, the overall running time of the simulator may not be bounded in general. Indeed, in the concentration step, finding a point x_j that significantly splits $\mathcal{K}_j$ may require super-polynomial time. Also, in the majority-separation step, while the sets $L_{\mathcal{K}_j}$ are small, computing them from $\mathcal{K}_j$ may also require super-poly time.

Nevertheless, we show that for certain classes of circuits, simulation can be done more efficiently, or even in polynomial time. Specifically, abstracting away from the above simulation process, we consider the notion of *learning via a majority-separation oracle*, where a given circuit C (or more generally a function) in a prescribed family is learned via oracle access to C and oracle access to the majority separation oracle $\mathbb{S}$, which takes as input (the description of a) a concentrated sub-family $\mathcal{K}$ that includes C and outputs a point x that separates C from $\text{maj}_\mathcal{K}$ (the majority of functions in $\mathcal{K}$).

The complexity of our simulator is then determined by how well can the class in question be learned by majority-separation oracles. While the strategy described above shows that any class of circuits can be learned by a majority-separation oracle with polynomially many queries to C and $\mathbb{S}$, the pattern of these queries and the way in which they are interleaved affects the complexity of the simulation. As a simple example, suppose that there is a constant number of oracle calls to either $\mathbb{S}$ or C. (This is the case in some classes for which worst-case VBB obfuscation was previously shown, such as point functions, constant-size

---

[1] We assume here (for simplicity and without loss of generality), that the distinguishing gap is always of the same sign.

set functions, and constant dimension hyper-planes.) In this case we can non-uniformly hardwire in advance a polynomial number of separating sets $\mathcal{L}_{\mathcal{K}_j}$ into our simulator, without having to compute them on the fly. Otherwise, the sets $\mathcal{L}_{\mathcal{K}_j}$ are determined *adaptively* and need to be computed on the fly.

Comparison to [BBC⁺14]. Barak et al. show that average-case VBB obfuscation for all *evasive* collections (these are collections that are concentrated around the constant zero function) implies *weak average-case* VGB for all collections, where weak VGB means that the simulator is allowed to make a slightly super-polynomial number of queries. The result is weaker in the sense that it only achieves average-case (rather than worst-case) simulation, and only weak VGB.

At a technical level, what allows us to get standard VGB, as opposed to weak VGB, is the fact that we assume IO for the family in question. More specifically, the level to which the simulator has to concentrate the candidate set is determined by the adversary and simulation quality. In the time of obfuscation, these parameters are not known. Relying on IO allows us to push the decision of how many iterations to make all the way to the simulation rather than having to make this decision at the time of obfuscation.

1.2 Semantically-Secure Graded Encoding Schemes: Background

Before describing how we get strong IO from semantically secure graded encoding schemes, we provide some background on the latter. A graded encoding scheme [GGH13a] consists of the following algorithms: InstGen that give a universe set $[k]$, outputs public parameters pp and secret parameters sp, where pp contains a description of a ring R; Encode that given sp, a set $S \subseteq [k]$ and $\alpha \in R$, generates an encoding $[\alpha]_S$; Add and Sub that, given encodings $[\alpha_1]_S$ and $[\alpha_2]_S$, generate encodings $[\alpha_1 + \alpha_2]_S$ and $[\alpha_1 - \alpha_2]_S$ respectively; Mult that, given encodings $[\alpha_1]_{S_1}$ and $[\alpha_2]_{S_2}$ such that $S_1 \cap S_2 = \emptyset$, generates an encoding $[\alpha_1 \cdot \alpha_2]_{S_1 \cup S_2}$; and isZero that given an encoding $[\alpha]_{[k]}$ outputs 1 if and only if $\alpha = 0$ (all the algorithms above also take as input pp).

[GGH13a, CLT13] consider standard versions of DDH-type security that can be conjectured to hold for their graded encoding schemes. Basing the security of obfuscation mechanisms on these assumptions seems at this point far out of reach, even if one considers only IO security. So which security properties of encoding schemes would suffice for this purpose? The high-level approach of Pass et al. [PTS13] is to devise a property that, not only hides "DDH-type relations" between encodings, but also any other relation that cannot be revealed using the admissible algebraic operations provided by the graded encoding interface. In other words, the encoding scheme should amount to an "ideal encoding scheme", where encodings are truly accessed only through admissible algebraic operations. This may, in particular, allow leveraging the existing proofs of VBB security in the ideal graded encoding model [BR13, BGK⁺13].

More specifically, Pass et al. take the following approach (described first in an oversimplified manner). Consider a *message sampler* $\mathbb{M}([k], R)$ that samples, from one of two distributions $\mathcal{D}_0$ or $\mathcal{D}_1$, a tuple $(S_1, m_1), \ldots, (S_\ell, m_\ell)$, where each $S_i \subseteq [k]$ and each $m_i \in R$, and ℓ is polynomial in the security parameter. We say that the sampler is *admissible* if no polynomially-bounded "algebraic adversary" that is given $\boldsymbol{S} = (S_1, \ldots, S_\ell)$, and can access the ring elements $\boldsymbol{m} = (m_1, \ldots, m_\ell)$ only via *an ideal encoding oracle*, is able to tell whether $(\boldsymbol{S}, \boldsymbol{m})$ were taken from $\mathcal{D}_0$ or $\mathcal{D}_1$. The ideal encoding oracle only allows the same algebraic manipulations allowed by the graded encoding interface, or put abstractly, it allows the adversary to choose any arithmetic circuit C that respects the set structure given by $\boldsymbol{S}$, and test whether $C(\boldsymbol{m}) = 0$. The requirement is that, for such an admissible sampler, an efficient adversary that obtains actual encodings $([m_i]_{S_i} : i \in [\ell])$, along with the corresponding public parameters pp, also cannot tell whether $(\boldsymbol{S}, \boldsymbol{m})$ was sampled from $\mathcal{D}_0$ or $\mathcal{D}_1$.

As noticed by Pass et al., the assumption formulated above is actually false—it is susceptible to a diagonalization attack in the spirit of the [BGI+01] impossibility result for general VBB obfuscation. More specifically, the unobfuscatable functions constructed in [BGI+01] can be directly used to obtain two distributions on circuits $\mathcal{C}_0$ and $\mathcal{C}_1$ which cannot be distinguished given only black-box access to C sampled from either $\mathcal{C}_0$ or $\mathcal{C}_1$, but given any circuit with the same functionality as the circuit C, it is easy to tell from which one of the two C was sampled from. This distinguishing attack could now be translated to our setting using any obfuscation scheme that uses ideal graded encoding, such as the ones of [BR13, BGK+13]. Indeed, we can define an admissible sampler that corresponds to distributions $\mathcal{D}_0$ and $\mathcal{D}_1$, which sample $(\boldsymbol{S}, \boldsymbol{m})$ by first sampling a circuit C taken from $\mathcal{C}_0$ or $\mathcal{C}_1$, respectively, and letting $(\boldsymbol{S}, \boldsymbol{m})$ correspond to the ideal obfuscation of C. Admissibility is guaranteed due to the VBB guarantee in the ideal encoding model, whereas in the real world, the actual encodings $([m_i]_{S_i})$ give a circuit that computes the same function of C, and thus allows determining from where the sample was taken.

Pass et al. get around this caveat by strengthening the admissibility requirement to require that $\mathcal{D}_0, \mathcal{D}_1$ are indistinguishable even to a *semi-bounded* algebraic adversary, namely an algebraic adversary that is computationally unbounded, but makes only a polynomial number of queries to the ideal graded encoding oracle. The above attack no longer applies since the circuit distributions $\mathcal{C}_0$ and $\mathcal{C}_1$ involve computational elements, such as encryption, which makes them completely distinguishable to unbounded attackers, even given only polynomially many oracle queries. More generally, as mentioned above, we do not have any analogous unobfuscatability results for VGB obfuscation, and thus there are no known attacks on this notion of semantic security.

Furthermore, Pass et al. show that even this relaxed assumption suffices for obtaining IO in the plain model. This is the case since for NC^1 circuits, the [BGK+13] obfuscator in the ideal-graded encoding model is VBB even against *semi-bounded* adversaries. (VBB, in this context, means that the simulator is poly-time, given oracle access to the algebraic adversary and the obfuscated program.) The eventual Pass et al. assumption is further relaxed in several ways, while still yielding their main application to IO.

1.3 Strong IO from Semantically-Secure Graded Encoding, and Back Again

We sketch our variant of the semantic security assumption, and explain how we obtain strong IO for NC^1 circuits from this variant. We also give evidence for the *necessity* of semantic security for obtaining strong IO.

To get strong IO for arbitrary circuit distributions (including distributions that are not efficiently samplable) we will need to rely on a somewhat stronger version of the semantic security assumption discussed above, that allows for computationally unbounded samplers. Some care has to be taken when formalizing this assumption.

Recall that the message sampler is given the description of a ring R. (This is required in order to sample obfuscations in the ideal graded encoding model that consist, for example, of random elements in R.) A computationally unbounded sampler that sees R may be able to recover information that compromises the security of the encodings (for example, the secret parameters). The sampler can produce encodings that reveal this secret information. Note that such a sampler may still be admissible since learning the secret parameters gives no advantage to an algebraic adversary. Luckily, however, we can do with a significantly weaker variant of semantic security where this attack is avoided.

Specifically, the sampling is done in two stages: first, an unbounded sampler S generates a poly-size auxiliary input string; second, an *efficient* encoder $\mathbb{M}$ gets the ring R and the auxiliary input string, and generates the final samples. We call this variant *strong-sampler semantic security*.

Strong-sampler semantic security is already sufficient for constructing strong IO for arbitrary circuit distributions. The idea is to have the unbounded sampler S sample the description of a circuit as an auxiliary input string, and then, the efficient encoder $\mathbb{M}$ samples an obfuscation of the auxiliary input circuit. We show, with a straightforward proof, that the following holds.

Theorem 2 (informal). *Let $\mathcal{O}$ be any obfuscator for a class $\mathcal{C}$ of circuits, that is VBB against semi-bounded adversaries in the ideal graded encoding model. Then instantiating the graded encoding oracle with a strong-sampler semantically-secure graded encoding scheme results in a strong indistinguishability obfuscator $\mathcal{O}'$ for $\mathcal{C}$, in the plain model.*

Then, relying on the Barak et al. obfuscation for NC^1 in the ideal graded encoding model [BGK+13] (which is indeed VBB against semi-bounded adversaries), we obtain the following corollary.

Corollary 1 (informal). *Assume there exists a strong-sampler semantically-secure encoding scheme. Then there exists a strong indistinguishability obfuscator for NC^1.*

We also give evidence for the *necessity* of semantically-secure graded encoding schemes for obtaining VGB. To this end, we focus on a version of graded encoding with restricted functionality called *mutlilinear jigsaw puzzles* [GGH+13b]. Unlike graded encodings, in mutlilinear jigsaw puzzles, encodings can only be generated together the the system parameters. We refer to the public parameters, together with the set of initialized encodings, as a puzzle. Instead of performing individual permitted operations over the encodings, all the jigsaw puzzle user can do is to specify an arithmetic circuit C that respects the set structure of the set of initialized encodings, and test whether C evaluates to 0 on these encodings or not. Semantic security of mutlilinear jigsaw puzzles is defined similarly to the graded encoding case. Despite their restricted functionality, semantically-secure mutlilinear jigsaw puzzles can replace graded encodings in our construction of strong IO for NC^1.

We observe that the existence of semantically-secure jigsaw puzzles is implied by VGB obfuscation for all circuits. To see why this is the case, consider the circuit P that has a set of ring elements $\boldsymbol{m} = (m_1, \ldots, m_\ell)$ hardwired into it, together with the corresponding sets $\boldsymbol{S} = (S_1, \ldots, S_\ell)$. The circuit P takes as input an arithmetic circuit C that respects the set structure given by $\boldsymbol{S}$, and tests whether $C(\boldsymbol{m}) = 0$. To initialize a puzzle from a set of encodings $(\boldsymbol{S}, \boldsymbol{m})$ we simply VGB obfuscate the circuit P.

1.4 More on Semantic Security

Next we discuss our results pertaining to the study of semantic security of graded encoding schemes. The negative results discussed below hold even for the basic notion of semantic security, where the message sampler is of polysize.

SAT lower bounds. As additional evidence to the power of semantically-secure graded encodings, we observe that they imply that there do no exist SAT solvers that run in time $2^{o(n)} \cdot \text{poly}(|C|)$, for a boolean circuit C with n input variables; namely, any worst-case SAT solver must be exponential in the number of variables. To show this lower bound, we rely on a result by Wee [Wee05], showing a similar lower from any point function obfuscation.

Efficiency limitations via a generic attack. We present an attack against any graded encoding scheme satisfying certain efficiency properties. Before specifying these efficiency properties, let us first describe the high-level idea behind the attack, from which they emerge.

Similarly to the attack described in Section 1.2, this attack is based on ideal graded encoding obfuscation. However, this attack holds even when admissibility is defined with respect to semi-bounded algebraic adversaries, rather than just bounded ones. More specifically, it relies on the fact that the [BGK+13] ideal obfuscation scheme is also VBB with respect to semi-bounded algebraic adversaries. Recall that this on its own is not enough to recover the attack from Section 1.2, since there is no general impossibility VGB obfuscation. The attack we describe now will indeed take a somewhat different approach, exploiting the particular interface of graded encoding schemes.

The idea is to construct two circuit distributions $\mathcal{C}_0, \mathcal{C}_1$, where any circuit $C_{b,r}$ sampled from $\mathcal{C}_b$ is associated with a random ring element $r \in R$. The circuit

$C_{b,r}$ reveals the hidden bit b only when given as input some public parameters pp and a proper encoding $[r]_{[k]}$ of the ring element r. The corresponding admissible sampler $\mathbb{M}$ would then sample from one of two distributions $\mathcal{D}_0, \mathcal{D}_1$, where sampling from $\mathcal{D}_b$ is done by sampling $C_{b,r}$ from $\mathcal{C}_b$, and outputting $(\boldsymbol{S}, \boldsymbol{m})$, that represents an ideal obfuscation of $C_{b,r}$, together with $([k], r)$. Intuitively, an ideal, even semi-bounded, algebraic adversary gains no more than oracle-access to $C_{b,r}$, together with the ability to evaluate low-degree arithmetic circuits on the random ring element r, thus it cannot learn the bit b.[2] In contrast, the real world distinguisher, which is given the public parameters pp and an actual encoding of r, can simply run the obfuscation on pp and the encoding of r, and learn b.

So what is needed to make the above attack applicable? First, since we only have ideal obfuscation against semi-bounded adversaries for NC^1, the circuit $C_{b,r}$ should be implementable in NC^1. Second, it is required that the size of the public parameters pp and the size of an encoding grows slower than k, the size of the universe for the sets that control the depth of allowed arithmetic computations. Indeed, in order to obfuscate $C_{b,r}$ in the ideal graded encoding model, it is required that the universe set $[k]$ is appropriately large (in particular, larger than the circuit's input). Thus, the public parameters and encoding received by $C_{b,r}$ as input must grow slower than k.[3]

Both of the above efficiency requirements are not satisfied by the candidate constructions of [GGH13a, CLT13] in their current forms. Indeed, for these schemes it is not known how to implement $C_{b,r}$ (or any procedure of equivalent effect) in NC^1. Also, in these schemes the size of the public parameters and encodings does grow with the maximal level k.

Still, it may be good to keep this attack in mind, pending potential improvements in the efficiency of obfuscation algorithms or graded encoding schemes. In fact, it appears prudent to weaken semantic security by requiring that it holds only given some a priory bound on either the level k, or on the number of elements ℓ output by an admissible sampler $\mathbb{M}$. This allows considering candidate schemes where certain parameters, such as the size of the ring (and induced size of encodings), are larger than these bound and are thus not susceptible to this type of attacks.

There may also exist certain tradeoffs in efficiency. For instance, in certain settings it may be more reasonable to let the size of parameters grow with ℓ, rather than with k. We define such a variant of *bounded semantic security*.

Relaxations of semantic security. As mentioned above, the notion of semantically-secure graded encodings lends itself to a number of variants along several axes. We study the relations among these relaxations and show that eventually they all

---

[2] This argument is a bit oversimplified; indeed, to argue that one cannot learn b given oracle access to r, we should also deal with the case that the adversary queries with improper public parameters and encodings. In the body, we deal with this by using a variant of the circuit $C_{b,r}$ that checks that any query corresponds to at most a small number of ring elements, and thus hits r only with negligible probability.

[3] We thank Rafael Pass for pointing this out.

suffice for obtaining the implications presented above for program obfuscation. Below we address two relaxations that were introduced by Pass et al. [PTS13] (and are already embedded into their main definition of semantic security).

First, Pass et al. consider *constant-message samplers* where the first $\ell - O(1)$ elements in the distributions $\mathcal{D}_0$ and $\mathcal{D}_1$ are required to be exactly the same, and are viewed as polynomially long "auxiliary-input" correlated to the last constant number of elements. (In later versions, Pass et al. limit the number of elements in the auxiliary distribution $\mathcal{Z}$ to a fixed polynomial, partially in light of the attacks described in this work.) Second, they strengthen the notion of admissibility where indistinguishability with respect to the algebraic adversary needs to hold in a strong *pointwise* sense. That is, for almost any two samples $(S_0, m_0), (S_1, m_1)$, taken from (potentially joint) distributions $(\mathcal{D}_0, \mathcal{D}_1)$, the algebraic adversary outputs the same bit when given (S_0, m_0) and when given (S_1, m_1). Finally, admissibility is further strengthened to only allow for "highly-entropic" samples. Indeed, this relaxation turns out to be essential in the context of the [GGH13a] graded encoding scheme (but not necessarily in the [CLT13] scheme).

From a technical perspective, the difference between the constant-message and multi-message definitions (in either the pointwise or non-pointwise case) is that the general transformation of Theorem 2 appears to require the seemingly stronger multi-message definition. In contrast, the specific construction of Pass et al. works even using only the single-message version.

We show that, in fact, the relaxed constant-message notion is equivalent to the multi-message notion. However, there are certain nuances to this equivalence:

- In the non-pointwise case, we show that even single-message (rather than constant-message) semantic-security implies multi-message semantic-security. Here the reduction essentially preserves the number of elements output by the attacker's message sampler. Specifically, any attacker against m-message semantic-security translates to an attacker against single-message semantic-security with auxiliary input of length m.

- In the pointwise case, this implication still holds, but with certain loss in parameters. Specifically, any attacker against m-message semantic-security with distinguishing advantage ϵ translates to an attacker against single-message semantic-security with auxiliary input of length $m \cdot \mathrm{poly}(n/\epsilon)$.

In conclusion, in the non-pointwise case, constant-message semantic-security does not constitute an actual relaxation, and would also lead to a generic construction of (strong) IO. In the pointwise case, however, the specific obfuscator of Pass et al. gives a quantitative security advantage compared with the generic construction. In particular, for **bounded** constant-message pointwise semantic-security, the generic transformation does not apply as far as we know, whereas the obfuscator of Pass et al. does. It would thus be interesting to come up with evidence as to whether moving to pointwise security amounts to a meaningful relaxation of the assumption.

Strong IO from new assumptions? Pass et al. also consider an alternative modification of semantic security, where instead of requiring indistinguishability with respect to *any* admissible sampler, it is only required for a *single* specific sampler. To get IO, however, they also require that indistinguishability holds against subexponential distinguishers. This assumption is incomparable to the assumption discussed here, and is not further studied in this work. Gentry et al. [GLSW14] recently constructed indistinguishability obfuscators based on a new assumption of a somewhat different flavor, regarding a more demanding variant of graded encoding. Whether these assumptions suffice for constructing strong indistinguishability obfuscators is an intriguing question.

Organization. Section 2 reviews the definitions of VBB, VGB and IO. Section 3 defines SIO and shows its equivalence to VBB for concentrated circuit distributions. Section 4 constructs worst case VGB and VBB obfsucators from strong IO. Section 5 constructs SIO from semantically-secure graded encoding schemes. The study of the semantically-secure graded encoding assumption appear in the full version of this work.

2   Obfuscation: VBB, VGB, Indistinguishability

We review three basic definitions of obfuscation that are used throughout the paper. We start by defining the functionality requirement, which all the notions share, and then define different security notions.

Definition 1 (Functionality). *A PPT algorithm $\mathcal{O}$ is an obfuscator for a collection of circuits $\mathcal{C} = \bigcup_{n \in \mathbb{N}} \mathcal{C}_n$, if for any $C \in \mathcal{C}$,*

$$\Pr_{\mathcal{O}} [\forall x : \mathcal{O}(C)(x) = C(x)] = 1 .$$

VBB and VGB Obfuscation. Virtual Black Box (VBB) obfuscation [BGI+01] guarantees that an obfuscated circuit $\mathcal{O}(C)$ does not reveal any predicate $\pi(C)$ that cannot be learned by an efficient simulator that is given only black-box access to C. The basic definition is *worst-case* in the sense that the simulator needs to be successful for any circuit in a given circuit collection. We later also address an *average-case* notion. In the definition below we use a slightly weaker definition than the standard one, and allow the simulator to depend on the distinguishing probability p.

Definition 2 (Worst-case VBB Obfuscation). *An obfuscator $\mathcal{O}$ for a collection of circuits $\mathcal{C} = \bigcup_{n \in \mathbb{N}} \mathcal{C}_n$ is worst-case VBB if for every poly-size adversary $\mathcal{A}$, and polynomial p, there exists a poly-size simulator $\mathcal{S}$, such that for every $n \in \mathbb{N}$, every predicate $\pi : \mathcal{C}_n \to \{0, 1\}$, and every $C \in \mathcal{C}_n$:*

$$\left| \Pr_{\mathcal{A}, \mathcal{O}} [\mathcal{A}(\mathcal{O}(C)) = \pi(C)] - \Pr_{\mathcal{S}} [\mathcal{S}^C(1^n) = \pi(C)] \right| \leq 1/p(n) .$$

Virtual Grey Box (VGB) obfuscation [BC10] relaxes VBB by allowing the simulator to have unbounded computational power, but still only a bounded number of oracle queries to C.

Definition 3 (Worst-case VGB Obfuscation). *An obfuscator $\mathcal{O}$ for a collection of circuits $\mathcal{C} = \bigcup_{n \in \mathbb{N}} \mathcal{C}_n$ is worst-case VGB if for every poly-size adversary $\mathcal{A}$, and polynomial p, there exists an unbounded simulator $\mathcal{S}$, and a polynomial q, such that for every $n \in \mathbb{N}$, every predicate $\pi : \mathcal{C}_n \to \{0,1\}$, and $C \in \mathcal{C}_n$:*

$$\left| \Pr_{\mathcal{A},\mathcal{O}}[\mathcal{A}(\mathcal{O}(C)) = \pi(C)] - \Pr_{\mathcal{S}}[\mathcal{S}^{C[q(n)]}(1^n) = \pi(C)] \right| \leq 1/p(n) \ ,$$

where $C[q(n)]$ is an oracle that allows at most $q(n)$ queries.

Indistinguishability Obfuscation. We next define the notion of indistinguishability obfuscation, introduced in [BGI+01].

Definition 4 (Indistinguishability obfuscation [BGI+01]). *An obfuscator for $\mathcal{C}$ is said to be an* indistinguishability obfuscator *for $\mathcal{C}$, denoted by $i\mathcal{O}$, if for any poly-size distinguisher $\mathcal{D}$, there exists a negligible function μ such that for all $n \in \mathbb{N}$, and any two circuits $C_0, C_1 \in \mathcal{C}_n$ of the same size and functionality,*

$$\Pr[b \leftarrow \{0,1\}; \mathcal{D}(C_0, C_1, i\mathcal{O}(C_b)) = b] \leq \frac{1}{2} + \mu(n) \ .$$

It can be readily seen that if an obfuscator $\mathcal{O}$ is VBB for a function collection $\mathcal{C}$ then it is also VGB for $\mathcal{C}$. Furthermore, if $\mathcal{O}$ is VGB for $\mathcal{C}$ then it is also an indistinguishability obfuscator for $\mathcal{C}$.

3   Strong Indistinguishability Obfuscation

In this section we define the notion of strong indistinguishability obfuscation (SIO). We start by defining the notion of concentrated distributions over circuits.

Concentrated Circuit Distributions. At a high-level, a distribution ensemble $\tilde{\mathcal{C}}$, over a circuit collection $\mathcal{C}$, is *concentrated*, if given polynomially many oracle queries to a random circuit C from the distribution, it is information theoretically hard to find an input x such that C does not agree with $\mathrm{maj}_{\tilde{\mathcal{C}}}$ on the point x, where $\mathrm{maj}_{\tilde{\mathcal{C}}}$ is the common output of circuits distributed according to $\tilde{\mathcal{C}}$. If $\tilde{\mathcal{C}}$ corresponds to the uniform distribution on some collection $\mathcal{C}$, $\mathrm{maj}_{\tilde{\mathcal{C}}}$ is simply the majority vote. Concentrated distributions naturally generalize the concept of *evasive distributions* studied in [BBC+14], in which the majority is always the all-zero function, i.e. $\mathrm{maj}_{\tilde{\mathcal{C}}} \equiv 0$.

Definition 5 (Concentrated circuit distributions)
Let $\mathcal{C} = \bigcup_{n \in \mathbb{N}} \mathcal{C}_n$ be a circuit collection, where $\mathcal{C}_n$ consists of circuits $C : \{0,1\}^n \to \{0,1\}$ of size $\mathrm{poly}(n)$, and let $\tilde{\mathcal{C}}_n$ be a distribution on $\mathcal{C}_n$. Let $\mathrm{maj}_{\tilde{\mathcal{C}}_n}(x) := \lfloor \mathbb{E}_{C \leftarrow \tilde{\mathcal{C}}_n} C(x) \rceil$ be the common output at point x of circuits drawn from $\tilde{\mathcal{C}}_n$.

1. *For any $\varepsilon \in [0, 1]$, $\tilde{C}_n$ is said to be ε-concentrated if*

$$\max_{x \in \{0,1\}^n} \Pr_{C \leftarrow \tilde{C}_n} \left[C(x) \neq \mathrm{maj}_{\tilde{C}_n}(x) \right] \leq \varepsilon \ .$$

2. *$\tilde{C}$ is said to be concentrated if for some negligible $\mu(\cdot)$, and any $n \in \mathbb{N}$, $\tilde{C}_n$ is $\mu(n)$-concentrated.*
3. *$\tilde{C}$ is said to be evasive if it is concentrated, and for any $n \in \mathbb{N}$ and any $x \in \{0, 1\}^n$, $\mathrm{maj}_{\tilde{C}_n}(x) = 0$.*
4. *We say that the collection C itself is concentrated (evasive) if the uniform distribution ensemble on circuits in C is concentrated (evasive).*

Strong Indistinguishability Obfuscation. Strong Indistinguishability Obfuscation requires that indistinguishability holds, even when C_0 and C_1 do not necessarily compute the exact same function, but are taken from two distributions $\tilde{C}_n^0$ and $\tilde{C}_n^1$ that are concentrated around the same function; namely, $\mathrm{maj}_{\tilde{C}_n^0} \equiv \mathrm{maj}_{\tilde{C}_n^1}$:

Definition 6 (Strong indistinguishability obfuscation). *An obfuscator for C is said to be a **strong** indistinguishability obfuscator for C, denoted by $i\mathcal{O}^*$, if for any two concentrated distribution ensembles $\tilde{C}^0, \tilde{C}^1$ on C, such that $\forall n \in \mathbb{N} : \mathrm{maj}_{\tilde{C}_n^0} \equiv \mathrm{maj}_{\tilde{C}_n^1}$, and any poly-size distinguisher $\mathcal{D}$, there exists a negligible function μ such that for all $n \in \mathbb{N}$,*

$$\Pr[b \leftarrow \{0, 1\}; (C_0, C_1) \leftarrow (\tilde{C}_n^0, \tilde{C}_n^1); \mathcal{D}(i\mathcal{O}^*(C_b)) = b] \leq \frac{1}{2} + \mu(n) \ .$$

Remark 1. Above, we do not require that the distributions $\tilde{C}^0, \tilde{C}^1$ are efficiently samplable. We can also consider a weaker definition where this restriction is added. In the full version of this work we show that this weaker version can be obtained from a weaker notion of semantic security.

We observe that any strong IO obfuscator for C is also an IO obfuscator for C. Indeed, for any two circuits C_0, C_1 of equivalent functionality, each of these circuits on its own is trivially concentrated around their common functionality.

4   Strong IO Is Equivalent to Worst-Case VGB

In this section, we prove that the notion of strong indistinguishability obfuscation (strong IO) is equivalent to VGB. Clearly, any VGB obfuscator for a class C is also a strong IO for C. We show that the converse is true as well. Namely, we show that any strong indistinguishability obfuscator $\mathcal{O}$ for a class C of circuits is a worst-case VGB obfuscator for C. In addition, we show that for classes C with some additional properties, $\mathcal{O}$ is in fact worst-case VBB. We refer the reader to Section 1.1 for an overview.

4.1 Definitions and Statement of Main Theorem

Notation and terminology. For a function $f : \{0,1\}^n \rightarrow \{0,1\}$, we say that a point $x \in \{0,1\}^n$ *separates* a circuit C from f if $C(x) \neq f(x)$. We say that a set $L \subseteq \{0,1\}^n$ separates C from f, if some $x \in L$ separates C from f. Given a circuit collection $\mathcal{K}$, we say that L separates $\mathcal{K}$ from f, if L separates any $C \in \mathcal{K}$ from f. Recall, that we say that a collection $\mathcal{K}$ is concentrated if the uniform distribution on $\mathcal{K}$ is concentrated around its majority function $\text{maj}_{\mathcal{K}}$.

Definition 7 (Majority-separating oracle). *Let $\mathcal{C}$ be a collection of boolean circuits defined over $\{0,1\}^n$, let $C \in \mathcal{C}$, and let $\varepsilon > 0$. An oracle $\mathbb{S}$ is said to be $(\mathcal{C}, C, \varepsilon)$-separating if given any ε-concentrated sub-collection $\mathcal{K} \subseteq \mathcal{C}$, represented by a circuit that samples uniform elements in $\mathcal{K}$, $\mathbb{S}(\mathcal{K})$ outputs a point $x \in \{0,1\}^n$ that separates C from $\text{maj}_{\mathcal{K}}$, or $\perp$ if no such point exists.*

Remark 2. In the above definition, and throughout this section, we often abuse notation and denote by $\mathcal{K}$ both the sub-collection and the circuit that samples uniform elements from the sub-collection.

Definition 8 (Learnability by majority-separating oracles). *A collection $\mathcal{C} = \bigcup_{n \in \mathbb{N}} \mathcal{C}_n$ of boolean circuits is said to be $(t, \mathbf{c}, s, \varepsilon)$-learnable by a majority-separation oracle if there exists a deterministic oracle-aided machine $\mathcal{L}$ such that, given oracle access to $C \in \mathcal{C}_n$ and a $(\mathcal{C}_n, C, \varepsilon(n))$-separating oracle $\mathbb{S}$, $\mathcal{L}^{C,\mathbb{S}}(1^n)$ outputs $\hat{C} \in \mathcal{C}_n$ of equivalent functionality to C, in time $t(n)$, using at most $s(n)$ queries to $\mathbb{S}$, and at most $\mathbf{c}_i(n)$ queries to C between the $i - 1$-st and the i-th calls to $\mathbb{S}$.*

Our main technical theorem shows that any strong indistinguishability obfuscator for a circuit collection $\mathcal{C}$ that is learnable via a majority separation oracle is also a worst-case simulation-based obfuscator. The size and query complexity of the worst-case simulator, in particular whether it is a VBB or VGB simulator, is determined by the learnability parameters $(t, \mathbf{c}, s, \varepsilon)$.

Theorem 3. *Let $\mathcal{C} = \bigcup_{n \in \mathbb{N}} \mathcal{C}_n$ be a circuit collection that is $(t, \mathbf{c}, s, \frac{1}{q})$-learnable by a majority-separating oracle, for some polynomial q. Let $\mathcal{O}$ be a strong indistinguishability obfuscator for $\mathcal{C}$, let $\mathcal{A}$ be a boolean poly-size adversary, and let p be a polynomial. Then $(\mathcal{A}, p)$ has a simulator $\mathcal{S}$ of size $O(|\mathcal{A}| + t \cdot s \cdot q^s \cdot \prod_{i=1}^{s} 2^{\mathbf{c}_i})$ with $O(\|\mathbf{c}\|_1 + q \cdot s)$ oracle queries. The simulator works in the worst-case for any $C \in \mathcal{C}$.*

In Section 4.2 we show that any circuit collection $\mathcal{C}$ is indeed $(t, \mathbf{c}, s, \frac{1}{q})$-learnable, for some setting of parameters (where $\|\mathbf{c}\|_1, q, s$ are polynomially bounded).

4.2 VGB and VBB by Majority-Separation Learning

In this section, we show that any class of circuits is learnable by a majority-separating oracle, with parameters that yield VGB simulation. In the full version of this work we discuss additional classes that can be learned with better parameters, yielding VBB simulation. This includes previously obfuscated classes as well as new ones.

VGB Obfuscation for All Circuits. We show

Theorem 4. *Let $\mathcal{C}$ be any circuit collection and let $\mathcal{O}$ be a strong indistinguishability obfuscator for $\mathcal{C}$. Then $\mathcal{O}$ is also a worst-case VGB obfuscator for $\mathcal{C}$.*

To prove Theorem 4, we show that any circuit collection is learnable by a majority-separating oracle, where the learner is of unbounded size, but only performs a polynomial number of queries to its oracles. Theorem 4 then follows from Theorem 3.

Lemma 1. *For any $q > 2$, any circuit collection $\mathcal{C} = \{\mathcal{C}_n\}_{n \in \mathbb{N}}$ is $(t, \mathbf{c}, s, \frac{1}{q})$-learnable by a majority-separating oracle for $t(n) = \infty$, $s(n) \leq \|\mathbf{c}(n)\|_1 \leq q(n) \cdot \log |\mathcal{C}_n|$.*

5 From Semantically-Secure Graded Encodings to Strong IO for NC^1

In this section we show that any semantically-secure graded encoding scheme, together with *any* ideal graded encoding obfuscation (i.e., any obfuscation that is virtual-black-box secure in the ideal encoding model) for a class $\mathcal{C}$ of circuits, implies *strong indistinguishability obfuscation* for $\mathcal{C}$.

Proposition 1. *Assume there exists a semantically-secure graded encoding scheme, and assume there exists an ideal graded encoding obfuscation for a circuit class $\mathcal{C}$. Then there exists a strong IO obfuscator for the circuit class $\mathcal{C}$, in the plain model (Definition 6).*

Our definition of semantically-secure graded encoding is a strengthening of the assumption of [PTS13]. One key difference from the assumption in [PTS13] is that here we consider semantic security even for distributions of messages that are not efficiently samplable. The definition can be found in the full version of this work where we also discuss several relaxations. The definition of VBB obfuscation in the ideal-graded-encoding model is also deferred to the full version of this work.

Proposition 1, combined with the recent VBB obfuscators for NC^1 in the ideal-graded-encoding model [BR13, BGK$^+$13], and with our results from Section 4, immediately yields strong IO and VGB obfuscation for NC^1.

Theorem 5. *Assume there exists a semantically-secure graded encoding scheme. Then there exist strong IO and worst-case VGB obfuscation for any collection in NC^1.*

Acknowledgements. We are grateful to Rafael Pass for enlightening discussions and valuable comments.

References

[BBC+14]   Barak, B., Bitansky, N., Canetti, R., Kalai, Y.T., Paneth, O., Sahai, A.: Obfuscation for evasive functions. In: Lindell, Y. (ed.) TCC 2014. LNCS, vol. 8349, pp. 26–51. Springer, Heidelberg (2014)

[BC10]   Bitansky, N., Canetti, R.: On strong simulation and composable point obfuscation. In: Rabin, T. (ed.) CRYPTO 2010. LNCS, vol. 6223, pp. 520–537. Springer, Heidelberg (2010)

[BCC+14]   Bitansky, N., Canetti, R., Cohn, H., Goldwasser, S., Kalai, Y.T., Paneth, O., Rosen, A.: The impossibility of obfuscation with auxiliary input or a universal simulator. CoRR, abs/1401.0348 (2014)

[BGI+01]   Barak, B., Goldreich, O., Impagliazzo, R., Rudich, S., Sahai, A., Vadhan, S.P., Yang, K.: On the (im)possibility of obfuscating programs. In: Kilian, J. (ed.) CRYPTO 2001. LNCS, vol. 2139, pp. 1–18. Springer, Heidelberg (2001)

[BGK+13]   Barak, B., Garg, S., Kalai, Y.T., Paneth, O., Sahai, A.: Protecting obfuscation against algebraic attacks. Cryptology ePrint Archive, Report 2013/631 (2013), http://eprint.iacr.org/

[BR13]   Brakerski, Z., Rothblum, G.N.: Virtual black-box obfuscation for all circuits via generic graded encoding. Cryptology ePrint Archive, Report 2013/563 (2013), http://eprint.iacr.org/

[CLT13]   Coron, J.-S., Lepoint, T., Tibouchi, M.: Practical multilinear maps over the integers. In: Canetti, R., Garay, J.A. (eds.) CRYPTO 2013, Part I. LNCS, vol. 8042, pp. 476–493. Springer, Heidelberg (2013)

[CV13]   Canetti, R., Vaikuntanathan, V.: Obfuscating branching programs using black-box pseudo-free groups. Cryptology ePrint Archive, Report 2013/500 (2013), http://eprint.iacr.org/

[GGH13a]   Garg, S., Gentry, C., Halevi, S.: Candidate multilinear maps from ideal lattices. In: Johansson, T., Nguyen, P.Q. (eds.) EUROCRYPT 2013. LNCS, vol. 7881, pp. 1–17. Springer, Heidelberg (2013)

[GGH+13b]   Garg, S., Gentry, C., Halevi, S., Raykova, M., Sahai, A., Waters, B.: Candidate indistinguishability obfuscation and functional encryption for all circuits. In: FOCS (2013)

[GK05]   Goldwasser, S., Kalai, Y.T.: On the impossibility of obfuscation with auxiliary input. In: FOCS, pp. 553–562 (2005)

[GLSW14]   Gentry, C., Lewko, A., Sahai, A., Waters, B.: Indistinguishability obfuscation from the multilinear subgroup elimination assumption. Cryptology ePrint Archive, Report 2014/309 (2014), http://eprint.iacr.org/

[Had00]   Hada, S.: Zero-knowledge and code obfuscation. In: Okamoto, T. (ed.) ASIACRYPT 2000. LNCS, vol. 1976, pp. 443–457. Springer, Heidelberg (2000)

[PTS13]   Pass, R., Telang, S., Seth, K.: Obfuscation from semantically-secure multilinear encodings. Cryptology ePrint Archive, Report 2013/781 (2013), http://eprint.iacr.org/

[SW13]   Sahai, A., Waters, B.: How to use indistinguishability obfuscation: Deniable encryption, and more. IACR Cryptology ePrint Archive 2013, 454 (2013)

[Wee05]   Wee, H.: On obfuscating point functions. IACR Cryptology ePrint Archive 2005, 1 (2005)

Breaking '128-bit Secure' Supersingular Binary Curves*
(Or How to Solve Discrete Logarithms in $\mathbb{F}_{2^{4 \cdot 1223}}$ and $\mathbb{F}_{2^{12 \cdot 367}}$)

Robert Granger[1], Thorsten Kleinjung[1], and Jens Zumbrägel[2]

[1] Laboratory for Cryptologic Algorithms, EPFL, Switzerland
[2] Institute of Algebra, TU Dresden, Germany
robbiegranger@gmail.com, thorsten.kleinjung@epfl.ch,
jens.zumbragel@ucd.ie

Abstract. In late 2012 and early 2013 the discrete logarithm problem (DLP) in finite fields of small characteristic underwent a dramatic series of breakthroughs, culminating in a heuristic quasi-polynomial time algorithm, due to Barbulescu, Gaudry, Joux and Thomé. Using these developments, Adj, Menezes, Oliveira and Rodríguez-Henríquez analysed the concrete security of the DLP, as it arises from pairings on (the Jacobians of) various genus one and two supersingular curves in the literature, which were originally thought to be 128-bit secure. In particular, they suggested that the new algorithms have no impact on the security of a genus one curve over $\mathbb{F}_{2^{1223}}$, and reduce the security of a genus two curve over $\mathbb{F}_{2^{367}}$ to 94.6 bits. In this paper we propose a new field representation and efficient general descent principles which together make the new techniques far more practical. Indeed, at the '128-bit security level' our analysis shows that the aforementioned genus one curve has approximately 59 bits of security, and we report a total break of the genus two curve.

Keywords: Discrete logarithm problem, supersingular binary curves, pairings, finite fields.

1 Introduction

The role of small characteristic supersingular curves in cryptography has been a varied and an interesting one. Having been eschewed by the cryptographic community for succumbing spectacularly to the subexponential MOV attack in 1993 [39], which maps the DLP from an elliptic curve (or more generally, the Jacobian of a higher genus curve) to the DLP in a small degree extension of

---

* The second author acknowledges the support of the Swiss National Science Foundation, via grant numbers 206021-128727 and 200020-132160, while the third author acknowledges the support of the Irish Research Council, grant number ELEVATEPD/2013/82.

J.A. Garay and R. Gennaro (Eds.): CRYPTO 2014, Part II, LNCS 8617, pp. 126–145, 2014.
© International Association for Cryptologic Research 2014

the base field of the curve, they made a remarkable comeback with the advent of pairing-based cryptography in 2001 [41,30,9]. In particular, for the latter it was reasoned that the existence of a subexponential attack on the DLP does not *ipso facto* warrant their complete exclusion; rather, provided that the finite field DLP into which the elliptic curve DLP embeds is sufficiently hard, this state of affairs would be acceptable.

Neglecting the possible existence of native attacks arising from the supersingularity of these curves, much research effort has been expended in making instantiations of the required cryptographic operations on such curves as efficient as possible [6,17,14,27,26,5,29,7,11,18,3,1], to name but a few, with the associated security levels having been estimated using Coppersmith's algorithm from 1984 [12,38]. Alas, a series of dramatic breakthrough results for the DLP in finite fields of small characteristic have potentially rendered all of these efforts in vain.

The first of these results was due to Joux, in December 2012, and consisted of a more efficient method – dubbed 'pinpointing' – to obtain relations between factor base elements [31]. For medium-sized base fields, this technique has heuristic complexity as low as $L(1/3, 2^{1/3}) \approx L(1/3, 1.260)^1$, where as usual $L(\alpha, c) = L_Q(\alpha, c) = \exp((c + o(1))(\log Q)^\alpha (\log \log Q)^{1-\alpha})$, with Q the cardinality of the field. This improved upon the previous best complexity of $L(1/3, 3^{1/3}) \approx L(1/3, 1.442)$ due to Joux and Lercier [36]. Using this technique Joux solved example DLPs in fields of bitlength 1175 and 1425, both with prime base fields.

Then in February 2013, Göloğlu, Granger, McGuire and Zumbrägel used a specialisation of the Joux-Lercier doubly-rational function field sieve (FFS) variant [36], in order to exploit a well-known family of 'splitting polynomials', i.e., polynomials which split completely over the base field [19]. For fields of the form $\mathbb{F}_{q^{kn}}$ with $k \geq 3$ fixed ($k = 2$ is even simpler) and $n \approx dq$ for a fixed integer $d \geq 1$, they showed that for binary (and more generally small characteristic) fields, relation generation for degree one elements runs in heuristic *polynomial time*, as does finding the logarithms of degree two elements (if q^k can be written as $q'^{k'}$ for $k' \geq 4$), once degree one logarithms are known. For medium-sized base fields of small characteristic a heuristic complexity as low as $L(1/3, (4/9)^{1/3}) \approx L(1/3, 0.763)$ was attained; this approach was demonstrated via the solution of example DLPs in the fields $\mathbb{F}_{2^{1971}}$ [21] and $\mathbb{F}_{2^{3164}}$.

After the initial publication of [19], Joux released a preprint [32] detailing an algorithm for solving the discrete logarithm problem for fields of the form $\mathbb{F}_{q^{2n}}$, with $n \leq q + d$ for some very small d, which was used to solve a DLP in $\mathbb{F}_{2^{1778}}$ [33] and later in $\mathbb{F}_{2^{4080}}$ [34]. For $n \approx q$ this algorithm has heuristic complexity $L(1/4 + o(1), c)$ for some undetermined c, and also has a heuristic polynomial time relation generation method, similar in principle to that in [19]. While the degree two element elimination method in [19] is arguably superior

---

[1] The original paper states a complexity of $L(1/3, (8/9)^{1/3}) \approx L(1/3, 0.961)$; however, on foot of recent communications the constant should be as stated.

– since elements can be eliminated on the fly – for other small degrees Joux's elimination method is faster, resulting in the stated complexity.

In April 2013 Göloğlu et al. combined their approach with Joux's to solve an example DLP in the field $\mathbb{F}_{2^{6120}}$ [22] and later demonstrated that Joux's algorithm can be tweaked to have heuristic complexity $L(1/4, c)$ [20], where c can be as low as $(\omega/8)^{1/4}$ [20], with ω the linear algebra constant, i.e., the exponent of matrix multiplication. Then in May 2013, Joux announced the solution of a DLP in the field $\mathbb{F}_{2^{6168}}$ [35].

Most recently, in June 2013, Barbulescu, Gaudry, Joux and Thomé announced a *quasi-polynomial time* for solving the DLP [4], for fields $\mathbb{F}_{q^{kn}}$ with $k \geq 2$ fixed and $n \leq q + d$ with d very small, which for $n \approx q$ has heuristic complexity

$$(\log q^{kn})^{O(\log \log q^{kn})}. \tag{1}$$

Since (1) is smaller than $L(\alpha, c)$ for any $\alpha > 0$, it is asymptotically the most efficient algorithm known for solving the DLP in finite fields of small characteristic, which can always be embedded into a field of the required form. Interestingly, the algorithmic ingredients and analysis of this algorithm are much simpler than for Joux's $L(1/4 + o(1), c)$ algorithm.

Taken all together, one would expect the above developments to have a substantial impact on the security of small characteristic parameters appearing in the pairing-based cryptography literature. However, all of the record DLP computations mentioned above used Kummer or twisted Kummer extensions (those with n dividing $q^k \mp 1$), which allow for a reduction in the size of the factor base by a factor of kn and make the descent phase for individual logarithms relatively easy. While such parameters are preferable for setting records (most recently in $\mathbb{F}_{2^{9234}}$ [25]), none of the parameters featured in the literature are of this form, and so it is not *a priori* clear whether the new techniques weaken existing pairing-based protocol parameters.

A recent paper by Adj, Menezes, Oliveira and Rodríguez-Henríquez has begun to address this very issue [2]. Using the time required to compute a single multiplication modulo the cardinality of the relevant prime order subgroup as their basic unit of time, which we denote by M_r, they showed that the DLP in the field $\mathbb{F}_{3^{6 \cdot 509}}$ costs at most $2^{73.7} \, M_r$. One can arguably interpret this result to mean that this field has 73.7 bits of security[2]. This significantly reduces the intended security level of 128 bits (or 111 bits as estimated by Shinohara *et al.* [42], or 102.7 bits for the Joux-Lercier FFS variant with pinpointing, as estimated in [2]). An interesting feature of their analysis is that during the descent phase, some elimination steps are performed using the method from the quasi-polynomial time algorithm of Barbulescu *et al.*, when one might have expected

---

[2] The notion of bit security is quite fuzzy; for the elliptic curve DLP it is usually intended to mean the logarithm to the base 2 of the expected number of group operations, however for the finite field DLP different authors have used different units, perhaps because the cost of various constituent algorithms must be amortised into a single cost measure. In this work we time everything in seconds, while to achieve a comparison with [2] we convert to M_r.

these steps to only come into play at much higher bitlengths, due to the high arity of the arising descent nodes.

In the context of binary fields, Adj *et al.* considered in detail the DLP in the field $\mathbb{F}_{2^{12 \cdot 367}}$, which arises via a pairing from the DLP on the Jacobian of a supersingular genus two curve over $\mathbb{F}_{2^{367}}$, first proposed in [3], with embedding degree 12. Using all of the available techniques they provided an upper bound of $2^{94.6} \, M_r$ for the cost of breaking the DLP in the embedding field, which is some way below the intended 128-bit security level. In their conclusion Adj *et al.* also suggest that a commonly implemented genus one supersingular curve over $\mathbb{F}_{2^{1223}}$ with embedding degree 4 [29,7,11,18,1], is not weakened by the new algorithmic advances, i.e., its security remains very close to 128 bits.

In this work we show that the above security estimates were incredibly optimistic. Our techniques and results are summarised as follows.

- **Field representation:** We introduce a new field representation that can have a profound effect on the resulting complexity of the new algorithms. In particular it permits the use of a smaller q than before, which not only speeds up the computation of factor base logarithms, but also the descent (both classical and new).

- **Exploit subfield membership:** During the descent phase we apply a *principle of parsimony*, by which one should always try to eliminate an element in the target field, and only when this is not possible should one embed it into an extension field. So although the very small degree logarithms may be computed over a larger field, the descent cost is *greatly reduced* relative to solving a DLP in the larger field.

- **Further descent tricks:** The above principle also means that elements can automatically be rewritten in terms of elements of smaller degree, via factorisation over a larger field, and that elements can be eliminated via Joux's Gröbner basis computation method [32] with $k = 1$, rather than $k > 1$, which increases its degree of applicability.

- **'128-bit secure' genus one DLP:** We show that the DLP in $\mathbb{F}_{2^{4 \cdot 1223}}$ can be solved in approximately 2^{40} s, or $2^{59} \, M_r$, with r a 1221-bit prime.

- **'128-bit secure' genus two DLP:** We report a total break of the DLP in $\mathbb{F}_{2^{12 \cdot 367}}$ (announced in [24]), which took about 52240 core-hours.

- **L(1/4, c) technique only:** Interestingly, using our approach the elimination steps à la Barbulesu *et al.* [4] were not necessary for the above estimate and break.

The rest of the paper is organised as follows. In §2 we describe our field representation and our target fields. In §3 we present the corresponding polynomial time relation generation method for degree one elements and degree two elements (although we do not need the latter for the fields targeted in the present paper), as well as how to apply Joux's small degree elimination method [32] with the new representation. We then apply these and other techniques to $\mathbb{F}_{2^{4 \cdot 1223}}$ in §4 and to $\mathbb{F}_{2^{12 \cdot 367}}$ in §5 . Finally, we conclude in §6.

2   Field Representation and Target Fields

In this section we introduce our new field representation and the fields whose DLP security we will address. This representation, as well as some preliminary security estimates, were initially presented in [19].

2.1   Field Representation

Although we focus on binary fields in this paper, for the purposes of generality, in this section we allow for extension fields of arbitrary characteristic. Hence let $q = p^l$ for some prime p, and let $\mathbb{K} = \mathbb{F}_{q^{kn}}$ be the field under consideration, with $k \geq 1$.

We choose a positive integer d_h such that $n \leq qd_h + 1$, and then choose $h_0, h_1 \in \mathbb{F}_{q^k}[X]$ with $\max\{\deg(h_0), \deg(h_1)\} = d_h$ such that

$$h_1(X^q)X - h_0(X^q) \equiv 0 \pmod{I(X)}, \tag{2}$$

where $I(X)$ is an irreducible degree n polynomial in $\mathbb{F}_{q^k}[X]$. Then $\mathbb{K} = \mathbb{F}_{q^k}[X]/(I(X))$. Denoting by x a root of $I(X)$, we introduce the auxiliary variable $y = x^q$, so that one has two isomorphic representations of $\mathbb{K}$, namely $\mathbb{F}_{q^k}(x)$ and $\mathbb{F}_{q^k}(y)$, with $\sigma : \mathbb{F}_{q^k}(y) \rightarrow \mathbb{F}_{q^k}(x) : y \mapsto x^q$. To establish the inverse isomorphism, note that by (2) in $\mathbb{K}$ we have $h_1(y)x - h_0(y) = 0$, and hence $\sigma^{-1} : \mathbb{F}_{q^k}(x) \rightarrow \mathbb{F}_{q^k}(y) : x \mapsto h_0(y)/h_1(y)$.

The knowledgeable reader will have observed that our representation is a synthesis of two other useful representations: the one used by Joux [32], in which one searches for a degree n factor $I(X)$ of $h_1(X)X^q - h_0(X)$; and the one used by Gölöğlu *et al.* [19,20], in which one searches for a degree n factor $I(X)$ of $X - h_0(X^q)$. The problem with the former is that it constrains n to be approximately q. The problem with the latter is that the polynomial $X - h_0(X^q)$ is insufficiently general to represent all degrees n up to qd_h. By changing the coefficient of X in the latter from 1 to $h_1(X^q)$, we greatly increase the probability of overcoming the second problem, thus combining the higher degree coverage of Joux's representation with the higher degree possibilities of [19,20].

The *raison d'être* of using this representation rather than Joux's representation is that for a given n, by choosing $d_h > 1$, one may use a smaller q. So why is this useful? Well, since the complexity of the new descent methods is typically a function of q, then subject to the satisfaction of certain constraints, one may use a smaller q, thus reducing the complexity of solving the DLP. This observation was our motivation for choosing field representations of the above form.

Another advantage of having an h_1 coefficient (which also applies to Joux's representation) is that it increases the chance of there being a suitable (h_1, h_0) pair with coefficients defined over a proper subfield of $\mathbb{F}_{q^k}$, which then permits one to apply the factor base reduction technique of [36], see §4 and §5.

2.2   Target Fields

For $i \in \{0, 1\}$ let $E_i/\mathbb{F}_{2^p} : Y^2 + Y = X^3 + X + i$. These elliptic curves are supersingular and can have prime or nearly prime order only for p prime, and have embedding degree 4 [16,6,17]. We focus on the curve

$$E_0/\mathbb{F}_{2^{1223}} : Y^2 + Y = X^3 + X, \tag{3}$$

which has a prime order subgroup of cardinality $r_1 = (2^{1223} + 2^{612} + 1)/5$, of bitlength 1221. This curve was initially proposed for 128-bit secure protocols [29] and has enjoyed several optimised implementations [7,11,18,1]. Many smaller p have also been proposed in the literature (see [5,16], for instance), and are clearly weaker.

For $i \in \{0, 1\}$ let $H_i/\mathbb{F}_{2^p} : Y^2 + Y = X^5 + X^3 + i$. These genus two hyperelliptic curves are supersingular and can have a nearly prime order Jacobian only for p prime (note that 13 is always a factor of $\#\mathrm{Jac}_{H_0}(\mathbb{F}_{2^p})$, since $\#\mathrm{Jac}_{H_0}(\mathbb{F}_2) = 13$), and have embedding degree 12 [5,16]. We focus on the curve

$$H_0/\mathbb{F}_{2^{367}} : Y^2 + Y = X^5 + X^3, \tag{4}$$

with $\#\mathrm{Jac}_H(\mathbb{F}_{2^{367}}) = 13 \cdot 7170258097 \cdot r_2$, and $r_2 = (2^{734} + 2^{551} + 2^{367} + 2^{184} + 1)/(13 \cdot 7170258097)$ is a 698-bit prime, since this was proposed for 128-bit secure protocols [3], and whose security was analysed in depth by Adj et $al.$ in [2].

3   Computing the Logarithms of Small Degree Elements

In this section we adapt the polynomial time relation generation method from [19] and Joux's small degree elimination method [32] to the new field representation as detailed in §2.1. Note that henceforth, we shall refer to elements of $\mathbb{F}_{q^{kn}} = \mathbb{F}_{q^k}[X]/(I(X))$ as field elements or as polynomials, as appropriate, and thus use x and X (and y and Y) interchangeably. We therefore freely apply polynomial ring concepts, such as degree, factorisation and smoothness, to field elements.

In order to compute discrete logarithms in our target fields we apply the usual index calculus method. It consists of a precomputation phase in which by means of (sparse) linear algebra techniques one obtains the logarithms of the factor base elements, which will consist of the low degree irreducible polynomials. Afterwards, in the individual logarithm phase, one applies procedures to recursively rewrite each element as a product of elements of smaller degree, in this way building up a $descent$ tree, which has the target element as its root and factor base elements as its leaves. This proceeds in several stages, starting with a continued fraction descent of the target element, followed by a special-Q lattice descent (referred to as degree-balanced classical descent, see [19]), and finally using Joux's Gröbner basis descent [32] for the lower degree elements. Details of the continued fraction and classical descent steps are given in §4, while in this section we provide details of how to find the logarithms of elements of small degree.

We now describe how the logarithms of degree one and two elements (when needed) are to be computed. We use the relation generation method from [19], rather than Joux's method [32], since it automatically avoids duplicate relations. For $k \geq 2$ we first precompute the set $\mathcal{S}_k$, where

$$\mathcal{S}_k = \{(a, b, c) \in (\mathbb{F}_{q^k})^3 \mid X^{q+1} + aX^q + bX + c \text{ splits completely over } \mathbb{F}_{q^k}\}.$$

For $k = 2$, this set of triples is parameterised by $(a, a^q, \mathbb{F}_q \ni c \neq a^{q+1})$, of which there are precisely $q^3 - q^2$ elements. For $k \geq 3$, $\mathcal{S}_k$ can also be computed very efficiently, as follows. Assuming $c \neq ab$ and $b \neq a^q$, the polynomial $X^{q+1} + aX^q + bX + c$ may be transformed (up to a scalar factor) into the polynomial $f_B(\overline{X}) = \overline{X}^{q+1} + B\overline{X} + B$, where $B = \frac{(b-a^q)^{q+1}}{(c-ab)^q}$, and $X = \frac{c-ab}{b-a^q}\overline{X} - a$. The set $\mathcal{L}$ of $B \in \mathbb{F}_{q^k}$ for which f_B splits completely over $\mathbb{F}_{q^k}$ can be computed by simply testing for each such B whether this occurs, and there are precisely $(q^{k-1} - 1)/(q^2 - 1)$ such B if k is odd, and $(q^{k-1} - q)/(q^2 - 1)$ such B if k is even [8]. Then for any (a, b) such that $b \neq a^q$ and for each $B \in \mathcal{L}$, we compute via $B = \frac{(b-a^q)^{q+1}}{(c-ab)^q}$ the corresponding (unique) $c \in \mathbb{F}_{q^k}$, which thus ensures that $(a, b, c) \in \mathcal{S}_k$. Note that in all cases we have $|\mathcal{S}_k| \approx q^{3k-3}$.

3.1   Degree 1 Logarithms

We define the factor base $\mathcal{B}_1$ to be the set of linear elements in x, i.e., $\mathcal{B}_1 = \{x - a \mid a \in \mathbb{F}_{q^k}\}$. Observe that the elements linear in y are each expressible in $\mathcal{B}_1$, since $(y - a) = (x - a^{1/q})^q$.

As in [36,19,20], the basic idea is to consider elements of the form $xy + ay + bx + c$ with $(a, b, c) \in \mathcal{S}_k$. The above two field isomorphisms induce the following equality in $\mathbb{K}$:

$$x^{q+1} + ax^q + bx + c = \frac{1}{h_1(y)}\left(yh_0(y) + ayh_1(y) + bh_0(y) + ch_1(y)\right). \quad (5)$$

When the r.h.s. of (5) also splits completely over $\mathbb{F}_{q^k}$, one obtains a relation between elements of $\mathcal{B}_1$ and the logarithm of $h_1(y)$. One can either adjoin $h_1(y)$ to the factor base, or simply use an $h_1(y)$ which splits completely over $\mathbb{F}_{q^k}$.

We assume that for each $(a, b, c) \in \mathcal{S}_k$ that the r.h.s. of (5) – which has degree $d_h + 1$ – splits completely over $\mathbb{F}_{q^k}$ with probability $1/(d_h + 1)!$. Hence in order for there to be sufficiently many relations we require that

$$\frac{q^{3k-3}}{(d_h + 1)!} > q^k, \quad \text{or equivalently} \quad q^{2k-3} > (d_h + 1)!. \quad (6)$$

When this holds, the expected cost of relation generation is $(d_h + 1)! \cdot q^k \cdot S_{q^k}(d_h + 1, 1)$, where $S_{q^k}(n, m)$ denotes the cost of testing whether a degree n polynomial is m-smooth, i.e., has all of its irreducible factors of degree $\leq m$. The cost of solving the resulting linear system using sparse linear algebra techniques is $O(q^{2k+1})$ arithmetic operations modulo the order r subgroup in which one is working.

3.2   Degree 2 Logarithms

For degree two logarithms, there are several options. The simplest is to apply the degree one method over a quadratic extension of $\mathbb{F}_{q^k}$, but in general (without any factor base automorphisms) this will cost $O(q^{4k+1})$ modular arithmetic operations. If $k \geq 4$ then subject to a condition on q, k and d_h, it is possible to find the logarithms of irreducible degree two elements on the fly, using the techniques of [19,20]. In fact, for the DLP in $\mathbb{F}_{2^{12 \cdot 367}}$ we use both of these approaches, but for different base fields, see §5.

Although not used in the present paper, for completeness we include here the analogue in our field representation of Joux's approach [32]. Since this approach forms the basis of the higher degree elimination steps in the quasi-polynomial time algorithm of Barbulescu et al., its analogue in our field representation should be clear.

We define $\mathcal{B}_{2,u}$ to be the set of irreducible elements of $\mathbb{F}_{q^k}[X]$ of the form $X^2 + uX + v$. For each $u \in \mathbb{F}_{q^k}$ one expects there to be about $q^k/2$ such elements[3]. As in [32], for each $u \in \mathbb{F}_{q^k}$ we find the logarithms of all the elements of $\mathcal{B}_{2,u}$ simultaneously. To do so, consider (5) but with x on the l.h.s. replaced with $Q = x^2 + ux$. Using the field isomorphisms we have that $Q^{q+1} + aQ^q + bQ + c$ is equal to

$$(y^2+u^qy)\left(\left(\tfrac{h_0(y)}{h_1(y)}\right)^2 + u\left(\tfrac{h_0(y)}{h_1(y)}\right)\right) + a(y^2+u^qy) + b\left(\left(\tfrac{h_0(y)}{h_1(y)}\right)^2 + u\left(\tfrac{h_0(y)}{h_1(y)}\right)\right) + c$$

$$= \tfrac{1}{h_1(y)^2}\left((y^2+u^qy)(h_0(y)^2+uh_0(y)h_1(y)+ah_1(y)^2) + b(h_0(y)^2+uh_0(y)h_1(y)) + ch_1(y)^2\right).$$

The degree of the r.h.s. is $2(d_h+1)$, and when it splits completely over $\mathbb{F}_{q^k}$ we have a relation between elements of $\mathcal{B}_{2,u}$ and degree one elements, whose logarithms are presumed known, which we assume occurs with probability $1/(2(d_h + 1))!$. Hence in order for there to be sufficiently many relations we require that

$$\frac{q^{3k-3}}{(2(d_h+1))!} > \frac{q^k}{2}, \quad \text{or equivalently} \quad q^{2k-3} > (2(d_h+1))!/2. \qquad (7)$$

Observe that (7) implies (6). When this holds, the expected cost of relation generation is $(2(d_h+1))! \cdot q^k \cdot S_{q^k}(2(d_h+1), 1)/2$. The cost of solving the resulting linear system using sparse linear algebra techniques is again $O(q^{2k+1})$ modular arithmetic operations, where now both the number of variables and the average weight is halved relative to the degree one case. Since there are q^k such u, the total expected cost of this stage is $O(q^{3k+1})$ modular arithmetic operations, which may of course be parallelised.

3.3 Joux's Small Degree Elimination with the New Representation

As in [32], let Q be a degree d_Q element to be eliminated, let $F(X) = \sum_{i=0}^{d_F} f_i X^i$, $G(X) = \sum_{j=0}^{d_G} g_j X^j \in \mathbb{F}_{q^k}[X]$ with $d_F + d_G + 2 \geq d_Q$, and assume without loss of generality $d_F \geq d_G$. Consider the following expression:

$$G(X) \prod_{\alpha \in \mathbb{F}_q} (F(X) - \alpha G(X)) = F(X)^q G(X) - F(X)G(X)^q \qquad (8)$$

---

[3] For binary fields there are precisely $q^k/2$ irreducibles, since X^2+uX+v is irreducible if and only if $\mathrm{Tr}_{\mathbb{F}_{q^k}/\mathbb{F}_2}(v/u^2) = 1$.

The l.h.s. is $\max(d_F, d_G)$-smooth. The r.h.s. can be expressed modulo $h_1(X^q)X - h_0(X^q)$ in terms of $Y = X^q$ as a quotient of polynomials of relatively low degree by using

$$F(X)^q = \sum_{i=0}^{d_F} f_i^q Y^i, \quad G(X)^q = \sum_{j=0}^{d_G} g_j^q Y^j \quad \text{and} \quad X \equiv \frac{h_0(Y)}{h_1(Y)}.$$

Then the numerator of the r.h.s. becomes

$$\Big(\sum_{i=0}^{d_F} f_i^q Y^i\Big)\Big(\sum_{j=0}^{d_G} g_j^q h_0(Y)^j h_1(Y)^{d_F-j}\Big) - \Big(\sum_{i=0}^{d_F} f_i^q h_0(Y)^i h_1(Y)^{d_F-i}\Big)\Big(\sum_{j=0}^{d_G} g_j^q Y^j\Big). \quad (9)$$

Setting (9) to be 0 modulo $Q(Y)$ gives a system of d_Q equations over $\mathbb{F}_{q^k}$ in the $d_F + d_G + 2$ variables $f_0, \ldots, f_{d_F}, g_0, \ldots, g_{d_G}$. By choosing a basis for $\mathbb{F}_{q^k}$ over $\mathbb{F}_q$ and expressing each of the $d_F + d_G + 2$ variables $f_0, \ldots, f_{d_F}, g_0, \ldots, g_{d_G}$ in this basis, this system becomes a bilinear quadratic system[4] of kd_Q equations in $(d_F + d_G + 2)k$ variables. To find solutions to this system, one can specialise $(d_F + d_G + 2 - d_Q)k$ of the variables in order to make the resulting system generically zero-dimensional while keeping its bilinearity, and then compute the corresponding Gröbner basis, which may have no solution, or a small number of solutions. For each solution, one checks whether (9) divided by $Q(Y)$ is $(d_Q - 1)$-smooth: if so then Q has successfully been rewritten as a product of elements of smaller degree; if no solutions give a $(d_Q - 1)$-smooth cofactor, then one begins again with another specialisation.

The degree of the cofactor of $Q(Y)$ is upper bounded by $d_F(1 + d_h) - d_Q$, so assuming that it behaves as a uniformly chosen polynomial of such a degree one can calculate the probability ρ that it is $(d_Q - 1)$-smooth using standard combinatorial techniques.

Generally, in order for Q to be eliminable by this method with good probability, the number of solutions to the initial bilinear system must be greater than $1/\rho$. To estimate the number of solutions, consider the action of $\mathrm{Gl}_2(\mathbb{F}_{q^k})$ on the set of pairs (F, G). The subgroups $\mathrm{Gl}_2(\mathbb{F}_q)$ and $\mathbb{F}_{q^k}^\times$ (via diagonal embedding) both act trivially on the set of relations, modulo multiplication by elements in $\mathbb{F}_{q^k}^\times$. Assuming that the set of (F, G) quotiented out by the action of the compositum of these subgroups (which has cardinality $\approx q^{k+3}$), generates distinct relations, one must satisfy the condition

$$q^{(d_F + d_G + 1 - d_Q)k - 3} > 1/\rho. \quad (10)$$

Note that while (10) is preferable for an easy descent, one may yet violate it and still successfully eliminate elements by using various tactics, as demonstrated in §5.

---

[4] The bilinearity makes finding solutions to this system easier [44], and is essential for the complexity analysis in [32] and its variant in [20].

4 Concrete Security Analysis of $\mathbb{F}_{2^{4 \cdot 1223}}$

In this section we focus on the DLP in the 1221-bit prime order r_1 subgroup of $\mathbb{F}_{2^{4 \cdot 1223}}^{\times}$, which arises from the MOV attack applied to the genus one supersingular curve (3). By embedding $\mathbb{F}_{2^{4 \cdot 1223}}$ into its degree two extension $\mathbb{F}_{2^{8 \cdot 1223}} = \mathbb{F}_{2^{9784}}$ we show that, after a precomputation taking approximately 2^{40} s, individual discrete logarithms can be computed in less than 2^{34} s.

4.1 Setup

We consider the field $\mathbb{F}_{2^{8 \cdot 1223}} = \mathbb{F}_{q^n}$ with $q = 2^8$ and $n = 1223$ given by the irreducible factor of degree n of $h_1(X^q)X - h_0(X^q)$, with

$$h_0 = X^5 + tX^4 + tX^3 + X^2 + tX + t, \quad h_1 = X^5 + X^4 + X^3 + X^2 + X + t,$$

where t is an element of $\mathbb{F}_{2^2} \setminus \mathbb{F}_2$. Note that the field of definition of this representation is $\mathbb{F}_{2^2}$.

Since the target element is contained in the subfield $\mathbb{F}_{2^{4 \cdot 1223}}$, we begin the classical descent over $\mathbb{F}_{2^4}$, we switch to $\mathbb{F}_q = \mathbb{F}_{2^8}$, i.e., $k = 1$, for the Gröbner basis descent, and, as explained below, we work over $\mathbb{F}_{q^k}$ with either $k = 1$ or a few $k > 1$ to obtain the logarithms of all factor base elements.

4.2 Linear Algebra Cost Estimate

In this precomputation we obtain the logarithms of all elements of degree at most four over $\mathbb{F}_q$. Since the degree 1223 extension is defined over $\mathbb{F}_{2^2}$ in our field representation, by the action of the Galois group $\mathrm{Gal}(\mathbb{F}_q/\mathbb{F}_{2^2})$ on the factor base, the number of irreducible elements of degree j whose logarithms are to be computed can be reduced to about $2^{8j}/(4j)$ for $j \in \{1, 2, 3, 4\}$.

One way to obtain the logarithms of these elements is to carry out the degree 1 relation generation method from §3.1, together with the elementary observation that an irreducible polynomial of degree k over $\mathbb{F}_q$ splits completely over $\mathbb{F}_{q^k}$. First, computing degree one logarithms over $\mathbb{F}_{q^3}$ gives the logarithms of irreducible elements of degrees one and three over $\mathbb{F}_q$. Similarly, computing degree one logarithms over $\mathbb{F}_{q^4}$ gives the logarithms of irreducible elements of degrees one, two, and four over $\mathbb{F}_q$. The main computational cost consists in solving the latter system arising from $\mathbb{F}_{q^4}$, which has size 2^{28} and an average row weight of 256.

However, we propose to reduce the cost of finding these logarithms by using $k = 1$ only, in the following easy way. Consider §3.3, and observe that for each polynomial pair (F, G) of degree at most d, one obtains a relation between elements of degree at most d when the numerator of the r.h.s. is d-smooth (ignoring factors of h_1). Note that we are not setting the r.h.s. numerator to be zero modulo Q or computing any Gröbner bases. Up to the action of $\mathrm{Gl}_2(\mathbb{F}_q)$ (which gives equivalent relations) there are about q^{2d-2} such polynomial pairs. Hence, for $d \geq 3$ there are more relations than elements if the smoothness probability of the r.h.s. is sufficiently high. Notice that $k = 1$ implies that the r.h.s. is divisible by $h_1(Y)Y - h_0(Y)$, thus increasing its smoothness probability and resulting in enough relations for $d = 3$ and for $d = 4$. After having solved the much smaller system for $d = 3$ we know the logarithms of all elements up to degree three, so that the average row weight for the system for $d = 4$ can be reduced to about $\frac{1}{4} \cdot 256 = 64$ (irreducible degree four polynomials on the l.h.s.). As above the size of this system is 2^{28}.

The cost for generating the linear systems is negligible compared to the linear algebra cost. For estimating the latter cost we consider Lanczos' algorithm to solve a sparse $N \times N$, $N = 2^{28}$, linear system with average row weight $W = 64$. As noted in [40,20] this algorithm can be implemented such that

$$N^2 \left(2\, W \, \mathrm{ADD} + 2\, \mathrm{SQR} + 3\, \mathrm{MULMOD}\right) \tag{11}$$

operations are used. On our benchmark system, an AMD Opteron 6168 processor at 1.9 GHz, using [28] our implementation of these operations took 62 ns, 467 ns and 1853 ns for an ADD, a SQR and a MULMOD, respectively, resulting in a linear algebra cost of 2^{40} s.

As in [2], the above estimate ignores communication costs and other possible slowdowns which may arise in practice. An alternative estimate can be obtained by considering a problem of a similar size over $\mathbb{F}_2$ and extrapolating from [37]. This gives an estimated time of 2^{42} s, or for newer hardware slightly less. Note that this computation was carried out using the block Wiedemann algorithm [13], which we recommend in practice because it allows one to distribute the main part of the computation. For the sake of a fair comparison with [2] we use the former estimate of 2^{40} s.

4.3 Descent Cost Estimate

We assume that the logarithms of elements up to degree four are known, and that computing these logarithms with a lookup table is free.

Small Degree Descent. We have implemented the small degree descent of §3.3 in Magma [10] V2.20-1, using Faugere's F4 algorithm [15]. For each degree from 5 to 15, on the same AMD Opteron 6168 processor we timed the Gröbner basis computation between 10^6 and 100 times, depending on the degree. Then using a bottom-up recursive strategy we estimated the following average running times in seconds for a full logarithm computation, which we present to two significant figures:

$$C[5, \ldots, 15] = [\, 0.038\, , 2.1\, , 2.1\, , 93\, , 95\, , 180\, , 190\, , 3200\, , 3500\, , 6300\, , 11000\,]\,.$$

Degree-Balanced Classical Descent. From now on, we make the conservative assumption that a degree n polynomial which is m-smooth, is a product of n/m degree m polynomials. In practice the descent cost will be lower than this, however, the linear algebra cost is dominating, so this issue is inconsequential for our security estimate. The algorithms we used for smoothness testing are detailed in the full version of the paper [23].

For a classical descent step with degree balancing we consider polynomials $P(X^{2^a}, Y) \in \mathbb{F}_q[X, Y]$ for a suitably chosen integer $0 \leq a \leq 8$. It is advantageous to choose P such that its degree in one variable is one; let d be the degree in the other variable. In the case $\deg_{X^{2^a}}(P) = 1$, i.e., $P = v_1(Y)X^{2^a} + v_0(Y)$, $\deg v_i \leq d$, this gives rise to the relation

$$L_v^{2^a} = \left(\frac{R_v}{h_1(X)^{2^a}} \right)^{2^8} \quad \text{where} \quad \begin{aligned} L_v &= \tilde{v}_1(X^{2^{8-a}})X + \tilde{v}_0(X^{2^{8-a}}), \\ R_v &= v_1(X)h_0(X)^{2^a} + v_0(X)h_1(X)^{2^a} \end{aligned}$$

in $\mathbb{F}_q[X]/(h_1(X^q)X - h_0(X^q))$ with $\deg L_v \leq 2^{8-a}d + 1$, $\deg R_v \leq d + 5 \cdot 2^a$, and $\tilde{v}_i$ being v_i with its coefficients powered by 2^{8-a}, for $i = 0, 1$. Similarly, in the case $\deg_Y(P) = 1$, i.e., $P = w_1(X^{2^a})Y + w_0(X^{2^a})$, $\deg w_i \leq d$, we have the relation

$$L_w^{2^a} = \left(\frac{R_w}{h_1(X)^{2^a d}} \right)^{2^8} \quad \text{where} \quad \begin{aligned} L_w &= \tilde{w}_1(X)X^{2^{8-a}} + \tilde{w}_0(X), \\ R_w &= h_1(X)^{2^a d} \left(w_1\left(\left(\frac{h_0(X)}{h_1(X)} \right)^{2^a} \right)X + w_0\left(\left(\frac{h_0(X)}{h_1(X)} \right)^{2^a} \right) \right) \end{aligned}$$

with $\deg L_w \leq d + 2^{8-a}$, $\deg R_w \leq 5 \cdot 2^a d + 1$ and again $\tilde{w}_i$ being w_i with its coefficients powered by 2^{8-a}, for $i = 0, 1$.

The polynomials v_i (respectively w_i) are chosen in such a way that either the l.h.s. or the r.h.s. is divisible by a polynomial $Q(X)$ of degree d_Q. Gaussian reduction provides a lattice basis $(u_0, u_1), (u'_0, u'_1)$ such that the polynomial pairs satisfying the divisibility condition above are given by $ru_i + su'_i$ for $i = 0, 1$, where $r, s \in \mathbb{F}_q[X]$. For nearly all polynomials Q it is possible to choose a lattice basis of polynomials with degree $\approx d_Q/2$ which we will assume for all Q appearing in the analysis; extreme cases can be avoided by look-ahead or backtracking techniques. Notice that a polynomial Q over $\mathbb{F}_{2^4} \subset \mathbb{F}_q$ can be rewritten as a product of polynomials which are also over $\mathbb{F}_{2^4}$, by choosing the basis as well as r and s to be over $\mathbb{F}_{2^4}$. This will be done in all steps of the classical descent. The polynomials r and s are chosen to be of degree four, resulting in 2^{36} possible pairs (multiplying both by a common non-zero constant gives the same relation).

In the final step of the classical eliminations (from degree 26 to 15) we relax the criterion that the l.h.s. and r.h.s. are 15-smooth, allowing also irreducibles of even degree up to degree 30, since these can each be split over $\mathbb{F}_q$ into two polynomials of half the degree, thereby increasing the smoothness probabilities. Admittedly, if we follow our worst-case analysis stipulation that all polynomials at this step have degree 26, then one could immediately split each of them into two degree 13 polynomials. However, in practice one will encounter polynomials of all degrees ≤ 26 and we therefore carry out the analysis without using the splitting shortcut, which will still provide an overestimate of the cost of this step.

In the following we will state the logarithmic cost (in seconds) of a classical descent step as $c_l + c_r + c_s$, where 2^{c_l} and 2^{c_r} denote the number of trials to get the left hand side and the right hand side m-smooth, and 2^{c_s} s is the time required for the corresponding smoothness test. Our smoothness tests were benchmarked on the AMD Opteron 6168 processor.

– **$d_Q = 26$ to $m = 15$:** We choose $\deg_{X^{2^a}} P = 1$, $a = 5$, Q on the right, giving $d = 17$ and $(\deg(L_v), \deg(R_v)) = (137, 151)$. On average the smoothness test $S_{2^8}(137, 30)$ takes 1.9 ms, giving a logarithmic cost of $13.4 + 15.6 - 9.0$, hence $2^{20.0}$ s. The expected number of factors is 19.2, so the subsequent cost will be less than $2^{17.7}$ s. Note that, as explained above, we use the splitting shortcut for irreducibles of even degree up to 30, resulting in the higher than expected smoothness probabilities.

– **$d_Q = 36$ to $m = 26$:** We choose $\deg_{X^{2^a}} P = 1$, $a = 5$, Q on the right, giving $d = 22$ and $(\deg(L_v), \deg(R_v)) = (177, 146)$. On average the smoothness test $S_{2^8}(146, 26)$ takes 1.9 ms, giving a logarithmic cost $18.7 + 13.6 - 9.0$, hence $2^{23.3}$ s. The expected number of factors is 12.4, so the subsequent cost will be less than $2^{23.9}$ s.

– **$d_Q = 94$ to $m = 36$:** We choose $\deg_Y P = 1$, $a = 0$, Q on the left, giving $d = 51$ and $(\deg(L_w), \deg(R_w)) = (213, 256)$. On average the smoothness test $S_{2^8}(213, 36)$ takes 5.1 ms, giving a logarithmic cost $15.0 + 20.3 - 7.5$, hence $2^{27.8}$ s. The expected number of factors is 13.0, so the subsequent cost will be less than $2^{28.4}$ s.

Continued Fraction Descent. For the continued fraction descent we multiply the target element by random powers of the generator and express the product as a ratio of two polynomials of degree at most 611. For each such expression we test if both the numerator and the denominator are 94-smooth. On average the smoothness test $S_{2^8}(611, 94)$ takes 94 ms, giving a logarithmic cost of $17.7 + 17.7 - 3.4$, hence $2^{32.0}$ s. The expected number of degree 94 factors on both sides will be 13, so the subsequent cost will be less than $2^{32.8}$ s.

Total Descent Cost. The cost for computing an individual logarithm is therefore upper-bounded by $2^{32.0}$ s $+ 2^{32.8}$ s $< 2^{34}$ s.

4.4   Summary

The main cost in our analysis is the linear algebra computation which takes about 2^{40} s, with the individual logarithm stage being considerably faster. In order to compare with the estimate in [2], we write the main cost in terms of M_r which gives $2^{59} M_r$, and thus an improvement by a factor of 2^{69}. Nevertheless, solving a system of cardinality 2^{28} is still a formidable challenge, but perhaps not so much for a well-funded adversary. For completeness we note that if one wants to avoid a linear algebra step of this size, then one can work over different fields, e.g., with $q = 2^{10}$ and $k = 2$, or $q = 2^{12}$ and $k = 1$. However, while this allows a partitioning of the linear algebra into smaller steps as described in §3.2 but at a slightly higher cost, the resulting descent cost is expected to be significantly higher.

5   Solving the DLP in $\mathbb{F}_{2^{12 \cdot 367}}$

In this section we present the details of our solution of a DLP in the 698-bit prime order r_2 subgroup of $\mathbb{F}_{2^{12 \cdot 367}}^{\times} = \mathbb{F}_{2^{4404}}^{\times}$, which arises from the MOV attack applied to the Jacobian of the genus two supersingular curve (4). Note that the prime order elliptic curve $E_1/\mathbb{F}_{2^{367}} : Y^2 + Y = X^3 + X + 1$ with embedding degree 4 also embeds into $\mathbb{F}_{2^{4404}}$, so that logarithms on this curve could have easily been computed as well.

5.1   Setup

To compute the target logarithm, as stated in §1 we applied a principle of parsi-mony, namely, we tried to solve all intermediate logarithms in $\mathbb{F}_{2^{12 \cdot 367}}$, considered as a degree 367 extension of $\mathbb{F}_{2^{12}}$, and only when this was not possible did we embed elements into the extension field $\mathbb{F}_{2^{24 \cdot 367}}$ (by extending the base field to $\mathbb{F}_{2^{24}}$) and solve them there.

All of the classical descent down to degree 8 was carried out over $\mathbb{F}_{2^{12 \cdot 367}}$, which we formed as the compositum of the following two extension fields. We defined $\mathbb{F}_{2^{12}}$ using the irreducible polynomial $U^{12} + U^3 + 1$ over $\mathbb{F}_2$, and defined $\mathbb{F}_{2^{367}}$ over $\mathbb{F}_2$ using the degree 367 irreducible factor of $h_1(X^{64})X - h_0(X^{64})$, where $h_1 = X^5 + X^3 + X + 1$, and $h_0 = X^6 + X^4 + X^2 + X + 1$. Let u and x be roots of the extension defining polynomials in U and X respectively, and let $c = (2^{4404} - 1)/r_2$. Then $g = x + u^7$ is a generator of $\mathbb{F}_{2^{4404}}^{\times}$ and $\bar{g} = g^c$ is a generator of the subgroup of order r_2. As usual, our target element was chosen to be $\bar{x}_\pi = x_\pi^c$ where

$$x_\pi = \sum_{i=0}^{4403} (\lfloor \pi \cdot 2^{i+1} \rfloor \bmod 2) \cdot u^{11-(i \bmod 12)} \cdot x^{\lfloor i/12 \rfloor}.$$

The remaining logarithms were computed using a combination of tactics, over $\mathbb{F}_{2^{12}}$ when possible, and over $\mathbb{F}_{2^{24}}$ when not. These fields were constructed as de-gree 2 and 4 extensions of $\mathbb{F}_{2^6}$, respectively. To define $\mathbb{F}_{2^6}$ we used the irreducible polynomial $T^6 + T + 1$. We then defined $\mathbb{F}_{2^{12}}$ using the irreducible polynomial $V^2 + tV + 1$ over $\mathbb{F}_{2^6}$, and $\mathbb{F}_{2^{24}}$ using the irreducible polynomial $W^4 + W^3 + W^2 + t^3$ over $\mathbb{F}_{2^6}$.

5.2   Degree 1 Logarithms

It was not possible to find enough relations for degree 1 elements over $\mathbb{F}_{2^{12}}$, so in accordance with our stated principle, we extended the base field to $\mathbb{F}_{2^{24}}$ to compute the logarithms of all 2^{24} degree 1 elements. We used the polynomial time relation generation from §3.1, which took 47 hours. This relative sluggishness was due to the r.h.s. having degree $d_h + 1 = 7$, which must split over $\mathbb{F}_{2^{24}}$. However, this was faster by a factor of 24 than it would have been otherwise, thanks to h_0 and h_1 being defined over $\mathbb{F}_2$. This allowed us to use the technique from [36] to reduce the size of the factor base via the automorphism $(x + a) \mapsto$

$(x + a)^{2^{367}}$, which fixes x but has order 24 on all non-subfield elements of $\mathbb{F}_{2^{24}}$, since $367 \equiv 7 \bmod 24$ and $\gcd(7, 24) = 1$. This reduced the factor base size to 699252 elements, which was solved in 4896 core hours on a 24 core cluster using Lanczos' algorithm, approximately 24^2 times faster than if we had not used the automorphisms.

5.3   Individual Logarithm

We performed the standard continued fraction initial split followed by degree-balanced classical descent as in §4.3, using Magma [10] and NTL [43], to reduce the target element to an 8-smooth product in 641 and 38224 core hours respectively. The most interesting part of the descent was the elimination of the elements of degree up to 8 over $\mathbb{F}_{2^{12}}$ into elements of degree one over $\mathbb{F}_{2^{24}}$, which we detail below. This phase was completed using Magma and took a further 8432 core hours. However, we think that the combined time of the classical and non-classical parts could be reduced significantly via a backwards-induction analysis of the elimination times of each degree.

Small Degree Elimination. As stated above we used several tactics to achieve these eliminations. The first was the splitting of an element of even degree over $\mathbb{F}_{2^{12}}$ into two elements of half the degree (which had the same logarithm modulo r_2) over the larger field. This automatically provided the logarithms of all degree 2 elements over $\mathbb{F}_{2^{12}}$. Similarly elements of degree 4 and 8 over $\mathbb{F}_{2^{12}}$ were rewritten as elements of degree 2 and 4 over $\mathbb{F}_{2^{24}}$, while we found that degree 6 elements were eliminable more efficiently by initially continuing the descent over $\mathbb{F}_{2^{12}}$, as with degree 5 and 7 elements.

The second tactic was the application of Joux's Gröbner basis elimination method from §3.3 to elements over $\mathbb{F}_{2^{12}}$, as well as elements over $\mathbb{F}_{2^{24}}$. However, in many cases condition (10) was violated, in which case we had to employ various recursive strategies in order to eliminate elements. In particular, elements of the same degree were allowed on the r.h.s. of relations, and we then attempted to eliminate these using the same (recursive) strategy. For degree 3 elements over $\mathbb{F}_{2^{12}}$, we even allowed degree 4 elements to feature on the r.h.s. of relations, since these were eliminable via the factorisation into degree 2 elements over $\mathbb{F}_{2^{24}}$.

In Figure 1 we provide a flow chart for the elimination of elements of degree up to 8 over $\mathbb{F}_{2^{12}}$, and for the supporting elimination of elements of degree up to 4 over $\mathbb{F}_{2^{24}}$. Nearly all of the arrows in Figure 1 were necessary for these field parameters (the exceptions being that for degrees 4 and 8 over $\mathbb{F}_{2^{12}}$ we could have initially continued the descent along the bottom row, but this would have been slower). The reason this 'non-linear' descent arises is due to q being so small, and d_H being relatively large, which increases the degree of the r.h.s. cofactors, thus decreasing the smoothness probability. Indeed these tactics were only borderline applicable for these parameters; if h_0 or h_1 had degree any larger than 6 then not only would most of the descent have been much harder, but it seems that one would be forced to compute the logarithms of degree 2 elements

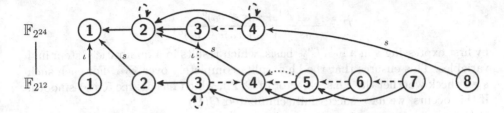

Fig. 1. This diagram depicts the set of strategies employed to eliminate elements over $\mathbb{F}_{2^{12}}$ of degree up to 8. The encircled numbers represent the degrees of elements over $\mathbb{F}_{2^{12}}$ on the bottom row, and over $\mathbb{F}_{2^{24}}$ on the top row. The arrows indicate how an element of a given degree is rewritten as a product of elements of other degrees, possibly over the larger field. Unadorned solid arrows indicate the maximum degree of elements obtained on the l.h.s. of the Gröbner basis elimination method; likewise dashed arrows indicate the degrees of elements obtained on the r.h.s. of the Gröbner basis elimination method, when these are greater than those obtained on the l.h.s. Dotted arrows indicate a fall-back strategy when the initial strategy fails. An s indicates that the element is to be split over the larger field into two elements of half the degree. An ι indicates that an element is promoted to the larger field. Finally, a loop indicates that one must use a recursive strategy in which further instances of the elimination in question must be solved in order to eliminate the element in question.

over $\mathbb{F}_{2^{24}}$ using Joux's linear system method from §3.2, greatly increasing the required number of core hours. As it was, we were able to eliminate degree 2 elements over $\mathbb{F}_{2^{24}}$ on the fly, as we describe explicitly below.

Finally, we note that our descent strategy is considerably faster than the alternative of embedding the DLP into $\mathbb{F}_{2^{24 \cdot 367}}$ and performing a full descent in this field, even with the elimination on the fly of degree 2 elements over $\mathbb{F}_{2^{24}}$, since much of the resulting computation would constitute superfluous effort for the task in hand.

Degree 2 Elimination over $\mathbb{F}_{2^{24}}$. Let $Q(Y)$ be a degree two element which is to be eliminated, i.e., written as a product of degree one elements. As in [19,20] we first precompute the set of 64 elements $B \in \mathbb{F}_{2^{24}}$ such that the polynomial $f_B(X) = X^{65} + BX + B$ splits completely over $\mathbb{F}_{2^{24}}$ (in fact these B's happen to be in $\mathbb{F}_{2^{12}}$, but this is not relevant to the method). We then find a Gaussian-reduced basis of the lattice $L_{Q(Y)}$ defined by

$$L_{Q(Y)} = \{(w_0(Y), w_1(Y)) \in \mathbb{F}_{2^{24}}[Y]^2 : w_0(Y)h_0(Y) + w_1(Y)h_1(Y) \equiv 0 \pmod{Q(Y)}\}.$$

Such a basis has the form $(u_0, Y + u_1), (Y + v_0, v_1)$, with $u_i, v_i \in \mathbb{F}_{2^{24}}$, except in rare cases, see Remark 1. For $s \in \mathbb{F}_{2^{24}}$ we obtain lattice elements $(w_0(Y), w_1(Y)) = (Y + v_0 + su_0, sY + v_1 + su_1)$.

Using the transformation detailed in §3, for each $B \in \mathbb{F}_{2^{24}}$ such that f_B splits completely over $\mathbb{F}_{2^{24}}$ we perform a Gröbner basis computation to find the set of $s \in \mathbb{F}_{2^{24}}$ that satisfy

$$B = \frac{(s^{64} + u_0 s + v_0)^{65}}{(u_0 s^2 + (u_1 + v_0)s + v_1)^{64}},$$

by first expressing s in a $\mathbb{F}_{2^{24}}/\mathbb{F}_{2^6}$ basis, which results in a quadratic system in 4 variables. This ensures that the l.h.s. splits completely over $\mathbb{F}_{2^{24}}$. For each such s we check whether the r.h.s. cofactor of $Q(Y)$, which has degree 5, is 1-smooth. If this occurs, we have successfully eliminated $Q(Y)$.

However, one expects on average just one s per B, and so the probability of $Q(Y)$ being eliminated in this way is $1 - (1 - 1/5!)^{64} \approx 0.415$, which was borne out in practice to two decimal places. Hence, we adopted a recursive strategy in which we stored all of the r.h.s. cofactors whose factorisation degrees had the form $(1, 1, 1, 2)$ (denoted type 1), or $(1, 2, 2)$ (denoted type 2). Then for each type 1 cofactor we checked to see if the degree 2 factor was eliminable by the above method. If none were eliminable we stored every type 1 cofactor of each degree 2 irreducible occurring in the list of type 1 cofactors of $Q(Y)$. If none of these were eliminable (which occurred with probability just 0.003), then we reverted to the type 2 cofactors, and adopted the same strategy just specified for each of the degree 2 irreducible factors. Overall, we expected our strategy to fail about once in every $6 \cdot 10^6$ such $Q(Y)$. This happened just once during our descent, and so we multiplied this $Q(Y)$ by a random linear polynomial over $\mathbb{F}_{2^{24}}$ and performed a degree 3 elimination, which necessitates an estimated 32 degree 2 polynomials being simultaneously eliminable by the above method, which thanks to the high probability of elimination, will very likely be successful for any linear multiplier.

5.4   Summary

Finally, after a total of approximately 52240 core hours (or $2^{48}\ M_{r_2}$), we found that $\bar{x}_\pi = \bar{g}^{\log}$, with (see [24] for a Magma verification script) $\log =$

40932089202142351640934477339007025637256140979451423541922853874473604
39015351684721408233687689563902511062230980145272871017382542826764695
59843114767895545475795766475848754227211594761182312814017076893242 .

Remark 1. During the descent, we encountered several polynomials $Q(Y)$ that were apparently not eliminable via the Gröbner basis method. We discovered that they were all factors of $h_1(Y) \cdot c + h_0(Y)$ for $c \in \mathbb{F}_{2^{12}}$ or $\mathbb{F}_{2^{24}}$, and hence $h_0(Y)/h_1(Y) \equiv c \pmod{Q(Y)}$. This implies that (9) is equal to $F(c)G^{(q)}(Y) + F^{(q)}(Y)G(c)$ modulo $Q(Y)$, where $G^{(q)}$ denotes the Frobenius twisted G and similarly for $F^{(q)}$. This cannot become 0 modulo $Q(Y)$ if the degrees of F and G are smaller than the degree of Q, unless F and G are both constants. However, thanks to the field representation, finding the logarithm of these $Q(Y)$ turns out to be easy. In particular, if $h_1(Y) \cdot c + h_0(Y) = Q(Y) \cdot R(Y)$ then $Q(Y) = h_1(Y)((h_0/h_1)(Y) + c)/R(Y) = h_1(Y)(X + c)/R(Y)$, and thus modulo r_2 we have $\log(Q(y)) \equiv \log(x + c) - \log(R(y))$, since $\log(h_1(y)) \equiv 0$. Since $(x + c)$ is in the factor base, if we are able to compute the logarithm of $R(y)$, then we are done. In all the cases we encountered, the cofactor $R(y)$ was solvable by the above methods.

6 Conclusion

We have introduced a new field representation and efficient descent principles which together make the recent DLP advances far more practical. As example demonstrations, we have applied these techniques to two binary fields of central interest to pairing-based cryptography, namely $\mathbb{F}_{2^{4\cdot1223}}$ and $\mathbb{F}_{2^{12\cdot367}}$, which arise as the embedding fields of (the Jacobians of) a genus one and a genus two supersingular curve, respectively. When initially proposed, these fields were believed to be 128-bit secure, and even in light of the recent DLP advances, were believed to be 128-bit and 94.6-bit secure. On the contrary, our analysis indicates that the former field has approximately 59 bits of security and we have implemented a total break of the latter.

References

1. Adikari, J., Hasan, M.A., Negre, C.: Towards faster and greener cryptoprocessor for eta pairing on supersingular elliptic curve over $\mathbb{F}_{2^{1223}}$. In: Knudsen, L.R., Wu, H. (eds.) SAC 2012. LNCS, vol. 7707, pp. 166–183. Springer, Heidelberg (2013)
2. Adj, G., Menezes, A., Oliveira, T., Rodríguez-Henríquez, F.: Weakness of $\mathbb{F}_{3^{6\cdot509}}$ for discrete logarithm cryptography. In: Cao, Z., Zhang, F. (eds.) Pairing 2013. LNCS, vol. 8365, pp. 20–44. Springer, Heidelberg (2014)
3. Aranha, D.F., Beuchat, J.-L., Detrey, J., Estibals, N.: Optimal eta pairing on supersingular genus-2 binary hyperelliptic curves. In: Dunkelman, O. (ed.) CT-RSA 2012. LNCS, vol. 7178, pp. 98–115. Springer, Heidelberg (2012)
4. Barbulescu, R., Gaudry, P., Joux, A., Thomé, E.: A heuristic quasi-polynomial algorithm for discrete logarithm in finite fields of small characteristic. In: Nguyen, P.Q., Oswald, E. (eds.) EUROCRYPT 2014. LNCS, vol. 8441, pp. 1–16. Springer, Heidelberg (2014)
5. Barreto, P.S.L.M., Galbraith, S.D., Héigeartaigh, C.Ó., Scott, M.: Efficient pairing computation on supersingular abelian varieties. Des. Codes Cryptography 42(3), 239–271 (2007)
6. Barreto, P.S.L.M., Kim, H.Y., Lynn, B., Scott, M.: Efficient algorithms for pairing-based cryptosystems. In: Yung, M. (ed.) CRYPTO 2002. LNCS, vol. 2442, pp. 354–369. Springer, Heidelberg (2002)
7. Beuchat, J.-L., López-Trejo, E., Martínez-Ramos, L., Mitsunari, S., Rodríguez-Henríquez, F.: Multi-core implementation of the Tate pairing over supersingular elliptic curves. In: Garay, J.A., Miyaji, A., Otsuka, A. (eds.) CANS 2009. LNCS, vol. 5888, pp. 413–432. Springer, Heidelberg (2009)
8. Bluher, A.W.: On $x^{q+1}+ax+b$. Finite Fields and Their Applications 10(3), 285–305 (2004)
9. Boneh, D., Franklin, M.: Identity-based encryption from the Weil pairing. In: Kilian, J. (ed.) CRYPTO 2001. LNCS, vol. 2139, pp. 213–229. Springer, Heidelberg (2001)
10. Bosma, W., Cannon, J., Playoust, C.: The Magma algebra system. I. The user language. J. Symbolic Comput. 24(3-4), 235–265 (1997)
11. Chatterjee, S., Hankerson, D., Menezes, A.: On the efficiency and security of pairing-based protocols in the type 1 and type 4 settings. In: Hasan, M.A., Helleseth, T. (eds.) WAIFI 2010. LNCS, vol. 6087, pp. 114–134. Springer, Heidelberg (2010)

12. Coppersmith, D.: Fast evaluation of logarithms in fields of characteristic two. IEEE Transactions on Information Theory 30(4), 587–593 (1984)
13. Coppersmith, D.: Solving homogeneous linear equations over GF(2) via block Wiedemann algorithm. Mathematics of Computation 62(205), 333–350 (1994)
14. Duursma, I., Lee, H.-S.: Tate pairing implementation for hyperelliptic curves $y^2 = x^p - x + d$. In: Laih, C.-S. (ed.) ASIACRYPT 2003. LNCS, vol. 2894, pp. 111–123. Springer, Heidelberg (2003)
15. Faugère, J.-C.: A new efficient algorithm for computing Gröbner bases (F_4). J. Pure Appl. Algebra 139(1-3), 61–88 (1999)
16. Galbraith, S.D.: Supersingular curves in cryptography. In: Boyd, C. (ed.) ASIACRYPT 2001. LNCS, vol. 2248, pp. 495–513. Springer, Heidelberg (2001)
17. Galbraith, S.D., Harrison, K., Soldera, D.: Implementing the Tate pairing. In: Fieker, C., Kohel, D.R. (eds.) ANTS 2002. LNCS, vol. 2369, pp. 324–337. Springer, Heidelberg (2002)
18. Ghosh, S., Roychowdhury, D., Das, A.: High speed cryptoprocessor for η_t pairing on 128-bit secure supersingular elliptic curves over characteristic two fields. In: Preneel, B., Takagi, T. (eds.) CHES 2011. LNCS, vol. 6917, pp. 442–458. Springer, Heidelberg (2011)
19. Göloğlu, F., Granger, R., McGuire, G., Zumbrägel, J.: On the function field sieve and the impact of higher splitting probabilities: Application to discrete logarithms in $\mathbb{F}_{2^{1971}}$ and $\mathbb{F}_{2^{3164}}$. In: Canetti, R., Garay, J.A. (eds.) CRYPTO 2013, Part II. LNCS, vol. 8043, pp. 109–128. Springer, Heidelberg (2013)
20. Göloğlu, F., Granger, R., McGuire, G., Zumbrägel, J.: Solving a 6120-bit DLP on a desktop computer. In: Lange, T., Lauter, K., Lisoněk, P. (eds.) SAC 2013. LNCS, vol. 8282, pp. 136–152. Springer, Heidelberg (2014)
21. Göloğlu, F., Granger, R., McGuire, G., Zumbrägel, J.: Discrete Logarithms in GF(2^{1971}). NMBRTHRY list (February 19, 2013)
22. Göloğlu, F., Granger, R., McGuire, G., Zumbrägel, J.: Discrete Logarithms in GF(2^{6120}). NMBRTHRY list (April 11, 2013)
23. Granger, R., Kleinjung, T., Zumbrägel, J.: Breaking '128-bit secure' supersingular binary curves (or how to solve discrete logarithms in $\mathbb{F}_{2^{4 \cdot 1223}}$ and $\mathbb{F}_{2^{12 \cdot 367}}$). Cryptology ePrint Archive, Report 2014/119
24. Granger, R., Kleinjung, T., Zumbrägel, J.: Discrete logarithms in the Jacobian of a genus 2 supersingular curve over GF(2^{367}). NMBRTHRY list (January 30, 2014)
25. Granger, R., Kleinjung, T., Zumbrägel, J.: Discrete Logarithms in GF(2^{9234}). NMBRTHRY list (January 31, 2014)
26. Granger, R., Page, D., Stam, M.: Hardware and software normal basis arithmetic for pairing-based cryptography in characteristic three. IEEE Trans. Computers 54(7), 852–860 (2005)
27. Granger, R., Page, D., Stam, M.: On small characteristic algebraic tori in pairing-based cryptography. LMS J. Comput. Math. 9, 64–85 (2006)
28. Granlund, T.: GNU MP: The GNU Multiple Precision Arithmetic Library, 5.0.5 edn. (2012), http://gmplib.org/
29. Hankerson, D., Menezes, A., Scott, M.: Software implementation of pairings. In: Identity-Based Cryptography. Cryptology and Information Security, vol. 2, pp. 188–206. IOS Press (2008)
30. Joux, A.: A one round protocol for tripartite Diffie-Hellman. In: Bosma, W. (ed.) ANTS 2000. LNCS, vol. 1838, pp. 385–393. Springer, Heidelberg (2000)
31. Joux, A.: Faster index calculus for the medium prime case. Application to 1175-bit and 1425-bit finite fields. In: Johansson, T., Nguyen, P.Q. (eds.) EUROCRYPT 2013. LNCS, vol. 7881, pp. 177–193. Springer, Heidelberg (2013)

32. Joux, A.: A new index calculus algorithm with complexity $L(1/4 + o(1))$ in very small characteristic. In: Lange, T., Lauter, K., Lisoněk, P. (eds.) SAC 2013. LNCS, vol. 8282, pp. 355–379. Springer, Heidelberg (2014)
33. Joux, A.: Discrete Logarithms in $GF(2^{1778})$. NMBRTHRY list (February 11, 2013)
34. Joux, A.: Discrete Logarithms in $GF(2^{4080})$. NMBRTHRY list (March 22, 2013)
35. Joux, A.: Discrete Logarithms in $GF(2^{6168})$. NMBRTHRY list (May 21, 2013)
36. Joux, A., Lercier, R.: The function field sieve in the medium prime case. In: Vaudenay, S. (ed.) EUROCRYPT 2006. LNCS, vol. 4004, pp. 254–270. Springer, Heidelberg (2006)
37. Kleinjung, T., et al.: Factorization of a 768-bit RSA modulus. In: Rabin, T. (ed.) CRYPTO 2010. LNCS, vol. 6223, pp. 333–350. Springer, Heidelberg (2010)
38. Lenstra, A.K.: Unbelievable security: Matching AES security using public key systems. In: Boyd, C. (ed.) ASIACRYPT 2001. LNCS, vol. 2248, pp. 67–86. Springer, Heidelberg (2001)
39. Menezes, A.J., Okamoto, T., Vanstone, S.A.: Reducing elliptic curve logarithms to logarithms in a finite field. IEEE Trans. Inform. Theory 39(5), 1639–1646 (1993)
40. Popovyan, I.: Efficient parallelization of lanczos type algorithms. Cryptology ePrint Archive, Report 2011/416 (2011), http://eprint.iacr.org/
41. Sakai, R., Mitsunari, S., Kasahara, M.: Cryptographic schemes based on pairing over elliptic curve. IEIC Technical Report 101(214), 75–80 (2001)
42. Shinohara, N., Shimoyama, T., Hayashi, T., Takagi, T.: Key length estimation of pairing-based cryptosystems using η_t pairing. In: Ryan, M.D., Smyth, B., Wang, G. (eds.) ISPEC 2012. LNCS, vol. 7232, pp. 228–244. Springer, Heidelberg (2012)
43. Shoup, V.: NTL: A library for doing number theory, 5.5.2 edn. (2009), http://www.shoup.net/ntl/
44. Spaenlehauer, P.-J.: Solving multihomogeneous and determinantal systems algorithms - complexity - applications. Ph.D. thesis, Université Pierre et Marie Curie, UPMC (2012)

Leakage-Tolerant Computation
with Input-Independent Preprocessing

Nir Bitansky[1,*], Dana Dachman-Soled[2], and Huijia Lin[3,**]

[1] Tel Aviv University, Tel Aviv, Israel
nirbitan@tau.ac.il
[2] University of Maryland, College Park, MD, USA
danadach@ece.umd.edu
[3] University of California, Santa Barbara, CA, USA
rachel.lin@cs.ucsb.edu

Abstract. Following a rich line of research on leakage-resilient cryptography, [Garg, Jain, and Sahai, CRYPTO11] and [Bitansky, Canetti, and Halevi, TCC12] initiated the study of *secure interactive protocols* in the presence of arbitrary leakage. They put forth notions of *leakage tolerance* for zero-knowledge and general secure multi-party computation that aim at capturing the best-possible security when the private inputs of honest parties are exposed to direct leakage. So far, only a handful of specific two-party functionalities have been successfully realized under the notion. General functionalities were only realized under weaker security notions [Boyle, Garg, Jain, Kalai, and Sahai, Crypto13], or relying on leakage-immune input-processing, which needs to be repeated for each and every execution [Boyle, Goldwasser, Jain, Kalai, STOC12].

We construct leakage-tolerant multi-party computation protocols for *general functions*, relying on *input-independent preprocessing* that is performed once and for-all. The protocols tolerate continual leakage, throughout an unbounded number of executions, provided that leakage is bounded within any particular execution. In the malicious setting, we also require a common reference string, and a constant fraction of honest parties.

At the core of our construction, is a tight connection between secure compilers in the Only-Computation-Leaks (OCL) model and leakage-tolerant protocols. In particular, we show that two-party leakage-tolerant protocols with input-independent preprocessing are essentially *equivalent* to two-component OCL compilers satisfying certain strong properties. We then show how to construct such *strong OCL compilers* in the plain model, with the help of $O(1)$ auxliary components.

1 Introduction

Secure Multiparty Computation (MPC) [Yao82, GMW87, BGW88, CCD88] is a central facet of modern cryptography. MPC protocols allows m mutually

* Supported by an IBM Ph.D. Fellowship, the Check Point Institute for Information Security, and the Israeli Ministry of Science and Technology.
** Part of this research was done when the third author was at CSAIL, MIT, supported by the following grants Eager - CNS-1347364, Simons – Agreement Dated 6-5-12, and Darpa - FA8750–11-2-0225.

J.A. Garay and R. Gennaro (Eds.): CRYPTO 2014, Part II, LNCS 8617, pp. 146–163, 2014.
© International Association for Cryptologic Research 2014

distrustful parties to securely compute any function $f(\bar{x})$ of their private inputs $\bar{x} = (x_1, \ldots, x_m)$. The security of such a protocol π, which guarantees privacy and correctness to honest parties, is captured through the *simulation paradigm* (also known as the *real-ideal paradigm*). The paradigm stipulates that the adversarial effect and view, in a "real-world" execution of π, can be simulated in an "ideal-world", where parties run an idealized protocol I_f. In the idealized protocol inputs are simply handed to a *trusted party*, often referred as an *ideal functionality*, that performs the computation for the parties.

Protocols in the traditional MPC model crucially rely on the assumption that *the internal computation state of honest parties is kept completely secret from the attackers,* and the sole way of affecting honest parties and gaining information regarding their secret state is through the communication interface. However, in reality, the attackers may design their own interfaces, via a myriads of side-channels (e.g. timing, radiation, etc., see [Sta09]), and learn information—termed *leakage*—about the secret state of honest parties. This growing threat has spurred a large body of research devoted to the development of *leakage-resilient cryptography* (see [ADW09] for a survey).

Following this line of research, the works of [GJS11, BCH12] initiated the study of secure interactive protocols in the presence of leakage. In this setting, the adversary can obtain leakage on honest parties' secret states (in addition to controlling the corrupted parties), modelled as the outputs of arbitrary leakage functions, chosen adaptively by the adversary during protocol execution.

A fundamental question concerns the level of security that can be achieved in this model. The common aim in leakage-resilient cryptography is to achieve the same security properties as in the traditional attack model, where there is no leakage. In the context of protocols, such a guarantee means that leakage on the state of honest parties causes no degradation of security; namely, the real world protocols retain the same security guarantees that the ideal world protocols have in a leakage-free environment, where honest parties' inputs are totally secret. However, such a strong guarantee is inherently *impossible* if the real world adversary can directly leak on parties' inputs.

The model in focus of this work: leakage-tolerance. Acknowledging that direct leakage on parties' inputs is often unavoidable, [GJS11, BCH12] put forth the model of *leakage-tolerance* (rather than resilience) that aims to achieve the "best-possible" security guarantee in this scenario. Intuitively, leakage-tolerance means that ℓ-bits of leakage on an honest party's internal state, including inputs, messages, and randomness, "translate" to at most ℓ-bits of leakage only on its private input and output. More precisely, under the real-ideal paradigm, it is required that a real-world executions subject to ℓ bits of leakage on honest parties' secret states, can be simulated by an ideal-world execution, subject to ℓ bits of leakage on the honest parties' ideal states.[1] Here, the ideal state of a party is specified

---

[1] [GJS11] proposes a weaker notion of leakage-tolerance for zero-knowledge protocols that allows a bigger leakage budget, $(1 + \varepsilon)\ell$ bits, in the ideal-world execution. In this work, we follow the more stringent notion, with $\varepsilon = 0$, proposed in [BCH12].

as a part of the description of the ideal functionality; in the most natural (and best possible) setting, it contains only the party's private input and output.

Prior Work. Boyle, Goldwasser, Jain and Kalai [BGJK12] circumvent the impossibility of leakage-resilient protocols, by relying on *leakage-free input-processing*. Informally, in every execution, a leakage-free input processing phase is executed first to encrypt each party's input, and only the ciphertexts are delivered to the parties; thus, the inputs themselves are never exposed to direct leakage. In addition, they rely on a leakage-free input-*independent* preprocessing phase that is *performed once for all executions*. In this setting, they construct MPC protocols for general m-party functions, which are resilient under continual leakage, if a constant fraction of parties is uncorrupted, the number of parties m is polynomial in the security parameter λ, and the leakage on any party within any single execution is a-priori bounded.

While leakage-free input processing leads to a strong guarantee of leakage-resilience; however, it significantly deviates from the regime of leakage-tolerance. Indeed, the main narrative behind leakage-tolerance is that leakage on inputs may be unavoidable; in particular, leakage-free input processing in each and every execution may be impossible or expensive to impose.

Boyle, Goldwasser, Jain, Kalai and Sahai [BGJ+13] construct MPC protocols for general deterministic functions that achieve *joint-state* leakage-tolerance (they do not rely on any leakage-free phase and do not require an a-priori bound on the amount of leakage). Specifically, they consider ideal functionalities where the ideal leaky state of each individual party includes the *joint inputs and outputs of all parties*. Roughly, this means that the effect of learning a leakage function on the *isolated* state of any single party can be "emulated" by a simulator that learns a leakage function of the *joint* inputs and outputs of all parties.

While certainly meaningful, the joint-state model does not seem to capture the best-possible tolerance in the face of leakage. Indeed, in the real world, parties maybe physically separated; thus allowing the real-world adversary only separate leakage on the isolated state of each parties. Ideally, we would expect that such separate leakage on the real state of a given party would translate to leakage on the inputs and outputs corresponding to this party alone. Joint-state leakage-tolerance, however, effectively means that, by leakage on any single party, the adversary may vicariously obtain leakage on the joint inputs and outputs of all the parties together. In this work, we shall aim at obtaining (separate-state) leakage-tolerance, and will refer to protocols achieving joint-state leakage tolerance as *weak leakage-tolerant protocols*.

Leakage-tolerant protocols with direct leakage on inputs (i.e., without leakage-free input-processing), in the separate-state model, are only known for specific two-party tasks, such as secure message transmission, commitment, oblivious transfer and zero knowledge [BCG+11, GJS11, BCH12, Pan14]. Determining the feasibility of such leakage-tolerant protocols for general tasks remains open.

1.1 Contributions

Our primary contribution is constructing multi-party leakage-tolerant protocols for general functions, relying only on an *input-independent* leakage-free preprocessing phase. The input-independent processing is done once and for all, and *continual* leakage-tolerance is maintained throughout any number of executions, provided that the leakage within each execution is bounded.

In more detail, in the model of input-independent processing, each party obtains an initial state to be used later in the computation. The initial states are sampled, without leakage, from a fixed (joint) distribution that is independent of the inputs (or function), which are determined online. The online phase proceeds in an arbitrary number of executions, where in each execution a multiparty function is computed on a new set of inputs. The entire state of each party, including its current inputs, randomness, and initial state, are subject to leakage at any point in the protocol's execution, with the restriction that between every two executions, the leakage on each party's private state is a-priori bounded (in length). The initial state is updated between the executions, under leakage, and previous states and inputs are erased (which is, in fact, necessary in the continual setting). The continual leakage-tolerance achieved in this model means that in every execution, the leakage from each honest party can be emulated by a simulator obtaining the same amount of leakage from its ideal state, consisting only of its own input and output in the current execution.

At the heart of our constructions, is a strong connection that we establish between Leakage-Tolerant Computation (LTC for short) and secure compilers in the *Only Computation Leaks (OCL) model*. We next recall the basics of OCL compilers, and overview the main results.

OCL vs. LTC. The OCL model [MR04] considers a setting where computation is performed with leaky memory, under the assumption that only the parts "touched" by the computation can leak information. The memory is initialized ahead of time and without leakage, typically with secret information associated with the computation. A (continual) OCL scheme, is meant to take any computation represented by a circuit $C(k, \cdot)$ with an associated secret k, and compile it, offline and without leakage, into a new computation $C'(k, \cdot)$ that fully protects the secret k when executed using leaky memory. The intuitive property that C' protects the secret k is formalized by the requirement that the adversary's view can be simulated given only the input and output of the computation.

To see the connection with the LTC setting, it is convenient to interpret the evaluation of an OCL-compiled circuit as a leaky distributed computation performed jointly by t honest parties (or components) [BCG+11, DF12, BGJK12]: The parties' memories are initiated with some preprocessed information about the secret k, and they communicate with each other via secret and authenticated secure channels; during the compuation, *bounded* leakage can be obtained from the different parties *separately*, but it is not possible to leak on the joint state of any two parties. Furthermore, in the basic OCL model, leakage is assumed to be *ordered*; namely, computations are done by the parties in a certain order, and at any point it is only possible to leak from the active party.

Thus, the differences between the models of OCL and LTC are the following: First, the secure communication assumption. This difference can be bridged using existing constructions of leakage-tolerant communication [BCG+11] to replace the secure channels. Second, the preprocessing of secret inputs: In OCL, a shared secret input k is preprocessed offline without leakage and split between the t parties; in contrast, in LTC, the parties receive their private inputs online under direct leakage. Another difference is that in the LTC model leakage is *unordered*; namely, it is possible to leak from any party at any time. Finally, the OCL model assumes that all parties are honest and only subject to bounded leakage, whereas in LTC, we must also deal with corruption of parties.

Bridging the gap: LTC with input-independent preprocessing and strong OCL. As discussed above, the LTC model is meant to model settings where inputs are unavoidably subject to leakage. Our first contribution is a generic transformation from a strengthened form of two-party OCL, referred to as *strong OCL*, to two-party LTC with input-independent preprocessing. Informally, the main feature of strong OCL schemes is that they allow simulating the internal states of the two parties without knowledge of the adversary's leakage functions, and moreover, simulation of the party that produces the output depends only on the output of the computation, *obliviously of the input*. In addition, strong OCL security is guaranteed even under *unordered* leakage.

The transformation yields (continual) LTC protocols for the case of two-party LTC with *no corruptions*, and is a crucial step towards achieving stronger forms of security. Furthermore, we show that strong OCL is *necessary* for LTC.

Theorem 1 (informal). *Any two-party strong continual OCL scheme implies two-party continual LTC relying on input-independent preprocessing (and secure channels), and vice versa. The LTC protocol is secure when no party is corrupted and can tolerate the same amount of leakage on every party as the OCL scheme.*

Obtaining Strong continual OCL. There are several known (continual) OCL schemes in the literature [JV10, GR10, DF12, GR12]; however, none satisfies the requirements of strong OCL as is. The OCL schemes of [JV10, GR10, DF12] can be rather directly augmented to satisfy the required strong properties; however, all of these schemes rely on a leakage-free hardware component. Thus far, the only scheme in the literature that does not rely on hardware is the Goldwasser and Rothblum scheme [GR12] (referred to as the GR scheme henthforth), which requires more than two components.[2]

To avoid any reliance on leakage-free hardware in our end result, we relax the requirement of strong 2-component OCL to *strong 2-component OCL with auxiliary parties*. Here the computation is carried out by two main components with the assistance of several auxiliary parties, where we require that the states of the auxiliary parties can be simulated obliviously of both the input and output

---

[2] In most part of [GR12], the scheme is described with polynomially many components. This same scheme can be reduced, however, to $O(1) > 2$ components at the cost of a worst leakage rate.

(the simulation guarantee for the two main parties remains unchanged). We construct such an OCL scheme, without any reliance on hardware. This, in particular, yields a multi-component OCL scheme without hardware.

Theorem 2 (informal). *There exists a continual strong 2-component OCL scheme with $O(1)$ auxiliary parties that does not rely on any hardware. Moreover, the scheme is unconditionally secure.*

Given a strong two-component OCL scheme with $O(1)$ auxiliary parties, Theorem 1 is then generalized to yield two-party LTC, with $O(1)$ *auxiliary parties*, whose ideal state is empty. These LTC protocol (assisted by the auxiliary parties) eventually lead to standard multi-party LTC, with no auxiliary parties.

Multiparty LTC and security against corruptions. We then leverage the two-party protocols, with $O(1)$ auxiliary parties, to obtain m-party (continual) LTC protocols that withstand up to $(1 - \epsilon)m$ corrupted parties, for any number of parties m that is polynomial in the security parameter.

We provide two transformations: The first is a generic transformation for the case of no party corruptions: it takes any m-party LTC protocol with (leakage-free) input-*dependent* preprocessing and obtains a new protocol relying only on input-*independent* preprocessing and two-party LTC (with auxiliary parties). The second achieves the same in the case of $(1 - \epsilon)m$ corruptions and is based on the specific protocol of Boyle et al. [BGJK12] in the common reference string. (The first transformation applies assuming that m is a large enough constant. The second requires that m is polynomial in the security paramter λ, which is inherited from [BGJK12].)

Theorem 3 (informal). *Any m-party LTC protocol with input-**dependent** preprocessing and two-party LTC with $O(1)$ auxiliary parties, both secure when no party is corrupted, imply an m-party LTC protocol with input-**independent** preprocessing when no party is corrupted (without additional auxiliary parties or hardware). Moreover, the [BGJK12] protocol, in the common reference string model, and any two-party LTC with $O(1)$ auxiliary parties as above, imply security under $(1 - \epsilon)m$ corruptions, for any constant ϵ, $m = \lambda^{\Omega(1)}$. The resulting protocols can tolerate the same amount of leakage as the original protocols.*

UNIVERSAL COMPOSABILITY AND OBLIVIOUS SIMULATION. All of our constructions are presented within the framework of universal composability (UC) with leakage [Can01, BCH12]. In particular, our protocols admit the strong form of emulation known as *leakage-oblivious simulation*. An oblivious simulator works obliviously of the actual leakage function that the adversary produces, and provides a way (more precisely, a state-translation function) that simulates the real world states of honest parties using their ideal state; namely, inputs and outputs. An essential feature of protocols with oblivious simulation (and thus of the protocols constructed in this work) is that they respect the (leaky) universal composition theorem [Can01, BCH12].

We note that [NVZ13] show how leakage-tolerant protocols with oblivious simulation imply protocols with a relaxed form adaptive security. Their result, however, does not address the input-independent processing model. Applying the same ideas to our leakage-tolerant protocols would naturally result in (relaxed) adaptively-secure protocols in the preprocessing model.

RANDOMIZED FUNCTIONALITIES. We can further extend the constructions of m-party LTC protocols from Theorem 3 to also support randomized functionalities. For this purpose we design a new leakage-tolerant m-party coin-tossing protocol (in the input-independent processing model). The protocol requires that the number of honest parties is as large as the number of parties in the two-party LTC protocol with auxiliary parties, for the no-corruption setting.

1.2 Techniques

We now present the main ideas and techniques behind out results. We begin by giving some intuition regarding the difficulty of constructing LTC protocols.

Why classical protocols are not leakage-tolerant. A common paradigm for 2PC and MPC protocols is for parties to first *secret share* their inputs, and then homomorphically compute a given boolean circuit over their shares. For example, in two-party GMW [GMW87], the invariant is that throughout the computation each one of the parties holds one random additive share for each wire in the circuit, where the two shares together encode the actual value of the wire; then, addition is done locally over shares, and multiplication is done with the help of *oblivious transfer.*

The additive secret sharing commonly used, however has very poor leakage-resilience properties. Indeed, it is possible to learn the value of any intermediate value in the circuit, by simply leaking a single bit from every party. In contrast, in the ideal world, where it is only possible to separately leak a single bit on the input and output of each party, learning the value of some intermediate wires might be impossible. This renders classical protocols entirely insecure.[3]

A plausible route towards circumventing this problem would be to use a *leakage-resilient secret sharing scheme* [BGK11, DLWW11, DF11], such as the inner product two-source extractor. The challenge is, however, to be able to compute the circuit gates over such shares in a leakage-resilient way. While this is not known in the plain model, this approach is successfully executed in existing OCL protocols (e.g. in [GR12, DF12] with the inner-product extractor), with the help of a leakage-free preprocessing phase. In the OCL setting, however, all secret inputs are preprocessed offline, while online inputs are public. Thus, a natural question is whether we can import the OCL techniques to the setting of LTC.

Before discussing how to bridge the gap between LTC an OCL, we first quickly cover some of the technical basics of the OCL compiler, which will be instrumental for our technical exposition.

---

[3] For example, the value of an intermediate wire might be the inner product of two uniformly random inputs, and thus statistically close to uniform, even under independent leakage as above. A rather similar problem also appears in other classical protocols (e.g., Yao [Yao82]), even if not as explicitly.

Strong OCL. It is convenient to consider two-party OCL schemes for universal circuits $U(k, \cdot)$ with a fixed secret input k. A continual strong OCL scheme Λ consists of a compiler algorithm Comp that preprocesses a secret k, and splits it into two shares, and a two party protocol between a left component P_L and a right component P_R whose memories are initiated with the two shares respectively; to evaluate a function f on k, the two components P_L and P_R interact with each other, where P_L receives the input f and P_R produces the output $y = f(k)$. The scheme may be assisted by additional *auxiliary parties* $P_{A_1}, \ldots, P_{A_a}$, who obtain an initial state from Comp and participate together with P_L, P_R in the protocol for computing $f(k)$.

The protocol proceeds in iterations: In each iteration i, the adversary may specify $f = f_i$ and obtain leakage from any one of the parties $P_L, P_R, P_{A_1}, \ldots, P_{A_a}$. Any leakage functions is evaluated only on the individual state of the leaking party; the number of bits leaked from any given party during a single iteration is bounded by some prefixed length function ℓ.

We require an oblivious simulator $\mathcal{S}$ that simulates the states of all parties $P_L, P_R, P_{A_1}, \ldots, P_{A_a}$ without knowledge of the leakage functions specified by the adversary; the leakage to the adversary is computed by evaluating these functions on the simulated states. We further require that $\mathcal{S}$ admits a special structure: The state of P_L in every iteration i can be simulated given the current input f_i and output $y_i = f_i(k)$. The state of P_R can be simulated given only the output y_i, and obliviously of f_i. The state of any auxiliary party P_{A_i} can be simulated obliviously of either f_i, y_i.

From strong OCL to LTC without corruption: To illustrate the idea behind our construction of LTC protocols secure without corruption, we focus in this technical overview, on the case where (P_0, P_1), assisted by $(P_{A_1}, \ldots, P_{A_a})$, jointly compute a single-output function f of their private inputs (x_0, x_1), and only one of them receives the output. Furthermore, we focus on the non-continual setting, where only a single execution is performed, and later on generalize to the continual setting.

Given a strong OCL scheme, obtaining a one-time leakage-tolerant protocol ρ is straightforward. An easy way to compute a function is to ask P_0 to send its input x_0 to P_1, who then computes $y = f(x_0, x_1)$ directly; however, this is obviously non-private. Instead, we may have P_0 encrypt its input x_0 using a one-time pad r and send the ciphertext $c = x_0 \oplus r$ to P_1. Now privacy is re-installed, but it becomes unclear how to perform the computation.

To remedy this, the OCL scheme provides a way for the two parties to jointly decrypt x_0 and compute $f(x_0, x_1)$. More precisely, the preprocessing phase samples the initial states of the OCL scheme with respect to a random string r, which will set as the OCL secret (referred to before as k, and distributes the left-component initial state to P_1 and the right-component initial state together with r to P_0. During the protocol execution, P_0 sends $c = r \oplus x_0$ to P_1; then, jointly with the auxiliary parties $P_{A_1}, \ldots, P_{A_a}$, they perform an OCL evaluation, where P_0 acts as the right component and P_1 acts the left component with input function $g((c, x_1), \cdot) = f(c \oplus (\cdot), x_1) = y$. The OCL evaluation computes the function g on the secret r, producing the desired output y at P_0.

Showing that the above protocol ρ is indeed leakage tolerant reduces to showing that the states of P_0 and P_1 can be simulated using their own input and output. By construction, P_0's state consists of x_0, r and the right-component state of OCL, while P_1's state contains x_1, c and the left-component state of OCL. A key observation is that since r is truly random, so is c. Therefore, the ciphertext c can be simulated directly using a random string $\tilde{c} \leftarrow U$, and later, the secret r can be simulated as $\tilde{c} \oplus x_0$; and the pair $(\tilde{c}, \tilde{r})$ is distributed identically to their counterparts (c, r) in the real execution. Next, it follows from the strong leakage resilience of the OCL scheme that the left-component state state_L in P_1 can be simulated using the input function $g(\tilde{c}, x_1), \cdot)$ and the output y, while the right-component state state_R in P_0 can be simulated using only y. Therefore, overall the simulated state $(x_0, \tilde{r}, \mathsf{state}_R)$ of P_0 and the simulated state of $(x_1, \tilde{c}, \mathsf{state}_L)$ of P_1 depend, respectively, on their own input and output. The state of the auxiliary parties is guaranteed, by strong OCL, to be simulatable independently of the input and output, as required. Thus, leakage-tolerance follows as required.

To generalize the above to the continual setting, requires a modification of the above protocol. We design a slightly more complicated protocol ρ', in which even the ciphertext c is computed using the OCL scheme by evaluating the function $g'(x_0, \cdot) = x_0 \oplus (\cdot)$ on the secret r; To do this, the preprocessing stage is modified to sample an additional set of OCL initial states with respect to the secure r, and to distribute the left-component initial state to P_0 who later acts as the left component when evaluating g'. The protocol ρ' is still a one-time protocol, but in which r is not fully revealed.

Moving to the continual case, instead of directly using r as the one-time pad, in the ith iteration, we use the pseudo-random string produced by $\mathsf{PRF}(r, i)$ as the one-time pad, where r is used as the seed. It follows from the continual strong leakage-resilience of the OCL scheme that the seed r is always kept secret, and thus all the one-time pads generated are pseudo-random.

From LTC with input-independent processing back to strong OCL. We briefly sketch how LTC with input-independent processing can be used to obtain strong OCL, thus implying that OCL is necessary for our goal. For simplicity, we describe the transformation with two parties, and with no auxiliary parties. It is not hard to see that by starting from an LTC with a auxiliary parties, we get strong OCL with a auxiliary parties.

The idea relies on the properties of inner product as a two-source extractor [CG88]. For an OCL secret $k \in \{0,1\}^n$ we consider a two-party function $g(f, \mathbf{L}_i, \mathbf{L}'), (\mathbf{R}, \mathbf{R}'))$ that takes as input a description of $f : \{0,1\}^n \to \{0,1\}^*$, matrices $\mathbf{L}_i, \mathbf{R}_i \in \mathbb{F}_2^{\kappa \times n}$, which will be inner product shares of the key k (that is, $\mathbf{L}_i[j], \mathbf{R}_i[j] \in \mathbb{F}_2^{\kappa}$ and $\langle \mathbf{L}_i[j], \mathbf{R}_i[j] \rangle = k_j$), and two random matrices $\mathbf{L}, \mathbf{R} \in \mathbb{F}_2^{\kappa' \times \kappa'}$, where $\kappa' = \mathrm{poly}(\kappa)$. The function computes $f(k) = f(\langle \mathbf{L}_i[1], \mathbf{R}_i[1] \rangle, \ldots, \langle \mathbf{L}_i[n], \mathbf{R}_i[n] \rangle)$, and in addition new random shares $\mathbf{L}_{i+1}[j], \mathbf{R}_{i+1}[j] \in \mathbb{F}_2^{\kappa}$ of the key k, which will be computed using randomness $\langle \mathbf{L}'[1], \mathbf{R}'[1] \rangle, \ldots, \langle \mathbf{L}'[\kappa'], \mathbf{R}'[\kappa'] \rangle$, derived by inner-product extraction.

At compilation, initial shares $\mathbf{L}_1, \mathbf{R}_1$ of the key k are sampled and distributed to the parties, and input-independent processing is done with respect to the function g. Then at each iteration i the parties compute the function where P_0 inputs $f, \mathbf{L}_i, \mathbf{L}'$, where $\mathbf{L}_i$ was produced in the previous iteration and $\mathbf{L}'$ was sampled uniformly at random by P_0 itself. P_1 accordingly inputs $\mathbf{R}_i, \mathbf{R}'$. The properties of the LTC ensure that throughout all the different shares $\mathbf{L}, \mathbf{L}', \mathbf{R}, \mathbf{R}'$ are only leaked on separately, within some small bound. Strong OCL simulation then follows directly by the LTC simulation guarantee.

Obtaining strong OCL. Intuitively, our construction combines the two-component OCL scheme of Dziembowski and Faust [DF12] (referred to as the DF scheme henceforth), which relies on a leakage-free hardware that samples random orthogonal vectors, with the key ciphertext bank module in the Goldwasser-Rothblum OCL scheme [GR12] (henceforth, the GR scheme). The ciphertext bank allows continual sampling of random orthogonal vectors at the presence of leakage using multiple components.

A natural idea is to use auxiliary parties to emulate the GR ciphertext bank in order to implement the hardware needed for the DF-scheme. However, combining the two schemes and showing that the joint scheme admits strong simulation turns out to be quite challenging: First, the GR-scheme is proven secure in a weaker model of OCL, where the leakage adversary can only obtain leakage from the component that is *currently activated*, implying that leakage occurs in the same order as the sequence of sub-computations. As a first step towards our construction, we argue that the GR-scheme is also secure against "unordered leakage". Second, we provide new simulation procedures for the DF-scheme and the GR ciphertext banks as required by strong OCL. We defer a more detailed description of the joint scheme to the full version.

From two-party LTC to multiparty LTC. We now briefly explain our transformations from m-party LTC protocols with input-*dependent* preprocessing to protocols, relying only on input-*independent preprocessing* and two-party LTC.

The high-level idea behind our transformations is as follows: The input processing of the multi-party LTC can be performed online, and under leakage, jointly by two parties. To process an input x_i of a given party P_{i_0}, it will use the help of another party P_{i_1}, and possibly of other auxiliary parties $P_{i'_1}, \ldots, P_{i'_a}$. The two parties would each sample independently a long enough random string r_{i_0} and r_{i_1}, respectively, and will use the LTC to compute the two-party function $g((x_i, r_{i_0}), r_{i_1})$ that computes the processing function $\bar{x}_i = \Pi(x_i; \mathsf{Ext}(r_{i_0}, r_{i_1}))$, where the randomness $r = \mathsf{Ext}(r_{i_0}, r_{i_1})$ is derived from the two random strings using a two-source extractor (e.g., inner product).

Once each party obtains this processed input, the parties then run the original protocol, no longer requiring leakage-free preprocessing. Intuitively, the two-source extraction guarantees—provided that there is only bounded separate leakage on each of the random strings—that the randomness $r = \mathsf{Ext}(r_{i_0}, r_{i_1})$ is statistically independent of the leakage, achieving the same effect as leakage-free input preprocessing.

The above intuition holds assuming that the party P_{i_1} assisting P_{i_0}, as well as the other assisting parties $P_{i_1'}, \ldots, P_{i_a'}$, are all honest. In particular, we achieve a protocol in the no corruption model. Indeed, assuming P_{i_0} is (even semi-honestly) corrupted, the adversary, who now knows r_{i_0} can learn any ℓ-bounded function of r, by leaking on r_{i_0}. Furthermore, a malicious party may even bias the result and hurt the correctness of the protocol.

An appealing approach towards overcoming this problem is to have each party P_i jointly process its input with all other parties P_j, and aggregate the processed inputs into a single input that is safe to use. We observe that the input-processing in the [BGJK12] protocol possesses additional properties, which give rise to such an approach. Specifically, in the [BGJK12] protocol the input processing function $\Pi(x_i, \mathsf{pk}, \mathsf{crs}) := (\mathsf{Enc}_{\mathsf{pk}}(x_i), \pi)$ samples an encryption of the input x_i under a public key pk for a fully-homomorphic encryption scheme, and a NIZK of knowledge π of the input x_i. Here the public key pk and the common reference string crs are determined as part of the input-independent processing (in particular, in [BGJK12], there is no leakage on the encryption's randomness).

We implement the above idea as follows: Let $a = O(1)$ be the number of auxiliary parties required for the two-party LTC. We let each P_i jointly compute with each coalition C of parties of size $a + 1$ an encryption c_C of **zero**, and a NIZK for it being an encryption of zero with respect to pk. The randomness for this computation is computed by a two-source extractor, as above. Then, P_i aggregates all these ciphers by adding them together to a new zero encryption $\mathsf{c} = \sum_{C \in \binom{[m]}{a+1}} \mathsf{c}_C$, and uses them to get a rerandomnized encryption c^{x_i} of his input x_i, by encrypting x_i under leakage (and thus non-securely) and then adding to it the aggregated zero encryption c. Also, P_i computes a NIZK of knowledge that it knows x_i and that c^{x_i} was generated by adding an encryption of x_i to ciphers c, and that NIZKs for the fact that they're zero ciphers were verified.

It can be shown that, in known fully homomorphic encryption schemes, the final encryption of x_i is semantically-secure provided that any one of the zero encryptions c_C, which is the case as long as there exists some non-corrupted coalition C of parties. Moreover, the NIZKs guarantee that malicious parties cannot bias the result of the computation.

The above transformation withstands the same number of corruptions as the [BGJK12] protocol: it allows $(1 - \epsilon)m$ corruptions, for $m > \lambda^{-\Omega(1)}$.

A note on universal composability and randomized functionalities. In the stand-alone setting, the security of our LTC protocols follows directly from the leakage resilience of [BGJK12] protocols in the stand-alone setting, which is shown in [BGJK12]. To show the full leakage tolerance defined in the UC setting [BCH12], we need to rely on protocols that are leakage-resilient in the UC setting, which is outside the scope of [BGJK12]. To bridge this gap, we modify the original [BGJK12] protocols to a UC variant, by replacing all building blocks in [BGJK12] with their corresponding UC counterparts. While most building blocks have standard UC version, there is no known leakage-tolerant m-party coin-tossing protocol in the UC setting. We construct such a protocol, in the input-independent pre-processing model, relying on our two-party LTC protocols with auxliary parties. This yields a coin-tossing protocol that us secure as long as sufficiently many parties are honest (the same as the number of parties in the LTC protocol). This coin-tossing protocol not only facilitates a UC variant of [BGJK12], but also allows implementing randomized functionalities.

1.3 Organization

In Section 2, we provide the formal definition of strong OCL compilers, and in Section 3 we construct two party LTC protocols secure with no corruption and only leakage from two component strong OCL compilers. Due to the lack of space, we leave the security proof of the two party LTC protocols, as well as the construction of strong OCL compilers and the final multiparty LTC protocols secure with corruptions to the full version.

2 Strong Only-Computation-Leaks Compilation

N-*component OCL.* A N-component OCL scheme for a circuit $C(k, \cdot)$, associated with a secret k, consists of an efficient compiler Comp and a N-party protocol $\Pi = (P_1^{\text{OCL}}, P_2^{\text{OCL}}, \cdots, P_N^{\text{OCL}})$. To compute $C(k, \cdot)$ in a leakage-resilient way, the circuit is *compiled* ahead of time by $\text{Comp}(C(k, \cdot))$ that produces an initial state $(\text{init}_1^{(k)}, \cdots, \text{init}_N^{(k)})$ for each one of the N parties, and this compilation is done "in the dark" without any leakage. Then, at computation time, the parties can compute together $y = C(k, x)$ for any input x by running the protocol Π.

Below we provide the formal definition of OCL schemes for *universal circuits* Since our end goal is constructing composable leakage tolerant protocols, where the simulator is oblivious of the leakage queries from the adversary, we consider strengthened OCL schemes which have obvious simulators.

OCL schemes with oblivious simulation: Let $\{U_T(k, f)\}_{T \in \mathbb{N}}$ denote the family of *universal circuits* where U_T takes two inputs k and f of length at most T, where f represents a T-step deterministic computation, and computes $f(k)$. (If the computation does not complete in T steps, we assume w.l.o.g. that the output of $U_T(k, f)$ is $\perp$).

Definition 1 (Continual N-component OCL schemes). *We say that* $\Lambda = (\text{Comp}, \Pi = \langle P_1^{ocl}, \cdots, P_N^{ocl} \rangle)$ *is a* continual, N-*component OCL scheme for the universal circuit family* $\{U_T(k, f)\}_{T \in \mathbb{N}}$ *if it satisfies the following properties.*

Initialization: *For every security parameter* λ *and* $T \in \mathbb{N}$, $k \in \{0, 1\}^T$, *the compiler* $\text{Comp}(1^\lambda, U_T, k)$ *runs in time* $\text{poly}(\lambda, T)$ *and outputs* N *initial states* $\text{init}_1, \text{init}_2, \cdots, \text{init}_N$.

Unbounded-time evaluation: *The evaluation procedure invokes the protocol* Π *between the components* $P_1^{ocl}(\text{init}_1)$, $P_2^{ocl}(\text{init}_2)$ *to* $P_N^{ocl}(\text{init}_N)$, *which interact in an arbitrary polynomial number of iterations: In the* i^{th} *iteration,* P_1^{ocl} *receives an input* $f_i \in \{0, 1\}^T$ *and* P_2^{ocl} *produces an output* y_i. *At the*

*end of the evaluation, an update procedure is carried out, producing the new
initial states for the next iteration; then all information other than the new
initial states are erased.*

*For every component $j \in [N]$, denote by $\text{init}_{i,j}$ the initial states of component
j at the onset of the ith iteration (in the first iteration, $\text{init}_{1,j} = \text{init}_j$), and
$\text{evl}_{i,j}$ the random coins tossed and messages exchanged by each P_j^{OCL} during
the ith iteration, including its state during the update phase.*

Correctness with adaptive input selection: *For every $\lambda \in \mathbb{N}$, $T \in \mathbb{N}$ polynomially related to λ, $k \in \{0,1\}^T$, auxiliary input $z \in \{0,1\}^{\text{poly}(\lambda)}$, and PPT
adversary $\mathcal{A}$, in the following real experiment $\text{RealExp}_{\mathcal{A}}^{\infty}(1^\lambda, T, k, z)$ where $\mathcal{A}$
initiates an arbitrary number of evaluations with adaptively chosen inputs,
it holds that with all but negligible probability, the outputs of all evaluations
are correct.*

We say that Λ has perfect correctness, if the above holds with probability 1.

We next describe the security experiments of OCL schemes. Λ is said to be
ℓ-leakage-resilient with oblivious simulation if there is a simulator $\mathcal{S}$, such that,
for every $\lambda \in \mathbb{N}$, $T \in \mathbb{N}$ polynomially related to λ, every $k \in \{0,1\}^T$, and auxiliary input $z \in \{0,1\}^{\text{poly}(\lambda)}$, the views of the adversary in the following real and
ideal experiments are indistinguishable. In the real world, the adversary obtains
leakage independently from each component during OCL evaluations (with inputs chosen adaptively by the adversary), whereas in the ideal world, it obtains
leakage from states of the components simulated by an oblivious simulator. More
formally,

$\text{RealExp}_{\mathcal{A}}^{\infty}(1^\lambda, T, k, z)$ (Real experiment): The adversary $\mathcal{A}(1^\lambda, T, k, z)$ proceeds
as follows:

1. The initial states $(\text{init}_1, \cdots, \text{init}_N) \leftarrow \text{Comp}(1^\lambda, U_T, k)$ are sampled.
2. $\mathcal{A}$ launches ℓ-bounded leakage attacks on an unbounded number of evaluations of its choice: In the ith iteration,
   (a) $\mathcal{A}$ submits an input function $f_i \in \{0,1\}^T$, which is evaluated on k by re-
       suming the protocol execution of Π between the components $P_1^{OCL}(\text{init}_{i,1})$,
       $\cdots, P_N^{OCL}(\text{init}_{i,N})$ with input f_i to the first component P_1^{OCL}.
   (b) $\mathcal{A}$ launches an ℓ-bounded leakage attack on the ith evaluation: It issues
       an arbitrary number of leakage queries (P_j^{OCL}, L) for $j \in [N]$ adaptively,
       and obtains leakage answers $L(\text{init}_{i,j}, \text{evl}_{i,j})$, as long as the total amount
       of leakage on each P_j^{OCL} in this iteration is smaller than $\ell(\lambda)$ bits.
   (c) $\mathcal{A}$ obtains the output of the evaluation, which is the output of P_2^{OCL}.

Denote by $\text{view}_{\mathcal{A}}^{\ell,\infty}(1^\lambda, T, k, z)$ the view of $\mathcal{A}$ in the above experiment.

$\text{IdealExp}_{\mathcal{S},\mathcal{A}}^{\infty}(1^\lambda, T, k, z)$ (Ideal experiment): The adversary $\mathcal{A}(1^\lambda, T, k, z)$ participates in the same experiment as above, except that during its ℓ-bounded leakage
attacks, it is given simulated answers: In the ith iteration,

(a) $\mathcal{A}$ submits an input function $f_i \in \{0,1\}^T$. $\mathcal{S}(1^\lambda, T, i, f_i, f_i(k); \mathbf{w}_i)$ is invoked,
    producing simulated states $(\widetilde{\text{intl}}_{i,1}, \cdots, \widetilde{\text{intl}}_{i,N}, \widetilde{\text{evl}}_{i,1}, \cdots, \widetilde{\text{evl}}_{i,N})$, where w_i

is the fresh random coins tossed for the simulation in iteration i and $\mathbf{w}_i = w_1, \cdots, w_i$ is all the random coins that have been tossed for simulation in the first i iterations.

(b) Whenever $\mathcal{A}$ issues a leakage query (P_j^{OCL}, L) for $j \in [N]$, it is given the simulated answer $L(\widetilde{intl}_{i,j}, \widetilde{evl}_{i,j})$, as long as the total amount of leakage on each P_j^{OCL} in this iteration is smaller than $\ell(\lambda)$ bits.

(c) $\mathcal{A}$ obtains the simulated output of the evaluation in $\widetilde{evl}_{i,2}$.

Denote by $\widetilde{view}_{S,\mathcal{A}}^{\ell,\infty}(1^\lambda, T, k, z)$ the view of $\mathcal{A}$ in the above experiment.

Definition 2 (Continual ℓ-Leakage-resilience with oblivious simulation). *We say that a continual OCL scheme Λ is continually ℓ-leakage-resilient with oblivious simulation if there is a* **PPT** *simulator S, such that, for every* **PPT** *adversary $\mathcal{A}$, the following two ensembles are indistinguishable.*

- $\{view_{\mathcal{A}}^{\ell,\infty}(1^\lambda, T, k, z)\}_{\lambda \in \mathbb{N}, T \in \mathbb{N}, k, z \in \{0,1\}^{\mathrm{poly}(n)}}$
- $\{\widetilde{view}_{S,\mathcal{A}}^{\ell,\infty}(1^\lambda, T, k, z)\}_{\lambda \in \mathbb{N}, T \in \mathbb{N}, k, z \in \{0,1\}^{\mathrm{poly}(n)}}$

Strong OCL schemes: In the above definition, the oblivious simulator simulates the states of all N components in each evaluation i depending on both the input f_i and output $f_i(k)$. We consider the following strengthening: Only the simulation of the first component depends on both the input and output, whereas the simulation of the second component depends solely on the output, and simulation of the rest components depends on neither the input nor the output.

Definition 3 (Continual strong OCL Schemes). *We say that $\Lambda = (\mathrm{Comp}, \Pi = (P_1^{OCL}, P_2^{OCL}, \cdots, P_N^{OCL}))$ is a* continually ℓ-leakage-resilient strong OCL scheme *if it satisfies the following property.*

Strong ℓ-leakage resilience: *Λ admits an oblivious simulator S satisfying Definition 2 with the following structure: S consists of three sub-algorithms (S_1, S_2, S_3) and on input $(1^\lambda, T, i, f_i, f_i(k) \; ; \; \mathbf{w}_i)$, S invokes these sub-algorithms as follows:*

- $S_1(1^\lambda, T, i, f_i, f_i(k); \mathbf{w}_i) = (\widetilde{intl}_{i,1}, \widetilde{evl}_{i,1})$
- $S_2(1^\lambda, T, i, f_i(k); \mathbf{w}_i) = (\widetilde{intl}_{i,2}, \widetilde{evl}_{i,2})$
- $S_3(1^\lambda, T, i; \mathbf{w}_i) = (\widetilde{intl}_{i,3}, \cdots, \widetilde{intl}_{i,N}, \widetilde{evl}_{i,3}, \cdots, \widetilde{intl}_{i,N})$

and outputs $(\widetilde{intl}_{i,1}, \cdots, \widetilde{intl}_{i,N}, \widetilde{evl}_{i,1}, \cdots, \widetilde{evl}_{i,N})$.

Strong two-component OCL with auxiliary components. In this work, we often consider the special case of a strong *two-component* OCL scheme, and refer to the two components as the left and right components, denoted by P_L^{OCL} and P_R^{OCL}. The strong oblivious simulation property ensures that the state of the left component in each evaluation can be simulated using both the input and output, whereas the state of the right component can be simulated using only the output. We sometimes view a strong $(N + 2)$-component OCL scheme as a strong 2-component OCL scheme using N auxiliary parties $P_{A_1}^{OCL}, \cdots, P_{A_N}^{OCL}$, whose states can be simulated independently of the input and output; in this case, we denote the strong oblivious simulator as $S = (S_L, S_R, S_A)$. Viewing strong N-component OCL as strong two-component OCL with auxiliary components is instrumental for our construction of leakage tolerant protocols.

3   Two-Party Leakage-Tolerant Protocols without Corruption

In this section, we show how to construct a two-party, a auxiliary-party, continual leakage-tolerant protocol ρ in the input-independent pre-processing model based on any strong, continual 2-component OCL scheme with a auxiliary parties. Our tranformation works for any number a of auxiliary parties, and, in particular works for the special case of $a = 0$. The protocol is secure against adversaries that leak a bounded amount of ℓ bits of information on the state of each honest party (separately) in each time period, but do not corrupt any of the parties.

Notation. By $\mathcal{F}^f_{2\text{LTC-AUX}}$ we denote the 2-party ideal leaky functionality computing function f with auxiliary parties. By $\mathcal{F}_{\text{LSC}}$ we denote the secure communication functionality and by $\mathcal{F}_{\text{LFS}}$ we denote the input-idependent leakage-free preprocessing functionality which provides the initial states for all parties.

We now state the main theorem of this section:

Theorem 4. *Assume the existence of a ℓ-continual-leakage-resilient strong OCL Λ scheme with some number, a, of auxiliary components for the universal circuit family and the existence of one-way functions. Then for every efficiently computable deterministic two-input two-output function $f : \{0,1\}^* \times \{0,1\}^* \to \{0,1\}^* \times \{0,1\}^*$, there is a protocol ρ that strongly UC-emulates the functionality $\mathcal{F}^f_{2\text{LTC-AUX}}$ under ℓ-bounded continual leakage, with a auxiliary parties, when no party is corrupted, in the $(\mathcal{F}_{\text{LSC}}, \mathcal{F}_{\text{LFS}})$-hybrid model (i.e. with secure communication and input-independent leakage-free preprocessing). Furthermore, if Λ has perfect correctness, ρ also has perfect correctness.*

Towards proving the theorem, we first observe that it suffices to consider only functions with a single output and design leakage-tolerant protocols where both parties obtain this output.

Proposition 1. *Assume the existence of a ℓ-continual-leakage-resilient strong OCL Λ scheme with a auxiliary components for the universal circuit family. Then, for every efficiently computable deterministic two-input function $f : \{0,1\}^* \times \{0,1\}^* \to \{0,1\}^*$, there is a protocol ρ that strongly UC-emulates the functionality $\mathcal{F}^{f,\infty}_{2\text{LTC-AUX}}$ under ℓ-bounded continual leakage, when no party is corrupted, in the $(\mathcal{F}_{\text{LSC}}, \mathcal{F}_{\text{LFS}})$-hybrid model (i.e. with secure communication and input-independent leakage-free preprocessing). Furthermore, if Λ has perfect correctness, ρ also has perfect correctness.*

Theorem 4 directly follows from Proposition 1 using standard techniques.

3.1 The Protocol ρ

Let λ be security parameter, and let f be an efficiently computable deterministic two-input function. Below we present a two-party leakage-tolerant protocol ρ that strongly emulates the functionality $\mathcal{F}_{2\text{LTC-AUX}}^{f,\infty}$ in the $(\mathcal{F}_{\text{LSC}}, \mathcal{F}_{\text{LFS}})$-hybrid model, where $\mathcal{F}_{\text{LSC}}$ is the secure communication functionality and $\mathcal{F}_{\text{LFS}}$ captures the leakage-free preprocessing functionality. The protocol assumes a ℓ-continual

---

The leakage tolerant protocol ρ

The input-independent preprocessing stage: The leakage-free sampling (LFS) functionality $\mathcal{F}_{\text{LFS}}^{\text{Comp}_\rho}$, on input $(1^\lambda, T)$, where T will be specified later, invokes a compilation algorithm Comp_ρ on $(1^\lambda, T)$, proceeding as follows[4]:

1. Sample $r \leftarrow U_\lambda$ uniformly at random.
2. Sample two pairs of initial states of the OCL scheme Λ w.r.t. secret r independently and randomly: $(\text{init}_L^1, \text{init}_R^1, \text{init}_A^1) \leftarrow \text{Comp}(1^\lambda, U_T, r)$ and $(\text{init}_L^2, \text{init}_R^2, \text{init}_A^2) \leftarrow \text{Comp}(1^\lambda, U_T, r)$.
3. Distribute $\Phi_0 = (\text{init}_L^1, \text{init}_R^2)$ to P_0, $\Phi_1 = (\text{init}_R^1, \text{init}_L^2)$ to P_1 and $\Phi_A = (\text{init}_A^1, \text{init}_A^2)$ to the auxiliary parties.

The online stage: For each iteration j, given the initial states Φ_0, Φ_1 and Φ_A sampled in the preprocessing stage, P_0, P_1 and $P_1^{\text{aux}}, \ldots, P_a^{\text{aux}}$ on common input $(1^\lambda, f, T)$, and private inputs $x_0^j \in \{0,1\}^n$ and $x_1^j \in \{0,1\}^n$, proceed in the following steps, where all messages are sent via the secure channel functionality $\mathcal{F}_{\text{LSC}}$:

1. *The first OCL evaluation—Compute an encryption $c^j = x_0^j \oplus \text{PRF}(r,j)$ of x_0^j:*

   P_0, P_1 and the auxiliary parties $P_1^{\text{aux}}, \ldots, P_a^{\text{aux}}$ compute $x_0^j \oplus \text{PRF}(r,j)$ using the OCL protocol Π: P_0 acts as the left component using initial state $\text{init}_L^{j,1}$, P_1 acts as the right component using initial state $\text{init}_R^{j,1}$ and $P_1^{\text{aux}}, \ldots, P_a^{\text{aux}}$ act as the auxiliary components using initial states $\text{init}_{A,1}^{j,1}, \ldots, \text{init}_{A,a}^{j,1}$. P_0 feeds the following function $g_1^j(r) = g_1^{(j,x_0^j)}(r) = x_0^j \oplus \text{PRF}(r,j)$ to the left component as input. At the end of the evaluation P_1 obtains $\tilde{c}^j$.

2. *The second OCL evaluation—Compute the output $f(x_0^j, x_1^j)$:*

   P_0, P_1 and $P_1^{\text{aux}}, \ldots, P_a^{\text{aux}}$ compute $y^j = f(x_0^j, x_1^j)$ by evaluating the function $g((\tilde{c}^j, x_1^j), \text{PRF}(r,j))$ using Π again: P_0 acts as the right component using initial state $\text{init}_R^{j,2}$, P_0 acts as the left component using initial state $\text{init}_L^{j,2}$ and parties $P_1^{\text{aux}}, \ldots, P_a^{\text{aux}}$ act as the auxiliary components using initial states $\text{init}_{A,1}^{j,2}, \ldots, \text{init}_{A,a}^{j,2}$, respectively. P_1 feeds the function $g_2^j(r) = g_2^{(j,\tilde{c}^j,x_1^j)}(r) = f(\tilde{c}^j \oplus \text{PRF}(r,j), x_1^j)$ to the left component as input. At the end of the evaluation P_0 obtains $\tilde{y}^j$.

3. P_0 sends $\tilde{y}^j$ to P_1. They both output $\tilde{y}^j$.

$T = T(n)$ is thus set to bound on the time for computing the functions (g_1^j, g_2^j) on two n-bit inputs.

Fig. 1. The Leakage Tolerant Protocol ρ

leakage-resilient strong 2-component OCL scheme with a auxiliary parties. $\Lambda = (\text{Comp}, \Pi = (P_L, P_R, P_{A_1}, \ldots, P_{A_a}))$ with an oblivious simulator $\mathcal{S} = (\mathcal{S}_L, \mathcal{S}_R, \mathcal{S}_A)$.

Let n be the length of the inputs $x_0^j, x_1^j \in \{0,1\}^n$ to be evaluated in the j-th iteration, which is polynomially related with the security parameter[5]. Our leakage-tolerant protocol below utilizes the OCL scheme to perform the evaluation of $f(x_0^j, x_1^j)$. To ensure input privacy, a party must avoid sharing its input in the clear with another party. Instead, in the j-th iteration, the parties first use the OCL scheme to allow P_1 to obtain an encrypted version $x_0^j \oplus \text{PRF}(r, j)$ of P_0's input, where PRF is a pseudorandom function and the PRF key r is encoded as the OCL secret. Then, instead of directly evaluating f, the OCL scheme is used again to evaluate the following function $g((c^j = (x_0^j \oplus \text{PRF}(r, j)), x_1), \text{PRF}(r, j)) = f(c^j \oplus \text{PRF}(r, j), x_1^j)$.

In the following, we simplify notation by denoting by $\text{init}_A^{j,b} = \text{init}_{A,1}^{j,b}, \ldots, \text{init}_{A,a}^{j,b}$, for $j \in [a]$ and $b \in \{1, 2\}$, the initial states of *all* auxiliary components of the b-th OCL in the j-th iteration. We similarly define $\text{evl}_A^{j,b}$. Moreover, by $\mathbf{x}_0^j = x_0^1, \ldots, x_0^j$ we denote the sequence of inputs of P_0 in the first j iterations, by $\mathbf{x}_1^j = x_1^1, \ldots, x_1^j$ we denote the sequence of inputs of P_1 in the first j iterations and by $\mathbf{y}^j = y_1^1, \ldots, y_1^j$ the sequence of outputs in the first j iterations.

We present the leakage-tolerant protocol ρ in detail in Figure 1:

Acknowledgement. We thank Elette Boyle and Abhishek Jain for valuable discussions and the anonymous reviewers for helpful comments and suggestions.

References

[ADW09]   Alwen, J., Dodis, Y., Wichs, D.: Survey: Leakage Resilience and the Bounded Retrieval Model. In: Kurosawa, K. (ed.) ICITS 2009. LNCS, vol. 5973, pp. 1–18. Springer, Heidelberg (2010)

[BCG+11] Bitansky, N., Canetti, R., Goldwasser, S., Halevi, S., Kalai, Y.T., Rothblum, G.N.: Program obfuscation with leaky hardware. In: Lee, D.H., Wang, X. (eds.) ASIACRYPT 2011. LNCS, vol. 7073, pp. 722–739. Springer, Heidelberg (2011)

[BCH12]   Bitansky, N., Canetti, R., Halevi, S.: Leakage-tolerant interactive protocols. In: Cramer, R. (ed.) TCC 2012. LNCS, vol. 7194, pp. 266–284. Springer, Heidelberg (2012)

[BGJ+13] Boyle, E., Garg, S., Jain, A., Kalai, Y.T., Sahai, A.: Secure computation against adaptive auxiliary information. In: Canetti, R., Garay, J.A. (eds.) CRYPTO 2013, Part I. LNCS, vol. 8042, pp. 316–334. Springer, Heidelberg (2013)

[BGJK12] Boyle, E., Goldwasser, S., Jain, A., Kalai, Y.T.: Multiparty computation secure against continual memory leakage. In: STOC 2012, pp. 1235–1254 (2012)

---

[5] The reason that we separate the security parameter from the length of the input is that the leakage-bound of the protocol only depends on the security parameter, but not the input length. Thus, by scaling up the security parameter, the absolute number of leakage bits that the protocol tolerates grows.

[BGK11]   Boyle, E., Goldwasser, S., Kalai, Y.T.: Leakage-resilient coin tossing. In: Peleg, D. (ed.) DISC 2011. LNCS, vol. 6950, pp. 181–196. Springer, Heidelberg (2011), http://eprint.iacr.org/2011/291

[BGW88]   Ben-Or, M., Goldwasser, S., Wigderson, A.: Completeness theorems for non-cryptographic fault-tolerant distributed computation (extended abstract). In: STOC, pp. 1–10 (1988)

[Can01]   Canetti, R.: Universally composable security: A new paradigm for cryptographic protocols. In: FOCS 2001, pp. 136–145 (2001)

[CCD88]   Chaum, D., Crépeau, C., Damgård, I.: Multiparty unconditionally secure protocols (extended abstract). In: STOC 1988, pp. 11–19 (1988)

[CG88]   Chor, B., Goldreich, O.: Unbiased bits from sources of weak randomness and probabilistic communication complexity. SIAM J. Comput. 17(2), 230–261 (1988)

[DF11]   Dziembowski, S., Faust, S.: Leakage-resilient cryptography from the inner-product extractor. In: Lee, D.H., Wang, X. (eds.) ASIACRYPT 2011. LNCS, vol. 7073, pp. 702–721. Springer, Heidelberg (2011)

[DF12]   Dziembowski, S., Faust, S.: Leakage-resilient circuits without computational assumptions. In: Cramer, R. (ed.) TCC 2012. LNCS, vol. 7194, pp. 230–247. Springer, Heidelberg (2012)

[DLWW11]   Dodis, Y., Lewko, A.B., Waters, B., Wichs, D.: Storing secrets on continually leaky devices. In: FOCS 2011, pp. 688–697 (2011)

[GJS11]   Garg, S., Jain, A., Sahai, A.: Leakage-resilient zero knowledge. In: Rogaway, P. (ed.) CRYPTO 2011. LNCS, vol. 6841, pp. 297–315. Springer, Heidelberg (2011)

[GMW87]   Goldreich, O., Micali, S., Wigderson, A.: How to play any mental game or a completeness theorem for protocols with honest majority. In: STOC 1987, pp. 218–229 (1987)

[GR10]   Goldwasser, S., Rothblum, G.N.: Securing computation against continuous leakage. In: Rabin, T. (ed.) CRYPTO 2010. LNCS, vol. 6223, pp. 59–79. Springer, Heidelberg (2010)

[GR12]   Goldwasser, S., Rothblum, G.N.: How to compute in the presence of leakage. In: FOCS 2012, pp. 31–40 (2012)

[JV10]   Juma, A., Vahlis, Y.: Protecting Cryptographic Keys against Continual Leakage. In: Rabin, T. (ed.) CRYPTO 2010. LNCS, vol. 6223, pp. 41–58. Springer, Heidelberg (2010)

[MR04]   Micali, S., Reyzin, L.: Physically observable cryptography. In: Naor, M. (ed.) TCC 2004. LNCS, vol. 2951, pp. 278–296. Springer, Heidelberg (2004)

[NVZ13]   Nielsen, J.B., Venturi, D., Zottarel, A.: On the connection between leakage tolerance and adaptive security. In: Kurosawa, K., Hanaoka, G. (eds.) PKC 2013. LNCS, vol. 7778, pp. 497–515. Springer, Heidelberg (2013)

[Pan14]   Pandey, O.: Achieving constant round leakage-resilient zero-knowledge. In: Lindell, Y. (ed.) TCC 2014. LNCS, vol. 8349, pp. 146–166. Springer, Heidelberg (2014)

[Sta09]   Standaert, F.-X.: Introduction to side-channel attacks. In: Verbauwhede, I.M.R. (ed.) Secure Integrated Circuits and Systems, pp. 27–44. Springer (2009)

[Yao82]   Yao, A.C.-C.: Protocols for secure computations (extended abstract). In: FOCS 1982, pp. 160–164 (1982)

Interactive Proofs under Continual Memory Leakage

Prabhanjan Ananth[1,*], Vipul Goyal[2], and Omkant Pandey[3,*]

[1] University of California, Los Angeles, USA
prabhanjan@cs.ucla.edu
[2] Microsoft Research, India
vipul@microsoft.com
[3] University of Illinois at Urbana Champaign, USA
omkant@uiuc.edu

Abstract. We consider the task of constructing interactive proofs for NP which can provide meaningful security for a prover even in the presence of continual memory leakage. We imagine a setting where an adversarial verifier participates in multiple sequential interactive proof executions for a fixed NP statement x. In every execution, the adversarial verifier is additionally allowed to leak a fraction of the (secret) memory of the prover. This is in contrast to the recently introduced notion of leakage-resilient zero-knowledge (Garg-Jain-Sahai'11) where there is only a single execution. Under multiple executions, in fact the entire prover witness might end up getting leaked thus leading to a complete compromise of prover security.

Towards that end, we define the notion of non-transferable proofs for all languages in NP. In such proofs, instead of receiving w as input, the prover will receive an "encoding" of the witness w such that the encoding is sufficient to prove the validity of x; further, this encoding can be "updated" to a fresh new encoding for the next execution. We then require that if (x, w) are sampled from a "hard" distribution, then no PPT adversary A^* can gain the ability to prove x (on its own) to an honest verifier, even if A^* has participated in polynomially many interactive proof executions (with leakage) with an honest prover whose input is (x, w). Non-transferability is a strong security guarantee which suffices for many cryptographic applications (and in particular, implies witness hiding).

We show how to construct non-transferable proofs for all languages in NP which can tolerate leaking a constant fraction of prover's secret-state during each execution. Our construction is in the *common reference string* (CRS) model. To obtain our results, we build a witness-encoding scheme which satisfies the following continual-leakage-resilient (CLR) properties:
- The encodings can be randomized to yield a fresh new encoding,
- There does not exist any efficient adversary, who receiving only a constant fraction of leakage on polynomially many fresh encodings of the same witness w, can output a valid encoding provided that the witness w along with its corresponding input instance x were sampled from a hard distribution.

Our encoding schemes are essentially re-randomizable non-interactive zero-knowledge (NIZK) proofs for circuit satisfiability, with the aforementioned CLR properties. We believe that our CLR-encodings, as well as our techniques to build them, may be of independent interest.

---

* Work done at Microsoft Research, India.

J.A. Garay and R. Gennaro (Eds.): CRYPTO 2014, Part II, LNCS 8617, pp. 164–182, 2014.
© International Association for Cryptologic Research 2014

1   Introduction

Traditionally, when defining security of a cryptographic primitive, the adversary is allowed to interact with the underlying cryptographic algorithms only in a black-box manner. Emergence of side channel attacks [39,5,6,32,25,31] has shown that enforcing only black-box access may be difficult in many practical settings. Such attacks exploit the physical characteristics of a cryptographic device, such as the time and electrical power taken by the device, to learn useful information about its secrets. This information is often sufficient to "break" the system completely.

Leakage resilient cryptography [16] focuses on the algorithmic aspects of this problem, by developing theoretical paradigms and security notions which can deliver meaningful security under such attacks. In the last few years, we have seen many exciting developments in this direction, resulting in interesting attack models for leakage, as well as cryptographic primitives that guard against leakage [26,16,3,35,38,15,17,22,34,33].

Continual Memory Leakage. To model leakage, one typically allows the adversary to submit leakage queries in the form of efficiently computable functions f, and provide it with $f(st)$, where st is the internal state st of the system during execution. The class of functions f and the length of their output determines the kind and the amount of leakage tolerated by the system. Initial works focused on the *bounded leakage* model which requires that the total amount of leakage throughout the life time of the system is a priori bounded. Construction of several cryptographic primitives such as public-key encryption, identity based encryption, signatures schemes, stream ciphers etc. are known in this model [30,9,4,12,2].

However, enforcing an a priori bound on the total leakage is somewhat unreasonable since usually it is not known how many times the system will be used. The emerging standard for leakage attacks, therefore, seems to be the *continual memory leakage* model. In this attack model, the secret state st of the cryptographic system is "updated" after each time-period *without changing its public parameters* (if any). The adversary is allowed queries f as before—we only require that the amount of leakage between any two *successive* updates is bounded. This a very natural and powerful attack model, allowing the adversary to learn an unbounded overall leakage. Many basic cryptographic primitives such as public-key encryption, signatures schemes, identity-based encryption etc. are now known in this model [14,29,9,8]. With few exceptions, almost all results in the continual leakage model have focused on *non-interactive* tasks such as public-key encryption.

Interactive Protocols under Leakage attacks. Modeling meaningful security for interactive protocols in the presence of leakage is a more delicate task. Very recent works focusing on zero-knowledge [20,37] and multi-party computation [7,8] have now emerged. While leakage-resilient zero-knowledge (LRZK) protocols of [20,37] do not put a bound on total leakage, the system still does not protect against *continual* leakage. Indeed, the LRZK notion as defined in [20,7] do not model updating a witness; the notion becomes meaningless as soon as the entire witness is leaked. In this work, we will focus on providing meaningful security for the prover in face of *continual* leakage.

Encoding-based Proofs under Continual Leakage. The goal of this paper is to construct interactive protocols in the continual leakage model for arbitrary NP relations. The setting for our interactive protocol, defined for an NP relation, is the following. There exists a prover, who has an instance and a witness that the instance belongs to that relation. The prover executes polynomially (in the security parameter) many times and in each execution he convinces a verifier (which can be different for different executions) that he has a valid witness corresponding to that instance without actually revealing the witness. Now that we have defined the model, we need to come up with a meaningful notion of security that will help us apply this in many practical scenarios especially identification schemes and witness hiding schemes which are of primary interest for us.

1. The first thing to observe here, irrespective of the security notion we consider, is that the prover cannot use the same witness in all the executions because by obtaining leakage on the witness bit-by-bit, the verifier can obtain the entire witness. To this end, we need a refreshing mechanism that updates the witnesses in between executions.
2. The first candidate security definition for our setting is the standard simulation based notion. This says that there exists a PPT simulator who can simulate (entirely on its own) the transcript of conversation between the prover and the verifier as well as simulate the leakage queries made by the verifier. However similar to an argument in [20], we observe that it is unlikely that such a security notion can be satisfied in our setting even if we allow a preprocessing phase[1]. An informal argument to rule out the simulation based notion in the preprocessing model for single execution leakage can be found in [20] (and the same argument applies directly for multiple executions with continual leakage as well). At a high level, the problem occurs when the verifier submits a leakage function that tests whether the memory state of the prover has a valid witness (or an encoding of the witness) or not. In this case, the simulator has no way of answering such a function query since the function query may be "obfuscated" in some sense. Hence, the simulator doesn't know what the right answer to such a query would be (the function may output a pre-encoded value on getting a valid witness which the simulator doesn't know). We refer the reader to [20] for more details.
3. Garg et. al. [20] overcame the above problem by giving more capability to the simulator. Their simulator is given access to a leakage oracle containing the witness. The security requirement is that the simulator, with access to a leakage oracle on the witness, should be able to output a transcript which is indistinguishable from the one obtained by the interaction between the honest prover and the verifier. One can try to adopt such a security notion to our setting as well. In each execution, the simulator is given access to a leakage oracle holding the witness. However note that this becomes meaningless under multiple executions. This is because, under multiple executions, the entire witness can be leaked from the leakage oracle by the simulator!
4. Next option we consider is a variant of the above model where in each execution, the simulator is given access to a leakage oracle containing *an encoding of a*

---

[1] In a preprocessing phase, a witness can be preprocessed before the execution of the protocol begins.

witness. In between executions, there is an update phase which refreshes the encodings. While this is a step towards our final definition, this does not quite give us anything meaningful yet. The reason is that there could be a protocol which satisfies this definition yet an adversarial verifier could obtain a valid encoded witness, by possibly combining leakages on many encodings obtained during its interaction with the prover! Thus, one needs to be more specific about the properties this witness encoding scheme satisfies.

5. From the above observation it follows that any security definition we consider should have the following property – given leakage on many encodings, an efficient adversary should not be able to compute a valid encoding. This is indeed the starting point of our work.

6. We introduce the security notion that we study in the paper. Suppose an efficient adversary after receiving leakage across many executions with the honest prover is able to convince an honest verifier, with non-negligible probability, that he has a valid encoding, then, we should be able to extract the encoding from the adversary with non-negligible probability. Not only does this definition imply that given leakage on many encodings, computing a valid encoding is hard but our definition is in fact stronger. This is because our definition rules out the possibility of obtaining even a partial encoding from the leakage on many encodings, using which membership of the instance in the language can be proven. We term proof systems that satisfy this definition as *non transferable proof systems* (NTP). We consider this as a simplified and a more direct way of obtaining the guarantees that notion in the previous bullet had (where we talk about obtaining leakage on encodings in the ideal world).

Our main result is to achieve the notion outlined above for all of NP in the common reference string (CRS) model. Achieving this notion in the plain model is left as an interesting open problem (and as we discuss later, even in the CRS model, achieving this notion already turns out to be quite intricate and non-trivial).

Note that most of the above discussions can be translated to the CRS setting with the exception of *simulatability of encodings* (see bullet 2). In other words, it is still possible to have a standard simulation based notion where the encodings are completely simulated. Towards that end, consider the following protocol. The prover receives a non interactive zero knowledge proof that the instance belongs to that language and then it sends the same proof to the verifier in every execution. Thus, this protocol is zero-knowledge for any polynomial number of executions even when the entire state of the prover is leaked in every execution. However, it is not clear how meaningful this protocol is: the adversary gets to learn a valid encoding of the witness which it can later use on its own. This is no different from the scenario where the prover gives out the witness to the verifier. Indeed, it cannot be used in the applications like identification schemes that we are interested in.

We believe that requiring the adversary to not be able to compute a valid encoding (which would allow him to prove on his own) is indeed a "right" security notion in the setting of interactive proofs with continuous leakage (indeed, there may be others notions). If the adversary can obtain the same witness (or the witness encoding) as being used by the prover, there is very little left in terms of the security guarantees that can be satisfied. Indeed, our notion of NTP is a formalization of this "intuition": not only it ensures that the adversary cannot get any valid witness encoding, but also, that it cannot compute any "partial" witness encoding which still allows him to complete the proof on his own.

To summarize, the main question that we are concerned with in this work is the following:

Do non-transferable proofs, under (non-trivial) continual memory leakage, exist for NP?

Our Results. In this work, we construct non-transferable CLR proof systems for all languages in NP assuming DDH is hard in pairing-based proofs (called eXternal Diffie Hellman assumption). Our results are in the CRS model. We obtain our result by constructing the so called *CLR-encodings* for all NP languages, and then use them to construct non-transferable (interactive) proofs for all of NP. We work in the CRS model; recall that, as we argued above, non-transferable proofs are non-trivial in the CRS model as well. In our continual leakage model, we do not allow leakage during the update phase. Further, we follow [20,37] and require that the randomness of a particular round of the protocol is not available to the prover until that round begins.[2]

To construct CLR-encodings, we first construct *re-randomizable* NIZK-proofs *for all NP-languages.* Prior to our work, such proofs were known only for specific, number-theoretic, relations. In addition, our re-randomizable NIZK proofs satisfy the stronger property of "controlled malleability": roughly, it means that the only way to construct new valid proofs is to either re-reandomize given proofs or create them from scratch (using the knowledge of an appropriate witness). These are of independent interest.

Finally, we find that the CLR-encodings are a cryptographic primitive of independent interest as well. We show how to use them to build other cryptographic primitives secure under continual memory leakage. In particular, we show how to construct continual leakage resistant public key encryption schemes, identity based encryption schemes as well as attribute based encryption schemes (for polynomial sized attribute space) using CLR-encodings and the recent new primitive of *witness encryption* [19] (see full version for more details). We hope that this will inspire future work, and our CLR-encoding primitive will be useful in obtaining more primitives resilient against continual leakage attacks. Even though some of these applications already exist under standard assumptions, ours is the first work that describes a generic tool that can be used to construct many continual-leakage-resilient primitives at once.

A brief overview of our approach. The starting point of our construction is the observation that we can use the "proof of a proof" approach which was used in [13]. That is, instead of receiving a witness w as input, the prover will receive a non-interactive proof π that $x \in L$. If such a proof is sound, then the prover can prove to the verifier

---

[2] It might be possible to avoid this model and allow leakage on all randomness *from the start*; however, it is usually difficult to ensure without making further assumptions on the statements. For example, in the context of LRZK, it might be impossible to support such leakage because a cheating verifier can first obtain a "short" commitment to the witness *and* randomness, and later try to check if the transcript is consistent with this state.

that "there exists a convincing non-interactive proof π for $x \in L$." Therefore, π acts as a different kind of witness for x, or, as an "encoding of w." A natural choice for such proofs is NIZK-proofs (which exist for all NP languages). These proofs have strong security properties, e.g., an adversary cannot decode w from a NIZK-proof for x (unless x is easy).

We start by constructing re-randomizable NIZKs for the NP-complete language of circuit satisfiability. However, re-randomization does not guarantee much as it might be possible to use leakage on many such NIZK-proofs and "stitch them together" to obtain a valid proof. To avoid this, we turn to the notion of *non-malleability*. More specifically, we require the following two properties:

- First, we require that each proof must have sufficient min-entropy of its own, say $\ell + \omega(\log n)$, where n is the security parameter. This guarantees that ℓ bits of leakage do not suffice to predict this proof exactly. In turn, given ℓ bits of leakage from polynomially many independent proofs will not suffice to guess *any* of those proofs exactly. This step guarantees some protection against leakage information theoretically.
- However, this leakage might still suffice to compute a *new* proof (that is different from all other proofs so far). To counter this, we require that each proof must be *simulation-extractable* [40,41]. This would essentially mean that the adversary cannot compute new proofs unless he knows a witness for x. But this is a bit too strong since it means that the proofs cannot even be re-randomized! Therefore, we need a "controlled" form of simulation extractability; one such notion is CM-SSE [10] but this falls short of what we need.

We formulate an appropriate notion called *decomposable NIZK with controlled malleability*. These are NIZK-proofs which can be re-randomized, and essentially "that is all" an adversary can do with such proofs. We construct such proofs for circuit satisfiability. To construct such proofs, at a very high level, we show how to combine the re-randomizable garbled circuits [21] with malleable proof-systems of Groth and Sahai [23,10].

Prior Work. A series of works [39,5,6,32,25,31], to name a few, studied a class of hardware-based attacks on various cryptosystems, which were later referred to as side channel attacks. Early theoretical works focused on increasingly sophisticated modeling of such attacks and protecting against them[3,15,38,26], resulting in constructions for public key encryption schemes [35,34], signature schemes [17], identity based encryption schemes [12] and so on. Early schemes in the bounded leakage model also appear in [4,30,12]. Later several works focused on continual leakage [14,29,8,22]. The works of [20,7,37,8] focused on interactive protocols in presence of leakage. Dodis et. a. [14] were the first to consider identification schemes in presence of continual leakage. To the best of our knowledge, ours is the first work to address the security of *general* interactive proofs in the presence of *continual leakage*.

The concept of non-transferrable proof systems appear in the works of [27,28,36], as well as early works on identification schemes [18,42,24,11]. The idea of "proof of a proof" has appeared, in a very different context, in the work of [13].

2   Notation and Definitions

We will use standard notation and assume familiarity with common cryptographic concepts. We use n as a security parameter. We say that a PPT sampling algorithm Sampler is a *hard distribution* over an NP relation R if Sampler outputs pairs $(x, w) \in R$, and for every PPT adversary A^* it holds that the probability, $\Pr[(x, w) \leftarrow \mathsf{Sampler}(1^n); w' \leftarrow A^*(x) \wedge (x, w') \in R]$ is negligible in n. For an NP-language L, we define a witness relation, denoted by R_L, such that $x \in L$ iff there exists w such that $(x, w) \in R_L$. We now define the notion of encoding schemes.

Encoding schemes. An encoding scheme offers a mechanism to not only encode the witnesses of an NP relation, say R_L, but also to refresh these encodings. Further, it can be verified whether x is in the language L or not by using the encodings of the witness(es) of x. Looking ahead, the prover in the continual leakage resistant system would not have the witness corresponding to the original relation. Instead he will possess the encoding of the witness, which he will consequently refresh in between executions. We now formally define the encoding scheme.

An encoding scheme for an NP-relation R_L, in the CRS model, is a tuple of PPT algorithms $\mathbb{E} = (\mathsf{PubGen}, \mathsf{Encode}, \mathsf{Update}, V_L)$[3]. The PubGen takes as input a security parameter and outputs a string CRS. The algorithm Encode takes as input $(x, w) \in R_L$ as well as CRS and it outputs a fresh "encoded" witness, say $\tilde{w}$. The update algorithm takes as input $(x, \tilde{w})$ as well as CRS and outputs an encoded witness $\tilde{w}'$ which may be different from $\tilde{w}$. The requirement from the update algorithm is that the distributions $\{\mathsf{Encode}(x, w, \mathsf{CRS})\}$ and $\{\mathsf{Update}(x, \mathsf{Encode}(x, w, \mathsf{CRS}))\}$ are computationally indistinguishable. The verifier V_L on input $(\mathsf{CRS}, x, \tilde{w})$ either decides to accept or reject.

Any encoding scheme for a relation R_L needs to satisfy two properties, namely completeness and soundness. These properties are defined the same way they are defined in an interactive proof system[4]. Intuitively, except with negligible error it should happen that the verifier V_L accepts an encoding (generated using either Encode or Update) of w iff w is a valid witness.

Encoding based proofs. Let (P, V) be a pair of PPT interactive Turing machines, L an NP language and $\mathbb{E} = (\mathsf{PubGen}, \mathsf{Encode}, \mathsf{Update}, V_L)$, a refreshable encoding scheme for L. We say that $(P, V, \mathbb{E})$ is an encoding-based proof for L if the following two conditions hold: firstly, (P, V) is an interactive proof (or argument) for L; and second, the prover algorithm P gets as input $x \in L$ along with an encoded witness $\tilde{w}$ such that $\tilde{w} = \mathsf{Encode}(x, w, \mathsf{CRS})$ (where CRS is the output of PubGen algorithm.), where w is

---

[3] The corresponding scheme in the standard model will not have the CRS generation algorithm. That is, it will consist of $(\mathsf{Encode}, \mathsf{Update}, V_L)$.

[4] *Completeness.* Let $(x, w) \in R_L$. And let CRS be the output of PubGen. Let $(x, \tilde{w}_i)$ be such that $\tilde{w}_i$ is either the output of (i) Encode on input (x, w, CRS) if $i = 1$ or (ii) it is the output of $\mathsf{Update}(x, \mathsf{CRS}, \tilde{w}_{i-1})$ for $i > 1$. Then, completeness property says that V_L accepts the input $(\mathsf{CRS}, x, \tilde{w})$ with probability 1.
Soundness. For every $x \notin L$, and every PPT algorithm P^*, and for sufficiently large n and advice $z \in \{0, 1\}^*$ we have that $\Pr[\mathsf{CRS} \leftarrow \mathsf{PubGen}(1^n); V_L(\mathsf{CRS}, x, P^*(x, \mathsf{CRS}, z)) = 1]$ is negligible in n.

a witness such that $(x, w) \in R_L$. Further, the prover can participate in many executions and in between executions he refreshes his encoded witness using the Update procedure by executing Update on an encoded witness $\tilde{w}$ to obtain a new encoding $\tilde{w}'$. In this work, we consider encoding based proof systems which satisfy extractabilty property, which is stronger than soundness.

We can also consider encoding based proofs in the CRS model. In this case, there is a Setup algorithm in addition to the algorithms $(P, V, \mathbb{E})$. Note that $\mathbb{E}$ has its own CRS generation algorithm PubGen. For simplicity we assume that the encoding based proof has a single CRS generation algorithm, denoted by PubGen, that internally runs the CRS generation algorithms of both $\mathbb{E}$ as well as the encoding based proof system (P, V).

Continual leakage attacks on encoding based proofs. Let $l := l(n)$ be a leakage parameter used for bounding the maximum leakage (in number of bits) allowed during a single execution of the proof. Let $(P, V, \mathbb{E})$ be an encoding-based proof for $L \in NP$. From now on, we will denote by R_L a witness relation for L. Let A^* be a PPT algorithm, called the *leakage adversary* and let $(x, w) \in R_L$. An adversary A^*, taking the role of a verifier, on input a statement $x \in L$ (and some advice $z \in \{0,1\}^*$), interacts with an honest prover P in polynomially many executions (or sessions). At the start of the first execution, the prover receives as auxiliary input an encoding of witness w, denoted by $\tilde{w}_1$. At the start of the i-th session, the prover receives as its auxiliary input a "refreshed" encoding of the encoded witness, namely $\tilde{w}_i = \mathsf{Update}(x, \tilde{w}_{i-1})$, and fresh randomness. Following prior work on LRZK [20,37], we will assume that the randomness required by P to execute round j of any given session i is not sampled *until that round begins*.

During the execution of session i, A^* is allowed to make adaptively chosen leakage queries by submitting an efficient leakage function f_j at the beginning of round j of the proof (P, V). In response, the adversary is given the value of $f_j(st_j)$ where st_j denotes the state of the honest prover in round j; st_j consists of the refreshed witness $\tilde{w}_i$ and the randomness of the prover in session i *up to round j* denoted $(r_1^i, \ldots, r_j^i)$ for that session. It is required that the total leakage bits received by A^* in session i — i.e., the sum of lengths of outputs of queries sent *during session i only*—is at most l. We say that A^* launches a continual leakage attack on the system $(\mathsf{PubGen}, P, V, \mathbb{E})$.

In this work, we study continual-leakage-resilient encoding based proofs under a specific security notion called *non-transferability*, and we term all the encoding based proofs that satisfy this security definition as non transferable proof systems.

Definition 1 (Continual-Leakage-Resilient Non-transferable Proofs). *Let $\Pi :=$ (PubGen, P, V, $\mathbb{E}$) be an encoding-based proof in the CRS model (as defined above) for an encoding scheme $\mathbb{E} =$ (PubGen, Encode, Update, V_L). We say that Π is a continuous-leakage-resilient non-transferable proof for L (CLR-NTP) with leakage-parameter $l(n)$, if for every PPT adversary Adv and every algorithm Sampler that is a hard distribution over R_L, and every advice string $z \in \{0,1\}^*$, the success probability of A^* in the following NTPGame, taken over the randomness of the game, is negligible in n.*

NTPGame(n, Sampler, Π, Adv, z). *The game proceeds in the following steps:*

1. **Initialize** Run PubGen(1^n)[5] *to obtain a CRS, say ρ; then run* Sampler(1^n) *to obtain* $(x, w) \in R_L$.

2. **Phase I:** *Initiate the adversary algorithm* Adv *on inputs* (x, z), *who launches a continual-leakage attack on the encoding-based proof system* $(P, V, \mathbb{E})$, *as described earlier. The prover P starts by receiving an encoded witness obtained using* Encode *and this is refreshed (using* Update*) at the start of every new session. At some point,* Adv *signals the end of this phase, at which point the next phase begins.*

3. **Phase II:** *The adversary* Adv *attempts to prove that $x \in L$ to the honest verifier algorithm V. The verifier V receives fresh random coins in this (stand alone) session. The game ends when this session ends.*

4. Adv *wins the game if V accepts the proof.*

In what follows, the leakage parameter $l(n)$ will often be implicit in the abbreviation CLR-NTP and clear from the context.

3   A Construction from CLR Encodings

In this section we describe a construction of non-transferable proof for all of NP. The construction will use encoding schemes (defined in Section 2) that are secure against continual leakage attacks and we term such encodings as CLR encodings. We first define CLR encodings and then assuming the existence of CLR encodings, we construct non-transferable proof systems. We defer the construction of the CLR encodings to the next section.

3.1   CLR Encodings for NP Languages

Continuous-Leakage-Resilient Encodings. Let $L \in NP$, $\mathbb{E} := ($PubGen, Encode, Update, $V_L)$ an encoding scheme defined for language L, $l := l(n)$ a leakage parameter, and Sampler a sampling algorithm for the relation R_L. Informally, we require that for every hard distribution Sampler for R_L, no PPT adversary Adv receiving at most l bits of leakage from polynomially many valid encodings $\tilde{w}_1, \tilde{w}_2, \ldots, \tilde{w}_{p(n)}$, generated by either Encode or Update, where p is a polynomial chosen by adversary Adv, can output a valid encoding that will be accepted by V_L.

Formally, we consider the following EncodingGame, parameterized by $l(n)$, played between a PPT adversary Adv and a challenger CH. At the start of the game, CH samples (x, w) by executing Sampler(1^n) and ρ by executing PubGen(1^n). It sends x to Adv and sets private$_1$ = Encode(ρ, x, w). The game now proceeds in sessions and a session must end before a new fresh session can be started. In each session s, Adv is allowed to make adaptively chosen leakage queries $f_1, f_2, \ldots, f_m$ such that their total output length, that is the sum of the output lengths of $f_1, \ldots, f_m$, is at most l. It receives $f_1($private$_s)$, $f_2($private$_s)$, $\ldots, f_m($private$_s)$ in response. At the beginning of session $s + 1$, the challenger sets private$_{s+1}$ = Update($x, \rho,$ private$_s$). In the end, Adv halts by outputting an encoding π^*. Adv wins the EncodingGame if $V_L(\rho, x, \pi^*) = 1$.

---

[5] Recall that PubGen includes the CRS generation algorithm of the proof system (P, V) as well as the CRS generation algorithm of the encoding scheme $\mathbb{E}$.

We say that the scheme $\mathbb{E} := (\mathsf{PubGen}, \mathsf{Encode}, \mathsf{Update}, V_L)$ is a *CLR Encoding* with leakage-parameter $l(n)$ if for every Sampler that is a hard distribution over R_L, there does not exist any PPT algorithm Adv that can win the EncodingGame with probability non-negligible in n.

In the next section, we will prove the following theorem.

Theorem 1. *There exists a CLR Encoding scheme for every language $L \in NP$ in the CRS model under the validity of eXternal Diffie Hellman assumption; the encoding scheme supports a constancy fraction of leakage, i.e., it supports $l(n) = \delta \cdot p(n)$ where δ is a constant and $p(n)$ is a polynomial bounding the size of the encodings for statements of length n.*

3.2 Our Construction

We are now ready to describe our construction. Given a CLR encoding, we can use standard tools and paradigms to build a CLR-NTP system for L. We only need to make sure that tools used have an appropriate "leakage resilient" property.

To this end, we use a leakage resilient zero-knowledge proof system (LRZK) for NP denoted by (P_{lr}, V_{lr}). Such proof systems were constructed in [20,37]. Recall that an interactive proof system (P_{lr}, V_{lr}) is said to be *leakage-resilient* if for every cheating verifier V^* there exists a simulator S_{lr} who produces a computationally indistinguishable view on input the statement x and access to a leakage-oracle $\mathcal{L}_w^n(\cdot)$. The leakage oracle takes as input efficient functions f and returns $f(w)$. While, in LRZK, S_{lr} must read no more leakage from $\mathcal{L}_w^n$ than it gives to V^* in the simulated view, for us a weaker requirement suffices as used by GJS. We use a leakage-parameter $l := l(n)$ which we wll fix later.

GJS construct a variant denoted $(1+\epsilon)$-LRZK in n/ϵ rounds for any constant ϵ based on general assumptions. This notion in fact suffices for our purpose. However, since we are willing to use a CRS, we can replace the "preamble" of the GJS protocol (whose sole purpose is to extract a secret string r from the verifier) by $E_{pk}(r)$ where E_{pk} is a binding public-key encryption scheme and pk is a part of the CRS. As a result we get a constant round protocol; the modified simulator now extracts the string r by decrypting the cipher-text and then continue as in GJS. Further, we now get standard LRZK instead of $(1+\epsilon)$-LRZK. Therefore, we let (P_{lr}, V_{lr}) denote this modified constant-round version of the GJS protocol.

The protocol. Let $\mathbb{E} := (\mathsf{PubGen}, \mathsf{Encode}, \mathsf{Update}, V_L)$ be a CLR L-Encoding for the *circuit-satisfiability* problem, and let (P_{lr}, V_{lr}) be the above mentioned LRZK protocol for an NP-complete language.

Our continual-leakage-resilient non-transferable proof system (CLR-NTP). denoted by (P_{ntp}, V_{ntp}), for a language $L \in NP$ is an encoding-based proof system which uses $\mathbb{E}$ as its encoding. It proceeds as follows:

1. A public CRS $\rho \leftarrow \mathsf{PubGen}(1^n)$ is sampled at the beginning.
   Let $(x, w) \in R_L$, and let $\mathsf{private}_1 \leftarrow \mathsf{Encode}(\rho, x, w)$ be an encoded witness. The prover P_{ntp} receives $(x, \mathsf{private}_1)$ as input; the verifier V_{ntp} receives x as its only

input. *(Note that there is no leakage during this step or during any of the update phases).*

2. The prover P_{lr} proves to the verifier V_{lr}) that there exists an input private* such that the (deterministic) algorithm V_L accepts the string $(\rho, x, \text{private}^*)$.

3. At the end of each session, the prover applies the Update algorithm of the encoding scheme $\mathbb{E}$ to its encoded witness private and receives a fresh encoding which is used for the next session.

We only state the following theorem. A formal proof of the theorem can be found in the full version.

Theorem 2. *The protocol (P_{ntp}, V_{ntp}) is a non-transferable proof system for circuit satisfiability in the continual memory leakage model (CLR-NTP), supporting a constant fraction of leakage under the validity of XDH assumption.*

4   Constructing CLR Encodings in the CRS Model

In this section, we provide a detailed overview of our construction of CLR encodings. We refer the reader to the full version of this work for full technical details and proofs of various theorems.

We construct CLR-encodings for the NP-complete language of *circuit-satisfiability* under standard assumptions. The language consists of circuits C as instances and w as witnesses such that $C(w) = 1$. From here on, L denotes the language of circuit-satisfiability and R is the corresponding relation. We will be working in the CRS model.

4.1   Step 1: NIZK Proofs as Encodings

As mentioned in the introduction, we start with the idea of "proof of a proof" [13], and encode a given witness as a NIZK-proof. The prover can prove to the verifier that "there exists a convincing non-interactive proof π for $x \in L$." Since we need to update the encodings in future executions, we need to find a way to update the NIZK-proofs. At this point, we have two options:

1. We can take the idea of "proof of a proof" even further. To update the encoding, the update algorithm computes a NIZK proof, say π_1 proving the statement that "there exists a valid NIZK proof π for $x \in L$." This process continues so that π_i uses π_{i-1}, and so on.

   Unfortunately, in this approach, the size of the encodings grows exponentially. We can attempt to reduce the size of π using more advanced techniques such as CS-proofs or SNARKs. However, these systems are usually based on either the random-oracle model or non-standard (knowledge-type) assumptions.

2. Another approach is to use NIZK proof systems which are **re-randomizable**: given π, it is possible to publicly compute a new proof π' of the *same size*; proof π' is computationally indistinguishable from a freshly computed NIZK proof. This is a promising approach, and also the starting point of our solution.

While re-randomization allows us to update the encodings π, it does little to "protect" the encoding under leakage. Consider an encoding which contains two (re-randomizable) proofs (π_1, π_2). Clearly, this new encoding is valid and re-randomizable; yet if one of the proofs is leaked, an adversary can obtain a full encoding from it and use it to prove that $x \in L$. Therefore, in addition to re-randomization, we need the property that it should be hard to obtain a valid encoding even given the leakage from polynomially many refreshed copies of an encoding.

We tackle this problem by considering two more properties of NIZKs:

- **Large min-entropy:** Suppose that each proof π has $\ell + \omega(\log n)$ bits of min-entropy. Then, even given ℓ bits of leakage on π, no adversary can predict π *exactly* with more than negligible probability. Consequently, if $\pi, \ldots, \pi_{k=\text{poly}(n)}$ are independently generated proofs, then no adversary receiving at most ℓ bits of leakage from each proof, can predict π_i exactly (for any $i \in [k]$) with more than negligible probability.

   At a high level, this property ensures that if an adversary computes a valid encoding, say π^*, it will be different from all of the encodings prover generates (via the update algorithm).[6]

- **Controlled malleability:** The previous bullet rules out the possibility that an adversary can output a π^* that is one of the valid encodings from $\pi_1, \ldots, \pi_k$; we now want to rule out the possibility that π^* can be a *re-randomization* of one of the encodings, say π_i. This is actually easily guaranteed if the proofs also have "controlled malleability" (CM) property. More details follow.

   • First, consider the simpler property of *simulation-extractability* (a strengthened version of *simulation soundness*) [40,41]. It states that if an adversary A, who receives polynomially many simulated proofs (possibly to false statements), outputs a new proof π^* for some statement x^*, and π^* is different from all proofs A has received, then there exists an extractor which extracts a witness for x^* from the proof π^*.[7]

      Clearly, if our proofs satisfy this notion, then A cannot output a new proof π^* (for x) that differs from all previous proofs. This is because if it does, the extractor will extract a witness w for x. This however, is not possible, since x is sampled from a hard distribution (in the CLR-encoding game). Unfortunately, this notion is too strong for us: it also rules out re-randomization which is essential to update the proofs.

   • We therefore turn to the notion of *controlled malleability* which is a generalization of simulation-extractability to allow for very limited type of mauling, specifically *re-randomization* [10].[8] In particular, suppose that the NIZK proof

---

[6] Note that this property by itself still does not guarantee much: most proofs already have this property, and if not, it is trivially satisfied by appending sufficiently many random bits (at the end of the proof) which will be ignored by the verifier. We therefore need the second property, which rules out the proofs of this kind.

[7] The extractor uses an extraction trapdoor corresponding to the system's CRS.

[8] The CKLM definition is more general: it talks about a general set of *allowable transformations* $\mathcal{T}$ instead of re-randomization.

system has the following extraction property. Consider an adversary A receiving several freshly computed proofs $\pi_1, \ldots, \pi_k$ for the statement x. If A outputs a convincing proof π^* for x that is different from all proofs it receives, then there exists an extractor which on input π^* outputs one of the following: (1) a witness for x, *or* (2) one of the previous proofs π_i and randomness r such that π^* is a re-randomization of π_i using randomness r. We will refer to this property as the **controlled malleability (CM)** property.[9]

Note that in our setting, these two properties would indeed suffice: it would be hard to extract one of the previous proofs π_i because A only receives ℓ bits of leakage but the proofs have large min-entropy property; further, it would also be hard to extract a witness for x since x would be sampled from a hard distribution in the security game of CLR-encodings. Therefore, if we have a NIZK satisfying these properties, we would indeed have CLR encodings.

4.2   Step 2: Decomposable NIZK Proofs

Unfortunately, the CM property defined above is a bit too strong, and we do not achieve it. Instead, we consider a slight variation of both *large min-entropy* and *CM* property, which will also suffice to construct CLR-encodings.

To do this, we consider NIZKs in which the proof Π for a statement x can be decomposed into a pair (y, π). We view y as an *encoding of the statement* x, and π as the actual proof. We require that $\Pi = (y, \pi)$ satisfy all usual properties of a re-randomizable NIZK proof. Let reRand $=$ (reRand$_1$, reRand$_2$) be re-rerandomization algorithm where the first and second parts re-randomize y and π respectively and appropriately.

Then, we modify the large min-entropy and CM properties, and require them to hold w.r.t. the first and second parts. More precisely, we require that:

- **Large min-entropy of *first part*:** We require that the first component y of an honestly generated proof $\Pi = (y, \pi)$ have sufficient min-entropy, e.g., $\ell + \omega(\log n)$ to tolerate ℓ bits of leakage. Note that this is a stronger property since if y has large min-entropy then so does Π. As before, it holds that given ℓ bits of leakage from polynomially many independent proof $\Pi_1, \ldots, \Pi_k$, it will be hard to compute y_i for any i where $\Pi = (y_i, \pi_i)$.
- **(Modified) Controlled malleability:** As before we consider an adversary A who receives independently generated proofs $\{\Pi_i = (y_i, \pi_i)\}_{i=1}^k$ and outputs a valid proof $\Pi^* = (y^*, \pi^*)$. All proofs are for the same statement x. Then, there exists an extractor which, on input Π^*, either outputs a witness for x, or it outputs (y_i, r) such that y^* is a re-randomization of y_i using r: i.e., $y^* = $ reRand$_1(y_i; r)$. Note that this property is strictly weaker than before: the extractor is only required to extract the first part of one of the proofs, namely y_i, but not the second part π_i.

Using the same arguments as before, it is not hard to see that NIZK-proofs satisfying these properties also imply CLR-encodings. We therefore formally define such

---

[9] We remark that this property is inspired by — but different from — the work of [10]. For simplicity, we are only presenting what is relevant to our context.

proofs, and call them *decomposable proofs with controlled malleability (CM) property*, or **decomposable ℓ-CM-NIZK** where ℓ is the parameter for large-min-entropy property. The construction of CLR-encodings from these proofs is straightforward: if $\mathbb{P}_{cm} = (\text{CRSSetup}_{cm}, P_{cm}, V_{cm}, \text{reRand}_{cm})$ is decomposable ℓ-CM-NIZK then the corresponding CLR encoding scheme $\mathbb{E} = (\text{PubGen}, \text{Encode}, \text{Update}, V)$ is obtained by setting $\text{PubGen} = \text{CRSSetup}_{cm}$, $\text{Encode} = P_{cm}$, $\text{Update} = \text{reRand}_{cm}$, and $V = V_{cm}$. We have the following theorem.

Theorem 3. *Suppose that* $\mathbb{P}_{cm} = (\text{CRSSetup}_{cm}, P_{cm}, V_{cm}, \text{reRand}_{cm})$ *is a decomposable ℓ-CM-NIZK for an NP relation R. Then, $\mathbb{E} = (\text{PubGen}, \text{Encode}, \text{Update}, V)$ is a CLR-encoding with leakage parameter ℓ w.r.t. every hard distribution D on relation R.*

4.3 Step 3: The Sub-proof Property

It is clear that in one of the the key challenge in our system is the (modified) controlled-malleability property (we will drop the word "modified" from here on). In this section, we will focus on this property only, and show that it can be achieved from a weaker property call the *sub-proof* property. More specifically, let $\mathbb{P}$ be a NIZK-PoK system which satisfies the first three properties — *decomposability, re-randomization*, and *large min-entropy* — and the following sub-proof property:

- **Sub-proof property**: There exist a special language L_{sub} and an (ordinary) NIZK-proof system $\mathbb{P}_{sub}$ for L_{sub} such that a pair $\Pi = (y, \pi)$ is a valid proof for a statement $x \in L$ *if and only if*:
  1. x is a prefix of y, i.e., $\exists\, \widetilde{y}$ such that $y = x \circ \widetilde{y}$;
  2. π is a valid proof that $y \in L_{sub}$ (according to the proof system $\mathbb{P}_{sub}$);

We refer to such a proof system as the *decomposable NIZK with sub-proof (SP) property*, or **decomposable ℓ-SP-NIZK**. We now show that the sub-proof property can be "boosted up" to achieve the CM-property using the techniques of Chase et al. [10].

To do so, we first observe that the re-randomization property of $\mathbb{P}$ imposes the following *malleability* requirement on $\mathbb{P}_{sub}$:

($\mathbb{P}_{sub}$ **must be malleable**): Suppose that $\Pi = (y, \pi)$ is a NIZK-proof that $x \in L$ according to the main system $\mathbb{P}$ (under some CRS ρ). By the sub-proof property, π is a NIZK-proof that $y \in L_{sub}$ according to the sub-proof system $\mathbb{P}_{sub}$. By re-randomization property, if we let $\Pi' = \text{reRand}\,(\Pi) = (\text{reRand}_1(y), \text{reRand}_2(\pi))$: $= (y', \pi')$, then Π' would also be a valid NIZK-proof for $x \in L$. This means that y' is a statement in L_{sub} and π' is a valid NIZK-proof for y'. Hence, the sub-proof system $\mathbb{P}_{sub}$ is *malleable* in the following sense: given a proof π for the statement $y \in L_{sub}$, it is possible to obtain a proof π' for a related statement $y' = \text{reRand}(y)$ (in the same language).

In the terminology of [10], reRand_1 is called an **allowable transformation** over the statements in L_{sub}, reRand_2 is the corresponding **mauling** algorithm, and $\mathbb{P}_{sub}$ is a reRand_1-*malleable* proof system. We now recall one of the main results from [10].

Result from [10]. Chase et al. formalize the general notion of a $\mathcal{T}$-malleable proof systems where $\mathcal{T}$ is an appropriate set of allowable transformations. They give a compiler which can convert such a $\mathcal{T}$-malleable proof system into a new proof system that achieves *controlled malleability* in the following sense. In the new proof system, given a proof π for a statement y, an adversary can obtain proof for another statement y' *if and only if* $y' = \tau(y)$ where τ is a transformation from the set $\mathcal{T}$. This property is formalized by defining the notion of *controlled-malleable simulation-soundness extractability* (CM-SSE) proofs. In our context, the set $\mathcal{T}$ corresponds to reRand$_1$, and a specific τ is equivalent to executing reRand$_1$ with a specific randomness. The notion of CM-SSE then amounts to the following (simpler) experiment:

- **(CM-SSE in our context)**: Let $(\rho, t_1, t_2) \leftarrow \mathsf{FakeCRS}(1^n)$ where ρ is a simulated CRS, and t_1 and t_2 are simulation and extraction trapdoors respectively. A PPT adversary A is allowed to interact with the simulator polynomially many times, say k times, where in i-th interaction A sends an (adaptively chosen) statement y_i and receives a simulated proof π_i (proving that $y_i \in L_{\mathsf{sub}}$). In the end, A outputs a pair (y^*, π^*) such that y^* is different from all previous statements $y_1, \ldots, y_k$ *and* π^* is a valid proof for y^*. We say that the system is CM-SSE if for every A who outputs (y^*, π^*) with noticeable probability, the extraction trapdoor t_2 either extracts a witness w^* for the statement y^*, or it extracts a previous statement (say) y_i and randomness r such that $y^* = \mathsf{reRand}_1(y_i; r)$.

Remark 1. To apply the CKLM transformation, the underlying proof system must satisfy some structural properties. In particular, the NP-relation $\mathcal{R}$ (for which the system works) and the set of transformations $\mathcal{T}$, both must be expressible as a system of bilinear equations over elements in bilinear groups. This property is called *CM-friendliness*. In our context, relation R_{sub} and algorithm reRand$_1$ must be CM-friendly, i.e., be expressible as a system of bilinear equations. We remark that the CKLM transformation can be based on the XDH assumption[10].

This essentially means that by suitably applying the CKLM compiler to the sub-proof system of a given (decomposable) ℓ-SP-NIZK, it should be possible to obtain the desired ℓ-CM-NIZK. This idea indeed works as proven in theorem below. One subtlety is that we need to assume that reRand$_1$ *does not change the prefix x of its input*. This is because, by definition, CM-SSE only extracts an instance y_i from which y^* has been obtained. To achieve ℓ-CM-NIZK we need extract a witness for x (which is same in all proofs). By requiring that reRand$_1$ does not change the prefix, we enforce that y_i and y^* will have the same prefix x, and this allows the reduction to go through.

Theorem 4. *Suppose that there exists a decomposable ℓ-SP-NIZK system $\mathbb{P}$ for an NP-relation R. Let $\mathsf{reRand} = (\mathsf{reRand}_1, \mathsf{reRand}_2)$ be the re-randomization algorithm, $\mathbb{P}_{\mathsf{sub}}$ be the sub-proof system, and R_{sub} be the sub-relation associated with $\mathbb{P}$. Then, under the validity of XDH assumption, there exists a decomposable ℓ-CM-NIZK for R provided that R_{sub} and reRand_1 are CM-friendly and reRand_1 does not change the prefix of its input.*

---

[10] CKLM transformation makes use of a structure-preserving signature scheme which can be based on DDH assumption in bilinear groups [1].

4.4 Step 4: Achieving the Sub-proof Property

We have seen that in order to construct a CLR L-encoding, it suffices to construct a proof-system $\mathbb{P}$ with the sub-proof property. In the final step, we construct such a proof system for the circuit satisfiability problem. Only an overview of the construction is provided here and the details appear in our full version.

From here on, let L represent the circuit satisfiability language, and R be the corresponding relation. The instances in L are circuits C and witness w is an input such that $C(w) = 1$. To construct $\mathbb{P}$, we will actually start by constructing the sub-proof system in question $\mathbb{P}_{\mathsf{sub}}$. The sub-proof system has to be malleable and must work for a "CM-friendly relation", say R_{sub}. The starting point of our construction is the observation instead of directly working with the given circuit C, we can actually work with a "garbled version" of C, denoted by GC. If we use re-randomizable garbled circuits, then we can randomize GC to yield a new garbled circuit GC$'$ for the same C. For the proofs, we need to construct an NIZK proof system which proves the correctness of GC and can also be re-randomized. One of the approaches is to use the malleable proof systems of Groth and Sahai [23]. However, such proof systems only exist for special languages involving number-theoretic statements in bilinear groups.

Therefore, our next task is to represent GC (and the fact that it is a garbled version of a given C) using a system of bilinear equations such that we can deploy GS-proofs to argue about the solutions to such a system. This is a rather complex step in the proof: we show in several smaller steps how to represent simple relations using bilinear equations, and use them to represent an entire circuit. By doing so, we actually hit two birds with one stone: on one hand, we obtain malleable-proofs for an NP-complete language, and on the other hand, the resulting system is *CM-friendly* (since we already start with equations in bilinear groups). This strategy eventually works out, although there are several other details to be filled in.

To summarize, the main steps in the construction are as follows:

1. The first step is to devise a method to represent a garbled circuit as a system of bilinear equations. We use a slight variant of the garbled-circuit construction in [21]. Thereafter, we show how to represent such a garbled circuit using a system of bilinear equations. We require that this representation of GC must re-randomizable (in a manner that is consistent with the randomization of garbled circuits).

2. Next, we define the InpGen algorithm (which will define the first part of the prover algorithm in the main system) as follows. InpGen first generates the garbled circuits and then outputs their representation as a system of bilinear equations. I.e., InpGen, on input (C, w), outputs (y, w') where $y := (C, \mathsf{GC}, \mathsf{w_{GC}})$, GC is the garbled circuit of C expressed as appropriate bilinear equations, $\mathsf{w_{GC}}$ are wire-keys for w expressed as appropriate group elements, and w' is the randomness (and wire-keys) used to generate everything so far. The sub-relation over (y, w') is essentially the garbled circuit relation, and denoted by R_{gc}.
   The re-randomization algorithm of the garbled-circuit construction will act as the algorithm reRand_1. Note that this algorithm will not change (the prefix) C of the input y. We also show that reRand_1 can also be appropriately expressed as a system of bilinear equations, and hence satisfy the condition of CM-friendliness.

3. Next, we show how to use GS-proofs to obtain a NIZK-proof that y is indeed consistent with C and that it is satisfiable. Using previous work [10], we can ensure that this system is malleable w.r.t. reRand$_1$. The prover algorithm at this step becomes the second component of the final prover; rest of the algorithms become the corresponding algorithms of the final system.

The resulting system is a decomposable NIZK with the sub-proof property, as stated in the theorem below. We defer the proof of the theorem to the full version.

Theorem 5. *Assuming the validity of XDH assumption, there exists a constant* $0 < \delta < 1$ *and a decomposable* ℓ-*SP-NIZK* $\mathbb{P}$ *for the relation* R_{gc} *such that* $\ell = \delta \cdot p(n)$ *where* $p(n)$ *is a fixed polynomial defining the length of proofs (for statements of length* n) *generated by* $\mathbb{P}$. *Further,* R_{gc} *and* $\mathbb{P}$ *are both CM-friendly.*

References

1. Abe, M., Groth, J., Haralambiev, K., Ohkubo, M.: Optimal structure-preserving signatures in asymmetric bilinear groups. In: Rogaway, P. (ed.) CRYPTO 2011. LNCS, vol. 6841, pp. 649–666. Springer, Heidelberg (2011)
2. Agrawal, S., Dodis, Y., Vaikuntanathan, V., Wichs, D.: On continual leakage of discrete log representations. In: Sako, K., Sarkar, P. (eds.) ASIACRYPT 2013, Part II. LNCS, vol. 8270, pp. 401–420. Springer, Heidelberg (2013)
3. Akavia, A., Goldwasser, S., Vaikuntanathan, V.: Simultaneous hardcore bits and cryptography against memory attacks. In: Reingold, O. (ed.) TCC 2009. LNCS, vol. 5444, pp. 474–495. Springer, Heidelberg (2009)
4. Alwen, J., Dodis, Y., Naor, M., Segev, G., Walfish, S., Wichs, D.: Public-key encryption in the bounded-retrieval model. In: Gilbert, H. (ed.) EUROCRYPT 2010. LNCS, vol. 6110, pp. 113–134. Springer, Heidelberg (2010)
5. Biham, E., Carmeli, Y., Shamir, A.: Bug attacks. In: Wagner, D. (ed.) CRYPTO 2008. LNCS, vol. 5157, pp. 221–240. Springer, Heidelberg (2008)
6. Biham, E., Shamir, A.: Differential fault analysis of secret key cryptosystems. In: Kaliski Jr., B.S. (ed.) CRYPTO 1997. LNCS, vol. 1294, pp. 513–525. Springer, Heidelberg (1997)
7. Bitansky, N., Canetti, R., Halevi, S.: Leakage-tolerant interactive protocols. In: Cramer, R. (ed.) TCC 2012. LNCS, vol. 7194, pp. 266–284. Springer, Heidelberg (2012)
8. Boyle, E., Goldwasser, S., Jain, A., Kalai, Y.T.: Multiparty computation secure against continual memory leakage. In: STOC, pp. 1235–1254. ACM (2012)
9. Brakerski, Z., Kalai, Y.T., Katz, J., Vaikuntanathan, V.: Overcoming the hole in the bucket: Public-key cryptography resilient to continual memory leakage. In: FOCS, pp. 501–510. IEEE (2010)
10. Chase, M., Kohlweiss, M., Lysyanskaya, A., Meiklejohn, S.: Malleable proof systems and applications. In: Pointcheval, D., Johansson, T. (eds.) EUROCRYPT 2012. LNCS, vol. 7237, pp. 281–300. Springer, Heidelberg (2012)
11. Cho, C., Ostrovsky, R., Scafuro, A., Visconti, I.: Simultaneously resettable arguments of knowledge. In: Cramer, R. (ed.) TCC 2012. LNCS, vol. 7194, pp. 530–547. Springer, Heidelberg (2012)
12. Chow, S., Dodis, Y., Rouselakis, Y., Waters, B.: Practical leakage-resilient identity-based encryption from simple assumptions. In: ACM CCS, pp. 152–161. ACM (2010)
13. De Santis, A., Yung, M.: Cryptographic applications of the non-interactive metaproof and many-prover systems. In: Menezes, A. J., Vanstone, S.A. (eds.) Advances in Cryptology - CRYPTO 1990. LNCS, vol. 537, pp. 366–377. Springer, Heidelberg (1991)

14. Dodis, Y., Haralambiev, K., Lopez-Alt, A., Wichs, D.: Cryptography against continuous memory attacks. In: FOCS, pp. 511–520. IEEE Computer Society (2010)
15. Dodis, Y., Pietrzak, K.: Leakage-resilient pseudorandom functions and side-channel attacks on feistel networks. In: Rabin, T. (ed.) CRYPTO 2010. LNCS, vol. 6223, pp. 21–40. Springer, Heidelberg (2010)
16. Dziembowski, S., Pietrzak, K.: Leakage-resilient cryptography. In: FOCS, pp. 293–302. IEEE (2008)
17. Faust, S., Kiltz, E., Pietrzak, K., Rothblum, G.N.: Leakage-resilient signatures. In: Micciancio, D. (ed.) TCC 2010. LNCS, vol. 5978, pp. 343–360. Springer, Heidelberg (2010)
18. Feige, U., Fiat, A., Shamir, A.: Zero-knowledge proofs of identity. J. Cryptology 1(2), 77–94 (1988)
19. Garg, S., Gentry, C., Sahai, A., Waters, B.: Witness encryption and its applications. In: STOC, pp. 467–476. ACM (2013)
20. Garg, S., Jain, A., Sahai, A.: Leakage-resilient zero knowledge. In: Rogaway, P. (ed.) CRYPTO 2011. LNCS, vol. 6841, pp. 297–315. Springer, Heidelberg (2011)
21. Gentry, C., Halevi, S., Vaikuntanathan, V.: i-hop homomorphic encryption and rerandomizable yao circuits. In: Rabin, T. (ed.) CRYPTO 2010. LNCS, vol. 6223, pp. 155–172. Springer, Heidelberg (2010)
22. Goldwasser, S., Rothblum, G.N.: Securing computation against continuous leakage. In: Rabin, T. (ed.) CRYPTO 2010. LNCS, vol. 6223, pp. 59–79. Springer, Heidelberg (2010)
23. Groth, J., Sahai, A.: Efficient non-interactive proof systems for bilinear groups. In: Smart, N.P. (ed.) EUROCRYPT 2008. LNCS, vol. 4965, pp. 415–432. Springer, Heidelberg (2008)
24. Guillou, L.C., Quisquater, J.-J.: A paradoxical indentity-based signature scheme resulting from zero-knowledge. In: Goldwasser, S. (ed.) Advances in Cryptology - CRYPTO 1988. LNCS, vol. 403, pp. 216–231. Springer, Heidelberg (1990)
25. Halderman, J.A., Schoen, S.D., Heninger, N., Clarkson, W., Paul, W., Calandrino, J.A., Feldman, A.J., Appelbaum, J., Felten, E.W.: Lest we remember: cold-boot attacks on encryption keys. Communications of the ACM 52(5), 91–98 (2009)
26. Ishai, Y., Sahai, A., Wagner, D.: Private circuits: Securing hardware against probing attacks. In: Boneh, D. (ed.) CRYPTO 2003. LNCS, vol. 2729, pp. 463–481. Springer, Heidelberg (2003)
27. Jakobsson, M.: Blackmailing using undeniable signatures. In: De Santis, A. (ed.) EUROCRYPT 1994. LNCS, vol. 950, pp. 425–427. Springer, Heidelberg (1995)
28. Jakobsson, M., Sako, K., Impagliazzo, R.: Designated verifier proofs and their applications. In: Maurer, U.M. (ed.) EUROCRYPT 1996. LNCS, vol. 1070, pp. 143–154. Springer, Heidelberg (1996)
29. Juma, A., Vahlis, Y.: Protecting cryptographic keys against continual leakage. In: Rabin, T. (ed.) CRYPTO 2010. LNCS, vol. 6223, pp. 41–58. Springer, Heidelberg (2010)
30. Katz, J., Vaikuntanathan, V.: Signature schemes with bounded leakage resilience. In: Matsui, M. (ed.) ASIACRYPT 2009. LNCS, vol. 5912, pp. 703–720. Springer, Heidelberg (2009)
31. Kocher, P.C.: Timing attacks on implementations of diffie-hellman, rsa, dss, and other systems. In: Koblitz, N. (ed.) CRYPTO 1996. LNCS, vol. 1109, pp. 104–113. Springer, Heidelberg (1996)
32. Kocher, P., Jaffe, J., Jun, B.: Differential power analysis. In: Wiener, M. (ed.) CRYPTO 1999. LNCS, vol. 1666, pp. 388–397. Springer, Heidelberg (1999)
33. Lewko, A., Lewko, M., Waters, B.: How to leak on key updates. In: Proceedings of the 43rd Annual ACM Symposium on Theory of Computing, pp. 725–734. ACM (2011)
34. Lewko, A., Rouselakis, Y., Waters, B.: Achieving leakage resilience through dual system encryption. In: Ishai, Y. (ed.) TCC 2011. LNCS, vol. 6597, pp. 70–88. Springer, Heidelberg (2011)

35. Naor, M., Segev, G.: Public-key cryptosystems resilient to key leakage. In: Halevi, S. (ed.) CRYPTO 2009. LNCS, vol. 5677, pp. 18–35. Springer, Heidelberg (2009)
36. Ostrovsky, R., Persiano, G., Visconti, I.: Constant-round concurrent non-malleable zero knowledge in the bare public-key model. In: Aceto, L., Damgård, I., Goldberg, L.A., Halldórsson, M.M., Ingólfsdóttir, A., Walukiewicz, I. (eds.) ICALP 2008, Part II. LNCS, vol. 5126, pp. 548–559. Springer, Heidelberg (2008)
37. Pandey, O.: Achieving constant round leakage-resilient zero-knowledge. In: Lindell, Y. (ed.) TCC 2014. LNCS, vol. 8349, pp. 146–166. Springer, Heidelberg (2014)
38. Pietrzak, K.: A leakage-resilient mode of operation. In: Joux, A. (ed.) EUROCRYPT 2009. LNCS, vol. 5479, pp. 462–482. Springer, Heidelberg (2009)
39. Quisquater, J.-J., Samyde, D.: Electromagnetic analysis (ema): Measures and counter-measures for smart cards. In: Attali, S., Jensen, T. (eds.) E-smart 2001. LNCS, vol. 2140, pp. 200–210. Springer, Heidelberg (2001)
40. Sahai, A.: Non-malleable non-interactive zero knowledge and adaptive chosen-ciphertext security. In: FOCS, pp. 543–553 (1999)
41. De Santis, A., Di Crescenzo, G., Ostrovsky, R., Persiano, G., Sahai, A.: Robust non-interactive zero knowledge. In: Kilian, J. (ed.) CRYPTO 2001. LNCS, vol. 2139, pp. 566–598. Springer, Heidelberg (2001)
42. Schnorr, C.-P.: Efficient signature generation by smart cards. Journal of Cryptology 4(3), 161–174 (1991)

Amplifying Privacy in Privacy Amplification

Divesh Aggarwal[1], Yevgeniy Dodis[1,*], Zahra Jafargholi[2,**], Eric Miles[2],
and Leonid Reyzin[3,***]

[1] New York University, New York, NY, USA
[2] Northeastern University, Boston, MA, USA
[3] Boston University, MA, USA

Abstract. We study the classical problem of privacy amplification,
where two parties Alice and Bob share a weak secret X of min-entropy
k, and wish to agree on secret key R of length m over a public communication channel completely controlled by a computationally unbounded
attacker Eve.

Despite being extensively studied in the literature, the problem of designing "optimal" efficient privacy amplification protocols is still open,
because there are several optimization goals. The first of them is (1)
minimizing the *entropy loss* $L = k - m$. Other important considerations include (2) minimizing the number of communication rounds, (3)
maintaining security even after the secret key is used (this is called *post-
application robustness*), and (4) ensuring that the protocol P does not
leak some "useful information" about the source X (this is called *source
privacy*). Additionally, when dealing with a very long source X, as happens in the so-called Bounded Retrieval Model (BRM), extracting as
long a key as possible is no longer the goal. Instead, the goals are (5)
to touch as little of X as possible (for efficiency), and (6) to be able to
run the protocol many times on the same X, extracting multiple secure
keys.

Achieving goals (1)-(4) (or (2)-(6) in BRM) simultaneously has remained open. In this work we improve upon the current state-of-the-art,
by designing a variety of new privacy amplification protocols, thereby
achieving the following goals for the first time:

- 4-round (resp. 2-round) *source-private* protocol with *optimal entropy
  loss* $L = O(\lambda)$, whenever $k = \Omega(\lambda^2)$ (resp. $k > \frac{n}{2}(1 - \alpha)$ for some
  universal constant $\alpha > 0$).
- 3-round *post-application-robust* protocols with *optimal entropy loss*
  $L = O(\lambda)$, whenever $k = \Omega(\lambda^2)$ or $k > \frac{n}{2}(1 - \alpha)$ (the latter is also
  source-private).
- The first BRM protocol capable of extracting the optimal number
  $\Theta(k/\lambda)$ of session keys, improving upon the previously best bound
  $\Theta(k/\lambda^2)$. (Additionally, our BRM protocol is post-application-
  robust, takes 2 rounds, and can be made source-private by increasing
  the number of rounds to 4.)

* Supported by NSF CNS Grants 1314568, 1319051, 1065288, 1017471, and Faculty
  Awards from Google and VMware.
** Supported by NSF grants CCF-0845003 and CCF-1319206.
*** Supported by NSF grants 1012798 and 1012910.

1   Introduction

We study the classical problem of *privacy amplification* [3,22,2,23] (PA), in which
two parties, Alice and Bob, share a weak secret X (of length n bits and min-
entropy $k < n$) and wish to agree on a close-to-uniform secret key R of length m
bits. We consider the active-adversary case, in which the communication channel
between Alice and Bob can be not only observed, but also fully controlled, by a
computationally unbounded attacker Eve. The most natural quantity to optimize
here is the *entropy loss* $L = k - m$ (for a given security level $\varepsilon = 2^{-\lambda}$), but several
other parameters (described below) are important as well.

Aside from being clean and elegant, this problem arises in a number of applications, such as biometric authentication, leakage-resilient cryptography, and
quantum cryptography. Additionally, the mathematical tools used to solve this
problem (such as randomness extractors [24]) have found many other applications in other areas of cryptography and complexity theory. Not surprisingly, PA
has been extensively studied in the literature, as we survey below.

In the easier "passive adversary" setting (in which Eve can observe, but not
modify), PA can be solved by applying a (strong) *randomness extractor* [24],
which uses a uniformly random nonsecret seed S to extract nearly uniform secret randomness from the weak secret X. A randomness extractor accomplishes
passive-adversary PA in one message: Alice sends the seed S to Bob, and both
parties compute the extracted key $R = \mathsf{Ext}(X; S)$. Moreover, it is known that
the optimal entropy loss of randomness extractors is $L = \Theta(\log(1/\varepsilon))$ [25], and
this bound can be easily achieved (e.g. using the Leftover Hash Lemma [16]).

ACTIVE EVE SETTING: NUMBER OF ROUNDS VS. ENTROPY LOSS. The situation is more complex in the "active Eve" setting. Existing one-message solutions [23,9] work for min-entropy $k > n/2$ and require large entropy loss
$L > n - k$. It was shown by [13,14] that $k > n/2$ is necessary, and that the
large entropy loss of $n - k$ is likely necessary, as well. Thus, we turn to protocols
of two or more rounds.

Two rounds were shown to be sufficient by [14], who proved, nonconstruc-
tively, the existence of two-round PA protocols with optimal entropy loss
$L = \Theta(\log(1/\varepsilon))$ for any k. (This was done using a strengthening of extractors,
called *non-malleable extractors*, whose existence was shown in [14].) Construc-
tively, no such protocols are known, and all known constructive results sacrifice
either the number of rounds, or the entropy loss, or the minimum entropy requirement. A protocol of [19, Theorem 1.9] (building on [27,17,6]) sacrifices the
number of rounds: it achieves $L = O(\log(1/\varepsilon))$, but only in $O(1 + \log(1/\varepsilon)/\sqrt{k})$
rounds. The protocol of [19, Theorem 1.6] (building on [14]) sacrifices the mini-
mum entropy requirement: it achieves $L = O(\log(1/\varepsilon))$ in two rounds, but only
when $k = \Omega(\log^2(1/\varepsilon))$. Protocols of [10,7,18,20] make an incomparable minimum entropy requirement: they also achieve $L = O(\log(1/\varepsilon))$ in two rounds,
but require that $k > n/2$ (with the exception of [20], who slightly relaxed it
to $k > \frac{n}{2}(1 - \alpha)$ for some tiny but positive constant α). These protocols also
built the first constructive non-malleable extractors when $k > n/2$. The result of

[19, Theorem 1.8] (building on [10,18]) further relaxes the entropy requirement to $k > \delta n$ for any constant $\delta > 0$. It also achieves $L = O(\log{(1/\varepsilon)})$ in two rounds, but the constant hidden in the O-notation is $g(\delta) = 2^{(1/\delta)^c}$ for some astronomical (and not even exactly known) constant c.[1] More generally, since some of the protocols mentioned above hide relatively large (or, as in the last example, even astronomical) constant factors, simpler protocols (such as [14] or [17]) may outperform asymptotically optimal ones for many realistic settings of parameters.

To summarize, the landscape of existing PA protocols is rather complex, even if we consider only the tradeoff between the min-entropy, the entropy loss, and the number of rounds. The situation becomes even more complex, if one adds additional highly desirable properties: *source privacy, post-application robustness*, and *local computability*. We consider those next.

SOURCE PRIVACY. Intuitively, this property demands that the transcript of the protocol (even together with the derived key R!) does not reveal any "useful information" about the source X; or, equivalently (as shown by [12]), that the transcript does not reveal any information at all about the *distribution* of X (beyond a lower bound k on its min-entropy). For the case of passive Eve, source privacy was considered by Dodis and Smith [12], who showed that randomness extractors are indeed source-private. For active Eve, the only work that considered this notion is the elegant paper [4], which constructed a 4-round private protocol with entropy loss $L = O(\log^2(1/\varepsilon))$. Thus, unlike for PA protocols without source privacy,

(A) *no source-private PA protocol is known which achieves either optimal entropy loss $L = O(\log{(1/\varepsilon)})$, or fewer than four rounds.*

POST-APPLICATION ROBUSTNESS. Informally, the basic authenticity notion of PA protocols, called *pre-application robustness* by [9], simply states that Eve cannot force Alice and Bob to agree on different keys $R_A \neq R_B$. While easy to define, this property is likely insufficient for most applications of PA protocols, because in any two-party protocol, one party (say, Bob) has to finish before the other party. In this case, Bob is not sure if Alice ever received his last message, and must somehow decide to use his derived key R_B. In doing so, he might leak some partial information about R_B (possibly all of it!), and Eve might now use this partial (or full) information to modify the last message that Bob originally sent to Alice. Motivated by these considerations, [9] defined a strong property called *post-application robustness*, which (intuitively) requires that Eve cannot modify Bob's last message and cause Alice to output $R_A \neq R_B$, even if given Bob's key R_B.

The only protocols known to achieve post-application robustness are in [9,14,10]. Of those, only the protocol of [10] achieves asymptotically optimally entropy loss: for entropy $k > \delta n$, it achieves entropy loss $O((1/\delta)^c \log{(1/\varepsilon)})$ in

---

[1] The value c depends on some existential results in additive combinatorics. However, it appears safe to conclude that it is astronomical, which translates into "triply astronomical" $g(\delta) = 2^{(1/\delta)^c}$, even for $\delta = 0.49$.

$O((1/\delta)^c)$ rounds for some astronomical constant c mentioned in Footnote 1. Most protocols in [27,9,14,6,10,7,18,20,19] are proven only for pre-application robustness (some works simply ignored the distinction). In particular,

(B) *no post-application robust, constant-round protocol with optimal entropy loss is known (with the exception of protocol of [10] using astronomical constants mentioned above).*

LOCAL COMPUTABILITY AND REUSABILITY. Local computability is of interest when the length and the min-entropy of the source X is much larger than the desired number of extracted bits m. In such a case, it is desirable to compute the output without having to read all of the source. This property is traditionally associated with the Bounded Retrieval Model (BRM) [15,8], where the random source X is made *intentionally huge*, so that X still has a lot of entropy k even after the attacker ("virus") managed to download a big fraction of X over time. For historical reasons, we will also use the term "BRM", but point out that local computability seems natural in any scenario where $k \gg m$, and not just the BRM application.

The right way to think about entropy loss in such a scenario is not via the formula $L = k - m$, because entropy from X is not "lost": much entropy remains in X even after the protocol execution, because most of X is not even accessed. In fact, the PA protocol may be run multiple times on the same X, to obtain multiple keys, until the entropy of X is exhausted. Specifically focusing on $m = \Theta(\log(1/\varepsilon))$ (so that the extracted key can be used to achieve ε security), "optimal" reusability means the ability to extract $\Theta(k/\log(1/\varepsilon))$ keys (assuming the entropy rate of X is constant).

In the passive adversary case, optimal reusability is achievable with locally computable randomness extractors [21,28]. In the active adversary case, however, the story is again more complicated. The only prior work to consider local computability in this setting is the work of [14]. Reusability has not been explicitly considered before, but it is easy to see that the locally computable protocol of [14] allows the extraction of $\Theta(k/\log^2(1/\varepsilon))$ keys. Thus,

(C) *no prior locally computable protocol achieves optimal reusability.*

1.1 Our Results

In this work, we solve open problems (A), (B), and (C), by designing several new techniques for building PA protocols. Many of our techniques are *general transformations* that convert a given protocol P into a "better" protocol P'. Given a wide variety of incomparable existing PA protocols (surveyed above), this modular approach will often allow us to obtain several improved protocols in "one go".

TWO METHODS OF ADDING SOURCE PRIVACY. Our first method (Section 3.2) maintains the number of rounds at 2, at the expense of using a strengthening of non-malleable extractors [14] (which we call *adaptive non-malleable extractors*) to derive a one-time pad to mask the "non-private" message which should be sent in the second round. (Given that we already use non-malleable extractors however, we might as well combine our protocol with the non-private protocol of [14] based on non-malleable extractors with similar parameters; this is what we do to keep things simple.) Our second method (Section 3.3), inspired by the specific protocol of [4], turns certain 2-round non-private protocols into 4-round private protocols, using standard extractors and XOR-universal hash functions. (The concrete protocol of [4] implicitly applied a very particular variant of our transformation to the two-round protocol of [14], but we get improved results using "newer" protocol [19].) In particular, either one of these transformations will provide (with different tradeoffs) a positive answer to Open Question (A). For completeness, we also observe (Section 3.1) that the 1-round PA protocols of [9] are already source-private.

PRE- TO POST-APPLICATION ROBUSTNESS. We make a very simple transformation which converts pre-application robust protocols to post-application robust protocols, at the cost of one extra round, but with almost no increase in the entropy loss. Although very simple, it immediately gives a variety of answers to Open Question (B) (and can also be combined with our first transformation, since it preserves source privacy).

Overall, by applying our transformations above to different protocols and in various orders, we get several improvements to existing protocols, summarized in Table 1 (which includes various solutions to Questions (A), (B), and more).

Table 1. Our improvement (also marked in RED) over prior PA protocols

Result	Entropy	Rounds	Entropy Loss		Source
			Pre-app	Post-app	Privacy
[14] (non-expl.)	$k = \Omega(\log(1/\varepsilon))$	2	$\Theta(\log(1/\varepsilon))$	$\Theta(\log(1/\varepsilon))$	NO
This work (non-expl.)	$k = \Omega(\log(1/\varepsilon))$	2	$\Theta(\log(1/\varepsilon))$	$\Theta(\log(1/\varepsilon))$	YES
[9]	$k > \frac{n}{2}$	1	$n - k - \Theta(\log(1/\varepsilon))$	$\frac{n}{2} + \Theta(\log(1/\varepsilon))$	YES[2]
[19]	$k = \Omega(\log^2(1/\varepsilon))$	2	$\Theta(\log(1/\varepsilon))$	$\Theta(\log^2(1/\varepsilon))$	NO
This work	$k = \Omega(\log^2(1/\varepsilon))$	3	$\Theta(\log(1/\varepsilon))$	$\Theta(\log(1/\varepsilon))$	NO
[4]	$k = \Omega(\log^2(1/\varepsilon))$	4	$\Theta(\log^2(1/\varepsilon))$	$\Theta(\log^2(1/\varepsilon))$	YES
This work	$k = \Omega(\log^2(1/\varepsilon))$	4	$\Theta(\log(1/\varepsilon))$	$\Theta(\log^2(1/\varepsilon))$	YES
This work	$k = \Omega(\log^2(1/\varepsilon))$	5	$\Theta(\log(1/\varepsilon))$	$\Theta(\log(1/\varepsilon))$	YES
[20]	$k > \frac{n}{2}(1-\alpha)$	2	$\Theta(\log(1/\varepsilon))$	$\frac{n}{2}(1-\alpha) + \Theta(\log(1/\varepsilon))$	NO
This work	$k > \frac{n}{2}(1-\alpha)$	2	$\Theta(\log(1/\varepsilon))$	$\frac{n}{2}(1-\alpha) + \Theta(\log(1/\varepsilon))$	YES
This work	$k > \frac{n}{2}(1-\alpha)$	3	$\Theta(\log(1/\varepsilon))$	$\Theta(\log(1/\varepsilon))$	YES

ACHIEVING LOCAL COMPUTABILITY AND OPTIMAL REUSABILITY. While only the work of [14] explicitly considered local computability, it is reasonable to ask if other existing protocols can be modified to be locally computable and reusable. To achieve optimal reusability, we focus on protocols with optimal entropy loss, because they have the property that the protocol transcript reduces the entropy of X by $O(\log(1/\varepsilon))$, leaving *residual* entropy of X high. They can be modified to extract a short key of length $\Theta(\log(1/\varepsilon))$, which will give optimal reusability.

To achieve local computability, extractors used within a protocol can be replaced with locally computable extractors. Indeed, the protocol of [6] seems

---

[2] We observe in this paper that this protocol is private.

to amenable to such modification. However, it is not constant-round. Most other constant-round protocols with optimal entropy loss [10,7,18,20] use non-malleable extractors, and this approach fails, because no locally computable (even non-constructive!) instantiations of non-malleable extractors are known.

However, we observe that the 2-round, optimal entropy loss protocol of [19, Theorem 1.6] does not use non-malleable extractors. Moreover, by making all extractors in that protocol locally computable, we get a locally computable, 2-round protocol. However, the security analysis of [19] uses a very delicate and interdependent setting of various parameters for the security proof to go through. Hence, it is not immediately clear if this intricate proof will go though if one uses locally computable extractors. Instead, we will develop a different, *modular* analysis underlying the key ideas of [19], which will give us a rigorous 2-round solution to open problem (C), as well as have other benefits we describe shortly. Specifically, we show a general transformation that turns certain (post-application) secure 2-round protocols into 2-round protocols *with optimal entropy loss $L = O(\log(1/\varepsilon))$ and residual min-entropy $k' = k - O(\log(1/\varepsilon))$* (Section 5). The transformation uses two-source extractor of [26] to compress the second message of the protocol to only $O(\log(1/\varepsilon))$ bits. By applying this transformation to the original (non-BRM) protocol of [14], we get a protocol very similar to the protocol of [19], but with a much more modular and easier-to-follow security analysis. On the other hand, by using the locally computable protocol of [14] instead (see Section 6), we get a 2-round locally computable protocol with optimal residual entropy (and, thus, reusability), solving open problem (C).[3] Furthermore, we can add source privacy by using our 2-to-4-round transformation mentioned earlier, which can be done via local computation as well.

These results are summarized in Table 2.

IMPROVING ENTROPY LOSS OF POST-APPLICATION ROBUST PROTOCOLS. As another advantage of our modular approach, we note that the transformation described in the previous paragraph is interesting not only in the context of local computability. It also allows one to turn *post-application* robust 2-round protocols with *sub-optimal* entropy loss L into 2-round *pre-application* robust protocols with *optimal* entropy loss, which then (using our pre-application to post-application transformation described above) can be turned into 3-round *post-application* robust protocols with *optimal* entropy loss. Namely, we can obtain optimal entropy loss at the expense of one extra round. (For the BRM setting, no extra round is needed, as we only extract "short" keys of length $O(\log(1/\varepsilon))$.)

---

[3] Interestingly, the main limitation of the non-BRM protocol of [19] — high min-entropy requirement $k = \Omega((\log(1/\varepsilon))^2)$ — is not an issue in the BRM model. Thus, we can view our result as finding a "practical application scenario" for the very elegant communication reduction technique developed by [19].

Table 2. Protocols in the Bounded Retrieval Model; each extracts $\Theta(\log(1/\varepsilon))$ bits per key, is post-application robust, and requires $k = \Omega(\log^2(1/\varepsilon))$. Entries in **RED** mark our improvements.

Result	Rounds	Residual Min-entropy	# Keys Extracted	Source Privacy
[14]	2	$k - \Theta(\log^2(1/\varepsilon))$	$\Theta(k/\log^2(1/\varepsilon))$	NO
This work	2	$k - \Theta(\log(1/\varepsilon))$	$\Theta(k/\log(1/\varepsilon))$	NO
This work	4	$k - \Theta(\log(1/\varepsilon))$	$\Theta(k/\log(1/\varepsilon))$	YES

2 Preliminaries

For a set S, we let U_S denote the uniform distribution over S. For an integer $m \in \mathbb{N}$, we let U_m denote the uniform distribution over $\{0,1\}^m$, the bit-strings of length m. For a distribution or random variable X we write $x \leftarrow X$ to denote the operation of sampling a random x according to X. For a set S, we write $s \leftarrow S$ as shorthand for $s \leftarrow U_S$.

ENTROPY AND STATISTICAL DISTANCE. The *min-entropy* of a random variable X is defined as $\mathbf{H}_\infty(X) \stackrel{\text{def}}{=} -\log(\max_x \Pr[X = x])$. We say that X is an (n,k)-*source* if $X \in \{0,1\}^n$ and $\mathbf{H}_\infty(X) \geqslant k$. For $X \in \{0,1\}^n$, we define the *entropy rate* of X to be $\mathbf{H}_\infty(X)/n$. We also define *average (aka conditional) min-entropy* of a random variable X conditioned on another random variable Z as

$$\mathbf{H}_\infty(X|Z) \stackrel{\text{def}}{=} -\log\left(\mathbb{E}_{z \leftarrow Z}\left[\ \max_x \Pr[X = x|Z = z]\ \right]\right)$$
$$= -\log\left(\mathbb{E}_{z \leftarrow Z}\left[2^{-\mathbf{H}_\infty(X|Z=z)}\right]\right),$$

where $\mathbb{E}_{z \leftarrow Z}$ denotes the expected value over $z \leftarrow Z$.

The *statistical distance* between two random variables W and Z distributed over some set S is

$$\Delta(W, Z) \stackrel{\text{def}}{=} \max_{T \subseteq S}(|W(T) - Z(T)|) = \frac{1}{2}\sum_{s \in S}|W(s) - Z(s)|.$$

Note that $\Delta(W, Z) = \max_D(\Pr[D(W) = 1] - \Pr[D(Z) = 1])$, where D is a probabilistic function. We say W is ε-close to Z, denoted $W \approx_\varepsilon Z$, if $\Delta(W, Z) \leq \varepsilon$. We write $\Delta(W, Z|Y)$ as shorthand for $\Delta((W, Y), (Z, Y))$.

We introduce some cryptographic primitives needed for our constructions.

EXTRACTORS. An extractor [24] can be used to extract uniform randomness out of a weakly-random value which is only assumed to have sufficient min-entropy. Our definition follows that of [11], which is defined in terms of conditional min-entropy.

Definition 1 (Extractors). *An efficient function* $\mathsf{Ext} : \{0,1\}^n \times \{0,1\}^d \to \{0,1\}^m$ *is an (average-case, strong)* (k, ε)-*extractor, if for all* X, Z *such that* X *is distributed over* $\{0,1\}^n$ *and* $\mathbf{H}_\infty(X|Z) \geq k$, *we get*

$$\Delta(\ (Z, Y, \mathsf{Ext}(X; Y))\ ,\ (Z, Y, U_m)\ ) \leqslant \varepsilon$$

where $Y \equiv U_d$ denotes the coins of Ext *(called the* seed*). The value $L = k - m$ is called the* entropy loss *of* Ext*, and the value d is called the* seed length *of* Ext*.*

MESSAGE AUTHENTICATION CODES. One-time message authentication codes (MACs) use a shared random key to authenticate a message in the information-theoretic setting.

Definition 2 (One-time MACs). *A function family* $\{\mathsf{MAC}_R : \{0,1\}^d \to \{0,1\}^v\}$ *is an ε-secure one-time MAC for messages of length d with tags of length v if for any $w \in \{0,1\}^d$ and any function (adversary) $A : \{0,1\}^v \to \{0,1\}^d \times \{0,1\}^v$,*

$$\Pr_R[\mathsf{MAC}_R(W') = T' \wedge W' \neq w \mid (W', T') = A(\mathsf{MAC}_R(w))] \leq \epsilon,$$

where R is the uniform distribution over the key space $\{0,1\}^\ell$.

XOR-UNIVERSAL HASH FUNCTIONS. We recall the definition of XOR-universal-hashing [5].

Definition 3 (ρ-XOR-Universal Hashing). *A family $\mathcal{H}$ of (deterministic) functions $h : \{0,1\}^u \to \{0,1\}^v$ is a called ρ-XOR-universal hash family, if for any $x_1 \neq x_2 \in \{0,1\}^u$ and any $a \in \{0,1\}^v$ we have $\Pr_{h \leftarrow \mathcal{H}}[h(x_1) \oplus h(x_2) = a] \leq \rho$. When $\rho = 1/2^v$, we say that $\mathcal{H}$ is (perfectly) XOR-universal. The value $\log |\mathcal{H}|$ is called the seed length of $\mathcal{H}$.*

2.1 Privacy Amplification

We define a privacy amplification protocol (P_A, P_B), executed by two parties Alice and Bob sharing a secret $X \in \{0,1\}^n$, in the presence of an active, computationally unbounded adversary Eve, who might have some partial information E about X satisfying $\mathbf{H}_\infty(X|E) \geqslant k$. Informally, this means that whenever a party (Alice or Bob) does not reject, the key R output by this party is random and statistically independent of Eve's view. Moreover, if both parties do not reject, they must output the same keys $R_A = R_B$ with overwhelming probability. The formal definition is given below.

Definition 4. *An interactive protocol (P_A, P_B), executed by Alice and Bob on a communication channel fully controlled by an active adversary Eve, is a (k, m, ϵ)-privacy amplification protocol if it satisfies the following properties whenever $\mathbf{H}_\infty(X|E) \geq k$:*

1. CORRECTNESS. *If Eve is passive, then $\Pr[R_A = R_B \wedge R_A \neq \perp \wedge R_B \neq \perp] = 1$.*

2. ROBUSTNESS. *We start by defining the notion of pre-application robustness, which states that even if Eve is active, $\Pr[R_A \neq R_B \wedge R_A \neq \perp \wedge R_B \neq \perp] \leqslant \epsilon$. The stronger notion of post-application robustness is defined similarly, except Eve is additionally given the key R_A the moment she completed the left execution (P_A, P_E), and the key R_B the moment she completed the right execution*

(P_E, P_B). For example, if Eve completed the left execution before the right execution, she may try to use R_A to force Bob to output a different key $R_B \notin \{R_A, \bot\}$, and vice versa.

3. EXTRACTION. Given a string $r \in \{0,1\}^m \cup \{\bot\}$, let $\mathsf{purify}(r)$ be $\bot$ if $r = \bot$, and otherwise replace $r \neq \bot$ by a fresh m-bit random string U_m: $\mathsf{purify}(r) \leftarrow U_m$. Letting E' denote Eve's view of the protocol, we require that

$$\Delta(R_A, \mathsf{purify}(R_A) \mid E') \leq \epsilon \quad \text{and} \quad \Delta(R_B, \mathsf{purify}(R_B) \mid E') \leq \epsilon$$

Namely, whenever a party does not reject, its key looks like a fresh random string to Eve.

The quantity $k - m$ is called the entropy loss and the quantity $\log(1/\epsilon)$ is called the security parameter of the protocol.

SOURCE PRIVACY. Following Bouman and Fehr [4], we now add the source privacy requirement for privacy amplification protocols. To define this property, we let $\mathsf{FullOutput}(X, E)$ denote the tuple (E', R_A, R_B), where Alice and Bob share a secret X and output keys R_A and R_B, respectively, and Eve starts with initial side information E and ends with final view E' at the end of the protocol.

Definition 5 (Source Privacy). *An interactive protocol (P_A, P_B), executed by Alice and Bob on a communication channel fully controlled by an active adversary Eve, is (k, ε)-private, if for any two distributions (X_0, E) and (X_1, E), where $\mathbf{H}_\infty(X_0|E) \geq k$ and $\mathbf{H}_\infty(X_1|E) \geqslant k$, we have*

$$\Delta(\mathsf{FullOutput}(X_0, E), \mathsf{FullOutput}(X_1, E)) \leq \varepsilon$$

Our definition is stronger than the definition of [4], who only required that the final transcript E' does not reveal any information about X.

3   New Private Protocols

3.1   One Round Private Protocol

Dodis et al [9] gave a construction of robust extractors using which they gave one-round (k, m, ε)-secure privacy amplification protocols for $k > n/2 + O(\log(1/\varepsilon))$. We argue the source privacy of their protocols in the full version[1], and thus get the following result.

Theorem 1. *For $k > n/2$, there is an explicit polynomial-time, one-round $(k, 2\varepsilon + 2^{-n/2})$-private, (k, m, ε)-secure privacy amplification protocol with pre-application robustness and entropy loss $k - m = n - k + O(\log(1/\varepsilon))$. We get post-application robustness at the cost of increasing the entropy loss to $n/2 + O(\log(1/\varepsilon))$.*

3.2   Two Round Private Protocol with Optimal Entropy Loss

In this section, we give a two round protocol that achieves optimal entropy loss $O(\log (1/\varepsilon))$ for pre-application robustness. For post-application robustness, the entropy loss is about $n/2$, but we show how to improve it to $O(\log (1/\varepsilon))$ in Section 4 at the cost of 1 additional round.

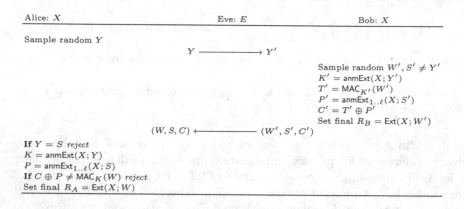

Protocol 1. New 2-round Source-Private Protocol for $\mathbf{H}_\infty(X|E) > n/2$

Our Two Round Private Protocol. Our protocol (Protocol 1) makes the protocol of [14] private, using the same idea as [4]: we apply a one-time pad P' to the tag sent by Bob in the second round, T', where the pad P' is derived from X. We make use of an adaptive non-malleable extractor, where the adversary $\mathcal{A}$ is allowed to see Y, Z, and additionally either $\mathsf{anmExt}(X; Y)$ or $R \equiv U_m$ before producing the modified seed Y', and still $\mathsf{anmExt}(X; Y)$ should be statistically close to R given $\mathsf{anmExt}(X; Y'), Y, Z$.

Using this, our protocol achieves the following result.

Theorem 2. *Let $2^{-n/4} < \varepsilon < 1/n$, and $\varepsilon' = \varepsilon/7$. Given a (τ, ε')-adaptive non-malleable extractor, for $k > \tau + \Theta(\log (1/\varepsilon))$ and output length $\Theta(\log (1/\varepsilon))$, there exists an explicit polynomial-time, two-round (k, ε)-private, (k, m, ε)-secure privacy amplification protocol with pre-application robustness and entropy loss $O(\log (1/\varepsilon))$. Furthermore, we get post-application robustness with entropy loss to $\tau + O(\log (1/\varepsilon))$.*

We can instantiate the above result using our construction (resp. existential proof) of adaptive non-malleable extractors to obtain the following results. The details can be found in the full version [1].

Corollary 1. *There exists a universal constant $\alpha > 0$, such that for $k > n/2(1 - \alpha)$, there exists an explicit polynomial-time, two-round (k, ε)-private, (k, m, ε)-secure privacy amplification protocol with pre-application robustness and entropy loss $O(\log (1/\varepsilon))$. We get post-application robustness at the cost of increasing the entropy loss to $n/2(1 - \alpha) + O(\log (1/\varepsilon))$.*

Corollary 2. *For $k = \Omega(\log(1/\varepsilon))$, there exists a two-round (k, ε)-private, (k, m, ε)-secure privacy amplification protocol with post-application robustness and entropy loss $k - m = O(\log(1/\varepsilon))$.*

3.3 Privacy Using Extractors and XOR-Universal Hashing

In this section, we use a ρ-XOR universal hash function family to construct a 4-round protocol for private privacy amplification, given any 2 round privacy amplification protocol of the form Protocol 2, where the string sent in the first round is sampled independent of X. We note that all known 2 round protocols in the literature are of this generic form.

Alice: X Eve: E Bob: X

Sample random Y Sample random W'

$\qquad\qquad\qquad Y \xrightarrow{\hspace{2cm}} Y'$

$\qquad\qquad\qquad\qquad\qquad\qquad\qquad\qquad K' = f_1(X, Y')$
$\qquad\qquad\qquad\qquad\qquad\qquad\qquad\qquad T' = f_2(K', W')$
$\qquad\qquad\qquad\qquad\qquad\qquad\qquad\qquad \text{Set final } R_B = g(X, W')$

$\qquad\qquad\qquad (W, T) \xleftarrow{\hspace{2cm}} (W', T')$

$K = f_1(X, Y)$
If $T \neq f_2(K, W)$ *reject*
Set final $R_A = g(X, W)$

Protocol 2. A Generic 2-round Privacy Amplification Protocol

Let $\ell = \log(1/\varepsilon)$. Let $\mathcal{H}$ be a ε-XOR universal family of hash functions from $\{0,1\}^{|T|}$ to $\{0,1\}^{2\ell}$, and let $\mathsf{Ext} : \{0,1\}^n \times \{0,1\}^d \mapsto \{0,1\}^{2\ell}$ be a $(k - 2\ell - 2|K| - |R_B|, \varepsilon)$ extractor. Using these, our protocol is depicted as Protocol 3.

Alice: X Eve: E Bob: X

Sample random Y, h Sample random W', S'

$\qquad\qquad\qquad Y \xrightarrow{\hspace{1.5cm}} Y'$
$\qquad\qquad\qquad W, S \xleftarrow{\hspace{1.5cm}} W', S'$
$\qquad\qquad\qquad h \xrightarrow{\hspace{1.5cm}} h'$

$\qquad\qquad\qquad\qquad\qquad\qquad\qquad\qquad K' = f_1(X, Y')$
$\qquad\qquad\qquad\qquad\qquad\qquad\qquad\qquad T' = f_2(K', W')$
$\qquad\qquad\qquad\qquad\qquad\qquad\qquad\qquad C' = h'(T') \oplus \mathsf{Ext}(X; S')$
$\qquad\qquad\qquad\qquad\qquad\qquad\qquad\qquad \text{Set final } R_B = g(X, W')$

$\qquad\qquad\qquad C \xleftarrow{\hspace{1.5cm}} C'$

$K = f_1(X, Y)$
$T = f_2(K, W)$
If $C \neq h(T) \oplus \mathsf{Ext}(X; S)$ *reject*
Set final $R_A = g(X, W)$

Protocol 3. A Generic 4-round Private Privacy Amplification Protocol

Theorem 3. *Let Protocol 2 be a 2-round $(k - u, m, \varepsilon)$-secure privacy amplification protocol with pre- (resp. post-) application robustness for $k - |T| - 2|K| - |R_B| \geqslant 2\ell$. Then Protocol 3 is a 4-round $(k, m, O(\sqrt{\varepsilon}))$-secure $(k, O(\sqrt{\varepsilon}))$-private privacy amplification protocol with pre- (resp. post-) application robustness.*

For a proof, refer to the full version. We apply this generic transformation to Li's recent 2 round (k, ε)-secure privacy amplification protocol for $k = \Omega(\log^2(1/\varepsilon))$, that achieves entropy loss $O(\log(1/\varepsilon))$ for pre-application robustness, and $O(\log^2(1/\varepsilon))$ for post-application robustness [19]. We get the following result.

Corollary 3. *For $k = \Omega(\log^2(1/\varepsilon))$, there exists an explicit polynomial-time, 4-round (k, ε)-private, (k, m, ε)-secure privacy amplification protocol with pre-application robustness and entropy loss $L = k - m = O(\log(1/\varepsilon))$. We get post-application robustness with entropy loss $O(\log^2(1/\varepsilon))$.*

In Section 4, we will see how to get a 5-round private privacy amplification protocol with post-application robustness and entropy loss $O(\log(1/\varepsilon))$.

4   From Pre-application to Post-application Robustness

In this section, we show a generic transformation from a t-round privacy amplification protocol $\mathcal{P}$ that achieves pre-application robustness to a $(t + 1)$-round protocol $\mathcal{P}'$ that achieves post-application robustness. The transformation can be described as follows.

Let $\ell = \log(1/\varepsilon)$. Without loss of generality, assume that the last message in $\mathcal{P}$ was sent from Bob to Alice. Let $\widetilde{R}_A, \widetilde{R}_B$ denote the first u bits of the keys computed by Alice and Bob, respectively (Set $\widetilde{R}_A = \bot$ if Alice rejects, and $\widetilde{R}_B = \bot$ if Bob rejects). We need a $(k - O(\ell), \varepsilon)$-extractor $\mathsf{Ext} : \{0,1\}^n \times \{0,1\}^d \to \{0,1\}^m$ and an ε-secure one-time MAC for d-bit messages, whose key length is u. Using these, the $(t + 1)$-round protocol is depicted as Protocol 4.

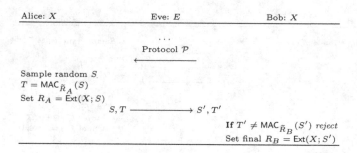

Protocol 4. $(t+1)$-round Privacy Amplification Protocol $\mathcal{P}'$ with post-application robustness.

Theorem 4. *If Protocol $\mathcal{P}$ is (k, m, ε)-secure privacy amplification protocol with pre-application robustness and residual entropy $k - O(\log(1/\varepsilon))$, then Protocol $\mathcal{P}'$ is a $(k, m - O(\log(1/\varepsilon)), O(\varepsilon))$ secure privacy amplification protocol with post-application robustness. Additionally, if $\mathcal{P}$ is (k, ε) private, then $\mathcal{P}'$ is $(k, O(\varepsilon))$ private.*

For a proof of this theorem, refer to the full version [1].

Using this result, we can get optimal entropy loss for post-application robustness for several protocols as described in the full version [1].

5 Increasing Residual Entropy

We now consider the task of preserving as much entropy as possible in the weak source X, which is a natural goal and has implications in the Bounded Retrieval Model (see section 6). Formally, the *residual entropy* of an interactive protocol is $\min_{E'} (\mathbf{H}_\infty(X \mid E'))$ where E' is the adversary's view after the protocol. We refer to $\mathbf{H}_\infty(X \mid E) - \min_{E'} (\mathbf{H}_\infty(X \mid E'))$ as the loss in residual entropy. Our main result is the following transformation achieving loss in residual entropy $O(\log(1/\varepsilon)$, i.e. linear in the security parameter, which is optimal up to constant factors.

Theorem 5. *Assume that there is a 2-round (k, m, ε)-secure privacy amplification protocol with post-application robustness in which the first message is independent of the (n, k)-source X and we have $\log n = O(\log(1/\varepsilon))$, $\varepsilon \geq 2^{-m/C}$, and $k \geq C \log(1/\varepsilon)$ for sufficiently large C.*

Then there is a 2-round (k', m', ε')-secure privacy amplification protocol with residual entropy $\geq k' - O(\log(1/\varepsilon'))$ provided that $k' \geq k + C' \log(1/\varepsilon)$ and $\varepsilon' \geq \varepsilon^{1/C'}$ for sufficiently large C', and $m' = k' - O(\log(1/\varepsilon'))$ for pre-application robustness or $m' = k' - k - O(\log(1/\varepsilon'))$ for post-application robustness.

To achieve the transformation of Theorem 5, we need the following notion of a *receipt* protocol, which is essentially a 2-round message authentication protocol in which the party who speaks first chooses the message. Such protocols can be defined as follows.

Definition 6. *A (k, ℓ, ε)-receipt protocol (for messages of length d) is a function* Receipt : $\{0, 1\}^d \times \{0, 1\}^r \times \{0, 1\}^n \to \{0, 1\}^\ell$ *that satisfies the following: for $Y \equiv U_r$, every $\mu \in \{0, 1\}^d$, every X such that $\mathbf{H}_\infty(X|E) \geq k$, and every $\mu' \neq \mu, Y'$ chosen by an adversary given μ, Y, E,*

$$\mathbf{H}_\infty(\mathsf{Receipt}(\mu, Y, X) \mid Y, \mathsf{Receipt}(\mu', Y', X)) \geq \log(1/\varepsilon).$$

The main ingredient in proving Theorem 5 is the following, the proof of which is deferred to the full version [1].

Theorem 6. *Assume that there exists a polynomial-time (k, ℓ, ε)-receipt protocol for d-bit messages such that Alice communicates $\leq \ell$ bits and $2^{-C\ell} \leq \varepsilon \leq 1/(C\ell)$ for sufficiently large C.*

Then for any $r \leq \log(1/\varepsilon)/100$, there exists a polynomial-time $(k, r, 2^{-\Omega(r)})$-receipt protocol for d-bit messages where Alice communicates $O(\ell)$ bits.

Finally, we obtain the following corollary by instantiating Theorem 5 using the 2-round privacy amplification protocol with post-application robustness due to Dodis and Wichs [14, Cor. 4].

Corollary 4. *For $k = \Omega(\log^2(1/\varepsilon))$, there exists an explicit polynomial-time 2-round (k, m, ε)-secure privacy amplification protocol with post-application robustness that achieves $m = \Omega(\log(1/\varepsilon))$ and residual entropy $k - O(\log(1/\varepsilon))$.*

6   Applications to the Bounded Retrieval Model

In the Bounded Retrieval Model (BRM) [8,15], Alice and Bob share an (intentionally) very large secret key X. The idea is that the size of X makes it infeasible for an attacker Eve to learn the entire string, even if she has infiltrated either Alice or Bob's storage device, because of limits on the amount of data that can be transmitted out of the device. Thus as in previous sections we assume that Eve has some adversarially chosen side information E about X, but that $k := \mathbf{H}_\infty(X|E)$ is not too small. Specifically here we think of $k = \alpha n$ for some constant $0 < \alpha < 1$.

Since reading the entire string X would be prohibitively inefficient, any function used by Alice or Bob that takes X as input must only read a small number of positions, i.e. it must be *locally computable*. Dodis and Wichs observe [14, Sec. 5] that their privacy amplification protocol has the property that each function taking X as input is a standard extractor. These can be replaced with the constructions of locally computable extractors due to Vadhan [28], and thus the protocol works in the BRM.

One downside of the [14] protocol is that the second message (which depends on X) has length $\Omega(\log^2(1/\varepsilon))$, and thus the loss in residual entropy is $\Omega(\log^2(1/\varepsilon)) = \Omega(m^2)$. It would be more desirable to have loss in residual entropy $O(m)$, as then Alice and Bob could derive a total of $\Omega(k/m)$ secret keys, as opposed to only $O(k/m^2)$ keys.

Corollary 4 shows that the loss in residual entropy can be reduced to $O(m)$. This protocol remains locally computable and thus applicable to the BRM, because still every function that takes X as input is a standard extractor and can be replaced by a locally computable extractor. In summary, we have the following.

Theorem 7. *For $k = \Omega(\log^2(1/\varepsilon))$, there exists an explicit polynomial-time 2-round $(k, m = \Omega(\log(1/\varepsilon)), \varepsilon)$-secure privacy amplification protocol in the BRM with post-application robustness and residual entropy $k - O(\log(1/\varepsilon))$, thus allowing a total of $\Omega(k/m)$ keys to be derived.*

By relaxing the number of rounds to four, we can obtain a BRM protocol that additionally has *source privacy* by instead plugging the [14, Cor. 4] protocol into the transformation of Theorem 3.

Theorem 8. *For $k = \Omega(\log^2(1/\varepsilon))$, there exists an explicit polynomial-time 4-round $(k, m = \Omega(\log(1/\varepsilon)), \varepsilon)$-secure (k, ε)-private privacy amplification protocol in the BRM with post-application robustness and residual entropy $k - O(\log(1/\varepsilon))$, thus allowing a total of $\Omega(k/m)$ keys to be derived.*

References

1. Aggarwal, D., Dodis, Y., Jafargholi, Z., Miles, E., Reyzin, L.: Amplifying privacy in privacy amplification. Cryptology ePrint Archive, Report 2013/723 (2014)
2. Bennett, C.H., Brassard, G., Crépeau, C., Maurer, U.M.: Generalized privacy amplification. IEEE Transactions on Information Theory 41(6), 1915–1923 (1995)
3. Bennett, C.H., Brassard, G., Robert, J.: Privacy amplification by public discussion. SIAM Journal on Computing 17(2), 210–229 (1988)
4. Bouman, N.J., Fehr, S.: Secure authentication from a weak key, without leaking information. In: Paterson, K.G. (ed.) EUROCRYPT 2011. LNCS, vol. 6632, pp. 246–265. Springer, Heidelberg (2011)
5. Carter, L., Wegman, M.N.: Universal classes of hash functions. J. Comput. Syst. Sci. 18(2), 143–154 (1979)
6. Chandran, N., Kanukurthi, B., Ostrovsky, R., Reyzin, L.: Privacy amplification with asymptotically optimal entropy loss. In: Proceedings of the 42nd Annual ACM Symposium on Theory of Computing (2010)
7. Cohen, G., Raz, R., Segev, G.: Non-malleable extractors with short seeds and applications to privacy amplification. In: IEEE Conference on Computational Complexity, pp. 298–308. IEEE (2012)
8. Di Crescenzo, G., Lipton, R., Walfish, S.: Perfectly secure password protocols in the bounded retrieval model. In: Halevi, S., Rabin, T. (eds.) TCC 2006. LNCS, vol. 3876, pp. 225–244. Springer, Heidelberg (2006)
9. Dodis, Y., Kanukurthi, B., Katz, J., Reyzin, L., Smith, A.: Robust fuzzy extractors and authenticated key agreement from close secrets. IEEE Transactions on Information Theory 58(9), 6207–6222 (2012)
10. Dodis, Y., Li, X., Wooley, T.D., Zuckerman, D.: Privacy amplification and nonmalleable extractors via character sums. In: Ostrovsky, R. (ed.) FOCS, pp. 668–677. IEEE (2011)
11. Dodis, Y., Ostrovsky, R., Reyzin, L., Smith, A.: Fuzzy extractors: How to generate strong keys from biometrics and other noisy data. SIAM Journal on Computing 38, 97–139 (2008)
12. Dodis, Y., Smith, A.: Entropic security and the encryption of high entropy messages. In: Kilian, J. (ed.) TCC 2005. LNCS, vol. 3378, pp. 556–577. Springer, Heidelberg (2005)
13. Dodis, Y., Spencer, J.: On the (non)universality of the one-time pad. In: Proceedings of the 43rd Annual IEEE Symposium on Foundations of Computer Science, p. 376. IEEE Computer Society (2002)
14. Dodis, Y., Wichs, D.: Non-malleable extractors and symmetric key cryptography from weak secrets. In: Mitzenmacher, M. (ed.) STOC, pp. 601–610. ACM (2009)
15. Dziembowski, S.: Intrusion-resilience via the bounded-storage model. In: Halevi, S., Rabin, T. (eds.) TCC 2006. LNCS, vol. 3876, pp. 207–224. Springer, Heidelberg (2006)
16. Håstad, J., Impagliazzo, R., Levin, L.A., Luby, M.: A pseudorandom generator from any one-way function. SIAM J. Comput. 28(4), 1364–1396 (1999)
17. Kanukurthi, B., Reyzin, L.: Key agreement from close secrets over unsecured channels. In: Joux, A. (ed.) EUROCRYPT 2009. LNCS, vol. 5479, pp. 206–223. Springer, Heidelberg (2009)
18. Li, X.: Design extractors, non-malleable condensers and privacy amplification. In: Karloff, H.J., Pitassi, T. (eds.) STOC, pp. 837–854. ACM (2012)

19. Li, X.: Non-malleable condensers for arbitrary min-entropy, and almost optimal protocols for privacy amplification. CoRR, abs/1211.0651 (2012)
20. Li, X.: Non-malleable extractors, two-source extractors and privacy amplification. In: FOCS, pp. 688–697. IEEE Computer Society (2012)
21. Lu, C.-J.: Encryption against storage-bounded adversaries from on-line strong extractors. J. Cryptology 17(1), 27–42 (2004)
22. Maurer, U.M.: Protocols for secret key agreement by public discussion based on common information. In: Brickell, E.F. (ed.) CRYPTO 1992. LNCS, vol. 740, pp. 461–470. Springer, Heidelberg (1993)
23. Maurer, U.M., Wolf, S.: Privacy amplification secure against active adversaries. In: Kaliski Jr., B.S. (ed.) CRYPTO 1997. LNCS, vol. 1294, pp. 307–321. Springer, Heidelberg (1997)
24. Nisan, N., Zuckerman, D.: Randomness is linear in space. J. Comput. Syst. Sci. 52(1), 43–52 (1996)
25. Radhakrishnan, J., Ta-Shma, A.: Bounds for dispersers, extractors, and depth-two superconcentrators. SIAM J. Discrete Math. 13(1), 2–24 (2000)
26. Raz, R.: Extractors with weak random seeds. In: Proceedings of the 37th Annual ACM Symposium on Theory of Computing, pp. 11–20 (2005)
27. Renner, R.S., Wolf, S.: Unconditional authenticity and privacy from an arbitrarily weak secret. In: Boneh, D. (ed.) CRYPTO 2003. LNCS, vol. 2729, pp. 78–95. Springer, Heidelberg (2003)
28. Vadhan, S.P.: Constructing locally computable extractors and cryptosystems in the bounded-storage model. J. Cryptology 17(1), 43–77 (2004)

On the Communication Complexity
of Secure Computation

Deepesh Data[1,*], Manoj M. Prabhakaran[2,**], and Vinod M. Prabhakaran[1,***]

[1] School of Technology and Computer Science,
Tata Institute of Fundamental Research, Mumbai, India
{deepeshd,vinodmp}@tifr.res.in
[2] Department of Computer Science,
University of Illinois, Urbana-Champaign, USA
mmp@illinois.edu

Abstract. Information theoretically secure multi-party computation (MPC) is a central primitive of modern cryptography. However, relatively little is known about the communication complexity of this primitive.

In this work, we develop powerful information theoretic tools to prove lower bounds on the communication complexity of MPC. We restrict ourselves to a concrete setting involving 3-parties, in order to bring out the power of these tools without introducing too many complications. Our techniques include the use of a data processing inequality for *residual information* — i.e., the gap between mutual information and Gács-Körner common information, a new *information inequality* for 3-party protocols, and the idea of *distribution switching* by which lower bounds computed under certain worst-case scenarios can be shown to apply for the general case.

Using these techniques we obtain tight bounds on communication complexity by MPC protocols for various interesting functions. In particular, we show concrete functions that have "communication-ideal" protocols, which achieve the minimum communication simultaneously on all links in the network. Also, we obtain the first *explicit* example of a function that incurs a higher communication cost than the input length in the secure computation model of Feige, Kilian and Naor [17], who had shown that such functions exist. We also show that our communication bounds imply tight lower bounds on the amount of randomness required by MPC protocols for many interesting functions.

1   Introduction

Information theoretically secure multi-party computation has been a central primitive of modern cryptography. The seminal results of Ben-Or, Goldwasser,

---

* Research supported in part by ITRA, Media Lab Asia, India.
** Research supported in part by NSF grants 1228856 and 0747027.
*** Research funded in part by ITRA, Media Lab Asia, India and a Ramanujan fellowship from DST, India.

J.A. Garay and R. Gennaro (Eds.): CRYPTO 2014, Part II, LNCS 8617, pp. 199–216, 2014.
© International Association for Cryptologic Research 2014

and Wigderson [3] and Chaum, Crépeau, and Damgård [9] showed that information theoretically secure function computation is possible between parties
connected by pairwise, private links as long as only a strict minority may collude in the honest-but-curious model (and a strictly less than one-third minority
may collude in the malicious model). Since then, several protocols have improved
the efficiency of these protocols.

However, relatively less is known about *lower bounds* on the amount of *communication* required by a secure multi-party computation protocol, with a few
notable exceptions [31,21,10,17]. In fact, [28] shows that establishing strong communication lower bounds (even with restrictions on the number of rounds) would
imply breakthrough lower bound results for other well-studied problems like
private-information retrieval and locally decodable codes. Further, due to the
standard upper bounds on the communication needed in a secure multi-party
computation protocol [3,9], such lower bounds would imply non-trivial circuit
complexity lower bounds — a notoriously hard problem in theoretical computer
science. The goal of this work is to develop tools to tackle the difficult problem
of lower bounds for communication in secure multi-party computation, even if
they do not immediately have direct implications to circuit complexity or locally
decodable codes.

In this work we develop novel *information-theoretic tools to obtain lower
bounds on the communication complexity of secure computation.* Our tools have
connections with information-complexity techniques developed in the context of
communication complexity and related problems. In particular, all these tools
are related to notions of "common information" introduced by Gács-Körner [22]
and Wyner [43].[1]

We shall restrict our study to a concrete setting that brings out the power of
these tools without introducing too many additional complications. Our setting
involves 3 parties (with security against corruption of any single party) of which
only two parties have inputs, X and Y, and only the third party produces an
output Z as a (possibly randomized) function of the inputs. This class of functions is similar to that studied in [17], but our protocol model is more general
(since it allows fully interactive communication), making it harder to establish
lower bounds. Also, our lower bounds apply to the semi-honest setting, where
security is required only against passive corruption.

Results and Techniques. We study the setting shown in Figure 1. We obtain
lower bounds on the expected number of bits that need to be exchanged between
each pair of parties when securely evaluating a (possibly randomized) function
of two inputs, so that Alice and Bob feed the inputs to the function, and Charlie
receives the output from the function. In fact, our bounds are on the entropy

---

[1] In communication complexity and related problems, the lower bound techniques relate to Wyner common information [39,6], whereas the tools in this work are more
directly related to Gács-Körner common information. Wyner common information
and Gács-Körner common information have been generalized to a measure of correlation represented as the "tension region" in [40].

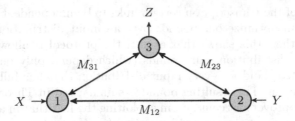

Fig. 1. A three-party secure computation problem. Alice (party-1) has input X and Bob (party-2) has Y. We require that (i) Charlie (party-3) obtains as output a randomized function of the other two parties' inputs, distributed as $p_{Z|XY}$, (ii) Alice and Bob learn no additional information about each other's inputs, and (iii) Charlie learns nothing more about X, Y than what is revealed by Z. All parties can talk to each other, over multiple rounds over bidirectional pairwise private links.

of the transcript between each pair,[2] and hence hold even when the protocol is amortized over several instances with independent inputs. Further, often these bounds do not depend on the input distribution (as long as the distribution has full support), and hold even if the protocol is allowed to depend on the input distribution.

At a high-level, the ingredients in deriving our lower bounds are the following:

- Firstly, we observe that, since Alice and Bob do not obtain any outputs, they are both forced to reveal their inputs fully (up to equivalent inputs) to the rest of the system, and further, Charlie's output depends on the inputs only through all the communication he has with the rest of the system. Combined with the privacy requirements, this immediately leads to a naïve lower bound: specifically, writing X, Y, Z as X_1, X_2, X_3, we have $H(M_{ij}) \geq H(X_i, X_j | X_k)$, where $\{i, j, k\} = \{1, 2, 3\}$.[3]
  We strengthen the naïve lower bounds by relying on a "data-processing inequality" for *residual information* — i.e., the gap between mutual-information and (Gács-Körner) common information — which lets us relate the residual information between the messages to the residual information between the inputs/outputs. This bound is given in Theorem 1.
- We can further improve the above lower bounds using a new tool, called *distribution switching*. The key idea is that the security requirement forces the

---

[2] The entropy bounds translate to bounds on the expected number of bits communicated, when we require that the messages on the individual links are encoded using (possibly adaptively chosen) prefix-free codes. See the full version [14] for details.

[3] We point out a simple example for which one can obtain a tight bound from this naïve bound: addition (in any group) requires one group element to be communicated between every pair of players, even with amortization over several independent instances. Previous lower bounds for secure evaluation of addition, while considering an arbitrary number of parties, either restricted themselves to bounding the *number of messages* required [21,10], or relied on non-standard security requirements (like "unstoppability" [21]).

distribution of the transcript on certain links to be independent of the inputs. Hence, we can optimize our bounds over all input distributions having full support. Further, this shows that even if the protocol is allowed to depend on the input distribution, our bounds (which depend only on the function being evaluated) hold for every input distribution that has full support over the input domain. The resulting bound is summarised in Theorem 2.

- A different improvement comes from exploiting the fact that in a protocol, the transcripts have to be generated by the parties interactively, rather than be created by an omniscient "dealer." An important technical contribution of this work is to provide a new tool towards this, in the form of a *new information inequality for 3-party interactive protocols* (Lemma 4). We use this along with the idea of *distribution switching* to significantly improve the above lower bounds by optimizing them using appropriate distributions of inputs. In fact, we can take the different terms in our bounds and *optimize each of them separately using different distributions over the inputs*. The resulting bounds (Theorem 3 and Theorem 4) are often stronger than what can be obtained by considering a single input distribution for the entire expression.

The resulting bounds are summarized in Theorem 1, Theorem 2, Theorem 3 and Theorem 4. We remark that unlike most of the existing results (for e.g. the bounds in [21] for summation (mod 2)), our lower bounds are not restricted to specific functions, but are applicable to all 3-party functions (except Theorem 2 and Theorem 4, which place some restrictions on the functions). To illustrate the use of our lower bounds, we apply them to several interesting example functions. In particular, we show the following:

- We analyze secure protocols for a few functions – GROUP-ADD, CONTROLLED-ERASURE and REMOTE-OT – and, applying our lower bounds, show that these protocols achieve *optimal communication complexity simultaneously on each link*. We call such a protocol a *communication-ideal* protocol. We leave it open to characterize which functions have communication-ideal protocols.

- We show an *explicit* deterministic function $f : \{0,1\}^n \times \{0,1\}^n \to \{0,1\}^{n-1}$ which has a communication-ideal protocol in which Charlie's total communication cost is (and must be at least) $3n - 1$ bits. In contrast, [17] showed that *there exist* functions $f : \{0,1\}^n \times \{0,1\}^n \to \{0,1\}$, for which Charlie must receive at least $3n - 4$ bits, if the protocol is required to be in their non-interactive model. (Note that our bound is incomparable to that of [17], since we require the output of our function to be longer; on the other hand, our bound uses an explicit function, and continues to hold even if we allow unrestricted interaction.)

In the full version of this paper [14], we extend the above results to lower bounds on a couple of related quantities. Firstly, we identify a *multi-secret sharing* primitive that is interesting on its own right, but also has the property that lower bounds on its share sizes serve as lower bounds for communication complexity of MPC protocols; some of our preliminary lower bounds are, in fact, bounds on the share sizes for such a multi-secret sharing scheme. Secondly, we show that our lower bounds for communication complexity also yield lower bounds on the amount of randomness needed in secure computation protocols. We analyze secure protocols for several natural functions, and prove that these protocols are *randomness-optimal*, i.e., they use the least amount of randomness.

Related Work. Communication complexity of multi-party computation without security requirements has been widely studied since [44] (see [33]), and more recently has seen the use of information-theoretic tools as well, in [7] and subsequent works. Independently, in the information theory literature communication requirements of interactive function computation have been studied (e.g. [37]).

In secure multi-party computation, there has been a vast literature on information-theoretic security, focusing on building efficient protocols, as well as characterizing various aspects like corruption models that admit secure protocols (e.g. [3,9,8,27,20,26]) and the number of rounds of interaction needed (e.g. [18,24,19,38,30]). Among other things, these results upper-bounded the communication complexity of multi-party secure computation in terms of the circuit complexity of the computation. Recently, [1] showed that, in general this upper bound is not tight by showing that all functions can be securely evaluated with sub-exponential communication (in our model of 3-party computation protocols), whereas most functions have exponential circuit complexity.

But *lower-bounding communication complexity* has received much less attention. For 2-party secure computation with security against passive corruption of one party (when the function admits such a protocol), communication complexity was combinatorially characterized in [31]. [21,10] gave tight lower and upper bounds on the number of messages needed for secure computation of addition (mod 2) by n parties. Further, relying on a stronger corruption model (fail-stop corruption), [21] also argued a lower bound for the amortized *communication complexity* of secure addition over any finite field. Feige et al. [17] obtained a lower bound on the communication complexity for a restricted class of 3-party protocols; along with positive results, they gave a modest lower bound for communication needed for evaluating random functions in this model. The difficulty of obtaining general lower bounds was pointed out by Ishai and Kushilevitz [28], who related such lower bounds to lower bounds for locally decodable codes and private information retrieval protocols. The connection to private information retrieval protocols was recently used in [1] to, among other things, derive the best known general upper bound on communication for Boolean functions in the model of [17]. The related question of how much randomness is required for secure computation seems to have received even less attention, but again, with some notable exceptions [32,5,23,34].

We remark that in a model with computational security, under computational hardness assumptions, the communication complexity of secure computation is linear in the input size, relying on fully homomorphic encryption ([25] and subsequent works) or exponential computation by the parties [36]. Also, in a model with exponential amount of correlated randomness shared among the parties, such a result was obtained in [29].

Information-theoretic tools have been successfully used in deriving bounds in various cryptographic problems like key agreement (e.g. [35,11]), secure 2-party computation (e.g. [15]) and secret-sharing and its variants (e.g. [2] and [4]). In this work, we rely on information-theoretic tools developed in [42,40], which also considered cryptographic problems. Some preliminary observations leading to this work appeared in [13].

2   Preliminaries

Notation. We write p_X to denote the distribution of a discrete random variable X; $p_X(x)$ denotes $\Pr[X = x]$. When clear from the context, the subscript of p_X will be omitted. The conditional distribution denoted by $p_{Z|U}$ specifies $\Pr[Z = z|U = u]$, for each value z that Z can take and each value u that U can take. A *randomized function* of two variables, is specified by a probability distribution $p_{Z|XY}$, where X, Y denote the two input variables, and Z denotes the output variable.

For random variables T, U, V, we write the *Markov chain* $T - U - V$ to indicate that T and V are conditionally independent conditioned on U: i.e., $I(T; V|U) = 0$. All logarithms are to the base 2 and entropies are in bits.

Protocols. A 3-party protocol Π is specified by a collection of "next message functions" (Π_1, Π_2, Π_3) which probabilistically map a *state* of the protocol to the next state (in a restricted manner), and output functions $(\Pi_1^{out}, \Pi_2^{out}, \Pi_3^{out})$ used to define the outputs of the parties as probabilistic functions of their views. We shall also allow the protocol to depend on the distribution of the inputs to the parties. (This would allow one to tune a protocol to be efficient for a suitable input distribution. Allowing this makes our lower bounds stronger; on the other hand, none of the protocols we give for our examples require this flexibility.)

Without loss of generality, the state of the protocol consists only of the inputs received by each party and the *transcript* of the messages exchanged so far.[4] We denote the final transcripts on the three links, after executing protocol Π on its specified input distribution by $M_{12}^{\Pi}, M_{23}^{\Pi}$ and M_{31}^{Π}. When Π is clear from the context, we simply write M_{12} etc. We define $M_1 = (M_{12}, M_{31})$ as the transcripts that party 1 can see; M_2 and M_3 are defined similarly. We define the view of the ith party, V_i to consist of M_i and that party's inputs and outputs (if any).

It is easy to see that a protocol, along with an input distribution, fully defines a joint distribution over all the inputs, outputs and the joint transcripts on all the links.

Secure Computation. We consider three party computation functionalities, in which Alice and Bob (parties 1 and 2) receive as inputs the random variables $X \in \mathcal{X}$ and $Y \in \mathcal{Y}$, respectively, and Charlie (party 3) produces an output $Z \in \mathcal{Z}$ distributed according to a specified distribution $p_{Z|XY}$. The sets $\mathcal{X}, \mathcal{Y}$

---

[4] Since the parties are computationally unbounded, there is no need to allow private randomness as part of the state; randomness for a party can always be resampled at every round conditioned on the inputs, outputs and messages in that party's view.

and $\mathcal{Z}$ are always finite. In secure computation, we shall consider the inputs to the computation to come from a distribution p_{XY} over $\mathcal{X} \times \mathcal{Y}$.

A (perfectly) secure computation protocol $\Pi(p_{XY}, p_{Z|XY}) = (\Pi_1, \Pi_2, \Pi_3, \Pi_3^{out})$ for $(p_{XY}, p_{Z|XY})$ is a protocol which satisfies the following conditions:

- Correctness: Output of Charlie, is distributed according to $p_{Z|X=x,Y=y}$, where x, y are the inputs to Alice and Bob
- Privacy: The privacy condition corresponds to "1-privacy", wherein at most one party is passively corrupt. Corresponding to security against Alice, Bob and Charlie, respectively, we have the following three Markov chains. $V_1 - X - (Y, Z)$, $V_2 - Y - (X, Z)$ and $V_3 - Z - (X, Y)$. Equivalently (see Footnote 4), $I(M_1; (Y, Z)|X) = I(M_2; (X, Z)|Y) = I(M_3; (X, Y)|Z) = 0$.

Intuitively, the privacy condition guarantees that even if one party (say Alice) is curious, and retains its view from the protocol (in particular, M_1), this view reveals nothing more to it about the inputs and outputs of the other parties (namely, Y, Z), than what its own inputs and outputs reveal (as long as the other parties erase their own views). In other words, a curious party may as well simulate a view for itself based on just its inputs and outputs, rather than retain the actual view it obtained from the protocol execution.

For simplicity, we prove all our results for *perfect security* as defined above; this is also the setting for classical positive results like that of [3]. But in fact, we expect all our bounds to extend to the setting of statistical security as well (following [41,40] who extend similar results to the statistical security case).[5] Also, the above security requirements are for an honest execution of the protocol (corresponding to honest-but-curious or passive corruption of at most one party). The lower bounds derived in this model typically continue to hold for active corruption as well (since for many functionalities, every protocol secure against active corruption is a protocol secure against passive corruption); in this case, when a party uses a broadcast channel (as would be necessary in our setting, where 1 out of 3 parties is corrupted), it is counted as sending individual messages to every other party.

Communication Complexity and Entropy. A standard approach to lower-bounding the number of bits in a string is to lower-bound its entropy. However, in an interactive setting, a party sees the messages in each round, rather than just a concatenation of all the bits sent over the entire protocol. In a setting where we allow variable length messages, this would seem to allow communicating more bits of information than the length of the transcript itself. But this allows the parties to learn when the message transmitted in a round ends, implicitly inserting an end-of-message marker into the bit stream. To account for this, one can require that the message sent in every round is a codeword in a prefix-free code. (The code itself can be dynamically determined based on previous

---

[5] We remark that our bounds do not apply to a relaxed security setting sometimes considered in the information theory literature: there the error in computation/security is only required to go to 0 as the size of the input grows to infinity. [12] gives an example where there is a strict gap between the communication complexity under this relaxed setting and the perfect security setting of this paper.

messages exchanged over the link.) It can be shown that, with this requirement, the number of bits communicated in each link is indeed lower-bounded by the entropy of the transcript in that link.

Normal Form. For a pair $(p_{XY}, p_{Z|XY})$, define the relations $x \cong x'$, $y \cong y'$ and $z \cong z'$ as follows.

1. For any $x, x' \in \mathcal{X}$, let $\mathcal{S}_{x,x'} = \{y \in \mathcal{Y} : p_{XY}(x, y) > 0, p_{XY}(x', y) > 0\}$. We say that $x \cong x'$, if $\forall y \in \mathcal{S}_{x,x'}$ and $z \in \mathcal{Z}$, we have $p_{Z|XY}(z|x, y) = p_{Z|XY}(z|x', y)$.
2. For any $y, y' \in \mathcal{Y}$, let $\mathcal{S}_{y,y'} = \{x \in \mathcal{X} : p_{XY}(x, y) > 0, p_{XY}(x, y') > 0\}$. We say that $y \cong y'$, if $\forall x \in \mathcal{S}_{y,y'}$ and $z \in \mathcal{Z}$, we have $p_{Z|XY}(z|x, y) = p_{Z|XY}(z|x, y')$.
3. Let $\mathcal{S} = \{(x, y) : p_{XY}(x, y) > 0\}$. For any $z, z' \in \mathcal{Z}$, we say that $z \cong z'$, if $\exists c \geq 0$ such that $\forall (x, y) \in \mathcal{S}$, we have $p_{Z|XY}(z|x, y) = c \cdot p_{Z|XY}(z'|x, y)$.

A pair $(p_{XY}, p_{Z|XY})$ is said to be in *normal form* if $x \cong x' \Rightarrow x = x'$, $y \cong y' \Rightarrow y = y'$, and $z \cong z' \Rightarrow z = z'$.

It is easy to show (as we do in the full version) that we may assume without loss of generality that $(p_{XY}, p_{Z|XY})$ is in normal form since any given $(p_{XY}, p_{Z|XY})$ can be transformed to a $(p_{X'Y'}, p_{Z'|X'Y'})$ in normal form so that any secure computation protocol for the former can be transformed to one for the latter with the same communication costs, and vice versa.

Communication-Ideal Protocol. We say that a protocol $\Pi(p_{XY}, p_{Z|XY})$ for securely computing a randomized function $p_{Z|XY}$, for a distribution p_{XY} is *communication-ideal* if for each $ij \in \{12, 23, 31\}$,

$$H(M_{ij}^{\Pi}) = \inf_{\Pi'(p_{XY}, p_{Z|XY})} H(M_{ij}^{\Pi'}),$$

where the infimum is over all secure protocols for $p_{Z|XY}$ with the same distribution p_{XY}. That is, a communication-ideal protocol achieves the least entropy possible for every link, simultaneously. We remark that it is not clear, *a priori*, how to determine if a given function $p_{Z|XY}$ has a communication-ideal protocol for a given distribution p_{XY}.

Common Information and Residual Information

Gács and Körner [22] introduced the notion of common information to measure a certain aspect of correlation between two random variables. The Gács-Körner common information of a pair of correlated random variables (U, V) can be defined as $H(U \sqcap V)$, where $U \sqcap V$ is a random variable with maximum entropy among all random variables Q that are determined both by U and by V (i.e., there are functions f and g such that $Q = f(U) = g(V)$). In [40], the gap between mutual information and common information was termed *residual information*: $RI(U; V) := I(U; V) - H(U \sqcap V)$.

In [42], Wolf and Wullschleger identified (among other things) the following important *data processing inequality* for residual information.

Lemma 1 ([42]). *If T, U, V, W are jointly distributed random variables such that the following two Markov chains hold: (i) $U - T - W$, and (ii) $T - W - V$, then*

$$RI(T; W) \leq RI((U, T); (V, W)).$$

The Markov chain conditions above correspond to the requirement that it is secure (against honest-but-curious adversaries) to require a pair of parties holding the views (U, T) and (V, W), to produce outputs T, W, respectively, because for the first party, the rest of its view, U, can be simulated based on the output T, independent of the output W (and similarly, for the second party). The lemma states that under such a secure transformation from views to outputs, the residual information can only decrease.

It is easy to see that the following is an equivalent definition of residual information (see [40]).

$$RI(U; V) = \min_{\substack{Q: \exists f, g \text{ s.t.} \\ Q = f(U) = g(V)}} I(U; V | Q). \tag{1}$$

The random variable Q which achieves the minimum is, in fact, $U \sqcap V$. Note that the residual information is always non-negative.

3   Lower Bounds on Communication Complexity

This section is divided into three parts. In Section 3.1, we derive preliminary lower bounds for secure computation. In each of the subsequent subsections, we give different improvements of the lower bounds derived in Section 3.1. Omitted proofs are available in the full version [14].

3.1   Preliminary Lower Bounds

We first state the following basic lemma for any protocol for secure computation. Similar results have appeared in the literature earlier (for instance, special cases of Lemma 2 appear in [16,41,13]).

Lemma 2. *Suppose $(p_{XY}, p_{Z|XY})$ is in normal form. Then, in any secure protocol $\Pi(p_{XY}, p_{Z|XY})$, the cut isolating Alice from Bob and Charlie must reveal Alice's input X, i.e., $H(X | M_{12}, M_{31}) = 0$. Similarly, $H(Y | M_{12}, M_{23}) = 0$ and $H(Z | M_{23}, M_{31}) = 0$.*

Lemma 2 states the simple fact that, for $(p_{XY}, p_{Z|XY})$ in normal form, the information about a party's input must flow out through the links she/he is part of, and the information about Charlie's output must flow in through the links he is part of. This crucially relies on the fact that Alice and Bob obtain no output, and Charlie has no input in our model.

We obtain a preliminary lower bound in Theorem 1 below by using the above lemma and the data-processing inequality for residual information in Lemma 1. Recall that the assumption below of $(p_{XY}, p_{Z|XY})$ being in normal form is without loss of generality.

Theorem 1. *Any secure protocol $\Pi(p_{XY}, p_{Z|XY})$, where $(p_{XY}, p_{Z|XY})$ is in normal form, should satisfy the following lower bounds on the entropy of the transcripts on each link.*

$$H(M_{23}) \geq \max\{RI(X;Z), RI(X;Y)\} + H(Y,Z|X), \tag{2}$$

$$H(M_{31}) \geq \max\{RI(Y;Z), RI(X;Y)\} + H(X,Z|Y), \tag{3}$$

$$H(M_{12}) \geq \max\{RI(X;Z), RI(Y;Z)\} + H(X,Y|Z). \tag{4}$$

Proof: We shall prove (2). The other two inequalities follow similarly.

$$H(M_{23}) \geq \max\{H(M_{23}|M_{31}), H(M_{23}|M_{12})\}$$
$$= \max\{I(M_{23}; M_{12}|M_{31}), I(M_{23}; M_{31}|M_{12})\} + H(M_{23}|M_{12}, M_{31}) \tag{5}$$

We can bound the last term of (5) as follows (to already get a naïve bound):

$$H(M_{23}|M_{12}, M_{31}) \stackrel{(a)}{=} H(M_{23}, Y, Z|M_{12}, M_{31}, X)$$
$$\geq H(Y, Z|M_{12}, M_{31}, X) \stackrel{(b)}{=} H(Y, Z|X),$$

where (a) follows from Lemma 2 and (b) follows from the privacy against Alice. Next, we lower bound the first term inside max of (5) by $RI(X;Z)$ as follows.

$$I(M_{23}; M_{12}|M_{31}) = I(M_{23}M_{31}; M_{12}M_{31}|M_{31}) \geq RI(M_{23}, M_{31}; M_{12}, M_{31}) \tag{6}$$

where the inequality follows from (1) by taking $Q = M_{31}$. Now, by privacy against Charlie, we have $(M_{23}, M_{31}) - Z - X$ and by privacy against Alice, we have $(M_{12}, M_{31}) - X - Z$. Applying Lemma 1 with the above Markov chains, together with Lemma 2, we get

$$RI(M_{23}, M_{31}; M_{12}, M_{31}) \geq RI(Z;X) = RI(X;Z).$$

Similarly, we can lower-bound the second term inside max of (5) by $RI(X;Y)$, completing the proof.   □

In the rest of the paper we will restrict our attention to p_{XY} which have full support. This will allow us to strengthen the preliminary bounds in Theorem 1.

3.2 Improved Lower Bounds via Distribution Switching

To improve the bounds in Theorem 1, we will use a technique we call *distribution switching*. This significantly improves the above bounds and leads to one of our main theorems.

The following lemma states that privacy requirements imply that the transcript M_{12} generated by a secure protocol computing $p_{Z|XY}$ is independent of both the inputs. Moreover, if the function $p_{Z|XY}$ satisfies some additional constraints, then the other two transcripts also become independent of the inputs. The *characteristic bipartite graph* of a distribution p_{XY} is defined as a bipartite graph on vertex set $\mathcal{X} \cup \mathcal{Y}$ such that $x \in \mathcal{X}$ and $y \in \mathcal{Y}$ have an edge between them whenever $p_{XY}(x,y) > 0$. The proof of the following lemma is along the lines of a similar lemma in [13].

Lemma 3. *Consider a function $p_{Z|XY}$.*

1. *Suppose that p_{XY} is such that the characteristic bipartite graph of p_{XY} is connected. Then, for any secure protocol $\Pi(p_{XY}, p_{Z|XY})$, we have $I(X, Y, Z; M_{12}) = 0$.*

2. *Suppose p_{XY} has full support and $p_{Z|XY}$ satisfies the following condition:*

   Condition 1. *There is no non-trivial partition $\mathcal{X} = \mathcal{X}_1 \cup \mathcal{X}_2$ (i.e., $\mathcal{X}_1 \cap \mathcal{X}_2 = \varnothing$ and neither $\mathcal{X}_1$ nor $\mathcal{X}_2$ is empty), such that if $\mathcal{Z}_k = \{z \in \mathcal{Z} : x \in \mathcal{X}_k, y \in \mathcal{Y}, p(z|x, y) > 0\}, k = 1, 2$, their intersection $\mathcal{Z}_1 \cap \mathcal{Z}_2$ is empty.*

   Then, for any secure protocol $\Pi(p_{XY}, p_{Z|XY})$, we have $I(X, Y, Z; M_{31}) = 0$.

3. *Suppose p_{XY} has full support and $p_{Z|XY}$ satisfies the following condition:*

   Condition 2. *There is no non-trivial partition $\mathcal{Y} = \mathcal{Y}_1 \cup \mathcal{Y}_2$ such that if $\mathcal{Z}_k = \{z \in \mathcal{Z} : x \in \mathcal{X}, y \in \mathcal{Y}_k, p(z|x, y) > 0\}, k = 1, 2$, their intersection $\mathcal{Z}_1 \cap \mathcal{Z}_2$ is empty.*

   Then, for any secure protocol $\Pi(p_{XY}, p_{Z|XY})$, we have $I(X, Y, Z; M_{23}) = 0$.

We note that p_{XY} will have a connected characteristic bipartite graph if it has full support.

We will now strengthen the lower bounds from Theorem 1. Specifically, we will argue that even if the protocol is allowed to depend on the input distribution (as we do here), correctness and privacy conditions will require that the lower bounds derived for when the distributions of the inputs are changed continue to hold for the original setting.

Theorem 2. *Consider any secure protocol $\Pi(p_{XY}, p_{Z|XY})$, where p_{XY} has full support and $(p_{XY}, p_{Z|XY})$ is in normal form.*

1. *We have the following strengthening of (4):*

$$H(M_{12}) \geq \max\{ \sup_{p_{X'Y'}} (RI(X'; Z') + H(X', Y'|Z')),$$

$$\sup_{p_{X'Y'}} (RI(Y'; Z') + H(X', Y'|Z'))\}, \quad (7)$$

   where the sup operations are over $p_{X'Y'}$ having full support and the objective functions are evaluated using $p_{X'Y'Z'}(x, y, z) = p_{X'Y'}(x, y)p_{Z|XY}(z|x, y)$.

2. *If $p_{Z|XY}$ satisfies Condition 1 of Lemma 3, we have the following strengthening of (3):*

$$H(M_{31}) \geq \max\{ \sup_{p_{X'Y'}} (RI(Y'; Z') + H(X', Z'|Y')),$$

$$\sup_{p_{X'Y'}} (RI(X'; Y') + H(X', Z'|Y'))\}, \quad (8)$$

   where the sup operations are over the same set of $p_{X'Y'}$ as in (7).

3. If $p_{Z|XY}$ satisfies Condition 2 of Lemma 3, we have the following strengthening of (2):

$$H(M_{23}) \geq \max\{ \sup_{p_{X'Y'}} (RI(X';Z') + H(Y',Z'|X')),$$

$$\sup_{p_{X'Y'}} (RI(X';Y') + H(Y',Z'|X'))\}, \quad (9)$$

where the sup operations are over the same set of $p_{X'Y'}$ as in (7).

Proof: Notice that any secure protocol $\Pi(p_{XY}, p_{Z|XY})$, where distribution p_{XY} has full support, continues to be a secure protocol even if we switch the input distribution to a different one $p_{X'Y'}$. This follows directly from examining the correctness and privacy conditions required for a protocol to be secure.

By Lemma 3, it follows that the transcript M_{12} of the protocol (under both the original and the switched input distributions) must remain independent of the input data X, Y. Furthermore, since $(p_{XY}, p_{Z|XY})$ is in normal form and $p_{X'Y'}$ has full support, $(p_{X'Y'}, p_{Z|XY})$ is also in normal form. Hence, (7) follows from (4) of Theorem 1. Similarly, if the function $p_{Z|XY}$ satisfies the condition 1 (resp. 2) of Lemma 3, we can show (8) (resp. (9)) as well. $\square$

3.3   An Information Inequality for Protocols and Improved Lower Bounds

We can give a different improvement to Theorem 1 by exploiting the fact that, in a protocol, transcripts are generated by the parties interactively rather than by an omniscient dealer. Towards this, we derive an information inequality relating the transcripts on different links in general 3-party protocols, in which parties do not share any common or correlated randomness or correlated inputs at the beginning of the protocol.

Lemma 4. In a 3-party protocol, if the inputs to the parties are independent of each other, then, for $\{\alpha, \beta, \gamma\} = \{1, 2, 3\}$,

$$I(M_{\gamma\alpha}; M_{\beta\gamma}) \geq I(M_{\gamma\alpha}; M_{\beta\gamma}|M_{\alpha\beta}).$$

Further, as in (6), $I(M_{\gamma\alpha}; M_{\beta\gamma}|M_{\alpha\beta}) \geq RI(M_{\gamma\alpha}, M_{\alpha\beta}; M_{\beta\gamma}, M_{\alpha\beta})$. Hence, if the inputs are independent of each other,

$$I(M_{\gamma\alpha}; M_{\beta\gamma}) \geq I(M_{\gamma\alpha}; M_{\beta\gamma}|M_{\alpha\beta}) \geq RI(M_{\gamma\alpha}, M_{\alpha\beta}; M_{\beta\gamma}, M_{\alpha\beta}). \quad (10)$$

This inequality provides us with a means to exploit the protocol structure behind the transcripts. For instance, consider a secure protocol $\Pi(p_X p_Y, p_{Z|XY})$, where p_X, p_Y have full support and $(p_X p_Y, p_{Z|XY})$ is in normal form. We have,

$$H(M_{12}) = I(M_{12}; M_{23}) + H(M_{12}|M_{23})$$
$$= I(M_{12}; M_{23}) + I(M_{12}; M_{31}|M_{23}) + H(M_{12}|M_{23}, M_{31})$$
$$\geq RI(X;Z) + RI(Y;Z) + H(X,Y|Z),$$

where the last inequality used $H(M_{12}|M_{23}, M_{31}) \geq H(X, Y|Z)$ and $I(M_{12}; M_{31}|M_{23}) \geq RI(Y; Z)$ (both as in the proof of Theorem 1) as well as $I(M_{12}; M_{23}) \geq RI(X; Z)$. Thus the term $\max\{RI(X; Z), RI(Y; Z)\}$ in (4) can be replaced by $RI(X; Z) + RI(Y; Z)$ for independent inputs.

In the full version we prove the following two theorems by making use of Lemma 4 and distribution switching:

Theorem 3. *The following communication complexity bounds hold for any secure protocol $\Pi(p_{XY}, p_{Z|XY})$, where p_{XY} has full support and $(p_{XY}, p_{Z|XY})$ is in normal form:*

$$H(M_{23}) \geq \left(\sup_{p_{X'}} RI(X'; Z')\right) + \left(\sup_{p_{X''}} H(Y, Z''|X'')\right), \tag{11}$$

$$H(M_{31}) \geq \left(\sup_{p_{Y'}} RI(Y'; Z')\right) + \left(\sup_{p_{Y''}} H(X, Z''|Y'')\right), \tag{12}$$

$$H(M_{12}) \geq \max\left\{\sup_{p_{X'}}\left(\sup_{p_{Y'}} RI(Y'; Z')\right) + \left(\sup_{p_{Y''}} RI(X'; Z'') + H(X', Y''|Z'')\right),\right.$$

$$\left.\sup_{p_{Y'}}\left(\sup_{p_{X'}} RI(X'; Z')\right) + \left(\sup_{p_{X''}} RI(Y'; Z'') + H(X'', Y'|Z'')\right)\right\}, \tag{13}$$

where the sup *operations are over distributions $p_{X'}, p_{X''}, p_{Y'}, p_{Y''}$ having full support. The terms in the right hand side of (11) are evaluated using the distribution p_Y of the data Y of Bob, i.e.,*

$$p_{X'YZ'}(x, y, z) = p_{X'}(x)p_Y(y)p_{Z|XY}(z|x, y),$$
$$p_{X''YZ''}(x, y, z) = p_{X''}(x)p_Y(y)p_{Z|XY}(z|x, y).$$

Similarly, the terms in (12) are evaluated using the distribution p_X of the data X of Alice. The lower bound in (13) does not depend on the distributions p_X and p_Y of the data. The terms on the top row of (13), for instance, are evaluated using

$$p_{X'Y'Z'}(x, y, z) = p_{X'}(x)p_{Y'}(y)p_{Z|XY}(z|x, y),$$
$$p_{X'Y''Z''}(x, y, z) = p_{X'}(x)p_{Y''}(y)p_{Z|XY}(z|x, y).$$

When the function satisfies certain additional constraints, we can strengthen the lower bounds on the $H(M_{23})$ and $H(M_{31})$ as follows:

Theorem 4. *Consider any secure protocol $\Pi(p_{XY}, p_{Z|XY})$, where p_{XY} has full support and $(p_{XY}, p_{Z|XY})$ is in normal form.*

1. *Suppose the function $p_{Z|XY}$ satisfies Condition 1 of Lemma 3. Then, we have the following strengthening of (12).*

$$H(M_{31}) \geq \sup_{p_{X'}}\left(\left(\sup_{p_{Y'}} RI(Y'; Z')\right) + \left(\sup_{p_{Y''}} H(X', Z''|Y'')\right)\right), \tag{14}$$

where the sup *operations are over distributions* $p_{X'}, p_{Y'}, p_{Y''}$ *having full support and the terms in the right hand side are evaluated using the distribution*

$$p_{X'Y'Z'Y''Z''}(x', y', z', y'', z'') =$$
$$p_{X'}(x')p_{Y'}(y')p_{Z|XY}(z'|x', y')p_{Y''}(y'')p_{Z|XY}(z''|x', y'').$$

2. *Suppose the function* $p_{Z|XY}$ *satisfies Condition 2 of Lemma 3. Then, we have the following strengthening of* (11).

$$H(M_{23}) \geq \sup_{p_{Y'}} \left(\left(\sup_{p_{X'}} RI(X'; Z') \right) + \left(\sup_{p_{X''}} H(Y', Z''|X'') \right) \right), \qquad (15)$$

where the sup *operations are over distributions* $p_{X'}, p_{X''}, p_{Y'}$ *having full support and the terms in the right hand side are evaluated using the distribution*

$$p_{X'Y'Z'X''Z''}(x', y', z', x'', z'') =$$
$$p_{X'}(x')p_{Y'}(y')p_{Z|XY}(z'|x', y')p_{X''}(x'')p_{Z|XY}(z''|x'', y').$$

Note that in Theorem 2, Theorem 3 and Theorem 4, any choice of $p_{X'Y'}$, $p_{X'}, p_{X''}$, $p_{Y'}, p_{Y''}$ (with full support) will yield a lower bound. For a given function, while all choices do yield valid lower bounds, one is often able to obtain the *best* lower bound analytically (as in Theorem 5, where it is seen to be the best as it matches an upper bound) or numerically (as in Theorem 6).

To summarize, for any secure computation problem $(p_{XY}, p_{Z|XY})$, expressed in the normal form, Theorem 1 gives lower bounds on entropies of transcripts on all three links. If, in addition, p_{XY} has full support, then for $H(M_{31})$, our best lower bound is the larger of (3) and (12); for $H(M_{23})$, it is the larger of (2) and (11); and for $H(M_{12})$, it is the larger of (7) and (13). Further, if $p_{Z|XY}$ satisfies condition 1 of Lemma 3, then for $H(M_{31})$, our best lower bound is the larger of (8) and (14); if $p_{Z|XY}$ satisfies condition 2 of Lemma 3, then for $H(M_{23})$, our best lower bound is the larger of (9) and (15).

Our communication lower bounds were developed for protocols whose designs may take into account the joint distribution of X and Y. However, the right hand sides of (7) and (13) do not depend on the distribution p_{XY} of the inputs. Thus, even though we allow the protocol to depend on the distributions, our lower bound on $H(M_{12})$ does not. The same is true for (8) and (14) for $H(M_{31})$ (resp. (9) and (15) for $H(M_{23})$), which apply when the function $p_{Z|XY}$ satisfies condition 1 (resp. 2) of Lemma 3. When these conditions are not satisfied, the communication complexity of the optimal protocol may indeed depend on the distribution of the input (see full version for an example).

4   Application to Specific Functions

In this section we consider a few important examples, and apply our generic lower bounds from above to these examples, to obtain interesting results. While many of these results are natural to conjecture, they are not easy to prove (see, for instance, Footnote 3).

Optimality of the FKN Protocol. Feige et al. [17] provided a generic (non-interactive) secure computation protocol for all 3-party functions in our model. This protocol uses a straight-forward (but "inefficient") reduction from an arbitrary function to a variant of the oblivious transfer problem, which we shall call the remote OT function (defined below), and then gives a simple protocol for this new function. While the resulting protocol is inefficient for most functions, one could ask whether the protocol that [17] used for REMOTE OT itself is optimal. We use our lower bounds from above to show that this is indeed the case.

The REMOTE $\binom{m}{1}$-OT_2^n function, is defined as follows: Alice's input $X = (X_0, X_1, \ldots, X_{m-1})$ is made up of m strings each of length n bits, and Bob has an input $Y \in \{0, 1, \ldots, m-1\}$. Charlie wants to compute $Z = f(X, Y) = X_Y$. The protocol of [17] requires nm bits to be exchanged over the Alice-Charlie (31) link, $n + \log m$ bits over the Bob-Charlie (23) link and $nm + \log m$ bits over the Alice-Bob (12) link. In the full version, we prove the following theorem, which shows that this protocol is optimal and in fact, a *communication-ideal* protocol.

Theorem 5. *Any secure protocol* $\Pi(p_{XY}, \text{REMOTE-OT})$ *for computing* REMOTE $\binom{m}{1}$-OT_2^n *for inputs* X *and* Y, *where* p_{XY} *has full support, must satisfy*

$$H(M_{31}) \geq nm, \quad H(M_{23}) \geq n + \log m, \quad and \quad H(M_{12}) \geq nm + \log m.$$

In the full version we also give two other examples (GROUP-ADD, CONTROLLED-ERASURE) which have communication-ideal protocols.

Separating Secure and Insecure Computation. A basic question of secure computation is whether it needs more bits to be communicated than the input-size itself (which suffices for insecure computation). While natural to expect, it is not easy to prove this. In their restricted model, [17] showed a non-explicit result, that for securely computing *most* Boolean functions on the domain $\{0,1\}^n \times \{0,1\}^n$, Charlie is required to receive at least $3n - 4$ bits, which is significantly more than the $2n$ bits sufficient for insecure computation.

REMOTE $\binom{2}{1}$-OT_2^n from above already gives us an explicit example of a function where this is true: the total input size is $2n + 1$, but the communication is at least $H(M_{31}) + H(M_{23}) \geq 3n + 1$. To present an easy comparison to the lower bound of [17], we can consider a symmetrized variant of REMOTE $\binom{2}{1}$-OT_2^n, in which two instances of REMOTE $\binom{2}{1}$-OT_2^n are combined, one in each direction. More specifically, $X = (A_0, A_1, a)$ where A_0, A_1 are of length $(n-1)/2$ (for an odd n) and a is a single bit; similarly $Y = (B_0, B_1, b)$; the output of the function is defined as an $(n-1)$ bit string $f(X, Y) = (A_b, B_a)$. Considering (say) the uniform input distribution over X, Y, the bounds for REMOTE $\binom{2}{1}$-OT_2^n add up to give us $H(M_{31}) \geq 3(n-1)/2 + 1$ and $H(M_{23}) \geq 3(n-1)/2 + 1$, so that the communication with Charlie is lower-bounded by $H(M_{31}) + H(M_{23}) \geq 3n - 1$.

This compares favourably with the bound of [17] in many ways: our lower bound holds even in a model that allows interaction; in particular, this makes the gap between insecure computation ($n-1$ bits in our case, $2n$ bits for [17]) and secure computation (about $3n$ bits for both) somewhat larger. More importantly, our lower bound is explicit (and tight for the specific function we use), whereas that of [17] is existential. However, our bound does not subsume that of [17], who considered *Boolean* functions. Our results do not yield a bound significantly larger than the input size, when the output is a single bit. It appears that this regime is more amenable to combinatorial arguments, as pursued in [17], rather than information theoretic arguments.

Communication Complexity of Securely Computing AND. We define the 3-party AND function as follows: Alice has an input bit X, Bob has an input bit Y and Charlie should obtain $Z = f(X, Y) = X \wedge Y$. In the full version, we compute the following lower bound.

Theorem 6. *Any secure protocol $\Pi(p_{XY}, \text{AND})$ for computing* AND *for inputs X and Y, where p_{XY} has full support over $\{0,1\}^n \times \{0,1\}^n$, must satisfy*

$$H(M_{31}) \geq n \log(3), \quad H(M_{23}) \geq n \log(3), \quad \text{and} \quad H(M_{12}) \geq n(1.826).$$

The best known protocol for AND (due to [17]) achieves $H(M_{12}) = 1 + \log(3)$, $H(M_{23}) = H(M_{31}) = \log(3)$. Our lower bounds on $H(M_{31})$ and $H(M_{23})$ match this, but there is a gap for $H(M_{12})$: an upper bound of $1+\log(3) \approx 2.585$ against a lower bound of 1.826. Closing this gap remains an open problem.

5   Conclusion

In this work we presented new tools to obtain lower bounds on the communication complexity of secure 3-party computation, and showed that they yield tight bounds for interesting examples. However, the general problem of obtaining tight lower bounds for communication complexity of secure computation is wide open; indeed, their implications to circuit lower bounds presents a "barrier" to obtaining super-linear bounds for explicit functions. We propose a possibly easier open problem: do there *exist* Boolean functions with super-linear communication complexity for secure computation? Note that lower bounds on circuit complexity do not directly translate to lower bounds on communication complexity of secure computation, as established by a sub-exponential upper bound of $2^{\tilde{O}(\sqrt{n})}$ for the latter [1]. Though it is plausible that for random Boolean functions, the actual communication cost is $2^{\Omega(n^\epsilon)}$ for some $\epsilon > 0$, none of the current techniques appear capable of delivering such a result.

References

1. Beimel, A., Ishai, Y., Kumaresan, R., Kushilevitz, E.: On the cryptographic complexity of the worst functions. In: Lindell, Y. (ed.) TCC 2014. LNCS, vol. 8349, pp. 317–342. Springer, Heidelberg (2014)
2. Beimel, A., Orlov, I.: Secret sharing and non-Shannon information inequalities. IEEE Transactions on Information Theory 57(9), 5634–5649 (2011)

3. Ben-Or, M., Goldwasser, S., Wigderson, A.: Completeness theorems for non-cryptographic fault-tolerant distributed computation. In: Proc. 20th STOC, pp. 1–10. ACM (1988)
4. Blundo, C., De Santis, A., Di Crescenzo, G., Gaggia, A.G., Vaccaro, U.: Multi-secret sharing schemes. In: Desmedt, Y.G. (ed.) Advances in Cryptology - CRYPTO 1994. LNCS, vol. 839, pp. 150–163. Springer, Heidelberg (1994)
5. Blundo, C., Santis, A.D., Persiano, G., Vaccaro, U.: Randomness complexity of private computation. Computational Complexity 8(2), 145–168 (1999)
6. Braun, G., Pokutta, S.: Common information and unique disjointness. In: FOCS, pp. 688–697 (2013)
7. Chakrabarti, A., Shi, Y., Wirth, A., Yao, A.C.-C.: Informational complexity and the direct sum problem for simultaneous message complexity. In: FOCS, pp. 270–278. IEEE (2001)
8. Chaum, D.: The spymasters double-agent problem: Multiparty computations secure unconditionally from minorities and cryptographically from majorities. In: Brassard, G. (ed.) Advances in Cryptology - CRYPTO 1989. LNCS, vol. 435, pp. 591–602. Springer, Heidelberg (1990)
9. Chaum, D., Crépeau, C., Damgård, I.: Multiparty unconditionally secure protocols. In: Proc. 20th STOC, pp. 11–19. ACM (1988)
10. Chor, B., Kushilevitz, E.: A communication-privacy tradeoff for modular addition. Inf. Process. Lett. 45(4), 205–210 (1993)
11. Csiszár, I., Narayan, P.: Secrecy capacities for multiple terminals. IEEE Transactions on Information Theory 50(12), 3047–3061 (2004)
12. Data, D., Dey, B.K., Mishra, M., Prabhakaran, V.M.: How to securely compute the modulo-two sum of binary sources, arXiv, 1405.2555 (preprint 2014)
13. Data, D., Prabhakaran, V.M.: Communication requirements for secure computation. In: Proc. 51st Annual Allerton Conference on Communication, Control, and Computing (2013)
14. Data, D., Prabhakaran, V.M., Prabhakaran, M.M.: On the communication complexity of secure computation, arXiv, 1311.7584 (preprint, 2014)
15. Dodis, Y., Micali, S.: Lower bounds for oblivious transfer reductions. In: Stern, J. (ed.) EUROCRYPT 1999. LNCS, vol. 1592, pp. 42–55. Springer, Heidelberg (1999)
16. Dodis, Y., Micali, S.: Parallel reducibility for information-theoretically secure computation. In: Bellare, M. (ed.) CRYPTO 2000. LNCS, vol. 1880, pp. 74–92. Springer, Heidelberg (2000)
17. Feige, U., Kilian, J., Naor, M.: A minimal model for secure computation (extended abstract). In: STOC, pp. 554–563. ACM (1994)
18. Fischer, M.J., Lynch, N.A.: A lower bound for the time to assure interactive consistency. Inf. Process. Lett. 14(4), 183–186 (1982)
19. Fitzi, M., Garay, J.A., Gollakota, S., Pandu Rangan, C., Srinathan, K.: Round-optimal and efficient verifiable secret sharing. In: Halevi, S., Rabin, T. (eds.) TCC 2006. LNCS, vol. 3876, pp. 329–342. Springer, Heidelberg (2006)
20. Fitzi, M., Hirt, M., Maurer, U.M.: General adversaries in unconditional multi-party computation. In: Lam, K.-Y., Okamoto, E., Xing, C. (eds.) ASIACRYPT 1999. LNCS, vol. 1716, pp. 232–246. Springer, Heidelberg (1999)
21. Franklin, M.K., Yung, M.: Communication complexity of secure computation (extended abstract). In: STOC, pp. 699–710. ACM (1992)
22. Gács, P., Körner, J.: Common information is far less than mutual information. Problems of Control and Information Theory 2(2), 149–162 (1973)
23. Gál, A., Rosén, A.: Omega(log n) lower bounds on the amount of randomness in 2-private computation. SIAM J. Comput. 34(4), 946–959 (2005)

24. Gennaro, R., Ishai, Y., Kushilevitz, E., Rabin, T.: The round complexity of verifiable secret sharing and secure multicast. In: Proceedings of the Thirty-third Annual ACM Symposium on Theory of Computing, pp. 580–589. ACM (2001)
25. Gentry, C.: Fully homomorphic encryption using ideal lattices. In: STOC, pp. 169–178. ACM (2009)
26. Hirt, M., Lucas, C., Maurer, U., Raub, D.: Passive corruption in statistical multiparty computation. In: Smith, A. (ed.) ICITS 2012. LNCS, vol. 7412, pp. 129–146. Springer, Heidelberg (2012), http://eprint.iacr.org/2012/272
27. Hirt, M., Maurer, U.M.: Complete characterization of adversaries tolerable in secure multi-party computation (extended abstract). In: PODC, pp. 25–34 (1997)
28. Ishai, Y., Kushilevitz, E.: On the hardness of information-theoretic multiparty computation. In: Cachin, C., Camenisch, J.L. (eds.) EUROCRYPT 2004. LNCS, vol. 3027, pp. 439–455. Springer, Heidelberg (2004)
29. Ishai, Y., Kushilevitz, E., Meldgaard, S., Orlandi, C., Paskin-Cherniavsky, A.: On the power of correlated randomness in secure computation. In: Sahai, A. (ed.) TCC 2013. LNCS, vol. 7785, pp. 600–620. Springer, Heidelberg (2013)
30. Katz, J., Koo, C.-Y., Kumaresan, R.: Improving the round complexity of vss in point-to-point networks. Inf. Comput. 207(8), 889–899 (2009)
31. Kushilevitz, E.: Privacy and communication complexity. In: FOCS, pp. 416–421. IEEE (1989)
32. Kushilevitz, E., Mansour, Y.: Randomness in private computations. SIAM J. Discrete Math. 10(4), 647–661 (1997)
33. Kushilevitz, E., Nisan, N.: Communication complexity. Cambridge University Press, New York (1997)
34. Lee, E.J., Abbe, E.: A Shannon approach to secure multi-party computations, arXiv, 1401.7360 (preprint, 2014)
35. Maurer, U.M., Wolf, S.: Secret-key agreement over unauthenticated public channels iii: Privacy amplification. IEEE Transactions on Information Theory 49(4), 839–851 (2003)
36. Naor, M., Nissim, K.: Communication preserving protocols for secure function evaluation. In: STOC, pp. 590–599 (2001)
37. Orlitsky, A., Roche, J.R.: Coding for computing. IEEE Transactions on Information Theory 47(3), 903–917 (2001)
38. Patra, A., Choudhary, A., Rabin, T., Rangan, C.P.: The round complexity of verifiable secret sharing revisited. In: Halevi, S. (ed.) CRYPTO 2009. LNCS, vol. 5677, pp. 487–504. Springer, Heidelberg (2009)
39. Prabhakaran, M.M., Prabhakaran, V.M.: Communication complexity lower bounds from assisted common information. Under Preparation
40. Prabhakaran, V.M., Prabhakaran, M.M.: Assisted common information with an application to secure two-party sampling. IEEE Transactions on Information Theory 60(6), 3413–3434 (2014)
41. Winkler, S., Wullschleger, J.: On the efficiency of classical and quantum oblivious transfer reductions. In: Rabin, T. (ed.) CRYPTO 2010. LNCS, vol. 6223, pp. 707–723. Springer, Heidelberg (2010); Full version, arXiv, 1205.5136
42. Wolf, S., Wullschleger, J.: New monotones and lower bounds in unconditional two-party computation. IEEE Transactions on Information Theory 54(6), 2792–2797 (2008)
43. Wyner, A.D.: The wire-tap channel. The Bell System Technical Journal 54(8), 1355–1387 (1975)
44. Yao, A.C.-C.: Some complexity questions related to distributive computing (preliminary report). In: STOC, pp. 209–213. ACM (1979)

Optimal Non-perfect Uniform Secret Sharing Schemes

Oriol Farràs[1],[*], Torben Hansen[2],[**], Tarik Kaced[3],[***], and Carles Padró[4],[†]

[1] Universitat Rovira i Virgili, Tarragona, Catalonia, Spain
[2] Aarhus University, Aarhus, Denmark
[3] The Chinese University of Hong Kong, Hong Kong
[4] Nanyang Technological University, Singapore
oriol.farras@urv.cat, torben.brandt.hansen@post.au.dk,
tarik@inc.cuhk.edu.hk, carlespl@ntu.edu.sg

Abstract. A secret sharing scheme is non-perfect if some subsets of participants that cannot recover the secret value have partial information about it. The information ratio of a secret sharing scheme is the ratio between the maximum length of the shares and the length of the secret. This work is dedicated to the search of bounds on the information ratio of non-perfect secret sharing schemes. To this end, we extend the known connections between polymatroids and perfect secret sharing schemes to the non-perfect case.

In order to study non-perfect secret sharing schemes in all generality, we describe their structure through their access function, a real function that measures the amount of information that every subset of participants obtains about the secret value. We prove that there exists a secret sharing scheme for every access function.

Uniform access functions, that is, the ones whose values depend only on the number of participants, generalize the threshold access structures. Our main result is to determine the optimal information ratio of the uniform access functions. Moreover, we present a construction of linear secret sharing schemes with optimal information ratio for the rational uniform access functions.

Keywords: Secret sharing, Non-perfect secret sharing, Information Ratio, Polymatroid.

* Supported by the European Commission under FP7 project "Inter-Trust", by the Gov. of Spain through Juan de la Cierva grant, projects TIN2011C27076-C03-01 and Consolider Ingenio 695 2010 CSD2007-00004, and by the Gov. of Catalonia under grant 2009 SGR 1135.
** Supported by a grant from the Oticon Foundation and an Erasmus Mobility grant.
*** Supported by a grant from University Grants Committee of the Hong Kong S.A.R. (Project No. AoE/E-02/08).
† Supported by the Singapore National Research Foundation under Research Grant NRF-CRP2-2007-03.

J.A. Garay and R. Gennaro (Eds.): CRYPTO 2014, Part II, LNCS 8617, pp. 217–234, 2014.
© International Association for Cryptologic Research 2014

1   Introduction

A *secret sharing scheme* is a method to protect a *secret value* by distributing it into *shares* among a set of *participants* in order to prevent both the disclosure and the loss of the secret. Only *information-theoretically secure* secret sharing schemes are considered in this paper. A set of participants is *authorized* if their shares determine the secret value, while the shares of the participants in a *forbidden* set do not contain any information on the secret value. The *access structure* $\Gamma = (\mathcal{A}, \mathcal{B})$ of a secret sharing scheme consists of the families $\mathcal{A}$ and $\mathcal{B}$ of the forbidden and, respectively, authorized sets of participants. A secret sharing scheme is *perfect* if every subset of participants is either authorized or forbidden.

Secret sharing was independently introduced by Shamir [35] and Blakley [6]. They presented constructions of perfect *threshold* secret sharing schemes, in which the authorized subsets are those having at least a certain number of participants. In these schemes, the shares have the same length as the secret, which is optimal for perfect secret sharing schemes [22].

Blakley and Meadows [7] introduced the *ramp* secret sharing schemes, the first proposed non-perfect secret sharing schemes. Their main purpose was to improve the efficiency of perfect threshold schemes by relaxing the security requirements. Namely, the shares can be shortened if some unauthorized sets are allowed to obtain *partial* information on the secret value. The access structure of a ramp scheme is described by means of two thresholds t and r. Every set with at most t participants is forbidden, while every set with at least r participants is authorized. In the ramp schemes proposed in [7], the length of every share is $1/(r - t)$ times the length of the secret, which is also optimal [29].

The threshold and ramp schemes proposed in those seminal works [6, 7, 35] are *linear*, that is, they can be described in terms of linear maps over a finite field [8, 23] or in terms of linear codes [27, 28]. Because of their efficiency and homomorphic properties, linear perfect secret sharing schemes play a fundamental role in several areas of cryptography.

Most of the works in the literature on secret sharing deal with perfect schemes. One of the main lines of research is the search for bounds on the length of the shares in perfect secret sharing schemes for general access structures. The main fundamental problems remain unsolved and, in particular, there is a huge gap between the known upper and lower bounds. Most of the known lower bounds are derived from bounds on the *information ratio*, that is, the ratio between the maximum length of the shares and the length of the secret. Such bounds can be found by using the entropy function, a method initiated Karnin et al. [22] and Capocelli et al. [10]. On the basis of the connections between information theory, matroid theory, and secret sharing found by Fujishige [16, 17], Brickell and Davenport [9], and Csirmaz [13], matroids and polymatroids have appeared to be a powerful tool, as it can be seen from several recent works [2–4, 25, 26]. Similar questions have been considered for non-perfect secret sharing schemes too [15, 18, 24, 29, 30, 33], but the research is much less developed in this direction. In particular, only basic bounds on the information ratio of non-perfect secret sharing schemes are known [29, 30].

This work deals with the search for bounds on the information ratio of non-perfect secret sharing schemes. Our main purpose is to further extend results and techniques on perfect secret sharing schemes to the non-perfect case, with a special stress on the use of polymatroids and the construction of efficient linear secret sharing schemes.

Our first step is to choose a suitable way to describe the security requirements of non-perfect secret sharing schemes. This description should be more precise than the access structure. That is, in addition to the forbidden and qualified sets, also the amount of information on the secret value that is obtained by the other sets should be taken into account. We introduce the *access function* of a secret sharing scheme (Definitions 1 and 5), which is a refinement of the *access hierarchies* that are used in [24, 30, 33]. The access function is defined in terms of the entropy function and it is a monotone increasing function on the power set of the set of participants. The forbidden and authorized sets are those in which the value of the access function is 0 and, respectively, 1. For all other sets, the access function measures the relative amount of information on the secret value given by the shares. A similar concept, *fractional access structure*, was introduced in [18], but the partial information on the secret is measured in a different way. The relation between these two approaches is discussed in Section 2.

Our first result deals with a fundamental question. Namely, given a real-valued access function, does there exist a secret sharing scheme realizing it? By answering this question in the affirmative in Theorem 1, we generalize the result by Ito, Saito and Nishizeki [19], who proved that there exists a perfect secret sharing scheme for every access structure. Our result is not entirely obvious since the usual approach of using linear schemes cannot work. Indeed, there are only countably many linear secret sharing schemes over finite fields, while there are uncountably many access functions. Therefore, some access functions are inherently non-linear or might only be realized in the limit by a sequence of linear schemes. Nevertheless, we prove that every rational-valued access function admits a linear secret sharing scheme. If the access function takes non-rational values, then our construction requires to take a non-uniform probability distribution on the set of possible values of the secret. Similarly to the known general constructions of perfect secret sharing schemes [5, 19], our general construction is inefficient because the length of the shares grows exponentially with the number of participants.

The main problem we consider in this work is the search for bounds on the information ratio of secret sharing schemes for general access functions. For the first time, we apply to non-perfect schemes the polymatroid-based techniques that have been so useful for the perfect case.

The well known connection between perfect secret sharing and polymatroids is extended to non-perfect schemes in Section 5. Our definition of access function appears to be most suitable for our purposes. This can be seen, for instance, in Proposition 3, in which the characterization by Csirmaz [13, Proposition 2.3] of the compatibility between polymatroids and access structures is easily extended to non-perfect secret sharing.

Two different lower bounds on the optimal information ratio of access functions are discussed in Section 6. The first one is the extension of the parameter κ [25] to the non-perfect case. The second one, which is denoted by ϵ, is introduced in this paper. It is a lower bound on κ, and hence a lower bound on the optimal information ratio. The value of ϵ is 1 on every perfect access function, and hence this new parameter is relevant only for the non-perfect case. As a consequence of Proposition 4, the parameter ϵ improves the previously known lower bound in Proposition 1 [29, 30]. We prove in Proposition 5 that $\epsilon \leq \kappa \leq n\epsilon$, where n is the number of participants. This generalizes the known bounds $1 \leq \kappa \leq n$ [13, 25] for perfect secret sharing. As in the perfect case, the upper bound on κ indicates the limitations of using only Shannon information inequalities in the search of lower bounds on the information ratio.

Our main result deals with *uniform* access functions, that is, the ones that take the same value on sets that have the same cardinality. They generalize the perfect threshold access structures. Our main result is presented in Section 8. Namely, we determine in Theorem 4 the exact value of the optimal information ratio of all uniform access functions. Moreover, our proof provides a method to construct a linear secret sharing scheme with optimal information ratio for every given rational uniform access function. This is done in several steps. First, we prove in Proposition 8 that every uniform access function is a suitable convex combination of ramp access functions. As a consequence, the values of κ and ϵ coincide for the uniform access functions. Moreover, combining Proposition 8 with the basic concatenation method described in Section 7, one can construct a linear secret sharing scheme with optimal information ratio (that is, equal to the lower bound ϵ) for every rational uniform access function.

Due to space restrictions, we only present the proofs of the main results. The reader can find the remaining proofs in the full version of the paper [14].

2   Related Work

Brickell and Davenport [9] proved that every perfect secret sharing scheme in which all shares have the same length as the secret defines a matroid. This connection between secret sharing schemes and matroids was first extended to non-perfect schemes by Kurosawa et al. [24], who characterized the non-perfect secret sharing schemes that define a matroid. Recently, a characterization with weaker conditions has been presented [15]. Similarly to the results in this paper, its proof is based on the connection between secret sharing and polymatroids.

The polymatroid-based method described in [13, 25] is applied here for the first time to find lower bounds on the optimal information ratio of non-perfect secret sharing schemes. Some lower bounds on the information ratio of non-perfect secret sharing schemes were found by that entropy-based method. Namely, the one given in Proposition 1 [29, 30] and a lower bound for a particular access function [30] that proves that the bound in Proposition 1 is not always attained.

The *almost-perfect secret sharing schemes* introduced in [21] are schemes whose access functions are close to a perfect access function. The possibility of improving the information ratio by realizing a perfect access structure with non-perfect secret sharing schemes with close access functions is explored in that work.

Ishai, Kushilevitz and Strulovich [18] introduced the notion of *fractional secret sharing*, which is a restriction of non-perfect secret sharing. The security requirements of a fractional secret sharing scheme are described in terms of its *fractional access structure*, which is a monotone decreasing function $F : \mathcal{P}(P) \to \{1,\dots,m\}$, where $\mathcal{P}(P)$ is the power set of the set P of participants. Given the shares of the participants in a set $X \subseteq P$, the secret is uniformly distributed over a set of $f(X)$ possible values. In particular, the secret value is uniformly distributed over a set of $m = F(\emptyset)$ elements. Observe that $F(X)$ measures the number of guessing attempts, and hence the amount of work, needed by the participants in X to find the secret value. The main results in [18] are the following: every fractional access structure is realizable, and every uniform (or *symmetric* in their terminology) fractional access structure is efficiently realizable.

The main difference between the approaches in [18] and in this paper is that a fractional access structure fixes the size of the set of possible values of the secret. The following observation illustrates this difference. Every fractional access structure determines a unique access function, but the converse is not true because an access function does not fix the size of the secret, but only the ratio with the amount of information obtained by the sets of participants. Being a more restrictive concept, the problems related to fractional secret sharing are more difficult. In particular, our results do not appear to have a direct application to fractional secret sharing. For example, no optimality result for uniform fractional access structures (an open problem posed in [18, Section 5]) can be directly obtained from our optimality results on uniform access functions. Another difference between the two approaches is the limited power of linear secret sharing schemes when dealing with fractional secret sharing. Indeed, a fractional access structure can be realized by a linear secret sharing scheme over a field of order q (see Definition 11) only if all its values are powers of q.

Our optimality result for uniform access functions (Theorem 4) is closely related to a recent result by Chen and Yeung [11]. They proved that every $(1, n-1)$-uniform polymatroid is almost entropic. By taking into account that $\kappa = \epsilon$ for the uniform access functions, that implies the result in Remark 6. Nevertheless, the other results in Section 8, namely the value of the optimal information ratio of all uniform access functions and the optimal construction for rational uniform access functions cannot be derived from the results in [11].

3   Secret Sharing Schemes

In this work we consider the definition of secret sharing scheme that is based on information theory, specifically, on the entropy function. For a complete introduction to secret sharing, see [1, 31], and for a textbook on information theory see [12]. We begin by introducing some notation. For a finite set Q, we use $\mathcal{P}(Q)$ to denote its *power set*, that is, the set of all subsets of Q. We use a compact notation for set unions, that is, we write XY for $X \cup Y$ and Xy for $X \cup \{y\}$. In addition, we write $X - Y$ for the set difference and $X - x$ for $X - \{x\}$. Let $X = \{1, \ldots, t\}$ be a set and let $(S_i)_{i \in X}$ be a tuple of discrete random variables. We write S_X for the random variable $S_1 \times \cdots \times S_t$, and $H(S_X)$ for its Shannon entropy. Recall that, for two such random variables S_X, S_Y, one can consider the *conditional entropy* $H(S_X|S_Y) = H(S_{XY}) - H(S_Y)$ and the *mutual information* $I(S_X:S_Y) = H(S_X) - H(S_X|S_Y)$. All through the paper, P and Q stand for finite sets with $Q = Pp_o$ for some $p_o \notin P$.

Definition 1 (Access function). *An* access function *on a set P is a monotone increasing function*

$$\Phi : \mathcal{P}(P) \to [0, 1]$$

with $\Phi(\emptyset) = 0$ and $\Phi(P) = 1$. An access function is said to be perfect *if its only values are 0 and 1. An access function is called* rational *if it only takes rational values.*

Definition 2 (Access structure). *If $\mathcal{A}, \mathcal{B} \subseteq \mathcal{P}(P)$ are nonempty families of subsets of P such that $\mathcal{A}$ is monotone decreasing, $\mathcal{B}$ is monotone increasing, and $\mathcal{A} \cap \mathcal{B} = \emptyset$, then the pair $\Gamma = (\mathcal{A}, \mathcal{B})$ is called an* access structure *on P. The sets in $\mathcal{A}$ and the sets in $\mathcal{B}$ are, respectively, the* forbidden *and the* authorized *sets of the access structure Γ. In a* perfect *access structure, every subset of P is either forbidden or authorized.*

Definition 3. *For an access function Φ on P, a set $X \subseteq P$ is* forbidden *for Φ if $\Phi(X) = 0$, and it is* authorized *for Φ if $\Phi(X) = 1$. Then every access function Φ on P determines an access structure $\Gamma(\Phi) = (\mathcal{A}(\Phi), \mathcal{B}(\Phi))$ on P, where $\mathcal{A}(\Phi) \subseteq \mathcal{P}(P)$ and $\mathcal{B}(\Phi) \subseteq \mathcal{P}(P)$ are the families of the forbidden and, respectively, the authorized subsets for Φ.*

Definition 4 (Secret sharing scheme). *Let Q be a finite set of participants, let $p_o \in Q$ be a distinguished participant, which is called* dealer, *and take $P = Q - p_o$. A secret sharing scheme Σ on the set P is a collection $(S_i)_{i \in Q}$ of discrete random variables such that $H(S_{p_o}) > 0$ and $H(S_{p_o}|S_P) = 0$. The random variable S_{p_o} corresponds to the* secret, *while the random variables $(S_i)_{i \in P}$ correspond to the* shares *of the secret that are distributed among the participants in P.*

Definition 5 (Access function and access structure of a secret sharing scheme). *The* access function Φ *of a secret sharing scheme $\Sigma = (S_i)_{i \in Q}$ is defined by*

$$\Phi(X) = \frac{I(S_{p_o}:S_X)}{H(S_{p_o})}.$$

In addition, $\Gamma(\Phi)$ is the access structure *of the secret sharing scheme Σ. A secret sharing scheme is* perfect *if its access function is perfect.*

If $X \subseteq P$ is an authorized set for Σ, then $H(S_{p_o}|S_X) = 0$, which implies that the secret values can be recovered from the shares of the participants in X. On the other hand, the random variables S_{p_o} and S_X are independent if X is a forbidden set for Σ. In this situation the shares of the participants in X do not provide any information on the secret value. In any other case, the value $\Phi(X)$ determines the amount of information on the secret that is provided by the shares of the participants in X.

Definition 6 (Gap and maximum increment). *The* gap $g(\Gamma)$ *of an access structure* $\Gamma = (\mathcal{A}, \mathcal{B})$ *is defined as* $g(\Gamma) = \min\{|B - A| : A \in \mathcal{A}, B \in \mathcal{B}\}$. *The gap* $g(\Phi)$ *of an access function* Φ *is defined as the gap of the associated access structure. The maximum value* $\Phi(Xy) - \Phi(X)$ *for* $X \subseteq P$ *and* $y \in P$ *is called the* maximum increment *of the access function* Φ *is denoted by* $\mu(\Phi)$.

Definition 7 (Least common denominator of a rational access function). *The* least common denominator $M(\Phi)$ *of a rational access function* Φ *is the least common denominator of the values of* Φ.

Definition 8 (Uniform access function). *An access function* Φ *on* P *is uniform if* $\Phi(A) = \Phi(B)$ *for every* $A, B \subseteq P$ *with* $|A| = |B|$. *Uniform secret sharing schemes are those with uniform access function.*

Definition 9 (Threshold access structure). *Let* t, r, n *be integers with* $0 \leq t < r \leq n$. *In the* (t, r, n)-*threshold access structure on a set* P *with* $|P| = n$, *the forbidden sets are those with at most* t *participants, and the authorized sets are those with at least* r *participants. The values* t *and* r *are called, respectively, the* privacy threshold *and the* reconstruction threshold.

A threshold access structure is perfect if and only if $r = t + 1$. Observe that every uniform access function defines a threshold access structure. The privacy and reconstruction thresholds of a uniform access function are those of the associated threshold access structure. Ramp access functions form an important class of uniform access functions.

Definition 10 (Ramp access function). *Given integers* t, r, n *with* $0 \leq t < r \leq n$, *the* (t, r, n)-*ramp access function on a set* P *with* $|P| = n$ *is defined by:* $\Phi(X) = 0$ *if* $|X| \leq t$, *and* $\Phi(X) = (|X| - t)/(r - t)$ *if* $t < |X| < r$, *and* $\Phi(X) = 1$ *if* $|X| \geq r$.

Example 1. A variant of Shamir's threshold scheme [35] provides a secret sharing scheme for every ramp access function. This construction was first presented in the seminal work on non-perfect secret sharing by Blakley and Meadows [7]. Consider the (t, r, n)-ramp access function on the set $P = \{1, \ldots, n\}$. Let $\mathbb{K}$ be a finite field of size $|\mathbb{K}| \geq n + g$, where $g = r - t$ is the gap of the access function, and take $n + g$ different elements $y_1, \ldots, y_g, x_1, \ldots, x_n \in \mathbb{K}$. By choosing uniformly at random a polynomial $f \in \mathbb{K}[X]$ with degree at most $r - 1$, one obtains random variables $S_{p_o} = (f(y_1), \ldots, f(y_g)) \in \mathbb{K}^g$ and $S_i = f(x_i) \in \mathbb{K}$ for every $i \in P$. It is not difficult to check that these random variables define a secret sharing scheme for the (t, r, n)-ramp access function on P.

The length of the shares is a measure for the efficiency of a secret sharing scheme. We use the Shannon entropy as an approximation of the shortest binary codification. The *information ratio* $\sigma(\Sigma)$ of a secret sharing $\Sigma = (S_i)_{i \in Q}$ is the ratio between the maximum length of the shares and the length of the secret value, that is,

$$\sigma(\Sigma) = \frac{\max_{i \in P} H(S_i)}{H(S_{p_o})}.$$

The *optimal information ratio* $\sigma(\Phi)$ *of an access function* Φ is defined as the infimum of the information ratios of the secret sharing schemes for Φ. A secret sharing scheme attaining $\sigma(\Phi)$ is called *optimal*. The following is a well known lower bound on the optimal information ratio. An alternative proof for this result is presented here in Propositions 4 and 5.

Proposition 1 ([22, 29, 30]). *Let* Φ *be an access function with maximum increment* $\mu(\Phi)$ *and gap* $g(\Phi)$*. Then its optimal information ratio* $\sigma(\Phi)$ *satisfies* $\sigma(\Phi) \geq \mu(\Phi) \geq 1/g(\Phi)$*. In particular, the optimal information ratio of every perfect access function is at least 1.*

Definition 11 (Linear secret sharing scheme). *Let* $\mathbb{K}$ *be a finite field and let* ℓ *be a positive integer. In a* $(\mathbb{K}, \ell)$*-linear secret sharing scheme, the random variables* $(S_i)_{i \in Q}$ *are given by surjective* $\mathbb{K}$*-linear maps* $S_i : E \to E_i$*, where the uniform probability distribution is taken on* E *and* $\dim E_{p_o} = \ell$*.*

In a $\mathbb{K}$-linear secret sharing scheme $(S_i)_{i \in Q}$, the random variable S_X is uniform on its support for every $X \subseteq Q$. Because of that, $H(S_X) = \text{rank}\, S_X \cdot \log |\mathbb{K}|$, and hence

$$I(S_{p_o} : S_X) = (\text{rank}\, S_{p_o} + \text{rank}\, S_X - \text{rank}\, S_{Xp_o}) \log |\mathbb{K}|.$$

This implies that the access function of every linear secret sharing scheme is rational and its information ratio is rational too. For a rational access function Φ, we define $\lambda(\Phi)$ as the infimum of the information ratios of the linear secret sharing schemes for Φ. Clearly, $\lambda(\Phi)$ is an upper bound of $\sigma(\Phi)$.

Remark 1. The secret sharing scheme presented in Example 1 is linear. As a consequence, the (t, r, n)-ramp access function admits a $(\mathbb{K}, g)$-linear secret sharing scheme for every finite field $\mathbb{K}$ with $|\mathbb{K}| \geq n+g$, where $g = r-t$. By Proposition 1, this linear scheme has optimal information ratio, equal to the lower bound $1/g$.

Remark 2. A $(\mathbb{K}, \ell)$-linear secret sharing scheme with information ratio σ is determined by linear maps $S_i : E \to E_i$ with $\dim E_i \leq \max\{\ell, \sigma\ell\}$ for every $i \in Q$ and $\dim E \leq \sum_{i \in Q} \dim E_i$. Therefore, the computation time for both the distribution phase (computing the secret value and the shares) and the reconstruction phase (partially or totally recovering the secret value from some shares) is polynomial in $\log |\mathbb{K}|$, ℓ, σ and the number of participants.

Remark 3. Let Φ be a rational access function on P and let $M = M(\Phi)$ be its least common denominator. Clearly, $\ell \geq M$ for every $(\mathbb{K}, \ell)$-linear secret sharing scheme for Φ. Therefore, by Remark 2, the efficiency of the linear secret sharing schemes for Φ depends on $M(\Phi)$.

4   A Secret Sharing Scheme for Every Access Function

It is well known that every perfect access function admits a perfect secret sharing scheme [19, 5]. We present in Theorem 1 an extension of this result to the general case.

Remark 4. Similarly to the construction in [19] for the perfect case, our general construction is based on a very simple perfect secret sharing scheme for which the only authorized set is the full set of participants. Let G be a finite abelian group (with additive notation). Let T_{p_o} be an arbitrary random variable with support G. Fix a participant $q \in P$ and take independent uniform random variables $(T_i)_{i \in P-q}$ with support G. Finally, take $T_q = T_{p_o} - \sum_{i \in P-q} T_i$. It is not difficult to see that $\mathbf{T} = (T_i)_{i \in Q}$ is a perfect secret sharing scheme whose only authorized set is P.

Theorem 1. *Every access function admits a secret sharing scheme. Moreover, every rational access function Φ admits a $(\mathbb{K}, M(\Phi))$-linear secret sharing schemes for every finite field $\mathbb{K}$.*

Proof. Let Φ be an access function on the set of participants P. Let M be the smallest positive integer such that $\lceil M\Phi(X) \rceil \neq \lceil M\Phi(Y) \rceil$ if $\Phi(X) \neq \Phi(Y)$. Consider the sets

- $\Omega = \{\lceil M\Phi(X) \rceil : X \subseteq P\} - \{0\} \subseteq \{1, \ldots, M\}$, and
- $\Omega_1 = \{\lceil M\Phi(X) \rceil : X \subseteq P, M\Phi(X) \notin \mathbb{Z}\} \subseteq \Omega$.

We construct in the following a secret sharing scheme $\Sigma = (S_i)_{i \in Q}$ for Φ.

We begin by describing the random variable S_{p_o} corresponding to the secret value. Specifically, we take $S_{p_o} = \prod_{k=1}^{M} S^k$, where $(S^k)_{1 \leq k \leq M}$ are independent random variables with entropy $H(S^k) = 1$ that are described next. Let $\mathbb{F}_2$ be the field with order 2 and let h be the binary entropy function. If $k = \lceil M\Phi(X) \rceil \in \Omega_1$, take $\epsilon_k = k - M\Phi(X)$, which satisfies $0 < \epsilon_k < 1$, and take $S^k = S_0^k \times S_1^k$, where S_0^k and S_1^k are independent random variables with support $\mathbb{F}_2$ such that $\Pr[S_0^k = 0] = \min h^{-1}(\epsilon_k)$ and $\Pr[S_1^k = 0] = \min h^{-1}(1 - \epsilon_k)$. If $k \in \{1, \ldots, M\} - \Omega_1$, then S^k is a uniform random variable with support $\mathbb{F}_2$.

Now, we proceed to describe the random variables corresponding to the shares of the participants. Take $k \in \Omega$. Let $\mathcal{C}_k \subseteq \mathcal{P}(P)$ be the family of the subsets $X \subseteq P$ with $\lceil M\Phi(X) \rceil = k$ that are minimal with this property. Consider the random variable

$$T_{p_o}^k = S^1 \times \cdots \times S^{k-1} \times \widehat{S}^k,$$

where $\widehat{S}^k = S_1^k$ if $k \in \Omega_1$ and $\widehat{S}^k = S^k$ otherwise. Observe that $H(T_{p_o}^k) = M\Phi(X)$ for every $X \in \mathcal{C}_k$. The support of $T_{p_o}^k$ is $\mathbb{F}_2^m$ for some integer $m \geq k$. For every $X \in \mathcal{C}_k$, take the secret sharing scheme $\mathbf{T}^{(X)} = (T_i^{(X)})_{i \in Xp_o}$ described in Remark 4 with $T_{p_o}^{(X)} = T_{p_o}^k$ and $G = \mathbb{F}_2^m$. The random variable $T_{p_o}^k$ is the same for all schemes $\mathbf{T}^{(X)}$ with $X \in \mathcal{C}_k$, that is, all these schemes distribute shares for the same secret value. The other random variables $T_i^{(X)}$ are instantiated independently for different sets X. For every participant $i \in P$ take the family of subsets

$$\mathcal{D}_i = \bigcup_{k \in \Omega} \{X \in \mathcal{C}_k : i \in X\} \subseteq \mathcal{P}(P).$$

Finally, the random variable S_i corresponding to the share of a participant $i \in P$ is defined by

$$S_i = \prod_{X \in \mathcal{D}_i} T_i^{(X)}.$$

That is, the share of every participant is composed of sub-shares from the schemes $\mathbf{T}^{(X)}$ corresponding to the sets $X \subseteq P$ such that $i \in X$ and $X \in \mathcal{C}_k$ for some $k \in \Omega$.

Clearly, $H(T_{p_o}^k | S_Y) = 0$ for every subset $Y \subseteq P$ with $k = \lceil M\Phi(Y) \rceil$. On the other hand, it is not difficult to prove that the shares of the participants in Y do not provide any information about the other components of the secret value, and hence $I(S_{p_o} : S_Y) = H(T_{p_o}^k) = M\Phi(Y)$. Since $H(S_{p_o}) = M$, this implies that the scheme $\Sigma = (S_i)_{i \in Q}$ has access function Φ.

Some modifications in the previous construction are needed to prove the second part of the theorem. If Φ is rational, take $M = M(\Phi)$, the least common denominator of Φ. The set Ω is defined analogously but in this case $\Omega_1 = \emptyset$. Given a finite field $\mathbb{K}$, take $S_{p_o} = \prod_{k=1}^M S^k$, where $(S^k)_{1 \leq k \leq M}$ are independent random variables and each S_k is a uniform random variable with support $\mathbb{K}$. At this point, a $(\mathbb{K}, M)$-linear secret sharing scheme with access function Φ can be constructed by using the same steps as in the previous construction. $\square$

The above construction is not efficient because the information ratio is exponential in the number of participants. The construction can be refined in order to slightly decrease the information ratio, but no constructions are known in which the information ratio is not exponential.

5   Polymatroids and Secret Sharing

The connection between perfect secret sharing schemes and polymatroids has been used in order to obtain bounds on the information ratio [13, 25]. It is derived from the connection between polymatroids and Shannon entropy that was discovered by Fujishige [16, 17] and is described here in Theorem 2. In this section, we discuss the extension of this connection to non-perfect secret sharing schemes. For a function $F : \mathcal{P}(Q) \to \mathbb{R}$, a subset $X \subseteq Q$ and $y, z \in Q$, we notate

$$\Delta_F(X; y, z) = F(Xy) + F(Xz) - F(Xyz) - F(X).$$

Definition 12. *A polymatroid is a pair $\mathcal{S} = (Q, f)$ formed by a finite set Q, the ground set, and a rank function $f : \mathcal{P}(Q) \to \mathbb{R}$ satisfying the following properties.*

- $f(\emptyset) = 0$.
- f *is* monotone increasing: *if $X \subseteq Y \subseteq Q$, then $f(X) \leq f(Y)$.*
- f *is* submodular: $f(X \cup Y) + f(X \cap Y) \leq f(X) + f(Y)$ *for every $X, Y \subseteq Q$.*

If $\mathcal{S} = (Q, f)$ is a polymatroid, then $\lambda\mathcal{S} = (Q, \lambda f)$ is also a polymatroid for every positive real number λ. We say that $\lambda\mathcal{S}$ is a *multiple* of $\mathcal{S}$. The following characterization of rank functions of polymatroids is a straightforward consequence of [34, Theorem 44.1].

Proposition 2. *A map* $f \colon \mathcal{P}(Q) \to \mathbb{R}$ *is the rank function of a polymatroid with ground set* Q *if and only if* $f(\emptyset) = 0$ *and* $\Delta_f(X; y, z) \geq 0$ *for every* $X \subseteq Q$ *and* $y, z \in Q - X$.

Theorem 2 (Fujishige [16, 17]). *If* $(S_i)_{i \in Q}$ *is a tuple of discrete random variables, then the map* $f \colon \mathcal{P}(Q) \to \mathbb{R}$ *defined by* $f(X) = H(S_X)$ *is the rank function of a polymatroid with ground set* Q.

Because of the connection between polymatroids and the Shannon entropy described in the previous theorem, and by analogy to the conditional entropy, we write $f(X|Y) = f(XY) - f(Y)$ for every $X, Y \subseteq Q$.

As a consequence of Theorem 2, every secret sharing scheme determines a polymatroid. For perfect secret sharing schemes, this connection was first used in [13]. This is a useful tool for the study of secret sharing schemes.

Definition 13. *Let* $\Sigma = (S_i)_{i \in Q}$ *be a secret sharing scheme on* P. *Every multiple of the polymatroid* (Q, f), *where* $f(X) = H(S_X)$ *for every* $X \subseteq Q$, *is called a* Σ-*polymatroid.*

Definition 14. *Let* Φ *be an access function on* P *and let* $\mathcal{S} = (Q, f)$ *be a polymatroid. Then* $\mathcal{S}$ *is an* Φ-*polymatroid if*

$$\Phi(X) = \frac{f(p_o) - f(p_o|X)}{f(p_o)}$$

for every $X \subseteq P$.

We say that a polymatroid $\mathcal{S} = (Q, f)$ is *normalized* if $f(p_o) = 1$. A polymatroid $\mathcal{S} = (P, f)$ is *compatible* with the access function Φ if $\mathcal{S}$ can be extended to a normalized Φ-polymatroid $\mathcal{S}' = (Q, f)$. The following is a generalization of a result by Csirmaz [13, Proposition 2.3] on perfect secret sharing.

Proposition 3. *A polymatroid* $\mathcal{S} = (P, f)$ *is compatible with an access function* Φ *on* P *if and only if* $\Delta_f(X; y, z) \geq \Delta_\Phi(X; y, z)$ *for every* $X \subseteq P$ *and* $y, z \in P - X$.

6   Lower Bounds on the Information Ratio

On the basis of the connection between secret sharing and polymatroids, we introduce in this section two parameters, $\kappa(\Phi)$ and $\epsilon(\Phi)$, that provide lower bounds on the optimal information ratio $\sigma(\Phi)$. The first one is a straightforward generalization of the corresponding parameter for perfect secret sharing that was introduced in [25]. The second one is only relevant for non-perfect secret sharing. It makes it possible to generalize a previous result by Csirmaz on the limitation of Shannon inequalities to find lower bounds on the information ratio and, more importantly, to find a tight lower bound on the optimal information ratio of uniform access functions.

For a polymatroid $\mathcal{S} = (Q, f)$ we define

$$\sigma_{p_o}(\mathcal{S}) = \frac{\max_{x \in P} f(x)}{f(p_o)}.$$

If $\mathcal{S}$ is a Σ-polymatroid, then $\sigma(\Sigma) = \sigma_{p_o}(\mathcal{S})$. In addition, we define

$$\kappa(\Phi) = \inf\{\sigma_{p_o}(\mathcal{S}) : \mathcal{S} \text{ is a } \Phi\text{-polymatroid}\}. \tag{1}$$

Observe that, if Σ is a secret sharing with access function Φ, then every Σ-polymatroid is a Φ-polymatroid. Because of that, $\kappa(\Phi) \leq \sigma(\Phi)$. It is not difficult to prove that $\kappa(\Phi) \geq \mu(\Phi) \geq 1/g(\Phi)$ for every access function Φ [15, 29, 30]. In particular, this implies the well known fact that the information ratio of every perfect secret sharing scheme is at least 1.

The search of $\kappa(\Phi)$ for an access function Φ can be restricted to the family of the normalized Φ-polymatroids. The value of $\kappa(\Phi)$, which is a lower bound on $\sigma(\Phi)$, can be computed by means of the linear programming program determined by the Shannon information inequalities and the access function. This approach has been used in several works on perfect secret sharing, as for instance [32]. The infimum in (1) is a minimum and, moreover, $\kappa(\Phi)$ is rational if Φ is rational.

For an ordering $\tau = (\tau_1, \ldots, \tau_n)$ of the participants in P, we take $A_0^\tau = \emptyset$ and $A_i^\tau = \{\tau_1, \ldots, \tau_i\}$ for every $i = 1, \ldots, n$. For a function $F : \mathcal{P}(P) \to \mathbb{R}$ and for $i = 1, \ldots, n$, consider $\delta_i^\tau(F) = \Delta_F(A_{i-1}^\tau; \tau_i, \tau_n)$. Observe that $\sum_{i=1}^n \delta_i^\tau(F) = F(\tau_n) - F(\emptyset)$.

Definition 15. *Let Φ be an access function on P, with $|P| = n$. We define $\epsilon(\Phi)$ as the maximum of $\sum_{i=1}^n \max\{0, \delta_i^\tau(\Phi)\}$ among all orderings τ of P.*

Observe that $\max\{0, \delta_i^\tau(\Phi)\} \leq \Phi(A_i^\tau) - \Phi(A_{i-1}^\tau)$, and hence $\epsilon(\Phi) \leq 1$. As a consequence of the next proposition, $\epsilon(\Phi) = 1$ if Φ is a perfect access function. In addition, this result provides an alternative proof for the previously known basic lower bounds [29, 33] (see Proposition 1).

Proposition 4. *Let Φ be an access function on P. Then $\epsilon(\Phi) \geq \Phi(Xy) - \Phi(X)$ for every $X \subseteq P$ and $y \in P - X$. In particular, $\epsilon(\Phi) \geq \mu(\Phi) \geq 1/g(\Phi)$.*

It is known that $1 \leq \kappa(\Phi) \leq |P|$ for every perfect access function [13, 22]. These bounds on κ are extended to the non-perfect case by proving that $\epsilon(\Phi) \leq \kappa(\Phi) \leq \epsilon(\Phi) \cdot |P|$ in Proposition 5. Combined with Proposition 4, this implies that $\epsilon(\Phi)$ is in general a better lower bound on $\kappa(\Phi)$ than $1/g(\Phi)$.

Proposition 5. *Let Φ be an access function on a set of n participants. Then $\epsilon(\Phi) \leq \kappa(\Phi) \leq n\epsilon(\Phi)$.*

7   Concatenating Secret Sharing Schemes

We analyze here a simple way to combine secret sharing schemes. For each $j = 1, \ldots, m$ consider a positive integer q_j and a secret sharing scheme $\Sigma_j = (S_{ji})_{i \in Q}$ with access function Φ^j. A secret sharing scheme $\Sigma = \prod_{j=1}^{m} \Sigma_j^{q_j}$ is obtained by concatenating m secret sharing schemes, each consisting of q_j instances of Σ_j. That is, $\Sigma = (S_i)_{i \in Q}$ with $S_i = (S_{1i})^{q_1} \times \cdots \times (S_{mi})^{q_m}$ for every $i \in Q$. Observe that $H(S_X) = \sum_{j=1}^{m} q_j H(S_{jX})$ for every $X \subseteq Q$. Because of that, the access function Φ of Σ is given by

$$\Phi(X) = \frac{I(S_{p_o} : S_X)}{H(S_{p_o})} = \frac{\sum_{j=1}^{m} q_j I(S_{jp_o} : S_{jX})}{\sum_{j=1}^{m} q_j H(S_{jp_o})}$$

for every $X \subseteq Q$. Therefore,

$$\Phi = \sum_{j=1}^{m} \rho_j \Phi^j, \quad \text{where} \quad \rho_j = \frac{q_j H(S_{jp_o})}{\sum_{k=1}^{m} q_k H(S_{kp_o})} \quad \text{for } j = 1, \ldots, m.$$

That is, Φ is a convex combination of the access functions $\Phi^1, \ldots, \Phi^m$. Moreover, if σ_j is the information ratio of Σ_j, then the information ratio σ of Σ satisfies $\sigma \leq \sum_{j=1}^{m} \rho_j \sigma_j$. For more details, see [14].

This leads to the following result, which will be used in our construction of optimal secret sharing schemes for rational uniform access functions.

Proposition 6. *For $j = 1, \ldots, m$, let Φ^j be an access function on P that admits a $(\mathbb{K}, \ell_j)$-linear secret sharing scheme with information ratio σ_j. Let $\rho_1, \ldots, \rho_n$ be rational numbers with $0 < \rho_j < 1$ and $\sum_{j=1}^{m} \rho_j = 1$. Let M be a positive integer such that $M \rho_j$ is integer for every $j = 1, \ldots, m$. Then the access function $\Phi = \sum_{j=1}^{m} \rho_j \Phi^j$ admits a $(\mathbb{K}, \ell)$-linear secret sharing scheme with information ratio $\sigma \leq \sum_{j=1}^{m} \rho_j \sigma_j$ and $\ell = M \ell_1 \cdots \ell_m$.*

8   Uniform Secret Sharing Schemes

Uniform access functions generalize the perfect threshold access structures. It is well known that these access structures admit a linear secret sharing scheme with optimal information ratio, namely Shamir's secret sharing scheme [35]. We extend here this fundamental result by determining in Theorem 4 the optimal information ratio of all uniform access functions and presenting in Theorem 3 a construction of linear secret sharing schemes with optimal information ratio for all rational uniform access functions.

A uniform access function Φ on a set P with $|P| = n$ is determined by the values

$$0 = \Phi_0 \leq \Phi_1 \leq \cdots \leq \Phi_n = 1,$$

where $\Phi(X) = \Phi_i$ for every $X \subseteq P$ with $|X| = i$. Therefore, a uniform access function is determined by its *increment vector*

$$\Phi' = (\Phi'_1, \ldots, \Phi'_n),$$

where $\Phi'_i = \Phi_i - \Phi_{i-1}$. Observe that $\Phi'_i \geq 0$ and $\sum_{i=1}^{n} \Phi'_i = 1$. We use the convention $\Phi'_{n+1} = 0$.

Proposition 7. *Every (rational) uniform access function is a (rational) convex combination of perfect ramp access functions.*

Similarly to the perfect case, every rational uniform access function admits a linear secret sharing scheme with information ratio equal to 1.

Corollary 1. *Let Φ be a rational uniform access function on a set P of n participants and let $M = M(\Phi)$ be the least common denominator of Φ. Then, for every finite field $\mathbb{K}$ with $|\mathbb{K}| \geq n+1$, the access function Φ admits a $(\mathbb{K}, M)$-linear secret sharing scheme with information ratio equal to 1.*

Remark 5. By Remark 2, the efficiency of this linear scheme depends on the least common denominator of the access function. Specifically, the computation time for both the distribution phase and the reconstruction phase is polynomial in $\log|\mathbb{K}|$, $M(\Phi)$ and n.

In the rest of this section we present a construction of optimal linear secret sharing schemes for all rational uniform access functions. Nevertheless, the schemes that are obtained in this way are not efficient in general because the size of the secret value is too large.

Clearly, $\delta_i^\tau(\Phi) = \Phi'_i - \Phi'_{i+1}$ for $i = 1, \ldots, n$ and for every ordering τ of P. Because of that, we notate $\delta_i(\Phi) = \Phi'_i - \Phi'_{i+1}$. In particular, the value of $\epsilon(\Phi)$ is determined by the increment vector.

Lemma 1. *Let Φ be a uniform access function on a set P of size n. Then*

$$\epsilon(\Phi) = \sum_{i=1}^{n} \max\{0, \delta_i(\Phi)\} = \sum_{i=1}^{n} \max\{0, \Phi'_i - \Phi'_{i+1}\}.$$

Example 2. Let Φ be the (t, r, n)-ramp access function, which is uniform and has gap $g = r - t$. The increment vector Φ' is given by $\Phi'_i = 0$ if $1 \leq i \leq t$ or $r + 1 \leq i \leq n + 1$, and $\Phi'_i = 1/g$ if $t + 1 \leq i \leq r$. Therefore, $\delta_t(\Phi) = -1/g$ and $\delta_r(\Phi) = 1/g$, and $\delta_i(\Phi) = 0$ if $i \neq r, t$, and hence $\epsilon(\Phi) = 1/g$.

We proved in Proposition 7 that every uniform access function is a convex combination of ramp access functions. The next proposition is a refinement of that result that makes it possible to find an optimal secret sharing scheme for every rational uniform access function.

Proposition 8. *Let Φ be a uniform access function on a set P. Then there exist ramp access functions $\Phi^1, \ldots, \Phi^m$ on P and positive real numbers $\rho_1, \ldots, \rho_m$ with $\sum_{j=1}^{m} \rho_i = 1$ such that*

$$\Phi = \rho_1\Phi^1 + \cdots + \rho_m\Phi^m$$

and $\epsilon(\Phi) = \rho_1\epsilon(\Phi^1) + \cdots + \rho_m\epsilon(\Phi^m)$. Moreover, if Φ is rational, then the values $\rho_1, \ldots, \rho_m$ are rational.

Proof. We use induction on the gap $g = g(\Phi)$. If $g = 1$, then Φ is a ramp access function and the result obviously holds. Suppose that $g > 1$. Take $n = |P|$ and let t and r be, respectively, the privacy and the reconstruction thresholds of Φ. Then $g = r - t$ and $\Phi'_i = 0$ if $1 \leq i \leq t$ or $r+1 \leq i \leq n+1$, while $\Phi'_{t+1}, \Phi'_r > 0$. Let ℓ be the smallest integer satisfying $t + 1 \leq \ell \leq r$ and $\Phi'_\ell = \min\{\Phi'_{t+1}, \ldots, \Phi'_r\}$. We distinguish two cases.

Case 1: $\Phi'_\ell = 0$. Then $t + 1 < \ell < r$ and $0 < \Phi_\ell < 1$. Take $\rho = \Phi_\ell$ and consider the uniform access functions Ψ^1 and Ψ^2 defined by

$$\Psi^1_i = \min\left\{\frac{\Phi_i}{\Phi_\ell}, 1\right\}, \quad \Psi^2_i = \max\left\{\frac{\Phi_i - \Phi_\ell}{1 - \Phi_\ell}, 0\right\}$$

for every $i = 0, 1, \ldots, n$. Clearly, $\Phi = \rho\Psi^1 + (1 - \rho)\Psi^2$. Since $\Phi'_\ell = \Phi_\ell - \Phi_{\ell-1} = 0$, we have that $\Psi^1_i = 1$ if $i \geq \ell - 1$, and hence $\delta_i(\Psi^1) = 0$ if $i \geq \ell$. In addition, $\Psi^2_i = 0$ if $i \leq \ell$, and hence $\delta_\ell(\Psi^2) \leq 0$ and $\delta_i(\Psi^2) = 0$ if $i \leq \ell - 1$. Therefore,

$$\epsilon(\Phi) = \sum_{i=1}^n \max\{0, \rho\delta_i(\Psi^1) + (1-\rho)\delta_i(\Psi^2)\}$$

$$= \rho\sum_{i=1}^{\ell-1} \max\{0, \delta_i(\Psi^1)\} + (1-\rho)\sum_{i=\ell+1}^n \max\{0, \delta_i(\Psi^2)\}$$

$$= \rho\epsilon(\Psi^1) + (1-\rho)\epsilon(\Psi^2).$$

Since $g(\Psi^1) \leq \ell - t < g(\Phi)$ and $g(\Psi^2) \leq r - \ell < g(\Phi)$ the theorem holds for Φ by the induction hypothesis.

Case 2: $\Phi'_\ell > 0$. Let Ψ^1 be the (t, r, n)-ramp access function on P and take $\rho = g\Phi'_\ell$. If $\rho = 1$, then $\Phi = \Psi^1$ and the proof is concluded. Suppose that $\rho < 1$ and take $\Psi^2 = (\Phi - \rho\Psi^1)/(1 - \rho)$. Observe that $\Psi^2_0 = 0$ and $\Psi^2_n = 1$. We claim that $(\Psi^2)'_i \geq 0$ for every $i = 1, \ldots, n$, and hence Ψ^2 is a uniform access function on P. Indeed, $(\Psi^2)'_i = 0$ if $1 \leq i \leq t$ or $r + 1 \leq i \leq n$, and $(\Psi^2)'_i = (\Phi'_i - \rho(\Psi^1)'_i)/(1-\rho) = (\Phi'_i - \Phi'_\ell)/(1-\rho) \geq 0$ if $t+1 \leq i \leq r$. Since Ψ_1 is a ramp access function, $\delta_t(\Psi^1) = -1/g$ and $\delta_r(\Psi^1) = 1/g$, and $\delta_i(\Psi^1) = 0$ if $i \neq r, t$. Then the three values $\delta_t(\Phi)$, $\delta_t(\Psi^1)$ and $\delta_t(\Psi^2)$ are non-positive, while $\delta_r(\Phi)$, $\delta_r(\Psi^1)$ and $\delta_r(\Psi^2)$ are non-negative. Therefore, $\Phi = \rho\Psi^1 + (1 - \rho)\Psi^2$ and $\epsilon(\Phi) = \rho\epsilon(\Psi^1) + (1 - \rho)\epsilon(\Psi^2)$. The proof is concluded by checking that Ψ^2 is a convex combination of ramp access functions in the required conditions. Observe that $(\Psi^2)'_\ell = 0$. If $\ell = t + 1$ or $\ell = r$, then $g(\Psi^2) < g(\Phi)$ and the result holds by the induction hypothesis. Finally, we can reduce to Case 1 if $t + 1 < \ell < r$. $\quad\square$

Corollary 2. *For every uniform access function* Φ, $\kappa(\Phi) = \epsilon(\Phi)$.

Theorem 3. *Let Φ be a rational uniform access function on a set of participants P. For every finite field $\mathbb{K}$ with $|\mathbb{K}| \geq |P| + g(\Phi)$, there exists a $\mathbb{K}$-linear secret sharing scheme with access function Φ and information ratio $\sigma = \epsilon(\Phi)$. As a consequence, every rational uniform access function admits a linear secret sharing scheme with optimal information ratio.*

Corollary 3. *For every rational uniform access function Φ, $\epsilon(\Phi) = \kappa(\Phi) = \sigma(\Phi) = \lambda(\Phi)$.*

The fact that $\kappa(\Phi) = \sigma(\Phi)$ for a rational uniform access function Φ, proved in Corollary 3, can also be derived from [11]. The result was obtained independently by means of different techniques. However, the computation of the explicit optimal information ratio, and the construction of the optimal scheme was an open problem.

The results presented in Theorem 3 and Corollary 3 deal with rational access functions. For some non-rational access functions, we can also apply the techniques used in the proof of Proposition 8 and construct optimal schemes (see [14]).

We do not have a general method to construct a scheme with optimal information ratio for every uniform access function but, as it is demonstrated in the following remark, we can find secret sharing schemes whose parameters are arbitrarily close to the required ones.

Remark 6. For every non-rational uniform access function Φ on a set P with n participants, there is a sequence of rational uniform access functions $(\Phi^k)_{k \in \mathbb{N}}$ such that $\lim_{k \to \infty} \sum_{i=0}^{n} |\Phi_i - \Phi_i^k| = 0$. Since $\lim_{k \to \infty} \epsilon(\Phi^k) = \epsilon(\Phi)$ and $\epsilon(\Phi^k) = \sigma(\Phi^k)$, there is a sequence of linear secret sharing schemes $(\Sigma_k)_{k \in \mathbb{N}}$ satisfying $\Phi(\Sigma_k) \to \Phi$ and $\sigma(\Sigma_k) \to \epsilon(\Phi)$.

Nevertheless, this is not enough to prove our main result, Theorem 4. Instead, the following proposition is needed.

Proposition 9. *For every uniform access function Φ, there exists a sequence of secret sharing schemes $(\Sigma^k)_{k \in \mathbb{N}}$ realizing Φ whose information ratios $\sigma(\Sigma^k)$ converge to $\epsilon(\Phi)$ as $k \to \infty$.*

Proof. By Theorem 3, the result is obvious for rational access functions. Let Φ be a non-rational uniform access function on a set P with n participants. By Proposition 8, there exist ramp access functions $\Phi^1, \ldots, \Phi^m$ on P and positive real numbers $\rho_1, \ldots, \rho_m$ with $\sum_{j=1}^{m} \rho_i = 1$ such that $\Phi = \rho_1 \Phi^1 + \cdots + \rho_m \Phi^m$ and $\epsilon(\Phi) = \rho_1 \epsilon(\Phi^1) + \cdots + \rho_m \epsilon(\Phi^m)$. For every $j = 1, \ldots, m$, there exists a sequence of rational numbers $(\rho_{jk})_{k \in \mathbb{N}}$ with $\lim_{k \to \infty} \rho_{jk} = \rho_j$ and $\rho_{jk} \leq \rho_j$ for every $k \in \mathbb{N}$. For every $k \in \mathbb{N}$, consider $\alpha_k = \sum_{j=1}^{m} \rho_{jk}$ and the uniform access functions

$$\Psi^k = \frac{\rho_{1k}}{\alpha_k} \Phi^1 + \cdots + \frac{\rho_{mk}}{\alpha_k} \Phi^m \quad \text{and} \quad \Upsilon^k = \frac{\rho_1 - \rho_{1k}}{1 - \alpha_k} \Phi^1 + \cdots + \frac{\rho_m - \rho_{mk}}{1 - \alpha_k} \Phi^m.$$

Let s be a positive integer with $2^s \geq n + g(\Phi)$ and let $\mathbb{K}$ be the finite field with order 2^s. Since Ψ^k is rational and $g(\Psi^k) \leq g(\Phi)$, by Theorem 3 there exists a $(\mathbb{K}, \ell_k)$-linear secret sharing scheme $\Sigma_1^k = (S_i^k)_{i \in Q}$ with access function Ψ^k and information ratio

$$\sigma(\Sigma_1^k) = \sum_{j=1}^{m} \frac{\rho_{jk}}{\alpha_k} \epsilon(\Phi^j) = \epsilon(\Psi^k).$$

Observe that $H(S^k_{p_o}) = s\ell_k$. Moreover, we can take ℓ_k large enough such that $\lceil s\ell_k \Upsilon^k_i \rceil \neq \lceil s\ell_k \Upsilon^k_{i+1} \rceil$ for every $0 \leq i \leq n-1$ with $\Upsilon^k_i \neq \Upsilon^k_{i+1}$. From the proof of Theorem 1, there exists a secret sharing scheme $\Sigma^k_2 = (T^k_i)_{i \in Q}$ with access function Υ^k and $H(T^k_{p_o}) = s\ell_k$. The information ratio of Σ^k_2 is upper bounded by a quantity ν_n that only depends on the number n of participants. Take positive integers q_k and q'_k such that $1 + q_k/q'_k = 1/\alpha_k$. Let Σ^k be the concatenation of q_k copies of Σ^k_1 and q'_k copies of Σ^k_2. Then the access function of Σ_k is $\alpha_k \Psi^k + (1 - \alpha_k) \Upsilon^k = \Phi$ and its information ratio satisfies $\epsilon(\Phi) \leq \sigma(\Sigma^k) \leq \alpha_k \epsilon(\Psi^k) + (1 - \alpha_k) \nu_n$. The proof is concluded by taking into account that $\lim_{k \to \infty} \epsilon(\Psi^k) = \epsilon(\Phi)$ and $\lim_{k \to \infty} 1 - \alpha_k = 0$. □

Theorem 4. *The optimal information ratio of every uniform access function Φ is equal to $\epsilon(\Phi)$.*

References

1. Beimel, A.: Secret-Sharing Schemes: A Survey. In: Chee, Y.M., Guo, Z., Ling, S., Shao, F., Tang, Y., Wang, H., Xing, C. (eds.) IWCC 2011. LNCS, vol. 6639, pp. 11–46. Springer, Heidelberg (2011)
2. Beimel, A., Ben-Efraim, A., Padró, C., Tyomkin, I.: Multi-linear Secret-Sharing Schemes. In: Lindell, Y. (ed.) TCC 2014. LNCS, vol. 8349, pp. 394–418. Springer, Heidelberg (2014)
3. Beimel, A., Livne, N., Padró, C.: Matroids Can Be Far From Ideal Secret Sharing. In: Canetti, R. (ed.) TCC 2008. LNCS, vol. 4948, pp. 194–212. Springer, Heidelberg (2008)
4. Beimel, A., Orlov, I.: Secret Sharing and Non-Shannon Information Inequalities. IEEE Trans. Inform. Theory 57, 5634–5649 (2011)
5. Benaloh, J., Leichter, J.: Generalized secret sharing and monotone functions. In: Goldwasser, S. (ed.) Advances in Cryptology - CRYPTO 1988. LNCS, vol. 403, pp. 27–35. Springer, Heidelberg (1990)
6. Blakley, G.R.: Safeguarding cryptographic keys. In: AFIPS Conference Proceedings, vol. 48, pp. 313–317 (1979)
7. Blakley, G.R., Meadows, C.: Security of Ramp Schemes. In: Blakely, G.R., Chaum, D. (eds.) Advances in Cryptology - CRYPTO 1984. LNCS, vol. 196, pp. 242–268. Springer, Heidelberg (1985)
8. Brickell, E.F.: Some ideal secret sharing schemes. J. Combin. Math. and Combin. Comput. 9, 105–113 (1989)
9. Brickell, E.F., Davenport, D.M.: On the classification of ideal secret sharing schemes. J. Cryptology 4, 123–134 (1991)
10. Capocelli, R.M., De Santis, A., Gargano, L., Vaccaro, U.: On the Size of Shares for Secret Sharing. J. Cryptology 6, 157–167 (1993)
11. Chen, Q., Yeung, R.W.: Two-Partition-Symmetrical Entropy Function Regions. In: ITW, pp. 1–5 (2013)
12. Cover, T.M., Thomas, J.A.: Elements of Information Theory, 2nd edn. Wiley, New York (2006)
13. Csirmaz, L.: The size of a share must be large. J. Cryptology 10, 223–231 (1997)
14. Farràs, O., Hansen, T., Kaced, T., Padró, C.: Optimal Non-Perfect Uniform Secret Sharing Schemes, Full version: https://eprint.iacr.org/2014/124

15. Farràs, O., Padró, C.: Extending Brickell–Davenport theorem to non-perfect secret sharing schemes. Des. Codes Cryptogr. (2013) (Online First)
16. Fujishige, S.: Polymatroidal Dependence Structure of a Set of Random Variables. Information and Control 39, 55–72 (1978)
17. Fujishige, S.: Entropy functions and polymatroids—combinatorial structures in information theory. Electron. Comm. Japan 61, 14–18 (1978)
18. Ishai, Y., Kushilevitz, E., Strulovich, O.: Lossy Chains and Fractional Secret Sharing. In: STACS 2013. LIPICS, vol. 20, pp. 160–171 (2013)
19. Ito, M., Saito, A., Nishizeki, T.: Secret sharing scheme realizing any access structure. In: Proc. IEEE Globecom 1987, pp. 99–102 (1987)
20. Jackson, W.-A., Martin, K.M.: Geometric secret sharing schemes and their duals. Des. Codes Cryptogr. 4, 83–95 (1994)
21. Kaced, T.: Almost-perfect secret sharing. In: Proceedings of 2011 IEEE International Symposium on Information Theory, ISIT 2011, pp. 1603–1607 (2011) Full version available at arXiv.org, http://arxiv.org/abs/1103.2544
22. Karnin, E.D., Greene, J.W., Hellman, M.E.: On secret sharing systems. IEEE Trans. Inform. Theory 29, 35–41 (1983)
23. Kothari, S.C.: Generalized Linear Threshold Scheme. In: Blakely, G.R., Chaum, D. (eds.) Advances in Cryptology - CRYPTO 1984. LNCS, vol. 196, pp. 231–241. Springer, Heidelberg (1985)
24. Kurosawa, K., Okada, K., Sakano, K., Ogata, W., Tsujii, S.: Nonperfect Secret Sharing Schemes and Matroids. In: Helleseth, T. (ed.) Advances in Cryptology - EUROCRYPT 1993. LNCS, vol. 765, pp. 126–141. Springer, Heidelberg (1994)
25. Martí-Farré, J., Padró, C.: On Secret Sharing Schemes, Matroids and Polymatroids. J. Math. Cryptol. 4, 95–120 (2010)
26. Martín, S., Padró, C., Yang, A.: Secret Sharing, Rank Inequalities and Information Inequalities. In: Canetti, R., Garay, J.A. (eds.) CRYPTO 2013, Part II. LNCS, vol. 8043, pp. 277–288. Springer, Heidelberg (2013)
27. Massey, J.L.: Minimal codewords and secret sharing. In: Proceedings of the 6th Joint Swedish-Russian Workshop on Information Theory, Molle, Sweden, pp. 269–279 (August 1993)
28. McEliece, R.J., Sarwate, D.V.: On Sharing Secrets and Reed-Solomon Codes. Commun. ACM 24, 583–584 (1981)
29. Ogata, W., Kurosawa, K., Tsujii, S.: Nonperfect Secret Sharing Schemes. In: Seberry, J., Zheng, Y. (eds.) Advances in Cryptology - AUSCRYPT 1992. LNCS, vol. 718, pp. 56–66. Springer, Heidelberg (1993)
30. Okada, K., Kurosawa, K.: Lower Bound on the Size of Shares of Nonperfect Secret Sharing Schemes. In: Pieprzyk, J.P., Safavi-Naini, R. (eds.) Advances in Cryptology - ASIACRYPT 1994. LNCS, vol. 917, pp. 33–41. Springer, Heidelberg (1995)
31. Padró, C.: Lecture Notes in Secret Sharing. Cryptology ePrint Archive 2012/674
32. Padró, C., Vázquez, L., Yang, A.: Finding Lower Bounds on the Complexity of Secret Sharing Schemes by Linear Programming. Discrete Appl. Math. 161, 1072–1084 (2013)
33. Paillier, P.: On ideal non-perfect secret sharing schemes. In: Christianson, B., Crispo, B., Lomas, M., Roe, M. (eds.) Security Protocols 1997. LNCS, vol. 1361, pp. 207–216. Springer, Heidelberg (1998)
34. Schrijver, A.: Combinatorial Optimization. Polyhedra and Efficiency. Springer, Berlin (2003)
35. Shamir, A.: How to share a secret. Commun. of the ACM 22, 612–613 (1979)

Proving the TLS Handshake Secure (As It Is)

Karthikeyan Bhargavan[1], Cédric Fournet[2], Markulf Kohlweiss[2],
Alfredo Pironti[1], Pierre-Yves Strub[3], and Santiago Zanella-Béguelin[1,2]

[1] INRIA, Paris, France
{firstname.name}@inria.fr
[2] Microsoft Research, Cambridge, UK
{fournet,markulf}@microsoft.com
[3] IMDEA Software Institute, Madrid, Spain
pierre-yves@strub.nu

Abstract. The TLS Internet Standard features a mixed bag of cryptographic algorithms and constructions, letting clients and servers negotiate their use for each run of the handshake. Although many ciphersuites
are now well-understood in isolation, their composition remains problematic, and yet it is critical to obtain practical security guarantees for
TLS, as all mainstream implementations support multiple related runs
of the handshake and share keys between algorithms.

We study the provable security of the TLS handshake, as it is implemented and deployed. To capture the details of the standard and its main
extensions, we rely on MITLS, a verified reference implementation of the
protocol. We propose new agile security definitions and assumptions for
the signatures, key encapsulation mechanisms (KEM), and key derivation algorithms used by the TLS handshake. To validate our model of key
encapsulation, we prove that both RSA and Diffie-Hellman ciphersuites
satisfy our definition for the KEM. In particular, we formalize the use
of PKCS#1v1.5 and build a 3,000-line EASYCRYPT proof of the security
of the resulting KEM against replayable chosen-ciphertext attacks under
the assumption that ciphertexts are hard to re-randomize.

Based on our new agile definitions, we construct a modular proof
of security for the MITLS reference implementation of the handshake,
including ciphersuite negotiation, key exchange, renegotiation, and resumption, treated as a detailed 3,600-line executable model. We present
our main definitions, constructions, and proofs for an abstract model of
the protocol, featuring series of related runs of the handshake with different ciphersuites. We also describe its refinement to account for the
whole reference implementation, based on automated verification tools.

1 Introduction

TLS is the most widely deployed protocol for securing communications and yet,
after two decades of attacks, patches and extensions, its practical security remains unresolved. One of the most troublesome aspects of the protocol is its
handling of a large number of cryptographic algorithms and constructions. New

J.A. Garay and R. Gennaro (Eds.): CRYPTO 2014, Part II, LNCS 8617, pp. 235–255, 2014.
© International Association for Cryptologic Research 2014

extensions are added to the protocol and its implementations, while older features remain for backward compatibility. Thus, TLS clients and servers offer many choices, and each run of the handshake involves a negotiation of the best protocol version, ciphersuite, and extensions available at both ends. Such a trade-off between flexibility and security creates several problems:

(1) It makes the security of TLS depend on its correct configuration, inasmuch as some versions (e.g. SSL2) and algorithms (e.g. MD5 and RC4) are much weaker than others, and may also suffer from different implementation flaws. In theory, only very restrictive configurations have been proved secure. In practice, dangerous mis-configurations of TLS are commonplace.

(2) It complicates the protocol logic, as the integrity of the negotiation itself relies on algorithms being negotiated; this is a persistent source of attacks, from protocol regression in SSL2 [27] to version fallback in current browsers [18].

(3) It demands stronger security assumptions, to reflect the fact that honest parties may use the same key materials with different algorithms. Intuitively, TLS *on its own* enables a range of chosen-protocol attacks whereby a weak algorithm (chosen by the attacker) may compromise the security of stronger algorithms (chosen by honest parties). We detail below several constructions of TLS that demand joint assumptions on collections of algorithms. Surprisingly, prior work on the provable security of TLS failed to make this observation or left it implicit.

Besides interference between multiple algorithms, TLS features dependencies between multiple runs of the handshake. For instance, a client connection may first run an RSA-based session to establish a master secret and keys for the record layer, then run a second session on the same connection, possibly with different algorithms and certificates. Using a parallel connection, the client may run a third *resumption* handshake, re-using the master secret of a prior session to derive new keys. At that point, the security of those keys depends on algorithms and constructions used in three runs of the handshake. (See for instance [5] for recent attacks involving three related handshakes.) This is in sharp contrast with prior work on the provable security of TLS [13, 16, 17], which focus on a fixed run of the protocol, for a fixed choice of algorithms.

1.1 Cryptographic Agility

Agile security considers families of schemes or protocols, all serving the same purpose, when the same keys are shared across members of the family. Acar et al. [1] propose agile definitions for pseudo-random functions (PRF) and encryption schemes, and advocate agility as a major practical concern for protocols like TLS. Instead, *combined,* or *joint security* [12] studies the sharing of keys between constructions serving different purposes, e.g. encryption and signing. TLS requires both agile and joint security; in the remainder we let the term *agility* encompass both concepts.

The agility mechanisms of TLS ares primarily driven by *ciphersuites* of the form TLS_*e*_*s*_WITH_*r*, which indicates a key encapsulation mechanism (KEM) *e* and signature scheme s for the handshake, and an authenticated encryption scheme r for the record layer. For instance, the commonly-used ciphersuite TLS_RSA_WITH_AES_256_CBC_SHA indicates an RSA handshake: the client sends a fresh premaster secret encrypted under the server public key; both parties use it to extract a master secret, used in turn as the seed of a SHA1-based PRF to derive 4 keys for SHA1-based MACs and AES encryption in CBC mode. TLS 1.2 currently has 314 registered ciphersuites. More precisely, the choice of algorithms depends on additional data exchanged during the handshake (hence subject to active attacks), including protocol versions, certificate requests, certificate chains, and various extensions in the first two messages of the handshake (e.g. for choosing hash functions and elliptic curves). Still, because of key reuse across algorithms, we stress that the security of TLS does not reduce to the security of a few thousand fixed-algorithm variants of the handshake.

1.2   Empirical Study of Web Servers and Browsers

Using an online analyzer [24], we gathered extended information on server configurations for 215 of the top 500 domains,[1] including the TLS versions, ciphersuites, certificates, and extensions they offer. These servers accept 64 ciphersuites, with an average of 12 and standard deviation of 6. They still widely deploy weak algorithms: 70% accept at least one ciphersuite with MD5 and 90% at least one with RC4. All servers but one offer several versions; 37% offer only SSL3 and TLS 1.0; 56% offer all 4 versions from SSL3 to TLS 1.2. Although now forbidden by the standard, 3% still accept SSL2.

We also tested 12 TLS clients, including major web browsers (Chrome, Firefox, Internet Explorer, Safari) and libraries (NSS, OpenSSL, SChannel, Secure Transport). These clients similarly propose a large number of ciphersuites, ranging from 19 to 36; they all propose weak hash (MD5) or encryption methods (RC4, or even no encryption).

1.3   Cross-Ciphersuite Attacks

As a first example, most TLS servers are configured to use the same RSA certificate both for signing handshake messages and for decrypting premaster secrets. Experimentally, 69% of the servers we tested propose at least one ciphersuite using RSA for encryption and one using it for signing, and *all* 138 of those use the same key for both purposes.

As a second example, Mavrogiannopoulos et al. [20] report a cross-protocol attack between plain Diffie-Hellman (DH) and Elliptic-Curve Diffie-Hellman (ECDH) ciphersuites, due to a mis-interpretation of the signed group description sent by the server. Each family of ciphersuites is (a priori) secure in isolation, but configurations enabling a DH client and an ECDH server are subject to their attack.

---

[1] http://www.alexa.com/topsites/global, as of January 2014, excluding domains with no valid HTTPS certificate.

Our third example concerns the record algorithms (the r in TLS_e_s_WITH_r). Recall that both parties derive keys for r immediately after the KEM phase, and start using them before verifying the Finished messages that confirm the integrity of the handshake. As an optimization, the optional False Start TLS extension [19] lets clients send private application data before key confirmation. Depending on r, the *same* key materials are split into IVs, MAC keys, and encryption keys of various lengths. Hence, the client and the server may start using the same bits with different algorithms r_C and r_S, for instance as an IV at the client and as a MAC key at the server. To our knowledge, we are the first to report this cross-algorithm attack against [19]. We do not have an exploit based on two standard record algorithms (r_C, r_S) but one can easily design a pair of schemes strong in isolation and subject to the attack, and key recovery attacks against any standard algorithm r_C could be used to attack strong r_S algorithms.

1.4 Multiple Sessions and Connections

Following the standard, we recall TLS terminology for multiple related handshakes; this differs from the key-exchange model of Bellare & Rogaway [3] with only one kind of sessions and no shared state between sessions. Local instances of the protocol provide a *connection* (concretely, taking ownership of a TCP connection), either as client or as server. Each connection goes through a sequence of *epochs*, each epoch running one *handshake*. For a given connection, we refer to additional handshakes in the sequence as *renegotiations*. We refer to epochs performing full handshakes as *sessions*, and to epochs performing abbreviated handshakes as *resumptions*. We have a transition from the current epoch to the next each time a handshake *completes* by successfully processing the last message of the handshake. Abstractly, the local instance never stops; it is then ready to send (or receive) the first message of the next handshake.

Sessions intend to establish a fresh *master secret*, associated with data extracted from the handshake messages that record its origin and purpose, and used to derive fresh keys for the record layer. *Resumptions* instead rely on a prior complete session to save the cost of public-key cryptography and directly derive fresh keys using the algorithms and master secret of the original session. For each epoch, the handshake consists of a series of messages exchanged using the current record-layer protection mechanisms, initially in the clear, then typically using authenticated encryption.

1.5 Proving the TLS Handshake Secure

The scope of this paper is the TLS handshake, as it is specified in the Internet Standard and (to a lesser extent) as it is commonly used. We model multiple, related sessions and connections, and the agility issues caused by multiple ciphersuites featuring RSA and DHE key exchanges. We also model unilateral and mutual authentication, based on RSA and (EC)DSA signatures. On the other hand, we do not cover static DH, PSK, and ECDHE key exchanges, and we do not investigate the joint usage of keys for signing and encryption. (See the full paper for their discussion.)

Our main result is provable security for a standard-compliant, reference implementation of the handshake, seen as a detailed cryptographic model of the protocol. Our provably-secure handshake code consists of 3,600 lines of F#. Its security relies on new agile assumptions, notably for its KEMs. We reduce them to lower-level assumptions on RSA encryption and Diffie-Hellman exchange, using a 3,000-line EASYCRYPT [2] proof. Working with a reference implementation, and testing it against mainstream implementations, forces us to handle the details of multiple handshakes and algorithms. Proving it secure requires both modularity and automation.

A feature of TLS that traditionally resists abstraction is that the handshake releases algorithms and derived keys to the record layer *before* the handshake completes, so that its last messages can be exchanged as TLS fragments protected by the new keys. We revisit the cryptographic folklore that the handshake can only be proved secure by including these encrypted messages. The kernel of the lore is that it cannot be proved using a Bellare & Rogaway-style key-exchange definition. To achieve modularity, we separate record-key generation from handshake completion: our main definition releases the record keys in the middle of the handshake, before signaling its completion a few messages later. Since the handshake does *not* rely on record-layer protection, we can safely let the handshake adversary control both the network and the record layer. Completion is still necessary to confirm that the record keys are secure before encrypting any application data—but not for encrypting handshake Finished messages.

We stress that this paper establishes the security of the *handshake*, seen as a component of TLS, not the full communications protocol. Our main construction provides key indistinguishability, and ensures agreement on parameters for the record layer. Our results complement those of Bhargavan et al. [4], who describe MITLS, an implementation of TLS verified in the computational model of cryptography; they focus on the main TLS API and application security, but rely on stronger, ad hoc assumptions for RSA and Diffie-Hellman ciphersuites. Our handshake is integrated with MITLS, which provides additional definitions and verified code for the record layer and the protocol logic. (Their security model ensures in particular that the record keys are used for protecting application data only after handshake completion [4].) By composing our results with theirs, we obtain security for a reference implementation of the TLS standard and the sample applications built and verified on top of MITLS.

1.6 Overview of the Paper

We see the use of a verified reference implementation and automated tools as essential to precisely account for multiple related epochs and algorithms in TLS; §6 briefly describes our use of high-level programming, type systems, and provers to carry out modular cryptographic verification at this scale. To present our result and explain its proof structure, however, we rely on more succinct definitions and constructions, given in §2–5 and outlined below. This more abstract treatment suffices to convey the main ideas, but it necessarily omits many aspects of the handshake, such as its message formats. We refer to the standard [9] or the implementation for the details. Also, for simplicity, we do not model forward secrecy and state reveal e.g. for master secrets, and we consider only static compromise for long-term keys.

Signatures (§2) and Certificates. We begin with a relatively simple agile definition. TLS supports three core signature algorithms, $s \in \{RSA, DSA, ECDSA\}$, used with a range of algorithms h to hash the text before signing. The hash algorithm depends on protocol versions, ciphersuites and extensions. TLS does not enforce any key-based hash algorithm policy, so we need a notion of security that tolerates *some* weak algorithms in the standard. For instance, a verifier tricked into using MD5 may remain secure, provided the signer only uses SHA1, and vice-versa. For each core algorithm s, we define h^*-H-security against an adversary that must forge a valid signature for algorithms (s, h^*), given access to signing oracles for any algorithms (s, h) with $h \in H$. We show that a family of secure schemes may not be jointly secure, but we leave open its concrete analysis for the range of algorithms used in TLS.

Our model excludes any validation rules for certificates and their PKI, an important problem outside the scope of the TLS standard. Our constructions simply authenticate the exchanged certificate chains, and use a specification function to extract from them the public keys used in the handshake.

Master Secrets, Key Encapsulation, and Key Derivation (§3). Following Krawczyk et al. [17], we use KEMs [8] to model key-exchange; this allows us to unify RSA and Diffie-Hellman within the same formalism. Instead of treating the whole handshake as a KEM, however, following Morrissey et al. [21], we decompose it into *premaster secret*, *master secret*, and *record-key derivation* phases; this yields the modularity we need e.g. for modeling the re-use of master secrets between handshakes. We show how to securely construct a master secret KEM from a premaster secret KEM for RSA and Diffie-Hellman ciphersuites (Theorem 1) and, independently, how to derive record keys and Finished messages from master secrets (see the full paper). We formalize the proof of Theorem 1 in EASYCRYPT. For RSA, this involves showing that countermeasures to Bleichenbacher's attacks [6, 15] provide enough protection against chosen-ciphertext attacks. We rely on the assumption that PKCS#1v1.5 ciphertexts are hard to re-randomize; we leave open the problem of further reducing this conjecture to standard RSA assumptions. Our result does not directly compare to the one of Krawczyk et al. as their KEM also includes key derivation and Finished messages, whereas we rely on this new assumption. To comply with the standard, we also support agility in the algorithm used to extract master secrets from a premaster secrets. As for agile signatures, we arrive at a definition parameterized by an algorithm for the encryptor and a set of algorithms for the decryptor.

Once established, the master secret is used to key a pseudo-random function (PRF) for multiple epochs for two purposes: (1) to derive the record-layer key materials for the epoch; and (2) to compute the MACs of all messages exchanged in an epoch to verify its integrity. Our corresponding security definition (in the full paper) requires that adversaries commit to a record-layer algorithm r before key derivation. This let us support the negotiation of r without having to make agile assumptions for the record layer, as discussed in §1.3.

Agile Security Model (§4) and Proof (§5) for Sequences of Handshakes. The main two goals of the handshake are to establish shared keys for the record layer, and to agree on many parameters, including those used in the handshake itself. To this end, we propose a new security definition that covers multiple epochs on different connections, related by resumptions and renegotiations. We equip our adversary (informally including the rest of TLS, the application, and the network) with oracles to create honest connections and long-term keys for clients and servers, to control their usage, and to exchange handshake messages. Each honest instance of the protocol represents a connection, and logs a sequence of *local assignments*, recording its view on the successive epochs of the connection. This enables us to capture TLS assignments in a generic manner. Our main integrity result is that, when a handshake completes, and under suitable conditions on algorithms and keys, honest clients and servers agree on all assignments for all epochs on the connection. More explicitly, for new sessions, both parties agree on a unique label; the negotiation algorithms, parameters, and key-exchange values; and the optional certificate chains for the client and the server. For resumptions, both parties agree on the label of the session being resumed, as well as a fresh unique label for key derivation.

We also provide secure key derivation, depending on distinguished exchange-value assignments for each ciphersuite. They are somewhat similar to session identifiers in Bellare-Rogaway models but are used to define both *safety*, akin to freshness, and partnering. A session is *safe* when honest client and server agree on these assignments, under suitable conditions on algorithms and long-term keys. As discussed above, our definition immediately releases all connection keys. We guarantee that the keys of safe sessions are indistinguishable from fresh random keys; this accounts for selective session key reveal and test queries in Bellare-Rogaway models. Additionally, we provide *verified safety*, that is, sufficient conditions on the recorded long-term keys that enable honest parties to infer that their session is safe.

Our main result (§5, Theorem 2) reduces the concrete security of the TLS handshake to agile assumptions on the constructions used for signatures, KEMs, and PRFs. Each epoch assigns a distinguished agility-parameter a, selecting all algorithms for the epoch. The theorem statement is parameterized by a predicate α on a that holds whenever all algorithms selected by a are (assumed to be) secure. Thus, it provides meaningful security only for epochs where $\alpha(a)$ holds, despite any other epochs. If α is always false, there is nothing to prove. If we care specifically about one ciphersuite, say TLS_DHE_DSS_WITH_3DES_EDE_CBC_SHA, we may apply our theorem with α set to true only when a selects that ciphersuite. This already improves on non-agile results for TLS that assume all honest parties agree *in advance* on a ciphersuite and reject any others.

Our model accounts for agility with respect to record algorithms, and yields channel security for MITLS without agile assumptions on the algorithms r used in the record layer. We thus validate the use of stateful LHAE [23] for clients and servers that negotiate r. We require, however, that no application data be sent before the Finished messages are verified. For implementations that violate this requirement [19], stronger agile assumptions seem unavoidable.

Code-Based Verified Implementation (§6). We finally present the reference implementation of the handshake we integrated into MITLS, and its verification against our security definition, based on the same modular proof structure but at a greater level of detail, relying on type-based verification for scalability. Our code supports the standard and commonly-used extensions; we tested it against various mainstream TLS clients and servers, using 4 versions ranging from SSL3 to TLS 1.2, 12 ciphersuites, and various subsets of extensions. It improves on the original MITLS code [4], which supported less features, and whose security relied on monolithic, TLS-specific assumptions for RSA and DH ciphersuites. The full paper reports experimental results showing that our code runs handshakes with reasonable performance. To enable its automated verification, our code is structured into small, independent modules (that is, program libraries) parameterized by algorithm descriptors. For instance, our library code for the HMAC-based PRF used in TLS implements agility before calling selected core algorithms, e.g. SHA1. In contrast, the code that implements SHA1 is outside the scope of our verification effort—we document our agile cryptographic assumption on it, and call a standard library. Each cryptographic construction used in the handshake corresponds to a separate library in the code. We define the security of libraries for multiple keys and multiple algorithms; the corresponding definitions and reductions to single-key security of individual algorithms appear in the full paper.

In summary, our work sheds light on important design and implementation issues of TLS. To our knowledge, we provide the first provable-security results for TLS that account for algorithm agility. We are also the first to give an abstract security model for handshakes related by resumption and renegotiation.

Further Reading. Our website http://www.mitls.org provides additional materials: the MITLS source code; the EASYCRYPT proof of Theorem 1; and a companion paper with empirical data on TLS handshakes, auxiliary definitions, constructions, and proofs, and extended discussions of attacks and related work.

2   Agile Signatures

An *agile signature scheme* consists of three algorithms: KeyGen is a standard key generation algorithm, while Sign and Verify take an extra agility parameter. For instance, given a core signature scheme $s = (\mathsf{keygen}, \mathsf{sign}, \mathsf{verify})$, the hash-then-sign scheme $S_s = (\mathsf{KeyGen}, \mathsf{Sign}, \mathsf{Verify})$ of TLS is defined as follows: $\mathsf{KeyGen} \triangleq \mathsf{keygen}$ generates a key pair for algorithm s; $\mathsf{Sign}(h, sk, m) \triangleq \mathsf{sign}(sk, h(m))$ computes a signature using the core scheme s and hash algorithm h; and $\mathsf{Verify}(h, pk, m, \sigma) \triangleq \mathsf{verify}(pk, h(m), \sigma)$ verifies a purported signature σ for message m

hashed with algorithm h. We define existential unforgeability under chosen-message attacks (EUF-CMA) for agile signatures.

Definition 1 (EUF-CMA). *Let* (KeyGen, Sign, Verify) *be an agile signature scheme, $p^\star$ a parameter, and P a set of parameters, and consider the following forgery game:*

Game EUF $\triangleq$
$pk, sk \leftarrow$ KeyGen(); $M := \varnothing$
$m', \sigma \leftarrow \mathcal{A}^{\mathsf{SIGN}}(pk)$
return $m' \notin M \wedge$ Verify$(p^\star, pk, m', \sigma)$

Oracle SIGN$(p, m) \triangleq$
if $p \notin P$ **then return** $\bot$
$M := M \cup \{m\}$
return Sign(p, sk, m)

The scheme is $(\epsilon, t, p^\star, P)$-secure against EUF-CMA if, for any $\mathcal{A}$ that runs in time t, the EUF *game returns* true *with probability at most ϵ.*

This definition generalizes plain EUF-CMA security; the two coincide for a scheme with fixed hash algorithm h, i.e. $(p^\star, P) = (h, \{h\})$. We do not require $p^\star \in P$; for instance, one may pragmatically assume that forging an MD5-based signature is hard when given only SHA1-based signatures. Indeed, the attacks of Stevens et al. [26] rule out (MD5, {MD5, ...})-security, but (MD5, {SHA1})-security may still hold. On the other hand, non-agile security does not imply agile security. Consider for instance the scenario where the pre-image security of MD5 is broken. Then the attacks described by Naccache and Shparlinski [22] are likely to break (SHA256, {MD5, SHA256})-security, even though (SHA256, {SHA256})-security would still hold.

The TLS standard features the following hash-then-sign schemes: prior to version 1.2, RSA PKCS#1v1.5 signatures use the concatenation of MD5 and SHA1 hashes and (EC)DSA signatures use SHA1. TLS 1.2 introduces additional agility to facilitate migration from MD5 and SHA1 to stronger algorithms. Designers are aware of agility problems, and prescribe ad hoc countermeasures [9, §7.4.3]. The standard still requires that (EC)DSA use SHA1, delaying the migration to stronger algorithms. It also adds an encoding of the hash algorithm identifier to guarantee that all hash algorithms have disjoint range.

Given algorithms h and h' with disjoint ranges, if the core signature scheme itself is (ϵ, t)-EUF-CMA secure on their joint range, then we have $(\epsilon', t', h, \{h, h'\})$-security for the corresponding agile hash-then-sign signature scheme, where the difference between ϵ, t and ϵ', t' depends on the reduction to the collision resistance of h. Sadly, the core signature schemes used in TLS are not EUF-CMA secure. The best we can do, for now, is thus to assume that the hash-then-sign signature scheme that uses them meets Definition 1.

3   Master Secrets and Key Encapsulation

Following [14, 17], we model the basic key-exchange functionality of TLS as different variations on KEMs. However, we separate the derivation of the master secret from the derivation of keys for the record-layer. We model the premaster secret phase for RSA and Diffie-Hellman exchanges as agile KEMs (keygen, !enc, dec) parameterized by a 2-byte protocol version string.

RSA. keygen generates a fresh RSA key pair (pk, sk); $enc(pv, pk)$ appends a randomly chosen 46-byte string to pv to obtain the premaster secret pms, and returns it with the ciphertext c resulting from its PKCS#1v1.5 encryption under pk; $dec(pv, sk, c)$ decrypts c with sk. If the padding is correct and the decrypted pms is exactly 48 bytes long, it returns pms with the first 2 bytes replaced by pv, otherwise it returns $\bot$; such errors are handled in our ms-KEM below.

Diffie-Hellman. keygen selects group parameters pp, generates a fresh pair of DH values (g^x, x), and returns $pk = (pp, g^x)$ and $sk = (pk, x)$ as public and private KEM keys; $enc(pv, (pp, g^x))$ samples y and returns $pms = g^{xy}$ and $c = g^y$; $dec(pv, (pk, x), c)$ returns $c^x = g^{xy}$. The ciphertext space guarantees that c is in a large prime-order subgroup specified by pk. In contrast to the RSA pms-KEM, neither enc nor dec depend on pv.

On their own, these two premaster secret KEMs are *not* secure under any indistinguishability notion, even under relatively weak active attacks such as, for instance, plaintext-checking attacks (PCA): recall the Bleichenbacher attack, and the lack of active security for basic Diffie-Hellman (e.g., querying a plaintext-checking oracle on c^r and pms^r for any $r \neq 1$, suffices to distinguish a random pms from the one encapsulated in c). Rather than using pms as a key, TLS feeds it through an agile *key extraction function* (KEF) parameterized by a hash algorithm, to compute the master secret ms.

We model this phase of the handshake as an *agile labeled KEM*, extending the labeled KEMs of [14, 17] with an agility parameter. Given an agile (unlabeled) pms-KEM $e = (\text{keygen}, \text{enc}, \text{dec})$ and an agile key extraction function family KEF, the master secret KEM $E_e = (\text{KeyGen}, \text{Enc}, \text{Dec})$ of TLS is defined as follows:

- $\text{KeyGen}() \triangleq \text{keygen}()$;

- $\text{Enc}(pv, h, pk, \ell) \triangleq pms, c \leftarrow \text{enc}(pv, pk); \; ms \leftarrow \text{KEF}(pv, h, pms, \ell)$;
$$\textbf{return } ms, c$$
generates a premaster secret pms and a ciphertext c using e, then derives a master secret ms for ℓ using KEF.

- $\text{Dec}(pv, h, sk, \ell, c) \triangleq pms \leftarrow \text{dec}(pv, sk, c); \; \textbf{if } pms = \bot \textbf{ then } pms \leftarrow pv \,\|\, \$$;
$$\textbf{return } \text{KEF}(pv, h, pms, \ell)$$
decrypts the ciphertext c to obtain pms. If decryption fails, it computes a fake pms by appending a random 46-byte string to pv (this is never the case for DH). It returns the value obtained from pms and ℓ using the agile KEF.

We assume sufficient checks to ensure that all arguments are well-formed before calling the master secret KEM algorithms; e.g., for Diffie-Hellman, our code validates group parameters and checks that pk and c belong to a large prime-order subgroup before calling Dec.

We define security for agile labeled KEMs as indistinguishability under replayable chosen-ciphertext attacks (IND-RCCA), a relaxation of CCA security, first introduced for public-key encryption by Canetti et al. [7].

Definition 2 (IND-RCCA). *Let* (KeyGen, Enc, Dec) *be an agile labeled KEM,* p^* *a parameter,* P *a set of parameters; and consider the following game:*

Game RCCA $\triangleq$	**Oracle ENC**$(\ell) \triangleq$	**Oracle DEC**$(p, \ell, c) \triangleq$
$pk, sk \leftarrow \text{KeyGen}()$	if $\ell \in L$ then return $\perp$	if $\ell \in L \vee p \notin P$ then return $\perp$
$K, L := \varnothing$	$k_0, c \leftarrow \text{Enc}(p^*, pk, \ell)$	$L := L \cup \{\ell\}$
$b \leftarrow \{0,1\}$	$k_1 \leftarrow \$$	$k \leftarrow \text{Dec}(p, sk, \ell, c)$
$b' \leftarrow \mathcal{A}^{\text{ENC,DEC}}(pk)$	$K(\ell) := K(\ell) \cup \{k_0, k_1\}$	if $k \in K(\ell)$ then return $\perp$
return $(b' = b)$	**return** k_b, c	**return** k

The RCCA advantage of $\mathcal{A}$, $\mathbf{Adv}_{p^*, P}^{\text{RCCA}}(\mathcal{A})$ *is defined as* $2 \Pr[\text{RCCA} : b' = b] - 1$. *The scheme is* (ϵ, t, p^*, P)-*secure against IND-RCCA-n when the advantage of any adversary* $\mathcal{A}$ *running in time* t *and making at most* n *queries to* ENC *is at most* ϵ. *We write IND-RCCA instead of IND-RCCA-1.*

The check $\ell \in L$ in the decryption oracle reflects a property of TLS: honest servers decrypt at most once for each nonce. The check $\ell \in L$ in the encryption oracle is analogous to the restriction of Krawczyk et al. [17] to define IND-CCCA security for non-agile KEMs.

The lemma below (proved by a standard hybrid argument in the full paper) enables us to prove security for a single query, then use the multi-query variant for reasoning about TLS in our main theorem.

Lemma 1. *If a KEM* (KeyGen, Enc, Dec) *is* $(\epsilon/n, t', p^*, P)$-*secure against IND-RCCA, then it is* (ϵ, t, p^*, P)-*secure against IND-RCCA-n, where* $t' = t + O(n \cdot t_{\text{Enc}})$ *and* t_{Enc} *is the worst-case cost of algorithm* Enc.

Next, we define the assumptions for our main theorem on the TLS master secret KEM: *non-randomizability under plaintext-checking attacks* (NR-PCA) and *one-wayness under plaintext-checking attacks* (OW-PCA).

Definition 3 (NR-PCA, OW-PCA). *Let* (keygen, enc, dec) *be an agile (unlabeled) KEM,* p^* *a parameter, and* P *a set of parameters. Consider the two games:*

Game OW-PCA $\triangleq$	**Game NR-PCA** $\triangleq$	**Oracle PCO**$(p, k, c) \triangleq$
$pk, sk \leftarrow \text{keygen}()$	$pk, sk \leftarrow \text{keygen}()$	if $p \notin P \vee k = \perp$ then
$k^*, c^* \leftarrow \text{enc}(p^*, pk)$	$k^*, c^* \leftarrow \text{enc}(p^*, pk)$	**return** $\perp$
$k \leftarrow \mathcal{A}^{\text{PCO}}(pk, c^*)$	$c \leftarrow \mathcal{A}^{\text{PCO}}(pk, c^*)$	$k' \leftarrow \text{Dec}(p, sk, c)$
return $(k = k^*)$	**return** $(c \neq c^* \wedge k^* = \text{dec}(p^*, sk, c))$	**return** $(k' = k)$

The NR-PCA advantage of $\mathcal{A}$, $\mathbf{Adv}_{p^*, P}^{\text{NR-PCA}}(\mathcal{A})$ *is the probability that the* NR-PCA *game returns* true. *The KEM is* (ϵ, t, p^*, P)-*secure against NR-PCA if the advantage of any adversary* $\mathcal{A}$ *running in time* t *is at most* ϵ. *OW-PCA advantage and security are defined analogously.*

The full paper gives preliminary theorems and conjectures on these assumptions, and relates our agile IND-RCCA KEMs to prior work and more standard assumptions. We hope this will stimulate further cryptanalytic work on TLS.

Our main result on KEMs is that the generic ms-KEM E_e of TLS is IND-RCCA secure if the underlying pms-KEM e is both NR-PCA and OW-PCA secure. The proof (in the full paper) has been formalized using EASYCRYPT. The proof is in the random oracle model for the agile KEF. As explained above, we consider the single challenge case.

Theorem 1 (RCCA from NR-PCA and OW-PCA). *Let $\mathcal{A}$ be a $(p^\star, P)$-RCCA adversary for E_e running in time $t_{\mathcal{A}}$ and making at most q_{KEF} and q_{DEC} queries to the random and decryption oracle, respectively. Let $p^\star = (pv^\star, h^\star)$ and $P' \triangleq \{pv \mid (pv, h) \in P\}$. There exist an OW-PCA adversary $\mathcal{B}$ and an NR-PCA adversary $\mathcal{C}$ against e, both running in time $t_{\mathcal{A}} + O(q_{\mathsf{DEC}} \cdot q_{\mathsf{KEF}})$, such that*

$$\mathbf{Adv}_{p^\star, P}^{\mathsf{RCCA}}(\mathcal{A}) \le 2\Big(\mathbf{Adv}_{pv^\star, P'}^{\mathsf{NR\text{-}PCA}}(\mathcal{B}) + \mathbf{Adv}_{pv^\star, P'}^{\mathsf{OW\text{-}PCA}}(\mathcal{C}) + 2^{|pv| - |pms|}\big(q_{\mathsf{KEF}} + q_{\mathsf{DEC}}\big)\Big).$$

The factor $2^{|pv| - |pms|}$ is the entropy of the value $pv \,\|\, \$$ used to derive the master secret when RSA decryption fails, as recommended by TLS 1.2 to mitigate Bleichenbacher attacks. With the DH pms-KEM, decryption never fails (as the ciphertext validation is done beforehand) so the last term above can be omitted.

4   Defining Agile Security for Sequences of Handshakes

Our security definition for handshakes is general enough to apply to TLS, as specified in the standard and coded in MITLS, while hiding implementation details like message formats and specific cryptographic constructions. The adversary creates and interacts with multiple instances i of a handshake protocol Π by calling Π's oracles, detailed below. Each instance has a fixed role $\mathcal{R}$, either $\mathcal{C}$ for Client or $\mathcal{S}$ for Server, and models a connection endpoint.

- KeyGen(v) creates and stores a new honest keypair for the long-term public-key algorithm v (in TLS, ranging over s for signing and e for key encapsulation) and returns the associated public key. Similarly, KeyInject(v, pk, sk) stores a dishonest keypair (assuming pk is not yet in the store).
- Init($\mathcal{R}, cfg_{\mathcal{R}}$) creates an instance with role $\mathcal{R}$ and local configuration $cfg_{\mathcal{R}}$; it returns a fresh handle i.
- Send$_i$(*frag*) lets an existing instance i process a fragment, depending on its current state. As a result, the instance may update its state, assign local variables, and return a response. (In TLS, responses range over sequences of handshake and CCS message fragments, intended to be sent to the peer, as well as error messages.)
- Control$_i$(*env*) changes the global, internal state of the handshake, e.g., enabling the adversary to control access to stored sessions and private keys by the protocol the next time Send will be called, or to trigger a renegotiation request. This single oracle accounts for many control functions in the MITLS handshake implementation. For example, Control provides the environment with means to reject certificates that it deems invalid.

Each instance maintains its private local state (e.g. using local variables). Each instance can go through a sequence of epochs (e.g. recording the number of cycles in the state machine). For each epoch, it records a sequence of *variable assignments*, extended as the result of calls to Send and Control. Each variable is assigned at most once in every epoch. The selection and ordering of assignments within an epoch depends on the protocol; for instance, a client epoch may assign its client-certificate variable, then send a message to the server, causing the server epoch to record the same assignment later in the protocol.

Our definition is based on local variable assignments, which summarize the view of clients and servers so far about each epoch. This is adequate to model the handshake as a component within TLS, but this differs from models based on *matching conversations* [3] that compare the (unparsed) messages they have sent and received so far. We use assignments to express the main goals of the protocol, for instance assigning a fresh random value to the record key variable k; and agreeing on all assignments as a session completes. We list below the main variables used in our presentation, but our definition can account for a more detailed model of the TLS handshake.

ℓ	epoch identifier; in TLS, the concatenation of the client and server random values.
$\ell_{session}$	resumption identifier; in TLS, the identifier of the epoch that completed the session being resumed. (The MITLS code also assigns the TLS *sessionId*, chosen by the server, but we do *not* use it as an identifier as it is not necessarily unique.)
a_C, a_S	client and server negotiation parameters; in TLS, they consist of protocol versions, ciphersuites, and extension messages.
a	agility parameter; in TLS, the protocol version, the negotiated ciphersuite, and data extracted from the first flight of messages sent by the server.
$cert_C$, $cert_S$	client and server certificate chains. In TLS, these certificates are optional; e.g. the assignment $cert_C := \bot$ denotes the absence of client certificate.
ex_C, ex_S	client and server exchange variables, possibly secret, used to specify safety.
k	record key for the epoch; in TLS, depending on a, this key is usually split into 4 keys for MAC & encrypt.
complete	successful completion flag, marking the end of the handshake for this epoch.

Unless explicitly mentioned for key-exchange materials, these variables are public: the adversary can read them, but not change them; the protocol can write them once in every epoch, but not read them. (This restriction matters only for the record key, as we replace it with a random value.) The agility-parameter variable a determines the algorithms and constructions used by the handshake. Our security properties are conditioned by a strength predicate $\alpha(a)$ that indicates whether those algorithms are strong enough to secure the epoch. When the role of an epoch is clear from the context, the *peer* refers to the opposite role, and the *peer-exchange variable* refers to the exchange variable of the opposite role (e.g. ex_C when $\mathcal{R}$ is $\mathcal{S}$).

We deliberately avoid modeling certificate validation. For the handshake, certificate chains are authenticated, uninterpreted bitstrings. We leave as future work supplementing our model with an application-level certificate infrastructure above the MITLS API. We assume given a public specification function pk(*cert*) that returns either the public key associated with a certificate chain, or ⊥. The session state does not need to explicitly mention public keys, but public keys can appear in exchange variables.

A security model for a protocol describes how queries are answered and how session variables are assigned. Next, we define properties of these models as they interact with an adversary.

Definition 4 (Honesty, Safety, Matching Algorithms and Completion).
For a handshake protocol Π and a strength predicate $\alpha(\cdot)$, an adversary that calls Π's oracles any number of times produces a trace of interleaved variable assignments for a series of epochs for each instance. In this trace:

- *As determined by its assigned agility parameter a: an epoch is either a session, with distinguished client- and server-exchange variables, or a resumption, with an $\ell_{session}$ variable; sessions (and their exchange variables) are either static or ephemeral; a static session has at least one static exchange variable; an ephemeral session has only ephemeral exchange variables.*
- *A (long-term) public key is honest for algorithm v if it was returned by a call to KeyGen(v). A session's ephemeral server-exchange variable assignment is honest if there is a server session with the same assignment to its server-exchange variable—and conversely for ephemeral client-exchange variables.*
- *A client session is safe if (i) $\alpha(a)$ holds; (ii) honest public keys for a's algorithms are assigned to all static exchange variables; and (iii) there is a server session with the same assignment to the ephemeral server-exchange variable. A server session is safe if the converse holds.*
- *A resumption is safe if $\alpha(a)$ holds and $\ell_{session}$ is the identifier of a safe and complete session.*
- *An epoch has matching algorithm $r = \text{record}(a)$ when there is a peer epoch with the same identifier ℓ and algorithm r.*
- *An epoch is complete when it includes the assignment complete := 1.*

Anticipating on §5, for TLS we define the client exchange value ex_C to be the master secret ms together with the KEM public key pk, and the server-exchange variable ex_S to be the public key pk of the KEM. The latter is static for TLS-RSA, but ephemeral for TLS-DHE. Here ms is explicitly secret and ephemeral.

Definition 5 (Handshake Security). *Let Π be a handshake protocol, $\alpha(\cdot)$ a strength predicate, and $\mathcal{A}$ an adversary that calls Π's oracles any number of times. Consider the following properties:*

(1) **Uniqueness:** *epoch identifiers are used at most once in each role.*
   Let $\mathbf{Adv}^U(\mathcal{A})$ be the probability that two different epochs with the same role assign the same value to ℓ when $\mathcal{A}$ terminates.
(2) **Verified Safety:** *if the peer of a session uses a strong signature algorithm to authenticate and the public-key for the peer signature is honest, then the peer-exchange variable assignment is honest.*

Let $\mathbf{Adv}^S(\mathcal{A})$ be the probability that, when $\mathcal{A}$ terminates, there is an epoch such that $\alpha(a)$ holds; the public key of the peer is honest; and the assignment to the peer exchange value is not honest (i.e. not assigned by any peer);

(3) **Agile Key Derivation:** depending on a random bit b, replace the record key assigned in safe epochs with matching algorithm r with a fresh $k \leftarrow$ KeyGen(r), assigning the same value to epochs that have the same identifier ℓ, algorithms kdf(a) and exchange variables or resumption identifier.
Let $\mathbf{Adv}^K(\mathcal{A}) = 2p - 1$ where p is the probability that $\mathcal{A}$ returns b.

(4) **Agreement:** for every safe and complete epoch, there is a safe epoch in the other role such that their two instances agree on all prior assignments.
Let $\mathbf{Adv}^I(\mathcal{A})$ be the probability that, when $\mathcal{A}$ terminates: an instance created by Init($\mathcal{R}, cfg$) assigns complete $:= 1$ in a safe epoch; and no instance created by Init($\overline{\mathcal{R}}, cfg'$) begins with a series of epochs with the same assignments to all variables (up to, but possibly excluding complete $:= 1$).

The handshake is (ϵ, t, α)-secure when for any adversary $\mathcal{A}$ running in time t, we have $\mathbf{Adv}^G(\mathcal{A}) \leq \epsilon$, for $G = U, S, K, I$.

Discussion. The properties above are given in chronological order: in TLS in particular, protocol instances first exchange fresh random values, then derive keys, and finally confirm the integrity of the session negotiation.

Property (1) simply ensures that ℓ provides a unique identifier, later authenticated using (4); we use these identifiers for matching client and server sessions.

Property (2) enables, for instance, a client that trusts both the negotiated algorithm and the server certificates to deduce that its server-exchange variable is honest, and conclude that its session is safe.

Property (3) idealizes the derived key; this is key indistinguishability. Recall that TLS uses the key before the two parties actually agree on the record algorithms. Conservatively, (3) idealizes the key only when the record algorithms match. As Krawczyk et al. [17], our model does not consider forward secrecy.

Property (4) guarantees agreement on all variable assignments at the client and server instances since their creation, not just the assignments of the current epoch. Hence, as soon as one epoch safely completes, the peers agree also on all prior epochs on that connection—even those that were not safe, or not verifiably safe. For TLS, this property holds only thanks to the (mandatory) secure renegotiation extension, which links each epoch to its predecessor. This property is closely related to the TLS renegotiation results of Giesen et al. [11]. They additionally propose an extension of TLS that would guarantee agreement on the full stream of application data, not just the handshake epochs. On the other hand, our model and security definition also cover resumptions and RSA ciphersuites, which are not covered by their results. Unlike previous analyses of TLS, our definition accounts for session resumptions. Property (4) guarantees agreement on the new epoch identifier ℓ and the identifier $\ell_{session}$ of the resumed session (and hence on the new record keys), as long as the original session is safe. The epochs of the original session may be on a different connection, between a different pair of instances; for those instances, safety for the original session independently guarantees agreement on all its original variable assignments.

TLS applications often group connections that use the same session or the same long-term key, allowing them to share resources and access rights. For example, web browsers allow all connections to the same server to share resources via the Same Origin Policy. It may seem desirable to guarantee a strong relationship between such connections, but our Property (4) guarantees agreement only for the sequence of epochs over a single connection. Indeed, the natural extension of this property to multiple connections does not hold for TLS, as shown by the triple handshake attack of Bhargavan et al. [5]. In this attack, an unsafe server-authenticated session is resumed on a new connection and then renegotiated with a new safe mutually-authenticated session. For the new safe epoch, Property (4) retroactively guarantees agreement on the prior resumption, but *not* on the original unsafe session that was resumed. Consequently, it is possible for a client and server instance to have a safe epoch but inconsistent variable assignments for the session associated with a prior resumed epoch; this leads to a variety of attacks, similar to the renegotiation attacks of Ray [25]. A stronger agreement can be achieved either at the application level, by checking agreement on prior connections, or by a protocol extension that includes a hash of the log of the original session in resumption handshakes [5]; we leave the modeling of this extension and its security for future work.

Compared with classic key exchange models [3] and the key exchange part of ACCE [13], our definition yields useful additional properties. Property (4) guarantees agreement on the negotiation parameters a_C and a_S for safe and complete epochs, thereby preventing version and ciphersuite rollback attacks.

Our definition also provides (some) security for anonymous connections, which can be composed with other authentication mechanisms to achieve application security. For example, renegotiation with client and server certificates may provide mutual authentication on top of an initial, safe, but anonymous handshake. Late application-level, client password authentication may also yield mutual authentication, as illustrated by MITLS [4].

5   Proving Agile Security for TLS Handshakes

We are now ready to reduce the security of TLS handshakes to the security of agile signatures, KEMs and PRFs. We structure the proof to apply simultaneously to the protocol, illustrated in Figures 1 and 2, and to its MITLS implementation.

Figure 1 shows the assignments performed by a client instance and a server instance that run two successive, matching handshakes on the same connection: for both instances, a static session, followed by a (renegotiated) resumption. Figure 2 similarly shows the assignments for an ephemeral session. The agility parameter a of the handshake indicates which algorithm to use for each underlying functionality. We write for instance $a := \mathsf{alg}_C(cfg_C, a_S)$ to retrieve a from the client configuration and the negotiation parameter of the server; $e, p := \mathsf{kem}(a)$ to retrieve the core algorithm e and public parameter of the master secret KEM from a; and $E_e.\mathsf{Enc}$ for encryption using the master-secret KEM for e.

Our second main theorem reduces the security of TLS handshakes to their underlying algorithms, depending on a *strength predicate* on their agility parameters.

Client		**Server**

$\ell_C \leftarrow \$;\ a_C := cfg_C.a_c$ —— $\mathtt{ClientHello}[\ell_C, a_C]$ ⟶ $\ell_S \leftarrow \$;\ \ell := \ell_C \| \ell_S;\ sid \leftarrow \$;$
$certs_S := cfg_S.cert;\ cert_C := \bot$
$pk := \mathsf{pk}(certs_S);\ ex_S := pk$

$\mathtt{ServerHello}[\ell_S, a_S, sid]$
$\mathtt{ServerCertificate}[certs_S]$ $sk := \text{lookup } sk \text{ using } pk$

$\ell := \ell_C \| \ell_S;\ a := \mathsf{alg}_C(cfg_C, a_S)$ ⟵ $\mathtt{ServerHelloDone}$ —— $a, a_S := \mathsf{alg}_S(cfg_S, a_C);$
$pk := \mathsf{pk}(certs_S)$
$c, ms \leftarrow \mathsf{E}_e.\mathsf{Enc}(p_E, pk, \ell)$
$ex_S := pk;\ ex_C := (pk, ms)$ —— $\mathtt{ClientKeyExchange}[c]$ ⟶ $ms \leftarrow \mathsf{E}_e.\mathsf{Dec}(p_E, sk, \ell, c)$
$k := \lfloor\mathsf{PRF}(p_D, ms, t_1 \| \ell_S \| \ell_C)\rfloor_r$ $ex_C := (pk, ms)$
$log_C := \langle\text{prior messages}\rangle$ $log_C := \langle\text{prior messages}\rangle$
$tag_C := \lfloor\mathsf{PRF}(p_D, ms, t_2 \| log_C)\rfloor_p$ —— $\mathtt{ClientFinished}[tag_C]$ ⟶ $tag_C \overset{?}{=} \lfloor\mathsf{PRF}(p_D, ms, t_2 \| log_C)\rfloor_p$
                                 $k := \lfloor\mathsf{PRF}(p_D, ms, t_1 \| \ell_S \| \ell_C)\rfloor_r$
$log_S := \langle\text{prior messages}\rangle$ $log_S := \langle\text{prior messages}\rangle$
$tag_S \overset{?}{=} \lfloor\mathsf{PRF}(p_D, ms, t_3 \| log_S)\rfloor_p$ ⟵ $\mathtt{ServerFinished}[tag_S]$ —— $tag_S := \lfloor\mathsf{PRF}(p_D, ms, t_3 \| log_S)\rfloor_p$
$complete := 1$ $complete := 1;\ \mathsf{store}(\ell, sid, ms)$

······· *Client resumes session (ℓ, sid, ms) using a_C and tag_C from the epoch above* ·······

$\ell_C \leftarrow \$;\ \ell_{\text{session}} := \ell$ —— $\begin{array}{c}\mathtt{ClientHello}\\ {}[\ell_C, a_C, sid, tag_C]\end{array}$ ⟶ $\text{lookup }(\ell', ms, tag_S)\text{ using } sid$
                                $\ell_S \leftarrow \$;\ \ell_{\text{session}} := \ell'$

$\ell := \ell_C \| \ell_S$ ⟵ $\begin{array}{c}\mathtt{ServerHello}\\ {}[\ell_S, a_S, sid, tag_C, tag_S]\end{array}$ — $\ell := \ell_C \| \ell_S$
$k := \lfloor\mathsf{PRF}(p_D, ms, t_1 \| \ell_S \| \ell_C)\rfloor_r$ $k := \lfloor\mathsf{PRF}(p_D, ms, t_1 \| \ell_S \| \ell_C)\rfloor_r$
$log_S := \langle\text{prior messages}\rangle$ $log_S := \langle\text{prior messages}\rangle$
$tag_S \overset{?}{=} \lfloor\mathsf{PRF}(p_D, ms, t_3 \| log_S)\rfloor_p$ ⟵ $\mathtt{ServerFinished}[tag_S]$ —— $tag_S := \lfloor\mathsf{PRF}(p_D, ms, t_3 \| log_S)\rfloor_p$
$log_C := \langle\text{prior messages}\rangle$ $log_C := \langle\text{prior messages}\rangle$
$tag_C := \lfloor\mathsf{PRF}(p_D, ms, t_2 \| log_C)\rfloor_p$ —— $\mathtt{ClientFinished}[tag_C]$ ⟶ $tag_C \overset{?}{=} \lfloor\mathsf{PRF}(p_D, ms, t_2 \| log_C)\rfloor_p$
$complete := 1$ $complete := 1$

Two epochs on the same connection: the first handshake establishes a session without client authentication using static keys; the second one resumes the session.
Conventions in the figure:
(1) We use $\overset{?}{=}$ for checks; a failed check stops the instance.
(2) We use $:=$ for assigning epoch variables; variables exchanged in messages are implicitly assigned, e.g. the server assigns ℓ_C and a_C after parsing the first message.
(3) We omit the extraction of the negotiated key exchange algorithm e and the parameters p_E, p_D from a; for instance, we write p_D for $\mathsf{prf}(a)$.
(4) We omit $\mathtt{ChangeCipherSpec}$ messages: they are not part of the handshake protocol.
(5) We write $\langle\text{prior messages}\rangle$ for the concatenation of all messages sent and received so far in the epoch, starting from the latest $\mathtt{ClientHello}$. (6) We let $\lfloor.\rfloor_r$ and $\lfloor.\rfloor_p$ be functions that truncate to record-key and MAC sizes.
(7) We let t_1, t_2, t_3 abbreviate the constant strings $\mathtt{"derive\ key"}$, $\mathtt{"client\ finished"}$, $\mathtt{"server\ finished"}$; we write $\|$ for bytestring concatenation.

Fig. 1. Abstract model of TLS handshake protocol (static handshake; resumption)

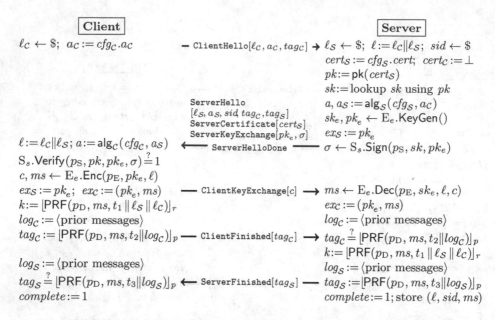

Fig. 2. Abstract model of TLS handshake protocol (ephemeral renegotiation)

Its proof (in the full paper) relies on intermediate definitions for multi-key libraries and, as a first step, uses hybrid arguments to lift security from our agile definitions to the multi-key setting.

Theorem 2 (TLS Handshake). *Let a, $a^\star$ range over the agility parameters supported by TLS. Let $P_s = \{p^\star \mid s, p^\star := \mathsf{sig}(a^\star)\}$, $P_e = \{p^\star \mid e, p^\star := \mathsf{kem}(a^\star)\}$, and $P = \{p^\star \mid p^\star := \mathsf{prf}(a^\star)\}$. Let α be a strength predicate (Definition 4) such that the following assumptions hold:*

(1) *If $\alpha(a)$ and $s, p := \mathsf{sig}(a)$ then S_s is EUF-CMA $(\epsilon_{s,p}, t_{s,p}, p, P_s)$-secure.*
(2) *If $\alpha(a)$ and $e, p := \mathsf{kem}(a)$ then E_e is IND-RCCA-n_{ms} $(\epsilon_{e,p}, t_{e,p}, p, P_e)$-secure.*
(3) *If $\alpha(a)$ and $p := \mathsf{prf}(a)$ then PRF is an (ϵ_p, t_p, p, P)-secure PRF.*

Let n_s bound the number of calls to $S_s.\mathsf{KeyGen}$. Let n and n_{ms} bound the number of epochs and sessions. Let n_e bound the number of calls to $E_e.\mathsf{KeyGen}$, both for ephemeral and static KEMs. The TLS handshake is (ϵ, t, α)-secure, where

$$\epsilon = \sum_s \sum_p n_s \epsilon_{s,p} + \sum_e \sum_p n_e \epsilon_{e,p} + n_{ms} \sum_p \epsilon_p + n^2(2^{-225} + 2^{-\min_p \lfloor \lfloor \cdot \rfloor_p \rfloor})$$

and where each t_* in the assumptions is at most t plus the cost of simulating Π in the reduction.

Discussion. In the theorem, the sets P_s, P_e, and P represent the worst case. Indeed, signers may, for those keys that they consider honest, stop using signature algorithm s together with weak hash functions, like MD5, while TLS may still support verification using such hash algorithms for backward compatibility. To model such scenarios, one could instead add P_s, P_e, and P to the state of the experiment to record which hash algorithms have been used so far for signing, decrypting and deriving keys to obtain a more precise statement.

6   Verified Reference Implementation

We jointly programmed the TLS handshake and developed its proof. We finally outline our code, and explain how its structure and automated verification relate to the cryptographic models of §2–5; we provide additional details and performance results in the full paper. Our handshake implementation for MITLS consists of 3,600 lines of F# code plus 2,050 lines of F7 specifications; it supports four protocol versions, three key exchange mechanisms, two signature algorithms, and four hash functions. It deals mostly with the protocol aspects; indeed, our cryptographic proof for Theorem 1, conducted with EASYCRYPT, concerns less than 200 lines of F#. Conversely, Theorem 2 involves the full codebase and proving it requires a modular design and automated program verification techniques.

We adopt the type-based cryptographic verification method of Fournet et al. [10], previously applied to MITLS by Bhargavan et al. [4, §2]. The MITLS library consists of 45 modules, not counting application code or platform libraries. Each module implements a single cryptographic functionality or protocol component and represents an abstraction boundary through its interface. A module is either trusted to be implemented correctly (e.g. the session database), or idealized under a cryptographic assumption (e.g. signatures) then verified, or perfectly verified (e.g. the protocol state machine). Each module interface specifies preconditions, postconditions, and type abstractions that govern the conditions under which secrets (keys, plaintexts, etc.) may be read or written by other modules.

We discuss the design of three important components that we modified during the course of this paper. *TLSInfo* defines agility parameters and logical predicates (corresponding to α in Definition 4) that specify algorithmic strength, honesty for both long-term-keys and ephemeral secrets, matching record algorithms, and handshake completion events. This new logical model is more detailed than the original one [4]; furthermore, we extended the session structure and logical model to provide a general treatment of protocol extensions. *HandshakeMessages* implements message formatting and parsing; agreement (Definition 5(4)) depends on its details, since only formatted data is cryptographically authenticated. This code is complicated but not especially deep, and best handled using automated verification. *Handshake* implements the handshake state machine (*Send* in §5). Its code is not as simple as suggested by the KEMs of §3, since the TLS standard employs different sequences of messages for (say) RSA and DHE handshakes. Hence, we have similar but separate code for them, each of their interfaces complying with the KEM abstraction of §3. Also, our code handles errors and warnings, omitted in this presentation but also verified.

Our new results on the handshake, composed with prior results on MiTLS [4] (the record layer, the top-level API, and various applications) yield agile, verified application security for TLS as it is.

References

1. Acar, T., Belenkiy, M., Bellare, M., Cash, D.: Cryptographic agility and its reation to circular encryption. In: Gilbert, H. (ed.) EUROCRYPT 2010. LNCS, vol. 6110, pp. 403–422. Springer, Heidelberg (2010)
2. Barthe, G., Grégoire, B., Heraud, S., Zanella-Béguelin, S.: Computer-aided security proofs for the working cryptographer. In: Rogaway, P. (ed.) CRYPTO 2011. LNCS, vol. 6841, pp. 71–90. Springer, Heidelberg (2011)
3. Bellare, M., Rogaway, P.: Entity authentication and key distribution. In: Stinson, D.R. (ed.) CRYPTO 1993. LNCS, vol. 773, pp. 232–249. Springer, Heidelberg (1994)
4. Bhargavan, K., Fournet, C., Kohlweiss, M., Pironti, A., Strub, P.-Y.: Implementing TLS with verified cryptograhic security. In: IEEE Symposium on Security and Privacy (2013)
5. Bhargavan, K., Delignat-Lavaut, A., Fournet, C., Pironti, A., Strub, P.-Y.: Triple handshakes and cookie cutters: Breaking and fixing authentication over TLS. In: IEEE Symposium on Security and Privacy (2014)
6. Bleichenbacher, D.: Chosen ciphertext attacks against protocols based on RSA encryption standard PKCS #1. In: Krawczyk, H. (ed.) CRYPTO 1998. LNCS, vol. 1462, pp. 1–12. Springer, Heidelberg (1998)
7. Canetti, R., Krawczyk, H., Nielsen, J.B.: Relaxing chosen-ciphertext security. In: Boneh, D. (ed.) CRYPTO 2003. LNCS, vol. 2729, pp. 565–582. Springer, Heidelberg (2003)
8. Cramer, R., Shoup, V.: Design and analysis of practical public-key encryption schemes secure against adaptive chosen ciphertext attack. SIAM J. Computing 33(1), 167–226 (2003)
9. Dierks, T., Rescorla, E.: The Transport Layer Security (TLS) Protocol Version 1.2 (2008)
10. Fournet, C., Kohlweiss, M., Strub, P.-Y.: Modular code-based cryptographic verification. In: ACM CCS 2011 (2011)
11. Giesen, F., Kohlar, F., Stebila, D.: On the security of TLS renegotiation. In: ACM CCS 2013 (2013)
12. Haber, S., Pinkas, B.: Securely combining public-key cryptosystems. In: ACM CCS 2001 (2001)
13. Jager, T., Kohlar, F., Schäge, S., Schwenk, J.: On the security of TLS-DHE in the standard model. In: Safavi-Naini, R., Canetti, R. (eds.) CRYPTO 2012. LNCS, vol. 7417, pp. 273–293. Springer, Heidelberg (2012)
14. Jonsson, J., Kaliski Jr., B.S.: On the security of RSA encryption in TLS. In: Yung, M. (ed.) CRYPTO 2002. LNCS, vol. 2442, pp. 127–142. Springer, Heidelberg (2002)
15. Klíma, V., Pokorný, O., Rosa, T.: Attacking RSA-based sessions in SSL/TLS. In: Walter, C.D., Koç, Ç.K., Paar, C. (eds.) CHES 2003. LNCS, vol. 2779, pp. 426–440. Springer, Heidelberg (2003)
16. Kohlar, F., Schäge, S., Schwenk, J.: On the security of TLS-DH and TLS-RSA in the standard model. Cryptology ePrint Archive, Report 2013/367 (2013)

17. Krawczyk, H., Paterson, K.G., Wee, H.: On the security of the TLS protocol: A systematic analysis. In: Canetti, R., Garay, J.A. (eds.) CRYPTO 2013, Part I. LNCS, vol. 8042, pp. 429–448. Springer, Heidelberg (2013)

18. Langley, A.: Unfortunate current practices for HTTP over TLS (2011), http://www.ietf.org/mail-archive/web/tls/current/msg07281.html

19. Modadugu, N., Langley, A., Moeller, B.: Transport Layer Security (TLS) False Start. Internet Draft (2010)

20. Mavrogiannopoulos, N., Vercauteren, F., Velichkov, V., Preneel, B.: A cross-protocol attack on the TLS protocol. In: ACM CCS 2012 (2012)

21. Morrissey, P., Smart, N.P., Warinschi, B.: A modular security analysis of the TLS handshake protocol. In: Pieprzyk, J. (ed.) ASIACRYPT 2008. LNCS, vol. 5350, pp. 55–73. Springer, Heidelberg (2008)

22. Naccache, D., Shparlinski, I.E.: Divisibility, Smoothness and Cryptographic Applications. ArXiv e-prints (October 2008)

23. Paterson, K.G., Ristenpart, T., Shrimpton, T.: Tag size does matter: Attacks and proofs for the TLS record protocol. In: Lee, D.H., Wang, X. (eds.) ASIACRYPT 2011. LNCS, vol. 7073, pp. 372–389. Springer, Heidelberg (2011)

24. Qualys SSL labs. SSL server test, https://www.ssllabs.com/ssltest/analyze.html

25. Ray, M.: Authentication gap in TLS renegotiation (2009), http://extendedsubset.com/Renegotiating_TLS.pdf

26. Stevens, M., Sotirov, A., Appelbaum, J., Lenstra, A., Molnar, D., Osvik, D.A., de Weger, B.: Short chosen-prefix collisions for MD5 and the creation of a rogue CA certificate. Cryptology ePrint Archive, Report 2009/111 (2009)

27. Wagner, D., Schneier, B.: Analysis of the SSL 3.0 protocol. In: 2nd USENIX Workshop on Electronic Commerce, WOEC 1996 (1996)

Memento: How to Reconstruct Your Secrets from a Single Password in a Hostile Environment

Jan Camenisch[1], Anja Lehmann[1], Anna Lysyanskaya[2], and Gregory Neven[1]

[1] IBM Research – Zurich, Rüschlikon, Switzerland
[2] Brown University, Providence, USA

Abstract. Passwords are inherently vulnerable to dictionary attacks, but are quite secure if guessing attempts can be slowed down, for example by an online server. If this server gets compromised, however, the attacker can again perform an offline attack. The obvious remedy is to distribute the password verification process over multiple servers, so that the password remains secure as long as no more than a threshold of the servers are compromised. By letting these servers additionally host shares of a strong secret that the user can recover upon entering the correct password, the user can perform further cryptographic tasks using this strong secret as a key such as encrypting data in the cloud. Threshold password-authenticated secret sharing (TPASS) protocols provide exactly this functionality. Unfortunately, the two only known schemes by Bagherzandi et al. (CCS 2011) and Camenisch et al. (CCS 2012) leak the password if a user mistakenly executes the protocol with malicious servers. Authenticating to the wrong servers is a common scenario when users are tricked in phishing attacks. We propose the first t-out-of-n TPASS protocol for any $n > t$ that does not suffer from this shortcoming. We prove our protocol secure in the UC framework, which for the particular case of password-based protocols offers important advantages over property-based definitions, e.g., by correctly modeling typos in password attempts.

1 Introduction

You wake up in a motel room. Where are you? How did you get here? You can't remember anything. Or perhaps you can. One word, a password, is engraved in your mind. You go outside and walk into the street. The first person you meet doesn't know you. The second seems to recognize you, or at least pretends to do so. He says he's your friend. He introduces you to other people who claim they are also your friends. They say they can help you reconstruct your memory—if you give the correct password. But why would you trust them? What if they are not your friends? What if they're trying to plant false memories into your brain? What if they're trying to learn your password, so they can retrieve your real memories from your real friends? How can you tell?

The above scenario, inspired by the movie "Memento" in which the main character suffers from short-term memory loss, leads to an interesting cryptographic problem that is also very relevant in practice. Namely, can a user securely recover his secrets from a set of servers, if all the user can or wants to remember is a single password and all of the servers may be adversarial? In particular, can he protect his precious password when

J.A. Garay and R. Gennaro (Eds.): CRYPTO 2014, Part II, LNCS 8617, pp. 256–275, 2014.
© International Association for Cryptologic Research 2014

accidentally trying to run the recovery with all-malicious servers? A solution for this problem can act as a natural bridge from human-memorizable passwords to strong keys for cryptographic tasks, all while avoiding offline dictionary attacks on the password. Practical applications include secure password managers (where the shared secret is a list of strongly random website passwords) and encrypting data in the cloud (where the shared secret is the encryption key) based on a single master password.

A single master password may seem a bad idea given that over the past few years, hundreds of millions of passwords have been stolen through server compromises, with major data breaches being reported at popular websites such as LinkedIn, Adobe, Yahoo!, and Twitter. Storing passwords in hashed form offers hardly any protection due to the efficiency of brute-force *offline attacks* using dictionaries. According to NIST [8], sixteen-character human-memorizable passwords have only 30 bits of entropy on average. With current graphical processors testing more than three hundred billion passwords per second [33], security must be considered lost as soon as an offline attack against the password data can be performed. Indeed, more than ninety percent of the 6.5 million password hashes pilfered from LinkedIn were cracked within six days [32]. Dedicated password hashes such as bcrypt [43] or PBKDF2 [37] only give a linear security improvement: a factor x more effort to verify passwords for an honest server makes offline dictionary attacks at a factor x harder.

However, as poorly as passwords stand their ground against offline attacks, they are actually fairly secure against *online* attacks, as long as attempts are slowed down or *throttled* by an honest server, e.g., by blocking accounts, presenting CAPTCHAs, or introducing time delays. The problem is that if a single server can check the correctness of a password, then that server—or any adversary breaking into it—must have access to some information that can be used for an offline attack. The obvious solution is to verify passwords through a distributed protocol involving multiple servers, in such a way that no single server, or no collusion up to a certain threshold, stores or obtains any information that can enable an offline attack.

Scenario. Recall our original goal that we don't just want to authenticate to a set of servers, we also want to store a (strong) secret that the user can later reconstruct from a subset of the servers using a single password, in such a way that the servers don't learn anything about the secret or the password. The secret can be used as a key for any other cryptographic purpose, for example, to encrypt and store a file in the cloud containing strong passwords and other credentials required for websites or online services. Those services thereby do not have to change their standard authentication mechanisms, ensuring a smooth deployment path. A commercial product along these lines called *RSA Distributed Credential Protection* [44] is already available.

When the user sets up his account, he carefully selects a list of names of servers that he will use in the protocol. He may make his selection based on the servers' reputation, perceived trust, or other criteria; the selection is important, because if too many of the selected servers are malicious, his password and secret are already compromised from the beginning. It is also clear that at setup the user must be able to authenticate the servers that he selected. In previous password-based schemes, setup is often assumed to take place out-of-band. Given the importance of the setup phase, we follow

Camenisch et al. [13] by explicitly modeling account setup and assuming that a public-key infrastructure (PKI) is in place to link server names to public keys.

When later the user wants to retrieve his secret, ideally, he should not need anything else than his username and password. In particular, he should not even have to remember the names of the servers he selected at setup. The list may be too long for the user to remember, and he can certainly not be expected to, at every retrieval, spend the same amount of thought on composing the list of names of the servers as during setup. Also, the user may retrieve his secret with a different device than the one that he used to create the account. For example, he may be logging in from his phone instead of his laptop, he may be installing a new device, or he may be borrowing a friend's tablet PC. Of course, we do have to assume that the device on which the user enters his single password is "clean", i.e., is not infected with malware, doesn't have a key-logger attached, etc. We make the minimal requirement that the user has a clean operating system and a clean web browser to work with, containing hardcoded keys of root certification authorities (CAs) and an implementation of our protocol. We explicitly do not want any user-specific state information from the setup phase to be needed on the device at the time of retrieval. Different users may select different server names, so the names of the selected servers cannot be hardcoded in the browser either. The list of servers that is used at retrieval may be different from the list used at setup: the user may forget some servers when authenticating, involve some servers that were not present at setup, mistype server URLs, or even be tricked into running the protocol with a set of all-malicious servers through a sort of phishing attack. Note that a PKI doesn't prevent this: malicious servers also have certified keys. Also note that users cannot rely on the servers to store user-specific state information that they later send back signed to the user, because the servers during retrieval may be malicious and lie about the content or wrongly pretend to have been part of the trusted setup set.

Existing Solutions. Threshold password-authenticated secret sharing (TPASS) schemes are the best fit for our problem: they allow a user to secret-share a secret K among n servers and protect it with a password p, so that the user can later recover K from any subset of $t + 1$ of the servers using p, but so that no coalition smaller than t learns anything about K or can mount an offline attack on p. Unfortunately, the two currently known TPASS protocols by Bagherzandi et al. [3] and Camenisch et al. [13] break down when the user tries to retrieve his secret from a set of all-malicious servers. In the former, the password is exposed to offline attacks, in the latter it is plainly leaked. We outline the attacks on both protocols in our full paper [12]. These attacks are of course quite devastating, as once the password is compromised, the malicious servers can recover the user's secret from the correct servers.

Our Contribution. We provide the first t-out-of-n TPASS protocol for any $n > t$ that does not require trusted, user-specific state information to be carried over from the setup phase. Our protocol requires the user to only remember a username and a password, assuming that a PKI is available; if he misremembers his list of servers and tries to retrieve his secret from corrupt servers, our protocol prevents the servers from learning anything about the password or secret, as well as from planting a different secret into the user's mind than the secret that he stored earlier.

Our construction is inspired by the protocol of Bagherzandi et al. [3] by relying on a homomorphic threshold cryptosystem, but the crucial difference is that in our retrieve protocol, the user never sends out an encryption of his password attempt. Instead, the user derives an encryption of the (randomized) quotient of the password used at setup and the password attempt. The servers then jointly decrypt the quotient and verify whether it yields "1", indicating that both passwords matched. In case the passwords were not the same, all the servers learn is a random value.

The Case for Universal Composability. We prove our protocol is secure in the universal composability (UC) framework [16]. The particular advantages of UC security notions for the special case of password-based protocols have been pointed out before [18,13]; we recall the main arguments here. First, all property-based security notions for threshold password-based protocols in the literature [40,47,38,3] assume that honest users choose their passwords from known, fixed, independent distributions. In reality, users share, reuse, and leak information related to their passwords outside of the protocol. Second, all known property-based notions allow the adversary to observe or even interact with honest users with their correct passwords, but not on incorrect yet related passwords—which is exactly what happens when a user makes a typo while entering his password. In the UC framework, this is modeled more naturally by letting the environment provide the passwords, so no assumptions need to be made regarding their distributions, dependencies, or leakages. Finally, property-based definitions consider the protocol in isolation, while the composition theorem of the UC framework guarantees secure composition with itself as well as with other protocols. Composition with other protocols is of particular importance in the considered TPASS setting, where a user shares and reconstructs a strong key K with multiple servers, and should be able to securely use that key in a different protocol, for instance to decrypt data kept in the cloud. Modeling secure composition of password-based protocols is particularly delicate given the inherent non-negligible success probability of the adversary when guessing the password. Following previous work [18,13], our UC notion absorbs the inherent guessing attacks into the ideal functionality itself. A secure protocol guarantees that the real world and ideal world are indistinguishable, thus the composition theorem continues to hold.

Building a UC secure protocol requires many additional tools, such as simulation-sound non-interactive zero-knowledge proofs with online witness extraction (which can be efficiently realized for discrete-logarithm based relations in the random-oracle model) and CCA2-secure encryption. It is all the more surprising that our final protocol is efficient enough for use in practice: It requires only $5n + 15$ and $14t + 24$ exponentiations from the user during setup and retrieval, respectively. Each server has to perform $n + 18$ and $7t + 28$ exponentiations in these respective protocols.

Related Work. In spite of their practical relevance, TPASS protocols only started to appear in the literature very recently. The t-out-of-n TPASS protocol by Bagherzandi et al. [3] was proved secure under a property-based security notion. As mentioned above, it relies on untamperable user memory and breaks down when the user retrieves its secret from all-corrupt servers. Our protocol can be seen as a strengthened version of the Bagherzandi et al. protocol; we refer to Section 4 for a detailed comparison. The 1-out-of-2 TPASS protocol by Camenisch et al. [13] was proved secure in the UC framework, but, by construction, leaks the password and secret if a user tries to retrieve his secret from all-corrupt servers.

Constructing TPASS protocols from generic multi-party computation (MPC) is possible but yields inefficient protocols. Our strong security requirements require public-key operations to be encoded in the to-be-evaluated circuit, while the state-of-the-art MPC protocols [23,24,22] require an expensive joint key-generation step to be performed at each retrieval. We refer to the full paper [12] for details.

The closely related primitive of threshold password-authenticated key exchange (TPAKE) lets the user agree on a fresh session key with each of the servers, but doesn't allow the user to store and recover a secret. Depending on the desired security properties, one can build a TPASS scheme from a TPAKE scheme by using the agreed-upon session keys to transmit the stored secret shares over secure channels [3].

The first TPAKE protocols due to Ford and Kaliski [29] and Jablon [36] were not proved secure. The first provably secure TPAKE protocol, a t-out-of-n protocol, was proposed by MacKenzie et al. [40]. The 1-out-of-2 protocol of Brainard et al. [9,47] is implemented in EMC's RSA Distributed Credential Protection [44]. Both protocols either leak the password or allow an offline attack when the retrieval is performed with corrupt servers (see the full paper [12]). The t-out-of-n TPAKE protocols by Di Raimondo and Gennaro [26] and the 1-out-of-2 protocol by Katz et al. [38] are proved secure under property-based (i.e., non-UC) notions. These protocols actually remain secure when executed with all-corrupt servers, but are restricted to the cases where $n > 3t$ and $(t, n) = (1, 2)$, respectively.

Boyen [7] presented a protocol related to TPASS, where a user can store a *random* value under a password with a *single* server. While being very efficient, this protocol fails to provide most of the security properties we require, i.e., the server can set up the user with a wrong secret, throttling is not possible, and no UC security is offered.

2   Definition of Security

Recall the goal of a TPASS scheme: at setup, a user secret-shares his data amongst n servers protected by a password p; at retrieval, he can recover his data from a subset of $t + 1$ of these n servers, assuming that at most t of them are corrupt. For the sake of simplicity, we assume that the user's data is a symmetric key K; the user can then always use K to encrypt and authenticate his actual data and store the resulting ciphertext in the cloud.

We want the user to be able to retrieve his data remembering only his username uid and his password, and perhaps the name of one or a couple of his trusted servers. The user has access to the PKI but cannot be assumed to store any additional information, cryptographic or other. In particular, the user does not have to remember the names or public keys of *all* of the servers among which he shared his key. Rather, in a step preceding the retrieval (that we don't model here), he can ask some servers to remind him of his full list of servers. Of course, these servers may lie if they are malicious, tricking the into retrieving his key from servers that weren't part of the original setup. We want to protect the user in this case and prevent the servers from learning the password.

Certain attacks are inherent and cannot be protected against. For example, a corrupt user can always perform an online attack on another user's password p by doing several retrieval attempts. It is therefore crucial that honest servers detect failed retrieval attempts, so that they can apply throttling mechanisms to stop or slow down the attack, such as blocking the user's account or asking the user to solve a CAPTCHA. The throttling mechanism should count retrieval attempts that remain pending for too long as failed attempts, since the adversary can always cut the communication before some of the servers were able to conclude.

A second inherent attack is that if at least $t + 1$ of the n servers at setup are corrupt, then these servers can mount an offline dictionary attack on the user's password p. Given the low entropy in human-memorizable passwords and the efficiency of offline dictionary attacks on modern hardware, one may conservatively assume that in this case the adversary simply learns p and K—which is how we model it here.

A somewhat subtle but equally unavoidable attack is that when an honest user makes a retrieval attempt with a set of all-corrupt servers, the servers can try to plant any key K^* of their choice into the user's output. This attack is unavoidable, because the corrupt servers can always pretend that they participated in a setup protocol for a "planted" password p^* and a "planted" key K^*, and then execute the retrieve protocol with the honest user using the information from this make-believe setup. If the planted password p^* matches the password p' the user is retrieving with, the user will retrieve the planted key K^* instead of his real key. Note that in the process, the adversary learns whether $p^* = p'$, thus he gets a guess at the password p'. This planting attack is even more critical if the user previously set up his account with at least $t + 1$ corrupted servers, because in that case the adversary already knows the real password p, which most likely is equal to the password p' with which the user runs the retrieval.

Finally, in our model, all participants are communicating over an adversarial network, which means that protocol failures are unavoidable: the adversary may block communication between honest servers and the user. As a result, we cannot guarantee that the user always succeeds in retrieving his data. In view of this, we chose to restrict the retrieval protocol to $t + 1$ servers: although this choice causes the retrieve protocol to fail if just one server refuses to (being adversarial), adversarial failures are already unavoidable in our network model. We could still try to guarantee some limited form of robustness (recall that, in the threshold cryptography literature, a protocol is *robust* if it can successfully complete its task despite malicious behavior from a fraction of participants) by requiring that, when $t + 1$ or more honest servers participate and the network does not fail, the user successfully recovers his data. However, while it seems not hard to add robustness to our protocols by applying the usual mechanisms found in the literature, it turns out that modeling robustness would considerably complicate our (already rather involved) ideal functionality.

2.1 Ideal Functionality

Assuming the reader is familiar with the UC framework [16], we now describe the ideal functionality $\mathcal{F}_{TPASS(t,n)}$ of t-out-of-n TPASS. For simplicity, we refer to $\mathcal{F}_{TPASS(t,n)}$ as $\mathcal{F}$ from now on. It interacts with a set of users $\{\mathcal{U}\}$, a set of servers $\{\mathcal{S}_i\}$ and an adversary $\mathcal{A}$. We consider static corruptions and assume that $\mathcal{F}$ knows which of the servers in $\{\mathcal{S}_i\}$ are corrupt.

The UC framework allows us to focus our analysis on a single protocol instance with a globally unique *session identifier sid*. Security for multiple sessions follows through the composition theorem [16] or, if different sessions are to share state, through the joint-state universal composition (JUC) theorem [19]. Here, we use the username *uid* as the session identifier *sid*, and let each setup and retrieve query be assigned a unique *sub-session identifier ssid* and *rsid* within the single-session functionality for $sid = uid$. When those sub-session identifiers are established through the functionality by Barak et al. [4], they have the form $ssid = (ssid', \mathbf{S})$ and $rsid = (rsid', \mathbf{S}')$, respectively, i.e., they consist of a globally unique string and the identifiers of the servers $\mathbf{S} = (\mathcal{S}_1, \ldots, \mathcal{S}_n)$ that agreed on that identifier. We will later motivate these choices; for now, it suffices to know that a session identifier $sid = uid$ corresponds to a single user account, and that the sub-session identifiers *ssid* and *rsid* refer to individual setup and retrieve queries for that account.

The functionality $\mathcal{F}$ has two main groups of interfaces, for setup and retrieve. For the sake of readability, we describe the behavior of those interfaces in a somewhat informal way here and provide their formal specification in the full paper [12].

Setup Interfaces: The SETUP-related interfaces allow a user $\mathcal{U}$ to instruct $\mathcal{F}$ to store the a key K, protected under a password p, among n servers $\mathbf{S} = (\mathcal{S}_1, \ldots, \mathcal{S}_n)$ of the user's choice.

1. A (SETUP, sid, $ssid$, p, K) message from a user $\mathcal{U}$ initiates the functionality for user name $uid = sid$. The sub-session identifier $ssid$ contains a list of n different server identities $\mathbf{S} = (\mathcal{S}_1, \ldots, \mathcal{S}_n)$ among which $\mathcal{U}$ wants to share his key K protected by the password p. If at least $t + 1$ servers in $\mathbf{S}$ are corrupt, $\mathcal{F}$ sends the password and the key to the adversary, otherwise it merely informs $\mathcal{A}$ that a setup sub-session is taking place. $\mathcal{F}$ also creates a record s where it stores $s.ssid$, $s.p$, $s.K$ and sets $s.\mathcal{R} \leftarrow \mathcal{U}$.
2. A (JOIN, sid, $ssid$, $\mathcal{S}_i$) message from the adversary $\mathcal{A}$ instructs $\mathcal{F}$ to let a server $\mathcal{S}_i$ join the setup. If $\mathcal{S}_i$ is honest, this means that $\mathcal{S}_i$ registers the setup and will not join any further setups for the same username $uid = sid$. The user is informed that $\mathcal{S}_i$ joined the setup.
3. A (STEAL, sid, $ssid$, $\hat{p}$, $\hat{K}$) message from $\mathcal{A}$ models a rather benign but unavoidable attack where the adversary "steals" the sub-session $ssid$ by intercepting and replacing the network traffic generated by $\mathcal{U}$, allowing $\mathcal{A}$ to replace the password and the key provided by $\mathcal{U}$ with his own choice $s.p \leftarrow \hat{p}$ and $s.K \leftarrow \hat{K}$. Note that this is not a very powerful attack, since the adversary could achieve essentially the same effect by letting a corrupt user initiate a separate setup session for $\hat{p}$, $\hat{K}$. Thus, the only difference is that here the adversary uses the $ssid$ generated by an honest user, and not a fresh one. Servers are unaware when such an attack takes place, but the user $\mathcal{U}$ cannot be made to believe that an honest server $\mathcal{S}_i$ has accepted his inputs. This is modeled by setting the recipient of server confirmations $s.\mathcal{R} \leftarrow \mathcal{A}$.

Retrieve Interfaces: The RETRIEVE-related interfaces allow $\mathcal{U}'$ to retrieve the key from $t + 1$ servers $\mathbf{S}'$ if $\mathbf{S}' \subseteq \mathbf{S}$ and $\mathcal{U}'$ furnishes the correct password; it also models the planting attack described earlier.

4. A (RETRIEVE, $sid, rsid, p'$) message from a user $\mathcal{U}'$ instructs $\mathcal{F}$ to initiate a retrieval for username $uid = sid$ with password p' from the set $\mathbf{S}' = \mathcal{S}_1, \dots, \mathcal{S}_{t+1}$ of $t + 1$ servers included in the sub-session identifier $rsid$. $\mathcal{F}$ then creates a retrieve record r, where it stores $r.rsid, r.p'$, sets $r.\mathcal{R} \leftarrow \mathcal{U}'$, and initially sets $r.ssid \leftarrow \perp$ and $r.K \leftarrow \perp$. If there was a setup sub-session $ssid$ that all honest servers in $\mathbf{S}'$ have joined and where all servers in $\mathbf{S}'$ also occur in $\mathbf{S}$, then $\mathcal{F}$ links this retrieve to $ssid$ by setting $r.ssid \leftarrow ssid$. $\mathcal{F}$ notifies the adversary and (with an adversarially determined delay) the honest servers in $\mathbf{S}'$ that a new retrieval is taking place. Note that the password attempt p' is *not* leaked to the adversary, even if all servers in $\mathbf{S}'$ are corrupt.

5. A (PLANT, $sid, rsid, p^*, K^*$) message from the adversary $\mathcal{A}$ allows him to perform the *planting attack* described earlier. Namely, if all $t + 1$ servers in the retrieval are corrupt, $\mathcal{A}$ can submit a password p^* and a key K^* to be planted. The functionality $\mathcal{F}$ tells $\mathcal{A}$ whether p^* matches the password attempt p'. If so, $\mathcal{F}$ also sets the key $r.K$ that will eventually be returned in this session to the to-be-planted key K^* provided by the adversary. Note that the adversary can perform only one planting attack per retrieval. So even if all $t + 1$ servers are corrupt, the adversary only obtains a single guess for the retrieval password p'.

6. A (STEAL, $sid, rsid, \hat{p}$) message from $\mathcal{A}$ allows the adversary to "steal" the sub-session identifier $rsid$, replacing the original password attempt $r.p'$ with $\hat{p}$ of his choice. Servers do not notice this attack taking place, but the originating user will conclude that the protocol failed, or not receive any output at all. This is modeled again by setting $r.\mathcal{R} \leftarrow \mathcal{A}$.

7. A (PROCEED, $sid, rsid, a$) message with $a \in \{\texttt{allow}, \texttt{deny}\}$ coming from an honest server $\mathcal{S}_i$ (after having been notified that a retrieval is taking place) indicates its (un)willingness to participate in the retrieval. This models the opportunity for an external throttling mechanism to refuse this retrieval attempt. Only when *all* honest servers have agreed to participate, the retrieval continues and the adversary learns whether the passwords matched (i.e., whether $r.p' = s.p$ with s being the setup record for $ssid$). If they matched, $\mathcal{F}$ also sets the key to be returned $r.K$ to the key shared during setup $s.K$.

8. A (DELIVER, $sid, rsid, \mathcal{P}, a$) message from $\mathcal{A}$ where $a = \texttt{allow}$ instructs $\mathcal{F}$ to output the final result of this retrieval to party $\mathcal{P}$, which can either be an honest server $\mathcal{S}_i$ or the user specified in $r.\mathcal{R}$. If $\mathcal{P} = r.\mathcal{R}$, the user will obtain the value $r.K$, where the result will signal a successful retrieval only if $r.K \neq \perp$, i.e., a key was assigned after the passwords matched. When $\mathcal{P} = \mathcal{S}_i$, the server will receive either a success or failure notification, indicating whether the passwords matched. Note that, in both cases, $\mathcal{A}$ can still turn a successful result into a failed one by passing $a = \texttt{deny}$ as input. This is because in the real world, the adversary can always make a party believe that a protocol ended unsuccessfully by simply dropping or invalidating correct messages. However, the inverse is not possible, i.e., the adversary can not make a mismatch of the passwords look like a match.

Session Identifiers. Our choice of (sub-)session identifiers merits some further explanation. In the UC framework, all machine instances participating in a protocol execution, including ideal functionalities, share a globally unique session identifier sid. Obviously, our SETUP and RETRIEVE interfaces must be called with the same sid to provide the expected functionality, because otherwise the instance cannot keep state between setup and retrieval. However, we insisted that a user can only be expected to remember a username and a password between setup and retrieve, but no further information such as public keys or random nonces. The sid therefore consists only of the username uid and thus cannot be used to uniquely identify different setup or retrieval sub-sessions for this username. To allow the functionality to refer to multiple simultaneous setup and retrieve sub-sessions, the participants of each sub-session establish a unique sub-session identifier $ssid$ or $rsid$ using the standard techniques mentioned earlier [4]. Therein, a unique identifier is created by simply concatenating the identities of the communicating parties and random nonces sent by all parties.

3   Preliminaries

In this section we introduce the building blocks for our protocols. These are three kinds of public-key encryption schemes, a signature scheme, and zero-knowledge proof protocols. We require two of the encryption schemes to be compatible, i.e., the message space to be the same algebraic group. To this end we make use of a probabilistic polynomial-time algorithm GGen that on input the security parameter 1^τ outputs the description of a multiplicative cyclic group $\mathbb{G}$, its prime order q, and a generator g, and require the key generation algorithms of the compatible encryption schemes to take $\mathbb{G}$ as input instead of the security parameter.

CPA-Secure Public-Key Encryption Scheme. Such a scheme consists of three algorithms (KGen, Enc, Dec). The key generation algorithm KGen on input $(\mathbb{G}, q, g)$ outputs a key pair (epk, esk). The encryption algorithm Enc, on input a public key epk and a message $m \in \mathbb{G}$, outputs a ciphertext C, i.e., $C \leftarrow \mathsf{Enc}_{epk}(m)$. The decryption algorithm Dec, on input the secret key esk and a ciphertext C, outputs a message $m \leftarrow \mathsf{Dec}_{esk}(C)$. We require this scheme to satisfy the standard CPA-security properties, with key generation defined as $\mathsf{KGen}(\mathsf{GGen}(1^\tau))$.

CCA2-Secure Labeled Public-Key Encryption Scheme. Any standard CCA2-secure scheme (KGen2, Enc2, Dec2) that supports labels [14] is sufficient fir our protocols. Therein, $(epk, esk) \leftarrow \mathsf{KGen2}(1^\tau)$ denotes the key generation algorithm. The encryption algorithm takes as input the public key epk, a message m, a label $l \in \{0,1\}^*$ and outputs a ciphertext $C \leftarrow \mathsf{Enc2}_{epk}(m, l)$. The decryption $\mathsf{Dec2}_{esk}(C, l)$ of C will either output a message m or a failure symbol $\bot$. The label l can be seen as context information which is non-malleably attached to a ciphertext C and restricts the decryption of C to that context, i.e., decryption with a label different from the one used for encryption will fail.

Semantically Secure (t, n)-Threshold Homomorphic Cryptosystem. Such a scheme consists of five algorithms (TKGen, TEnc, PDec, VfDec, TDec). The key generation algorithm TKGen, on input $(\mathbb{G}, q, g, t, n)$, outputs a public key tpk and n partial key pairs $(tpk_1, tsk_1), \ldots, (tpk_n, tsk_n)$.

The encryption algorithm TEnc, on input a public key tpk and a message $m \in \mathbb{G}$, outputs a ciphertext C. The partial decryption algorithm PDec, on input (tsk_i, C), outputs a decryption share d_i and a proof π_{d_i}. The decryption share verification algorithm VfDec, on input $(tpk_i, C, d_i, \pi_{d_i})$, verifies that d_i is correct w.r.t. C and tpk_i. The threshold decryption algorithm TDec, on input C and $k \geq t + 1$ decryption shares $d_{i_1}, \ldots, d_{i_k}$, outputs a plaintext m or $\bot$.

Our protocol will require that the threshold scheme has an appropriate *homomorphic property*, namely that there is an efficient operation $\odot$ on ciphertexts such that, if $C_1 \in \text{TEnc}_{tpk}(m_1)$ and $C_2 \in \text{TEnc}_{tpk}(m_2)$, then $C_1 \odot C_2 \in \text{TEnc}_{tpk}(m_1 \cdot m_2)$. We will also use exponents to denote the repeated application of $\odot$, e.g., C_1^2 to denote $C_1 \odot C_1$.

Further, the scheme needs to be *sound* and *semantically secure*. In a nutshell, the former means that for a certain set of public keys $tpk, tpk_1, \ldots, tpk_n$ a ciphertext C can be opened only in an unambiguous way. The latter property of semantic security can be seen as an adaptation of the normal semantic security definition to the threshold context, where the adversary can now have up to t of the partial secret keys. In our full paper [12], we provide a detailed description of those properties, which are an adaption of the definitions by Cramer, Damgård, and Nielsen [20] for semantically secure threshold homomorphic encryption. The full paper further contains a construction based on the ElGamal cryptosystem that achieves our security notion.

Existentially Unforgeable Signature Scheme. By (SKGen, Sign, Vf) we denote such schemes, with $(spk, ssk) \leftarrow \text{SKGen}(1^\tau)$ being the key generation algorithm. For signing of a message $m \in \{0, 1\}^*$, we write $\sigma \leftarrow \text{Sign}_{ssk}(m)$, and for verification we write $b \leftarrow \text{Vf}_{spk}(m, \sigma)$, where the output b will be either 1 or 0, indicating success or failure.

Simulation-Sound Zero-Knowledge Proof System. We require a non-interactive zero-knowledge (NIZK) proof system to prove relations among different ciphertexts. We use an informal notation for this proof system, e.g., $\pi \leftarrow \text{NIZK}\{(m) : C_1 = \text{TEnc}_{tpk}(m) \wedge C_2 = \text{Enc}_{epk}(m)\}$ *(ctxt)* denotes the generation of a non-interactive zero-knowledge proof that is bound to a certain context *ctxt* and proves that C_1 and C_2 are both proper encryptions of the same message m under the public key tpk and epk for the encryption scheme TEnc and Enc, respectively. We require the proof system to be simulation-sound [45] and zero-knowledge. In the full paper [12], we give concrete realizations of the NIZK proofs that we require in our protocols assuming specific instantiations of the encryption schemes.

4 Our TPASS Protocol

The core of our construction bears a lot in common with that of Bagherzandi et al. [3], which however does rely on trusted user storage and is not proven to be UC secure. We first summarize the idea of their construction and then explain the changes and extensions we made to remove the trusted storage assumption and achieve UC security according to our TPASS functionality.

The high-level idea of Bagherzandi et al. [3] is depicted in Figure 1 and works as follows: In the setup protocol, the user generates keys for a threshold encryption scheme,

---

Setup : $\mathcal{U}(p, K, \mathbf{S})$ with public parameters $\mathbb{G}, q, g, t, n$

User generates threshold keys $(tpk, (tpk_i, tsk_i)_{i=1,\ldots,n}) \leftarrow \mathsf{TKGen}(\mathbb{G}, q, g, t, n)$, encrypts p and K: $C_p \leftarrow \mathsf{TEnc}_{tpk}(p)$, $C_K \leftarrow \mathsf{TEnc}_{tpk}(K)$, and sends (C_p, C_K, tpk, tsk_i) to each server $\mathcal{S}_i$ in $\mathbf{S}$.

Retrieve : $\mathcal{U}(p', \mathbf{S}, tpk) \rightleftharpoons (\mathcal{S}_1(C_p, C_K, tpk, tsk_1), \ldots, \mathcal{S}_n(C_p, C_K, tpk, tsk_n))$

User $\mathcal{U}$: $C_{p'} \leftarrow \mathsf{TEnc}_{tpk}(p')$, send $C_{p'}$ to each server in $\mathbf{S}$
Server $\mathcal{S}_i$: compute $C_{\mathrm{test},i} \leftarrow (C_p \odot (C_{p'})^{-1})^{r_i}$ for random r_i, send $C_{\mathrm{test},i}$ to $\mathcal{U}$
User $\mathcal{U}$:   compute $C_{\mathrm{test}} \leftarrow \odot_{i=1}^n C_{\mathrm{test},i}$, send C_{test} to each server in $\mathbf{S}$
Server $\mathcal{S}_i$: compute $d_i \leftarrow \mathsf{PDec}_{tsk_i}(C_{\mathrm{test}} \odot C_K)$, send d_i to $\mathcal{U}$
User $\mathcal{U}$:   output $K' \leftarrow \mathsf{TDec}(C_{\mathrm{test}} \odot C_K, d_1, \ldots, d_n)$

Fig. 1. Construction outline of the Bagherzandi et al. protocol. For the sake of simplicity, we slightly deviate from the notation introduced in Section 3 and omit the additional output of π_{d_i} of PDec.

encrypts both the password p and the key K using the generated public key, and sends these encryptions and generated decryption key shares to all n servers in $\mathbf{S}$. In addition to his username and password, the user here needs to remember the main public key tpk of the threshold scheme and the servers he ran the setup with. In the retrieve protocol, the user encrypts his password attempt p' under tpk and sends the ciphertext to all the servers in $\mathbf{S}$. The servers now compute an encryption of the password quotient p/p' and combine it with the encryption of the key K. With their help, the user decrypts this combined encryption. If $p = p'$, this will decrypt to 1 the original key K, otherwise it will decrypt to a random value.

It is easy to see that the user *must* correctly remember tpk and the exact set of servers, as he sends out an encryption of his password attempt p' under tpk. If tpk can be tampered with and changed so that the adversary knows the decryption key, then the adversary can decrypt p'. (Bagherzandi et al. [3] actually encrypt $g^{p'}$, so that the malicious servers must still perform an offline attack to obtain p' itself. However, given the typical low entropy of passwords, the password p' can be considered as leaked.)

Retrieve : $\mathcal{U}(p', \mathbf{S}') \rightleftharpoons (\mathcal{S}_1(C_p, C_K, tpk, tsk_1), \ldots, \mathcal{S}_n(C_p, C_K, tpk, tsk_n))$

User $\mathcal{U}$:   request ciphertexts and threshold public key from all servers in $\mathbf{S}'$
Server $\mathcal{S}_i$: send $(C_p, C_K, tpk)_i$ to $\mathcal{U}$
User $\mathcal{U}$:   if all servers sent the same (C_p, C_K, tpk), compute $C_{\mathrm{test}} \leftarrow (C_p \odot \mathsf{TEnc}_{tpk}(1/p'))^r$
       for random r and send C_{test} to each server in $\mathbf{S}'$
Server $\mathcal{S}_i$: compute $C_{\mathrm{test},i} \leftarrow (C_{\mathrm{test}})^{r_i}$ for random r_i, send $C_{\mathrm{test},i}$ to $\mathcal{U}$
User $\mathcal{U}$:   compute $C'_{\mathrm{test}} \leftarrow \odot_{i=1}^n C_{\mathrm{test},i}$, send C'_{test} to each server in $\mathbf{S}'$
Server $\mathcal{S}_i$: compute $d_i \leftarrow \mathsf{PDec}_{tsk_i}(C'_{\mathrm{test}})$, send d_i to $\mathcal{U}$
User $\mathcal{U}$:   if $\mathsf{TDec}(C'_{\mathrm{test}}, d_1, \ldots, d_n) = 1$, send $d_1, \ldots, d_n$ to each server in $\mathbf{S}'$
Server $\mathcal{S}_i$: if $\mathsf{TDec}(C'_{\mathrm{test}}, d_1, \ldots, d_n) = 1$, compute $d'_i \leftarrow \mathsf{PDec}_{tsk_i}(C_K)$, send d'_i to $\mathcal{U}$
User $\mathcal{U}$:   output $K' \leftarrow \mathsf{TDec}(C_K, d'_1, \ldots, d'_n)$

Fig. 2. Construction outline of our retrieval protocol (setup idea as in Figure 1).

Removing the Trusted User-Storage Requirement. Roughly, we change the retrieval protocol such that the user never sends out an encryption of his password attempt p', but instead sends an encryption of the randomized quotient p/p'. Thus, if the user mistakenly talks to adversarial servers instead of his true friends, these servers can try a guess at p', but will not be able to learn anything more. Our retrieval protocol begins with the user requesting the servers in $\mathbf{S}'$ (which may or may not be a subset of $\mathbf{S}$) to send him the ciphertexts and threshold public key he allegedly used in setup. If all servers respond with the same information, the user takes the received encryption of p and uses the homomorphism to generate a randomized encryption of p/p'. The servers then jointly decrypt this ciphertext. If it decrypts to 1, i.e., the two passwords match, then the servers send the user their decryption shares for the ciphertext encrypting the key K. By separating the password check and the decryption of K, the user can actually double-check whether his password was correct and whether he reconstructed his real key K.

Making the Protocol UC-Secure. The second main difference of our protocol is its UC security, which requires further mechanisms and steps added to the construction outlined in Figure 2. We briefly summarize the additional changes, the detailed description of our protocol follows later. First, in the security proof we need to extract p, p', and K from the protocol messages. This is achieved through a common reference string (CRS) that contains the public key PK of a semantically secure encryption scheme and the parameters for a non-interactive zero-knowledge (NIZK) proof system. Values that need be extractable are encrypted under PK and NIZK proofs are added to ensure that the correct value is encrypted. Further, all $t + 1$ servers explicitly express their consent with previous steps by signing all messages. The user collects, verifies, and forwards these signatures, so that all servers can verify the consent of all other servers. Some of these ideas were discussed by Bagherzandi et al., but only for a specific instantiation of ElGamal encryption and without aiming for full-blown UC security. Our protocol, on the other hand, is based on generic building blocks and securely implements the UC functionality presented in Section 2.

How to Remember the Servers. For the retrieve protocol, we assume that the input of the user contains $t + 1$ server names. In practice, however, the user might not remember these names. This is an orthogonal issue and there are a number of ways to deal with it. For instance, if the user remembers a single server name, he can contact that server and ask to be reminded of the names of all n servers. The user can then decide with which $t + 1$ of these servers to run the retrieve protocol. The user could even query more than one server and see whether they agree on the full server list. Again, the crucial point is that the security of our protocol does not rely on remembering the $t + 1$ server names correctly, as the security of the password p' is not harmed, even when the user runs the retrieve protocol with $t + 1$ malicious servers.

A Note on Robustness. As discussed in Section 2, the restriction to run the retrieve protocol with exactly $t + 1$ servers rather stems from the complexity that robustness would add to our ideal functionality, than from an actual protocol limitation. With asynchronous communication channels, one can achieve only a very limited form of robustness where the protocol succeeds if there are enough honest players *and* the adversary, who controls the network, lets the honest players communicate. Conceptually, one could add such limited robustness by running the retrieve protocol with all n servers and in each step continue the protocol only with the first k servers that sent valid response, where $t + 1 \leq k \leq n$. Bagherzandi et al. [3] handle robustness similarly by running the protocol with all n servers, mark servers that cause the protocol to fail as corrupt, and restart the protocol with at least $t + 1$ servers that appear to be good. To obtain better robustness guarantees, one must impose stronger requirements on the network such as assuming synchronous and broadcast channels, as is often done in the threshold cryptography literature [1,2,20]. With synchronous channels, protocols can achieve a more meaningful version of robustness, where it is ensured that inputs of all honest parties will be included in the computation and termination of the protocol is guaranteed when sufficient honest parties are present [39]. However, in practice, networks are rarely synchronous, and it is known that the properties guaranteed in a synchronous world cannot simultaneously be ensured in an asynchronous environment [21,6]. Thus, given the practical setting of our protocol, we prefer the more realistic assumptions over modeling stronger (but unrealistic) robustness properties.

4.1 Detailed Description of Our TPASS Protocol

In our protocol description, when we say that a party sends a message m as part of the setup or retrieve protocol, the party actually sends a message (SETUP, sid, $ssid$, i, m) or (RETRIEVE, sid, $rsid$, i, m), respectively, where i is a sequence number corresponding to the step number in the respective part of the protocol. Each party will only accept the first message that it receives for a specific (sub-)session identifier and sequence number. All subsequent messages from the same party for the same step of the protocol will be ignored.

Each party locally maintains state information throughout the different steps of one protocol execution; servers $\mathcal{S}_i$ additionally maintain a persistent state variable $st_i[sid]$ associated with the username $sid = uid$ that is common to all executions. Before starting a new execution of the setup or retrieve protocol, we assume that the parties use standard techniques [16,4] to agree on a fresh and unique sub-session identifier $ssid'$ and $rsid'$, respectively, that is given as an input to the protocol. Each party then only accepts messages that include a previously established sub-session identifier, messages with unknown identifiers will be ignored. We also assume that the sub-session identifiers $ssid$ and $rsid$ explicitly contain the identities of the communicating servers $\mathbf{S}$ and $\mathbf{S}'$, respectively. Using the techniques described in [4], the sub-session identifier would actually also contain the identifier of the user. However, as we do not assume that users have persistent public keys, we could not verify whether a certain user indeed belongs to a claimed identifier, and thus we discard that part of the output.

Setup Protocol. We assume that the system parameters contain a group $\mathbb{G} = \langle g \rangle$ of order q that is a τ-bit prime, and that the password p and the key K can be mapped into $\mathbb{G}$. In the following we assume that p and K are indeed elements of $\mathbb{G}$. We further assume that each server $\mathcal{S}_i$ has a public key (epk_i, spk_i), where epk_i is a public encryption key for the CCA2-secure encryption scheme generated by KGen2 and spk_i is a signature verification key generated by SKGen. We also assume a public-key infrastructure where servers can register their public keys, modeled by the ideal functionality

$\mathcal{F}_{CA}$ by Canetti [17]. Moreover, we require a common reference string, retrievable via functionality $\mathcal{F}_{CRS}$, containing a public key $PK \in \mathbb{G}$ of the CPA-secure public-key encryption scheme, distributed as if generated through KGen, but to which no party knows the corresponding secret key.

The user $\mathcal{U}$, on input (SETUP, sid, $ssid$, p, K) with $ssid = (ssid', \mathbf{S})$, runs the following protocol with all servers in $\mathbf{S}$. Whenever a check fails for a party (either the user or one of the servers), the party aborts the protocol without any output.

Step S1. The user $\mathcal{U}$ sets up secret key shares and note

(a) Query functionality $\mathcal{F}_{CRS}$ to obtain PK and, for each $\mathcal{S}_i$ occurring in $\mathbf{S}$, query $\mathcal{F}_{CA}$ to obtain $\mathcal{S}_i$'s public keys (epk_i, spk_i).

(b) Run $(tpk, tpk_1, \ldots, tpk_n, tsk_1, \ldots, tsk_n) \leftarrow \mathsf{TKGen}(\mathbb{G}, q, g, t, n)$ and encrypt the password p and the key K under both tpk and PK, i.e., compute

$$C_p \leftarrow \mathsf{TEnc}_{tpk}(p), \ C_K \leftarrow \mathsf{TEnc}_{tpk}(K), \ \widetilde{C}_p \leftarrow \mathsf{Enc}_{PK}(p), \ \widetilde{C}_K \leftarrow \mathsf{Enc}_{PK}(K).$$

(c) Generate a non-interactive zero-knowledge proof π_0 that the ciphertexts encrypt the same password and key, bound to $\mathtt{ctxt} = (sid, ssid, tpk, \mathbf{tpk}, C_p, C_K, \widetilde{C}_p, \widetilde{C}_K)$, where $\mathbf{tpk} = (tpk_1, \ldots, tpk_n)$:

$$\pi_0 \leftarrow \mathrm{NIZK}\{(p, K) : C_p = \mathsf{TEnc}_{tpk}(p) \wedge C_K = \mathsf{TEnc}_{tpk}(K) \wedge$$
$$\widetilde{C}_p = \mathsf{Enc}_{PK}(p) \wedge \widetilde{C}_K = \mathsf{Enc}_{PK}(K)\} \, (\mathtt{ctxt}) \, .$$

(d) Set $note = (ssid, tpk, \mathbf{tpk}, C_p, C_K, \widetilde{C}_p, \widetilde{C}_K, \pi_0)$.

(e) Compute $C_{S,i} \leftarrow \mathsf{Enc2}_{epk_i}(tsk_i, (sid, note))$ and send a message $(note, C_{S,i})$ to server $\mathcal{S}_i$ for $i = 1, \ldots, n$.

Step S2. Each server $\mathcal{S}_i$ checks & confirms user message

(a) Receive $(note, C_{S,i})$ with $note = (ssid, tpk, \mathbf{tpk}, C_p, C_K, \widetilde{C}_p, \widetilde{C}_K, \pi_0)$. Check that the variable $st_i[sid]$ has not been initiated yet. Check that the note is valid, i.e., that the proof π_0 is correct and that the sets $\mathbf{tpk}$ and $\mathbf{S}$ have the same cardinality (recall that $\mathbf{S}$ is included in $ssid$). Further, check that $\mathsf{Dec2}_{esk_i}(C_{S,i}, (sid, note))$ decrypts to a valid threshold decryption key tsk_i w.r.t. the received public keys.

(b) Sign sid and note as $\sigma_{1,i} \leftarrow \mathsf{Sign}_{ssk_i}(sid, note)$ and send the signature $\sigma_{1,i}$ to $\mathcal{U}$.

Step S3. The user $\mathcal{U}$ verifies & forwards server signatures

(a) When valid signatures $(\sigma_{1,1}, \ldots, \sigma_{1,n})$ are received from all servers $\mathcal{S}_i$ in $\mathbf{S}$, forward them to all servers in $\mathbf{S}$.

Step S4. Each server $\mathcal{S}_i$ verifies & confirms server consent

(a) Upon receiving a message $(\sigma_{1,1}, \ldots, \sigma_{1,n})$ from $\mathcal{U}$, check that all signatures $\sigma_{1,i}$ for $i = 1, \ldots, n$ are valid w.r.t. the local $note$.

(b) Store necessary information in the state $st_i[sid] \leftarrow (note, tsk_i)$.

(c) Compute $\sigma_{2,i} \leftarrow \mathsf{Sign}_{ssk_i}((sid, note), \mathtt{success})$, send $\sigma_{2,i}$ to $\mathcal{U}$, and output the tuple (SETUP, sid, $ssid$).

Step S5. The user $\mathcal{U}$ outputs the servers' acknowledgments

(a) Whenever receiving a valid signature $\sigma_{2,i}$ from a server $\mathcal{S}_i$ in $\mathbf{S}$, output (SETUP, sid, $ssid$, $\mathcal{S}_i$).

Retrieval Protocol. The user $\mathcal{U}'$ on input (RETRIEVE, $sid, rsid, p'$) where $rsid = (rsid', \mathbf{S}')$ runs the following retrieval protocol with the list of $t+1$ servers specified in $\mathbf{S}'$. Whenever a check fails for a party, the party sends a message (RETRIEVE, $sid, rsid,$ fail) to all other parties and aborts with output (DELIVER2S, $sid, rsid,$ fail) if the party is a server, or with output (DELIVER2U, $sid, rsid, \perp$) if it is the user. Further, whenever a party receives a message (RETRIEVE, $sid, rsid,$ fail), it aborts with the same respective outputs.

Step R1. The user $\mathcal{U}'$ creates ephemeral encryption key & requests notes

(a) Query $\mathcal{F}_{CRS}$ to obtain PK and, for each $\mathcal{S}_i$ in $\mathbf{S}'$, query $\mathcal{F}_{CA}$ to obtain $\mathcal{S}_i$'s public keys (epk_i, spk_i).

(b) Generate a key pair $(epk_U, esk_U) \leftarrow \mathsf{KGen2}(1^\tau)$ for the CCA2-secure encryption scheme that will be used to securely obtain the shares of the key K from the servers.

(c) Encrypt the password attempt p' under the CRS as $\widetilde{C}_{p'} \leftarrow \mathsf{Enc}_{PK}(p')$.

(d) Request the note from each server by sending $(epk_U, \widetilde{C}_{p'})$ to each server $\mathcal{S}_i \in \mathbf{S}'$.

Step R2. Each server $\mathcal{S}_i$ retrieves & sends signed note

(a) Upon receiving a retrieve request $(epk_U, \widetilde{C}_{p'})$, check if a record $st_i[sid] = (note, tsk_i)$ exists. Parse $note = (ssid, tpk, \mathbf{tpk}, C_p, C_K, \widetilde{C}_p, \widetilde{C}_K, \pi_0)$ and check that all servers in $\mathbf{S}'$ also occur in $\mathbf{S}$. (Recall, that sid and $rsid$ are contained in the header of the message, $\mathbf{S}'$ is included in $rsid$ and $\mathbf{S}$ in $ssid$.)

(b) Query $\mathcal{F}_{CA}$ to obtain the public keys (epk_j, spk_j) of all the other servers $\mathcal{S}_j$ in $\mathbf{S}'$.

(c) Compute $\sigma_{4,i} \leftarrow \mathsf{Sign}_{ssk_i}(sid, rsid, epk_U, \widetilde{C}_{p'}, note)$ and send $(note, \sigma_{4,i})$ back to the user.

Step R3. The user $\mathcal{U}'$ verifies & distributes signatures

(a) Upon receiving the first message $(note_i, \sigma_{4,i})$ from a server $\mathcal{S}_i \in \mathbf{S}'$, verify the validity of $\sigma_{4,i}$ w.r.t. the previously sent values, and parse $note_i$ as $(ssid, tpk, \mathbf{tpk}, C_p, C_K, \widetilde{C}_p, \widetilde{C}_K, \pi_0)$. Check that all servers in $\mathbf{S}'$ occur in $\mathbf{S}$, that the lists $\mathbf{tpk}$ and $\mathbf{S}$ are of equal length, and that the proof π_0 is valid w.r.t. sid. If all checks succeed, set $note \leftarrow note_i$.

(b) Upon receiving any subsequent message $(note_j, \sigma_{4,j})$ from $\mathcal{S}_j$ in $\mathbf{S}'$, check that $\sigma_{4,j}$ is valid for the same note the first server had sent, i.e., verify $note_j = note$. Proceed only after $(note_j, \sigma_{4,j})$ messages from all servers $\mathcal{S}_j \in \mathbf{S}'$ have been received and processed.

(c) Send $(\sigma_{4,j})_{\mathcal{S}_j \in \mathbf{S}'}$ to all servers in $\mathbf{S}'$.

Step R4. Each server $\mathcal{S}_i$ proceeds or halt

(a) Upon receiving a message $(\sigma_{4,j})_{\mathcal{S}_j \in \mathbf{S}'}$ from the user, verify the validity of every signature $\sigma_{4,j}$ w.r.t. to the locally stored $note$. Output (RETRIEVE, $sid, rsid$) to the environment.

(b) Upon input (PROCEED, $sid, rsid, a$) from the environment, check that $a =$ allow, otherwise abort the protocol.

(c) Compute a signature $\sigma_{5,i} \leftarrow \mathsf{Sign}_{ssk_i}(rsid,$ allow$)$ and send $\sigma_{5,i}$ to $\mathcal{U}'$.

Step R5. The user $\mathcal{U}'$ computes the encrypted password quotient

(a) Upon receiving a message $\sigma_{5,i}$ from a server $\mathcal{S}_i$ in $\mathbf{S}'$, check that $\sigma_{5,i}$ is a valid signature on $(rsid, \mathtt{allow})$. Proceed only after a valid signature $\sigma_{5,i}$ has been received from all servers $\mathcal{S}_i \in \mathbf{S}'$.

(b) Use the homomorphic encryption scheme to encrypt p' and entangle it with the ciphertext C_p from $note$, which supposedly encrypts the password p. That is, select a random $r \leftarrow_{\mathtt{R}} \mathbb{Z}_q$ and compute $C_{\mathrm{test}} \leftarrow (C_p \odot \mathsf{TEnc}_{tpk}(1/p'))^r$.

(c) Generate a proof that C_{test} and $\widetilde{C}_{p'}$ are based on the same password attempt p'. To prevent man-in-the-middle attacks, the proof is bound to $\mathtt{ctxt} = (sid, rsid, note, epk_U, C_{\mathrm{test}}, \widetilde{C}_{p'})$ which in particular includes the values epk_U, C_{test}, and $\widetilde{C}_{p'}$ provided by the user so far:

$$\pi_1 \leftarrow \mathrm{NIZK}\{(p',r) : C_{\mathrm{test}} = (C_p \odot \mathsf{TEnc}_{tpk}(1/p'))^r \wedge \widetilde{C}_{p'} = \mathsf{Enc}_{PK}(p')\}(\mathtt{ctxt})$$

(d) Send a message $(C_{\mathrm{test}}, \pi_1, (\sigma_{5,j})_{\mathcal{S}_j \in \mathbf{S}'})$ to all servers in $\mathbf{S}'$.

Step R6. Each server $\mathcal{S}_i$ re-randomizes the quotient encryption

(a) Upon receiving a message $(C_{\mathrm{test}}, \pi_1, (\sigma_{5,j})_{\mathcal{S}_j \in \mathbf{S}'})$, verify the proof π_1 and validate all signatures $\sigma_{5,j}$.

(b) Choose $r_i \leftarrow_{\mathtt{R}} \mathbb{Z}_q$, compute the re-randomized ciphertext $C'_{\mathrm{test},i} \leftarrow (C_{\mathrm{test}})^{r_i}$ and the proof of correctness $\pi_{2,i} \leftarrow \mathrm{NIZK}\{(r_i) : C'_{\mathrm{test},i} = (C_{\mathrm{test}})^{r_i}\}$. Sign the ciphertext together with the session information as $\sigma_{6,i} \leftarrow \mathsf{Sign}_{ssk_i}(sid, rsid, C_{\mathrm{test}}, \widetilde{C}_{p'}, C'_{\mathrm{test},i})$. Send the message $(C'_{\mathrm{test},i}, \pi_{2,i}, \sigma_{6,i})$ to $\mathcal{U}'$.

Step R7. The user $\mathcal{U}'$ verifies & distributes the re-randomized quotient encryptions

(a) Upon receiving $(C'_{\mathrm{test},j}, \pi_{2,j}, \sigma_{6,j})$ from all servers $\mathcal{S}_j$ in $\mathbf{S}'$, where the proof $\pi_{2,i}$ and the signature $\sigma_{6,i}$ are valid w.r.t. the previously sent C_{test}, send $(C'_{\mathrm{test},j}, \pi_{2,j}, \sigma_{6,j})_{\mathcal{S}_j \in \mathbf{S}'}$ to all servers in $\mathbf{S}'$.

Step R8. Each server $\mathcal{S}_i$ computes combined quotient encryption & sends its decryption share

(a) Upon receiving $t+1$ tuples $(C'_{\mathrm{test},j}, \pi_{2,j}, \sigma_{6,j})_{\mathcal{S}_j \in \mathbf{S}'}$ from the user, verify all proofs $\pi_{2,j}$ and all signatures $\sigma_{6,j}$.

(b) Derive $C'_{\mathrm{test}} \leftarrow \odot_{\mathcal{S}_j \in \mathbf{S}'} C'_{\mathrm{test},j}$ and compute the verifiable decryption share of C'_{test} as $(d_i, \pi_{d_i}) \leftarrow \mathsf{PDec}_{tsk_i}(C'_{\mathrm{test}})$.

(c) Sign the share as $\sigma_{7,i} \leftarrow \mathsf{Sign}_{ssk_i}(rsid, C'_{\mathrm{test}}, d_i)$ and send $(d_i, \pi_{d_i}, \sigma_{7,i})$ to $\mathcal{U}'$.

Step R9. The user $\mathcal{U}'$ checks if $p = p'$ & distributes shares

(a) When receiving a tuple $(d_i, \pi_{d_i}, \sigma_{7,i})$ from a server $\mathcal{S}_i$ in $\mathbf{S}'$, verify that the signature $\sigma_{7,i}$ and the proof π_{d_i} for the decryption share are valid w.r.t. the locally computed $C'_{\mathrm{test}} \leftarrow \odot_{\mathcal{S}_j \in \mathbf{S}'} C'_{\mathrm{test},j}$.

(b) After having received correct decryption shares from all $t+1$ servers in $\mathbf{S}'$, check whether the passwords match by verifying that $\mathsf{TDec}(C'_{\mathrm{test}}, \{d_j\}_{\mathcal{S}_j \in \mathbf{S}'}) = 1$.

(c) Send all decryption shares, proofs, and signatures, $(d_j, \pi_{d_j}, \sigma_{7,j})_{\mathcal{S}_j \in \mathbf{S}'}$ to all servers $\mathcal{S}_i$ in $\mathbf{S}'$.

Step R10. Each servers S_i checks if $p = p'$ & sends decryption share for K

(a) Upon receiving $t + 1$ tuples $(d_j, \pi_{d_j}, \sigma_{7,j})_{S_j \in \mathbf{S}'}$, verify that all proofs π_{d_j} and signatures $\sigma_{7,j}$ are valid w.r.t. the locally computed C'_{test}.
(b) Check whether $\mathsf{TDec}(C'_{\text{test}}, \{d_j\}_{S_j \in \mathbf{S}'}) = 1$.
(c) Compute the decryption share for the key K as $(d'_i, \pi_{d'_i}) \leftarrow \mathsf{PDec}_{tsk_i}(C_K)$.
(d) Compute the ciphertext $C_{R,i} \leftarrow \mathsf{Enc2}_{epk_U}((d'_i, \pi_{d'_i}), (epk_U, spk_i))$ using the user's public key and the own signature public key as label, generate $\sigma_{8,i} \leftarrow \mathsf{Sign}_{ssk_i}(rsid, C_{R,i})$, and send $(C_{R,i}, \sigma_{8,i})$ to the user. Output $(\texttt{DELIVER2S}, sid, rsid, \texttt{success})$.

Step R11. The user $\mathcal{U}'$ reconstructs K

(a) Upon receiving a pair $(C_{R,i}, \sigma_{8,i})$ from a server S_i in $\mathbf{S}'$, check that $\sigma_{8,i}$ is valid and, if so, decrypt $C_{R,i}$ to $(d'_i, \pi_{d'_i}) \leftarrow \mathsf{Dec2}_{esk_U}(C_{R,i}, (epk_U, spk_i))$. Verify the validity of d'_i by verifying the proof $\pi_{d'_i}$ w.r.t. C_K taken from $note$.
(b) Once all $t + 1$ valid shares have been received, restore the key $K' \leftarrow \mathsf{TDec}(C_K, \{d'_j\}_{S_j \in \mathbf{S}'})$ and output $(\texttt{DELIVER2U}, sid, rsid, K')$.

4.2   Security and Efficiency

We now provide the results of our security analysis. The proof of Theorem 1 is given in the full paper [12].

Theorem 1. *If* $(\mathsf{TKGen}, \mathsf{TEnc}, \mathsf{PDec}, \mathsf{VfDec}, \mathsf{TDec})$ *is a semantically secure* (t, n)-*threshold homomorphic cryptosystem,* $(\mathsf{KGen}, \mathsf{Enc}, \mathsf{Dec})$ *is a CPA-secure encryption scheme,* $(\mathsf{KGen2}, \mathsf{Enc2}, \mathsf{Dec2})$ *is a CCA2-secure labeled encryption scheme, the signature scheme* $(\mathsf{SKGen}, \mathsf{Sign}, \mathsf{Vf})$ *is existentially unforgeable, and a simulation-sound concurrent zero-knowledge proof system is deployed, then our* Setup *and* Retrieve *protocols described in Section 4 securely realize* $\mathcal{F}$ *in the* $\mathcal{F}_{CA}$ *and* $\mathcal{F}_{CRS}$-*hybrid model.*

When instantiated with the ElGamal based encryption scheme for $(\mathsf{TKGen}, \mathsf{TEnc}, \mathsf{PDec}, \mathsf{VfDec}, \mathsf{TDec})$ and $(\mathsf{KGen}, \mathsf{Enc}, \mathsf{Dec})$ (as described in the full version [12]), with the ElGamal encryption scheme with Fujisaki-Okamoto padding [27,30] for $(\mathsf{KGen2}, \mathsf{Enc2}, \mathsf{Dec2})$, with Schnorr signatures [46,42] for $(\mathsf{SKGen}, \mathsf{Sign}, \mathsf{Vf})$, and with the Σ-protocols described in the full paper [12], then by the UC composition theorem and the security of the underlying building blocks we have the following corollary:

Corollary 1. *The* Setup *and* Retrieve *protocols described in Section 4 and instantiated as described above, securely realize* $\mathcal{F}$ *under the DDH-assumption for the group generated by* GGen *in the random-oracle and the* $\mathcal{F}_{CA}$, $\mathcal{F}_{CRS}$-*hybrid model.*

Efficiency Analysis: With the primitives instantiated as for Corollary 1, the user has to do $5n + 15$ exponentiations in $\mathbb{G}$ for the Setup protocol and $14t + 24$ exponentiations in the Retrieve protocol. The respective figures for each server are $n + 18$ and $7t + 28$. Counting hash values as half a group element, setup requires four rounds of communication with $n(2.5n + 18.5)$ total transmitted group elements, while retrieval takes ten rounds with $(t + 1)(36.5 + 2.5n + 10.5(t + 1))$ elements.

Acknowledgements. This research was supported by the European Community's Seventh Framework Programme through grant PERCY (agreement no. 321310). Anna Lysyanskaya is supported by NSF awards 0964379 and 1012060 and by IBM and Google faculty awards.

References

1. Abe, M., Fehr, S.: Adaptively Secure Feldman VSS and Applications to Universally-Composable Threshold Cryptography. In: Franklin, M. (ed.) CRYPTO 2004. LNCS, vol. 3152, pp. 317–334. Springer, Heidelberg (2004)
2. Almansa, J.F., Damgård, I.B., Nielsen, J.B.: Simplified Threshold RSA with Adaptive and Proactive Security. In: Vaudenay, S. (ed.) EUROCRYPT 2006. LNCS, vol. 4004, pp. 593–611. Springer, Heidelberg (2006)
3. Bagherzandi, A., Jarecki, S., Saxena, N., Lu, Y.: Password-protected secret sharing. In: ACM CCS 2011 (2011)
4. Barak, B., Lindell, Y., Rabin, T.: Protocol initialization for the framework of universal composability. Cryptology ePrint Archive, Report 2004/006 (2004)
5. Bellare, M., Rogaway, P.: Random oracles are practical: A paradigm for designing efficient protocols. In: ACM CCS 1993(1993)
6. Ben-Or, M., Canetti, R., Goldreich, O.: Asynchronous secure computation. In: STOC 1993 (1993)
7. Boyen, X.: Hidden credential retrieval from a reusable password. In: ASIACCS 2009 (2009)
8. Burr, W.E., Dodson, D.F., Newton, E.M., Perlner, R.A., Polk, W.T., Gupta, S., Nabbus, E.A.: Electronic authentication guideline. NIST Special Publication 800-63-1 (2011)
9. Brainard, J., Juels, A., Kaliski Jr., B.S., Szydlo, M.: A new two-server approach for authentication with short secrets. In: USENIX 2003 (2003)
10. Camenisch, J., Kiayias, A., Yung, M.: On the portability of generalized Schnorr proofs. In: Joux, A. (ed.) EUROCRYPT 2009. LNCS, vol. 5479, pp. 425–442. Springer, Heidelberg (2009)
11. Camenisch, J., Krenn, S., Shoup, V.: A framework for practical universally composable zero-knowledge protocols. In: Lee, D.H., Wang, X. (eds.) ASIACRYPT 2011. LNCS, vol. 7073, pp. 449–467. Springer, Heidelberg (2011)
12. Camenisch, J., Lehmann, A., Lysyanskaya, A., Neven, G.: Memento: How to reconstruct your secrets from a single password in a hostile environment. Cryptology ePrint Archive, Report 2014/429 (2014)
13. Camenisch, J., Lysyanskaya, A., Neven, G.: Practical yet universally composable two-server password-authenticated secret sharing. In: ACM CCS 2012 (2012)
14. Camenisch, J., Shoup, V.: Practical verifiable encryption and decryption of discrete logarithms. In: Boneh, D. (ed.) CRYPTO 2003. LNCS, vol. 2729, pp. 126–144. Springer, Heidelberg (2003)
15. Camenisch, J., Stadler, M.A.: Efficient group signature schemes for large groups (extended abstract). In: Kaliski Jr., B.S. (ed.) CRYPTO 1997. LNCS, vol. 1294, pp. 410–424. Springer, Heidelberg (1997)
16. Canetti, R.: Universally composable security: A new paradigm for cryptographic protocols. In: FOCS 2001 (2001)
17. Canetti, R.: Universally composable signature, certification, and authentication. In: 17th Computer Security Foundations Workshop. IEEE Computer Society (2004)
18. Canetti, R., Halevi, S., Katz, J., Lindell, Y., MacKenzie, P.: Universally composable password-based key exchange. In: Cramer, R. (ed.) EUROCRYPT 2005. LNCS, vol. 3494, pp. 404–421. Springer, Heidelberg (2005)

19. Canetti, R., Rabin, T.: Universal composition with joint state. In: Boneh, D. (ed.) CRYPTO 2003. LNCS, vol. 2729, pp. 265–281. Springer, Heidelberg (2003)

20. Cramer, R., Damgård, I.B., Nielsen, J.B.: Multiparty computation from threshold homomorphic encryption. In: Pfitzmann, B. (ed.) EUROCRYPT 2001. LNCS, vol. 2045, pp. 280–300. Springer, Heidelberg (2001)

21. Chor, B., Moscovici, L.: Solvability in asynchronous environments. In: FOCS 1989 (1989)

22. Damgård, I., Keller, M., Larraia, E., Pastro, V., Scholl, P., Smart, N.P.: Practical covertly secure MPC for dishonest majority—or: Breaking the SPDZ limits. In: Crampton, J., Jajodia, S., Mayes, K. (eds.) ESORICS 2013. LNCS, vol. 8134, pp. 1–18. Springer, Heidelberg (2013)

23. Damgård, I.B., Nielsen, J.B.: Universally composable efficient multiparty computation from threshold homomorphic encryption. In: Boneh, D. (ed.) CRYPTO 2003. LNCS, vol. 2729, pp. 247–264. Springer, Heidelberg (2003)

24. Damgård, I., Pastro, V., Smart, N., Zakarias, S.: Multiparty computation from somewhat homomorphic encryption. In: Safavi-Naini, R., Canetti, R. (eds.) CRYPTO 2012. LNCS, vol. 7417, pp. 643–662. Springer, Heidelberg (2012)

25. Desmedt, Y., Frankel, Y.: Threshold cryptosystems. In: Brassard, G. (ed.) Advances in Cryptology - CRYPTO 1989. LNCS, vol. 435, pp. 307–315. Springer, Heidelberg (1990)

26. Di Raimondo, M., Gennaro, R.: Provably secure threshold password-authenticated key exchange. In: Biham, E. (ed.) EUROCRYPT 2003. LNCS, vol. 2656, pp. 507–523. Springer, Heidelberg (2003)

27. El Gamal, T.: A public key cryptosystem and a signature scheme based on discrete logarithms. In: Blakely, G.R., Chaum, D. (eds.) Advances in Cryptology - CRYPTO 1984. LNCS, vol. 196, pp. 10–18. Springer, Heidelberg (1985)

28. Fiat, A., Shamir, A.: How to prove yourself: Practical solutions to identification and signature problems. In: Odlyzko, A.M. (ed.) Advances in Cryptology - CRYPTO 1986. LNCS, vol. 263, pp. 186–194. Springer, Heidelberg (1987)

29. Ford, W., Kaliski Jr., B.S.: Server-assisted generation of a strong secret from a password. WETICE 2000. In: IEEE Computer Society 2000 (2000)

30. Fujisaki, E., Okamoto, T.: Secure integration of asymmetric and symmetric encryption schemes. In: Wiener, M. (ed.) CRYPTO 1999. LNCS, vol. 1666, pp. 537–554. Springer, Heidelberg (1999)

31. Garay, J.A., MacKenzie, P.D., Yang, K.: Strengthening zero-knowledge protocols using signatures. In: Biham, E. (ed.) EUROCRYPT 2003. LNCS, vol. 2656, pp. 177–194. Springer, Heidelberg (2003)

32. Goodin, D.: Why passwords have never been weaker—and crackers have never been stronger. Ars Technica (2012)

33. Gosney, J.: Password Cracking HPC. In: Passwords 2012 Conference (2012)

34. Herley, C., van Oorschot, P.C.: A research agenda acknowledging the persistence of passwords. IEEE Security & Privacy (2012)

35. Herley, C., van Oorschot, P.C., Patrick, A.S.: Passwords: If we're so smart, why are we still using them? In: Dingledine, R., Golle, P. (eds.) FC 2009. LNCS, vol. 5628, pp. 230–237. Springer, Heidelberg (2009)

36. Jablon, D.P.: Password authentication using multiple servers. In: Naccache, D. (ed.) CT-RSA 2001. LNCS, vol. 2020, pp. 344–360. Springer, Heidelberg (2001)

37. Kaliski, B.: PKCS #5: Password-Based Cryptography Specification. IETF RFC 2898 (2000)

38. Katz, J., MacKenzie, P., Taban, G., Gligor, V.: Two-server password-only authenticated key exchange. In: Ioannidis, J., Keromytis, A., Yung, M. (eds.) ACNS 2005. LNCS, vol. 3531, pp. 1–16. Springer, Heidelberg (2005)

39. Katz, J., Maurer, U., Tackmann, B., Zikas, V.: Universally Composable Synchronous Computation. In: Sahai, A. (ed.) TCC 2013. LNCS, vol. 7785, pp. 477–498. Springer, Heidelberg (2013)
40. MacKenzie, P., Shrimpton, T., Jakobsson, M.: Threshold password-authenticated key exchange. In: Yung, M. (ed.) CRYPTO 2002. LNCS, vol. 2442, pp. 385–400. Springer, Heidelberg (2002)
41. MacKenzie, P., Yang, K.: On simulation-sound trapdoor commitments. In: Cachin, C., Camenisch, J.L. (eds.) EUROCRYPT 2004. LNCS, vol. 3027, pp. 382–400. Springer, Heidelberg (2004)
42. Pointcheval, D., Stern, J.: Security proofs for signature schemes. In: Maurer, U.M. (ed.) Advances in Cryptology - EUROCRYPT 1996. LNCS, vol. 1070, pp. 387–398. Springer, Heidelberg (1996)
43. Provos, N., Mazières, D.: A Future-Adaptable Password Scheme. In: USENIX 1999 (1999)
44. RSA, The Security Division of EMC. New RSA innovation helps thwart "smash-and-grab" credential theft. Press release (2012)
45. Sahai, A.: Non-malleable non-interactive zero knowledge and adaptive chosen-ciphertext security. In: FOCS 1999 (1999)
46. Schnorr, C.-P.: Efficient signature generation by smart cards. J. Cryptol. 4(3), 161–174 (1991)
47. Szydlo, M., Kaliski, B.: Proofs for two-server password authentication. In: Menezes, A. (ed.) CT-RSA 2005. LNCS, vol. 3376, pp. 227–244. Springer, Heidelberg (2005)

Scalable Zero Knowledge via Cycles of Elliptic Curves

Eli Ben-Sasson[1], Alessandro Chiesa[2], Eran Tromer[3], and Madars Virza[2]

[1] Technion, Haifa, Israel
eli@cs.technion.ac.il
[2] MIT, Cambridge, MA, USA
{alexch,madars}@csail.mit.edu
[3] Tel Aviv University, Tel Aviv, Israel
tromer@cs.tau.ac.il

Abstract. Non-interactive zero-knowledge proofs of knowledge for general NP statements are a powerful cryptographic primitive, both in theory and in practical applications. Recently, much research has focused on achieving an additional property, *succinctness*, requiring the proof to be very short and easy to verify. Such proof systems are known as *zero-knowledge succinct non-interactive arguments of knowledge* (zk-SNARKs), and are desired when communication is expensive, or the verifier is computationally weak.

Existing zk-SNARK implementations have severe scalability limitations, in terms of space complexity as a function of the size of the computation being proved (e.g., running time of the NP statement's decision program). First, the size of the proving key is quasilinear in the upper bound on the computation size. Second, producing a proof requires "writing down" all intermediate values of the entire computation, and then conducting global operations such as FFTs.

The bootstrapping technique of Bitansky et al. (STOC '13), following Valiant (TCC '08), offers an approach to scalability, by recursively composing proofs: proving statements about acceptance of the proof system's own verifier (and correctness of the program's latest step). Alas, recursive composition of known zk-SNARKs has never been realized in practice, due to enormous computational cost.

Using new elliptic-curve cryptographic techniques, and methods for exploiting the proof systems' field structure and nondeterminism, we achieve the first zk-SNARK implementation that practically achieves recursive proof composition. Our zk-SNARK implementation runs random-access machine programs and produces proofs of their correct execution, on today's hardware, for any program running time. It takes constant time to generate the keys that support *all* computation sizes. Subsequently, the proving process only incurs a constant multiplicative overhead compared to the original computation's time, and an essentially-constant additive overhead in memory. Thus, our zk-SNARK implementation is the first to have a well-defined, albeit low, clock rate of "verified instructions per second".

Keywords: computationally-sound proofs, proof-carrying data, zero-knowledge, elliptic curves.

J.A. Garay and R. Gennaro (Eds.): CRYPTO 2014, Part II, LNCS 8617, pp. 276–294, 2014.
© International Association for Cryptologic Research 2014

1   Introduction

Non-interactive zero-knowledge proofs of knowledge [BFM88, NY90, BDSMP91] are a powerful tool, studied extensively both in theoretical and applied cryptography. Recently, much research has focused on achieving an additional property, *succinctness*, that requires the proof to be very short and easy to verify. A proof system with this additional property is called a *zero-knowledge Succinct Non-interactive ARgument of Knowledge* (zk-SNARK). Because succinctness is a desirable, sometimes critical, property in numerous security applications, prior work has investigated zk-SNARK implementations. Unfortunately, all implementations to date suffer from severe scalability limitations, due to high space complexity, as we now explain.

1.1   Scalability Limitations of Prior zk-SNARK Implementations

Expensive Preprocessing.   As in any non-interactive zero-knowledge proof, a zk-SNARK requires a one-time trusted setup of public parameters: a *key generator* samples a proving key (used to generate proofs) and a verification key (used to check proofs); the key pair is then published as the proof system's parameters.

Most zk-SNARK constructions [Gro10, Lip12, BCI+13, GGPR13, PGHR13, BCG+13a, Lip13, BCTV14b], including all published implementations [PGHR13, BCG+13a, BCTV14b], require *expensive preprocessing* during key generation. Namely, the key generator takes as input an upper bound on the computation size, e.g., in the form of an explicit NP decision circuit C output by a *circuit generator*; then, the key generator's space complexity, as well as the size of the output proving key, depends at least linearly on this upper bound. Essentially, the circuit C is explicitly laid out and encoded so as to produce the proof system's parameters.

One way to mitigate the costs of expensive preprocessing is to make C universal, i.e., design C so that it can handle more than one choice of program [BCTV14b]. Yet, C *still* depends on upper bounds on the program size and number of execution steps. Moreover, even if key generation is carried out only once per circuit C, the resulting large proving key must be stored, and accessed, *each time a proof is generated*. Prior implementations of zk-SNARKs quickly become space-bound already for modest computation sizes, e.g., with proving keys of over 4 GB for circuits of only 16 million gates [BCTV14b].[1]

Thus, expensive preprocessing severely limits scalability of a zk-SNARK.

Space-Intensive Proof Generation.   Related in part to the aforementioned expensive preprocessing, the prover in all published zk-SNARK implementations has large space complexity. Essentially, the proving process requires writing down the *entire* computation (e.g., the evaluation of the circuit C) all at once, and then conduct a global computation (such as Fast Fourier transforms, or multi-exponentiations) based on it. In particular, if C expresses the execution of a program, then proving requires writing down the full trace of intermediate states throughout the program execution.

---

[1] Even worse, the reported numbers are for "data at rest": the proving key consists of a list of elliptic-curve points, which are *compressed* when not in use. However, when the prover uses the proving key to produce a proof, the points are uncompressed (and represented via projective or Jacobian coordinates), and take about three times as much space in memory.

Tradeoffs are possible, using block-wise versions of the global algorithms, and repeating the computation to reproduce segments of the trace. These decrease the prover's space complexity but significantly increase its time complexity, and thus do not adequately address scalability.

Remark 1. Even when relaxing the goal (by allowing interaction, "theorem batching", or non-zero-knowledge proofs), all published implementations of proof systems for outsourcing NP computations [SBW11, SMBW12, SVP+12, SBV+13, BFR+13] also suffer from both of the above scalability limitations.[2]

1.2   What We Know from Theory

Ideally, we would like to implement a zk-SNARK that does not suffer from either of the scalability limitations mentioned in the previous section, i.e., a zk-SNARK where:

- Key generation is *cheap* (i.e., its running time only depends on the security parameter) and *suffices for all computations* (of polynomial size). Such a zk-SNARK is called **fully succinct**.
- Proof generation is carried out *incrementally*, alongside the original computation, by updating, at each step, a proof of correctness of the computation so far. Such a zk-SNARK is called **incrementally computable**.

Work in cryptography tells us that the above properties can be achieved in theoretical zk-SNARK constructions. Namely, building on the work of Valiant on incrementally-verifiable computation [Val08] and the work of Chiesa and Tromer on proof-carrying data [CT10, CT12], Bitansky et al. [BCCT13] showed how to construct zk-SNARKs that are fully-succinct and incrementally-computable.

Concretely, the approach of [BCCT13] consists of a transformation that takes as input a *preprocessing* zk-SNARK (such as one from existing implementations), and *bootstraps* it, via recursive proof composition, into a new zk-SNARK that is fully-succinct and incrementally-computable. In recursive proof composition, a prover produces a proof about an NP statement that, among other checks, also ensures the accepting computation of the proof system's own verifier. In a zk-SNARK, proof verification is asymptotically cheaper than merely verifying the corresponding NP statement; so recursive proof composition is viable, in theory. In practice, however, this step introduces concretely enormous costs: even if zk-SNARK verifiers can be executed in just a few milliseconds on a modern desktop [PGHR13, BCTV14b], zk-SNARK verifiers still take millions of machine cycles to execute. Hence, known zk-SNARK implementations cannot achieve *even one step* of recursive proof composition in practical time. Thus, whether recursive proof composition can be realized in practice, with any reasonable efficiency, has so far remained an intriguing open question.

Remark 2 (PCPs). Suitably instantiating Micali's "computationally-sound proofs" [Mic00] yields fully-succinct zk-SNARKs. However, it is not known how to also

---

[2] In contrast, when outsourcing P computations, there are implementations without expensive preprocessing: [CMT12, TRMP12, Tha13] consider low-depth circuits, and [CRR11] consider outsourcing to multiple provers at least one of which is honest.

achieve incremental computation with this approach (without also invoking the aforementioned approach of Bitansky et al. [BCCT13]). Indeed, [Mic00] requires probabilistically-checkable proofs (PCPs) [BFLS91], where one can achieve a prover that runs in quasilinear-time [BCGT13b], but only by requiring space-intensive computations — again due to the need to write down the entire computation and conducting global operations on it.

1.3 Contributions

We present the first prototype implementation that practically achieves recursive composition of zk-SNARKs. This enables us to achieve the following results:

(i) Scalable zk-SNARKs. We present the first implementation of a zk-SNARK that is fully succinct and incrementally computable. Our implementation follows the approach of Bitansky et al. [BCCT13].

Our zk-SNARK works for proving/verifying computations on a general notion of random-access machine. The key generator takes as input a *machine specification*, consisting of settings for random-access memory (number of addresses and number of bits at each address) and a CPU circuit, defining the machine's behavior. The keys sampled by the key generator support proving/verifying computations, of any polynomial length, on this machine. Thus, our zk-SNARK implementation directly supports many architectures (e.g., floating-point processors, SIMD-based processors, etc.) — one only needs to specify memory settings and a CPU circuit.

Compared to the original machine computation, our zk-SNARK only imposes a constant multiplicative overhead in time and an essentially-constant additive overhead in space. Indeed, the proving process steps through the machine's computation, each time producing a new proof that the computation is correct so far, by relying on the prior proof; each proof asserts the satisfiability of a constant-size circuit, and requires few resources in time and space to produce. Our zk-SNARK scales, on today's hardware, to any computation size.

(ii) Proof-Carrying Data. The main tool in [BCCT13]'s approach is *proof-carrying data* (PCD) [CT10, CT12], a cryptographic primitive that encapsulates the security guarantees provided by recursive proof composition. Thus, as a stepping stone towards the aforementioned zk-SNARK implementation, we also achieve the first implementation of PCD, for arithmetic circuits.

(iii) Evaluation on vnTinyRAM. We evaluate our zk-SNARK on a specific choice of random-access machine: vnTinyRAM, a simple RISC von Neumann architecture that is supported by the most recent preprocessing zk-SNARK implementation [BCTV14b]. The evaluation confirms our expectations that our approach is slower for small computations but achieves scalability to large computations.

We evaluated our prototype on 16-bit and 32-bit vnTinyRAM with 16 registers (as in [BCTV14b]). For instance, for 32-bit vnTinyRAM, our prototype incrementally proves correct program execution at the cost of 35.5 seconds per program step, using a 64.4 MB proving key and 1,008 MB of additional memory. In contrast, for a T-step program, the system of [BCTV14b] requires roughly $0.05 \cdot T$ seconds, *provided* that roughly $3.1 \cdot T$ MB of main memory is available. Thus for $T > 326$ our system is more space-efficient, and the savings in space continue to grow as T increases. (These numbers are for an 80-bit security level.)

The Road Ahead. Obtaining scalable zk-SNARKs is but one application of PCD. More generally, PCD enables efficient "distributed theorem proving", which has applications ranging from securing the IT supply chain, to information flow control, and to distributed programming-language semantics [CT10, CT12, CTV13]. Now that a first prototype of PCD has been achieved, these applications are waiting to be explored in practice.

1.4 Summary of Challenges and Techniques

As we recall in Section 2, bootstrapping zk-SNARKs involves two main ingredients: a collision-resistant hash function and a preprocessing zk-SNARK. Practical implementations of both ingredients exist. So one may conclude that "practical bootstrapping" is merely a matter of stitching together implementations of these two ingredients. As we now explain, this conclusion is mistaken, because bootstrapping a zk-SNARK in practice poses several challenges that must be tackled in order to obtain any reasonable efficiency.

Common Theme: Leverage Field Structure. The techniques that we employ to overcome efficiency barriers leverage the fact that the "native" NP language whose membership is proved/verified by the zk-SNARK is the satisfiability of $\mathbb{F}$-arithmetic circuits, for a certain finite field $\mathbb{F}$. While any NP statement can be reduced to $\mathbb{F}$-arithmetic circuits, the proof system is most efficient for statements expressible as $\mathbb{F}$-arithmetic circuits of small size. Prior work only partially leveraged this fact, by using circuits that conduct large-integer arithmetic or "pack" bits into field elements for non-bitwise checks (e.g., equality) [PGHR13, BCG+13a, BFR+13, BCTV14b]. In this paper, we go further and, for improved efficiency, use circuits that conduct *field operations*.

Challenge: How to Efficiently "Close the Loop"? By far the most prominent challenge is efficiently "closing the loop". In the bootstrapping approach, each step requires proving a statement that (i) verifies the validity of previous zk-SNARK proofs; and (ii) checks another execution step. For recursive composition, this statement needs to be expressed as an $\mathbb{F}$-arithmetic circuit C_{pcd}, so that it can be proved using the very same zk-SNARK. In particular, we need to *implement the verifier V as an $\mathbb{F}$-arithmetic circuit C_V* (a subcircuit of C_{pcd}).

In principle, constructing C_V is possible, because circuits are a universal model of computation. In fact, not just in principle: much research has been devoted to improve the efficiency and functionality of circuit generators in practice [SVP+12, BCGT13a, SBV+13, PGHR13, BCG+13a, BCTV14b]. Hence, a reasonable approach to construct C_V is to apply a suitable circuit generator to a suitable software implementation of V.

However, such an approach is likely to be inefficient. Circuit generators strive to support complex program computations, by providing ways to efficiently handle data-dependent control flow, memory accesses, and so on. Instead, verifiers in preprocessing zk-SNARK constructions are "circuit-like" programs, consisting of few pairing-based arithmetic checks that do not use complex data-dependent control flow or memory accesses.

Thus, we want to avoid circuit generators, and somehow directly construct C_V so that its size is not huge. As we shall explain (see Section 3), this is not merely a programmatic difficulty, but there are *mathematical obstructions* to constructing C_V efficiently.

Main Technique: PCD-Friendly Cycles of Elliptic Curves.   In our underlying preprocessing zk-SNARK, the verifier V consists mainly of operations in an elliptic curve over a field $\mathbb{F}'$, and is thus expressed, most efficiently, as a $\mathbb{F}'$-arithmetic circuit. We observe that if this field $\mathbb{F}'$ is the same as the aforementioned native field $\mathbb{F}$ of the zk-SNARK's statement, then recursive composition can be orders of magnitude more efficient than otherwise. Unfortunately, as we shall explain, the "field matching" $\mathbb{F} = \mathbb{F}'$ is mathematically impossible.

In contrast, we show how to circumvent this obstruction by using multiple, suitably-chosen elliptic curves, that lie on a *PCD-friendly cycle*. For example, a PCD-friendly 2-cycle consists of two curves such that the (prime) size of the base field of one curve equals the group order of the other curve, and vice versa. Our implementation uses a PCD-friendly cycle of elliptic curves (found at a great computational expense) to attain zk-SNARKs that are *tailored* for recursive proof composition.

Additional Technique: Nondeterministic Verification of Pairings.   The zk-SNARK verifier involves, more specifically, several pairing-based checks over its elliptic curve. Yet, each pairing evaluation is very expensive, if not carefully performed. To further improve efficiency, we exploit the fact that the zk-SNARK supports NP statements, and provide a hand-optimized circuit implementation of the zk-SNARK verifier that leverages nondeterminism for improved efficiency. For instance, in our construction, we make heavy use of *affine* coordinates for both curve arithmetic and divisor evaluations [LMN10], because these are particularly efficient to *verify* (as opposed to *computing*, for which projective or Jacobian coordinates are known to be faster).

Challenge: How to Efficiently Verify Collision-Resistant Hashing?   Bootstrapping zk-SNARKs uses, at multiple places, a collision-resistant hash function H and an arithmetic circuit C_H for verifying computations of H. If not performed efficiently, this would be another bottleneck.

For instance, the aforementioned circuit C_{pcd}, besides verifying prior zk-SNARK proofs, is also tasked with verifying one step of machine execution. This involves not only checking the CPU execution but also the validity of loads and stores to random-access memory, done via memory-checking techniques based on Merkle trees [BEG+91, BCGT13a]. Thus C_{pcd} also needs to have a subcircuit to check Merkle-tree authentication paths. Constructing such circuits is straightforward, given a circuit C_H for verifying computations of H. But the main question here is how to pick H so that C_H can be small. Indeed, if random-access memory consists of A addresses, then checking an authentication path requires at least $\lceil \log A \rceil \cdot |C_H|$ gates. If C_H is large, this subcircuit *dwarfs* the CPU, and "wastes" most of the size of C_{pcd} for a single load/store.

Merely picking some standard choice of hash function H (e.g., SHA-256 or Keccak) yields C_H with tens of thousands of gates [PGHR13, BCG+14], making hash verifications very expensive. Is this inherent?

Additional Technique: Field-Specific Hashes.   We select a hash H that is tailored to efficient verification in the field $\mathbb{F}$. In our setting, $\mathbb{F}$ has prime order p, so its additive group is isomorphic to $\mathbb{Z}_p$. Thus, a natural approach is to let H be a *modular subset-sum* function over $\mathbb{Z}_p$. For suitable parameter choices and for random coefficients, subset-sum functions are collision-resistant [Ajt96, GGH96]. In this paper we base all of our collision-resistant hashing on suitable subset sums, and thereby greatly reduce the burden of hashing.[3]

1.5   Roadmap

The rest of this paper is organized as follows. In Section 2 we recall the main ideas of [BCCT13]'s approach. In Section 3, we discuss our construction of preprocessing zk-SNARKs that are tailored for efficient recursive composition of proofs; due to space constraints, we leave the other discussions (construction of proof-carrying data and scalable zk-SNARK) to the full version of this paper [BCTV14a]. In Section 4, we evaluate our system on the random-access machine vnTinyRAM.

2   Preliminaries

2.1   Preprocessing zk-SNARKs for Arithmetic Circuits

Given a field $\mathbb{F}$, the *circuit satisfaction problem* of an $\mathbb{F}$-arithmetic circuit $C : \mathbb{F}^n \times \mathbb{F}^h \to \mathbb{F}^l$ is defined by the relation $\mathcal{R}_C = \{(x, a) \in \mathbb{F}^n \times \mathbb{F}^h : C(x, a) = 0^l\}$; its language is $\mathcal{L}_C = \{x \in \mathbb{F}^n : \exists\, a \in \mathbb{F}^h, C(x, a) = 0^l\}$.

A **preprocessing zk-SNARK** for $\mathbb{F}$-arithmetic circuit satisfiability (see, e.g., [BCI$^+$13]) is a triple of polynomial-time algorithms (G, P, V), called *key generator*, *prover*, and *verifier*. The key generator G, given a security parameter λ and an $\mathbb{F}$-arithmetic circuit $C : \mathbb{F}^n \times \mathbb{F}^h \to \mathbb{F}^l$, samples a *proving key* pk and a *verification key* vk; these are the proof system's public parameters, which need to be generated only once per circuit. After that, anyone can use pk to generate non-interactive proofs for the language $\mathcal{L}_C$, and anyone can use the vk to check these proofs. Namely, given pk and any $(x, a) \in \mathcal{R}_C$, the honest prover $P(\text{pk}, x, a)$ produces a proof π attesting that $x \in \mathcal{L}_C$; the verifier $V(\text{vk}, x, \pi)$ checks that π is a valid proof for $x \in \mathcal{L}_C$. A proof π is a proof of knowledge, as well as a (statistical) zero-knowledge proof. The succinctness property requires that π has length $O_\lambda(1)$ and V runs in time $O_\lambda(|x|)$, where O_λ hides a (fixed) polynomial in λ.

See the full version of this paper for details [BCTV14a].

2.2   Proof-Carrying Data

Proof-carrying data (PCD) [CT10, CT12] is a cryptographic primitive that encapsulates the security guarantees obtainable via recursive composition of proofs. Since recursive proof composition naturally involves multiple (physical or virtual) parties, PCD is phrased in the language of a dynamically-evolving *distributed computation* among mutually-untrusting computing nodes, who perform local

---

[3] We note that subset-sum functions were also used in [BFR$^+$13], but, crucially, they were *not* tailored to the field. This is a key difference in usage and efficiency. (E.g., our hash function can be verified in ≤ 300 gates, while [BFR$^+$13] report 13,000.)

computations, based on local data and previous messages, and then produce output messages. Given a *compliance predicate* Π to express local checks, the goal of PCD is to ensure that any given message z in the distributed computation is Π-*compliant*, i.e., is consistent with a history in which each node's local computation satisfies Π. This formulation includes as special cases incrementally-verifiable computation [Val08] and targeted malleability [BSW12].

Concretely, a proof-carrying data (PCD) system is a triple of polynomial-time algorithms $(\mathbb{G}, \mathbb{P}, \mathbb{V})$, called *key generator*, *prover*, and *verifier*. The key generator $\mathbb{G}$ is given as input a predicate Π (specified as an arithmetic circuit), and outputs a proving key pk and a verification key vk; these keys allow anyone to prove/verify that a piece of data z is Π-compliant. This is achieved by attaching a short and easy-to-verify proof to each piece of data. Namely, given pk, received messages z_{in} with proofs π_{in}, local data z_{loc}, and a claimed outgoing message z, $\mathbb{P}$ computes a new proof π to attach to z, which attests that z is Π-compliant; the verifier $\mathbb{V}(\mathsf{vk}, z, \pi)$ verifies that z is Π-compliant. A proof π is a proof of knowledge, as well as a (statistical) zero-knowledge proof; succinctness requires that π has length $O_\lambda(1)$ and $\mathbb{V}$ runs in time $O_\lambda(|z|)$.

Finally, note that since Π is expressed as an $\mathbb{F}$-arithmetic circuit for a given field $\mathbb{F}$, the size of messages and local data are fixed; we denote these sizes by $n_{\mathsf{msg}}, n_{\mathsf{loc}} \in \mathbb{N}$. Similarly, the number of input messages is also fixed; we call this the *arity*, and denote it by $s \in \mathbb{N}$. Moreover, for convenience, Π also takes as input a flag $b_{\mathsf{base}} \in \{0,1\}$ denoting whether the node has no predecessors (i.e., b_{base} is a "base-case" flag). Overall, Π takes an input $(z, z_{\mathsf{loc}}, z_{\mathsf{in}}, b_{\mathsf{base}}) \in \mathbb{F}^{n_{\mathsf{msg}}} \times \mathbb{F}^{n_{\mathsf{loc}}} \times \mathbb{F}^{s \cdot n_{\mathsf{msg}}} \times \mathbb{F}$.

See the full version of this paper for details [BCTV14a].

2.3   The Bootstrapping Approach

Our implementation follows [BCCT13], which we now review. The approach consists of a transformation that, on input a preprocessing zk-SNARK and a collision-resistant hash function, outputs a scalable zk-SNARK. Thus, the input zk-SNARK is *bootstrapped* into one with improved scalability properties.

So fix a preprocessing zk-SNARK (G, P, V) and collision-resistant function H. The goal is to construct a fully-succinct incrementally-computable zk-SNARK (G^*, P^*, V^*) for proving/verifying the correct execution on a given random-access machine $\mathbf{M}$. Informally, we describe the transformation in four steps.

Step 1: from zk-SNARKs to PCD. The first step, independent of $\mathbf{M}$, is to construct a PCD system $(\mathbb{G}, \mathbb{P}, \mathbb{V})$, by using the zk-SNARK (G, P, V). This step involves recursive composition of zk-SNARK proofs.

Step 2: Delegate the Machine's Memory. The second step is to reduce the footprint of the machine $\mathbf{M}$, by delegating its random-access memory to an untrusted storage, via standard memory-checking techniques based on Merkle trees [BEG+91, BCGT13a]. We thus modify $\mathbf{M}$ so that its "CPU" receives values loaded from memory as nondeterministic guesses, along with corresponding authentication paths that are checked against the root of a Merkle tree based on the hash function H. Thus, $\mathbf{M}$'s state only consists of a (short) CPU state, and a (short) root of the Merkle tree that "summarizes" memory.[4]

Step 3: Design a Predicate $\Pi_{M,H}$ for Step-Wise Verification. The third step is to design a compliance predicate $\Pi_{M,H}$ that ensures that the only $\Pi_{M,H}$-compliant messages z are the ones that result from the correct execution of the (modified) machine M, one step at a time; this is analogous to the notion of incremental computation [Val08]. Crucially, because $\Pi_{M,H}$ is only asked to verify one step of execution at a time, we can implement $\Pi_{M,H}$'s requisite checks with a circuit of merely constant size.

Step 4: Construct New Proof System. The new zk-SNARK $(G^\star, P^\star, V^\star)$ is constructed as follows. The new key generator $G^\star$ is set to the PCD generator $\mathbb{G}$ invoked on $\Pi_{M,H}$. The new prover $P^\star$ uses the PCD prover $\mathbb{P}$ to prove correct execution of M, one step at a time and conducting the incremental distributed computation "in his head". The new verifier $V^\star$ simply uses the PCD verifier $\mathbb{V}$ to verify $\Pi_{M,H}$-compliance. In sum, since $\Pi_{M,H}$ is small and suffices for all computations, the new zk-SNARK is scalable: it is fully succinct; moreover, because the new prover computes a proof for each new step based on the previous one, it is also incrementally computable. (See the full version of this paper for definitions of these properties [BCTV14a].)

Our goal is to realize the above approach in a practical implementation.

Security of Recursive Proof Composition. Security in [BCCT13] is proved by using the *proof-of-knowledge property* of zk-SNARKs; we refer the interested reader to [BCCT13] for details. One aspect that must be addressed from a theoretical standpoint is the *depth* of composition. Depending on assumption strength, one may have to recursively compose proofs in "proof trees above the message chain", rather than along the chain. From a practical perspective we make the heuristic assumption that depth of composition does not affect security of the zk-SNARK, because no evidence suggests otherwise for the constructions that we use.

3 PCD-Friendly Preprocessing zk-SNARKs

We first construct preprocessing zk-SNARKs that are tailored for efficient recursive composition of proofs.

3.1 PCD-Friendly Cycles of Elliptic Curves

Let $\mathbb{F}$ be a finite field, and (G, P, V) a preprocessing zk-SNARK for $\mathbb{F}$-arithmetic satisfiability. The idea of recursive proof composition is to prove/verify satisfiability of an $\mathbb{F}$-arithmetic circuit C_{pcd} that checks the validity of previous proofs (among other things). Thus, we need to implement the verifier V as an $\mathbb{F}$-arithmetic circuit C_V, to be used as a sub-circuit of C_{pcd}.

---

[4] Similarly to [BCCT13] and our realization thereof, Braun et al. [BFR+13] leverage memory-checking techniques based on Merkle trees [BEG+91] for enabling a circuit to "securely" load from and store to an untrusted storage. However, the systems' goals (batched verification of MapReduce computations in a 2-move protocol) and techniques are different (cf. Footnote 3).

How to write C_V depends on the algorithm of V, which in turn depends on which elliptic curve is used to instantiate the pairing-based zk-SNARK. For prime r, in order to prove statements about $\mathbb{F}_r$-arithmetic circuit satisfiability, one instantiates (G, P, V) using an elliptic curve E defined over some finite field $\mathbb{F}_q$, where the group $E(\mathbb{F}_q)$ of $\mathbb{F}_q$-rational points has order $r = \#E(\mathbb{F}_q)$ (or, more generally, r divides $\#E(\mathbb{F}_q)$). Then, all of V's arithmetic computations are over $\mathbb{F}_q$, or extensions of $\mathbb{F}_q$ up to degree k, where k is the embedding degree of E with respect to r (i.e., the smallest integer k such that r divides $q^k - 1$). We motivate our approach by first describing two "failed attempts".

Attempt #1: Pick Curve with $q = r$. Ideally, we would like to select a curve E with $q = r$, so that V's arithmetic is over the *same field* for which V's native NP language is defined. Unfortunately, this cannot happen: the condition that E has embedding degree k with respect to r implies that r divides $q^k - 1$, which implies that $q \neq r$. The same implication holds even if $E(\mathbb{F}_q)$ has a non-prime order n and the prime r (with respect to which k is defined) only divides n. So, while appealing, this idea cannot even be instantiated.

Attempt #2: Long Arithmetic. Since we are stuck with $q \neq r$, we may consider doing "long arithmetic": simulating $\mathbb{F}_q$ operations via $\mathbb{F}_r$ operations, by working with bit chunks to perform integer arithmetic, and modding out by q when needed. Alas, having to work at the "bit level" implies a blowup on the order of $\log q$ compared to native arithmetic. So, while this approach can at least be instantiated, it is very expensive.

Our Approach: Cycle through Multiple Curves. We formulate, and instantiate, a new property for elliptic curves that enables us to completely circumvent long arithmetic, even with $q \neq r$. In short, our idea is to base recursive proof composition, not on a single zk-SNARK, but on *multiple* zk-SNARKs, each instantiated on a different elliptic curve, that *jointly* satisfy a special property.

For the simplest case, suppose we have two primes q_α and q_β, and elliptic curves $E_\alpha/\mathbb{F}_{q_\alpha}$ and $E_\beta/\mathbb{F}_{q_\beta}$ such that $q_\alpha = \#E_\beta(\mathbb{F}_{q_\beta})$ and $q_\beta = \#E_\alpha(\mathbb{F}_{q_\alpha})$, i.e., the size of the base field of one curve equals the group order of the other curve, and vice versa. We then construct two preprocessing zk-SNARKs $(G_\alpha, P_\alpha, V_\alpha)$ and $(G_\beta, P_\beta, V_\beta)$, respectively instantiated on the two curves $E_\alpha/\mathbb{F}_{q_\alpha}$ and $E_\beta/\mathbb{F}_{q_\beta}$.

Now note that $(G_\alpha, P_\alpha, V_\alpha)$ works for $\mathbb{F}_{q_\beta}$-arithmetic circuit satisfiability, but all of V_α's arithmetic computations are over $\mathbb{F}_{q_\alpha}$ (or extensions thereof); while $(G_\beta, P_\beta, V_\beta)$ works for $\mathbb{F}_{q_\alpha}$-arithmetic circuits, but V_β's arithmetic computations are over $\mathbb{F}_{q_\beta}$ (or extensions thereof). Instead of having each zk-SNARK handle statements about its *own* verifier, as in the prior attempts (i.e., writing V_α as a $\mathbb{F}_{q_\beta}$-arithmetic circuit, or V_β as a $\mathbb{F}_{q_\alpha}$-arithmetic circuit), we instead let each zk-SNARK handle statements about the verifier of the *other* zk-SNARK. That is, we write V_α as a $\mathbb{F}_{q_\alpha}$-arithmetic circuit C_{V_α}, and V_β as a $\mathbb{F}_{q_\beta}$-arithmetic circuit C_{V_β}.

We can then perform recursive proof composition by *alternating* between the two proof systems. Roughly, one can use P_α to prove successful verification of a proof by C_{V_β} and, conversely, P_β to prove successful verification of a proof by C_{V_α}. Doing so in alternation ensures that fields "match up", and no long arithmetic is needed. (This sketch omits key technical details; see the full version of this paper [BCTV14a].)

Since E_α and E_β facilitate constructing PCD, we say that (E_α, E_β) is a *PCD-friendly 2-cycle of elliptic curves*. More generally, the idea extends to cycling through ℓ curves satisfying this definition:

Definition 1. *Let $E_0, \ldots, E_{\ell-1}$ be elliptic curves, respectively defined over finite fields $\mathbb{F}_{q_0}, \ldots, \mathbb{F}_{q_{\ell-1}}$, with each q_i a prime. We say that $(E_0, \ldots, E_{\ell-1})$ is a **PCD-friendly cycle** of length ℓ if each E_i is pairing friendly and, moreover, $\forall\, i \in \{0, \ldots, \ell-1\}$, $q_i = \#E_{i+1 \bmod \ell}(\mathbb{F}_{q_{i+1 \bmod \ell}})$.*

To our knowledge this notion has not been explicitly sought before. Though, fortunately, a family that satisfies this notion is already known, as discussed in the next subsection.

3.2 Two-Cycles Based on MNT Curves

We construct pairs of elliptic curves, E_4 and E_6, that form PCD-friendly 2-cycles (E_4, E_6). These are MNT curves [MNT01] of embedding degrees 4 and 6. Our construction also ensures that E_4 and E_6 are sufficiently *2-adic* (see below), a desirable property for efficient implementations of preprocessing zk-SNARKs.

MNT Curves and the KT Correspondence.   Miyaji, Nakabayashi, and Takano [MNT01] characterized prime-order elliptic curves with embedding degrees $k = 3, 4, 6$; such curves are now known as *MNT curves*. Given an elliptic curve E defined over a prime field $\mathbb{F}_q$, they gave necessary and sufficient conditions on the pair (q, t), where t is the *trace* of E over $\mathbb{F}_q$, for E to have embedding degree $k = 3, 4, 6$. We refer to an MNT curve with embedding degree k as an MNTk curve. Karabina and Teske [KT08] proved an explicit 1-to-1 correspondence between MNT4 and MNT6 curves:

Theorem 1 ([KT08]). *Let $n, q > 64$ be primes. Then the following two conditions are equivalent: (i) n and q represent an elliptic curve $E_4/\mathbb{F}_q$ with embedding degree $k = 4$ and $n = \#E(\mathbb{F}_q)$; (ii) n and q represent an elliptic curve $E_6/\mathbb{F}_n$ with embedding degree $k = 6$ and $q = \#E(\mathbb{F}_n)$.*

PCD-Friendly 2-Cycles on MNT Curves.   The above theorem implies that:

Each MNT6 curve lies on a PCD-friendly 2-cycle with the corresponding MNT4 curve (and vice versa).

Thus, a PCD-friendly 2-cycle can be obtained by constructing an MNT4 curve and its corresponding MNT6 curve. Next, we explain at high level how this can be done.

Constructing PCD-Friendly 2-Cycles.   First, we recall the only known method to construct MNTk curves [MNT01]. It consists of two steps:

- Step I: *curve discovery*. Find suitable $(q, t) \in \mathbb{N}^2$ such that there exists an ordinary elliptic curve $E/\mathbb{F}_q$ of prime order $n := q + 1 - t$ and embedding degree k.
- Step II: *curve construction*. Starting from (q, t), use the *Complex-Multiplication method* (CM method) [AM93] to compute the equation of E over $\mathbb{F}_q$.

The complexity of Step II depends on the *discriminant* D of E, which is the square-free part of $4q - t^2$. At present, the CM method is feasible for discriminants D up to size 10^{16} [Sut12]. Thus, Step I is conducted in a way that results in candidate parameters

(q, t) inducing relatively-small discriminants, to aid Step II. (Instead, "most" (q, t) induce a discriminant D of size $\sqrt{q}$, which is too large to handle.) Concretely, [MNT01] derived, for $k \in \{3,4,6\}$ and discriminant D, Pell-type equations whose solutions yield candidate parameters (q, t) for MNTk curves $E/\mathbb{F}_q$ of trace t and discriminant D. So Step I can be performed by iteratively solving the *MNTk Pell-type equation*, for increasing discriminant size, until a suitable (q, t) is found.

The above strategy can be extended, in a straightforward way, to construct PCD-friendly 2-cycles. First perform Step I to obtain suitable parameters (q_4, t_4) for an MNT4 curve $E_4/\mathbb{F}_{q_4}$; the parameters (q_6, t_6) for the corresponding MNT6 curve $E_6/\mathbb{F}_{q_6}$ are $q_6 := q_4 + 1 - t_4$ and $t_6 := 2 - t_4$. Then perform Step II for (q_4, t_4) to compute the equation of E_4, and then also for (q_6, t_6) to compute that of E_6. The complexity in both cases is the same: one can verify that E_4 and E_6 have the same discriminant. The two curves E_4 and E_6 form a PCD-friendly 2-cycle (E_4, E_6).

Suitable Cycle Parameters. We now explain what "suitable (q_4, t_4)" means in our context, by specifying a list of additional properties that we wish a PCD-friendly cycle to satisfy.

- *Bit lengths.* In a 2-cycle (E_4, E_6), the curve E_4 is "less secure" than E_6, because E_4 has embedding degree 4 while E_6 has embedding degree 6. Thus, we use E_4 to set lower bounds on bit lengths. Since we aim at a security level of 80 bits, we need $r_4 \geq 2^{160}$ and $q_4 \geq 2^{240}$ (so that $\sqrt{r_4} \geq 2^{80}$ and $q_4^4 \geq 2^{960}$ [FST10]). Since $\log r_4 \approx \log q_4$ for MNT4 curves, we only need to ensure that q_4 has 240 bits.[5]
- *Towering friendliness.* We restrict our focus to moduli q_4 and q_6 that are *towering friendly* (i.e., congruent to 1 modulo 6) [BS10]; this improves the efficiency of arithmetic in $\mathbb{F}_{q_4}^4$ and $\mathbb{F}_{q_6}^6$ (and their subfields).
- *2-adicity.* As discussed in [BCG$^+$13a, BCTV14b], if a pairing-based preprocessing zk-SNARK (G, P, V) is instantiated with an elliptic curve $E/\mathbb{F}_q$ of prime order r (or with $\#E(\mathbb{F}_q)$ divisible by a prime r), it is important, for efficiency reasons, that $r - 1$ is divisible by a large power of 2, i.e., $\nu_2(r - 1)$ is large. (Recall that $\nu_2(n)$, the 2-adic order of n, is the largest power of 2 dividing n.) Concretely, if G is invoked on an $\mathbb{F}_r$-arithmetic circuit C, it is important that $\nu_2(r - 1) \geq \lceil \log |C| \rceil$. We call $\nu_2(r - 1)$ the *2-adic order of E*, or the *2-adicity of E*.
  So let ℓ_4 and ℓ_6 be the target values for $\nu_2(r_4 - 1)$ and $\nu_2(r_6 - 1)$. One can verify that, for any MNT-based PCD-friendly 2-cycle (E_4, E_6), it holds that $\nu_2(r_4 - 1) = 2 \cdot \nu_2(r_6 - 1)$; in other words, E_4 is always "twice as 2-adic" as E_6. Thus, to achieve the target 2-adic orders, it suffices to ensure that $\nu_2(r_4 - 1) \geq \max\{\ell_4, 2\ell_6\}$ (where, as before, $r_4 := q_4 + 1 - t_4$). As we shall see it will suffice to take $\nu_2(r_4 - 1) \geq 34$.

Of the above properties, the most restrictive one is 2-adicity, because it requires seeing enough curves until, "by sheer statistics", one finds (q_4, t_4) with a high-enough value for $\nu_2(r_4 - 1)$. Collecting enough samples is costly because, as discriminant size increases, the density of MNT curves decreases: empirically, one finds that the number MNT curves with discriminant $D \leq N$ is (approximately) less than $\sqrt{N}$ [KT08].

---

[5] Alas, since E_4 has a low embedding degree, the ECDLP in $E(\mathbb{F}_{q_4})$ and DLP in $\mathbb{F}_{q_4}^4$ are "unbalanced": the former provides 120 bits of security, while the latter only 80. Moreover, the same is true for E_6: the ECDLP in $E(\mathbb{F}_{q_6})$ provides 120 bits of security, while the DLP in $\mathbb{F}_{q_4}^6$ only 80. Finding PCD-friendly cycles without these inefficiencies is an open problem.

An Extensive Computation for a Suitable Cycle.   Overall, finding and constructing a suitable cycle required a substantial computational effort.

- *Cycle discovery.* In order to find suitable parameters for a cycle, we explored a large space: all discriminants up to $1.1 \cdot 10^{15}$, requiring about 610,000 core-hours on a large cluster of modern x86 servers. Our search algorithm is a modification of [KT08, Algorithm 3]. Among all the 2-cycles that we found, we selected parameters (q_4, t_4) and (q_6, t_6) for a 2-cycle (E_4, E_6) of curves such that: (i) q_4, q_6 each have 298 bits; (ii) q_4, q_6 are towering friendly; and (iii) $\nu_2(r_4 - 1) = 34$ and $\nu_2(r_6 - 1) = 17$. The bit length of q_4, q_6 is higher than the lower bound of 240; we entail this cost so to pick a rare cycle with high 2-adicity, which helps the zk-SNARK's efficiency more than the slowdown incurred by the higher bit length.
- *Cycle construction.* Both E_4 and E_6 have discriminant 614144978799019, whose size requires state-of-the-art techniques in the CM method [Sut11, ES10, Sut12] in order to explicitly construct the curves.[6]

Below, we report the parameters and equations for the 2-cycle (E_4, E_6) that we selected.

$E_4/\mathbb{F}_{q_4} : y^2 = x^3 + A_4 x + B_4$ where

$A_4 = 2$,

$B_4 = 423894536526684178289416011533888240029318103673896002803341544124054745019340795360841685$,

$q_4 = 475922286169261325753349249653048451545124879242694725395555128576210262817955800483758081$.

$E_6/\mathbb{F}_{q_6} : y^2 = x^3 + A_6 x + B_6$ where

$A_6 = 11$,

$B_6 = 106700080510851735677967319632585352256454251201367587890185989362936000262606668469523074$,

$q_6 = 475922286169261325753349249653048451545124878552823515553267735739164647307408490559963137$.

Security.   One may wonder if curves lying on PCD-friendly cycles are weak (e.g., in terms of DL hardness). Yet, MNT4 and MNT6 curves of suitable parameters are widely believed to be secure, and they *all* fall in PCD-friendly 2-cycles. The additional requirement of high 2-adicity is not known to cause weakness either.

3.3 A Matched Pair of Preprocessing zk-SNARKs

Based on the cycle (E_4, E_6), we designed and constructed two preprocessing zk-SNARKs for arithmetic circuit satisfiability: (G_4, P_4, V_4) based on the curve E_4, and (G_6, P_6, V_6) on E_6. The software implementation follows [BCTV14b], the fastest preprocessing zk-SNARK implementation for circuits at the time of writing. We thus adapt the techniques in [BCTV14b] to our algebraic setting, which consists of the two MNT curves E_4 and E_6, and achieve efficient implementations of (G_4, P_4, V_4) and (G_6, P_6, V_6).

The implementation itself entails many algorithmic and engineering details, and we refer the reader to [BCTV14b] for a discussion of these techniques. We only provide a

---

[6] The authors are grateful to Andrew V. Sutherland for generous help in running the CM method on such a large discriminant.

high-level efficiency comparison between the preprocessing zk-SNARK of [BCTV14b] based on Edwards curves (also at 80-bit security), and our implementations of (G_4, P_4, V_4) and (G_6, P_6, V_6); see the full version of this paper. Our implementation is slower, because of two main reasons: (i) MNT curves do not enjoy advantageous properties that Edwards curves do; and (ii) the modulus sizes are larger (298 bits in our case vs. 180 bits in [BCTV14b]). On the other hand, the fact that MNT curves lie on a PCD-friendly 2-cycle is crucial for the PCD construction described next.

4   Evaluation on vnTinyRAM

We evaluate our scalable zk-SNARK when the given random-access machine M equals vnTinyRAM, a simple RISC von Neumann architecture [BCTV14b, BCG+13b]. For comparison, we also compare [BCTV14b]'s preprocessing zk-SNARK (which also supports vnTinyRAM) with our scalable zk-SNARK.

We ran our experiments on a desktop PC with a 3.40 GHz Intel Core i7-4770 CPU and 16 GB of RAM available. Unless otherwise specified, all times are in single-thread mode; as for our multi-core experiments, we enabled one thread for each of the CPU's 4 cores (for a total of 4 threads).

Recalling vnTinyRAM. The architecture vnTinyRAM is parametrized by the *word size*, denoted w, and the *number of registers*, denoted k. In terms of instructions, vnTinyRAM includes load and store instructions for accessing random-access memory (in byte or word blocks), as well as simple integer, shift, logical, compare, move, and jump instructions. Thus, vnTinyRAM can efficiently implement control flow, loops, subroutines, recursion, and so on. Complex instructions (e.g., floating-point arithmetic) are not directly supported and can be implemented "in software". See the full version of this paper for how vnTinyRAM can be expressed in our random-access machine formalism (i.e., given w, k, how to construct M to express w-bit vnTinyRAM with k registers).

Costs on vnTinyRAM. The performance of our zk-SNARK $(G^\star, P^\star, V^\star)$ on vnTinyRAM is easy to characterize, because it is determined by few quantities. For the key generator $G^\star$, the relevant quantities are:
- the constant time and space complexity of $G^\star$, when given as input a description of vnTinyRAM; and
- the constant sizes of the generated proving key pk and verification key vk.

For the proving algorithm $P^\star$, which proceeds step by step alongside the original computation, they are:
- the constant time necessary to incrementally compute the new (constant-size) proof at each step; and
- the constant space needed to compute the new proof (on top of the space needed by the original program).[7]

---

[7] The prover also needs to store the Merkle tree's intermediate hashes, which incurs a linear overhead in the program's space complexity. Since this overhead is small, and can even be reduced by saving only the high levels of the Merkle tree (and recomputing, "on demand", the local neighborhood of accessed leaves), we focus on the additive overhead needed for proving.

Finally, the verifier $V^\star$ takes as input a program $\mathcal{P}$ and a time bound T, and runs in time $O(|\mathcal{P}| + \log T)$; in our implementation, we fix $T \leq 2^{300}$ (plenty enough), so that $V^\star$ runs in time $O(|\mathcal{P}|)$.

In Figure 1, we report our measurements for two settings of vnTinyRAM: $(w,k) = (16,16)$ and $(w,k) = (32,16)$, i.e., 16-bit and 32-bit vnTinyRAM with 16 registers. (The same settings as in [BCTV14b].)

	16-bit vnTinyRAM $(w,k) = (16,16)$		32-bit vnTinyRAM $(w,k) = (32,16)$	
	key generator $G^\star$			
	TIME			
	1 thread	4 threads	1 thread	4 threads
total	44.5 s	15.9 s	53.8 s	19.4 s
	SPACE			
memory	1.0 GB	1.2 GB	1.1 GB	1.4 GB
pk size		51.5 MB		64.4 MB
vk size		1.3 kB		1.3 kB

	prover $P^\star$			
	TIME			
	1 thread	4 threads	1 thread	4 threads
per step	33.1 s	11.5 s	35.5 s	12.1 s
	SPACE			
memory	0.9 GB	1.2 GB	1.1 GB	1.3 GB
proof		374 B		374 B

	verifier $V^\star$	
	TIME	
$|\mathcal{P}| = 10$	23.6 ms	24.3 ms
$|\mathcal{P}| = 10^2$	24.1 ms	24.9 ms
$|\mathcal{P}| = 10^3$	30.1 ms	31.1 ms
$|\mathcal{P}| = 10^4$	91.0 ms	94.1 ms
in general	$\approx (23.48 + 0.00674|\mathcal{P}|)$ ms	$\approx (24.17 + 0.00698|\mathcal{P}|)$ ms

Fig. 1. Performance of our scalable zk-SNARK on 16-bit and 32-bit vnTinyRAM

Comparison with [BCTV14b].   In Figure 2, we compare the efficiency of [BCTV14b]'s preprocessing zk-SNARK and our scalable zk-SNARK, for a (random) program $\mathcal{P}$ of 10^4 instructions, as a function of T (the number of vnTinyRAM computation steps).

		key generator		key sizes		prover		verifier
		TIME	SPACE	|pk|	|vk|	TIME	SPACE	TIME
16-bit vnTinyRAM	[BCTV14b]	$0.09 \cdot T$ s	$1.8 \cdot T$ MB	$0.3 \cdot T$ MB	40.4 kB	$0.04 \cdot T$ s	$1.9 \cdot T$ MB	24.2 ms
$(w,k) = (16,16)$	**this work**	44.5 s	933 MB	51.5 MB	1.3 kB	$33.1 \cdot T$ s	873 MB	91.0 ms
32-bit vnTinyRAM	[BCTV14b]	$0.14 \cdot T$ s	$3.0 \cdot T$ MB	$0.4 \cdot T$ MB	80.3 kB	$0.05 \cdot T$ s	$3.1 \cdot T$ MB	41.0 ms
$(w,k) = (32,16)$	**this work**	53.8 s	1,082 MB	64.4 MB	1.3 kB	$35.5 \cdot T$ s	1,008 MB	94.1 ms

Fig. 2. Comparison between [BCTV14b]'s preprocessing zk-SNARK and our scalable zk-SNARK

The (approximate) asymptotic efficiency for [BCTV14b] was obtained by linearly interpolating [BCTV14b]'s measurements (which were collected on a machine with similar characteristics as our benchmarking machine). As for our measurements, we use the relevant numbers from Figure 1.

Conclusion. Our experiments demonstrate that, as expected, our approach is slower for small computations but, on the other hand, offers scalability to large computations by avoiding any space-intensive computations.

Indeed, [BCTV14b] (as well as other preprocessing zk-SNARK implementations [PGHR13, BCG$^+$13a]) require space-intensive computations to maintain their efficiency. As T grows, such approaches simply run out of memory, and must resort to "computing in blocks", sacrificing time complexity.

In contrast, our zk-SNARK, while requiring more time per execution step, merely requires a constant amount of memory to prove any number of execution steps. In particular, our zk-SNARK becomes more space-efficient than [BCTV14b]'s zk-SNARK when $T > 460$ for 16-bit vnTinyRAM, and when $T > 326$ for 32-bit vnTinyRAM; moreover, these savings in space grow unbounded as T increases.

Finally, being scalable, our zk-SNARK implementation is the first to achieve a well-defined clock rate of *verified instructions per second* (VIPS). Concretely, for vnTinyRAM, we obtain the following VIPS values:

	16-bit vnTinyRAM $(w, k) = (16,16)$	32-bit vnTinyRAM $(w, k) = (32,16)$
1 thread	$\text{VIPS} = \frac{1}{33.1}\text{Hz}$	$\text{VIPS} = \frac{1}{35.5}\text{Hz}$
4 threads	$\text{VIPS} = \frac{1}{11.5}\text{Hz}$	$\text{VIPS} = \frac{1}{12.1}\text{Hz}$

While perhaps too slow for most applications, our prototype empirically demonstrates the feasibility of the bootstrapping approach as a way to achieve scalability of zk-SNARKs and, more generally, to achieve the rich functionality of proof-carrying data.

Acknowledgments. We thank Andrew V. Sutherland for generous help in running the CM method on elliptic curves with large discriminants. We thank Damien Stehlé and Daniele Micciancio for discussions about the security of subset-sum functions. We thank Koray Karabina for answering questions about algorithms in [KT08].

This work was supported by: the Broadcom Foundation and Tel Aviv University Authentication Initiative; the Center for Science of Information (CSoI), an NSF Science and Technology Center, under grant agreement CCF-0939370; the Check Point Institute for Information Security; the European Community's Seventh Framework Programme (FP7/2007-2013) under grant agreement number 240258; the Israeli Centers of Research Excellence I-CORE program (center 4/11); the Israeli Ministry of Science and Technology; the Leona M. & Harry B. Helmsley Charitable Trust; the Simons Foundation, with a Simons Award for Graduate Students in Theoretical Computer Science; and the Skolkovo Foundation with agreement dated 10/26/2011.

References

[Ajt96]   Ajtai, M.: Generating hard instances of lattice problems. In: STOC 1996 (1996)

[AM93]   Atkin, A.O.L., Morain, F.: Elliptic curves and primality proving. Math. Comp (1993)

[BCCT13]   Bitansky, N., Canetti, R., Chiesa, A., Tromer, E.: Recursive composition and bootstrapping for SNARKs and proof-carrying data. In: STOC 2013 (2013)

[BCG$^+$13a]   Ben-Sasson, E., Chiesa, A., Genkin, D., Tromer, E., Virza, M.: SNARKs for C: Verifying program executions succinctly and in zero knowledge. In: Canetti, R., Garay, J.A. (eds.) CRYPTO 2013, Part II. LNCS, vol. 8043, pp. 90–108. Springer, Heidelberg (2013)

[BCG$^+$13b]   Ben-Sasson, E., Chiesa, A., Genkin, D., Tromer, E., Virza, M.: TinyRAM architecture specification v2.00 (2013), URL: http://scipr-lab.org/tinyram

[BCG$^+$14]   Ben-Sasson, E., Chiesa, A., Garman, C., Green, M., Miers, I., Tromer, E., Virza, M.: Zerocash: Decentralized anonymous payments from Bitcoin. In: SP 2014 (2014)

[BCGT13a]   Ben-Sasson, E., Chiesa, A., Genkin, D., Tromer, E.: Fast reductions from RAMs to delegatable succinct constraint satisfaction problems. In: ITCS 2013 (2013)

[BCGT13b]   Ben-Sasson, E., Chiesa, A., Genkin, D., Tromer, E.: On the concrete efficiency of probabilistically-checkable proofs. In: STOC 2013 (2013)

[BCI$^+$13]   Bitansky, N., Chiesa, A., Ishai, Y., Paneth, O., Ostrovsky, R.: Succinct non-interactive arguments via linear interactive proofs. In: Sahai, A. (ed.) TCC 2013. LNCS, vol. 7785, pp. 315–333. Springer, Heidelberg (2013)

[BCTV14a]   Ben-Sasson, E., Chiesa, A., Tromer, E., Virza, M.: Scalable zero knowledge via cycles of elliptic curves. Cryptology ePrint Archive (2014)

[BCTV14b]   Ben-Sasson, E., Chiesa, A., Tromer, E., Virza, M.: Succinct non-interactive zero knowledge for a von Neumann architecture. In: Security 2014 (2014), http://eprint.iacr.org/2013/879

[BDSMP91]   Blum, M., De Santis, A., Micali, S., Persiano, G.: Non-interactive zero-knowledge. SIAM J. Comp. (1991)

[BEG$^+$91]   Blum, M., Evans, W., Gemmell, P., Kannan, S., Naor, M.: Checking the correctness of memories. In: FOCS 1991 (1991)

[BFLS91]   Babai, L., Fortnow, L., Levin, L.A., Szegedy, M.: Checking computations in polylogarithmic time. In: STOC 1991 (1991)

[BFM88]   Blum, M., Feldman, P., Micali Non-interactive, S.: zero-knowledge and its applications. In: STOC 1988 (1988)

[BFR$^+$13]   Braun, B., Feldman, A.J., Ren, Z., Setty, S., Blumberg, A.J., Walfish, M.: Verifying computations with state. In: SOSP 2013 (2013)

[BS10]   Benger, N., Scott, M.: Constructing tower extensions of finite fields for implementation of pairing-based cryptography. In: Hasan, M.A., Helleseth, T. (eds.) WAIFI 2010. LNCS, vol. 6087, pp. 180–195. Springer, Heidelberg (2010)

[BSW12]   Boneh, D., Segev, G., Waters, B.: Targeted malleability: Homomorphic encryption for restricted computations. In: ITCS 2012 (2012)

[CMT12]   Cormode, G., Mitzenmacher, M., Thaler, J.: Practical verified computation with streaming interactive proofs. In: ITCS 2012 (2012)

[CRR11]   Canetti, R., Riva, B., Rothblum, G.N.: Practical delegation of computation using multiple servers. In: CCS 2011 (2011)

[CT10]   Chiesa, A., Tromer, E.: Proof-carrying data and hearsay arguments from signature cards. In: ICS 2010 (2010)

[CT12]   Chiesa, A., Tromer, E.: Proof-carrying data: Secure computation on untrusted plat-
            forms (high-level description). In: The Next Wave: The National Security Agency's
            Review of Emerging Technologies (2012)

[CTV13]   Chong, S., Tromer, E., Vaughan, J.A.: Enforcing language semantics using proof-
            carrying data. ePrint 2013/513 (2013)

[ES10]   Enge, A., Sutherland, A.V.: Class invariants by the CRT method. In: Hanrot, G.,
            Morain, F., Thomé, E. (eds.) ANTS-IX. LNCS, vol. 6197, pp. 142–156. Springer,
            Heidelberg (2010)

[FST10]   Freeman, D., Scott, M., Teske, E.: A taxonomy of pairing-friendly elliptic curves.
            Journal of Cryptology (2010)

[GGH96]   Goldreich, O., Goldwasser, S., Halevi, S.: Collision-free hashing from lattice prob-
            lems. Technical report, ECCC TR95-042 (1996)

[GGPR13]   Gennaro, R., Gentry, C., Parno, B., Raykova, M.: Quadratic span programs and suc-
            cinct NIZKs without PCPs. In: Johansson, T., Nguyen, P.Q. (eds.) EUROCRYPT
            2013. LNCS, vol. 7881, pp. 626–645. Springer, Heidelberg (2013)

[Gro10]   Groth, J.: Short pairing-based non-interactive zero-knowledge arguments. In: Abe,
            M. (ed.) ASIACRYPT 2010. LNCS, vol. 6477, pp. 321–340. Springer, Heidelberg
            (2010)

[KT08]   Karabina, K., Teske, E.: On prime-order elliptic curves with embedding degrees k
            = 3, 4, and 6. In: van der Poorten, A.J., Stein, A. (eds.) ANTS-VIII 2008. LNCS,
            vol. 5011, pp. 102–117. Springer, Heidelberg (2008)

[Lip12]   Lipmaa, H.: Progression-free sets and sublinear pairing-based non-interactive zero-
            knowledge arguments. In: Cramer, R. (ed.) TCC 2012. LNCS, vol. 7194, pp. 169–
            189. Springer, Heidelberg (2012)

[Lip13]   Lipmaa, H.: Succinct non-interactive zero knowledge arguments from span pro-
            grams and linear error-correcting codes. In: Sako, K., Sarkar, P. (eds.) ASIACRYPT
            2013, Part I. LNCS, vol. 8269, pp. 41–60. Springer, Heidelberg (2013)

[LMN10]   Lauter, K., Montgomery, P.L., Naehrig, M.: An analysis of affine coordinates for
            pairing computation. In: Joye, M., Miyaji, A., Otsuka, A. (eds.) Pairing 2010.
            LNCS, vol. 6487, pp. 1–20. Springer, Heidelberg (2010)

[Mic00]   Micali, S.: Computationally sound proofs. SIAM J. Comp. (2000)

[MNT01]   Miyaji, A., Nakabayashi, M., Takano, S.: New explicit conditions of elliptic curve
            traces for FR-reduction. IEICE Transactions on Fundamentals of Electronics, Com-
            munications and Computer Sciences (2001)

[NY90]   Naor, M., Yung, M.: Public-key cryptosystems provably secure against chosen ci-
            phertext attacks. In: STOC 1990 (1990)

[PGHR13]   Parno, B., Gentry, C., Howell, J., Raykova, M.: Pinocchio: Nearly practical verifi-
            able computation. In: Oakland 2013 (2013)

[SBV+13]   Setty, S., Braun, B., Vu, V., Blumberg, A.J., Parno, B., Walfish, M.: Resolving the
            conflict between generality and plausibility in verified computation. In: EuroSys
            2013 (2013)

[SBW11]   Setty, S., Blumberg, A.J., Walfish, M.: Toward practical and unconditional verifi-
            cation of remote computations. In: HotOS 2011 (2011)

[SMBW12]   Setty, S., McPherson, M., Blumberg, A.J., Walfish, M.: Making argument systems
            for outsourced computation practical (sometimes). In: NDSS 2012 (2012)

[Sut11]   Sutherland, A.V.: Computing Hilbert class polynomials with the Chinese remainder
            theorem. Math. Comp. (2011)

[Sut12]   Sutherland, A.V.: Accelerating the CM method. LMS Journal of Computation and
            Mathematics (2012)

[SVP+12]   Setty, S., Vu, V., Panpalia, N., Braun, B., Blumberg, A.J., Walfish, M.: Taking proof-based verified computation a few steps closer to practicality. In: Security 2012 (2012)

[Tha13]   Thaler, J.: Time-optimal interactive proofs for circuit evaluation. In: Canetti, R., Garay, J.A. (eds.) CRYPTO 2013, Part II. LNCS, vol. 8043, pp. 71–89. Springer, Heidelberg (2013)

[TRMP12]   Thaler, J., Roberts, M., Mitzenmacher, M., Pfister, H.: Verifiable computation with massively parallel interactive proofs. CoRR (2012)

[Val08]   Valiant, P.: Incrementally verifiable computation or proofs of knowledge imply time/space efficiency. In: Canetti, R. (ed.) TCC 2008. LNCS, vol. 4948, pp. 1–18. Springer, Heidelberg (2008)

Switching Lemma for Bilinear Tests and Constant-Size NIZK Proofs for Linear Subspaces

Charanjit S. Jutla[1] and Arnab Roy[2]

[1] IBM T.J. Watson Research Center, Yorktown Heights, NY 10598, USA
[2] Fujitsu Laboratories of America, Sunnyvale, CA 94085, USA

Abstract. We state a switching lemma for tests on adversarial responses involving bilinear pairings in hard groups, where the tester can effectively switch the randomness used in the test from being given to the adversary at the outset to being chosen after the adversary commits its response. The switching lemma can be based on any k-linear hardness assumptions on one of the groups. In particular, this enables convenient information theoretic arguments in the construction of sequence of games proving security of cryptographic schemes, mimicking proofs and constructions in the random oracle model.

As an immediate application, we show that the computationally-sound quasi-adaptive NIZK proofs for linear subspaces that were recently introduced [JR13b] can be further shortened to *constant*-size proofs, independent of the number of witnesses and equations. In particular, under the XDH assumption, a length n vector of group elements can be proven to belong to a subspace of rank t with a quasi-adaptive NIZK proof consisting of just a single group element. Similar quasi-adaptive aggregation of proofs is also shown for Groth-Sahai NIZK proofs of linear multi-scalar multiplication equations, as well as linear pairing-product equations (equations without any quadratic terms).

Keywords: NIZK, bilinear pairings, quasi-adaptive, Groth-Sahai, Random Oracle, IBE, CCA2.

1 Introduction

Testing pairing equations in bilinear groups is a fundamental component of numerous cryptographic schemes spanning public key encryption schemes, signatures, zero knowledge proofs and so on. We state and prove a *switching lemma* for testing pairing equations in bilinear groups, where an adversary is given some random group elements from one of the groups, and the pairing test (of equality and/or inequality) is performed on adversary's output and the same random group elements. We show that the tester can replace the random group elements in the test with a new set of fresh random group elements, effectively mimicking the behavior of a random oracle. This switching lemma can be based on any k-linear hardness assumptions on one of the groups. This not only enables convenient information theoretic arguments in the construction of sequence of

J.A. Garay and R. Gennaro (Eds.): CRYPTO 2014, Part II, LNCS 8617, pp. 295–312, 2014.
© International Association for Cryptologic Research 2014

games proving security of cryptographic schemes, but also allows more efficient protocols reminiscent of the Fiat-Shamir paradigm using random oracles [FS86].

Fiat-Shamir paradigm is best illustrated by the conversion of 3-round sigma protocol [Dam] for proof of knowledge (PoK) of discrete logarithms to a random oracle based NIZK. Consider an example where the prover is trying to prove possession of the discrete logarithm x of a public value g^x. In the first round the prover commits to a random value r by sending g^r. In response, the verifier generates a fresh random value c and sends to the prover. The prover then responds with $r + cx$. This constitutes an honest verifier zero-knowledge PoK. In transforming this to a NIZK, a public random oracle H is used and the prover just transmits $(g^r, r + H(g^r, g^x) \cdot x)$. Essentially the random oracle induces the effect of a 'fresh' randomness that can be used for verification and is not under any effective control of the prover. In this paper we create an analogous effect in the standard model using the hardness of k-linear problems (such as DDH and DLIN) in bilinear groups. We show that even if the random testing values are public and hence known to the prover, during verification one can switch to freshly generated testing values with negligible change in the probability of success of the verification.

As an immediate application, we show that the computationally-sound quasi-adaptive NIZK (QA-NIZK) proofs for linear subspaces that we recently introduced in [JR13b] can be further shortened to *constant*-size proofs, independent of the number of variables and equations. In [JR13b], it was shown that for languages that are linear subspaces of vector spaces of the bilinear groups, one can obtain more efficient computationally-sound NIZK proofs compared to [GS08] in a slightly different *quasi-adaptive* setting, which suffices for many cryptographic applications. In the quasi-adaptive setting, a class of parametrized languages $\{L_\rho\}$ is considered, parametrized by ρ, and the CRS generator is allowed to generate the CRS based on the language parameter ρ. However, the CRS simulator in the zero-knowledge setting is required to be a single efficient algorithm that works for the whole parametrized class or probability distributions of languages, by taking the parameter as input. This property was referred to as *uniform simulation*.

The main idea underlying the construction in [JR13b] can be summarized as follows. Consider the language L_ρ (over a cyclic group $\mathbb{G}$ of order q, in additive notation) defined as

$$L_\rho = \left\{ \langle l_1, l_2, l_3 \rangle \in \mathbb{G}^3 \mid \exists x_1, x_2 \in \mathbb{Z}_q : l_1 = x_1 \cdot \mathbf{g}, \, l_2 = x_2 \cdot \mathbf{f}, \, l_3 = (x_1 + x_2) \cdot \mathbf{h} \right\}$$

where $\rho \overset{\text{def}}{=} (\mathbf{g}, \mathbf{f}, \mathbf{h})$ is the parameter defining the language. Suppose that the CRS can be set to be a basis for the null-space $L_\rho^\perp$ of the language L_ρ. Then, just pairing a potential language candidate with $L_\rho^\perp$ and testing for all-zero suffices to prove that the candidate is in L_ρ, as the null-space of $L_\rho^\perp$ is just L_ρ. However, efficiently computing null-spaces in hard bilinear groups is itself hard. Thus, an efficient CRS simulator cannot generate $L_\rho^\perp$. However, it was shown that it suffices to give as CRS a (hiding) commitment that is computationally indistinguishable from a binding commitment to $L_\rho^\perp$.

Our contributions. Utilizing the switching lemma, for n equations in t variables, our computationally-sound quasi-adaptive NIZK proofs for linear subspaces require only k group elements under the k-linear decisional assumption [HK07, Sha07]. Thus, under the XDH[1] assumption for bilinear groups, our proofs require only *one* group element. In contrast, the Groth-Sahai system requires $(n + 2t)$ group elements and our previous system required $(n - t)$ group elements. Similarly, under the decisional linear assumption (DLIN), our proofs require only 2 group elements, whereas the Groth-Sahai system requires $(2n + 3t)$ group elements and our previous system required $(2n - 2t)$ group elements. These parameters are summarized in Table 1. While our CRS size grows linearly with n, the number of pairing operations is competitive and could be significantly less compared to earlier schemes for appropriate n and t.

Table 1. Comparison with Groth-Sahai, Jutla-Roy (2013) and Schnorr-NIZKs for Linear Subspaces. Parameter t is the number of unknowns and n is the dimension of the vector space, i.e. the number of equations. *See text for recent independent work.*

	XDH			DLIN		
	Proof	CRS	#Pairings	Proof	CRS	#Pairings
Groth-Sahai	$n + 2t$	4	$2n(t + 2)$	$2n + 3t$	9	$3n(t + 3)$
Jutla-Roy '13	$n - t$	$2t(n - t) + 2$	$(n - t)(t + 2)$	$2n - 2t$	$4t(n - t) + 3$	$2(n - t)(t + 2)$
Schnorr (RO)	$t + 2$	–	–	–	–	–
This paper	1	$n + t + 1$	$n + 1$	2	$2(n + t + 2)$	$2(n + 2)$

Note that Schnorr proofs of multiple equations in the random oracle model can also be combined into a proof consisting of only two group elements (by taking random linear combinations employing the random oracle), but it still requires commitments to all the variables. Thus, our proofs are even shorter than Schnorr proofs. On the other hand, Schnorr proofs are proof of knowledge (as opposed to ours or Groth-Sahai), and can be somewhat faster to verify as they only use exponentiation instead of pairings. We also show that proofs of multiple linear scalar-multiplication equations, as well as multiple *linear* pairing product equations (i.e. without any quadratic terms) can be aggregated into a single proof in the Groth-Sahai system. This can lead to significant shortening of proofs of multiple linear pairing product equations. The comparisons are tabulated in Table 2. We remark that this is in contrast to the batching of Groth-Sahai proof verification [BFI+10], where the proofs were not aggregated, but multiple pairing equations were batched together during the verification step. We can use similar batching techniques to improve the verification step; therefore, we skip taking these optimizations into consideration. A recent work [LPJY14] has also obtained constant-size QA-NIZK proofs under DLIN (but not under XDH). While our proofs are marginally shorter (2 against 3 in DLIN), they additionally show constant-size unbounded simulation-sound QA-NIZK proofs.

---

[1] XDH is the assumption that DDH is hard in one of the pairing groups. Also note that DDH is same as the k-linear assumption for $k = 1$.

Table 2. Comparison with (1) Groth-Sahai for n number of linear Scalar Multiplication Equations: $\vec{y} \cdot \vec{\mathbf{a}}_j + \vec{b}_j \cdot \vec{\mathbf{x}} = \mathbf{u}_j$, with $j \in [1, n]$, $\vec{y} \in \mathbb{Z}_q{}^s$, $\vec{\mathbf{x}} \in \mathbb{G}^t$ and $\mathbf{u}_j \in \mathbb{G}$. and (2) Groth-Sahai for n number of linear Pairing Product Equations: $e(\vec{\mathbf{y}}, \vec{\mathbf{a}}_j) + e(\vec{b}_j, \vec{\mathbf{x}}) = \mathbf{u}_j$, with $j \in [1, n]$, $\vec{\mathbf{y}} \in \mathbb{G}^s$, $\vec{\mathbf{x}} \in \mathbb{G}^t$ and $\mathbf{u}_j \in \mathbb{G}_T$.

	DLIN Linear Multi-Scalar and Linear Pairing-Product		
	Proof	CRS	#Pairings
Groth-Sahai	$3(s+t) + 9n$	9	$9n(s+t+3) + n$
This paper	$3(s+t) + 18$	$9 + 4n$	$18(s+t+3) + n$

While the cryptographic literature is replete with applications using NIZK proofs of algebraic languages over bilinear groups, and many examples were given in [JR13b] involving NIZK proofs of linear subspaces, we focus on two particular cases where aggregation of proofs of linear subspaces lead to interesting results. We consider a construction of [CCS09] to convert key-dependent message (KDM) CPA secure encryption scheme [BHHO08] into a KDM-CCA2 secure scheme which involved proving $O(N)$ linear equations, where N is the security parameter. With our aggregation of proofs, the size of this proof (in the quasi-adaptive setting) is reduced to just 2 group elements (under the DLIN assumption) from the earlier $O(N)$ sized quasi-adaptive proofs and Groth-Sahai proofs. It is also easy to see that the quasi-adaptive setting for proving the NIZK suffices, as is the case for most applications. As another application we reduce the size of the publicly-verifiable CCA2-IBE scheme obtained in [JR13b] by another group element to just five group elements plus a tag. This makes it shorter than the highly-optimized CCA2-IBE scheme obtained using the [CHK04] paradigm from hierarchical-IBE (HIBE) and in addition is publicly-verifiable.

Organization of the paper. We begin the rest of the paper with the switching lemma for bilinear tests in hard groups in Section 2. We recall the quasi-adaptive NIZK definitions in Section 3 and develop constant-size quasi-adaptive NIZKs for linear subspaces in Section 4. In Section 5, we apply our switching lemma to aggregate Groth-Sahai NIZKs. Finally, we provide application examples in Section 6. We defer detailed proofs, formal descriptions and a summary of standard hardness assumptions that we use to the full paper [JR13a].

2　Switching Lemma for Bilinear Tests in Hard Groups

Notations. Consider bilinear groups $\mathbb{G}_1$ and $\mathbb{G}_2$ with pairing e to a target group $\mathbb{G}_T$, all of prime order q, and random generators $\mathbf{g}_1 \in \mathbb{G}_1$ and $\mathbf{g}_2 \in \mathbb{G}_2$. Let $\mathbf{0}_1, \mathbf{0}_2$ and $\mathbf{0}_T$ be the identity elements in the three groups $\mathbb{G}_1, \mathbb{G}_2$ and $\mathbb{G}_T$ respectively. We will use additive notation for group operations, with $\mathbb{G}_1, \mathbb{G}_2$ and $\mathbb{G}_T$ viewed as $\mathbb{Z}_q$-vector spaces. The scalar product by $\mathbb{Z}_q$ elements naturally extends to vectors and matrices of group elements. The pairing operation also naturally extends to vectors of elements (by summation) and correspondingly to matrices of elements. Column vectors will be denoted by an arrow over the letter, e.g. $\vec{r}$ for (column) vector of $\mathbb{Z}_q$ elements, and $\vec{d}$ as (column) vector of group elements. Thus, $e(\vec{f}^\top, \vec{h}) = \sum_i e(f_i, h_i)$.

Switching Lemma Usage Example. We demonstrate the usage of the Switching Lemma by way of a toy example. Suppose we are given three elements $g, f\,(= a \cdot g), h(= b \cdot g)$ in the group $\mathbb{G}_1$ and we need a proof system, not necessarily ZK, for tuples of the form $(x \cdot g, x \cdot f, x \cdot h)$. Towards that end, suppose the following CRS is published: $((ar_1 + br_2) \cdot g_2, -r_1 \cdot g_2, -r_2 \cdot g_2)$. So the pairing test $e(x \cdot g, (ar_1 + br_2) \cdot g_2) + e(x \cdot f, -r_1 \cdot g_2) + e(x \cdot h, -r_2 \cdot g_2) = 0_T$, satisfies completeness, i.e., it holds for valid tuples.

However, how do we know that it is sound? A look at the pairing equation shows that there is a fair degree of freedom to satisfy it, without being a valid tuple. So we definitely have to resort to a computational hardness assumption to argue soundness. This is where we invoke the switching lemma, which is based on a hardness assumption. Thus even though we publish a CRS that uses r_1, r_2, during verification we can switch them with fresh r'_1, r'_2 chosen randomly and independently.

This means if a candidate tuple (l_1, l_2, l_3) satisfies the original test with a certain probability, then it also satisfies the switched test: $e(l_1, (ar'_1 + br'_2) \cdot g_2) + e(l_2, -r'_1 \cdot g_2) + e(l_3, -r'_2 \cdot g_2) = 0_T$ with almost the same probability. Rearranging, we get: $r'_1 \cdot e(a \cdot l_1 - l_2, g_2) + r'_2 \cdot e(b \cdot l_1 - l_3, g_2) = 0_T$. Now, observe that the r'_1, r'_2 were chosen *after* the tuple was given. So with high probability, both of $e(a \cdot l_1 - l_2, g_2)$ and $e(b \cdot l_1 - l_3, g_2)$ must be 0_T. Therefore, $l_2 = a \cdot l_1$ and $l_3 = b \cdot l_1$, thus proving soundness.

Another way to look at this is that we produced a single CRS by random linear combination of CRS'es to prove the individual languages $\{(x \cdot g, x \cdot f) \mid x \in \mathbb{Z}_q\}$ and $\{(x \cdot g, x \cdot h) \mid x \in \mathbb{Z}_q\}$. Since the combined CRS is given to the adversary, we cannot resort to information-theoretic arguments to separate the individual equations. However with the switching lemma in play, the separation follows.

Switching Lemma Intuition. First consider the asymmetric bilinear group setting, where DDH holds in each individual group, and there is no easy isomorphism from $\mathbb{G}_1$ to $\mathbb{G}_2$ and vice-versa. If an Adversary $\mathcal{A}$ is given two random group elements, say r_1, r_2, from $\mathbb{G}_2$, then one would like to claim that it is highly improbable for $\mathcal{A}$ to produce non-zero f_1, f_2 in $\mathbb{G}_1$ such that $e(f_1, r_1) + e(f_2, r_2) = 0_T$. First, note that if the groups were symmetric, then this is easy to achieve by setting $f_1 = r_2$ and $f_2 = -r_1$. But, since we are in the asymmetric setting, the improbability is proven under DDH holding for $\mathbb{G}_2$ as follows: We will show that an adversary $\mathcal{A}$ that can produce a non-zero f_1, f_2 satisfying the above pairing equation can be used to produce an adversary B that can break DDH. So, given a DDH challenge, $g_2, x \cdot g_2, y \cdot g_2$ and h which is either $xy \cdot g_2$ or $z \cdot g_2$, adversary B passes $g_2, x \cdot g_2$ to $\mathcal{A}$ (note they are random and independent). Since $\mathcal{A}$ produces non-zero f_1, f_2 such that $e(f_1, g_2) + e(f_2, x \cdot g_2) = 0_T$, it follows that $f_1 = -x \cdot f_2$. Then comparing $e(f_1, y \cdot g_2)$ with $-e(f_2, h)$ allows B to distinguish the two versions of h.

Surprisingly, a similar claim holds when the adversary $\mathcal{A}$ is given an arbitrary long, say length n vector $\vec{r}$ of random group elements from $\mathbb{G}_2$, and $\mathcal{A}$ is required to produce a length n vector $\vec{f}$ (from $\mathbb{G}_1$) such that $e(\vec{f}^\top, \vec{r}) = \mathbf{0}_T$. This is proven using a hybrid argument, and for that purpose it is useful to restate the above claim of improbability as a switching lemma: *Given* $\mathbf{r}_1, \mathbf{r}_2$, *if an adversary has probability* Δ *success in producing non-zero* $\mathbf{f}_1, \mathbf{f}_2$ *such that* $e(\mathbf{f}_1, \mathbf{r}_1) + e(\mathbf{f}_2, \mathbf{r}_2) = \mathbf{0}_T$, *then the probability of* $e(\mathbf{f}_1, \mathbf{r}_1') + e(\mathbf{f}_2, \mathbf{r}_2') = \mathbf{0}_T$ *holding is also negligibly close to* Δ, *where* $\mathbf{r}_1', \mathbf{r}_2'$ *are chosen after* $\mathcal{A}$ *commits* $\mathbf{f}_1, \mathbf{f}_2$.

Moving on to the symmetric bilinear groups, and assuming that the k-linear (commonly called DLIN for 2-linear) assumption holds for the groups, one can show that if $\mathcal{A}$ is now given k independent pairs of random group elements, then it is highly improbable for $\mathcal{A}$ to produce non-zero $\mathbf{f}_1, \mathbf{f}_2$ such that the above pairing test holds for all k pairs (with the same $\mathbf{f}_1, \mathbf{f}_2$). Again, a switching lemma variant is more useful for proving the general lemma for n-vectors. Further, the reduction to the k-linear assumption is achieved by embedding the k-linear challenge in a single pairing test which is a random linear combination of the k pairing tests.

We now state the switching lemma in its full generality and later remark on interesting special cases.

Lemma 1 (Switching Lemma). *Let $\mathcal{D}$ be an arbitrary efficiently samplable distribution over $n \times m$ matrices from $\mathbb{Z}_q$. For any PPT adversary $\mathcal{A}$ producing a vector of n elements from group $\mathbb{G}_1$, let $\Delta_{\mathcal{A}}$ be the following probability*

$$\Pr \left[\begin{array}{c} \mathbf{R} \xleftarrow{\$} \mathbb{G}_2^{m \times k}, \ \mathsf{C}^{n \times m} \leftarrow \mathcal{D}, \ \vec{\mathbf{f}}^{n \times 1} \leftarrow \mathcal{A}(\mathbf{g}_1, \mathbf{g}_2, \mathbf{R}, \mathsf{C}) : \\ \vec{\mathbf{f}} \neq \vec{\mathbf{0}}_1^{n \times 1} \ \text{and} \ e\left(\vec{\mathbf{f}}^\top, \mathsf{C} \cdot \mathbf{R}\right) = \vec{\mathbf{0}}_T^{1 \times k} \end{array} \right]$$

Then, under the k-linear assumption for group $\mathbb{G}_2$, the following probability is negligibly close to $\Delta_{\mathcal{A}}$.

$$\Pr \left[\begin{array}{c} \mathbf{R} \xleftarrow{\$} \mathbb{G}_2^{m \times k}, \ \mathsf{C}^{n \times m} \leftarrow \mathcal{D}, \ \vec{\mathbf{f}}^{n \times 1} \leftarrow \mathcal{A}(\mathbf{g}_1, \mathbf{g}_2, \mathbf{R}, \mathsf{C}), \ \mathbf{R}' \xleftarrow{\$} \mathbb{G}_2^{m \times k} : \\ \vec{\mathbf{f}} \neq \vec{\mathbf{0}}_1^{n \times 1} \ \text{and} \ e\left(\vec{\mathbf{f}}^\top, \mathsf{C} \cdot \mathbf{R}'\right) = \vec{\mathbf{0}}_T^{1 \times k} \end{array} \right]$$

The absolute value of the difference in the probabilities is bounded by $m \cdot \mathrm{ADV}(klin)$.

Remarks. If we assume that the distribution $\mathcal{D}$ overwhelmingly produces full ranked matrices, then observe that the later probability is information theoretically close to 0. Hence we can state:

$$\Pr \left[\begin{array}{c} \mathbf{R} \xleftarrow{\$} \mathbb{G}_2^{m \times k}, \ \mathsf{C}^{n \times m} \leftarrow \mathcal{D}, \ \vec{\mathbf{f}}^{n \times 1} \leftarrow \mathcal{A}(\mathbf{g}_1, \mathbf{g}_2, \mathbf{R}, \mathsf{C}) : \\ \vec{\mathbf{f}} \neq \vec{\mathbf{0}}_1^{n \times 1} \ \text{and} \ e\left(\vec{\mathbf{f}}^\top, \mathsf{C} \cdot \mathbf{R}\right) = \vec{\mathbf{0}}_T^{1 \times k} \end{array} \right] \approx_{k-linear} 0$$

If, however, $\mathcal{D}$ produces singular matrices non-negligibly often, then there is an efficient adversary that can induce the event $\vec{\mathbf{f}} \neq \vec{\mathbf{0}}_1$ and $e(\vec{\mathbf{f}}^\top, \mathsf{C} \cdot \mathbf{R}) = \vec{\mathbf{0}}_T$.

The switching lemma still stands since the same adversary can induce the event $\vec{\mathbf{f}} \neq \vec{\mathbf{0}}_1$ and $e\left(\vec{\mathbf{f}}^{\top}, \mathsf{C} \cdot \mathsf{R}'\right) = \vec{\mathbf{0}}_T$ just as easily.

Although the presence of the C matrix is not strictly essential for the settings that we consider in this paper, we leave it in this form for generalizations to groups where the scalar field or ring is not easily invertible.

Instead of using the k-linear assumption directly, we use a related assumption which we call the *k-lifted linear assumption* and is implied by the k-linear assumption with a perfect reduction (see [JR13a] for a proof).

Definition 1 (k-lifted linear assumption). *For a constant $k \geq 1$, assuming a generation algorithm $\mathcal{G}$ that outputs a tuple $(q, \mathbb{G})$ such that $\mathbb{G}$ is of prime order q, the k-lifted linear assumption asserts that it is computationally infeasible to distinguish between*

$$\mathrm{TUPLE}_0 = (b_1 \cdot \mathbf{g}, \cdots, b_k \cdot \mathbf{g}, r_1 \cdot \mathbf{g}, \cdots, r_k \cdot \mathbf{g}, \; b_1 s_1 \cdot \mathbf{g}, \cdots, b_k s_k \cdot \mathbf{g}, \sigma \cdot \mathbf{g})$$

and

$$\mathrm{TUPLE}_1 = (b_1 \cdot \mathbf{g}, \cdots, b_k \cdot \mathbf{g}, r_1 \cdot \mathbf{g}, \cdots, r_k \cdot \mathbf{g}, \; b_1 s_1 \cdot \mathbf{g}, \cdots, b_k s_k \cdot \mathbf{g}, s_{k+1} \cdot \mathbf{g})$$

where $\mathbf{g}$ is chosen randomly from $\mathbb{G}$, b_i, r_i and s_i are chosen randomly from $\mathbb{Z}_q$, and $\sigma = \sum_{i=1}^{n} r_i s_i$.

Note that the k-linear assumption is a variant of the k-lifted linear assumption with all $r_1, ..., r_k$ equal to one. Now we prove the switching lemma under this weaker assumption.

Proof. (of Lemma 1) When $m \leq k$, the lemma follows information-theoretically (although the proof for $m > k$ also works for this case) by noting that in this case R will have rank m with high probability. Now we focus on the case that $m > k$. Consider the following inductive hypothesis (over j):

$$\Pr\left[\begin{array}{c} \mathsf{R} \xleftarrow{\$} \mathbb{G}_2^{m \times k}, \; \mathsf{C}^{n \times m} \leftarrow \mathcal{D}, \; \vec{\mathbf{f}}^{n \times 1} \leftarrow \mathcal{A}(\mathbf{g}_1, \mathbf{g}_2, \mathsf{R}, \mathsf{C}), \; \mathsf{R}' \xleftarrow{\$} \mathbb{G}_2^{m \times k} : \\ \vec{\mathbf{f}} \neq \vec{\mathbf{0}}^{m \times 1} \;\; \text{and} \;\; e\left(\vec{\mathbf{f}}^{\top}, \mathsf{C} \cdot \mathsf{R}''\right) = \vec{\mathbf{0}}_T^{1 \times k} \end{array}\right]$$

differs from $\Delta_{\mathcal{A}}$ by at most $j \cdot \mathrm{ADV}(klin)$, where R'' has the first $(m - j)$ rows same as first $(m - j)$ rows of R and the last j rows same as the last j rows of R'. In the base case, i.e., when $j = 0$, this is same as the hypothesis (antecedent) in the lemma, and when $j = m$, this induction hypothesis is same as the claim (consequent) in the lemma. Thus, we just need to prove the induction step.

For an adversary $\mathcal{A}$, suppose the difference in the two probabilities corresponding to (induction hypothesis for) $j = t$ and $j = t + 1$ be δ. More precisely, denote the probability for adversary $\mathcal{A}$ corresponding to $j = t$ by Δ^t. Thus, we are supposing that $|\Delta^t - \Delta^{t+1}| \geq \delta$. Using $\mathcal{A}$ as a black box we will demonstrate an adversary $\mathcal{S}$ that will have advantage at least (negligibly close to) δ to break the k-lifted linear assumption.

So, let a *k-lifted linear* challenger produce: $(b_1 \cdot \mathbf{g}_2, \cdots, b_k \cdot \mathbf{g}_2, r_1 \cdot \mathbf{g}_2, ..., r_k \cdot \mathbf{g}_2, b_1 s_1 \cdot \mathbf{g}_2, ..., b_k s_k \cdot \mathbf{g}_2, \chi)$ in the group $\mathbb{G}_2$, where χ is either $\left(\sum_{i=1}^{n} r_i s_i\right) \cdot \mathbf{g}_2$ or random. Note that b_i, r_i and s_i are chosen randomly and independently by the challenger.

Let vectors $\vec{r}$ and $\vec{s}$ be defined component-wise as $(\vec{r})_i = r_i \cdot g_2$ and $(\vec{s})_i = s_i$, respectively. Define the k by k matrix B as the diagonal matrix with the i-th diagonal element set to b_i. Further, let $\mathbf{B} = B \cdot g_2$.

S samples $C^{n \times m} \leftarrow \mathcal{D}$, and chooses g_1 at random. It next samples an $(m - t - 1)$ by k matrix R_1 at random from $\mathbb{Z}_q$ (i.e. all elements of the matrix chosen randomly and independently from $\mathbb{Z}_q$). It sets $\mathbf{R_1} = R_1 \cdot \mathbf{B}$. It further samples a t by k matrix $\mathbf{R_2}$ at random from $\mathbb{G}_2$ (i.e. all elements chosen randomly and independently from $\mathbb{G}_2$). Finally S sets $\mathbf{R}$ to be the rows of $\mathbf{R_1}$, the row $\vec{r}^{\top}$ and the rows of $\mathbf{R_2}$ combined (in that order) to form an m by k matrix. Observe that all of $\mathbf{R}$'s entries are independently random. The adversary $\mathcal{A}$ is then given g_1, g_2, $\mathbf{R}$ and C. The adversary $\mathcal{A}$ in response produces $\vec{f}$. Now, S first checks that $\vec{f}$ is non-zero. It then chooses another t by k matrix R_2 at random from $\mathbb{Z}_q$ and sets $\mathbf{R_2} = R_2 \cdot \mathbf{B}$. Noting that S has access to $B \cdot \vec{s} \cdot g_2$, S (efficiently) performs the following bilinear test

$$e\left(\vec{f}^{\top}, C \cdot \begin{bmatrix} R_1 \cdot B \cdot \vec{s} \cdot g_2 \\ \chi \\ R_2 \cdot B \cdot \vec{s} \cdot g_2 \end{bmatrix} \right) \stackrel{?}{=} \mathbf{0}_T \tag{1}$$

S outputs 1 if the test succeeds, and otherwise outpus 0.

Note that the above experiment has two games, one corresponding to real k-lifted linear challenge TUPLE_0 choice, and one corresponding to fake k-lifted linear TUPLE_1 challenge choice. We will call these games $\mathbf{G}_0$ (the real game) and $\mathbf{G}_0'$ (the fake game). Our aim is to show that the probability of S outputting 1 in the real game $\mathbf{G}_0$ differs from the probability of its outputting 1 in the fake game $\mathbf{G}_0'$ by (negligibly close to) δ. To prove this, we first modify the above two games. In the modified games $\mathbf{G}_1$ and $\mathbf{G}_1'$, S itself chooses the k-lifted linear challenges according to the same distribution as in $\mathbf{G}_0$ and $\mathbf{G}_0'$. However, it defers the choice of $\vec{s}$ to after $\mathcal{A}$ has responded (noting that $\mathcal{A}$ is not given anything related to $\vec{s}$). After $\mathcal{A}$ responds, S chooses $\vec{s}$ at random, and also sets χ as $\vec{r}^{\top} \cdot \vec{s}$ in $\mathbf{G}_1$ and as $\vec{r}'^{\top} \cdot \vec{s}$ in $\mathbf{G}_1'$, where $\vec{r}'$ is another random k-tuple independent of $\vec{r}$. Adversary S then performs the same test (1) as above, and outputs 1 if the test succeeds, and otherwise it outputs 0. Since the distributions in games $\mathbf{G}_1$ and $\mathbf{G}_1'$ are identical to the distributions in $\mathbf{G}_0$ and $\mathbf{G}_0'$ (resp.), the probabilities of S outputting 1 remains the same in the respective games.

Now, note that in the (real) game $\mathbf{G}_1$ the above test (1) is equivalent to testing

$$e\left(\vec{f}^{\top}, C \cdot \begin{bmatrix} R_1 \\ \vec{r}^{\top} \\ R_2 \end{bmatrix} \cdot \vec{s} \right) \stackrel{?}{=} \mathbf{0}_T \tag{2}$$

and in the (fake) game $\mathbf{G}_1'$ the test (1) is equivalent to testing (2) but with $\vec{r}$ replaced by $\vec{r}'$. Now define games $\mathbf{G}_2$ and $\mathbf{G}_2'$ which are identical to games $\mathbf{G}_1$ and $\mathbf{G}_1'$ (resp.) except that instead of (1) the final test performed by S in $\mathbf{G}_2$ is

$$e\left(\vec{\mathbf{f}}^{\top}, \mathbf{C} \cdot \begin{bmatrix} \mathbf{R}_1 \\ \vec{\mathbf{r}}^{\top} \\ \mathbf{R}_2 \end{bmatrix}\right) \stackrel{?}{=} \vec{\mathbf{0}}_T^{1 \times k} \tag{3}$$

and the final test performed by $\mathcal{S}$ in $\mathbf{G}_2'$ is same but with $\vec{\mathbf{r}}$ replaced by $\vec{\mathbf{r}}'$. Going through the details of games $\mathbf{G}_2$ and $\mathbf{G}_2'$, it is clear that probability of $\mathcal{S}$ outputting 1 in $\mathbf{G}_2$ ($\mathbf{G}_2'$) is exactly Δ^t (resp. Δ^{t+1}). Moreover, since the distributions in $\mathbf{G}_1$ and $\mathbf{G}_2$ are identical, Δ^t is also the probability of (3) holding in $\mathbf{G}_1$. Thus, the probability of (2) holding in $\mathbf{G}_1$ is at least Δ^t, and no more except for the probability of (3) not holding and yet (2) holding. Since in game $\mathbf{G}_1$, $\vec{\mathbf{s}}$ is chosen after $\mathcal{A}$ responds, this additional probability is at most the probability (over random choice of $\vec{\mathbf{s}}$) of (2) holding for any fixed choice of rest of the coins in the game for which (3) does not hold. This probability is at most $1/q$. It follows that the probability of $\mathcal{S}$ outputting 1 in $\mathbf{G}_1$ (and hence in $\mathbf{G}_0$) differs from Δ^t by at most $1/q$. A similar argument shows that the probability of $\mathcal{S}$ outputting 1 in $\mathbf{G}_1'$ (and hence in game $\mathbf{G}_0'$) differs from Δ^{t+1} by at most $1/q$. Since, by hypothesis $|\Delta^{t+1} - \Delta^t| \geq \delta$, this completes the proof.

3   Quasi-Adaptive NIZK Proofs

We recall here the definitions from [JR13b] and provide a summary. Instead of considering NIZK proofs for a (witness-) relation R, the authors consider Quasi-Adaptive NIZK proofs for a probability distribution $\mathcal{D}$ on a collection of (witness-) relations $\mathcal{R} = \{R_\rho\}$. The quasi-adaptiveness allows for the common reference string (CRS) to set based on R_ρ after the latter has been chosen according to $\mathcal{D}$. However the simulator generating the CRS (in the simulation world) is required to be a single probabilistic polynomial time algorithm that works for the whole collection of relations $\mathcal{R}$.

To be more precise, they consider ensemble $\{\mathcal{D}_\lambda\}$ of distributions on collection of relations $\mathcal{R}_\lambda$, where each $\mathcal{D}_\lambda$ specifies a probability distribution on $\mathcal{R}_\lambda = \{R_{\lambda,\rho}\}$. When λ is clear from context it can be dropped. Since in the quasi-adaptive setting the CRS could depend on the relation, an associated *parameter language* $\mathcal{L}_{\mathrm{par}}$ is considered such that a member of this language is enough to characterize a particular relation, and this language member is provided to the CRS generator.

A tuple of algorithms $(\mathsf{K}_0, \mathsf{K}_1, \mathsf{P}, \mathsf{V})$ is called a *QA-NIZK* proof system for witness-relations $\mathcal{R}_\lambda = \{R_\rho\}$ with parameters sampled from a distribution $\mathcal{D}$ over associated parameter language $\mathcal{L}_{\mathrm{par}}$, if there exists a probabilistic polynomial time simulator $(\mathsf{S}_1, \mathsf{S}_2)$, such that for all non-uniform PPT adversaries $\mathcal{A}_1, \mathcal{A}_2, \mathcal{A}_3$ we have:

Quasi-Adaptive Completeness:

$$\Pr[\lambda \leftarrow \mathsf{K}_0(1^m); \rho \leftarrow \mathcal{D}_\lambda; \psi \leftarrow \mathsf{K}_1(\lambda, \rho); (x, w) \leftarrow \mathcal{A}_1(\lambda, \rho, \psi);$$
$$\pi \leftarrow \mathsf{P}(\psi, x, w) : \mathsf{V}(\psi, x, \pi) = 1 \text{ if } R_\rho(x, w)] = 1$$

Quasi-Adaptive Soundness:

$$\Pr[\lambda \leftarrow \mathsf{K}_0(1^m); \rho \leftarrow \mathcal{D}_\lambda; \psi \leftarrow \mathsf{K}_1(\lambda, \rho);$$
$$(x, \pi) \leftarrow \mathcal{A}_2(\lambda, \rho, \psi) : \; \mathsf{V}(\psi, x, \pi) = 1 \text{ and } \neg(\exists w : R_\rho(x, w))] \approx 0$$

Quasi-Adaptive Zero-Knowledge:

$$\Pr[\lambda \leftarrow \mathsf{K}_0(1^m); \rho \leftarrow \mathcal{D}_\lambda; \psi \leftarrow \mathsf{K}_1(\lambda, \rho) : \mathcal{A}_3^{\mathsf{P}(\psi, \cdot, \cdot)}(\lambda, \rho, \psi) = 1] \approx$$
$$\Pr[\lambda \leftarrow \mathsf{K}_0(1^m); \rho \leftarrow \mathcal{D}_\lambda; (\psi, \tau) \leftarrow \mathsf{S}_1(\lambda, \rho) : \mathcal{A}_3^{\mathsf{S}(\psi, \tau, \cdot, \cdot)}(\lambda, \rho, \psi) = 1],$$

where $\mathsf{S}(\psi, \tau, x, w) = \mathsf{S}_2(\psi, \tau, x)$ for $(x, w) \in R_\rho$ and both oracles (i.e. P and S) output failure if $(x, w) \notin R_\rho$.

Note that ψ is the CRS in the above definitions.

4   Aggregating Quasi-Adaptive Proofs of Linear Subspaces

We summarize the linear-subspace QA-NIZK setting of [JR13b] here and refer the reader to that paper for details.

Linear Subspace Languages. We consider languages that are linear subspaces of vectors of $\mathbb{G}_1$ elements. In other words, the languages we are interested in can be characterized as languages parametrized by $\mathbf{A}$ as below:

$$L_{\mathbf{A}} = \{\vec{\mathbf{x}}^\top \cdot \mathbf{A} \in \mathbb{G}_1^n \mid \vec{\mathbf{x}} \in \mathbb{Z}_q^{\;t}\}, \text{ where } \mathbf{A} \text{ is a } t \times n \text{ matrix of } \mathbb{G}_1 \text{ elements.}$$

Here $\mathbf{A}$ is an element of the associated *parameter language* $\mathcal{L}_{\mathrm{par}}$, which is all $t \times n$ matrices of $\mathbb{G}_1$ elements. The parameter language $\mathcal{L}_{\mathrm{par}}$ also has a corresponding witness relation $\mathcal{R}_{\mathrm{par}}$, where the witness is a matrix of $\mathbb{Z}_q$ elements : $\mathcal{R}_{\mathrm{par}}(\mathbf{A}, \mathsf{A})$ iff $\mathbf{A} = \mathsf{A} \cdot \mathbf{g}_1$.

Robust and Efficiently Witness-Samplable Distributions. Let the $t \times n$ dimensional matrix $\mathbf{A}$ be chosen according to a distribution $\mathcal{D}$ on $\mathcal{L}_{\mathrm{par}}$. The distribution $\mathcal{D}$ is called *robust* if with probability close to one the left-most t columns of $\mathbf{A}$ are full-ranked. A distribution $\mathcal{D}$ on $\mathcal{L}_{\mathrm{par}}$ is called *efficiently witness-samplable* if there is a probabilistic polynomial time algorithm such that it outputs a pair of matrices $(\mathbf{A}, \mathsf{A})$ that satisfy the relation $\mathcal{R}_{\mathrm{par}}$ (i.e., $\mathcal{R}_{\mathrm{par}}(\mathbf{A}, \mathsf{A})$ holds), and further the resulting distribution of the output $\mathbf{A}$ is same as $\mathcal{D}$. For example, the uniform distribution on $\mathcal{L}_{\mathrm{par}}$ is efficiently witness-samplable, by first picking A at random, and then computing $\mathbf{A}$.

QA-NIZK Construction. We now describe a computationally sound quasi-adaptive NIZK $(\mathsf{K}_0, \mathsf{K}_1, \mathsf{P}, \mathsf{V})$ for linear subspace languages $\{L_{\mathbf{A}}\}$ with parameters sampled from a robust and efficiently witness-samplable distribution $\mathcal{D}$ over the associated parameter language $\mathcal{L}_{\mathrm{par}}$. As a conceptual starting point, we first develop a construction which has k^2 element proofs, demonstrating a single application of the Switching Lemma. Later, we give a k element construction which linearly combines the first construction proofs and uses an additional layer of Switching Lemma application. Our description here is self sufficient and relates to the scheme in [JR13b] in that we linearly combine proofs of multiple elements yielding constant-size proofs.

Algorithm K_1: The algorithm K_1 generates the CRS as follows. Let $\mathbf{A}^{t\times n}$ be the parameter supplied to K_1. Let $s \stackrel{\text{def}}{=} n - t$: this is the number of equations in excess of the unknowns. It generates a matrix $\mathsf{D}^{t\times k^2}$ with all elements chosen randomly from $\mathbb{Z}_q$ and k elements $\{b_v\}_{v\in[1,k]}$ and sk elements $\{r_{iu}\}_{i\in[1,s],u\in[1,k]}$ all chosen randomly from $\mathbb{Z}_q$. Define matrices $\mathsf{R}^{s\times k^2}$ and $\mathsf{B}^{k^2\times k^2}$ component-wise as follows:

$$(\mathsf{R})_{i,k(u-1)+v} = r_{iu}, \text{ with } i \in [1,s], u,v \in [1,k].$$

$$(\mathsf{B})_{ij} = \begin{cases} b_v & \text{if } i = j = k(u-1)+v, \text{ with } u,v \in [1,k] \\ 0 & \text{if } i \neq j, \text{ with } i,j \in [1,k^2] \end{cases}$$

Intuitively, the matrix R is a k times column-wise repetition of the r_{ij}'s, and if we denote $\{b_v\}_{v\in[1,k]}$ by $\vec{b}$, then the diagonal matrix B is just the vector $\vec{b}$ repeated k times along the diagonal (i.e. $\mathsf{B}_{k(u-1)+v,k(u-1)+v}$ is b_v and not b_u).

The common reference string (CRS) has two parts $\mathbf{CRS}_p$ and $\mathbf{CRS}_v$ which are to be used by the prover and the verifier respectively.

$$\mathbf{CRS}_p^{t\times k^2} := \mathbf{A} \cdot \begin{bmatrix} \mathsf{D} \\ \mathsf{R}\,\mathsf{B}^{-1} \end{bmatrix} \qquad \mathbf{CRS}_v^{(n+k^2)\times k^2} = \begin{bmatrix} \mathsf{D}\,\mathsf{B} \\ \mathsf{R} \\ -\mathsf{B} \end{bmatrix} \cdot \mathbf{g}_2$$

Prover P: Given candidate $\vec{l}^{1\times n} = \vec{x}^\top \cdot \mathbf{A}$ with witness vector $\vec{x}^{t\times 1}$, the prover generates the following proof consisting of k^2 elements in $\mathbb{G}_1$: $\vec{p}^{1\times k^2} := \vec{x}^\top \cdot \mathbf{CRS}_p$

Verifier V: Given candidate $\vec{l}^{1\times n}$, and proof $\vec{p}^{1\times k^2}$, the verifier checks the following (k^2 equations) :

$$e\left(\left[\vec{l}^{1\times n} \mid \vec{p}^{1\times k^2}\right], \mathbf{CRS}_v\right) \stackrel{?}{=} \vec{0}_T^{1\times k^2}$$

Theorem 1. *The above algorithms* $(\mathsf{K}_0, \mathsf{K}_1, \mathsf{P}, \mathsf{V})$ *constitute a computationally sound quasi-adaptive NIZK proof system for linear subspace languages* $\{L_\mathbf{A}\}$ *with parameters* $\mathbf{A}$ *sampled from a robust and efficiently witness-samplable distribution* $\mathcal{D}$ *over the associated parameter language* $\mathcal{L}_{par}$, *given any group generation algorithm for which the k-linear assumption holds for group* $\mathbb{G}_2$.

Proof Intuition. We now give a proof sketch for soundness and defer the full proof, including completeness and zero knowledge, to the full paper [JR13a].

Soundness: We prove soundness by transforming the system over a sequence of games. Consider an adversary $\mathcal{A}$ that *wins* if it can produce a "proof" $\vec{\mathbf{p}}$ for a candidate $\vec{l}$ that is not in $L_{\mathbf{A}}$ and yet the pairing test $e\left(\left[\vec{l}^{1 \times n} \mid \vec{\mathbf{p}}^{1 \times k^2}\right], \mathbf{CRS}_v\right)$ $\overset{?}{=} \vec{\mathbf{0}}_T^{1 \times k^2}$ holds. Game $\mathbf{G}_0$ just replicates the soundness security definition. In Game $\mathbf{G}_1$ the CRS is generated using parameter witness A and its null-space, and this can be done efficiently by the challenger as the parameter distribution is efficiently witness samplable. After this transformation, we show that in the case of a certain event, a verifying proof of a non-language member implies breaking the k-linear assumption in group $\mathbb{G}_2$, while in the case of the event not occurring we can apply the Switching Lemma to bound the probability of the adversary winning.

In Game $\mathbf{G}_1$, the challenger efficiently samples $\mathbf{A}$ according to distribution $\mathcal{D}$, along with witness A (since $\mathcal{D}$ is an efficiently witness samplable distribution). Since A is a $t \times (t+s)$ dimensional rank t matrix, there is a rank s matrix $\begin{bmatrix} W^{t \times s} \\ I^{s \times s} \end{bmatrix}$ of dimension $(t+s) \times s$ whose columns form a complete basis for the null-space of A, which means $A \cdot \begin{bmatrix} W^{t \times s} \\ I^{s \times s} \end{bmatrix} = 0^{t \times s}$. In this game, the NIZK CRS is computed as follows: Generate matrix $D'^{\,t \times k^2}$ with elements randomly chosen from $\mathbb{Z}_q$ and the matrices $R^{s \times k^2}$ and $B^{k^2 \times k^2}$ as in the real CRS. Implicitly set: $D = D' + W\,R\,B^{-1}$. Therefore we have,

$$\mathbf{CRS}_p^{t \times k^2} = A \cdot \begin{bmatrix} D \\ R\,B^{-1} \end{bmatrix} = A \cdot \left(\begin{bmatrix} D' \\ 0^{s \times k^2} \end{bmatrix} + \begin{bmatrix} W \\ I^{s \times s} \end{bmatrix} \cdot R\,B^{-1} \right) = A \cdot \begin{bmatrix} D' \\ 0^{s \times k^2} \end{bmatrix}$$

$$\mathbf{CRS}_v^{(n+k^2) \times k^2} = \begin{bmatrix} D\,B \\ R \\ -B \end{bmatrix} \cdot g_2 = \begin{bmatrix} D'\,B + W\,R \\ R \\ -B \end{bmatrix} \cdot g_2$$

Suppose that $\mathcal{A}$ wins $\mathbf{G}_1$. Now, let us partition the $\mathbb{Z}_q$ matrix A as $\begin{bmatrix} A_0^{t \times t} \mid A_1^{t \times s} \end{bmatrix}$ and the candidate vector $\vec{l}$ as $\begin{bmatrix} \vec{l}_0^{1 \times t} \mid \vec{l}_1^{1 \times s} \end{bmatrix}$. Note that, since A_0 has rank t, the elements of $\vec{l}_0$ are 'free' elements and $\vec{l}_0$ can be extended to a unique n element vector $\vec{l}\,'$, which is a member of $L_{\mathbf{A}}$. This member vector $\vec{l}\,'$ can be computed as $\vec{l}\,'^{1 \times n} := \begin{bmatrix} \vec{l}_0 \mid -\vec{l}_0 \cdot W \end{bmatrix}$, where $W = -A_0^{-1}A_1$. The proof of $\vec{l}\,'$ is computed as $\vec{\mathbf{p}}\,'^{1 \times k^2} := \vec{l}_0 \cdot D'$. Since $\mathcal{A}$ wins $\mathbf{G}_1$, then $(\vec{l}, \vec{\mathbf{p}})$ passes the verification test, and further by design $(\vec{l}\,', \vec{\mathbf{p}}\,')$ passes the verification test. Thus, we obtain: $(\vec{l}_1' - \vec{l}_1) \cdot R = (\vec{\mathbf{p}}\,' - \vec{\mathbf{p}}) \cdot B$, where $\vec{l}_1'^{1 \times s} = -\vec{l}_0 \cdot W$. This gives us a set of equalities, for all $u \in [1, k]$:

$$\sum_{i=1}^{s} (l_{1i}' - l_{1i}) \cdot r_{iu} = (\mathbf{p}_{k(u-1)+1}' - \mathbf{p}_{k(u-1)+1}) \cdot b_1 = \cdots = (\mathbf{p}_{k(u-1)+k}' - \mathbf{p}_{k(u-1)+k}) \cdot b_k$$

$$(4)$$

Note that since $\vec{l}$ is not in the language, there exists an $i \in [1, s]$, such that $\vec{l}'_{1i} - \vec{l}_{1i} \neq 0$. Now consider the event E defined as follows:

$$\text{Event } E \equiv \text{ For some } u \in [1, k]: \sum_{i=1}^{s}(l'_{1i} - l_{1i}) \cdot r_{iu} \neq \mathbf{0}_1 \qquad (5)$$

Our strategy now is to show that the probability of $\mathcal{A}$ winning in both the events E and $\neg E$ is negligible. Under the event $\neg E$, we apply the Switching Lemma to switch the r_{iu}'s to a fresh set of random values r'_{iu}'s while verifying. After that, we argue information theoretically that the probability of winning the switched game is negligible. Under the event E, we show that one can build a k-linear challenge adversary using $\mathcal{A}$, such that if $\mathcal{A}$ wins then this new adversary can efficiently compute the (least) u in Event E, and using the multiple equalities in Equation 4 it can break the k-linear challenge.   □

We now show that the proof system described above with k^2 group elements can be further shortened to just k group elements. The main idea is to observe that Equation 4 is again several sets of equations, and we can carefully set up the system so that the prover only shows random linear combinations of Equation 4. Then resorting to Switching Lemma we conclude that the individual equations must be true. We now describe this optimized Quasi-Adaptive NIZK proof system in detail.

QA-NIZK construction with k elements. In this construction the **Algorithm K₁** generates the CRS as follows. It generates a matrix $\mathsf{D}^{t \times k}$ with all elements chosen randomly from $\mathbb{Z}_q$ and k elements $\{b_v\}_{v \in [1,k]}$ and k^3 elements $\{t_{uvw}\}_{u,v,w \in [1,k]}$ and sk elements $\{r_{iu}\}_{i \in [1,s], u \in [1,k]}$ all chosen randomly from $\mathbb{Z}_q$. Define matrices $\mathsf{R}^{s \times k}$ and $\mathsf{B}^{k \times k}$ component-wise as follows:

$$(\mathsf{R})_{iw} = \sum_{u=1}^{k}\sum_{v=1}^{k} r_{iu}t_{uvw}, \text{ with } i \in [1, s], w \in [1, k].$$

$$(\mathsf{B})_{vw} = \sum_{u=1}^{k} b_v t_{uvw}, \text{ with } v, w \in [1, k].$$

The construction of $\mathbf{CRS}_p$ and $\mathbf{CRS}_v$ remain algebraically the same, although now they use lesser elements. The prover and verifier also retain the same algebraic form. The set of equalities for this construction corresponding to the equation $(\vec{l}'_1 - \vec{l}_1) \cdot \mathsf{R} = (\vec{p}' - \vec{p}) \cdot \mathsf{B}$, is for all $w \in [1, k]$:

$$\sum_{i=1}^{s}\left[(l'_{1i} - l_{1i}) \cdot \left(\sum_{u=1}^{k}\sum_{v=1}^{k} r_{iu}t_{uvw}\right)\right] - \sum_{v=1}^{k}\left[(\mathbf{p}'_v - \mathbf{p}_v) \cdot \left(\sum_{u=1}^{k} b_v t_{uvw}\right)\right] = \mathbf{0}_1$$

$$(6)$$

Rearranging, we get for all $w \in [1, k]$:

$$\sum_{u=1}^{k}\sum_{v=1}^{k} t_{uvw}\left(\sum_{i=1}^{s}[(l'_{1i} - l_{1i}) \cdot r_{iu}] - (\mathbf{p}'_v - \mathbf{p}_v) \cdot b_v\right) = \mathbf{0}_1 \qquad (7)$$

Now, using the Switching Lemma and after applying information theoretic arguments, we transition to a game where the adversary wins if it wins the original game and the following event occurs:

$$\text{For all } u \in [1,k]: \sum_{i=1}^{s}(l'_{1i} - l_{1i}) \cdot r_{iu} = (\mathbf{p}'_1 - \mathbf{p}_1) \cdot b_1 = \cdots = (\mathbf{p}'_k - \mathbf{p}_k) \cdot b_k \quad (8)$$

After this point, the proof is analogous to the previous QA-NIZK construction. Detailed proof is given in [JR13a]. We also give a more optimized construction in [JR13a] which uses less randomness and enjoys a better security reduction.

5   Aggregating Groth-Sahai Proofs

We show that proofs of multiple linear scalar-multiplication equations, as well as multiple *linear* pairing product equations can be aggregated into a single proof in the Groth-Sahai system. We will focus on describing the aggregation for the scalar-multiplication equations, as the results for the linear pairing product equations are obtained in almost an identical manner.

Consider bilinear groups $\mathbb{G}_1$ and $\mathbb{G}_2$ with pairing e into a third group $\mathbb{G}_T$. Consider equations of the type

$$\sum_{i=1}^{n} y_i \cdot \mathbf{a}_i + \sum_{i=1}^{m} b_i \cdot \mathbf{x}_i = \mathbf{t}_1 \quad (9)$$

where the variables y_i are to take values in $\mathbb{Z}_q$, the variables $\mathbf{x}_i$ are to take values in $\mathbb{G}_1$. The constants $\mathbf{a}_i$ are in $\mathbb{G}_1$, and scalar constants b_i are in $\mathbb{Z}_q$. Moreover, $\mathbf{t}_1$ is in $\mathbb{G}_1$.

When the bilinear group is symmetric, i.e. $\mathbb{G}_1 = \mathbb{G}_2$, and under the DLIN assumption, the Groth-Sahai NIZK proof of the above equation requires commitments to the variables, each commitment being of size *three* group elements (for both y_i or $\mathbf{x}_i$). In addition it requires a proof of *nine* group elements. When there are multiple equations of the above kind in the same variables, the commitments to the variables remain the same, but each equation requires nine group elements. In other words, if there are $m + n$ variables and k equations, the full proof of the k equations has size $3 \cdot (m + n) + 9k$ group elements.

We will now show that in the quasi-adaptive setting, the full proof of the k equations can be obtained with size $3 \cdot (m + n) + 9$ group elements. We first describe how the proof is done in the Groth-Sahai system, and then we will point out the relevant changes. The proofs and commitments actually belong to the $\mathbb{Z}_q$-module $\mathbb{G}^3$ (where $\mathbb{G} = \mathbb{G}_1 = \mathbb{G}_2$).

We will write these groups in additive notation, and the bilinear pairing operation $e(A,B)$ written in infix notation as $A \otimes B$, with the pairing operation defining a tensor product $\mathbb{G} \otimes \mathbb{G}$ over $\mathbb{Z}_q$. Without loss of generality (see e.g. A2.2 in [Eis95]), we can assume that $\mathbb{G}_T = \mathbb{G} \otimes \mathbb{G}$. Further, this naturally extends to a tensor product $\mathbb{G}^3 \otimes \mathbb{G}^3$. One can also define a tensor product $\mathbb{Z}_q \otimes \mathbb{G}$, but since $\mathbb{G}$ is a $\mathbb{Z}_q$-module, this tensor product is just $\mathbb{G}$.

Let $\iota_1 : \mathbb{Z}_q \to \mathbb{G}^3$, $\iota_2 : \mathbb{G} \to \mathbb{G}^3$, $p_1 : \mathbb{G}^3 \to \mathbb{Z}_q$, $p_2 : \mathbb{G}^3 \to \mathbb{G}$ be group homomorphisms s.t. $\iota_1 \circ p_1$, and $\iota_2 \circ p_2$ are identity maps in $\mathbb{Z}_q$ and $\mathbb{G}$ resp. Note that the maps ι_1 and ι_2 naturally define a group homomorphism ι_T from $\mathbb{Z}_q \otimes \mathbb{G}$ $(= \mathbb{G})$ to $\mathbb{G}^3 \otimes \mathbb{G}^3$, and similarly p_1 and p_2 define a group homomorphism p_T from $\mathbb{G}^3 \otimes \mathbb{G}^3$ to $\mathbb{Z}_q \otimes \mathbb{G}$ $(= \mathbb{G})$.

The NIZK common reference string (CRS) consists of three elements from $\mathbb{G}^3$, i.e. $\boldsymbol{u}_1, \boldsymbol{u}_2, \boldsymbol{u}_3 \in \mathbb{G}^3$. They are chosen as follows: $\boldsymbol{u}_1 = (\alpha \cdot \mathbf{g}, \mathcal{O}, \mathbf{g})$, and $\boldsymbol{u}_2 = (\mathcal{O}, \beta \cdot \mathbf{g}, \mathbf{g})$, and $\boldsymbol{u}_3 = r\boldsymbol{u}_1 + s\boldsymbol{u}_2$, for random $\alpha, \beta, r, s \in \mathbb{Z}_q$, and random $\mathbf{g} \in \mathbb{G}\backslash\mathcal{O}$. This real-world CRS $\vec{\boldsymbol{u}}$ is sometimes also referred to as the *binding* CRS.

The map $\iota_2(\mathcal{Z})$ is just $(\mathcal{O}, \mathcal{O}, \mathcal{Z})$, and $p_2(\mathcal{Z}_1, \mathcal{Z}_2, \mathcal{Z}_3) = \mathcal{Z}_3 - \alpha^{-1}\mathcal{Z}_1 - \beta^{-1}\mathcal{Z}_2$, which shows that $\iota_2 \circ p_2$ is an identity map. It also shows that $p_2(\boldsymbol{u}_1) = p_2(\boldsymbol{u}_2) = p_2(\boldsymbol{u}_3) = \mathcal{O}$. Now, the **commitments** to elements $\mathcal{Z}$ in $\mathbb{G}$ are made by picking r_1, r_2, r_3 at random from $\mathbb{Z}_q$, and setting $c_2(\mathcal{Z}) = \iota_2(\mathcal{Z}) + r_1\boldsymbol{u}_1 + r_2\boldsymbol{u}_2 + r_3\boldsymbol{u}_3$. Thus, $p_2(c_2(\mathcal{Z})) = \mathcal{Z}$, and hence the name binding CRS.

The map $\iota_1(z)$ is $\iota_2(z \cdot \mathbf{g})$, and hence commitment to $z \in \mathbb{Z}_q$ is $c_1(z) = c_2(z \cdot \mathbf{g})$.

For equations of the form (9) , i.e. $\vec{y} \cdot \vec{\mathbf{a}} + \vec{b} \cdot \vec{\mathbf{x}} = \mathbf{t}_1$, a proof $\vec{\boldsymbol{\pi}}$ (along with commitments to variables) is obtained by setting $\vec{\boldsymbol{\pi}} = S^\top \iota_2(\vec{\mathbf{a}}) + R^\top \iota_1(\vec{b}) + \vec{\boldsymbol{\theta}}$, where R is the matrix of rows (r_1, r_2, r_3), coming from $c_2(\mathbf{x}_i)$, one for each committed variable $\mathbf{x}_i$, and S is the matrix of rows (r_1, r_2, r_3), coming from $c_1(y_i)$, one for each committed variable y_i. Note, $\vec{\boldsymbol{\pi}}$ is vector of three $\mathbb{G}^3$ elements. The vector $\vec{\boldsymbol{\theta}}$ is set to be a random linear combination of $H_i\vec{\boldsymbol{u}}$, where H_i are finitely many matrices, and form a basis for the solutions to $\vec{\boldsymbol{u}} \bullet H\vec{\boldsymbol{u}} = 0$. It turns out that these matrices H_i are independent of the ZK simulator trapdoors α and β.

Let "$\bullet$" denote the dot product of vectors of elements from $\mathbb{G}^3$ and $\mathbb{G}^3$ w.r.t. product $\otimes$. The commitments $\vec{c}_1, \vec{c}_2$ and the proof are **verified** by the following equation:

$$\iota_1(\vec{b}) \bullet \vec{c}_2 + \vec{c}_1 \bullet \iota_2(\vec{\mathbf{a}}) = \iota_1(1) \bullet \iota_2(\mathbf{t}_1) + \vec{\boldsymbol{u}} \bullet \vec{\boldsymbol{\pi}}.$$

Quasi-Adaptive Aggregation. In the quasi-adaptive setting [JR13b], the NIZK CRS is allowed to depend on the language parameters, but with the further requirement that the ZK simulation be uniform. In the above context, the language parameters are $\vec{\mathbf{a}}$ and $\vec{b}$. Note $\mathbf{t}_1$ is *not* a language parameter, as it is a quantity produced by the prover.

So, let there be k equations in the same variables, with the j-th equation being

$$\vec{y} \cdot \vec{\mathbf{a}}^j + \vec{b}^j \cdot \vec{\mathbf{x}} = \mathbf{t}_1^j \tag{10}$$

In the above setting the prover produces k proofs, $\vec{\boldsymbol{\pi}}^j$. We would like the prover to give a random linear combination of these proofs, where the randomness is fixed in the CRS setup. In the DLIN setting, we need two different linear combinations. Thus, let the CRS generator choose two random $\mathbb{Z}_q$-vectors $\vec{\rho}$ and $\vec{\psi}$. The prover is required to produce $\vec{\boldsymbol{\pi}}_\rho = \sum_{j \in [1,k]} \rho_j \cdot \vec{\boldsymbol{\pi}}^j$ and $\vec{\boldsymbol{\pi}}_\psi = \sum_{j \in [1,k]} \psi_j \cdot \vec{\boldsymbol{\pi}}^j$. To be able to do so, the prover needs $\sum_j \rho_j \cdot \iota_2(\vec{\mathbf{a}}^j)$, $\sum_j \rho_j \cdot \iota_1(\vec{b}^j)$ (and similar terms using ψ_j). The $\vec{\theta}$ terms in the proofs need not be linearly combined, and the prover can just add one such term to each of $\vec{\pi}_\rho$ and $\vec{\pi}_\psi$, as its purpose is only to allow zero-knowledge simulation (i.e. witness hiding). The CRS generator can certainly produce these elements and give them as part of the CRS. The CRS generator also needs to give as part of the verification CRS the terms $\langle \iota_1(\rho_j) \rangle_j$ and $\langle \iota_1(\psi_j) \rangle_j$. In order to apply the switching lemma, we show in the proof of the theorem below that if $\vec{\mathbf{a}}^j$ are efficiently witness samplable, then the CRS generator can also simulate this verification CRS given $\rho_j \cdot \mathbf{g}$ and $\psi_j \cdot \mathbf{g}$.

The verification is now done as follows:

$$(\sum_j \rho_j \cdot \iota_1(\vec{b}^j)) \bullet \vec{c_2} + \vec{c_1} \bullet (\sum_j \rho_j \cdot \iota_2(\vec{\mathbf{a}}^j)) = \sum_j (\iota_1(\rho_j) \bullet \iota_2(\mathbf{t}_1^j)) + \vec{u} \bullet \vec{\pi}_\rho \quad (11)$$

$$(\sum_j \psi_j \cdot \iota_1(\vec{b}^j)) \bullet \vec{c_2} + \vec{c_1} \bullet (\sum_j \psi_j \cdot \iota_2(\vec{\mathbf{a}}^j)) = \sum_j (\iota_1(\psi_j) \bullet \iota_2(\mathbf{t}_1^j)) + \vec{u} \bullet \vec{\pi}_\psi$$

$$(12)$$

Theorem 2. *The above system constitutes a computationally-sound quasi-adaptive NIZK proof system for equations (10) with parameters $\langle \vec{\mathbf{a}}^j \rangle_j$, $\langle \vec{b}^j \rangle_j$, whenever $\langle \vec{\mathbf{a}}^j \rangle_j$ are chosen according to an efficiently witness-samplable distribution, and given any group generation algorithm for which the DLIN assumption holds.*

Proof of the theorem can be found in [JR13a]. Since Groth-Sahai proofs of more general equations (involving quadratic terms) require pairing of adversarially supplied commitments with each other, the switching lemma is not directly applicable. It remains an open problem to aggregate such NIZK proofs.

6   Extensions and Applications

Tags. We extend the system of Section 4 to include tags mirroring [JR13b]. The tags are elements of $\mathbb{Z}_q$, are included as part of the proof and are used as part of the defining equations of the language. We still get k element proofs based on the k-linear assumption. Details are in [JR13a].

KDM-CCA2 Encryption [CCS09]. In the paper [CCS09], the authors construct a public key encryption scheme simultaneously secure against key dependent chosen plaintext (KDM) and adaptive chosen ciphertext attacks (CCA2). They apply a Naor-Yung "double encryption" paradigm to combine any KDM-CPA secure scheme with any IND-CCA2 secure scheme along with an appropriate NIZK proof, to obtain a KDM-CCA2 secure scheme. In a particular construction, they obtain short ciphertexts by combining the KDM-CPA secure scheme of [BHHO08] with the IND-CCA2 scheme of [CS98], along with a Groth-Sahai NIZK proof. We show that the NIZK proof required in this construction can be considerably shortened. We defer the reader to [CCS09] for details of the scheme, and just describe the equations to be proved here. Consider bilinear groups $\mathbb{G}_1$ and $\mathbb{G}_2$ in which the K-linear and L-linear assumptions hold, respectively.

Let $\vec{\mathbf{g}}_1, \cdots, \vec{\mathbf{g}}_K, \mathbf{h}_1, \cdots, \mathbf{h}_K$ be part of the public key of the KDM-CPA secure encryption scheme and let $\vec{\mathbf{f}}_1, \cdots, \vec{\mathbf{f}}_K, \mathbf{c}_1, \cdots, \mathbf{c}_K, \mathbf{d}_1, \cdots, \mathbf{d}_K, \mathbf{e}_1, \cdots, \mathbf{e}_K$ be part of the public key of the IND-CCA2 secure encryption scheme. Let $(\vec{\mathbf{g}}, \mathbf{h}) \in \mathbb{G}_1^N \times \mathbb{G}_1$ be a ciphertext from the KDM-CPA secure encryption scheme and $(\vec{\mathbf{f}}, \mathbf{a}, \mathbf{b}) \in \mathbb{G}_1^{K+1} \times \mathbb{G}_1 \times \mathbb{G}_1$ be a ciphertext from the IND-CCA2 secure encryption scheme, with label l. Let $t = H(\vec{\mathbf{f}}, \mathbf{a}, l)$, where H is a collision resistant hash. The purpose of the NIZK proof is to establish that they encrypt the same plaintext. This translates to the following statement:

$$\exists r_1, \cdots, r_K, w_1, \cdots, w_K : \begin{pmatrix} \vec{\mathbf{g}} = \sum_{i=1}^{K} r_i \cdot \vec{\mathbf{g}}_i \ \wedge \ \vec{\mathbf{f}} = \sum_{i=1}^{K} w_i \cdot \vec{\mathbf{f}}_i \ \wedge \\ \mathbf{b} = \sum_{i=1}^{K} w_i \cdot (\mathbf{d}_i + t \cdot \mathbf{e}_i) \ \wedge \\ \mathbf{h} - \mathbf{a} = \sum_{i=1}^{K} r_i \cdot \mathbf{h}_i - \sum_{i=1}^{K} w_i \cdot \mathbf{c}_i \end{pmatrix}$$

This translates into $N + K + 3$ equations in $2K$ variables. Using the Groth-Sahai NIZK scheme, this requires $(2K)(L+1)$ elements of $\mathbb{G}_2$ and $(N + K + 3)L$ elements of $\mathbb{G}_1$. In our scheme this requires L elements of $\mathbb{G}_1$ in the proof - **1** under DDH and **2** under DLIN assumptions in $\mathbb{G}_2$.

CCA2-IBE Scheme [JR13b]. The definition of CCA2-secure encryption [BDPR98] naturally extends to the Identity-Based Encryption setting [CHK04]. In [JR13b], the authors construct a fully adaptive CCA2-secure IBE, which also allows public verification of the assertion that a ciphertext is valid for the particular claimed identity. The IBE scheme has four group elements (and a tag), where one group element serves as one-time pad for encrypting the plaintext. The remaining three group elements form a linear subspace with one variable as witness and three integer tags corresponding to: (a) the identity, (b) the tag needed in the IBE scheme, and (c) a 1-1 (or universal one-way) hash of some of the elements. It was shown that if these three group elements can be QA-NIZK proven to be consistent, and given the unique proof property of the QA-NIZKs, then the IBE scheme can be made CCA2-secure. Since, there are three components, and one variable the QA-NIZK required only two group elements under SXDH. We slightly shorten the proof to one element under SXDH. We defer the reader to [JR13b] for details of the original system, and just describe the Key Generation and Encryption steps in [JR13a].

References

[BDPR98] Bellare, M., Desai, A., Pointcheval, D., Rogaway, P.: Relations among notions of security for public-key encryption schemes. In: Krawczyk, H. (ed.) CRYPTO 1998. LNCS, vol. 1462, pp. 26–45. Springer, Heidelberg (1998)

[BFI+10] Blazy, O., Fuchsbauer, G., Izabachène, M., Jambert, A., Sibert, H., Vergnaud, D.: Batch Groth-Sahai. In: Zhou, J., Yung, M. (eds.) ACNS 2010. LNCS, vol. 6123, pp. 218–235. Springer, Heidelberg (2010)

[BHHO08] Boneh, D., Halevi, S., Hamburg, M., Ostrovsky, R.: Circular-Secure Encryption from Decision Diffie-Hellman. In: Wagner, D. (ed.) CRYPTO 2008. LNCS, vol. 5157, pp. 108–125. Springer, Heidelberg (2008)

[CCS09]   Camenisch, J., Chandran, N., Shoup, V.: A public key encryption scheme
            secure against key dependent chosen plaintext and adaptive chosen cipher-
            text attacks. In: Joux, A. (ed.) EUROCRYPT 2009. LNCS, vol. 5479, pp.
            351–368. Springer, Heidelberg (2009)
[CHK04]   Canetti, R., Halevi, S., Katz, J.: Chosen-ciphertext security from identity-
            based encryption. In: Cachin, C., Camenisch, J.L. (eds.) EUROCRYPT
            2004. LNCS, vol. 3027, pp. 207–222. Springer, Heidelberg (2004)
[CS98]   Cramer, R., Shoup, V.: A practical public key cryptosystem provably se-
            cure against adaptive chosen ciphertext attack. In: Krawczyk, H. (ed.)
            CRYPTO 1998. LNCS, vol. 1462, pp. 13–25. Springer, Heidelberg (1998)
[Dam]   Damgård, I.: On Σ protocols, http://www.daimi.au.dk/~ivan/Sigma.pdf
[Eis95]   Eisenbud, D.: Commutative algebra with a view toward algebraic geome-
            try. Graduate Texts in Mathematics, vol. 150. Springer (1995)
[FS86]   Fiat, A., Shamir, A.: How to prove yourself: Practical solutions to iden-
            tification and signature problems. In: Odlyzko, A.M. (ed.) Advances in
            Cryptology - CRYPTO 1986. LNCS, vol. 263, pp. 186–194. Springer, Hei-
            delberg (1987)
[GS08]   Groth, J., Sahai, A.: Efficient non-interactive proof systems for bilinear
            groups. In: Smart, N.P. (ed.) EUROCRYPT 2008. LNCS, vol. 4965, pp.
            415–432. Springer, Heidelberg (2008)
[HK07]   Hofheinz, D., Kiltz, E.: Secure hybrid encryption from weakened key en-
            capsulation. In: Menezes, A. (ed.) CRYPTO 2007. LNCS, vol. 4622, pp.
            553–571. Springer, Heidelberg (2007)
[JR13a]   Jutla, C., Roy, A.: Switching lemma for bilinear tests and constant-size
            NIZK proofs for linear subspaces. Cryptology ePrint Archive, Report
            2013/670 (2013), http://eprint.iacr.org/2013/670
[JR13b]   Jutla, C.S., Roy, A.: Shorter quasi-adaptive NIZK proofs for linear sub-
            spaces. In: Sako, K., Sarkar, P. (eds.) ASIACRYPT 2013, Part I. LNCS,
            vol. 8269, pp. 1–20. Springer, Heidelberg (2013)
[LPJY14]   Libert, B., Peters, T., Joye, M., Yung, M.: Non-malleability from Mal-
            leability: Simulation-Sound Quasi-Adaptive NIZK Proofs and CCA2-
            Secure Encryption from Homomorphic Signatures. In: Nguyen, P.Q.,
            Oswald, E. (eds.) EUROCRYPT 2014. LNCS, vol. 8441, pp. 514–532.
            Springer, Heidelberg (2014)
[Sha07]   Shacham, H.: A Cramer-Shoup encryption scheme from the linear as-
            sumption and from progressively weaker linear variants. Cryptology ePrint
            Archive, Report 2007/074 (2007), http://eprint.iacr.org/

Physical Zero-Knowledge Proofs
of Physical Properties

Ben Fisch[1,*], Daniel Freund[2,*], and Moni Naor[3,**]

[1] Columbia University, New York, NY, USA
benafisch@gmail.com
[2] Cornell University, Ithaca, NY, USA
freund90@mac.com
[3] Weizmann Institute of Science, Rehovot, Israel
moni.naor@weizmann.ac.il

Abstract. Is it possible to prove that two DNA-fingerprints match, or that they do not match, without revealing *any* further information about the fingerprints? Is it possible to prove that two objects have the same design without revealing the design itself? In the digital domain, *zero-knowledge* is an established concept where a prover convinces a verifier of a statement without revealing any information beyond the statement's validity. However, zero-knowledge is not as well-developed in the context of problems that are *inherently physical*. In this paper, we are interested in protocols that prove physical properties of physical objects without revealing further information. The literature lacks a unified formal framework for designing and analyzing such protocols. We suggest the first paradigm for formally defining, modeling, and analyzing *physical zero-knowledge* (PhysicalZK) protocols, using the Universal Composability framework. We also demonstrate applications of physical zero-knowledge to DNA profiling and neutron radiography. Finally, we explore *public observation proofs*, an analog of public-coin proofs in the context of PhysicalZK.

1 Introduction

Zero-knowledge proofs are protocols that prove an assertion without revealing any information beyond that assertion's validity. Zero-knowledge proofs were first introduced by Goldwasser, Micali, and Rackoff in 1985 [16]. The power of zero-knowledge proofs is quite remarkable: anything that can be proved efficiently can be proved with a zero-knowledge protocol, under the cryptographic assumption that one-way functions exist (see Goldreich [9]).

Zero-knowledge proofs have also been considered in a physical setting. A number of works have explored constructions of zero-knowledge protocols that can be physically implemented [26,19,24,23]. One goal of those works was to design

---

* Most of the work was done while the authors were visiting the Weizmann Institute of Science. Research in part supported by the Kupcinetz-Getz Summer Science School.
** Incumbent of the Judith Kleeman Professorial Chair.

J.A. Garay and R. Gennaro (Eds.): CRYPTO 2014, Part II, LNCS 8617, pp. 313–336, 2014.
© International Association for Cryptologic Research 2014

protocols with simple procedures and security arguments that the participating parties could easily understand. An added advantage of simple physical protocols is that humans can implement them without the aid of computers. Moran and Naor [24] give methods for polling people on sensitive issues using physical envelopes as an alternative to electronic polling, where humans might not trust computers to behave honestly. Many works have also addressed the incorporation of physical hardware into broader cryptographic schemes. In some cases, these hybrid protocols achieve efficiency or security gains that are unachievable in a standard computation model. Examples of physically realizable functionalities that have been suggested for aiding general cryptographic protocols include tamper-evidence [23], tamper-proof tokens [12,7,21,22,25,20,18], one-time programs [15], and physically uncloneable functions [3].

Previous literature on zero-knowledge in a physical setting addressed physical protocols for tasks that could otherwise be solved digitally. There is comparatively little formal work on protocols for inherently physical tasks that cannot be solved digitally. One example that has been studied rigorously is *distance bounding protocols*, introduced by Brands and Chaum in 1993 [2], in which a verifier party determines or verifies an upper bound on its physical distance to a prover party. In 2012, Glaser, Barak, and Goldston [8] suggested applying zero-knowledge concepts to the task of proving that a nuclear weapon is authentic without revealing sensitive information about its actual design, a problem that arises in the context of nuclear disengagement treaties. They presented an ϵ-knowledge protocol for this task, but did not have a rigorous framework for formally defining and analyzing the protocol's ϵ-knowledge security.

Our Contributions. We present the first formal treatment of *physical zero-knowledge* (PhysicalZK) proofs for inherently physical claims. In our setting, a prover convinces a verifier that an input object satisfies a given physical property. Our framework for designing and analyzing PhysicalZK protocols uses the *Universally Composable* (UC) security framework [4], popularly applied in analysis of hybrid protocols involving physical hardware.

Expanding on Glaser et al., we present the first PhysicalZK protocols for the nuclear verification problem, or the general task of verifying object neutron radiograph equality. We also demonstrate an application of PhysicalZK proofs to DNA profiling in which a prover (e.g. a suspect) convinces a verifier (e.g. the police) that its DNA profile does not match a target profile (e.g. obtained from a crime scene) without revealing to the verifier any further information about the profiles, and discuss a protocol for parental testing.

A further goal of our work is to initiate a rigorous study into the foundations of physical zero-knowledge. We point out both differences and similarities between physical and standard ZK where they arise. In particular, Section 3 compares the UC properties of physical vs. digital ZK, and Section 6 explores a physical analog of public coin proofs.

2   What Is Physical Zero-Knowledge (PhysicalZK)?

A standard zero-knowledge proof involves a binary relation R and an input x. A prover convinces a verifier that there exists a witness w such that $(x, w) \in R$. The verifier "learns nothing" from the protocol except the existence of w, and possibly the fact that the prover "knows" w. (See Goldreich [9] for formal definitions, classical theorems, and variants of zero-knowledge).

Previously, the term *physical zero-knowledge* was used for physically implemented ZK protocols, involving physical tools such as scissors, playing cards, envelopes, or pez dispensers. However, the underlying tasks in those protocols were still logical in nature (e.g. solving a Sudoku puzzle [19], finding Waldo [26]).

In our definition of physical zero-knowledge (PhysicalZK), a prover convinces a verifier that a physical input object has a physical property Π. The verifier should "learn nothing" except the validity of the statement "X satisfies Π." A physical measurement M verifies Π, possibly requiring the assistance of a measurement device D. Asymmetry between the prover and the verifier arises not from secret knowledge or computational power, but from *access permissions* to the object and measurement device. Since a verifier might forcefully break its restricted access, the threat model we consider only addresses adversaries that avoid being caught (similar to the *covert adversary* model [1]). Before proceeding, we give a few simple examples.

1. **Coke vs. Pepsi "blind test":** Alice demonstrates to Bob her ability to distinguish between the tastes of Coke and Pepsi using the classic blind test. However, the simplest test is not zero-knowledge. Bob might give Alice a cup of Sprite, and gain information from her response. One fix is to use indistinguishable coffee lids. Alice observes that Bob prepares cups of Coke and Pepsi. Bob then supplies Alice with the lids, Alice marks the inside of each lid with her secret signature, and covers the cups. After the blind test, Alice commits her response on a piece of paper. But before handing the paper to Bob, she will remove the lid, and check for her signature.

2. **Bins and Balls Equality:** Alice proves to Bob that two bins X and Y (of capacity n) contain the same number of balls. The following ϵ-knowledge protocol was given in [8]. Alice chooses $N > n$, and prepares two new pairs of bins, each of capacity $N + n$, labelled B_0 and B_1 respectively. Alice chooses two independent random values r_0 and r_1 uniformly distributed in $[0, N)$. Concealing the bins from Bob, she adds r_0 balls to each bin in B_0, and r_1 to each bin in B_1. Bob randomly selects $i \in \{0, 1\}$, Alice hands Bob the pair B_i, and Bob checks that both bins in the pair have equal numbers of balls. Alice then pours the contents of X into one bin in the remaining pair, and the contents of Y into the other. Finally, Bob checks the final contents of the bins to verify that they contain equal numbers of balls. Alice's success of cheating is at most $1/2$. (Appendix A contains a full analysis).

3. **Litmus Test:** Alice proves to Bob that her solution is basic/acidic without revealing the actual pH. Blue litmus paper turns red in acidic solution, and red litmus paper turns blue in basic solution. First, Bob tests Alice's litmus paper in known basic/acidic solutions to check that it operates correctly. After the protocol is

complete, the litmus paper must be completely destroyed (to prevent Bob from later examining traces of the solution remaining on the paper).

3   PhysicalZK in the UC Security Framework

The UC Framework. The Universally Composable security framework (UC) of Canetti [4] defines two worlds: the "real" world in which the real protocol is executed, and the "ideal" world in which an ideal process is implemented with the help of a trusted third party. A protocol *environment machine* $\mathcal{Z}$ interacts with the protocols in both worlds, setting each party's inputs, and reading their outputs. Although $\mathcal{Z}$ does not see internal communication between parties, it communicates freely with an adversary $\mathcal{A}$. When $\mathcal{A}$ *corrupts* a party, it assumes the party's identity, and takes control of its communication. A real protocol *UC-emulates* an ideal process if for every real adversary $\mathcal{A}$ there exists an ideal world adversary $\mathcal{S}$ such that no environment $\mathcal{Z}$ can distinguish between its interactions with $\mathcal{A}$ in the real protocol and $\mathcal{S}$ in the ideal process. The *universal composition* theorem states that if π is a protocol involving sub-protocol calls to an ideal functionality $\mathcal{F}$, ρ is a protocol that UC-emulates $\mathcal{F}$, and $\pi^{\rho/\mathcal{F}}$ is the *hybrid protocol* obtained by replacing calls to $\mathcal{F}$ in π with calls to ρ, then $\pi^{\rho/\mathcal{F}}$ UC-emulates π.

Modeling Physical Protocols. We separate physical protocols into a *logical layer* and a *physical layer*. All the physical operations of the protocol belong to the physical layer. Every physical operation serves an ideal function, and can be modeled by an ideal process in an abstract computation model with interactive turing machines (ITMs). This translation is based on physical assumptions. The logical layer is the hybrid world protocol obtained by replacing all physical operations with oracle calls to their ideal functionalities.

For example, consider the operation of pouring x balls into a bin and sealing the bin. We can define an ideal functionality $\mathcal{T}$ and an ideal process for this operation as follows. $\mathcal{T}$ stores tuples of the form $(value, id, creator, holder, state)$. Upon receiving the two commands **Create**(x, id) and **Seal**(id) from party P_i, $\mathcal{T}$ stores the tuple $(x, id, P_i, P_i, sealed)$, and will deny requests to view the value x that come from any party other than P_i. However, any party P_j may send a special command **Force**(id) to $\mathcal{T}$, and $\mathcal{T}$ will respond by sending the entire tuple to P_j and broadcasting to all other participating parties that P_j issued the **Force** command. This emulates the real behavior of a party who forcefully breaks open the sealed bin without permission, and is labeled a cheater.

Rigorous analysis can be applied to the hybrid world logical layer. We can then interpret the universal composition property of our model as formally reducing security to the most basic physical assumptions necessary: if the hybrid world logical layer UC-emulates $\mathcal{F}$, then any real world physical protocol emulating the hybrid protocol also realizes $\mathcal{F}$.

UC Physical Commitments. Bit-commitment is impossible to UC-realize in the standard computation model without trusted setup assumptions [5]. However, physical assumptions change matters. Consider the following trivial protocol in which the parties continuously observe each other. Alice commits to her bit by placing it in a sealed container, and de-commits by opening the container. To prevent Bob from forcibly cheating, Alice could run the protocol behind a secure glass screen (see Section 6 on *public observation protocols*). There is also a more sophisticated UC secure bit-commitment protocol using tamper-evident envelopes [23], which does not require continuous observation.

Likewise, ZK is not UC-realizable without setup assumptions, but there are UC-secure ZK proofs for any NP relation given UC bit-commitment [5,6]. Therefore, it is possible to implement UC-secure ZK protocols for any NP relation using UC physical commitments.

Ideal Functionality $\mathcal{F}_{ZK}^{\Pi}$. The ideal functionality $\mathcal{F}_{ZK}^{\Pi}$ is described in Figure 1, running with parties Prover, Verifier, and an oracle $\mathcal{F}_C$ that compiles the ideal functionalities for a collection C of physical operations in the real world.

Functionality $\mathcal{F}_{ZK}^{\Pi}$

Π is a unary predicate representing a physical property Π. If id_X uniquely identifies a physical set X, the statement $id_X \in \Pi$ translates the physical world statement "X satisfies property Π." $\mathcal{F}_C$ includes an ideal functionality $\mathcal{F}_{M^\Pi}$ for the physical measurement operation M^π required to verify Π, which outputs $\Pi(id_X) \in \{0, 1\}$. The parameter $\mathsf{leak}(id_X)$ represents information that is leaked when Verifier forcefully cheats.

- Upon receiving $(id_X, pid_C, \mathsf{Prover}, \mathsf{Verifier})$ from the party Prover, $\mathcal{F}_{ZK}^{\Pi}$ queries the $\mathcal{F}_C$ specified by the process identifier pid_C to compute $\Pi(id_X)$, and sends $(\Pi(id_X), id_X, \Pi)$ to Verifier.
- Upon receiving the instruction cheat from Verifier, send $(id_X, \mathsf{leak}(id_X))$ to Verifier, and send $(\mathtt{Cheater}, \mathsf{Verifier})$ to Prover. If Prover sends cheat, send $(\mathtt{Cheater}, \mathsf{Prover})$ to Verifier. Upon receiving the instruction fail from either party, send $\mathtt{Failed}$ to both parties.

Fig. 1. Ideal world PhysicalZK

Let $\rho^{\mathcal{F}_C/C}$ denote the $\mathcal{F}_C$-hybrid model translation of a physical protocol ρ with physical operation collection C. A proof that $\rho^{\mathcal{F}_C/C}$ UC-emulates $\mathcal{F}_{ZK}^{\Pi}$ captures (up to physical assumptions) that ρ is secure against any adversary in the real physical world whose behavior is restricted to operations in C.[1] A generic

---

[1] Ideally, C should define a sufficient set of operations such that any action outside this set will either be recognized as malicious or irrelevant to the protocol. This is not a formal mathematical notion, but a physical assumption.

procedure for this analysis is outlined in Figure 2. Appendix A includes a full UC modeling and security proof for the Bins and Balls Equality protocol of [8]. Let HYBRID$_{\rho,\mathcal{F}_C,\mathcal{A},\mathcal{Z}}$ and IDEAL$_{\mathcal{F}^{\Pi}_{ZK},\mathcal{F}_C,\mathcal{S},\mathcal{Z}}$ respectively denote the random variables describing the output of environment $\mathcal{Z}$ after interacting with $\mathcal{A}$ in $\rho^{\mathcal{F}_C/\mathcal{C}}$ and $\mathcal{S}$ in the ideal process for $\mathcal{F}^{\Pi}_{ZK}$.

Definition 1. [2] *A physical protocol ρ is a **physical zero-knowledge** protocol for property Π with respect to the physical operation set $\mathcal{C}$ if for any $\mathcal{A}$ there exists $\mathcal{S}$ such that for all environments $\mathcal{Z}$ outputting a single bit:*

$$\text{HYBRID}_{\rho,\mathcal{F}_C,\mathcal{A},\mathcal{Z}} \approx \text{IDEAL}_{\mathcal{F}^{\Pi}_{ZK},\mathcal{F}_C,\mathcal{S},\mathcal{Z}}$$

Main Differences from Standard $\mathcal{F}^{R}_{ZK}$. One difference is to allow the verifier to obtain leakage by overtly cheating; however, $\mathcal{F}^{R}_{ZK}$ could be extended similarly. A more fundamental difference is the way $\mathcal{F}^{\Pi}_{ZK}$ verifies Π. $\mathcal{F}^{R}_{ZK}$ requires the prover to submit a witness w along with the input x so that $\mathcal{F}^{R}_{ZK}$ may efficiently verify $(x, w) \in R$. $\mathcal{F}^{R}_{ZK}$ cannot find a witness w on its own since UC requires the trusted party to be computationally efficient. In contrast, $\mathcal{F}^{\Pi}_{ZK}$ verifies $id_X \in \Pi$ on its own, as it only needs the prover to transfer *access permissions*, not secret knowledge.

This difference has significant consequences. $\mathcal{F}^{R}_{ZK}$ cannot be realized in UC without trusted setup because the simulator must straight-line extract a witness from its interaction with the real prover, implying that the real verifier could do so as well. UC-emulation of $\mathcal{F}^{\Pi}_{ZK}$ does not require extraction. Standard ZK proofs in UC are *zero-knowledge proofs-of-knowledge* (ZKPoK), whereas $\mathcal{F}^{\Pi}_{ZK}$ is not. Thus, although UC protocols for $\mathcal{F}^{\Pi}_{ZK}$ may rely on physical assumptions, they do not fundamentally require trusted setup assumptions.

4   Neutron Radiography

Glaser, Barak, and Goldston [8] were the first to suggest applying zero-knowledge proofs to the problem of authenticating nuclear warheads without revealing sensitive information about their design. One approach to authentication is "template-matching." The inspecting party possesses a template warhead, presumably confirmed to be authentic. The opposing party must prove that each warhead brought to the dismantlement queue is identical (in design) to the template.

Neutron radiography can be used to compare objects. An object is bombarded with neutrons, and the intensity of neutron scattering is measured over a range of angles. Glaser et al. suggested using *passive bubble detectors*[3] to physically

---

[2] To differentiate *statistical*, *computational*, and *perfect* PhysicalZK, we can easily extend the definition to depend on the type of indistinguishability (statistical/computational/perfect) that the relation $\approx$ describes.

[3] A passive bubble detector contains droplets of superheated liquid dispersed throughout a clear gel. When a neutron hits a droplet, it vaporizes the droplet producing a visible air bubble trapped in the gel.

$\mathcal{F}_C$-hybrid protocol UC-emulation of $\mathcal{F}_{ZK}^{\Pi}$

We can assume that $\mathcal{A}$ acts as a proxy for the environment $\mathcal{Z}$ [4].

- When $\mathcal{A}$ corrupts Verifier, $\mathcal{Z}$ either sees a successful run of $\rho^{\mathcal{F}_C/\mathcal{C}}$, or receives $(0, id_X, \Pi)$. Since $\mathcal{S}$ only receives a receipt $(\Pi(id_X), id_X, \Pi)$, it must simulate the hybrid world proof, invoking an instance of $\mathcal{F}_C$, and dummy parties P and V. $\mathcal{S}$ plays the role of prover (P), and uses $\mathcal{Z}$'s messages to play the verifier (V), whose messages $\mathcal{S}$ forwards to $\mathcal{Z}$. If failure occurs or cheating is detected in the simulation, $\mathcal{S}$ sends either fail or cheat to $\mathcal{F}_{ZK}^{\Pi}$.
- $\mathcal{S}$'s simulation must be straight-line (it cannot rewind $\mathcal{Z}$), but $\mathcal{S}$ can extract any hybrid world "physical commitments" that the dummy party V makes with $\mathcal{F}_C$. ($\mathcal{S}$ omits from its messages to $\mathcal{Z}$ any notification of the command **Force** it internally uses with $\mathcal{F}_C$ to force open the commitments).
- When $\mathcal{A}$ corrupts Prover, $\mathcal{S}$ simulates the hybrid world protocol with $\mathcal{Z}$, but now playing the verifier's role (no secret input needed). If the simulation succeeds, $\mathcal{S}$ forwards $\mathcal{Z}$'s input to $\mathcal{F}_{ZK}^{\Pi}$. If failure or cheating occurs, $\mathcal{S}$ sends either fail or cheat to $\mathcal{F}_{ZK}^{\Pi}$.

Fig. 2. The $\mathcal{F}_C$-hybrid model security proof

record the neutron counts at randomly selected angles. The task of comparing the physically recorded counts essentially reduces to Bins and Balls Equality. The GBG protocol for Bins and Balls Equality (see Section 2) only achieves ϵ-*knowledge* with $\epsilon = n/N$ (security is broken with $O(N)$ repetitions). We present a modified protocol that achieves perfect PhysicalZK.

Protocol 4.1 guarantees that the number of balls the verifier eventually counts is uniformly distributed in $[N, 2N)$. Instead of preparing bin pair j containing $r_j \in [0, N)$ balls, the prover prepares a quadruple j of bins: one pair of bins with $r_j \in [0, N)$ balls each, and a second with $N + r_j$ balls each. If the number of balls in the prover's original bins is $x < N$, then exactly one of $N + r_j + x$ and $r_j + x$ lies in the interval $[N, 2N)$. Only this bin pair is retained and displayed.

Soundness: The soundness error is at most $\frac{1}{2k}$. The verifier would accept a false claim (when $x \neq y$) only if it selects a quadruple $j \in [k]$ and labeling of the bins in the final pair so that $x + r_j = y + r_j'$, where r_j and r_j' are the initializations of the bins labelled bin_x and bin_y respectively. If more than one quadruple contains an incorrect initialization such that $r_j \neq r_j'$, then the verifier catches the prover. If one labeling results in $x + r_j = y + r_j'$, then the opposite labeling does not. Therefore, this event occurs with probability at most $\frac{1}{2k}$.

Perfect Zero-Knowledge: We show that the distribution of balls in the final pair of bins (bin_x and bin_y) is the uniform distribution over $[N, 2N)$. Fix an arbitrary input value $0 \leq a < N$ for the number of balls that bins X and Y each hold. Let Z denote the number of balls in bin_x and bin_y at the end of the protocol. $Z = r_j + a$ when $r_j + a \geq N$, and $Z = r_j + a + N$ when $r_j + a < N$. Consider $t \in [N, 2N)$. If $t \in [N, N + a)$, then $Pr[Z = t] = Pr[r_j = t - a] = 1/N$. Otherwise, if $t \in [N + a, 2N)$, then $Pr[Z = t] = Pr[r_j = t - a - N] = 1/N$.

Protocol 4.1: Bins and Balls Quadruples

Input: Two bins X and Y, which both contain x and y balls respectively. The maximum capacity of each bin is N.

1. Prover prepares and seals k "quadruples" of bins $Q_1, ..., Q_k$, where each "quadruple" Q_i consists of two pairs of bins, $pair_{i,0}$ and $pair_{i,1}$. Each bin has capacity at least $2N$. For all $1 \leq i \leq k$, Prover randomly selects uniformly distributed values $r_i \in [0, N)$, and prepares each Q_i such that each bin in $pair_{i,0}$ contains r_i balls, and each bin in $pair_{i,1}$ contains $r_i + N$ balls.
2. Verifier randomly selects j uniformly distributed in $[1, k]$, and requests to view all $Q_{i \neq j}$. Prover reveals all quadruples $Q_{i \neq j}$, and Verifier checks that these quadruples were initialized correctly.
3. Prover selects a final pair out of Q_j: if $x + r_j \geq N$, then Prover chooses $pair_{j,0}$, and if $x + r_j < N$, then Prover chooses $pair_{j,1}$. Prover destroys the other pair.
4. Verifier labels the bins in the remaining pair as "bin_x" and "bin_y" (he can do this randomly to add a $1/2$ factor to the soundness error).
5. Prover pours the contents of X into bin_x, and the contents of Y into bin_y. Prover reveals the contents of bin_x and bin_y, and Verifier accepts the proof if and only if the two bins contain the same number of balls.

The complete formal proof is very similar to the proof in Appendix A.2. Roughly, since the distribution in the final pair is uniform and independent of the input, the simulator can run the protocol on an empty input.

4.1   From Bins and Balls to Neutron Bombardment

We adapt Protocol 4.1 to the problem of proving object radiograph equivalence. Neutron detectors are placed at a finite number of angles around each object, and a neutron source is fired at both objects for the same duration of time. A measurement device is used to measure the counts of neutrons that each detector has physically recorded.

Measurement Devices. The parties mutually possess a neutron source with a known flow rate, and physical neutron detectors. Each party has its own measurement device D for obtaining the physically recorded neutron count of any neutron detector. In the hybrid world, D is modeled as an ideal functionality $\mathcal{F}_D$. When given the input id_X corresponding to an object X, $\mathcal{F}_D$ records a measurement value, and outputs a function of the measured value.

Operation Init(d, r). This initializes the given neutron detector d to the integer value r. We assume that the prover and verifier can perform this operation without the other party knowing the value r.[4]

---

[4] The appropriate time to run the neutron source at a detector during initialization is calculated from the flow rate of the source. However, the initialization value should be hidden from the other party. The initialization can be done privately, or using a concealed on/off switch on the neutron source.

Prover Types. We consider two types of provers. Prover Type I has prior knowledge of the exact neutron counts x_θ and y_θ at any angle θ that the verifier chooses to examine, and Prover Type II does not possess this knowledge.

Drawbacks. A Type I prover is required to know the values of x and y for any angle θ. A Type II prover is allowed to re-handle the detectors after the neutron collection, possibly giving her the opportunity to dishonestly meddle with the results. In Appendix B we include a different zero-knowledge protocol for ORE that avoids both of these issues. The protocol uses a measurement device that outputs neutron counts modulo N.

Protocol 4.2: Perfect PhysicalZK protocol for Object Radiograph Equivalence (ORE)

Input: Two objects X and Y with equal ORE's, denoted $X \sim Y$.

1. Prover: Prepare k "quadruples" $Q_1, ..., Q_k$ of neutron detectors (as in Protocol 4.1), selecting random values $r_i \in [0, N)$ for $1 \leq i \leq k$, and using the **Init**(d, r_i) operation on each.
2. Verifier: For each Q_i, $1 \leq i \leq k$, randomly select $b_i \leftarrow \{0, 1\}$.
   - If $b_i = 0$: examine the detectors in Q_i, and check that the neutron count initialization is valid. **Fail** if invalid.
   - If $b_i = 1$: run collection test on Q_i.

Collection Test on Q_i:

1. Verifier: For each pair of the quadruple Q_i, randomly choose one detector to label d_x^i, and label the other d_y^i. Select a random angle θ, and send this to the prover.
2. Prover: Run the neutron source on X and Y, collecting at the angle θ, using detectors labeled d_x^i for X and d_y^i for Y.
   - *Type I Prover*: Choose the unique detector pair that has a count in the range $[N, 2N)$, and discard the other pair. Hand this pair to the verifier.
   - *Type II Prover*: Examine the contents of both detector pairs, and proceed as a Type I prover.
3. Verifier: Check that the detectors received from the prover have equal neutron counts. **Fail** if the counts are not equal.

Tolerance δ. The verifier could accept if and only if $|x_\theta - y_\theta| < \delta$. Two changes are necessary. First, the prover should choose which pair to discard based on the lower of the two values x and y. Unfortunately, the difference $|x - y|$ is still revealed. Second, the verifier must ensure that $N \geq max_\theta\{|x_\theta - y_\theta| + \delta\}$. Otherwise the prover could fool the verifier into accepting that $|x_\theta - y_\theta| < \delta$ when $|x_\theta - y_\theta| > N - \delta$. Verifier can incorporate checking the size of N into the cut-and-choose protocol, but needs to know some loose upper bound on $x_\theta - y_\theta$.

Soundness and Completeness: Protocol 4.2 has perfect completeness and soundness error at most $(\frac{1+\beta}{2})^k$, where $\beta < 1$ is the probability that $x_\theta = y_\theta$

at a uniformly distributed angle θ (when $X \not\sim Y$). Suppose Prover cheats on c out of k detector quadruples. The probability that Verifier doesn't check any of the c bad quadruples is 2^{-c}. The probability Prover passes on all the $k - c$ good quadruples is $\frac{1}{2} + \frac{1}{2} \cdot \beta$ because it passes always if Verifier chooses to check and with probability β if Verifier chooses to run a collection test. By independence, the soundness error is thus $(\frac{1}{2})^c \cdot (\frac{1}{2} + \frac{\beta}{2})^{k-c}$. Since $\beta \geq 0$, an optimal (cheating) strategy is to set $c = 0$, giving error at most $(\frac{1}{2} + \frac{\beta}{2})^k$.

Perfect Zero-Knowledge: Follows from the analysis of Protocol 4.1.

5   DNA Profiling

In recent years, genetic privacy in DNA profiling has become the subject of wide debate. Privacy issues obstruct criminal investigations, deterring non-guilty suspects from otherwise providing DNA samples, and giving guilty suspects legitimate excuses to refuse testing. We present a zero-knowledge protocol through which a suspect can prove to the police that his DNA profile does not match a crime scene profile. We also sketch an adaptation of the BBQ (Protocol 4.1) primitive to DNA testing. One potential application is a zero-knowledge protocol for parental testing.

STR Analysis. DNA profiling uses STR analysis. STR stands for "Short Tandem Repeats," which are short nucleotide sequences that repeat in tandem. In certain locations of the human genome, although all humans posses the same repeating sequence, the exact number of repeat units is highly variable from person to person. The variations of a gene or genetic locus in the human population are called *alleles*. Every individual has two alleles of each gene, one from each parent.

CODIS Profiles. In the United States, all forensic laboratories share CODIS (the Combined DNA Index System), which uses 13 specific STR loci to identify individuals. A CODIS DNA profile vector consists of 13 pairs of STR sequence lengths, one pair for each loci.

DNA Primers. A genetic profile is generated through STR analysis. PCR (the polymerase chain reaction) is run with oligonucleotide primers to isolate and amplify each STR repeat sequence. Primers determine the specific start and end nucleotides of the sequence to be amplified, and thus control the lengths of the flanking regions that are cut out along with the STR sequences (see Figure 3). We will use the notation $P_{i,j}$ to denote a primer pair that isolates the ith locus STR sequence, and produces a pair of fragments of sizes $m_{i,1} + j$ and $m_{i,2} + j$, when $m_{i,1}$ and $m_{i,2}$ are the sizes of the ith locus alleles.

Electrophoresis. In capillary electrophoresis, the most popular technique for DNA profiling, DNA fragments are fluorescently labelled during PCR, and passed through a capillary tube. Smaller fragments pass faster than larger ones. A laser detects the fragments as they pass by. The length of a fragment is deduced from the time the fragment takes to reach the laser.

δ-CE device. We imagine a slightly modified capillary electrophoresis apparatus in which the laser can only be operated for a limited time window δ, effectively limiting the range of DNA fragment sizes that will be detected.[5]

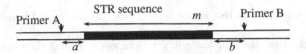

Fig. 3. STR sequence of length m cut out with flanking region $a + b$

5.1   DNA Inequality

At a basic level, the *police* (POL) will give the *defender* (DEF) one of the two DNA samples at random, and DEF must correctly identify the sample received. In general, DEF is not the suspect, but a public defender forensic team representing the suspect. There are two main challenges in proving zero-knowledge. First, the protocol simulator needs to extract the verifier's challenges without rewinding. Second, we must prevent POL from substituting a third auxiliary DNA sample mid-protocol. (Similarly, in the ZK protocol for *graph non-isomorphism* (GNI), the verifier proves it knows an isomorphism between its challenge graph and one of the two input graphs [11]). Additionally, we rely on the physical assumption that two DNA samples from the same person are always indistinguishable, whereas samples from individuals with distinct profiles are always distinguishable.

Random Seals. We require non-forgeable tamper-evident seals. We imagine that tamper-evident seals could be manufactured with a "random" pattern that is uniformly and independently distributed in a sufficiently large domain. A similar random pattern trick was suggested in [26].

Seal Covers. Our protocol also requires seal covers that function as physical commitments. The covers should be designed to hide any identifiable markings on the tamper-evident seals, and it should be possible to open the seals without removing their covers.

Completeness and Soundness. As DEF is able to differentiate between two distinct DNA samples, the protocol has perfect completeness. If the DNA samples are the same, DEF guesses the challenge sample correctly with probability 1/2. (The error is exponentially reduced by repetition).

Hybrid Model. The compiler holds tuples for each DNA sample input. The value attribute of each tuple is the DNA profile vector. The *random seal* operations queries an RO for a value r, tags r to the tuple, locks read/write access,

---

[5] Only the owner of the δ-CE can trust its operations. The party operating the δ-CE can ensure the limited-laser functionality by using a charged capacitor to power the laser so that the laser retains power for at most time δ. The operation can be repeated by recharging the capacitor.

Protocol 5.1: Zero-knowledge proof for DNA inequality

Preparation: Two test tubes are jointly prepared, one with DNA sample C, and one with S. DEF places identifiable tamper-evident seals on each test tube: one identifies S, and the other identifies C. POL covers the seals.

1. POL conceals the two test tubes, selects one randomly, and hands it back to DEF.
2. DEF checks that the seal has not yet been opened, and then opens the seal without removing its cover. DEF determines the profile of the sample in the test tube, and physically commits to the identity of the sample in the test tube.
3. POL "de-commits" to the challenge test tube by removing the cover on its seal.
4. DEF checks and identifies the uncovered seal to see that it wasn't replaced. If it is not the original seal, then DEF terminates the protocol. Otherwise, DEF opens its commitment from Step 2.
5. POL checks that DEF committed to the correct identity of the challenge sample.

---

and outputs r to the calling party. The *scramble* functionality swaps the *id* attributes of two tuples with probability $1/2$. The environment machine initializes and locks tuples for each DNA input sample. This emulates the "joint preparation" for the physical reason that no information is revealed to either party until they use analysis tools, such as electrophoresis, to examine the products of the PCR preparation.

Zero-Knowledge. The ideal functionality DNA Inequality is denoted $\mathcal{F}_{ZK}^{DI}$. The job of the simulator $\mathcal{S}$ in the case that $\mathcal{A}$ corrupts the hybrid world prover (DEF) was handled generically in Section 3, Figure 2. The case that $\mathcal{A}$ corrupts the hybrid world verifier (POL) is more interesting. First, $\mathcal{S}$ learns from $\mathcal{F}_{ZK}^{DI}$ if DEF passes or fails the ideal execution on the environment Z's input. Next, $\mathcal{S}$ simulates the hybrid model protocol, playing the role of DEF while Z uses $\mathcal{A}$ to control POL. Recall that $\mathcal{S}$ can straight-line extract physical commitments in the hybrid world (see Figure 2). Thus, $\mathcal{S}$ always knows the identity of POL's challenge sample, or that POL is cheating. $\mathcal{S}$ sets DEF's commitment in Step 2 to the correct identity of the challenge if DEF should pass, and the incorrect if DEF should fail. $\mathcal{S}$ only de-commits in Step 4 if POL did not cheat. If POL did not cheat, both $\mathcal{S}$ and the hybrid world DEF supply identical responses to Z, namely the identity of POL's challenge. If POL did cheat on its challenge, then $\mathcal{S}$ terminates its simulation. The hybrid world DEF will also terminate unless it fails to catch POL cheating, which only occurs with negligible probability (POL must guess the secret RO tag in order to fool DEF).

Testing a Village. There are cases where entire villages have been tested to see if the DNA profile of anyone in the population matches the crime scene DNA profile. Protocol 5.1 can be naturally extended for proving that a DNA profile does not exist in a population. Consider a population of 400 people. The verifier DEF receives 401 DNA samples, one from every individual in the population, plus a crime scene sample, all delivered in a set of 401 identical, covered, sealed, and randomly permuted test-tubes. DEF is required to find the crime scene sample C and hand it back to POL. If another individual has the same profile as C, then DEF fails with probability at least $\frac{1}{2}$.

5.2   Parental Testing

To prove a parent-child relationship using DNA profiling, it is necessary to show that DNA samples from the parent and child share at least one allele in each STR locus. We construct a zero-knowledge protocol for this task using an analogous technique to the Bins and Balls Quadruples (BBQ) scheme from Section 4.

Recall that a *primer* $P_{i,j}$ is used to cut out the ith locus STR alleles with flanking regions of total length j. Performing STR analysis with randomized primers P_{i,r_i} for r_i uniformly distributed in $[0, N)$ is analogous to adding a random number of balls to a bin. The quadruples of bins in BBQ translate to quadruples of test tubes running reactions with randomized primers. However, a technical caveat arises: each ith locus actually contains a pair of alleles that will be amplified with the same P_{i,r_i}, producing a pair of fragment lengths whose joint distribution is *not* necessarily uniform!

In the special case of paternity testing, this issue can be easily avoided by choosing to compare STR regions on the Y chromosome, which is uniquely passed from father to son. In more general circumstances, the prover can choose to reveal only one allele from each locus (using δ-CE), which is sufficient for showing that the two DNA samples share at least one allele in each STR locus.

6   Public Coin and Public Observation Proofs

A *private coin protocol* is one in which the verifier's random bits ("coin flips") must be kept private during the protocol. In contrast, the verifier's messages in a *public coin protocol* only consist of the outcomes of its coin flips. Public-coin physical protocols are *publicly observable* in the sense that the verifier can sit behind a glass screen throughout the protocol's execution, sending messages to the prover, and observing the prover's physical operations. Thus, unlike general physical zero-knowledge protocols, public observation physical zero-knowledge protocols do not rely on tamper-evident functionalities or (as heavily) on a covert threat model. While all public-coin protocols are publicly observable, not all publicly observable protocols need to be public-coin. For instance, the protocol may involve private-coin computational subprotocols.

In this section, we present an example of a public observation protocol for a special case of *DNA Inequality* in which the suspect's DNA S should pass if in at least one of the 13 CODIS loci it has an allele that is not present in the crime scene DNA C (notated $S \not\subset C$). The protocol becomes a perfectly complete test for DNA Inequality when only homozygotic gene regions are compared (e.g. X or Y chromosomes in males). The construction of the protocol involves a reduction to standard cryptography, using bit-commitment and generic ZK proofs for NP statements. Our protocol takes advantage of the fact that when $S \not\subset C$, the total number of distinct gene alleles in $S \cup C$ increases. Our protocol is closely related to the well-known public-coin protocol for GNI, making use of a classical set lower-bound protocol [17,10].

In the digital setting, public-coin $\mathcal{ZK}$ equals private-coin $\mathcal{ZK}$ [27,13,14,28]. In contrast, we don't know of a general method for converting any physical zero-knowledge protocol into a public observation zero-knowledge protocol. The known constructions of public-coin $\mathcal{ZK}$ proofs from private-coin $\mathcal{ZK}$ proofs involve simulating the private-coin verifier and applying universal hash functions to its messages. We do not know of any general analogous method for hashing physical messages. Furthermore, the public-coin verifier must be able to check set containment in the private-coin verifier's messages. In the physical setting, it is unclear whether the public-coin verifier can always assess the physical content of the private-coin verifier's messages, particularly when they involve physical concealment.f

Protocol 6.1: Public Observation ZK Protocol for DNA Inequality

- **Allele vector sets.** $V(C)$ is the set of *distinct* vectors of length 13 that can be formed by choosing one allele from each STR locus of C. $1 \leq |V(C)| \leq 2^{13}$. Similarly, we define $V(S \cup C)$, where each allele can be chosen from either S or C. If S introduces a new allele in at least one locus, then $|V(S \cup C)| \geq \frac{3}{2}|V(C)|$, and otherwise, $|V(S \cup C)| = |V(C)|$.
- **Equipment.** DEF has a *mod δ-CE device*, which is a δ-CE device that displays the lengths of DNA fragments modulo N, where N is a power of 2 greater than the longest possible fragment that will be measured. DEF also has access to a collection of DNA primers $\{P_{i,j}\}$.
- **Parameters ℓ and m.** Choose the smallest integers ℓ and m that satisfy the following conditions: $(\frac{3}{2})^{\ell} \geq 2m$ and $2^{m-\frac{1}{2}} \leq (\frac{3}{2})^{\ell} \cdot |V(C)|^{\ell} \leq 2^{m}$.[6]
- **Allele length vector set Z.** $Z \subset (\mathbb{Z}_N)^{13}$ is the set of possible vectors whose ith component is the length mod N of an allele taken from the ith locus of either S or C. It is the set of vectors of allele lengths mod N corresponding to $V(S \cup C)$.
- **Randomized allele length vector set $\mathbf{r}(Z^{\ell})$.** Let $\mathbf{r}$ be a random uniformly distributed vector in $(\mathbb{Z}_N)^{13\ell}$. Define $\mathbf{r}(Z^{\ell}) = \{\mathbf{r}+\mathbf{z} \mid \mathbf{z} \in Z^{\ell}\}$ where addition is over $(\mathbb{Z}_N)^{13\ell}$.

---

[6] Let $K = |V(C)|$. Note that $K = 2^r$ for $r \leq 13$. Choose the smallest integer x such that $(\frac{3}{2})^x \geq 2x \cdot log(\frac{3}{2}K) + 1$. One can verify that $x \leq 15$. Now find the smallest $\ell \geq x$ such that the fractional part of $\ell \cdot log(\frac{3}{2}K)$ is in $[\frac{1}{2}, 1) \cup \{0\}$. This will hold for either $x, x+1$, or $x+2$. Finally, set m to be the unique integer such that $m - \frac{1}{2} \leq \ell \cdot log(\frac{3}{2}K) \leq m$. Now ℓ and m satisfy $(\frac{3}{2})^{\ell} \geq 2\ell \cdot log(\frac{3}{2}K) + 1 \geq 2m$ and $2^{m-\frac{1}{2}} \leq (\frac{3}{2}K)^{\ell} \leq 2^m$.

- **Hash function family.** Set $k = log(N) \cdot 13\ell$, and choose a canonical encoding of $(\mathbb{Z}_N)^{13\ell}$ in $GF(2^k)$. We will use a family $\mathcal{H}_{k,m} := \{h_{a,b}\}$ of universal hash functions from $GF(2^k) \to GF(2^m)$ where $a, b \in GF(2^k)$, $a \neq 0$, and $h_{a,b}$ maps $x \mapsto ax + b$ and truncates the last $k - m$ bits.
- **Hash function shift.** For any vector $\mathbf{r} \in (\mathbb{Z}_N)^{13\ell}$ and $h \in \mathcal{H}_{k,m}$, define $\mathbf{r}(h) = h_{a,b-r\cdot a}$, where r is the encoding of $\mathbf{r}$ in $GF(2^k)$. Note that $\mathbf{r}(h)(\mathbf{r} + \mathbf{z}) = h(\mathbf{z})$.
- **Preparing $\mathbf{r}(Z^\ell)$.** DEF prepares ℓ sets of test tubes $T_1, ..., T_\ell$. Each T_i consists of test tubes $\{C_{i,j}\}_{j\in[13]}$ containing C and $\{S_{i,j}\}_{j\in[13]}$ containing S. For all (i, j), DEF selects an independent random value $r_{i,j}$ uniformly distributed in $[0, N)$, and then runs PCR on $C_{i,j}$ and $S_{i,j}$ with the primer pair $P_{j,r_{i,j}}$.[7] To display a vector $\mathbf{z} \in \mathbf{r}(Z^\ell)$ to POL, DEF chooses from every T_i the appropriate set of 13 test tubes containing the target fragments contained in $\mathbf{z}$, and then chooses time windows $t_{i,1}, ..., t_{i,13}$ to run the $mod\ \delta$-CE device on each test tube in order to only detect the target fragments.

Protocol 6.1 - Public observation ZK for DNA Inequality

POL observes DEF throughout the rounds. We assume that DEF cannot change the behavior of the $mod\ \delta$-CE device while under observation. Let $(\mathsf{com}, \mathsf{dec})$ denote a commitment scheme.

1. DEF: Choose a random uniformly distributed $\mathbf{r} \in (\mathbb{Z}_N)^{13\ell}$ and prepare $\mathbf{r}(Z^\ell)$. Find a set of m hash functions $H = \{h_i\}_{i=1}^m \subset \mathcal{H}$ so that $\bigcup_{i=1}^m h_i(Z^\ell) = \{0,1\}^m$. Compute commitments to the hash functions $\mathbf{r}(h_i)$ for each i, denoted $Com_H = \{\mathsf{com}(\mathbf{r}(h_1)), ..., \mathsf{com}(\mathbf{r}(h_m))\}$. Send Com_H to POL.
2. POL: Pick a uniformly distributed $y \in \{0,1\}^m$.
3. DEF: Find an $h_i \in H$ and $\mathbf{z} \in Z^\ell$ such that $h_i(\mathbf{z}) = y$. Display the allele vector $\mathbf{v} = \mathbf{z} + \mathbf{r}$ from the set $\mathbf{r}(Z^\ell)$.
4. DEF and POL: Execute a UC-secure ZK proof of the NP statement "there exists $x \in Com_H$ such that $\mathsf{dec}(x) = h$ and $h(\mathbf{v}) = y$."

Perfect Completeness: We use the fact that for any set $\mathcal{A} \subseteq \{0,1\}^m$ of size $|\mathcal{A}| \geq 2^{m-\frac{1}{2}}$, there there exists an (expected polynomial time computable) set of m hash functions $h_1, ..., h_m \in \mathcal{H}_{k,m}$ such that $\cup_i h_i(\mathcal{A}) = \{0,1\}^m$. (We include a proof of this fact using the Probabilistic Method in the full version of this paper). When S has at least one distinct allele from C, then $|Z^\ell| = |V(S \cup C)^\ell| \geq (\frac{3}{2})^\ell \cdot |V(C)^\ell| \geq 2^{m-\frac{1}{2}}$. Given any $\mathbf{z}$ there is at least one h_i among the m preselected functions that satisfies $h_i(\mathbf{z}) = y$. Since $\mathbf{r}(h_i)(\mathbf{z} + \mathbf{r}) = h_i(\mathbf{z}) = y$, there exists $h \in Com_\mathcal{H}$ such that $h(\mathbf{v}) = y$.

Soundness error 1/2: When S does not contain any distinct alleles from C, then $|Z| = |V(C)|$. For any set of m hash functions $\{h_i\}_{i=1}^m$, the size of

---

[7] POL observes that the same primer is applied to $C_{i,j}$ and $S_{i,j}$.

$\cup_{i=1}^{m} h_i(Z^\ell)$ is at most $m \cdot |V(C)|^\ell \leq m \cdot (\frac{3}{2})^{-\ell} \cdot 2^m \leq \frac{m}{2m} \cdot 2^m = 2^{m-1}$. Thus, the probability that a uniformly selected target $y \in \{0,1\}^m$ is in the image $\cup_i h_i(Z^\ell)$ is at most $\frac{1}{2}$. Note that this soundness error bound is independent of the prover's mod δ-CE device behavior, whether randomized or deterministic. The device output ensemble consists of at most $2^{m-1}/m$ distinct random variables over $\{0,1\}^m$, i.e. one variable X_z for each input configuration $z \in Z^\ell$. Given the uniformly selected target y, the probability that $h_i(X_z) = y$ for some i and z is (by a union bound) at most $1/2$.

Zero-Knowledge: We separately analyze the physical (DEF reveals $\mathbf{v}$ to POL) and computational (DEF proves there exists $x \in Com_{\mathcal{H}}$ such that $\mathtt{dec}(x) = h$ and $h(\mathbf{v}) = y$) stages of the protocol. In the physical stage, POL only sees the indices of $\mathbf{v} = \mathbf{z} + \mathbf{r}$, and since $\mathbf{r}$ is uniformly distributed independent of $\mathbf{z}$, $\mathbf{v}$ is also uniformly distributed. In the computational stage, DEF and POL execute a ZK protocol that is UC-secure under either physical assumptions or computational setup assumptions. The composed protocol securely realizes PhysicalZK by the universal composition theorem.

7 Conclusion and Future Directions

The need for privacy pervades not only the world of digital information, but physical information as well. Privacy in nuclear disengagement treaties and DNA profiling are just two examples of pertinent real world problems requiring inherently physical rather than digital solutions, and motivate the importance of developing a better theoretical foundation for physical cryptography.

A starting point is the rigorous analysis of protocols. The approach presented in this work separates the logical and physical components of a protocol using the language of modern cryptography, formally reducing mathematical claims of security and correctness to the underlying physical assumptions theory cannot address. Beyond that, there are structural questions: are there ZK proofs for every physical property, or secure computation protocols for every physical task? We noted parallels between public observation physical ZK protocols and public coin digital ZK protocols. Can any physical ZK protocol be made into a publicly observable one?

In the physical world, opposite to the digital, general theories and impossibility results seem difficult or impossible to achieve with only the tools of mathematics. Nonetheless, investigating general theories is an interesting direction for future work, perhaps beginning with restricted classes of physical operations. An orthogonal direction is to explore other models. In subsequent work, we show several techniques for solving generic physical tasks using a *disposable circuits model* in which digital information can be destroyed.

Acknowledgements. We are grateful to Tamir Biezuner, who helped us navigate the procedures involved in DNA profiling, and in particular, helped us adapt the *bins and balls quadruples* technique to DNA profiling using randomized primer selection. We also thank the anonymous reviewers for their helpful suggestions and comments.

References

1. Aumann, Y., Lindell, Y.: Security against covert adversaries: Efficient protocols for realistic adversaries. In: Vadhan, S.P. (ed.) TCC 2007. LNCS, vol. 4392, pp. 137–156. Springer, Heidelberg (2007)
2. Brands, S., Chaum, D.: Distance-bounding protocols (extended abstract). In: Helleseth, T. (ed.) EUROCRYPT 1993. LNCS, vol. 765, pp. 344–359. Springer, Heidelberg (1994)
3. Brzuska, C., Fischlin, M., Schröder, H., Katzenbeisser, S.: Physically uncloneable functions in the universal composition framework. In: Rogaway, P. (ed.) CRYPTO 2011. LNCS, vol. 6841, pp. 51–70. Springer, Heidelberg (2011)
4. Canetti, R.: Universally composable security: A new paradigm for cryptographic protocols. In: 42nd FOCS. Full version at Cryptology ePrint Archive, Report 2001/055, pp. 136–145. IEEE (October 2001), http://eprint.iacr.org/2001/055
5. Canetti, R., Fischlin, M.: Universally composable commitments. In: Kilian, J. (ed.) CRYPTO 2001. LNCS, vol. 2139, pp. 19–40. Springer, Heidelberg (2001)
6. Canetti, R., Kushilevitz, E., Lindell, Y.: On the limitations of universally composable two-party computation without set-up assumptions. In: Biham, E. (ed.) EUROCRYPT 2003. LNCS, vol. 2656, pp. 68–86. Springer, Heidelberg (2003)
7. Gennaro, R., Lysyanskaya, A., Malkin, T., Micali, S., Rabin, T.: Algorithmic tamper-proof (ATP) security: Theoretical foundations for security against hardware tampering. In: Naor, M. (ed.) TCC 2004. LNCS, vol. 2951, pp. 258–277. Springer, Heidelberg (2004)
8. Glaser, A., Barak, B., Goldston, R.J.: A new approach to nuclear warhead verification using a zero-knowledge protocol. In: 53rd Annual INMM Meeting. Institute of Nuclear Materials Management (2012)
9. Goldreich, O.: Foundations of Cryptography: Basic Tools, vol. 1. Cambridge University Press, Cambridge (2001)
10. Goldreich, O., Mansour, Y., Sipser, M.: Interactive proof systems: Provers that never fail and random selection. In: 28th IEEE Symposium on Foundations of Computer Science, pp. 449–461 (1987)
11. Goldreich, O., Micali, S., Wigderson, A.: Proofs that yield nothing but their validity or all languages in NP have zero-knowledge proof systems. J. ACM 38(3), 691–729 (1991)
12. Goldreich, O., Ostrovsky, R.: Software protection and simulation on oblivious rams. J. ACM 43(3), 431–473 (1996)
13. Goldreich, O., Sahai, A., Vadhan, S.: Honest verifier statistical zero-knowledge equals general statistical zero-knowledge. In: 30th Annual ACM Symposium on Theory of Computing, pp. 399–408 (May 1998)
14. Goldreich, O., Vadhan, S.: Comparing entropies in statistical zero-knowledge with applications to the structure of szk. In: 14th Annual IEEE Conference on Computational Complexity, pp. 54–73 (May 1999)
15. Goldwasser, S., Kalai, Y.T., Rothblum, G.N.: One-time programs. In: Wagner, D. (ed.) CRYPTO 2008. LNCS, vol. 5157, pp. 39–56. Springer, Heidelberg (2008)
16. Goldwasser, S., Micali, S., Rackoff, C.: The knowledge complexity of interactive proof systems. SIAM Journal on Computing 18(1), 186–208 (1989)
17. Goldwasser, S., Sipser, M.: Private coins versus public coins in interactive proof systems. In: STOC, pp. 59–68 (1986)
18. Goyal, V., Ishai, Y., Sahai, A., Venkatesan, R., Wadia, A.: Founding cryptography on tamper-proof hardware tokens. In: Micciancio, D. (ed.) TCC 2010. LNCS, vol. 5978, pp. 308–326. Springer, Heidelberg (2010)

19. Gradwohl, R., Naor, M., Pinkas, B., Rothblum, G.N.: Cryptographic and physical zero-knowledge proof systems for solutions of sudoku puzzles. Theory Comput. Syst. 44(2), 245–268 (2009)
20. Hazay, C., Lindell, Y.: Constructions of truly practical secure protocols using standardsmartcards. In: ACM CCS 2008, pp. 491–500. ACM Press (2008)
21. Hofheinz, D., Müller-Quade, J., Unruh, D.: Universally composable zero-knowledge arguments and commitments from signature cards. In: 5th Central European Conference on Cryptology MoraviaCrypt 2005 (2005)
22. Katz, J.: Universally composable multi-party computation using tamper-proof hardware. In: Naor, M. (ed.) EUROCRYPT 2007. LNCS, vol. 4515, pp. 115–128. Springer, Heidelberg (2007)
23. Moran, T., Naor, M.: Basing cryptographic protocols on tamper-evident seals. In: Caires, L., Italiano, G.F., Monteiro, L., Palamidessi, C., Yung, M. (eds.) ICALP 2005. LNCS, vol. 3580, pp. 285–297. Springer, Heidelberg (2005)
24. Moran, T., Naor, M.: Polling with physical envelopes: A rigorous analysis of a human-centric protocol. In: Vaudenay, S. (ed.) EUROCRYPT 2006. LNCS, vol. 4004, pp. 88–108. Springer, Heidelberg (2006)
25. Moran, T., Segev, G.: David and goliath commitments: UC computation for asymmetric parties using tamper-proof hardware. In: Smart, N.P. (ed.) EUROCRYPT 2008. LNCS, vol. 4965, pp. 527–544. Springer, Heidelberg (2008)
26. Naor, M., Naor, Y., Reingold, O.: Applied kid cryptography or how to convince your children you are not cheating. Journal of Craptology (1) (April 1999)
27. Okamoto, T.: On relationships between statistical zero-knowledge proofs. In: 28th ACM STOC, pp. 649–658 (May 1996)
28. Vadhan, S.P.: An unconditional study of computational zero knowledge. SIAM Journal on Computing, 176–185 (2004)

A   Full Security Proof Example: GBG ϵ-Knowledge

The basic Bins and Balls Equality (BBE) protocol of [8] is described in Section 2. The protocol does not actually achieve PhysicalZK, but it does achieve the closely related notion of ϵ-knowledge. Here $\epsilon = n/N$, where n is the capacity of the input bins, and N is the range of the random number of balls added.

Definition 2. *A protocol ρ UC ϵ-emulates an ideal functionality $\mathcal{T}$ if for any $\mathcal{A}$ there exists $\mathcal{S}$ such that for all environments $\mathcal{Z}$ outputting a single bit: $\Delta(\mathrm{REAL}_{\rho,\mathcal{A},\mathcal{Z}}, \mathrm{IDEAL}_{\mathcal{T},\mathcal{S},\mathcal{Z}}) \leq \epsilon$, where Δ denotes statistical variation distance. Equivalently, $|Pr[\mathrm{REAL}_{\rho,\mathcal{A},\mathcal{Z}} = 1] - Pr[\mathrm{IDEAL}_{\mathcal{T},\mathcal{S},\mathcal{Z}} = 1]| \leq \epsilon$*

*A physical protocol π is a **physical ϵ-knowledge** protocol for property Π with respect to the physical operation set $\mathcal{C}$ if its $\mathcal{F}_{\mathcal{C}}$-hybrid translation ρ UC ϵ-emulates $\mathcal{F}_{ZK}^{\Pi}$.*

Hybrid World Modeling of the GBG Protocol. The operations of the hybrid model compiler $\mathcal{F}_{\mathsf{BB}}$ are listed in Figure 4. A bin in the hybrid world is represented by a tuple of the form $(id, value, holder, state)$ stored by $\mathcal{F}_{\mathsf{BB}}$. Every operation listed models a real world operation used in the GBG protocol. In the real world protocol, parties can monitor each other by seeing who is holding or operating on a bin. To model this in the hybrid world, $\mathcal{F}_{BB}$ allows any party to request the $(id, holder, state)$ of any tuple. Additionally, $\mathcal{F}_{BB}$ notifies all parties of any operation executed, and its status (accept or reject). The hybrid world GBG protocol is described in Figure 5.

Hybrid model compiler $\mathcal{F}_{BB}$

For the following descriptions, each command is sent from party P_i to $\mathcal{F}_{BB}$.

- **Create** $(value)$: *Initializes a new bin with value number of balls.* Generate a unique id for the bin, store the tuple $(id, value, P_i, "open")$, and send the receipt $(id, value, P_i, "open")$ to P_i.
- **Seal** (id): *Seals a bin.* Check in the tuple $(id, value, holder, state)$ that $holder = P_i$ and $state = "open"$. If yes, updates the tuple to $(id, value, P_i, "sealed")$. Otherwise reject.
- **Break** (id): *Unseals a bin.* Check in the tuple $(id, value, holder, state)$ that $holder = P_i$. If yes, updates the tuple to $(id, value, P_i, "open")$. Otherwise reject.
- **Combine** (id_1, id_2): *Combines the contents of two bins.* Check that $holder_1 = holder_2 = P_i$, $state_1 = state_2 = "open"$. If yes, then update the tuples to $(id_1, 0, P_i, "open")$ and $(id_2, value_1 + value_2, P_i, "open")$. Otherwise reject.
- **Open** (id): *Opens and returns the bin contents count.* Check in $(id, value, holder, state)$ that $state = "open"$ and $holder = P_i$. If yes, send $(id, value)$ to P_i. Otherwise reject.
- **Send** (m, P_j): *Relays message to P_j.* Relay the message m to P_j. In the physical setting this could be implemented using a number of equivalent forms of communication (speech, writing, etc.)
- **Transfer** (id, P_j): *Transfers bin possession to P_j.* Check in $(id, value, holder, state)$ that $holder = P_i$. If yes, update the tuple to $(id, value, P_j, state)$.
- **Force** (id): *Forcefully opens a bin.* Send the entire tuple $(id, value, holder, state)$ to P_i.

Fig. 4. The $\mathcal{F}_{BB}$-hybrid model ideal functionalities for BB

Ideal Functionality Modeling for BBE. Recall the definition of the general PhysicalZK ideal functionality $\mathcal{F}_{ZK}^{\Pi}$ in Figure 1 of Section 3. The corresponding ideal functionality of a PhysicalZK protocol for BBE is $\mathcal{F}_{ZK}^{BBE}$, where BBE denotes the physical property of bins containing equal numbers of balls. The GBG protocol, however, has a non-negligible soundness error of $1/2$. While normally the error would be exponentially reduced by repetition, this protocol cannot be repeated on the same physical input. The input is consumed when new balls are added to the input bins. Thus, we will use a modified functionality $\mathcal{F}_{ZK,\delta}^{BBE}$ that allows the prover to cheat with probability at most δ (Figure 6).

$\mathcal{F}_{ZK,\delta}^{BBE}$ runs with a party Prover, a party Verifier, and an instance of $\mathcal{F}_{BB}$ specified by a process identifier pid_{BB}. If $(id_X, x, holder, state)$ and $(id_Y, y, holder, state)$

$\mathcal{F}_{\mathsf{BB}}$-**hybrid protocol** π

Input: $\mathcal{F}_{\mathsf{BB}}$ has the input stored as $(id_X, x, \mathsf{Prover}, \text{``sealed''})$ and $(id_Y, y, \mathsf{Prover}, \text{``sealed''})$ where $0 \leq x, y \leq n$.

If $x \neq y$, Prover sends $(\mathtt{Reject}, (id_X, id_Y), \mathsf{BBE})$ to Verifier. If $x = y$:

1. Prover: Randomly select r_0, r_1 in $[0, N)$. **Create** (r_0) twice and **Create** (r_1) twice, receiving receipts from $\mathcal{F}_{\mathsf{BB}}$ with four unique id values $B_{0,0}, B_{0,1}, B_{1,0}$, and $B_{1,1}$. For each $B_{i,j}$, **Seal**$(B_{i,j})$ and **Send**$(B_{i,j}, \mathsf{Verifier})$.
2. Verifier: Select a random choice bit $\sigma \in \{0, 1\}$. **Send** $(\sigma, \mathsf{Prover})$.
3. Prover: **Break**$(B_{1-\sigma,0})$, and **Break**$(B_{1-\sigma,1})$. **Transfer** $(B_{1-\sigma,0}, \mathsf{Verifier})$ and **Transfer** $(B_{1-\sigma,1}, \mathsf{Verifier})$.
4. Verifier: **Open** $(B_{1-\sigma,0})$ and **Open** $(B_{1-\sigma,1})$, receiving from $\mathcal{F}_{\mathsf{BB}}$ two receipts $(B_{1-\sigma,0}, v)$ and $(B_{1-\sigma,1}, v')$. Check that $v = v'$. If $v \neq v'$, output (**Cheater**, Prover).
5. Prover: **Combine** $(id_X, B_{\sigma,0})$, and **Combine**$(id_Y, B_{\sigma,1})$. **Transfer** $(B_{\sigma,0}, \mathsf{Verifier})$ and **Transfer** $(B_{\sigma,1}, \mathsf{Verifier})$.
6. Verifier: **Open** $(B_{\sigma,0})$ and **Open** $(B_{\sigma,1})$, receiving $(B_{\sigma,0}, v)$ and $(B_{\sigma,1}, v')$ from $\mathcal{F}_{\mathsf{BB}}$. Check that $v = v'$. If yes, output (**Accept**, $(id_X, id_Y), \mathsf{BBE}$). Otherwise, output (**Reject**, $(id_X, id_Y), \mathsf{BBE}$).

If the protocol prematurely fails due to an invalid message or operation, both parties output **Failed**. If Prover ever receives a receipt (**Force**, Verifier, accept) from $\mathcal{F}_{BB}$, it outputs (**Cheater**, Verifier).

Fig. 5. The $\mathcal{F}_{\mathsf{BB}}$-hybrid model protocol for BBE

are two tuples stored by $\mathcal{F}_{\mathsf{BB}}$, the relation $(id_X, id_Y) \in \mathsf{BBE}$ holds if and only if $x = y$.

Hybrid Protocol Experiment. $\mathcal{Z}$ activates $\mathcal{F}_{\mathsf{BB}}$ and sets the input to π, the $\mathcal{F}_{\mathsf{BB}}$-hybrid protocol, by initializing the tuples $(id_X, x, \mathsf{Prover}, \text{``sealed''})$ and $(id_Y, y, \mathsf{Prover}, \text{``sealed''})$. Next, $\mathcal{Z}$ activates Prover and Verifier, sending them the process identifier pid_{BB} for $\mathcal{F}_{\mathsf{BB}}$. Prover and Verifier execute π as described in Figure 5. $\mathcal{A}$ corrupts one, both, or none of the parties. $\mathcal{A}$ controls the communication of any party it corrupts, but only acts as a proxy for $\mathcal{Z}$. After π has completed, each party sends its output to $\mathcal{Z}$, who's final view of the experiment consists of its communication tape with $\mathcal{A}$ and the outputs of Prover and Verifier. Finally, $\mathcal{Z}$ outputs a single decision bit $\mathrm{HYBRID}_{\pi, \mathcal{F}_{\mathsf{BB}}, \mathcal{A}, \mathcal{Z}}$.

Ideal Process Experiment. $\mathcal{Z}$ activates $\mathcal{F}_{\mathsf{BB}}$ with process identifier pid_{BB} and initializes the tuples $(id_X, x, \mathsf{Prover}, \text{``sealed''})$ and $(id_Y, y, \mathsf{Prover}, \text{``sealed''})$. $\mathcal{Z}$ sends $(id_X, id_Y, pid_{\mathsf{BB}})$ to both Prover and Verifier. Prover will activate $\mathcal{F}_{ZK,\delta}^{\mathsf{BBE}}$, whose PID we denote as $\perp$. First, Prover transfers input tuple access to $\mathcal{F}_{ZK,\delta}^{\mathsf{BBE}}$, executing **Transfer**$(id_X, \perp)$ and **Transfer**$(id_Y, \perp)$. Next, it sends pid_{BB} and the

---

Functionality $\mathcal{F}_{ZK,\delta}^{\mathsf{BBE}}$

- Upon receiving $(id_X, id_Y, pid, \mathsf{Prover}, \mathsf{Verifier})$ from the party Prover, query the instance of $\mathcal{F}_{\mathsf{BB}}$ specified by pid to obtain the tuples $(id_X, x, \bot, \text{``}sealed\text{''})$ and $(id_Y, y, \bot, \text{``}sealed\text{''})$. If $x = y$, send $(\mathtt{Accept}, (id_X, id_Y), \mathsf{BBE})$ to the party $\mathsf{Verifier}$. If $x \neq y$, send $(\mathtt{Reject}, (id_X, id_Y), \mathsf{BBE})$.
- Upon receiving the instruction $(\mathbf{Cheat}, \mu)$ from Prover for $0 \leq \mu \leq \delta$, with probability μ send $(\mathtt{Accept}, (id_X, id_Y), \mathsf{BBE})$ to $\mathsf{Verifier}$, and otherwise send $(\mathbf{Cheater}, \mathsf{Prover})$. Upon receiving the instruction $\mathbf{Cheat}$ from $\mathsf{Verifier}$, send $(\mathbf{Cheater}, \mathsf{Verifier})$ to Prover, and $(\mathbf{Cheater}, x, y)$ to $\mathsf{Verifier}$.
- If the queries to $\mathcal{F}_{\mathsf{BB}}$ are unsuccessful, the initial input is invalid, or upon receiving the instruction $\mathbf{Fail}$ from either party, send $\mathtt{Failed}$ to both parties.

Fig. 6. Ideal functionality for a ZK proof of BBE with soundness error δ

input $(id_X, id_Y, pid_{\mathsf{BB}}, \mathsf{Prover}, \mathsf{Verifier})$ to $\mathcal{F}_{ZK,\delta}^{\mathsf{BBE}}$, which in turn proceeds according to Figure 6. As in [4], $\mathcal{F}_{ZK,\delta}^{\mathsf{BBE}}$ will also mediate the ideal adversary corruption mechanism. $\mathcal{S}$ corrupts a party P by sending the command $(\mathbf{corrupt}\ P)$ to $\mathcal{F}_{ZK,\delta}^{\mathsf{BBE}}$. Since $\mathcal{S}$ cannot corrupt P before $\mathcal{F}_{ZK,\delta}^{\mathsf{BBE}}$ is activated, $\mathcal{S}$ cannot modify the environment's input before $\mathcal{F}_{ZK,\delta}^{\mathsf{BBE}}$ receives it. Thus, while $\mathcal{S}$ may modify outputs to its corrupted parties, it cannot compromise the ideal process's output to an uncorrupted party. Prover and Verifier both forward the outputs they receive to $\mathcal{Z}$, and $\mathcal{Z}$ outputs a single decision bit $\mathrm{IDEAL}_{\mathcal{F}_{ZK}^{\mathsf{BBE}}, \mathcal{F}_{\mathsf{BB}}, \mathcal{S}, \mathcal{Z}}$.

Lemma 1. *The $\mathcal{F}_{\mathsf{BB}}$-hybrid protocol π of Figure 5 UC ϵ-emulates the ideal functionality $\mathcal{F}_{ZK,\delta}^{\mathsf{BBE}}$ with $\delta = 1/2$ and $\epsilon = n/N$. In other words, for any $\mathcal{A}$ there exists $\mathcal{S}$ such that for any $\mathcal{Z}$:*

$$|Pr[\mathrm{HYBRID}_{\pi, \mathcal{F}_{\mathsf{BB}}, \mathcal{A}, \mathcal{Z}} = 1] - Pr[\mathrm{IDEAL}_{\mathcal{F}_{ZK,\delta}^{\mathsf{BBE}}, \mathcal{F}_{\mathsf{BB}}, \mathcal{S}, \mathcal{Z}} = 1]| \leq \epsilon$$

Proof. We use the following simple fact. Given any distinguisher algorithm D and random variables X and Y with finite range Ω:

$$|Pr[D(X) = 1] - Pr[D(Y) = 1]| \leq \Delta(X, Y)$$

We will show that for all $\mathcal{A}$ there exists $\mathcal{S}$ such that for any environment $\mathcal{Z}$, the environment's respective views in the hybrid protocol experiment with $\mathcal{A}$ and the ideal process experiment with $\mathcal{S}$ are ϵ-close in statistical distance. We consider separately the four cases in which $\mathcal{A}$ corrupts the Prover, the Verifier, both parties, or neither parties. (We continue to write $\mathcal{F}_{ZK,\delta}^{\mathsf{BBE}}$, but it should be understood that $\delta = 1/2$).

$\mathcal{A}$ **corrupts Prover.** $\mathcal{S}$ obtains (id_X, id_Y, pid_{BB}) from $\mathcal{F}_{ZK,\delta}^{\mathsf{BBE}}$. $\mathcal{S}$ runs a separate instance of $\mathcal{F}_{\mathsf{BB}}$, and simulates the hybrid protocol π using empty entries for id_X and id_Y, and two dummy parties P (for Prover) and V (for Verifier). $\mathcal{S}$ plays P using the messages coming from $\mathcal{Z}$. $\mathcal{S}$ sends back to $\mathcal{Z}$ any receipts that P receives from $\mathcal{F}_{\mathsf{BB}}$. However, $\mathcal{S}$ does not send P's output from the simulation $\mathcal{Z}$. Instead, it sends one of the following messages to $\mathcal{F}_{ZK,\delta}^{\mathsf{BBE}}$:

- If either P or V output `Failed`, then S sends **Fail**.
- If P cheated in the initialization of only one pair, then S sends (**Cheat**, $1/2$).
- If P cheated in the initialization of both pairs, then S sends (**Cheat**, 0).
- If none of the above apply, then S sends $(id_X, id_Y, pid_{BB}, \text{Prover}, \text{Verifier})$.

Verifier writes the output received from $\mathcal{F}_{ZK,\delta}^{BBE}$ to its local output tape. S receives Prover's output from $\mathcal{F}_{ZK,\delta}^{BBE}$, and writes it to Prover's output tape. Z's view is identical to its view in the hybrid protocol experiment with $\mathcal{A}$ corrupting Prover.

$\mathcal{A}$ **corrupts Verifier.** S obtains Z's input (id_X, id_Y, pid_{BB}), and receives output from $\mathcal{F}_{ZK,\delta}^{BBE}$: either $(\text{Accept}, (id_X, id_Y), BBE)$ in the case that $x = y$, or $(\text{Reject}, (id_X, id_Y), BBE)$ in the case that $x \neq y$. If `Reject`, then S writes the output to Verifier's output tape. If `Accept`, then S must simulate $\mathcal{A}$'s view of the hybrid protocol π.

S runs a separate instance of $\mathcal{F}_{BB}$, creating empty entries for id_X and id_Y, and dummy parties P and V. S uses messages coming from Z to play V, and forwards any receipts that V receives to Z. The only possible receipts V receives that could be *statistically different* in this simulated π and the hybrid experiment π are the receipts $(B_{\sigma,0}, v_\pi)$ and $(B_{\sigma,1}, v_\pi')$ obtained in Step 6. Since the output is `Accept`, we know that $v_\pi = v_\pi'$. v_π is uniformly distributed in $[x, x + N)$, and the simulation output v_{sim} is uniformly distributed in $[0, N)$. The statistical difference is $\Delta(v_\pi, v_{sim}) = x/N \leq n/N = \epsilon$.

Next, S determines what message to send $\mathcal{F}_{ZK,\delta}^{BBE}$. If P outputs **Cheater**, it sends **Cheat** to $\mathcal{F}_{ZK,\delta}^{BBE}$, receives (**Cheater**, x, y) as output, and writes to Verifier's output tape whatever $\mathcal{A}$ would. If P outputs `Failed`, S sends **Fail**, and writes `Failed` to Verifier's output tape. Otherwise, S does not send anything, and simply writes $(\text{Accept}, (id_X, id_Y), BBE)$ to Verifier's output tape.

The outputs are identical to the outputs in the hybrid protocol experiment given the same inputs. Therefore, the statistical difference between Z's views of the hybrid protocol experiment and ideal process experiment on the same inputs is precisely the statistical difference in its communication with $\mathcal{A}$ and S during π and the simulated π, which is at most $\epsilon = n/N$.

$\mathcal{A}$ **Corrupts Both or Neither Parties.** If both, then S also corrupts both. Since there are no secrets kept from S, it can run the hybrid world experiment without help. If neither, S does nothing. For the same inputs, the outputs of Prover and Verifier are identical in the hybrid protocol experiment and the ideal process experiment when there is no corruption.

By Lemma 1, the GBG protocol is a physical ϵ-knowledge protocol for BBE with respect to the operation set BB.

B   ORE with a Mod-Counter

We present an alternative zero-knowledge protocol for ORE using a *mod-counter*, a measurement device that outputs neutron counts modulo N. When N is greater than the maximum possible neutron count, the neutron counts are equal if and only if they are congruent modulo N.

Who brings the mod-counter $\mathcal{M}$? The verifier cannot trust a prover's device to output correct values. Likewise, the prover cannot trust the verifier, who might program the device to secretly store actual integer count values. Thus, our solution is to have the verifier *program check* the prover's mod-counter. We model $\mathcal{M}$ as an adaptive program computing a sequence of functions $\{f_i\}$ on a sequence of inputs d_i, where each f_i is a function of $d_1, ..., d_i$, and each d_i is a physical neutron detector. We assume that the prover cannot remotely change the device's state once the verifier takes possession of it.

1. **Prover's preparation:** The prover prepares and labels m detector pairs. In each pair $1 \leq i \leq m$, the prover initializes the count of both detectors, d_i^1 and d_i^2, to a random $r_i \in [0, N)$. (Only a dishonest prover may set d_i^1 to r_i^1 and d_i^2 to r_i^2 such that $r_i^1 \neq r_i^2$).
2. **Verifier's tests:** We assume the verifier has a device to obtain the neutron count of any detector. For each ith detector pair, the verifier performs *RandomCompute* with probability 1/5, and otherwise runs *RandomTest*:
   - *RandomCompute.* Choose random values $y_i^1, y_i^2 \in [0, N)$. Increase the count of d_i^1 by y_i^1 to $r_i^1 + y_i^1$, and increase the count of d_i^2 to $r_i^2 + y_i^2$. Select random θ_i, and run the neutron source, recording the scattering at θ_i from X using d_i^1 and from X' using d_i^2. Query $\mathcal{M}$ with d_i^1 and d_i^2. The test passes if and only if $\mathcal{M}(d_i^1) - y_i^1 = \mathcal{M}(d_i^2) - y_i^2$.
   - *RandomTest.* Measure the detectors to uncover r_i^1 and r_i^2, and reject the protocol if $r_i^1 \neq r_i^2$. Select random $t_i^1, t_i^2 \in [0, 2N)$. Increase the count of d_i^1 to $r_i^1 + t_i^1$, and increase the count of d_i^2 to $r_i^2 + t_i^2$. Query $\mathcal{M}$ with d_i^1 and d_i^2. The test passes if and only if $\mathcal{M}(d_i^1) \equiv r_i^1 + t_i^1 \pmod{N}$ and $\mathcal{M}(d_i^2) \equiv r_i^2 + t_i^2 \pmod{N}$.

Completeness: When $x_{\theta_i} = x'_{\theta_i}$ for all i, then an honest prover will pass all rounds. Since $\mathcal{M}$ behaves correctly, it will pass every RandomTest. For all i, the prover sets $r_i^1 = r_i^2 = r_i$ so that in RandomCompute $\mathcal{M}(x_{\theta_i} + r_i + y_i^1) - y_i^1 = \mathcal{M}(x'_{\theta_i} + r_i + y_i^2) - y_i^2 \equiv x_{\theta_i} + r_i \pmod{N}$.

Zero-Knowledge: If the prover is honest, then on every round the initialization values are $r_i^1 = r_i^2 = r_i$ and the neutron counts collected from X and X' and angle θ_i are $x_{\theta_i} = x'_{\theta_i} = x_i$. Since r_i is uniformly distributed in $[0, N)$, $x_i + r_i \pmod{N}$ is also uniformly distributed in $[0, N)$, independent of x_i.

Soundness: The success probability of a cheating prover is bounded by $(\frac{4+\beta}{5})^m$ where $\beta < 1$ is an upper bound on the probability that $x_\theta = x'_\theta$ at a uniformly distributed angle θ when $X \not\sim X'$.

Proof: To simplify, assume $\mathcal{M}$'s input on each round is the *pair* of detectors. In reality, $\mathcal{M}$ is actually weaker, since it cannot examine the second detector before outputting a response for the first. We'll work with the hybrid model $\mathcal{M}$, which is an oracle taking integer pair inputs. Denote the output $\mathcal{M}(a,b) = (\mathcal{M}_1(a,b), \mathcal{M}_2(a,b))$. Further, assume that the prover knows the values of x_θ and x'_θ at every angle θ, and can predict the sequence of angles $\{\theta_i\}_{i=1}^{2m}$ that the verifier selects. (Soundness with a stronger prover holds against a weaker one). In calculating our upper bound we only consider deterministic $\mathcal{M}$ strategies since the verifier's strategy is independent and for any fixed protocol input there is a deterministic strategy for $\mathcal{M}$ that maximizes the soundness error.

Case 1: $x_{\theta_i} = x'_{\theta_i}$, occurs with probability β. The prover doesn't benefit from cheating. Case 2: $x_{\theta_i} \neq x'_{\theta_i}$, the prover's strategies are to either cheat on the initializations so that $x_{\theta_i} + r_i^1 = x'_{\theta_i} + r_i^2$, or instead to keep $r_i^1 = r_i^2$, and program $\mathcal{M}$ to cheat. In the latter case, the prover preselects the $k_i \leq N^2$ pairs for which $\mathcal{M}$ should pass RandomCompute. Setting $\mathcal{M}(a,b) = (a \bmod N, b + x_{\theta_i} - x'_{\theta_i} \bmod N)$ results in $k_i = N^2$.

If γ_i is the conditional probability that the prover passes RandomCompute on the *i*th round given that $x_{\theta_i} \neq x'_{\theta_i}$, then Lemma 2 implies the prover succeeds in round i with probability at most: $\beta + (1-\beta)\left(\frac{1}{5}\gamma_i + \frac{4}{5}(1 - \frac{\gamma_i}{4})\right) = \beta + (1-\beta)\frac{4}{5} = \frac{4+\beta}{5}$.

Lemma 2. *The prover's probability of passing RandomTest on the ith round given that $x_{\theta_i} \neq x'_{\theta_i}$ is at most $1 - \gamma_i/4$, and the optimal prover strategy achieves this probability exactly.*

Proof. Consider first the case where $r_i^1 = r_i^2 = r_i$. In this case, $\gamma_i = \frac{k_i}{N^2}$, where k_i is the number of pairs (y_i^1, y_i^2) in $[0,N) \times [0,N)$ for which $\mathcal{M}_1(x_{\theta_i} + r_i^1 + y_i^1) - y_i^1 = \mathcal{M}_2(x'_{\theta_i} + r_i^2 + y_i^2) - y_i^2$. For each of these k_i pairs, let $t_i^1 = x_{\theta_i} + y_i^1$ and $t_i^2 = x'_{\theta_i} + y_i^2$. Either $\mathcal{M}_1(t_i^1 + r_i) \neq t_i^1 \pmod{N}$ or $\mathcal{M}_2(t_i^2 + r_i) \neq t_i^2 \pmod{N}$, and so $\mathcal{M}$ would fail RandomTest if the verifier chooses the pair (t_i^1, t_i^2). Thus, there are at least k_i distinct pairs in $[0,2N) \times [0,2N)$, that cause $\mathcal{M}$ to fail RandomTest, and the prover's conditional probability of passing RandomTest is at most $1 - \frac{k_i}{(2N)^2} = 1 - \frac{\gamma_i}{4}$. The prover can program $\mathcal{M}$ to pass (i.e. behave normally) on all other $4N^2 - k_i$ pairs, so there is a prover strategy that passes RandomTest with probability exactly $1 - \frac{\gamma_i}{4}$. We will show that this strategy is optimal. Consider the second case where the prover cheats by setting r_i^1 and r_i^2 so that $x_{\theta_i} + r_i^1 = x'_{\theta_i} + r_i^2$. With this strategy, $\gamma_i = 1$, but since $r_i^1 \neq r_i^2$, the prover always fails RandomTest, failing the overall round with probability $1/2$. This strategy is suboptimal because the previous strategy passes with probability $3/4$ when setting $k_i = N^2$ so that $\gamma_i = 1$.

Client-Server Concurrent Zero Knowledge with Constant Rounds and Guaranteed Complexity*

Ran Canetti[1,**], Abhishek Jain[2], and Omer Paneth[3,***]

[1] Boston University, MA, USA and Tel-Aviv University, Israel
canetti@bu.edu
[2] Boston University, MA, USA and MIT, Cambridge, MA, USA
abhishek@csail.mit.edu
[3] Boston University, MA, USA
omer@bu.edu

Abstract. The traditional setting for concurrent zero knowledge considers a server that proves a statement in zero-knowledge to multiple clients in multiple concurrent sessions, where the server's actions in a session are *independent* of all other sessions. Persiano and Visconti [ICALP 05] show how keeping a limited amount of global state across sessions allows the server to significantly reduce the overall complexity while retaining the ability to interact concurrently with an unbounded number of clients. Specifically, they show a protocol that has only slightly super-constant number of rounds; however the communication complexity in each session of their protocol depends on the number of other sessions and has no a-priori bound. This has the drawback that the client has no way to know in advance the amount of resources required for completing a session of the protocol up to the moment where the session is completed.

We show a protocol that does not have this drawback. Specifically, in our protocol the client obtains a bound on the communication complexity of each session at the start of the session. Additionally the protocol is *constant-rounds*. Our protocols is fully concurrent, and assumes only collision-resistant hash functions. The proof requires considerably different techniques than those of Persiano and Visconti. Our main technical tool is an adaptation of the "committed-simulator" technique of Deng et. al [FOCS 09].

1 Introduction

Concurrent security of a protocol means that security is preserved even when many copies of the protocol may be executed concurrently with each other and with other, potentially unknown protocols. Concurrent security is essential for

---

* This paper is supported by the NSF EAGER grant, and NSF Algorithmic Foundations grant no. 1218461.
** Supported by the Check Point Institute for Information Security.
*** Supported by the Simons award for graduate students in theoretical computer science.

J.A. Garay and R. Gennaro (Eds.): CRYPTO 2014, Part II, LNCS 8617, pp. 337–350, 2014.
© International Association for Cryptologic Research 2014

protocols designed for modern networks, such as the Internet. However, it often imposes a cost on the complexity of the protocol. For example, stand-alone zero-knowledge protocols can be implemented in a constant number of rounds based on any one way function, while constant-round concurrent zero-knowledge protocols are not known without relying on non-standard assumptions or trusted setup.

Concurrent Zero Knowledge. The *concurrent zero knowledge* task [8] considers a natural and special case of concurrent security. Here there is a server that wants to prove theorems in zero-knowledge [9] to multiple clients (verifiers). For that purpose, the server runs an instance (i.e., a *session*) of a protocol with each client. There may be an unbounded (albeit polynomial) number of sessions, and sessions may execute concurrently with adversarially controlled delay and ordering of messages. Furthermore, the prover side of each session should execute without knowledge of any other session. This simplifies the design for the server and allows the prover side to be executed on separate machines without coordination. Still, for security we only consider two cases: one where all or some provers are corrupted, and one where all or some of the verifiers are corrupted. While the concurrent zero-knowledge setting is a substantial restriction of general composition it distills an important aspect of the general challenge of concurrent security. Indeed, this setting was extensively studied, with special attention to minimizing the number of rounds [18,13,17,3,12,6,14]. Furthermore, techniques developed for concurrent zero knowledge have been found useful in the study of more general concurrent systems (see e.g., [5]).

The state of the art for protocols based on standard assumptions is $\Omega(\log n)$ rounds, where n is the security parameter. Furthermore, for protocols with black-box simulation we know that $\tilde{\Omega}(\log n)$ is the best possible.

Correlated Provers. Persiano and Visconti [16] consider a relaxed variant of the classic concurrent zero knowledge model, where the server is allowed to somewhat correlate its strategies in the different sessions. Here one has to make sure that the correlation is on the one hand effectively implementable by the server, and on the other hand preserves the overall efficiency and performance from the point of view of the client. Specifically, they present a zero-knowledge protocol where the server keeps track of the *number* of currently open sessions at any time. It then starts off each session to have a constant number of messages whose length depends polynomially on the number of currently open sessions. If the number of sessions increases beyond some threshold before the session is over, the session has to be "re-done" with longer messages. Overall, it is guaranteed that if n^c sessions are executed concurrently to a session, then the protocols of [16] requires $O(c)$ rounds and $n^{O(c)}$ communication for that session.

The global state to be kept by the server in this protocol is indeed minimal and reasonable. Additionally, the number of rounds in every session grows very slowly with the number of sessions, significantly improving the best known "pure" concurrent zero-knowledge protocols (as long as the total number of sessions is polynomial). However, this protocol has the strong disadvantage that a client has no way of knowing, at any point during the protocol execution, how much communication it will need in order to complete the session.

This Work. We present a new concurrent zero-knowledge protocol where, like the [16] protocol, the server keeps track of the number of sessions currently open. Our protocol improves upon the protocol of [16] in two ways:

- **Constant rounds.** Our protocol takes six messages, regardless of the number of concurrent sessions.
- **Guaranteed complexity.** In our protocol, the server announces in the beginning of every session the communication complexity of the session. The server cannot dynamically increase the communication complexity of a session to accommodate new clients that arrive during the session's execution.

The Importance of Guaranteed Complexity. The advantage of having guaranteed complexity is best explained by an analogy: Consider a customer that is placing a call to a call center and is being put on hold. The customer's waiting is likely to become more endurable and efficient if the call center commits to (or estimates) the required waiting time at the beginning of the call. In our setting, the client's resource is communication rather than waiting time. Clearly, clients benefit from knowing ahead of time how much communication is required from them to participate in the protocol. For example, a client with limited communication resources would prefer to learn ahead of time that its resources are insufficient to complete the protocol, rather than during the session after all its resources have already been spent.

The Protocol of [16]. The protocol of [16] is based on the *bounded concurrent* protocol of Barak [1]. Barak's protocol is secure as long the number of concurrent sessions does not exceed some bound that depends on the communication complexity of the protocol. Very roughly, Persiano and Visconti show that it is possible to add rounds to the protocol and increase its communication "on-the-fly" as new occurrent sessions start. However, as a result, the round complexity of their protocol must depend on the number of sessions, and the server cannot guarantee the complexity of any session ahead of time.

It may seem that bounded concurrency is of no use for designing protocols with guaranteed complexity. Indeed, when the server commits the communication complexity of, say, the first session, it has no bound on the number of sessions that will be started concurrently to the first session.

Our Protocol. Counter to the above intuition, our protocol does leverage bounded concurrency techniques of Barak. However, our approach departs from [16] in the following manner: we set the communication complexity of every session only based on the order in which the sessions *start*. The first n sessions to start execute a bounded concurrent protocol that is secure for n concurrent session. The following $n^2 - n$ sessions execute a bounded concurrent protocol that is secure for n^2 session, and so on. Importantly, the communication complexity of a session is not affected by sessions that start *after* it. This in particular means that the [1,16] simulation technique is inadequate in our setting. Indeed, our security proof differs significantly from that in [1,16].

1.1 Our Techniques

We start by recalling Barak's zero-knowledge protocol and its simulation. Barak's protocol starts with a preamble phase where the prover sends a commitment c and the verifier responds with a random challenge r. Any prover that can commit to a program that predicts r can obtain a "trapdoor" and cheat in the proof phase. The zero-knowledge simulator will be able to obtain a trapdoor by committing to the code of the verifier itself. Next we discuss two approaches for extending Barak's protocol to the concurrent setting.

Bounded Concurrency. In the concurrent setting, the simulator cannot simply commit to the code of the verifier. Indeed, the verifier's code eventually predicts r, but might only do so after receiving convincing proofs in other sessions. Furthermore, when the simulator sends the commitment c in some session, it did not yet compute the proofs in upcoming sessions (in fact, these proofs might depend on c); therefore it cannot commit to such proof together with the verifier's code.

The approach in [1] is to change the protocol as follows: to obtain a trapdoor, the simulator must commit to a program that predicts r given some auxiliary information z (that may be chosen after r is sent). To maintain soundness, z must be much shorter then r. The simulator can now encode the simulated proofs in a bounded number of other sessions into z. This results in a bounded concurrent protocol. As argued above, this technique, on its own, is inadequate for our setting.

Committed Simulator. A different approach, that we will refer to as the "committed simulator" approach, is as follows: even if the number of concurrent sessions is unbounded, the simulated proofs in all these sessions still have a short description, which is the code of the simulator itself. Concretely, in every session, the simulator will commit to a version of itself that simulates the interaction with the verifier in all other sessions until the verifier sends the challenge r in that session.

The problem with this approach is bounding the running time of the simulator. If the simulator commits to itself in the preamble phase, then in the proof phase the simulator will prove a statement on its own execution. This execution might contain the proof phase of in some other sessions where the simulator also proved a statement on its own execution. For some adversarial schedules, such recursive construction of proof becomes too expensive. Nonetheless, variants of the committed simulator approach were successfully applied in many different settings [7,4,11,15,10,6].

Our Approach. Our simulation combines these two approaches to obtain a protocol with constant rounds and guaranteed complexity, assuming only collision resistant hashing. In a nutshell, we leverage the bounded concurrent simulation technique to "flatten" the recursion tree, avoiding the blowup in the simulator's running time. A more detailed description follows.

We start by assigning a *level* to each session. All sessions that execute a bounded concurrent protocol for n^i sessions, are assigned level i. Our protocol is defined such that for every i, the total number of sessions at all levels $\leq i$ is at most n^i. It follows that in every session at level i, the verifier's challenge is long enough to account for all the messages received by the verifier in sessions at levels at most i.

To deal with the messages sent in sessions at levels larger than i, we turn to the committed simulator approach. The main idea is that we can avoid the exponential blowup in the running time of the simulator by committing only to specific *parts* of the simulator that are in charge of simulating the sessions at levels larger than i rather than the entire code of the simulator.

The Simulator. The simulator Sim is divided into multiple components $\{\mathsf{Sim}_i\}$ where the i'th component Sim_i is in charge of simulating sessions at level i. To simulate a session at level i, Sim_i will commit to a program Π_i that contains the verifier's code together with the code of all the simulator's components Sim_j for $j > i$. We can think of the program Π_i as a new verifier that simulates all sessions at levels $> i$ internally and forwards externally the messages in sessions at level $\leq i$. Since sessions at level i execute a bounded concurrent protocol for n^i sessions, and the total number of sessions at levels $< i$ is at most n^i, we have that Sim_i can encode all the messages sent to Π_i as auxiliary input.

Finally, we argue that the running time of the simulator is polynomial. Using the analysis of the bounded concurrent protocol, we have that the running time of the component Sim_i is polynomial in the running time of the program Π_i. Since Π_i simply emulates all the simulator components Sim_j for $j > i$, we have that the running time of Sim_i is only polynomially larger than the total running time of all the components Sim_j for $j > i$. Since the total number of concurrent sessions started by an efficient adversary is bounded by some polynomial n^c, we get that the total number of levels is constant and therefore the running time of all the simulator's components is bounded by a polynomial.

Avoiding Circular Use of Randomness. We note that by using the above leveled simulation strategy we do not only avoid the blowup in the simulator's running time, but also avoid some of the technical complications that arise when the simulator commits to its own code. For example, in [4,10,6], the simulator needs to commit to its own code together with the randomness that it will use to simulate the rest of the protocol. The aforementioned works develop additional techniques to deal with this problem. In our setting, since every component only commits to the randomness used by the *higher* level component, such circular use of randomness is avoided, resulting in simpler protocol and analysis.

Taking Advantage of Terminating Sessions. It is natural to require that, as existing sessions terminate and the load on the server decreases, the complexity of the protocol in new sessions decreases as well. We note that extending our simulation strategy to satisfy this requirement is not straight-forward. The problem is that our simulation strategy assumes that for every session at level i, the total number of concurrent sessions at levels $\leq i$ is bounded by n^i. However, consider the scheduling where all sessions at levels $\leq i$ terminate and a new session starts. If we choose to decrease the protocol complexity in the new session, then the total number of sessions at levels $\leq i$ may exceed n^i. We demonstrate a slightly more complicated server strategy where the complexity of new sessions does decrease as old sessions terminate (while preserving overall simulatability).

1.2   Related Work

Concurrent Zero-Knowledge in the Plain Model. Improving the round-complexity of concurrent zero-knowledge proofs in the plain model has been an active area of research. The round complexity of concurrent zero-knowledge with black-box simulation was studied in [18,13,17], resulting in protocols with logarithmic round-complexity (which is essentially optimal [3]). Constant-round protocols with non-black-box simulation where constructed based on different non-standard assumption such as interactive knowledge assumptions [12], statistically sound P-certificates [6] and differing input (or extractable) obfuscation [14].

Optimistic Concurrent Zero-Knowledge. The work of Rosen and Shelat [19] also studies the round complexity of concurrent zero-knowledge proofs with a correlated prover in the client-server setting. Their focus is on improving the round complexity of concurrent zero-knowledge with respect to "optimistic" adversarial schedules. That is, the round complexity of their protocol significantly decreases when the scheduling of messages does not include too many nested sessions. However, for a worst-case adversarial schedules, [19] give no improvement over logarithmic round-complexity of [17] while our protocol has constant rounds in the worst cast. However, unlike in our protocol, the communication complexity in [19] has a fixed upper bound that is independent of the adversary.

2   The Guaranteed Complexity Model

In this section we formally define a protocol in the guaranteed complexity model. We start by describing the general syntax and the model of communication. We then consider the specific case of zero-knowledge proof systems in the guaranteed complexity model and present a security definition for the same.

Let Server be interactive PPT machine that interacts with multiple clients in concurrent sessions and let $\{\langle \mathcal{S}_\ell, \mathcal{C}_\ell \rangle\}_{\ell \in \mathbb{N}}$ be a family of protocols parameterized by a *load parameter* ℓ where for every $\ell \in \mathbb{N}$, $\mathcal{S}_\ell$ and $\mathcal{C}_\ell$ are PPT machines. A protocol in the guaranteed complexity model is defined by the tuple $\Pi = (\mathsf{Server}, \{\langle \mathcal{S}_\ell, \mathcal{C}_\ell \rangle\})$.

(Honest) Protocol Execution. The execution of a protocol $\Pi = (\mathsf{Server}, \{\langle \mathcal{S}_\ell, \mathcal{C}_\ell \rangle\})$ consists of a single server executing the algorithm Server while interacting with multiple clients concurrently. To initiate a new session a client sends a special *session initiation message* to the server. In response to the session initiation message, the server chooses a load parameter ℓ for the session and sends it to the client. In the rest of the session we require that the algorithm Server follows the strategy $\mathcal{S}_\ell$ while the client follows the strategy $\mathcal{C}_\ell$.

An execution of the protocol Π with $p(n)$ sessions is defined by the randomness of all the clients and the schedule of messages across all the sessions. Even though for every fixed load parameter ℓ, the strategies $\mathcal{S}_\ell, \mathcal{C}_\ell$ are efficient, the server algorithm may choose ℓ to be very large, increasing the running time of the concurrent execution. Therefore we explicitly require the efficiency of a concurrent execution.

Definition 1. *A protocol* (Server, $\{\langle \mathcal{S}_\ell, \mathcal{C}_\ell \rangle\}$) *in the guaranteed complexity model is efficient if for every polynomial p there exists another polynomial q such that the running time of Server in every execution with $p(n)$ sessions is bounded by $q(n)$.*

Zero Knowledge in the Guaranteed Complexity Model. Let $\Pi = ($Server, $\{\langle \mathcal{S}_\ell, \mathcal{C}_\ell \rangle\})$ be a protocol in the guaranteed complexity model and let $\mathcal{L}$ be an NP language with witness relation $\mathcal{R}_\mathcal{L}$. We say that Π is an interactive proof system for $\mathcal{L}$ if for every $\ell \in \mathbb{N}$, the protocol $\langle \mathcal{S}_\ell, \mathcal{C}_\ell \rangle$ is an interactive proof for $\mathcal{L}$. Next we define the zero-knowledge property.

Let Π be an interactive proof for language $\mathcal{L}$ in the guaranteed complexity model. Let n be the security parameter. Consider a concurrent adversary V^* that start $m(n)$ concurrent session with the server for some polynomial m. Let $x \in \mathcal{L}^m$ be the vector of instances used in the different session and let w be a vector of the corresponding witnesses used by the server. We allow V^* to control the scheduling of the messages across all the sessions. Let $\mathsf{View}_{V^*}(x, w, z)$ be the random variable describing the output of V^* in the above experiment when executed with auxiliary input z.

Definition 2 (Concurrent Zero-Knowledge in the Guaranteed Complexity Model). *Let $\Pi = ($Server, $\{\langle \mathcal{S}_\ell, \mathcal{C}_\ell \rangle\})$ be an interactive proof system for language $\mathcal{L}$ in the guaranteed complexity model. We say that Π is zero knowledge if for every polynomial m, and for every PPT concurrent adversary V^* starting $m(n)$ sessions there exists a PPT algorithm $\mathcal{S}$, such that for every instances vector $x \in \mathcal{L}^{m(n)}$, every witnesses vector w such that $(x_i, w_i) \in \mathcal{R}_\mathcal{L}$ for all $i \in [m(n)]$, and for every auxiliary input $z \in \{0,1\}^{\mathrm{poly}(n)}$ the following ensembles are computationally indistinguishable,*

$$\{\mathsf{View}_{V^*}(x, w, z)\}_{n \in \mathbb{N}} \approx_c \{\mathcal{S}(x, z)\}_{n \in \mathbb{N}}.$$

3 Constant-Round Zero-Knowledge in the Guaranteed Complexity Model

In this section we describe a constant-round ZK protocol $\Pi_{\mathsf{zk}} = ($Server, $\{\langle P_\ell, V_\ell \rangle\})$ in the guaranteed complexity model. We start by defining a family of protocols $\{\langle P_\ell, V_\ell \rangle\}_{\ell \in \mathbb{N}}$ where, roughly speaking, the protocol $\langle P_\ell, V_\ell \rangle$ is simply Barak's bounded-concurrent ZK protocol [1] with n^ℓ as the a priori bound on the number of sessions. We then define the server algorithm Server to complete the description of Π_{zk}.

The Protocol $\langle P_\ell, V_\ell \rangle$. The protocol will make use of the following primitives: a statistically binding commitment Com, a family $\mathcal{H} = \{\mathcal{H}_n\}_{n \in \mathbb{N}}$ of collision-resistant hash functions such that $h \in \mathcal{H}_n$ maps strings in $\{0,1\}^*$ to strings in $\{0,1\}^n$, and a witness-indistinguishable universal argument UA for an **NTIME**$(T(n))$-complete language where $T : \mathbb{N} \to \mathbb{N}$ is a "slightly" superpolynomial function, for example $T(n) = n^{\log \log n}$ [2]. In the description of the protocol, the length of the verifier's messages will depend on a parameter m that denotes the total length of the *prover's* messages in the protocol.

Common Input: $x \in \mathcal{L}$.
Auxiliary Input to P: A witness w for $x \in \mathcal{L}$.

Initiation Stage:
V_ℓ samples $h \leftarrow \mathcal{H}_n$ and sends h to P_ℓ.
Preamble Stage:
    1. P_ℓ sends $c = \mathsf{Com}(h(0^n))$ to V_ℓ.
    2. V_ℓ samples $r \leftarrow \{0,1\}^{n^\ell \cdot m + n}$ and sends r to P_ℓ.
Proof Stage:
    P_ℓ and V_ℓ execute the protocol UA where P_ℓ proves that $x \in \mathcal{L} \vee (h, c, r) \in \mathcal{L}_U$.

The language $\mathcal{L}_U$ is defines as follows: $(h, c, r) \in \mathcal{L}_U$ iff there exist a program $\Pi \in \{0,1\}^*$, a string $y \in \{0,1\}^*$, and randomness s for Com such that:

1. $|y| \leq |r| - n$.
2. $c = \mathsf{Com}(h(\Pi); s)$.
3. $\Pi(y)$ outputs r within $T(n)$ steps.

Fig. 1. Protocol Family $\langle P_\ell, V_\ell \rangle$ for ZK in the Guaranteed Complexity Model (Protocol 1)

Remark 1. The relation $\mathcal{L}_U$ presented in Protocol 1 is slightly oversimplified. For this relation, we can prove the security of Protocol 1 when $\mathcal{H}$ is collision-resistant against "slightly" super-polynomial sized circuits. For simplicity of exposition, in this manuscript, we will work with this assumption. We stress, however, that as discussed in several prior works (see e.g., [2]), this assumption can be removed by using an appropriate error-correcting code.

The Server Algorithm Server. We start by describing a simple server algorithm that only assigns monotonically increasing values of the load parameter to new sessions. In Section 3.2, we describe a better server algorithm that decreases the load parameter when some of the concurrent sessions terminate.

The algorithm Server maintains a variable SessionCount that counts the number of concurrent sessions started so far. Whenever a client initiates a new session, Server increases the value of SessionCount. When a new clients sends a session initiation message to the server, Server sets the load parameter ℓ for that session such that $n^{\ell-1} \leq$ SessionCount $\leq n^{\ell}$.

In the next section we prove the following theorem:

Theorem 1. *Assuming h is a hash function ensemble that is collision-resistant against circuits of size $n^{\log n}$, Com is a statistically binding commitment, and UA is a witness-indistinguishable universal argument for $\mathbf{NTIME}(n^{\log \log n})$, the protocol $\Pi_{zk} = ($Server, $\{\langle P_\ell, V_\ell \rangle\})$ is concurrent zero-knowledge in the guaranteed complexity model.*

3.1 Proof of Theorem 1

The proof that for every $\ell \in \mathbb{N}$, the protocol $\langle P_\ell, V_\ell \rangle$ is complete and sound, follows directly from the analysis of the bounded-concurrent ZK protocol in [1]. In this section we first show that for every $\ell \in \mathbb{N}$, Protocol 1 is efficient according to Definition 1. We then show that Π_{zk} is ZK in the guaranteed complexity model.

Protocol 1 Is Efficient. Let p be a polynomial and let ℓ_{max} be such that for large enough values of n, $p(n) < n^{\ell_{max}}$. By the definition of the server algorithm Server, in an execution with $p(n)$ sessions, the load parameter of every session is at most ℓ_{max}. Since the running time of P_ℓ only grows with ℓ, we have that the running time of Server in every session is at most the running time of $P_{\ell_{max}}$ and therefore the total running time of Server is bounded by a polynomial that depends only on p.

Protocol Π_{zk} Is ZK in the Guaranteed Complexity Model. Let V^* be a malicious verifier that starts at most $n^{\ell_{max}}$ sessions for some constant ℓ_{max}. By the definition of the server algorithm Server, the load parameter of every session in an honest execution is at most ℓ_{max}. We construct a simulator Sim $=$ (Sim$_{load}$, $\{$Sim$_\ell\})$ consisting of Sim$_{load}$ and ℓ_{max} other components $\{$Sim$_\ell\}_{\ell \in [\ell_{max}]}$. Roughly speaking, the component Sim$_{load}$ simulates the servers responses to the clients session initiation message in all sessions. The component Sim$_\ell$ simulates all the executions of $\langle P_\ell, V_\ell \rangle$ in sessions with load parameter ℓ. We now give more details.

The Component Sim$_{load}$. This component simulates the server's responses to the clients session initiation message in all sessions. This simulation involves assigning a load parameter for every session started by V^*. Since the honest server Server selects the load parameter in each session based only on the (public) adversarial scheduling, Sim$_{load}$ can use the exact same algorithm as Server, resulting in a perfect simulation of these messages.

The Component Sim$_\ell$. This component simulates the interaction of $\langle P_\ell, V_\ell \rangle$ in all the sessions with load parameter ℓ. At a high-level, the simulation will follow the simulation strategy of Barak's bounded-concurrent ZK protocol [1]. According to this strategy, the simulator sends a commitment c to the code of the verifier and then uses this code as a trapdoor witness, proving that c is commitment to a code Π that outputs the random string r sent by the verifier. All the messages simulated in concurrent sessions are given to Π as auxiliary input. The main problem is that in order to guarantee that the protocol is sound, the program Π is only allowed to get an auxiliary input of bounded length; however, the number of concurrent sessions in our setting are not bounded.

We fix this problem in the following manner. Instead of simply committing to the code of V^*, Sim_ℓ will commit to a program V_ℓ^* that includes the code of V^* as well as the code of the simulation components $\mathsf{Sim}_{\mathsf{load}}$ and $\mathsf{Sim}_{\ell+1}, \ldots, \mathsf{Sim}_{\ell_{\max}}$. Roughly speaking, the program V_ℓ^* will simulate all the sessions with load parameter $\ell' > \ell$ internally, and therefore Sim_ℓ will need to provide as auxiliary input only the messages of concurrent sessions where the load parameter is at most ℓ. It follows from the description of Server that the number of concurrent sessions where the load parameter is at most ℓ is bounded by some polynomial (that depends on ℓ). Therefore, it is possible to include all of these messages as an auxiliary input to V_ℓ^*.

Next we formally describe the simulator component Sim_ℓ, starting with the definition of the program V_ℓ^*.

The Program V_ℓ^*. V_ℓ^* is an interactive algorithm that includes the code of V^* together with the code of the simulation components $\mathsf{Sim}_{\mathsf{load}}$ and $\mathsf{Sim}_{\ell+1}, \ldots, \mathsf{Sim}_{\ell_{\max}}$. V_ℓ^* uses the same randomness as Sim to execute V^* and all the other simulation components. V_ℓ^* will emulate the execution of V^*, and will use the mentioned simulator components to internally simulate the responses to the session initiation messages in all sessions as well the prover messages of the protocols $\langle P_{\ell'}, V_{\ell'} \rangle$ executed in the sessions with load parameter $\ell' > \ell$. In the sessions with load parameter $\ell' \leq \ell$, V_ℓ^* will forward the messages of the protocol $\langle P_{\ell'}, V_{\ell'} \rangle$ externally.

In every session with load parameter ℓ, Sim_ℓ will simulate the execution of $\langle P_\ell, V_\ell \rangle$ as follows:

1. Sim_ℓ receives the description of a hash function h from V^*.
2. Sim_ℓ sends a commitment c to the hash of the code of a program Π that given auxiliary input $y = (m_1, \ldots, m_t)$, emulates an execution of V_ℓ^* when receiving the messages $m_1, \ldots, m_t$, and outputs V_ℓ^*'s next message.
3. Sim_ℓ receives the the random string r from V^*.
4. Sim_ℓ sends a UA proof using a trapdoor witness that contains the code of the program Π and an appropriate auxiliary input string y. The string y is a list of all the prover messages that were simulated by Sim in all sessions with load parameter at most ℓ and sent before V^* sent the random string r in the present session.

This completes the description of the simulator. Next, we turn to its analysis.

Analysis of Sim. We start by showing that Sim_ℓ constructs a valid witness for the statement $(h, c, r) \in \mathcal{L}_U$. This amounts to proving that $\Pi(y)$ outputs r and that $|y| \leq |r| - n$. We also need to show that the running time of $\Pi(y)$ is at most $T(n)$. We will show that the last statement is correct when we analyze the running time of the simulation. Finally, we will prove the indistinguishability of the adversary's view in the real and ideal world.

Proof That $\Pi(y)$ Outputs r. The program $\Pi(y)$ outputs the next message of V_ℓ^* given the external messages in y. V_ℓ^* emulates V^* using the same randomness as Sim. It is left to show that the messages sent to V^* emulated by V_ℓ^* and by Sim are identical. Recall that the messages sent to V^* in the execution emulated by V_ℓ^* are as follows: in sessions with load parameter larger than ℓ, the messages are generated by the internal simulation of V_ℓ^*, and the messages sent in sessions with load parameter at most ℓ are specified in y. For sessions with load parameter larger than ℓ, the messages sent to V^* in the emulation of V_ℓ^* and of Sim are identical since they are generated using the same simulation algorithm and using the same randomness (by the construction of V_ℓ^*). For sessions with load parameter at most ℓ, the messages sent to V^* in the emulation of V_ℓ^* and of Sim are identical by the way the auxiliary input string y is constructed.

Proof That $|y| \leq |r| - n$. The auxiliary input string y constructed by Sim_ℓ contains only prover messages in sessions with load parameter at most ℓ. By the definition of the server algorithm Server there could be at most n^ℓ such sessions, and the total length of all the prover messages in every session is bounded by the parameter m. Therefore we have $|y| \leq n^\ell \cdot m$. Since V_ℓ samples $r \in \{0, 1\}^{n^\ell \cdot m + n}$ we have that $|y| \leq |r| - n$.

Proof That the Simulation Is Polynomial Time. It is enough to show that all components of Sim are polynomial time. Since $\mathsf{Sim}_{\mathsf{load}}$ just follows the honest server algorithm, the efficiency of $\mathsf{Sim}_{\mathsf{load}}$ follows from the efficiency of the protocol. For every $\ell \in [\ell_{\mathsf{max}}]$ we show that the running time of Sim_ℓ is bounded by a polynomial in the security parameter (that depends on ℓ and on V^*). Since Sim_ℓ constructs the program V_ℓ^*, commits to its code, and provides a UA proof of its execution, the running time of Sim_ℓ is polynomial in the size and running time of V_ℓ^*. Additionally, since Sim_ℓ reads the entire transcript of the execution and uses it to construct the auxiliary input y in every session it simulates, the running time of Sim_ℓ is polynomial in the total length of the transcript. Note that the total length of the transcript is always bounded by the running time of V^* which is polynomial in the security parameter.

We start by bounding the running time of $\mathsf{Sim}_{\ell_{\mathsf{max}}}$. The program $V_{\ell_{\mathsf{max}}}^*$ only consists of the code of V^* and the code of $\mathsf{Sim}_{\mathsf{load}}$ and therefore, the running time of $V_{\ell_{\mathsf{max}}}^*$ is a polynomial. It follows that the running time of $\mathsf{Sim}_{\ell_{\mathsf{max}}}$ is also a polynomial. Now, for every $\ell \in [\ell_{\mathsf{max}}]$, the program V_ℓ^* only consists of the code of V^*, the code of $\mathsf{Sim}_{\mathsf{load}}$, and the code of $\mathsf{Sim}_{\ell'}$ for every $\ell \leq \ell' < \ell_{\mathsf{max}}$. Since ℓ_{max} is a constant depending only on V^*, and assuming that for all $\ell \leq \ell' < \ell_{\mathsf{max}}$ the running time of every $\mathsf{Sim}_{\ell'}$ is polynomial, the running time of V_ℓ^* and therefore also of Sim_ℓ must be polynomial. By induction we have that for every $\ell \in [\ell_{\mathsf{max}}]$

the running time of Sim_ℓ is bounded by a polynomial, and therefore the entire simulation is polynomial time.

Using the above proof, we complete the proof that Sim_ℓ constructs a valid trapdoor witness. Sim_ℓ constructs a program Π and auxiliary input y, and we need to show that the running time of $\Pi(y)$ is bounded by some super-polynomial function $T(n)$. The running time analysis above implies that for every $\ell \in [\ell_{\max}]$, the running time of V_ℓ^* and the size of the auxiliary input y constructed by Sim_ℓ are polynomial. The simulator component Sim_ℓ constructs a program Π that simulates V_ℓ^* sending it messages from y. It follows that the running time of $\Pi(y)$ is polynomial and therefore bounded by $T(n)$.

Proof That the Simulated View and the Real View Are Indistinguishable. For $0 \leq \ell \leq \ell_{\max}$, let H_i be the hybrid experiment that is identical to the execution of Sim except that every session executing the the protocol $\langle P_{\ell'}, V_{\ell'} \rangle$ for $\ell' \leq \ell$ follow the honest prover strategy using the valid witness w_j for the statement $x_j \in \mathcal{L}$ in that session. of that session. Since $\mathsf{Sim}_{\mathsf{load}}$ simulates the responses to the sessions initiation messages perfectly we have that:

$$H_{\ell_{\max}} = \mathsf{View}_{V^*}(\boldsymbol{x}, \boldsymbol{w}, z), \quad H_0 = \mathcal{S}(\boldsymbol{x}, z) .$$

It is therefore sufficient to prove that for every $0 \leq \ell < \ell_{\max}$, $H_\ell \approx_c H_{\ell+1}$. By the definition of the server algorithm Server, the number of sessions with load parameter ℓ is at most n^ℓ. For $0 \leq i \leq n^\ell$, let $H_{\ell,i}$ be the hybrid experiment that is identical to H_ℓ except that the first i sessions executing of the protocol $\langle P_\ell, V_\ell \rangle$ follow the honest prover strategy using a the valid witness w_j for the statement $x_j \in \mathcal{L}$ in that session. It follows that:

$$H_{\ell,n^\ell} = H_{\ell+1}, \quad H_{\ell,0} = H_\ell .$$

It is therefore sufficient to prove that for every $0 \leq i < n^\ell$, $H_{\ell,i} \approx_c H_{\ell,i+1}$.

Let $H'_{\ell,i}$ be the hybrid experiment that is identical to the $H_{\ell,i}$ except that the execution of the witness-indistinguishable universal argument UA in the proof stage of the ith execution of the protocol $\langle P_\ell, V_\ell \rangle$ uses a valid witness w_j for the session's statement $x_j \in \mathcal{L}$ instead of the trapdoor witness. Note that in an execution of Sim, the randomness of the component Sim_ℓ used for the UA prover executed in the proof stage of the protocol $\langle P_\ell, V_\ell \rangle$ is also used by the components $\mathsf{Sim}_{\ell'}$ for $\ell' < \ell$ in the construction of the program $V_{\ell'}^*$. However, in the experiment $H_{\ell,i}$, all the simulator components $\mathsf{Sim}_{\ell'}$ for $\ell' < \ell$ are replaced by executions of the honest prover. Since the randomness of the component Sim_ℓ used for the simulation of the UA prover in the protocol $\langle P_\ell, V_\ell \rangle$ is not used in any other part of the simulation, it follows from the indistinguishability property of UA that $H_{\ell,i} \approx_c H'_{\ell,i}$.

Note that the experiment $H_{\ell,i+1}$ is identical to the experiment $H'_{\ell,i}$ except that in the experiment $H_{\ell,i+1}$, the prover commitment c given in the preamble stage of the i'th execution of the protocol $\langle P_\ell, V_\ell \rangle$ is a commitment to the all zero string, following the honest prover strategy. As before, the randomness of the component Sim_ℓ used for the simulation of c sent in the protocol $\langle P_\ell, V_\ell \rangle$ is not used in any other part of the simulation and therefore it follows from the computational-hiding property of Com that $H_{\ell,i+1} \approx_c H'_{\ell,i}$.

Overall we have that for every $0 \leq \ell \leq \ell_{\max}, 0 \leq i \leq n^\ell$, $H_{\ell,i} \approx_c H_{\ell,i+1}$. Since $\ell \leq \ell_{\max}$ is a constant, n^ℓ is a polynomial and therefore we have that for every $0 \leq \ell \leq \ell_{\max}$, $H_{\ell+1} \approx_c H_\ell$ and also that $H_{\ell_{\max}} \approx_c H_0$ as required.

3.2   Decreasing the Load Parameter

In this section, we describe a different server algorithm that takes into account the termination of sessions and decreases the load parameter for new sessions accordingly. We start by describing the new server algorithm Server', and then describe the required changes to the simulation.

We identify the technical condition required for the simulation to work, and design a server algorithm Server' that always gives new sessions the lowest possible load parameter such that the technical condition still satisfies. The validity of our simulation relies on the validity of the following technical condition: for a session with load parameter ℓ_i, the number of sessions concurrent to it with load parameters at most ℓ_i is bounded by n^i. Before describing the algorithm Server' let us first introduce some notation. Let t be the number of open sessions at the moment a new client sends its session initiation message. For $i \in [t]$, let ℓ_i be the load parameter for the i'th open session. For $i \in [t]$, let t_i be the total number of sessions with load parameters at most i that are concurrent to session i. First note that if we set the load parameter of the new session to ℓ then for every session i such that $\ell_i \geq \ell$, the value t_i increases by 1. This will contradict the technical condition only if the value of t_i was already at its maximal allowed value n^{ℓ_i}.

Using the above notation, the algorithm Server' is easy to describe: Server' will set the load parameter of a new session to be the minimal value ℓ such that for every session i with $\ell_i \geq \ell$ we have $t_i < n^{\ell_i}$. While the behavior of the server algorithm Server' is not obvious, we can prove that it satisfies some natural conditions. For example we can show that if no sessions with load parameter ℓ are currently active, then the load parameter assigned to the next session to start cannot exceed ℓ.

Modifying the Simulator. Next we discuss the necessary changes to the simulator. In the current description of the simulator, every program V_ℓ^* that Sim commits to, internally emulates V^* starting from its initial state. As a result, we must give V_ℓ^* auxiliary input z that consists of the messages in all concurrent sessions with load parameter at most ℓ starting from the beginning of the concurrent execution. The problem is that the definition of the server algorithm Server' does not guarantee that such auxiliary input z is sufficiently short. Instead it only gives a bound on the number sessions with load parameter at most ℓ that are executed *concurrently* to the current session. In particular, Server' does not guarantee anything about the number of sessions that terminated before the current session had started. The solution is based on the observation that providing V_ℓ^* auxiliary input z that contains messages sent before the current session had started is wasteful. Instead, Sim can commit the a program $\tilde{V}^*{}_\ell$ that already contains these messages hardwired into it.

References

1. Barak, B.: How to go beyond the black-box simulation barrier. In: FOCS, pp. 106–115 (2001)
2. Barak, B., Goldreich, O.: Universal arguments and their applications. SIAM J. Comput. 38(5), 1661–1694 (2008)
3. Canetti, R., Kilian, J., Petrank, E., Rosen, A.: Black-box concurrent zero-knowledge requires (almost) logarithmically many rounds. SIAM J. Comput. 32(1), 1–47 (2002)
4. Canetti, R., Lin, H., Paneth, O.: Public-coin concurrent zero-knowledge in the global hash model. In: Sahai, A. (ed.) TCC 2013. LNCS, vol. 7785, pp. 80–99. Springer, Heidelberg (2013)
5. Canetti, R., Lindell, Y., Ostrovsky, R., Sahai, A.: Universally composable two-party and multi-party secure computation. In: STOC, pp. 494–503 (2002)
6. Chung, K.M., Lin, H., Pass, R.: Constant-round concurrent zero knowledge from p-certificates. In: FOCS (2013)
7. Deng, Y., Goyal, V., Sahai, A.: Resolving the simultaneous resettability conjecture and a new non-black-box simulation strategy. In: FOCS, pp. 251–260 (2009)
8. Dwork, C., Naor, M., Sahai, A.: Concurrent zero-knowledge. In: STOC, pp. 409–418 (1998)
9. Goldwasser, S., Micali, S., Rackoff, C.: The knowledge complexity of interactive proof systems. SIAM J. Comput. 18(1), 186–208 (1989)
10. Goyal, V.: Non-black-box simulation in the fully concurrent setting. In: STOC, pp. 221–230 (2013)
11. Goyal, V., Jain, A., Ostrovsky, R., Richelson, S., Visconti, I.: Concurrent zero knowledge in the bounded player model. In: Sahai, A. (ed.) TCC 2013. LNCS, vol. 7785, pp. 60–79. Springer, Heidelberg (2013)
12. Gupta, D., Sahai, A.: On constant-round concurrent zero-knowledge from a knowledge assumption. IACR Cryptology ePrint Archive 2012, 572 (2012)
13. Kilian, J., Petrank, E.: Concurrent and resettable zero-knowledge in polyloalgorithm rounds. In: STOC, pp. 560–569 (2001)
14. Pandey, O., Prabhakaran, M., Sahai, A.: Obfuscation-based non-black-box simulation and four message concurrent zero knowledge for np. IACR Cryptology ePrint Archive 2013, 754 (2013)
15. Pass, R., Rosen, A., Tseng, W.L.D.: Public-coin parallel zero-knowledge for np. J. Cryptology 26(1), 1–10 (2013)
16. Persiano, G., Visconti, I.: Single-prover concurrent zero knowledge in almost constant rounds. In: Caires, L., Italiano, G.F., Monteiro, L., Palamidessi, C., Yung, M. (eds.) ICALP 2005. LNCS, vol. 3580, pp. 228–240. Springer, Heidelberg (2005)
17. Prabhakaran, M., Rosen, A., Sahai, A.: Concurrent zero knowledge with logarithmic round-complexity. In: FOCS, pp. 366–375 (2002)
18. Richardson, R., Kilian, J.: On the concurrent composition of zero-knowledge proofs. In: Stern, J. (ed.) EUROCRYPT 1999. LNCS, vol. 1592, pp. 415–431. Springer, Heidelberg (1999)
19. Rosen, A., Shelat, A.: Optimistic concurrent zero knowledge. In: Abe, M. (ed.) ASIACRYPT 2010. LNCS, vol. 6477, pp. 359–376. Springer, Heidelberg (2010)

Round-Efficient Black-Box Construction of Composable Multi-Party Computation

Susumu Kiyoshima

NTT Secure Platform Laboratories, Japan
kiyoshima.susumu@lab.ntt.co.jp

Abstract. We present a *round-efficient* black-box construction of a general MPC protocol that satisfies composability in the plain model. The security of our protocol is proven in angel-based UC framework under the minimal assumption of the existence of semi-honest oblivious transfer protocols. When the round complexity of the underlying oblivious transfer protocol is $r_{OT}(n)$, the round complexity of our protocol is $\max(\widetilde{O}(\log^2 n), O(r_{OT}(n)))$. Since constant-round semi-honest oblivious transfer protocols can be constructed under standard assumptions (such as the existence of enhanced trapdoor permutations), our result gives $\widetilde{O}(\log^2 n)$-round protocol under these assumptions. Previously, only an $O(\max(n^\epsilon, r_{OT}(n)))$-round protocol was shown, where $\epsilon > 0$ is an arbitrary constant.

We obtain our MPC protocol by constructing a $\widetilde{O}(\log^2 n)$-round CCA-secure commitment scheme in a black-box way under the assumption of the existence of one-way functions.

1 Introduction

Protocols for *secure multi-party computation* (MPC) enable mutually distrustful parties to compute a functionality without compromising the correctness of the outputs and the privacy of their inputs. In the seminal work of Goldreich et al. [11], a general MPC protocol was constructed in a model with malicious adversaries and a dishonest majority.[1] (By "a general MPC protocol," we mean a protocol that can be used to securely compute any functionality.)

In this paper, we consider a *black-box construction* of a general MPC protocol that guarantees *composable security*. Before stating our result, we explain black-box constructions and composable security.

Black-Box Constructions. A construction of a protocol is *black-box* if it uses the underlying cryptographic primitives only in a black-box way (that is, only through their input/output interfaces). In contrast, if a construction uses the codes of the underlying primitives, it is *non-black-box*.

As argued in [17], constructing black-box constructions is important for both theoretical and practical reasons. Theoretically, it is important because understanding whether non-black-box use of cryptographic primitives is necessary for a

---

[1] In the following, we consider only such a model.

J.A. Garay and R. Gennaro (Eds.): CRYPTO 2014, Part II, LNCS 8617, pp. 351–368, 2014.
© International Association for Cryptologic Research 2014

cryptographic task is of great interest. Practically, it is important because black-box constructions are typically more efficient than non-black-box ones in terms of both communication complexity and computational complexity. In fact, since known non-black-box constructions of general MPC protocols compute general NP reductions to execute zero-knowledge proofs (this is where the codes of the primitives are used), they are highly inefficient and hard to implement. Thus, constructing black-box constructions of general MPC protocols is an important step toward practical general MPC protocols.

Recently, a series of works studied black-box constructions of general MPC protocols. Ishai et al. [17] showed the first construction of a general MPC protocol that uses the underlying low-level primitives (such as enhanced trapdoor permutations and homomorphic public-key encryption schemes) in a black-box way. Combined with the subsequent work of Haitner [15], which showed a black-box construction of a (malicious) oblivious transfer protocol based on a semi-honest oblivious transfer protocol, their work gives a black-box construction of a general MPC protocol based on a semi-honest oblivious transfer protocol [16]. Subsequently, Wee [30] reduced the round complexity of [17] to $O(\log^* n)$, and Goyal [12] further reduced the round complexity to $O(1)$.

These black-box protocols are proven to be secure in the *stand-alone setting*. That is, the protocols of [17,30,12] are secure in the setting where a single instance of the protocol is executed at a time.

Composable Security. Compared with the stand-alone setting, the *concurrent setting* is more general and realistic. In the concurrent setting, many instances of many different protocols are concurrently executed in an arbitrary schedule. Thus, in the concurrent setting, adversaries can perform a coordinated attack in which they choose messages in each instance based on the executions of the other instances.

As a strong and realistic security notion in the concurrent setting, Canetti [2] proposed *universally composable (UC) security*. The main advantage of UC security is *composability*, which guarantees that when we compose many UC-secure protocols, we can prove the security of the resultant protocol from the security of its components. Thus, UC security enables us to construct secure protocols in a modular way. Composability also guarantees that a protocol remains secure even when it is concurrently executed with any other protocols in any schedule. Thus, UC-secure protocols are secure in the concurrent setting. Canetti et al. [6] constructed a UC-secure general MPC protocol in the *common reference string (CRS) model* (i.e., in a model in which all parties are given a common public string that is chosen by a trusted third party). Black-box constructions of UC-secure general MPC protocols were shown in the $\mathcal{F}_{OT}$-hybrid model [18] and in the $\mathcal{F}_{COM}$-hybrid model [8] (i.e., in a model with the ideal oblivious transfer functionality and in a model with the ideal commitment functionality).

UC security, however, turned out to be too strong to achieve in the *plain model*. That is, even with non-black-box use of cryptographic primitives, we cannot construct UC-secure general MPC protocols in a model with no trusted setup [3,4].

To achieve composable security in the plain model, Prabhakaran and Sahai [29] proposed a variant of UC security called *angel-based UC security*. Roughly speaking, angel-based UC security is the same as UC security except that the adversary and the simulator have access to an additional entity—the *angel*—that allows some judicious use of super-polynomial-time resources. Although angel-based UC security is weaker than UC security, angel-based UC security guarantees meaningful security in many cases. (For example, angel-based UC security implies *super-polynomial-time simulation (SPS) security* [26,1,10,27]. In SPS security, we allow the simulator to run in super-polynomial time; thus SPS security guarantees that whatever an adversary can do in the real world can also be done in the ideal world *in super-polynomial time*.) Furthermore, it was proven that, like UC security, angel-based UC security guarantees composability. Prabhakaran and Sahai [29] presented a general MPC protocol that satisfies angel-based UC security in the plain model based on new assumptions. Subsequently, Malkin et al. [24] constructed another general MPC protocol that satisfies angel-based UC security in the plain model based on a new number-theoretic assumption.

Recently, several works constructed general MPC protocols with angel-based UC security under standard assumptions. Canetti et al. [5] constructed a polynomial-round general MPC protocol in angel-based UC security assuming the existence of enhanced trapdoor permutations. Subsequently, Lin [20] and Goyal et al. [14] reduced the round complexity to $\widetilde{O}(\log n)$ under the same assumption. They also showed that with enhanced trapdoor permutations that are secure against quasi-polynomial-time adversaries, the round complexity of their protocols can be reduced to $O(1)$.

The construction of these MPC protocols are, however, non-black-box. That is, in the protocols of [5,20,14], the underlying primitives are used in a non-black-box way.

Black-Box Constructions of Composable Protocols. Lin and Pass [22] showed the first black-box construction of a general MPC protocol that guarantees composable security in the plain model. The security of their protocol is proven under angel-based UC security and based on the minimal assumption of the existence of semi-honest oblivious transfer (OT) protocols. The round complexity of their protocol is $O(\max(n^\epsilon, r_{\mathrm{OT}}(n)))$, where $\epsilon > 0$ is an arbitrary constant and $r_{\mathrm{OT}}(n)$ is the round complexity of the underlying semi-honest OT protocols. Thus, with enhanced trapdoor permutations (from which we can construct constant-round semi-honest OT protocols), their result gives an $O(n^\epsilon)$-round protocol. Subsequently, Kiyoshima et al. [19] constructed a constant-round protocol from constant-round semi-honest OT protocols that are secure against quasi-polynomial-time adversaries and one-way functions that are secure against subexponential-time adversaries.

Summarizing the state-of-the-art, for composable protocols in the plain model, we have

- logarithmic-round non-black-box constructions under a standard polynomial-time hardness assumption [20,14],

- a polynomial-round black-box construction under a standard polynomial-time hardness assumption [22], and
- constant-round black-box or non-black-box constructions under standard super-polynomial-time hardness assumptions [20,14,19].

Thus, for composable protocols based on standard polynomial-time hardness assumptions, there exists a gap between the round complexity of the non-black-box protocols (logarithmic rounds [20,14]) and that of the black-box protocols (polynomial rounds [22]). The following is therefore an important open question.

> Does there exist a **round-efficient** black-box construction of a general MPC protocol that guarantees composability in the plain model under polynomial-time hardness assumptions?

1.1   Our Result

In this paper, we greatly narrow the gap between the round complexity of black-box composable general MPC protocols and the round complexity of non-black-box ones.

Main Theorem (Informal). *Assume the existence of $r_{\mathrm{OT}}(n)$-round semi-honest oblivious transfer protocols. Then, there exists a $\max(\widetilde{O}(\log^2 n), O(r_{\mathrm{OT}}(n)))$-round black-box construction of a general MPC protocol satisfying angel-based UC security in the plain model.*

Recall that, assuming the existence of enhanced trapdoor permutations, we have a constant-round semi-honest OT protocol. Thus, under this assumption, our main theorem gives a $\widetilde{O}(\log^2 n)$-round protocol.

We prove our main theorem by constructing a $\widetilde{O}(\log^2 n)$-round black-box construction of a *CCA-secure commitment scheme* [5,20,22,14,19] from one-way functions.

Theorem (Informal). *Assume the existence of one-way functions. Then, there exists a $\widetilde{O}(\log^2 n)$-round black-box construction of a CCA-secure commitment scheme.*

Roughly speaking, a CCA-secure commitment scheme is a tag-based commitment scheme (i.e., a commitment scheme that takes an n-bit string—a *tag*—as an additional input) such that the hiding property holds even against adversaries that interact with the *committed-value oracle* during the interaction with the challenger. The committed-value oracle interacts with the adversary as an honest receiver in many concurrent sessions of the commit phase. At the end of each session, if the commitment of this session is invalid or has multiple committed values, the oracle returns $\perp$ to the adversary. Otherwise, the oracle returns the unique committed value to the adversary.

Lin and Pass [22] showed that in angel-based UC security, an $O(\max(r_{\mathrm{CCA}}(n), r_{\mathrm{OT}}(n)))$-round general MPC protocol can be obtained in a black-box way from a $r_{\mathrm{CCA}}(n)$-round CCA-secure commitment scheme and a $r_{\mathrm{OT}}(n)$-round semi-honest OT protocol. Thus, we can prove our main theorem by combining the above theorem with the result of [22].

1.2   Outline

In Section 2, we give an overview of our CCA secure commitment scheme. Due
to lack of space, we defer formal proofs to the full version.

2   Overview of Our CCA-Secure Commitment Scheme

Key elements for obtaining CCA-secure commitment schemes are *concurrent
extractability* and *non-malleability*. With these elements, we can show that
the committed-value oracle is useless for breaking the hiding property. Non-
malleability is used to show that the sessions between the adversary and the
oracle are independent of the session between the adversary and the challenger.
Then, concurrent extractability is used to show that the committed-value oracle
can be emulated in polynomial time by extracting the committed values from
the adversary.

Before constructing our CCA-secure commitment scheme, we first construct
two building blocks: (i) a commitment scheme CECom' that is *concurrently extractable without over-extraction* and (ii) a *one-one CCA-secure* commitment
scheme CCACom$^{1:1}$. The former guarantees concurrent extractability and the
latter guarantees (slightly strong) non-malleability.

2.1   Building Block 1: Concurrently Extractable Commitment
Scheme without Over-Extraction

A commitment scheme is *concurrently extractable* if a rewinding extractor can
extract the committed values from any committer even in the concurrent setting,
and a concurrently extractable commitment scheme is *concurrently extractable
without over-extraction* if the extractor outputs $\perp$ whenever the commitment
is invalid.[2] (Basic extractability, in contrast, allows the extractor to output an
arbitrary value when the commitment is invalid.) There exists a commitment
scheme CECom that is concurrently extractable *with* over-extraction based on
the existence of one-way functions [25].

To construct a commitment scheme that is concurrently extractable without
over-extraction, we start from the following scheme (in which the cut-and-choose
technique is used in the same way as in the previous works of black-box protocols
[7,8,30,22,19]).

1. Let v be the value to be committed. Then, the committer computes an
   $(n + 1)$-out-of-$10n$ Shamir's secret sharing $s = (s_1, \ldots, s_{10n})$ of value v and
   commits to each s_j in parallel by using CECom.
2. Then, the receiver sends a random subset $\Gamma \subset [10n]$ of size n.
3. The committer reveals s_j for every $j \in \Gamma$ and decommits the corresponding
   commitments.

---

[2] A commitment is *valid* if there exists a valid decommitment of this commitment;
otherwise, it is *invalid*. A commitment is *accepted* if the receiver does not abort in
the commit phase; otherwise, it is *rejected*.

4. The receiver accepts the commitment if and only if the decommitments are valid for every $j \in \Gamma$.

For $j \in [10n]$, let the j-th *column* be the j-th CECom commitment. The use of the cut-and-choose technique guarantees that when the receiver accepts a commitment, the CECom commitments are valid in "most" columns. Then, since we can extract the committed value of CECom whenever the CECom commitment is valid, we can extract s_j in most columns on an accepted commitment. We can therefore recover v from the extracted values of the CECom commitments by using the error-correcting property of Shamir's secret sharing scheme.[3]

Unfortunately, although the above scheme is concurrently extractable without over-extraction, we cannot prove its hiding property. This is because the receiver requests the committer to open adaptively-chosen CECom commitments (in other words, the receiver performs a selective opening attack).

We therefore modify the scheme in the following way. At the beginning of the scheme, we let the receiver commit to Γ by using a statistically binding commitment scheme Com. Now, since the receiver no longer choose the subset adaptively, we can prove the hiding property by a standard technique. Furthermore, at first sight, the hiding property of Com seems to guarantee that the scheme remains to be concurrently extractable without over-extraction.

In the modified scheme, however, we cannot prove that the scheme is concurrently extractable without over-extraction. This is because we can no longer show that most of the CECom commitments are valid in an accepted commitment. Consider, for example, that there exists a cheating committer C^* such that receiving a Com commitment to Γ at the beginning, C^* somehow generates an invalid CECom commitment in the j-th column for every $j \notin \Gamma$ and commits to 0 in the j-th column for every $j \in \Gamma$. Then, although C^* seems to break the hiding property of Com, we do not know how to use C^* to break the hiding property of Com. To see this, observe the following. Recall that since CECom is an extractable commitment scheme *with* over-extraction, the extractor of CECom may output an arbitrary value when the CECom commitment is invalid. Thus, when we extract the committed values of CECom from C^*, the extracted value may be 0 in every column. Hence, although C^* behaves differently in CECom based on the value of Γ, we cannot detect it.

To overcome this problem, we use the commitment scheme wExtCom that was introduced by Goyal et al. [13]. The commit phase of wExtCom consists of three stages: commit, challenge, and reply. In the commit stage, the committer commits to random $a_0, a_1 \in \{0,1\}^n$ such that $a_0 \oplus a_1 = v$; in the challenge stage, the receiver sends a random bit $ch \in \{0,1\}$; in the reply stage, the committer reveals a_{ch} and decommits the corresponding commitment. We note that wExtCom is extractable only in a weak sense—extractions may fail with probability at most $1/2$—but wExtCom is extractable without over-extraction. That is, the extractor may output $\perp$ with probability at most $1/2$, but when the extractor outputs $v \neq \perp$, the commitment is valid and its committed value is v. We also note that wExtCom satisfies the following property: After the commit

---

[3] Recall that Shamir's secret sharing is also a codeword of Reed-Solomon code.

stage, if the committer returns a valid reply with probability $1/\text{poly}(n)$ for both $ch = 0$ and $ch = 1$, then the committed value can be extracted with probability 1 in expected polynomial time.

With wExtCom, we modify our scheme as follows: After committing to s with CECom, the committer commits to (s_j, d_j) for each $j \in [10n]$ in parallel with wExtCom, where (s_j, d_j) is a decommitment of the j-th CECom commitment. Then, we show that in most columns on an accepted commitment, the wExtCom commitment is valid and its committed value is a valid decommitment of the corresponding CECom commitment. Toward this end, we observe the following.

- If a cheating committer generates an accepting commitment with non-negligible probability, then in wExtCom of more than $9n$ columns, the cheating committer returns a valid reply with non-negligible probability for both $ch = 0$ and $ch = 1$. (If the cheating committer returns a valid reply with non-negligible probability for both $ch = 0$ and $ch = 1$ in wExtCom of at most $9n$ columns, then there are n columns in which the wExtCom commitment is accepted with probability at most $1/2 + \text{negl}(n)$. Thus, the probability that all wExtCom commitments are accepted is negligible, and therefore the commitment is accepted with at most negligible probability.)
- Thus, from the property of wExtCom, we can extract the committed values of wExtCom without over-extraction in most columns.
- Then, from the property of the cut-and-choose technique, we can show that in most columns of an accepted commitment, the wExtCom commitment is valid and its committed value is a valid decommitment of the corresponding CECom commitment. Note that since the committed values of wExtCom commitments can be extracted without over-extraction, we can show that the cheating committer cannot give invalid wExtCom commitments in many columns.

Then, since this implies that most of the CECom commitments are valid whenever the commitment is accepted, we can extract the committed value of the scheme without over-extraction as before, i.e., by extracting the committed values of CECom commitments and using the error-collecting property of Shamir's secret sharing scheme.

A formal description of our concurrently extractable commitment scheme CECom' is shown in Fig. 1. (For technical reasons, we set the number of columns to $40n$.) In Appendix A, we give a formal proof for the fact that in most columns on an accepted commitment, the wExtCom commitment is valid and its committed value is a valid decommitment of the CECom commitment. The formal proof is more complicated than the above proof sketch because we execute the wExtCom commitments in parallel and thus the columns are not independent of each other. The proof of this fact is the most complicated part of the analysis of CECom': Given this fact, we can show the concurrent extractability by using the technique used in the previous works [7,8,30,22,19].

To commit to $v \in \{0,1\}^n$, the committer C does the following with the receiver R.

Step 1. R commits to a random sublet $\Gamma \subset [40n]$ of size n by using Com.

Step 2. C computes an $(n + 1)$-out-of-$40n$ Shamir's secret sharing $s = (s_1, \ldots, s_{40n})$ of value v. Then, for each $j \in [40n]$ in parallel, C commits to s_j by using CECom. Let (s_j, d_j) be the decommitment of the j-th commitment.

Step 3. For each $j \in [40n]$ in parallel, C commits to (s_j, d_j) by using wExtCom.

Step 4. R decommits the Step 1 commitment to Γ.

Step 5. For each $j \in \Gamma$, C decommits the j-th Step 3 commitment to (s_j, d_j). Then, for each $j \in \Gamma$, R checks whether the decommitment is valid and whether the decommitted value (s_j, d_j) is a valid decommitment of the j-th Step 2 commitment.

Fig. 1. A concurrently commitment scheme CECom′

2.2   Building Block 2: One-One CCA-Secure Commitment Scheme

A *one-one CCA-secure commitment scheme*, which is closely related to a *non-malleable commitment scheme*, is one that is CCA secure w.r.t. a restricted class of adversaries that execute only a single session with the committed-value oracle and immediately receive the answer from the oracle at the end of the session.[4]

We construct a black-box $O(\log n)$-round one-one CCA-secure commitment scheme by simplifying the CCA-secure commitment scheme of [22] and using the *DDN* $\log n$ *trick* [9,23], which transforms a concurrent non-malleable commitment scheme for tags of length $O(\log n)$ to a non-malleable commitment scheme for tags of length $O(n)$ without increasing the round complexity. In the following, we assume the familiarity to the scheme of [22]. Roughly speaking, the scheme of [22] consists of polynomially-many *rows*—each row is a parallel execution of (a part of) the trapdoor commitment scheme of [28]—and a cut-and-choose phase, which forces the committer to give valid and consistent trapdoor commitments in every row. If we reduce the number of rows from $\mathsf{poly}(n)$ to $\ell(n)$ in the scheme of [22], where $\ell(n)$ is the length of the tags, the resultant scheme is no longer CCA secure. It is easy to verify, however, that the scheme is *parallel CCA secure*, i.e., it is CCA secure w.r.t. a restricted class of adversaries that give a single parallel query to the oracle and receive the answers immediately. (This is because when the adversaries give only a single parallel query, the recursive rewinding does not occur in the extraction and thus we require only a single rewinding opportunity.) Then, we set $\ell(n) := O(\log n)$ and apply the DDN $\log n$ trick to

---

[4] In contrast, a non-malleable commitment scheme is one that is CCA secure w.r.t. a restricted class of adversaries that execute a single session with the oracle and receive the answer *after completing the interactions with the challenger and the oracle.*

the above parallel CCA-secure commitment scheme. It is not hard to see that the resultant scheme is one-one CCA secure.

2.3   CCA-Secure Commitment Scheme from the Building Blocks

Given CECom' and CCACom$^{1:1}$, we construct a CCA-secure commitment scheme CCACom roughly as follows, where the committer commits to a value v with tag tag.

1. The receiver commits to a random subset $\Gamma \subset [10n]$ of size n by using CCACom$^{1:1}$ with tag tag.
2. The committer computes an $(n + 1)$-out-of-$10n$ Shamir's secret sharing $s = (s_1, \ldots, s_{10n})$ of value v and commits to each s_j in parallel by using a normal statistically binding commitment scheme Com.
3. For $\eta(n) := r_{\mathrm{CEC}}(n) + 1$ times in sequence (where $r_{\mathrm{CEC}}(n)$ is the round complexity of CECom'), the committer does the following: the committer commits to s_j for every $j \in [10n]$ by using CECom' in parallel. Each parallel commitment is called a *row*.
4. The receiver decommits the commitment of the first step and reveals Γ.
5. For every $j \in \Gamma$, the committer decommits all of the $\eta(n)$ commitments whose committed values are s_j.

Our scheme differs from the previous CCA-secure commitment schemes [5,22,20,14] in that it uses a one-one CCA-secure commitment scheme instead of a non-malleable commitment scheme; furthermore, our scheme uses a one-one CCA-secure commitment scheme in the reverse order. That is, whereas the previous schemes (implicitly or explicitly) use non-malleable commitment schemes from the committer to the receiver, our scheme uses a one-one CCA secure commitment scheme from the receiver to the committer. (Very recently, the same strategy is used in [19].)

Using a one-one CCA-secure commitment scheme in the reverse order is crucial in showing the *simulation-soundness* of the cut-and-choose phase. We say that the adversary (or the challenger) *cheats* if in an accepted commitment there exists a row whose committed shares disagree with s in more than n indexes. Using the one-one CCA security of CCACom$^{1:1}$, we can show that the adversary cannot cheat in every session of the *right interaction* (i.e., the interaction between the adversary and the oracle) even when the adversary receives a commitment in which the challenger cheats in the *left interaction* (i.e., the interaction between the adversary and the challenger). Roughly speaking, this is because the adversary can emulate the cheating challenger in polynomial time by making a single query to the committed-value oracle of CCACom$^{1:1}$ and receiving Γ; therefore, from one-one CCA security of CCACom$^{1:1}$, the commitment that the adversary receives on the left is useless for breaking the hiding property of CCACom$^{1:1}$ on the right, and thus the adversary cannot cheat on the right from the property of the cut-and-choose technique. Note that non-malleability is insufficient for this argument since the hiding property of CCACom$^{1:1}$ need to hold even when the adversary receives the answer from the oracle immediately after completing the query to the oracle. We also note that CECom′ must be concurrently extractable *without* over-extraction since otherwise the adversary may give invalid commitments in more than n indexes without being detected in the cut-and-choose phase. (As explained in Section 2.1, the existence of such an adversary does not contradict the one-one CCA security of CCACom$^{1:1}$ if over-extraction can occur.)

Given the simulation-soundness of the cut-and-choose phase, we can show the CCA security of CCACom by, as in the analysis of previous CCA-secure commitment schemes [5,22,20], rewinding the adversary and emulating the committed-value oracle in polynomial time. Toward this end, we consider a series of hybrid experiments in which the commitment that the adversary receives on the left is gradually changed as follows: In the i-th hybrid experiment ($i \in [\eta(n)]$), we switch the committed value from s_j to 0 for every $j \notin \Gamma$ in the i-th row, where Γ is extracted by brute force. Note that the $(i-1)$-st hybrid and the i-th hybrid differ only in the i-th row. The problem is that the adversary accesses the committed-value oracle, which runs in super-polynomial time. Then, to show the indistinguishability between the $(i-1)$-st hybrid and the i-th hybrid, we observe the following. Since there are $r_{\text{CEC}} + 1$ rows (in particular, the number of rows is bigger than the number of rounds in CECom′), we can extract the committed shares in a row on every right session without disturbing the hiding property of CECom′ in the i-th row on the left. (Here, we use a technique used in [21]. Roughly speaking, we extract the committed shares from a row that contains no message of the CECom′ commitment of the i-th row on the left.) Recall that, since CECom′ is concurrently extractable without over-extraction, we can extract the committed shares without over-extraction. Then, since the simulation-soundness guarantees that these shares agree with s in at least $9n$ indexes, we can compute v from these shares by using the error-correcting property of Shamir's secret sharing. Therefore we can emulate the oracle in polynomial time by rewinding the adversary (without disturbing the hiding property of CECom′ in the i-th row) and computing v as above. Thus, the indistinguishability of the $(i-1)$-st hybrid and the i-th hybrid follows from the hiding property of CECom′. Then, we consider another hybrid experiment: This experiment is the same as the $\eta(n)$-th hybrid except that the committed value of the j-th Com commitment in Step 2 is switched from s_j to 0 for every $j \notin \Gamma$. From the same argument as above, this hybrid is indistinguishable from the $\eta(n)$-th hybrid. Then, since in this hybrid the adversary does not receive any information about v, the CCA security follows.

We note that the actual argument is more complicated. For example, we need to show the simulation-soundness even for the adversary accessing the committed-value oracle. To solve this problem, we increase the number of rows (i.e., $\eta(n)$) and emulate the oracle in polynomial time without disturbing the one-one CCA security of CCACom$^{1:1}$. To show that the oracle can be emulated, we require the simulation soundness; thus, there seems to be a circular argument, i.e., we require the simulation soundness to show the simulation soundness. In the formal analysis, we show that this issue can be avoided. For details, see the full version.

Comparison with the CCA-secure commitment scheme of [19]. The above CCA-secure commitment scheme is based on the CCA-secure commitment scheme of [19], which is constructed from one-way functions that are secure against subexponential-time adversaries. The scheme of [19] is the same as the above scheme except for the following.

- There is only a single row, and CECom is used instead of CECom' (i.e., a concurrently extractable scheme *with* over-extraction is used).
- The underlying commitment schemes Com, CECom, and CCACom$^{1:1}$ are secure against subexponential-time adversaries. In particular, Com is hiding against T_1-time adversaries but is completely broken in time $o(T_2)$, CECom is hiding against T_2-time adversaries but is completely broken in time $o(T_3)$, and CCACom$^{1:1}$ is one-one CCA secure against T_3-time adversaries, where (T_1, T_2, T_3) is a hierarchy of running times such that $T_3 \gg T_2 \gg T_1 \gg n^{\omega(1)}$. This is where subexponentially hard one-way functions are required.

The high-level strategy for proving CCA security is the same, i.e., showing the simulation soundness from one-one CCA security of CCACom$^{1:1}$ and then considering hybrid experiments in which committed values of CECom and Com are gradually switched. The proof of [19] is, however, different from ours in the following.

- In the proof of the simulation soundness, the issue of over-extraction is solved by extracting the committed values of CECom by brute force. (Note that even when the committed values of CECom are extracted by brute force, the one-one CCA security of CCACom$^{1:1}$ still holds since the committed values of CECom extractable in time $o(T_3)$ and one-one CCA security of CCACom$^{1:1}$ holds against T_3-time adversaries.)
- When the committed values of CECom are switched, the indistinguishability follows immediately from the fact that CECom is hiding against T_2-time adversaries and the running time of the committed-value oracle is $o(T_2)$. (The committed-value oracle computes its output by extracting the committed values of Com by brute force. Thus, its running-time is $o(T_2)$.)

Thus, the proof of [19] heavily depends on the subexponentially hard security of the underlying commitment schemes. Roughly speaking, we weaken the assumption of [19] by doing the following.

- To show the simulation soundness without subexponentially hard security, we replace CECom with CECom', which is concurrently extractable *without* over-extraction.
- To show the indistinguishability when we switch the committed values of CECom', we increase the number of rows so that the committed-value oracle can be emulated in polynomial time by rewinding the adversary while preserving the hiding property of CECom'.

Overall, despite of the similarity of the high-level structure between the scheme of [19] and ours, the details of the security proofs have a lot of difference.

References

1. Barak, B., Sahai, A.: How to play almost any mental game over the net - concurrent composition via super-polynomial simulation. In: FOCS, pp. 543–552 (2005)
2. Canetti, R.: Universally composable security: A new paradigm for cryptographic protocols. In: FOCS, pp. 136–145 (2001)
3. Canetti, R., Fischlin, M.: Universally composable commitments. In: Kilian, J. (ed.) CRYPTO 2001. LNCS, vol. 2139, pp. 19–40. Springer, Heidelberg (2001)
4. Canetti, R., Kushilevitz, E., Lindell, Y.: On the limitations of universally composable two-party computation without set-up assumptions. In: Biham, E. (ed.) EUROCRYPT 2003. LNCS, vol. 2656, pp. 68–86. Springer, Heidelberg (2003)
5. Canetti, R., Lin, H., Pass, R.: Adaptive hardness and composable security in the plain model from standard assumptions. In: FOCS, pp. 541–550 (2010)
6. Canetti, R., Lindell, Y., Ostrovsky, R., Sahai, A.: Universally composable two-party and multi-party secure computation. In: STOC, pp. 494–503 (2002)
7. Choi, S.G., Dachman-Soled, D., Malkin, T., Wee, H.: Black-box construction of a non-malleable encryption scheme from any semantically secure one. In: Canetti, R. (ed.) TCC 2008. LNCS, vol. 4948, pp. 427–444. Springer, Heidelberg (2008)
8. Choi, S.G., Dachman-Soled, D., Malkin, T., Wee, H.: Simple, black-box constructions of adaptively secure protocols. In: Reingold, O. (ed.) TCC 2009. LNCS, vol. 5444, pp. 387–402. Springer, Heidelberg (2009)
9. Dolev, D., Dwork, C., Naor, M.: Nonmalleable cryptography. SIAM J. Comput. 30(2), 391–437 (2000)
10. Garg, S., Goyal, V., Jain, A., Sahai, A.: Concurrently secure computation in constant rounds. In: Pointcheval, D., Johansson, T. (eds.) EUROCRYPT 2012. LNCS, vol. 7237, pp. 99–116. Springer, Heidelberg (2012)
11. Goldreich, O., Micali, S., Wigderson, A.: How to play any mental game or a completeness theorem for protocols with honest majority. In: STOC. pp. 218–229 (1987)
12. Goyal, V.: Constant round non-malleable protocols using one way functions. In: STOC, pp. 695–704 (2011)
13. Goyal, V., Lee, C.K., Ostrovsky, R., Visconti, I.: Constructing non-malleable commitments: A black-box approach. In: FOCS, pp. 51–60 (2012)
14. Goyal, V., Lin, H., Pandey, O., Pass, R., Sahai, A.: Round-efficient concurrently composable secure computation via a robust extraction lemma. Cryptology ePrint Archive, Report 2012/652 (2012), http://eprint.iacr.org/
15. Haitner, I.: Semi-honest to malicious oblivious transfer—the black-box way. In: Canetti, R. (ed.) TCC 2008. LNCS, vol. 4948, pp. 412–426. Springer, Heidelberg (2008)
16. Haitner, I., Ishai, Y., Kushilevitz, E., Lindell, Y., Petrank, E.: Black-box constructions of protocols for secure computation. SIAM J. Comput. 40(2), 225–266 (2011)
17. Ishai, Y., Kushilevitz, E., Lindell, Y., Petrank, E.: Black-box constructions for secure computation. In: STOC, pp. 99–108 (2006)
18. Ishai, Y., Prabhakaran, M., Sahai, A.: Founding cryptography on oblivious transfer – efficiently. In: Wagner, D. (ed.) CRYPTO 2008. LNCS, vol. 5157, pp. 572–591. Springer, Heidelberg (2008)
19. Kiyoshima, S., Manabe, Y., Okamoto, T.: Constant-round black-box construction of composable multi-party computation protocol. In: Lindell, Y. (ed.) TCC 2014. LNCS, vol. 8349, pp. 343–367. Springer, Heidelberg (2014)
20. Lin, H.: Concurrent Security. Ph.D. thesis, Cornell University (2011)

21. Lin, H., Pass, R.: Concurrent non-malleable zero knowledge with adaptive inputs. In: Ishai, Y. (ed.) TCC 2011. LNCS, vol. 6597, pp. 274–292. Springer, Heidelberg (2011)

22. Lin, H., Pass, R.: Black-box constructions of composable protocols without set-up. In: Safavi-Naini, R., Canetti, R. (eds.) CRYPTO 2012. LNCS, vol. 7417, pp. 461–478. Springer, Heidelberg (2012)

23. Lin, H., Pass, R., Venkitasubramaniam, M.: Concurrent non-malleable commitments from any one-way function. In: Canetti, R. (ed.) TCC 2008. LNCS, vol. 4948, pp. 571–588. Springer, Heidelberg (2008)

24. Malkin, T., Moriarty, R., Yakovenko, N.: Generalized environmental security from number theoretic assumptions. In: Halevi, S., Rabin, T. (eds.) TCC 2006. LNCS, vol. 3876, pp. 343–359. Springer, Heidelberg (2006)

25. Micciancio, D., Ong, S.J., Sahai, A., Vadhan, S.P.: Concurrent zero knowledge without complexity assumptions. In: Halevi, S., Rabin, T. (eds.) TCC 2006. LNCS, vol. 3876, pp. 1–20. Springer, Heidelberg (2006)

26. Pass, R.: Simulation in quasi-polynomial time, and its application to protocol composition. In: Biham, E. (ed.) EUROCRYPT 2003. LNCS, vol. 2656, pp. 160–176. Springer, Heidelberg (2003)

27. Pass, R., Lin, H., Venkitasubramaniam, M.: A unified framework for UC from only OT. In: Wang, X., Sako, K. (eds.) ASIACRYPT 2012. LNCS, vol. 7658, pp. 699–717. Springer, Heidelberg (2012)

28. Pass, R., Wee, H.: Black-box constructions of two-party protocols from one-way functions. In: Reingold, O. (ed.) TCC 2009. LNCS, vol. 5444, pp. 403–418. Springer, Heidelberg (2009)

29. Prabhakaran, M., Sahai, A.: New notions of security: achieving universal composability without trusted setup. In: STOC, pp. 242–251 (2004)

30. Wee, H.: Black-box, round-efficient secure computation via non-malleability amplification. In: FOCS, pp. 531–540 (2010)

A   Formal Proof

In this section, we give a formal proof for the fact that in most columns on an accepted commitment of CECom', the wExtCom commitment is valid and its committed value is a valid decommitment of the CECom commitment. This is the most complicated part of the analysis of CECom': Given this fact, we can show the concurrent extractability by using the technique used in the previous works [7,8,30,22,19].

Lemma 1. *Let C^* be any cheating committer that concurrently executes many sessions of the commit phase of* CECom'. *Then, the following holds except with negligible probability: In more than $38n$ columns on every accepted session, the* wExtCom *commitment is valid and its committed value is a valid decommitment of the* CECom *commitment.*

Proof. First, we give some definitions. In each session, for $j \in [40n]$, the j-th column is the pair of the j-th CECom commitment in Step 2 and the j-th wExtCom commitment in Step 3. We say that a column is *consistent* if in the column the committed value of the wExtCom commitment is a valid decommitment of the CECom commitment; otherwise, the column is *inconsistent*. We say that C^* *cheats* in a session if (i) every wExtCom commitment is accepted, (ii) the j-th column is consistent for every $j \in \Gamma$, and (iii) there exist at least $2n$ inconsistent columns.

To prove the lemma, it suffices to show that in every session the probability that C^* cheats is negligible.

Assume for contradiction that for infinitely many n, there is a session in which C^* cheats with probability at least $1/\text{poly}(n)$. In the following, we fix any such n. Then, since the number of sessions is at most $\text{poly}(n)$, there is an $i^* \in [\text{poly}(n)]$ such that in the i^*-th session, C^* cheats with probability at least $1/n^c$ for a constant c.

Then, let us consider an adversary $\mathcal{B}$ against the hiding property of Com. For random subsets $\Gamma_0, \Gamma_1 \subset [40n]$ of size n, $\mathcal{B}$ tries to distinguish a Com commitment to Γ_0 from a Com commitment to Γ_1 as follows. $\mathcal{B}$ internally invokes C^* and honestly emulates the interaction between C^* and honest receivers except that in the i^*-th session, $\mathcal{B}$ does the following.

- In Step 1, $\mathcal{B}$ receives a Com commitment from the external committer (the committed value is either Γ_0 or Γ_1) and forwards the commitment to C^* as the Step 1 commitment.
- When Step 3 is accepted (i.e., all the wExtCom commitments are accepted), $\mathcal{B}$ does the following repeatedly: $\mathcal{B}$ rewinds C^* to the point that the next-message is the challenge bits of wExtCom in the i^*-th session; then $\mathcal{B}$ sends new random challenge bits and honestly interacts with C^* until the end of Step 3 (i.e., until receiving the replies in wExtCom). After collecting other n^{c+3} accepted transcripts of Step 3, $\mathcal{B}$ outputs 1 if the following hold:
  (i) from these $n^{c+3}+1$ accepted transcript (the first one and the subsequent n^{c+3} ones), $\mathcal{B}$ can extract the committed values of wExtCom in at least $39n$ columns,
  (ii) in at least n columns of these columns, the extracted values are not valid decommitments of the corresponding CECom commitments, and
  (iii) for every $j \in \Gamma_1$, either the extraction of the j-th column fails or the extracted value of the j-th column is a valid decommitment of the corresponding CECom commitment.
  Otherwise, $\mathcal{B}$ outputs 0. In the following, the first transcript that $\mathcal{B}$ generates in Step 3 is called the *main thread* and other n^{c+3} accepted transcripts are called the *look-ahead threads*.

If $\mathcal{B}$ rewinds C^* more than n^{3c+4} times, $\mathcal{B}$ terminates and outputs fail.

First, we show that an expected polynomial-time adversary $\mathcal{B}'$ successfully distinguishes Com commitments, where $\mathcal{B}'$ is the same as $\mathcal{B}$ except that $\mathcal{B}'$ does not terminate after $\mathcal{B}'$ rewinds C^* more than n^{3c+4} times. When $\mathcal{B}'$ receives a commitment to Γ_0, since the internal C^* receives no information of Γ_1, the probability that $\mathcal{B}'$ outputs 1 is exponentially small. (This is because when Condition (i) and Condition (ii) hold, the probability that Condition (iii) holds is exponentially small.) Thus, it remains to show that when $\mathcal{B}'$ receives a commitment to Γ_1, the probability that $\mathcal{B}'$ outputs 1 is at least $1/\text{poly}(n)$. Let extract be the event that $\mathcal{B}'$ extracts the committed values of wExtCom commitments from at least $39n$ columns, and let cheat be the event that C^* cheats in the i^*-th session on the main thread. Then, to show that $\mathcal{B}'$ outputs 1 with probability at least $1/\mathrm{poly}(n)$, it suffices to show that

$$\Pr\left[\text{cheat} \wedge \text{extract}\right] \geq \frac{1}{\mathrm{poly}(n)} . \tag{1}$$

(Recall the we can extract the committed values of wExtCom without over-extraction.) Let ρ be a prefix of a transcript between C^* and honest receivers such that after ρ, a honest receiver sends challenge bits of wExtCom in the i^*-th session. Let prefix_ρ be the event that a prefix of the main thread is ρ. Then, since the probability that C^* cheats in the i^*-th session is at least $1/n^c$, from an average argument, we have $\Pr\left[\text{cheat} \mid \text{prefix}_\rho\right] \geq 1/2n^c$ with probability at least $1/2n^c$ over the choice of ρ (i.e., when we obtain ρ by emulating the interaction between C^* and honest receivers). Let Δ be the set of prefixes such that $\Pr\left[\text{cheat} \mid \text{prefix}_\rho\right] \geq 1/2n^c$ holds. Then, since we have $\sum_{\rho \in \Delta} \Pr\left[\text{prefix}_\rho\right] \geq 1/2n^c$, we have

$$\Pr\left[\text{cheat} \wedge \text{extract}\right] \geq \sum_{\rho \in \Delta} \Pr\left[\text{cheat} \wedge \text{extract} \mid \text{prefix}_\rho\right] \cdot \Pr\left[\text{prefix}_\rho\right]$$

$$\geq \min_{\rho \in \Delta} \left(\Pr\left[\text{cheat} \wedge \text{extract} \mid \text{prefix}_\rho\right]\right) \cdot \sum_{\rho \in \Delta} \Pr\left[\text{prefix}_\rho\right]$$

$$\geq \frac{1}{2n^c} \min_{\rho \in \Delta} \left(\Pr\left[\text{cheat} \wedge \text{extract} \mid \text{prefix}_\rho\right]\right) . \tag{2}$$

Thus, to show Equation (1), it suffices to show that for any $\rho \in \Delta$, we have

$$\Pr\left[\text{cheat} \wedge \text{extract} \mid \text{prefix}_\rho\right] \geq \frac{1}{\mathrm{poly}(n)} . \tag{3}$$

In the following, we fix any $\rho^* \in \Delta$. Then, we have

$$\Pr\left[\text{cheat} \mid \text{prefix}_{\rho^*}\right] \geq \frac{1}{2n^c} . \tag{4}$$

Thus, from Equation (4), we have

$$\Pr\left[\text{cheat} \wedge \text{extract} \mid \text{prefix}_{\rho^*}\right] = \Pr\left[\text{cheat} \mid \text{prefix}_{\rho^*}\right] \cdot \Pr\left[\text{extract} \mid \text{prefix}_{\rho^*} \wedge \text{cheat}\right]$$

$$\geq \frac{1}{2n^c} \Pr\left[\text{extract} \mid \text{prefix}_{\rho^*} \wedge \text{cheat}\right] \tag{5}$$

Thus, to show Equation (3), it suffices to show that

$$\Pr\left[\text{extract} \mid \text{prefix}_{\rho^*} \wedge \text{cheat}\right] \geq \frac{1}{\mathrm{poly}(n)} . \tag{6}$$

Recall that when cheat occurs, Step 3 of the i^*-th session is accepted on the main thread. Thus, for any $j \in [40n]$, when cheat occurs and the challenge bit of wExtCom in the j-th column is $b \in \{0,1\}$ on the main thread, we can extract the committed value of the the j-th column if in the n^{c+3} look-ahead threads there is an accepted transcript of wExtCom such that the challenge bit of the j-th column is $1 - b$. Then, to show Equation (6), we show that when Step 3 of the i^*-th session is accepted on the main thread with prefix ρ^*, the probability that the challenge bit of wExtCom is b is "high" for any $b \in \{0,1\}$ in "most" columns. Let ch_j be a random variable for the challenge bit of wExtCom in the j-th column of the i^*-th session on the main thread, and let accept be the event that every wExtCom commitment is accepted in the i^*-th session on the main thread. (We have $\Pr[\mathsf{accept}] \geq \Pr[\mathsf{cheat}]$ from the definitions.) Then, for any $j \in [40n]$ and $b \in \{0,1\}$,

$$
\begin{aligned}
&\Pr\left[ch_j = b \mid \mathsf{accept} \wedge \mathsf{prefix}_{\rho^*}\right] \\
&= \frac{\Pr\left[ch_j = b \wedge \mathsf{accept} \wedge \mathsf{prefix}_{\rho^*}\right]}{\Pr\left[\mathsf{accept} \wedge \mathsf{prefix}_{\rho^*}\right]} \\
&\geq \frac{\Pr\left[ch_j = b \wedge \mathsf{cheat} \wedge \mathsf{prefix}_{\rho^*}\right]}{\Pr\left[\mathsf{prefix}_{\rho^*}\right]} \\
&= \frac{\Pr\left[\mathsf{cheat} \mid ch_j = b \wedge \mathsf{prefix}_{\rho^*}\right] \Pr\left[ch_j = b \wedge \mathsf{prefix}_{\rho^*}\right]}{\Pr\left[\mathsf{prefix}_{\rho^*}\right]} \\
&= \Pr\left[\mathsf{cheat} \mid ch_j = b \wedge \mathsf{prefix}_{\rho^*}\right] \Pr\left[ch_j = b\right] \ .
\end{aligned}
\tag{7}
$$

(Here, we use $\Pr\left[ch_j = b \wedge \mathsf{prefix}_{\rho^*}\right] = \Pr[ch_j = b] \cdot \Pr\left[\mathsf{prefix}_{\rho^*}\right]$.) Below, we show that in at least $39n$ columns of the i^*-th session, for any $b \in \{0,1\}$ we have

$$
\Pr\left[\mathsf{cheat} \mid ch_j = b \wedge \mathsf{prefix}_{\rho^*}\right] \geq \frac{1}{160n^{c+1}} \ .
\tag{8}
$$

Let

$$
A := \left\{ j \in [40n] \ \middle| \ \exists b_j \in \{0,1\} \ \text{s.t.} \ \Pr\left[\mathsf{cheat} \mid ch_j = b_j \wedge \mathsf{prefix}_{\rho^*}\right] < \frac{1}{160n^{c+1}} \right\} \ .
$$

Then we have

$$
\begin{aligned}
\Pr\left[\mathsf{cheat} \mid \mathsf{prefix}_{\rho^*}\right] &\leq \Pr\left[\bigwedge_{j \in A} ch_j = 1 - b_j\right] + \Pr\left[\mathsf{cheat} \bigwedge \left(\bigvee_{j \in A} ch_j = b_j\right) \ \middle| \ \mathsf{prefix}_{\rho^*}\right] \\
&\leq 2^{-|A|} + \sum_{j \in A} \Pr\left[\mathsf{cheat} \wedge ch_j = b_j \mid \mathsf{prefix}_{\rho^*}\right] \\
&= 2^{-|A|} + \sum_{j \in A} \Pr\left[\mathsf{cheat} \mid ch_j = b_j \wedge \mathsf{prefix}_{\rho^*}\right] \Pr\left[ch_j = b_j\right] \\
&\leq 2^{-|A|} + \sum_{j \in A} \Pr\left[\mathsf{cheat} \mid ch_j = b_j \wedge \mathsf{prefix}_{\rho^*}\right] \\
&< 2^{-|A|} + 40n \cdot \frac{1}{160n^{c+1}} \\
&\leq 2^{-|A|} + \frac{1}{4n^c} \ .
\end{aligned}
\tag{9}
$$

Then, from Equations (4) and (9), we have $|A| = O(\log n)$ and therefore $|A| \leq n$. Thus, in at least $39n$ columns, for any $b \in \{0,1\}$ we have Equation (8). Then, from Equations (7) and (8) and from $\Pr[ch_j = b] = 1/2$, for any $j \in [40n] \setminus A$ and any $b \in \{0,1\}$, we have

$$\Pr\left[ch_j = b \mid \mathsf{accept} \wedge \mathsf{prefix}_{\rho^*}\right] \geq \frac{1}{320n^{c+1}} \ .$$

Then, since the distributions of the look-ahead threads are the same as that of the main thread, we have that under the condition that prefix_{ρ^*} and cheat occur, for any $j \in [40n] \setminus A$, the adversary $\mathcal{B}'$ requires another $320n^{c+1}$ accepted transcripts on average to extract the committed value of $\mathsf{wExtCom}$ in the j-th columns. Since $\mathcal{B}'$ collects n^{c+3} accepted transcripts, for any $j \in [40n] \setminus A$ the adversary $\mathcal{B}'$ extracts the committed value of $\mathsf{wExtCom}$ in the j-th column except with probability $320n^{c+1}/n^{c+3} = 320/n^2$ under the condition that prefix_{ρ^*} and cheat occur. (Here, we use Markov's inequality.) Then, from the union bound, except with probability $39n \cdot 320/n^2 = 12480/n$, for every $j \in [40n] \setminus A$ the adversary $\mathcal{B}'$ extracts the committed value of $\mathsf{wExtCom}$ in the j-th column. Thus, we have

$$\Pr\left[\mathsf{extract} \mid \mathsf{prefix}_{\rho^*} \wedge \mathsf{cheat}\right] \geq 1 - \frac{12480}{n} \ . \tag{10}$$

Then, from Equations (5) and (10), we have

$$\Pr\left[\mathsf{cheat} \wedge \mathsf{extract} \mid \mathsf{prefix}_{\rho^*}\right] \geq \frac{1}{2n^c} \cdot \left(1 - \frac{12480}{n}\right) \geq \frac{1}{4n^c} \ . \tag{11}$$

Then, since ρ^* is any prefix in Δ, from Equations (2) and (11) we have

$$\Pr\left[\mathsf{cheat} \wedge \mathsf{extract}\right] \geq \frac{1}{2n^c} \cdot \frac{1}{4n^c} = \frac{1}{8n^{2c}} \ .$$

Thus, we have Equation (1). We therefore conclude that $\mathcal{B}'$ outputs 1 with probability at least $1/8n^{2c}$ when $\mathcal{B}'$ receives a commitment to Γ_1. Thus, $\mathcal{B}'$ successfully distinguishes a commitment to Γ_1 from a commitment to Γ_0.

Now, we are ready to show that $\mathcal{B}$ breaks the hiding property of Com. Clearly, the running time of $\mathcal{B}$ is at most $\mathsf{poly}(n)$. Note that, to show that $\mathcal{B}$ can distinguish Com commitments, it suffices to show that the output of $\mathcal{B}$ is the same as that of $\mathcal{B}'$ except with probability $1/n^{2c+1}$. (This is because $\mathcal{B}'$ outputs 1 with negligible probability when $\mathcal{B}'$ receives a commitment to Γ_0 whereas $\mathcal{B}'$ outputs 1 with with probability $1/8n^{2c}$ when $\mathcal{B}'$ receives a commitment to Γ_1.) Recall that the output of $\mathcal{B}$ differs from that of $\mathcal{B}'$ if and only if $\mathcal{B}'$ rewinds C^* more than n^{3c+4} times. Let ρ be any prefix of a transcript between C^* and honest receivers such that after ρ, the next message is the challenge bits of $\mathsf{wExtCom}$ in the i^*-th session. Let $T(n)$ be a random variable for the number of rewinding in $\mathcal{B}'$. Then, we have

$$\mathrm{E}\left[T(n) \mid \mathsf{prefix}_{\rho}\right] \leq \Pr\left[\mathsf{accept} \mid \mathsf{prefix}_{\rho}\right] \cdot \frac{n^{c+3}}{\Pr\left[\mathsf{accept} \mid \mathsf{prefix}_{\rho}\right]} = n^{c+3} \ .$$

Thus, we have

$$E\left[T(n)\right] = \sum_{\rho} \Pr\left[\mathsf{prefix}_\rho\right] E\left[T(n) \mid \mathsf{prefix}_\rho\right]$$

$$\leq n^{c+3} \sum_{\rho} \Pr\left[\mathsf{prefix}_\rho\right] \leq n^{c+3} .$$

Then, from Markov's inequality, $\mathcal{B}'$ rewinds C^* more than n^{3c+4} times with probability at most $n^{c+3}/n^{3c+4} = 1/n^{2c+1}$. Thus, the output of $\mathcal{B}$ is the same as that of $\mathcal{B}'$ except with probability $1/n^{2c+1}$, and therefore $\mathcal{B}$ distinguishes a commitment to Γ_1 from a commitment to Γ_0.

$\square$

Secure Multi-Party Computation
with Identifiable Abort

Yuval Ishai[1,*], Rafail Ostrovsky[2,**], and Vassilis Zikas[3,***]

[1] Computer Science Department, Technion, Haifa, Israel
yuvali@cs.technion.ac.il
[2] Computer Science Department, UCLA, Los Angeles, CA, USA
rafail@cs.ucla.edu
[3] Computer Science Department, ETH Zurich, Switzerland
vzikas@inf.ethz.ch

Abstract. Protocols for secure multi-party computation (MPC) that resist a dishonest majority are susceptible to "denial of service" attacks, allowing even a single malicious party to force the protocol to abort. In this work, we initiate a systematic study of the more robust notion of *security with identifiable abort*, which leverages the effect of an abort by forcing, upon abort, at least one malicious party to reveal its identity.

We present the first *information-theoretic* MPC protocol which is secure with identifiable abort (in short ID-MPC) using a correlated randomness setup. This complements a negative result of Ishai et al. (TCC 2012) which rules out information-theoretic ID-MPC in the OT-hybrid model, thereby showing that *pairwise* correlated randomness is insufficient for information-theoretic ID-MPC.

In the standard model (i.e., without a correlated randomness setup), we present the first computationally secure ID-MPC protocol making *black-box* use of a standard cryptographic primitive, namely an (adaptively secure) oblivious transfer (OT) protocol. This provides a more efficient alternative to existing ID-MPC protocols, such as the GMW protocol, that make a non-black-box use of the underlying primitives.

As a theoretically interesting sidenote, our black-box ID-MPC provides an example for a natural cryptographic task that can be realized using a *black-box access* to an OT protocol but cannot be realized unconditionally using an ideal OT oracle.

* Supported by the European Union's Tenth Framework Programme (FP10/2010-2016) under grant agreement no. 259426 ERC-CaC, ISF grant 1361/10, and BSF grant 2012378.
** Work supported in part by NSF grants 09165174, 1065276, 1118126 and 1136174, US-Israel BSF grant 2008411, OKAWA Foundation Research Award, IBM Faculty Research Award, Xerox Faculty Research Award, B. John Garrick Foundation Award, Teradata Research Award, and Lockheed-Martin Corporation Research Award. This material is based upon work supported by the Defense Advanced Research Projects Agency through the U.S. Office of Naval Research under Contract N00014 -11 -1-0392. The views expressed are those of the author and do not reflect the official policy or position of the Department of Defense or the U.S. Government.
*** Portions of this work were done at UCLA. Work supported in part by the Swiss National Science Foundation (SNF) Ambizione grant PZ00P2_142549.

J.A. Garay and R. Gennaro (Eds.): CRYPTO 2014, Part II, LNCS 8617, pp. 369–386, 2014.
© International Association for Cryptologic Research 2014

1   Introduction

Recent advances in secure multiparty computation have led to protocols that compute large circuits in a matter of seconds. Most of these protocols, however, are restricted to provide security against semi-honest adversaries, or alternatively assume an honest majority. A notable exception is the SPDZ line of work [3,16,14,15,34] which tolerates a majority of malicious parties. SPDZ is optimized for the pre-processing model and demonstrates a remarkably fast on-line phase, largely due to the fact that it uses information-theoretic techniques and, thus, avoids costly cryptographic operations. Unfortunately, all these efficient MPC protocols for the case of a dishonest majority are susceptible to the following *denial-of-service (DoS)* attack: even a single malicious party can force an abort without any consequences (i.e., without even being accused of cheating). Although classical impossibility results for MPC prove that abort-free computation is impossible against dishonest majorities, vulnerability to DoS attacks is an issue that should be accounted for in any practical application.

Summary of Known Results. The seminal works on MPC [47,21,2,9,42] establish tight feasibility bounds on the tolerable number of corruptions for perfect, statistical (aka information-theoretic or unconditional), and computational (aka cryptographic) security. For semi-honest adversaries, unconditionally secure protocols exist if there is an honest majority, or if the parties have access to a complete functionality oracle or other types of setup. An arguably minimal setup is giving the parties (appropriately) correlated random strings before the inputs are known. We refer to this as the *correlated randomness model*.

When there is no honest majority and the adversary is malicious, full security that includes fairness cannot be achieved [12]. Instead, one usually settles for the relaxed notion of *security with abort*: Either the protocol succeeds, in which case every party receives its output, or the protocol aborts, in which case all honest parties learn that the protocol aborted. (Because of the lack of fairness, the adversary can learn its outputs even when the protocol aborts.) The GMW protocol [21,19] realizes this notion of security under standard cryptographic assumptions. Interestingly, this protocol also satisfies the following useful *identifiability* property: upon abort every party learns the identity of some corrupted party. This property is in the focus of our work.

To the best of our knowledge, all protocols achieving this notion of security (e.g., [21,7]) are based on the same paradigm of using public zero-knowledge proofs to detect deviation from the protocol. While elegant and conceptually simple, this approach leads to inefficient protocols that make a non-black-box use of the underlying cryptographic primitives.[1] The situation is even worse in the information-theoretic setting, where an impossibility result from [31] (see

---

[1] Alternatively, protocols such as the CDN protocol [13] make a use of ad-hoc zero-knowledge proofs based on specific number theoretic intractability assumptions. The disadvantage of these protocols is that they require public-key operations for each gate of the circuit being evaluated, and cannot get around this by using optimization techniques such as efficient OT extension [28].

also [44, Section 3.7]) proves that information-theoretic MPC with identifiable abort is impossible even in the OT-hybrid model, i.e., where parties can make ideal calls to an oblivious transfer (OT) functionality [41].

Our Contributions. We initiate a systematic study of this more robust and desirable notion of *secure MPC with identifiable abort* (ID-MPC). An ID-MPC protocol leverages the effect of an abort by forcing, upon abort, at least one malicious party to reveal its identity. This feature discourages cheaters from aborting, and in many applications allows for full recovery by excluding the identified cheater and restarting the protocol. We provide formal security definitions both in the setting of Universal Composition (UC) [5] and in the stand-alone setting [21,19,4]. Furthermore, we study feasibility and efficiency of ID-MPC in both the information-theoretic and the computational security models.

For the information-theoretic model, we present a general compiler that transforms any MPC protocol which uses correlated randomness to achieve security against semi-honest adversaries into a similar protocol which is secure with identifiable abort against malicious adversaries. As a corollary, we get the first information-theoretic ID-MPC protocol in the correlated randomness model. This protocol complements an impossibility result from [31], which rules out information-theoretic ID-MPC in the OT-hybrid model. Indeed, the insufficiency of OT implies that *pairwise* correlated randomness is not sufficient for information-theoretic ID-MPC, but leaves open the question of whether or not n-wise correlations are, which is answered affirmatively here.

In the computational security model, we present an ID-MPC protocol for realizing sampling functionalities, namely ones that sample and distribute correlated random strings, which only makes a *black-box* use of an (adaptively secure) *OT protocol* and ideal calls to a commitment functionality.[2] Using this protocol for realizing the setup required by the information-theoretic protocol yields the first ID-MPC protocol which makes a black-box use of standard cryptographic primitives. This holds both in the UC framework [5], under standard UC-setups, and in the plain stand-alone model [21,19,4]. Combined with the abovementioned impossibility result from [31], this provides an interesting example for a natural cryptographic task that can be realized using a *black-box* access to an OT protocol but cannot be unconditionally realized using an ideal OT oracle.

Our results demonstrate that ID-MPC is not only the most desirable notion from a practical point of view, but it also has the potential to be efficiently implemented. To this end, one can instantiate our construction with efficient OT protocols from the literature [39,10,36,17].[3] Furthermore, pre-computing the

---

[2] The ideal commitments can be replaced by a black-box use of a commitment protocol, or alternatively realized by making a black-box use of OT [27,38]. The OT protocol can be secure against either semi-honest or malicious adversaries, as these two flavors are equivalent under black-box reductions [24,11].

[3] Our analysis requires the underlying OT to be adaptively secure. Proving the same statement for a static OT protocol is a theoretically interesting open problem. From a practical point of view, however, many instances of adaptively secure OT can be efficiently implemented from few such instances in the (programmable) random oracle model [28,36].

randomness in an off-line phase yields a protocol in the pre-processing model which, similarly to SPDZ-style protocols, has an information-theoretic online phase. Investigating how our methodology can be fine-tuned towards practice remains an interesting direction for future work. Finally, our protocols can be used to improve the efficiency of a number of protocols in the fairness-related literature, e.g., [29,18,26,37,48,22,1], as these works implicitly use ID-MPC (typically instantiated by GMW) to realize a sampling functionality.

Comparison to Existing Work. Our information-theoretic protocol can be seen as a new *feasibility* result, since the current literature contains no (efficient or inefficient) information-theoretic ID-MPC protocol from correlated randomness. Similarly, our computational protocol can also be seen as a "second-order" feasibility result, since this is the first ID-MPC protocol making *black-box* use of a standard cryptographic primitive. Notwithstanding, much of the motivation for considering black-box constructions in cryptography is derived from the goal of practical efficiency, and indeed the most practical protocols today (whether Yao-based or GMW-based) are black-box protocols that do not need to know the "code" of the underlying cryptographic primitives.

2 The Model

We prove our security statements in the universal composition (UC) framework of Canetti [5]: in a nutshell, a protocol π (securely) UC realizes a functionality $\mathcal{F}$ if for any adversary $\mathcal{A}$ attacking π there exists an ideal adversary, the simulator $\mathcal{S}$, that makes an ideal evaluation of $\mathcal{F}$ indistinguishable from a protocol execution with $\mathcal{A}$ in the eyes any environment $\mathcal{Z}$. When $\mathcal{Z}$, $\mathcal{A}$, and $\mathcal{S}$ are polynomially bounded we say that the protocol realizes $\mathcal{F}$ (with computational security); otherwise, when $\mathcal{Z}$, $\mathcal{A}$, and $\mathcal{S}$ are unbounded, we say that the protocol *unconditionally* realizes $\mathcal{F}$ (with information-theoretic security).

For simplicity we restrict our description to computation of non-reactive functionalities, also known as *secure function evaluation (SFE)*. (The general case can be reduced to this case by using a suitable form of secret sharing [31] for maintaining the secret state of the reactive functionality.) Moreover, we describe our protocols as *synchronous protocols*, i.e., round-based protocols where messages sent in some round are delivered by the beginning of the next round; such protocols can be executed in UC as demonstrated in [33,35]. The advantage of such a "synchronous" description is dual: first, it yields simpler descriptions of functionalities and protocols; indeed, because the parties are aware of the round in which each message should be sent/received, we can avoid always explicitly writing all the message/protocol IDs in the descriptions. Second, it is compatible with the protocol description in the stand-alone model of computation [20,4], which allows us to directly translate our results into that model.

Our protocols assume n parties from the set $\mathcal{P} = \{p_1, \ldots, p_n\}$. We prove our results for a non-adaptive adversary who actively corrupts parties *at the beginning* of the protocol execution, but our results can be extended to the adaptive case.[4] Our results are with respect to an (often implicit) security parameter k, where we use the standard definition of negligible and overwhelming from [19].

Correlated Randomness as a Sampling Functionality. Our protocols are in the *correlated randomness* model, i.e., they assume that the parties initially, before receiving their inputs, receive appropriately correlated random strings. In particular, the parties jointly hold a vector $\boldsymbol{R} = (R_1, \ldots, R_n) \in (\{0,1\}^*)^n$, where p_i holds R_i, drawn from a given efficiently samplable distribution $\mathcal{D}$. This is, as usual, captured by giving the parties initial access to an ideal functionality $\mathcal{F}_{\mathsf{Corr}}^{\mathcal{D}}$, known as a *sampling functionality*, which, upon receiving a default input from any party, samples $\boldsymbol{R}$ from $\mathcal{D}$ and distributes it to the parties. Hence, a protocol in the correlated randomness model is formally an $\mathcal{F}_{\mathsf{Corr}}^{\mathcal{D}}$-hybrid protocol.

Information-Theoretic Signatures. Our protocols use information-theoretic (i.t.) signatures [45,43,46] to commit a party to messages it sends. Roughly speaking, these are information-theoretic analogues to standard digital signatures, i.e., they allow some party p_i, the *signer*, to send a message m to a party p_j, the *receiver*, along with a string σ that we refer to as the *signature*, such that the receiver can at a later point publicly open σ and prove to every party that the message m was indeed sent from p_i. Note that in order to achieve i.t. security the verification key cannot be publicly known. Rather, in i.t. signatures, the signer has a signing key sk and every party $p_i \in \mathcal{P}$ holds a different private verification key vk_i corresponding to sk.

In our protocols different (independent) signing keys are used for each signature. In this case, i.t. signatures provide the following guarantees with overwhelming probability (against an unbounded adversary): *(completeness)* A signature with the correct singing key will be accepted by any honest verifier in $\mathcal{P}$; *(unforgeability)* the adversary cannot come up with a signature that will be accepted by some (honest) verifier without knowing the signing key; *(consistency)* an adversarial signer cannot come up with a signature that will be accepted by some honest verifier and rejected by another.

3   Security with Identifiable Abort

We put forward the notion of *secure multi-party computation with identifiable abort*, also referred to as *Identifiable MPC (ID-MPC)* which allows the computation to fail (abort), but ensures that when this happens every party is informed about it, and they also agree on the index i of some corrupted party $p_i \in \mathcal{P}$ (we say then that *the parties abort with p_i*). More concretely, for an arbitrary functionality $\mathcal{F}$, we define $[\mathcal{F}]_{\perp}^{\mathrm{ID}}$ to be the corresponding functionality with identifiable abort, which behaves as $\mathcal{F}$ with the following modification: upon receiving from the simulator a special command (abort, p_i), where $p_i \in \mathcal{P}$ is a corrupted party (if p_i is not corrupted then $[\mathcal{F}]_{\perp}^{\mathrm{ID}}$ ignores the message), $[\mathcal{F}]_{\perp}^{\mathrm{ID}}$ sets the output of all (honest) parties to (abort, p_i).

---

[4] In fact, some of our protocols use optimizations tailored to proving adaptive security.

Definition 1. *Let $\mathcal{F}$ be a functionality and $[\mathcal{F}]_{\perp}^{\mathrm{ID}}$ be the corresponding functionality with identifiable abort. We say that a protocol π securely realizes $\mathcal{F}$ with identifiable abort if π securely realizes the functionality $[\mathcal{F}]_{\perp}^{\mathrm{ID}}$.*

The UC composition theorem extends in a straightforward manner to security with identifiable abort. The concrete composition statement can be found in the full version.

4 Unconditional ID-MPC from Correlated Randomness

In this section we describe our unconditionally secure identifiable MPC protocol in the correlated randomness model. In fact, our result is more general, as we provide a compiler that transforms any given unconditionally secure protocol in the semi-honest correlated randomness model into an unconditionally secure ID-MPC protocol in the (malicious) correlated randomness model. Although the correlated randomness provided by the setup in the malicious protocol is different than the semi-honest, the latter can be obtained from the former by an efficient transformation. Informally, our statement can be phrased as follows:

Let π_{sh} be an $\mathcal{F}_{\mathsf{Corr}}^{\mathcal{D}}$-hybrid protocol (for an efficiently computable distribution $\mathcal{D}$), which *unconditionally* UC realizes a functionality $\mathcal{F}$ in the presence of a *semi-honest* adversary. Then there exists a compiler turning π_{sh} into an $\mathcal{F}_{\mathsf{Corr}}^{\mathcal{D}'}$-hybrid protocol (for an appropriate efficiently computable distribution $\mathcal{D}'$), which *unconditionally* UC realizes $\mathcal{F}$ *with identifiable abort* (in the malicious model).

Overview of the Compiler. We start by providing a high-level overview of our compiler. As is typical, the semi-honest protocol π_{sh} which we compile works over standard point-to-point (insecure) channels. Furthermore, without loss of generality (see Section 4.3) we assume that π_{sh} is *deterministic*.

Let $\boldsymbol{R}^{\mathrm{sh}} = (R_1^{\mathrm{sh}}, \ldots, R_n^{\mathrm{sh}})$ denote the setup used by the semi-honest protocol π_{sh} (i.e., each p_i holds string R_i^{sh}). The setup for the compiled protocol distributes $\boldsymbol{R}^{\mathrm{sh}}$ to the parties, and commits every party to its received string. Subsequently, the parties proceed by, first, committing to their inputs and, then, executing their π_{sh}-instructions in a publicly verifiable manner: whenever, p_i would send a message m in π_{sh}, in the compiled protocol p_i broadcasts m and publicly proves, in zero-knowledge, that the broadcasted message is consistent with his committed input and setup string R_i^{sh}. For the above approach to work for unbounded adversaries and allow for identifiability, we need the commitment scheme and the associated zero-knowledge proofs to be unconditionally secure and failures to be publicly detectable. We construct such primitives relying on appropriately correlated randomness in Sections 4.1 and 4.2, respectively.

4.1 Commitments with Identifiable Abort

In this section we provide a protocol which unconditionally UC realizes the standard (one-to-many) multi-party commitment functionality $\mathcal{F}_{\mathrm{COM}}$ with identifiable abort. $\mathcal{F}_{COM}$ allows party $p_i \in \mathcal{P}$, the *committer*, to commit to a message m and later on publicly open m while guaranteeing the following properties: (hiding) no party in $\mathcal{P} \setminus \{p_i\}$ receives any information on m during the commit phase; (binding) at the end of the commit phase a message m' is fixed (where $m' = m$ if the committer is honest), such that only m' might be accepted in the reveal phase (and m' is always accepted when the committer is honest).

Our protocol Π_{COM} which i.t. securely realizes $\mathcal{F}_{COM}$ with identifiable abort assumes the following correlated-randomness setup: for p_i to commit to a value $m \in \{0,1\}^*$, p_i needs to hold a uniformly random string $r \in \{0,1\}^{|m|}$ along with an information-theoretic signature σ on r, where every party in $\mathcal{P}$ holds his corresponding verification key (but no party, not even p_i, gets to learn the signing key). Given the above setup, p_i can commit to m by broadcasting $y = m \oplus r$. To, later on, open the commitment y, p_i broadcasts r along with the signature σ, where every party verifies the signature and outputs $m = y \oplus r$ if it is valid, otherwise aborts with p_i (i.e., outputs (abort, p_i)).

The hiding property of Π_{COM} follows from the fact that r is uniformly random. Moreover, the unforgeability of the signature scheme ensures that the commitment is binding and publicly verifiable. Finally, the completeness of the scheme ensures that the protocol aborts only when the committer p_i is corrupted. Additionally, same as all UC commitments, the above scheme is *extractable*, i.e., the simulator of a corrupted committer can learn, already in the commit phase, which message will be opened so that he can input it to the functionality, and *equivocal*, i.e., the simulator of a corrupted receiver can open a commitment to any message of his choice.[5] Taking a glimpse at the proof both properties follow from the fact that the simulator controls the setup: knowing r allows the simulator to extract m from the broadcasted message, whereas knowing the signing key sk allows him to generate a valid signature/opening to any message.

Theorem 1. *The protocol Π_{COM} unconditionally UC realizes the functionality $\mathcal{F}_{COM}$ with identifiable abort.*

4.2 Setup-Commit-Then-Proof

Next we present a protocol which allows the parties receiving random strings (drawn from some joint distribution $\mathcal{D}$) to publicly prove, in zero-knowledge, that they use these strings in a protocol. Our protocol implements the *Setup-Commit-then-Prove* functionality $\mathcal{F}_{SCP}$ which can be viewed as a modification of the Commit-then-Prove functionality from [7] restricting the committed witnesses to be distributed by the setup instead of being chosen by the provers. More concretely $\mathcal{F}_{SCP}$ works in two phases: in a first phase, it provides a string/witness R_i to each $p_i \in \mathcal{P}$, where $\boldsymbol{R} = (R_1, \ldots, R_n)$ is drawn from $\mathcal{D}$; in a second phase, $\mathcal{F}_{SCP}$ allows every party p_i to prove q-many NP statements of the type

---

[5] In [31] a primitive called *unanimously identifiable commitments* (UIC) was introduced for this purpose, but the definition of UIC does not guarantee all the properties we need for UC secure commitments.

$\mathcal{R}(x, R_i) = 1$ for the same publicly known NP relation $\mathcal{R}_i$ and the witness R_i received from the setup, but for potentially different (public) strings x.

In the remainder of this section we describe a protocol which unconditionally securely realizes the setup-commit-then-proof functionality $\mathcal{F}_{\text{SCP}}$ in the correlated randomness model. To this direction, we first show how to realize the sigle-use version of $\mathcal{F}_{\text{SCP}}$, denoted as $\mathcal{F}_{\text{1SCP}}$, and then use the UC composition with joint state theorem (JUC) [8] to derive a protocol for $\mathcal{F}_{\text{SCP}}$. The functionality $\mathcal{F}_{\text{1SCP}}$ works exactly as $\mathcal{F}_{\text{SCP}}$ with the restriction that it allows a prover $p \in \mathcal{P}$ to do a *single* (instead of q-many) proofs for a witness w of a given NP relation $\mathcal{R}$.

Our protocol for realizing the functionality $\mathcal{F}_{\text{1SCP}}$ with identifiable abort uses the idea of "MPC in the head" [25,30,32]. In particular, let $\mathcal{F}_{\text{D}}$ denote the $(n+1)$-party (reactive) functionality among the players in $\mathcal{P}$ and a special player p_D, the *dealer*, which works as follows: In a first phase, $\mathcal{F}_{\text{D}}$ receives a message $w \in \{0,1\}^{\text{poly}(k)}$ from p_D and forwards w to $p \in \mathcal{P}$. In a second phase, p sends x to $\mathcal{F}_{\text{D}}$, which computes $b := \mathcal{R}(x, w)$ and outputs (b, x) to every $p_j \in \mathcal{P} \setminus \{p_D\}$. Clearly, any protocol in the plain model which unconditionally realizes $\mathcal{F}_{\text{D}}$ with an honest dealer p_D, where p_D does not participate in the second phase, can be turned into a protocol which securely realizes $\mathcal{F}_{\text{1SCP}}(\mathcal{P}, \mathcal{D}, \mathcal{R}, p)$ in the correlated randomness model. Indeed, one needs to simply have the corresponding sampling functionality play the role of p_D (where w is drawn from $\mathcal{D}$). In the following we show how to design such a protocol using the idea of player-simulation [25].

Let $\Pi_{(n+1,m),t}$ be a protocol which perfectly securely (and robustly) realizes $\mathcal{F}_{\text{D}}$ in the client-server model [25,30,32], among the clients $\mathcal{P} \cup \{p_D\}$ and an additional m servers. Such a protocol exists assuming $t < m/3$ servers are corrupted [2]. For simplicity, assume that $\Pi_{(n+1,m),t}$ has the following properties, which are consistent to how protocols from the literature, e.g., [2], would realize functionality $\mathcal{F}_{\text{D}}$ in the client-server setting: (i) for computing the first phase of $\mathcal{F}_{\text{D}}$, $\Pi_{(n+1,m),t}$ has p_D share his input w among the m servers with a secret sharing scheme that is perfectly t-*private* (the shares of any t servers leak no information on w) and perfectly t-*robust* (the sharing can be reconstructed even when up to t cheaters modify their shares), and, also p_D hands *all* the shares to p (ii) p_D does not participate in the second phase of $\Pi_{(n+1,m),t}$ (this is wlog as p_D is a client with no input or output in this second phase), and (iii) the output $(\mathcal{R}(x, w), x)$ is publicly announced (i.e., is in the view of every server at the end of the protocol).

Assuming p_D is honest, a protocol Π_{n+1} for unconditionally realizing $\mathcal{F}_{\text{D}}$ with identifiable abort (among only the players in $\mathcal{P} \cup \{p_D\}$) can be built based on the above protocol $\Pi_{(n+1,m),t}$ as follows: for the first phase, p_D generates shares of a t-robust and t-private sharing of w as he would do in $\Pi_{(n+1,m),t}$ and sends them to p. In addition to sending the shares, p_D commits p to each share by sending him an i.t. signature on it and distributing the corresponding verification keys to the players in $\mathcal{P}$. For the second phase, p emulates in his head the second phase of the execution of $\Pi_{(n+1,m),t}$ among m virtual servers $\hat{p}_1, \ldots, \hat{p}_m$ where each server has private input his share, as received from p_D in the first phase, and a public input x (the same for all clients); p publicly commits to the view of each server. Finally, the parties in $\mathcal{P} \setminus \{p\}$ challenge p to open a random subset $\mathcal{J} \subseteq [m]$ of size t of the committed views and announce the corresponding input-signatures which p received from p_D. If the opened views are inconsistent with an accepting execution of $\Pi_{(n+1,m),t}$ on input x and the committed shares—i.e., some output is 0, or some opening fails, or some signature does not verify for the corresponding (opened) private input, or for some pair of views the incoming messages do not match the outgoing messages—then the parties abort with p.

The security of the protocol Π_{n+1} is argued similarly to [30, Theorem 4.1]: on the one hand, when p is honest then we can use the simulator for $\Pi_{(n+1,m),t}$ to simulate that views of the parties in $\mathcal{J}$. The perfect t-security of $\Pi_{(n+1,m),t}$ and the t-privacy of the sharing ensures that this simulation is indistinguishable from the real execution. On the other hand, when p is corrupted, then we only need to worry about correctness. Roughly, correctness is argued as follows: if there are at most $t < m/3$ incorrect views, then the t-robustness of $\Pi_{(n+1,m),t}$ and of the sharing ensures that the output in any of the other views will be correct; by a standard counting argument we can show that the probability that some of these views is opened is overwhelming when $m = O(k)$. Otherwise, (i.e., if there are more than t-incorrect views) then with high probability a pair of such views will be opened and the inconsistency will be exposed.

To derive, from Π_{n+1}, a protocol for $\mathcal{F}_{1\text{SCP}}(\mathcal{P}, \mathcal{D}, \mathcal{R}, p)$ in the correlated randomness model, we have the sampling functionality, $\mathcal{F}_{\text{Corr}}^{1\text{SCP}}$ play the role of the dealer p_D. In addition to the committed shares, $\mathcal{F}_{\text{Corr}}^{1\text{SCP}}$ generates the necessary setup enabling any prover $p \in \mathcal{P}$ to commit to the m (virtual) servers' views in the second phase of the protocol Π_{n+1}. Furthermore, to simplify the description, we also have $\mathcal{F}_{\text{Corr}}^{1\text{SCP}}$ create a "coin-tossing setup" which players in $\mathcal{P}$ can use to sample the random subset $\mathcal{J} \in [m]$ of views to be opened: $\mathcal{F}_{\text{Corr}}^{1\text{SCP}}$ hands to each $p_j \in \mathcal{P}$ a random string c_j and commits p_j to it; the coin sequence c for choosing $\mathcal{J}$ is then computed by every p_j opening c_j and taking $c = \oplus_{j=1}^{n} c_j$. In the following we give a detailed description of the protocol $\Pi_{1\text{SCP}}$ for implementing $\mathcal{F}_{1\text{SCP}}$, where we denote by $\langle w \rangle = (\langle w \rangle^1, \ldots, \langle w \rangle^m)$ a perfectly t-private and t-robust secret sharing of a given value w among players in some $\hat{\mathcal{P}} = (\hat{p}_1, \ldots, \hat{p}_m)$ (e.g., the sharing from [2] which is based on bivariate polynomials), where $\langle w \rangle^i$ denotes the ith share of $\langle w \rangle$, i.e., the state of the (virtual) server $\hat{p}_i$ after the sharing is done.

Theorem 2. *Let $\Pi_{(n+1,m),t}$ be a protocol as described above among $n + 1$ clients and $m = O(k)$ servers which perfectly securely (and robustly) realizes the functionality $\mathcal{F}_D$ in the presence of $t < m/3$ corrupted servers. The $(\mathcal{F}_{\text{Corr}}^{1\text{SCP}}(\mathcal{P}, \mathcal{D}, m, t, \mathcal{R})$-hybrid) protocol $\Pi_{1\text{SCP}}(\mathcal{P}, \mathcal{D}, \mathcal{R}, m, t, p)$ unconditionally securely realizes the functionality $\mathcal{F}_{1\text{SCP}}(\mathcal{P}, \mathcal{D}, \mathcal{R}, p)$ with identifiable abort.*

The Multiple-Proof Extension of $\mathcal{F}_{1\text{scp}}$. In order to realize functionality $\mathcal{F}_{\text{SCP}}$ we need to extend $\mathcal{F}_{1\text{SCP}}$ to distribute a vector $\boldsymbol{R} = (R_1, \ldots, R_n)$ of witnesses, one for each party, (instead of only one witness) sampled from some efficient distribution $\mathcal{D}$, and allow every $p_i \in \mathcal{P}$ to prove up to q statements of the type

---

Protocol $\Pi_{1\text{SCP}}(\mathcal{P}, \mathcal{D}, m, t, \mathcal{R}, p)$

Setup-Commit Phase: To obtain the appropriate setup, prover p sends $(\text{CorrRand}, p)$ to the sampling functionality $\mathcal{F}_{\text{Corr}}^{\text{1SCP}}(\mathcal{P}, \mathcal{D}, m, t, \mathcal{R})$, which distributes the following random strings and signatures (where every $p_j \in \mathcal{P}$ receives the corresponding verification keys):

- The prover p receives a sharing $\langle w \rangle = (\langle w \rangle^1, \ldots, \langle w \rangle^m)$ of w along with corresponding signatures $\sigma(\langle w \rangle^1), \ldots, \sigma(\langle w \rangle^m)$ and (privately) outputs $(\text{witness}, w)$.
- Every $p_i \in \mathcal{P}$ receives the challenge-string c_i along with a corresponding signature $\sigma(c_i)$.
- The prover also receives random strings $v_1, \ldots, v_m$ along with corresponding signatures $\sigma(v_1), \ldots, \sigma(v_m)$ to use for committing to the server's views in $\Pi_{(n+1,m),t}$.

Prove Phase: Upon p receiving his input $(\text{ZK-prover}, x)$ the following steps are executed:

1. If $\mathcal{R}(x, w) = 0$ then p broadcasts $(\text{not-verified}, p)$ and every party halts with output $(\text{not-verified}, p)$. Otherwise, p broadcasts $(\mathcal{R}, x)$.

2. p emulates in his head the second phase of protocol $\Pi_{(n+1,m),t}$ where each server $\hat{p}_j \in \hat{\mathcal{P}} = \{\hat{p}_1, \ldots, \hat{p}_m\}$ has private input $\langle w \rangle^j$ and public input x.

3. For each $\hat{p}_j \in \hat{\mathcal{P}}$, p commits, by invocation of protocol $\Pi_{\text{COM}}(\mathcal{P})$, to the view $\text{VIEW}_j \in \{0,1\}^{V_j}$ of $\hat{p}_j$ in the above emulated execution using v_j from his setup.

4. For each $p_i \in \mathcal{P}$: p_i announces the random string c_i and the corresponding signature $\sigma(c_i)$ and every $p_j \in \mathcal{P}$ verifies, using his corresponding verification keys, validity of the signatures and aborts with p_i in case the check fails.

5. The parties compute $c = \sum_{i=1}^{n} c_i$ and use it as random coins to sample a random t-size set $\mathcal{J} \subseteq [m]$.

6. For each $j \in \mathcal{J}$: p opens the commitment to VIEW_j and announces the signature $\sigma(\langle w \rangle^j)$. If any of the openings fails or any of the announced signatures is not valid for the input-share appearing in the corresponding view, then the protocol aborts with p_i.

7. Otherwise, the parties check that the announced views are consistent with an execution of protocol $\Pi_{(n+1,m),t}$ with the announced inputs in which the (global) output is 1, i.e., they check that in all the announced views the output equals 1 and all signatures are valid, and that for all pairs $(j, k) \in \mathcal{J}^2$: the incoming messages in $\hat{p}_j$'s view match the outgoing messages in $\hat{p}_k$'s view. If any of these checks fails then the protocol aborts with p_i, otherwise, every party outputs $(\text{verified}, x, p)$.

$\mathcal{R}(R_i, x)$ for potentially different public inputs x. The corresponding sampling functionality is derived as follows: it first samples $\boldsymbol{R}$ and subsequently it emulates, for each $p_i \in \mathcal{P}$, q independent setups for $\Pi_{1\text{SCP}}$ (for the same random value R_i and relation $\mathcal{R}_i$). Given such a sampling functionality the protocol Π_{SCP} for unconditionally securely realizing $\mathcal{F}_{\text{SCP}}$ with identifiable abort is straight-forward: The parties receive the random strings $R_1, \ldots, R_n$ along with q proof setups for each party. Then, for each invocation of the prove phase, party p_i executes the prove phase of protocol Π_{1SCP} using the corresponding proof setup.[6]

Theorem 3. *Protocol* $\Pi_{SCP}(\mathcal{P}, \mathcal{D}, \mathcal{R}, q)$ *unconditionally securely realizes the functionality* $\mathcal{F}_{SCP}(\mathcal{P}, \mathcal{D}, \mathcal{R}, q)$ *with identifiable abort.*

The proof follows from the security of Π_{1SCP} by a direct application of the universal composition with joint state (JUC) theorem [8].

4.3 The "Semi-honest to Malicious with Abort" Compiler

We are now ready to describe our main compiler, denoted as $C(\cdot)$ which compiles any given protocol π_{sh} secure in the semi-honest model using (only) correlated randomness into a protocol $C(\pi_{sh})$ which is secure with abort in the (malicious) correlated randomness model.[7]

We make the following simplifying assumptions on the semi-honest protocol π_{sh} which are without loss of generality, since all existing semi-honest protocols in the correlated randomness model can be trivially turned to satisfy them:

- We assume that π_{sh} has a known (polynomial) upper bound $\mathrm{Rnd}_{\pi_{sh}}$ on the number of rounds, where in each round every party sends a single message.
- We assume that π_{sh} is *deterministic*. Any π_{sh} can be turned into such by having the setup include for each $p_i \in \mathcal{P}$ a uniformly random and independent string r_i that p_i uses as his coins.
- Finally, we assume that π_{sh} starts off by having every party send to all parties a one-time pad encryption of his input x_i using as key the first $|x_i|$ bits from r_i (those bits are not reused). Clearly, this modification does not affect the security of π_{sh} as the simulator can easily simulate this step by broadcasting a random string. Looking ahead in the proof, this will allow the simulator to extract the corrupted parties' inputs.

The compiler $C(\pi_{sh})$ uses the protocol Π_{SCP} as follows: Denote by $R^{sh} = (R_1^{sh}, \ldots, R_n^{sh})$ the setup used by π_{sh} and by $\mathcal{D}^{sh}$ the corresponding distribution. Let also $\mathcal{R}_{\pi_{sh}, i}$ denote the relation corresponding to p_i's next message function. More concretely, if $h_{\pi_{sh}, i} \in \{0, 1\}^*$ denotes the history of messages seen by p_i and m is a message, then $\mathcal{R}_{\pi_{sh}, i}((h_{\pi_{sh}, i}, m), R_i) = 1$ if m is the next message of p_i in an execution with history $h_{\pi_{sh}}$, and setup R_i, otherwise $\mathcal{R}_{\pi_{sh}, i}((h_{\pi_{sh}, i}, m), R_i) = 0$. The compiled protocol $C(\pi_{sh})$ starts by executing the setup-commit phase of protocol $\Pi_{SCP}(\mathcal{P}, \mathcal{D}^{sh}, \mathcal{R} = (\mathcal{R}_{\pi_{sh}, 1}, \ldots, \mathcal{R}_{\pi_{sh}, n}), \mathrm{Rnd}_{\pi_{sh}})$. Subsequently, every $p_i \in \mathcal{P}$ executes his π_{sh} instructions, where in each round instead of sending its message m over the point-to-point channel, p_i broadcasts m and proves, using the proof phase of protocol Π_{SCP}, that $\mathcal{R}_{\pi_{sh}, n}((h_{\pi_{sh}}, m), R_i) = 1$. If Π_{SCP} aborts with some p_i then our compiler also aborts with p_i. Otherwise, the security of Π_{SCP} ensures that every p_i followed π_{sh} for the given setup; therefore, security of our

---

[6] Recall that we implicitly assume that all messages generated from the setup have unique identifiers so that the parties know which ones to use for which proof.

[7] Note that $C(\pi_{sh})$ uses broadcast which can be trivially realized by a protocol assuming appropriate correlated randomness, e.g., [40].

compiler follows from the security of π_{sh}. Note that the corresponding sampling functionality for $C(\pi_{sh})$ is computable in time polynomial in the running time of the sampling functionality $\mathcal{F}_{\mathsf{Corr}}^{\mathcal{D}^{sh}}$ for protocol π_{sh}.

Theorem 4. *Let π_{sh} be a protocol as above which unconditionally UC realizes a functionality $\mathcal{F}$ in the presence of a semi-honest adversary in the $\mathcal{F}_{\mathsf{Corr}}^{\mathcal{D}^{sh}}$-hybrid (correlated randomness) model. Then the compiled protocol $C(\pi_{sh})$ unconditionally UC realizes the functionality $\mathcal{F}$ with identifiable abort in the presence of a malicious adversary in the $\mathcal{F}_{\mathsf{Corr}}^{\mathrm{SCP}}$-hybrid (correlated randomness) model.*

Note that any (semi-honest) OT-hybrid protocol can be cast as a protocol in the correlated randomness model by precomputing the OT. Hence, by instantiating π_{sh} with any semi-honest OT hybrid protocol. e.g., [20], we obtain the following corollary.

Corollary 1. *There exists a protocol which unconditionally UC realizes any well-formed [7] multi-party functionality with identifiable abort.*

The question of feasibility of unconditional security with identifiable abort from correlated randomness has been open even in the simpler *standalone* model [21,19,4]. As a corollary of Theorem 4 one can derive a positive statement also for that model.

Corollary 2 (Stand-alone security with identifiable abort). *There exists a protocol which unconditionally securely evaluates any given function f with identifiable abort in the stand alone correlated randomness model.*

5   SFE Using Black-Box OT

In this section, we provide a generic MPC protocol which is (computationally) secure with identifiable abort making black-box use of an (adaptively) secure UC protocol for one-out-of two oblivious transfer $\mathcal{F}_{\mathsf{OT}}$ (see [39] for a formal description) in the Common Reference String (CRS) model.

The high-level idea of our construction is the following: as we have already provided an unconditional implementation of ID-MPC based (only) on correlated randomness, it suffices to provide a protocol $\Pi_{\mathsf{Corr}}^{\mathsf{CSP}}$ with the above properties for implementing the corresponding sampling functionality $\mathcal{F}_{\mathsf{Corr}}^{\mathsf{SCP}}$. Indeed, given such a protocol $\Pi_{\mathsf{Corr}}^{\mathsf{CSP}}$, we can first use it to compute the setup needed for $C(\pi_{sh})$ (for any appropriate semi-honest protocol π_{sh}, e.g., the one from [21]) and then use π_{sh} to evaluate any given functionality; if either the setup generation or π_{sh} aborts with some p_i then the construction also aborts with p_i.

In the remainder of this section we describe $\Pi_{\mathsf{Corr}}^{\mathcal{F}_{\mathsf{Corr}}^{\mathsf{SCP}}}$. In fact, we provide a protocol $\Pi_{\mathsf{Corr}}^{\mathcal{D}}$ which allows to implement any sampling functionality $\mathcal{F}_{\mathsf{Corr}}^{\mathcal{D}}$ for a given efficiently computable distribution $\mathcal{D}$. The key idea behind our construction in the following: as the functionality $\mathcal{F}_{\mathsf{Corr}}^{\mathcal{D}}$ receives no (private) inputs from the parties, we can have every party commit to its random tape, and then attempt to realize $\mathcal{F}_{\mathsf{Corr}}^{\mathcal{P}}$ by a protocol which is secure with (non-identifiable) abort; if the evaluation aborts then the parties open the commitments to their random tapes and use these tapes to detect which party cheated. Note that, as the parties have no private inputs, announcing their views does not violate privacy of the computation.

For the above idea to work we need to ensure that deviation from the honest protocol can be consistently detected by every party (upon opening the committed random coins). Therefore, we define the following $\mathcal{P}$-*verifiability* property. For any given execution of a protocol Π, we say that a party p_i *correctly executed* Π *with respect to* (x_i, r_i) *(up to round ρ) in the CRS model* if p_i sent all his messages as instructed by Π on this input x_i, random coins r_i and the common reference string C. Let Π be a protocol in the CRS model which starts by having every party commit to its random tape. Π is $\mathcal{P}$-*verifiable* if there exists a deterministic polynomial algorithm $\mathcal{D}$, called *the detector*, with the following property: given the CRS, the inputs of the parties, their committed randomness, and the view of any honest p_j, $\mathcal{D}$ outputs the identity of a party $p_i \in \mathcal{P}$ who did not correctly execute Π (if such a party exists).

In the remainder of this section we provide the details of our protocol. As our protocols makes black-box use of a UC secure 12OT protocol in the CRS model, for it to be $\mathcal{P}$-verifiable the underlying 12OT protocol needs to also be $\mathcal{P}$-verifiable. Therefore, in the following, first, we show how to obtain from any given OT protocol Π_{OT} a $\mathcal{P}$-verifiable OT protocol Π_{VOT} (making black-box use of Π_{OT}), and, subsequently, we show how to use Π_{VOT} to transform an OT-hybrid SFE protocol into a $\mathcal{P}$-verifiable SFE protocol in the CRS model. Finally, at the end of the current section, we show how to use our $\mathcal{P}$-verifiable SFE protocol to implement any sampling functionality $\mathcal{F}_{\mathsf{Corr}}^{\mathcal{P}}$ with identifiable abort making black-box use of Π_{OT}.

$\mathcal{P}$-Verifiable OT. Let Π_{OT} be a (two-party) protocol which adaptively UC securely realizes $\mathcal{F}_{\mathsf{OT}}$, among parties p_1 and p_2 in the CRS model (e.g., [10,39]). For $i \in \{1, 2\}$ denote by $f_{\Pi_{\mathsf{OT}}}^i$ the next message function of p_i defined as follows: let VIEW_i be the view of party p_i at the beginning of round ρ in an execution of Π_{OT};[8] then $f_{\Pi_{\mathsf{OT}}}^i(\mathrm{VIEW}_i) = m$ is the message which p_i sends in round ρ of protocol Π_{OT}, given that his current view is VIEW_i (if ρ is the last round, then, by default, $m = (\mathsf{out}, y)$, where y is p_i's output). Observe that $f_{\Pi_{\mathsf{OT}}}$ is a deterministic function. Without loss of generality, assume that protocol Π_{OT} has a known number of rounds $\mathrm{Rnd}_{\Pi_{\mathsf{OT}}}$, where in each round only one of the parties p_1 and p_2 sends a message (from $\{0, 1\}^k$). Let, also, $\mathcal{F}_{\mathsf{OT}}^{\mathcal{P}}$ denote the multi-party extension of $\mathcal{F}_{\mathsf{OT}}$, in which parties other than p_1 and p_2 provide a default input and receive a default output, i.e., $\mathcal{F}_{\mathsf{OT}}^{\mathcal{P}}$ corresponds to the function $f_{\mathsf{OT}}^{\mathcal{P}}((x_0, x_1), b, \lambda, \ldots, \lambda) = (\bot, x_b, \bot, \ldots, \bot)$. We describe a multi-party $\mathcal{P}$-verifiable protocol Π_{VOT} which securely realizes the functionality $\mathcal{F}_{\mathsf{OT}}^{\mathcal{P}}$.

The protocol Π_{VOT} works as follows: initially, every party commits to its random tape. Subsequently, the parties execute their Π_{OT} instructions with the

---

[8] Recall that VIEW_i consists of the inputs and randomness of p_i along with all messages received up to round r.

following modification: whenever, for $i, j \in \{1, 2\}$, p_i is to send a message $m \in \{0, 1\}^k$ to p_j, he chooses the first k unused bits from his random tape (denote by K the string resulting by concatenating these bits), broadcast a one-time pad encryption $c = m \oplus K$ of m with key K, and *privately* opens the corresponding commitments towards p_j. If the opening fails then p_j publicly complains and p_i replies by broadcasting K; p_j recovers m by decrypting c. Clearly, the above modification does not affect the security of Π_{OT} (as all keys are chosen using fresh and independent randomness), therefore Π_{VOT} securely realizes $\mathcal{F}_{\text{OT}}^{\mathcal{P}}$. Additionally, the above protocol is $\mathcal{P}$-verifiable: indeed, because the entire transcript is broadcasted, the view of any party contains all information needed to check whether or not the transcript is consistent with any given set of inputs and committed randomness. For simplicity, in the following we state the security in the $\{CRS, \hat{\mathcal{F}}_{\text{COM}}\}$-*hybrid model* i.e., where, in addition to the CRS the protocol can make ideal calls to a (one-to-many) commitment functionality $\hat{\mathcal{F}}_{\text{COM}}$ which behaves exactly as $\mathcal{F}_{\text{COM}}$ but allows both public and private opening of the committed value. [9] We point out that all security statements in the lemma are with respect to an adaptive adversary.

Lemma 1. *Assuming Π_{OT} UC securely realizes the two-party 12OT functionality $\mathcal{F}_{\text{OT}}$ in the CRS model, the protocol Π_{VOT} (defined above) satisfies the following properties: (security) Π_{VOT} UC securely realizes the multi-party extension $\mathcal{F}_{\text{OT}}^{\mathcal{P}}$ of $\mathcal{F}_{\text{OT}}$ (defined above) in the $\{CRS, \hat{\mathcal{F}}_{\text{COM}}\}$-hybrid model; ($\mathcal{P}$-verifiability) Π_{VOT} is $\mathcal{P}$-verifiable. Furthermore, Π_{VOT} makes black-box use of (the next-message function of) Π_{OT}.*

$\mathcal{P}$-Verifiable MPC with (Non-identifiable) Abort. The next step is to add verifiability to a given adaptively UC secure OT-hybrid MPC protocol $\Pi^{\mathcal{F}_{\text{OT}}}$. Wlog, we assume that $\Pi^{\mathcal{F}_{\text{OT}}}$ only makes calls to $\mathcal{F}_{\text{OT}}$ and to a broadcast channel. (Indeed, $\mathcal{F}_{\text{OT}}$ can be used to also implement secure bilateral communication as follows: to send message x, the sender inputs (x, x) and the receiver input $b = 1$.)

Denote by $\Pi^{\Pi_{\text{VOT}}}$ the version of $\Pi^{\mathcal{F}_{\text{OT}}}$ which starts off by having every party publicly commit to its random tape and has all calls to $\mathcal{F}_{\text{OT}}$ replaced by invocations of protocol Π_{VOT} instantiated with fresh/independent randomness. More precisely, $\Pi^{\Pi_{\text{VOT}}}$ is derived from $\Pi^{\mathcal{F}_{\text{OT}}}$ as follows:

– Initially every party commits to its random tape using one-to-many commitments.

– All calls to $\mathcal{F}_{\text{OT}}$ (including the ones used as above to implement bilateral communication) are replaced by invocations of protocol Π_{VOT}. (The random coing do not need to be committed again; the above commitments are used in the invocations of Π_{VOT}.)

– For each party p_i a specific part of p_i's random tape is associated with each invocation of Π_{VOT}. This part is used only in this invocation and nowhere else in the protocol.

---

[9] We can use any of the CRS-based commitment protocols [6,7] to instantiate $\hat{\mathcal{F}}_{\text{COM}}$.

The following lemma states the achieved security, where as in Lemma 1 all security statements are with respect to an adaptive adversary. The proof follows from the security of $\Pi^{\mathcal{F}_{OT}}$ and the security/$\mathcal{P}$-verifiability of Π_{VOT}.

Lemma 2. *Let $\mathcal{F}$ be a UC functionality and $\Pi^{\mathcal{F}_{OT}}$ be a protocol which unconditionally UC securely realizes $\mathcal{F}$ in the $\mathcal{F}_{OT}$-hybrid model with (non-identifiable) abort, and for a protocol Π_{OT} which UC securely realizes $\mathcal{F}_{OT}$ in the CRS model, let Π_{VOT} be the corresponding $\mathcal{P}$-verifiable protocol (as in Lemma 1). Then protocol $\Pi^{\Pi_{VOT}}$, defined above, satisfies the following properties: (security) $\Pi^{\Pi_{VOT}}$ UC securely realizes $\mathcal{F}$ with (non-identifiable) abort in the $\{CRS, \hat{\mathcal{F}}_{COM}\}$-hybrid model; ($\mathcal{P}$-verifiability) Protocol $\Pi^{\Pi_{VOT}}$ is $\mathcal{P}$-verifiable. Furthermore, $\Pi^{\Pi_{VOT}}$ makes black-box use of (the next-message function of) Π_{OT}.*

The Setup Compiler. We next describe the protocol $\Pi^{\mathcal{D}}_{Corr}$ which securely realizes any given sampling functionality $\mathcal{F}^{\mathcal{D}}_{Corr}$ (for an efficiently computable distribution $\mathcal{D}$), while making black-box use of a UC secure OT-protocol in the CRS model and ideal calls to $\hat{\mathcal{F}}_{COM}$. The idea is to, first, have every party commits to its random coins and then invoke $\Pi^{\Pi_{VOT}}$ to securely realize functionality $\mathcal{F}^{\mathcal{D}}_{Corr}$ using these coins; if the evaluation aborts, then the parties open their committed randomness and use the detector $\mathcal{D}$ to figure out which party cheated. Because the parties have no inputs, opening their randomness does not violate privacy.

Unfortunately, the above over-simplistic protocol is not simulatable. Intuitively, the reason is that $\Pi^{\Pi_{VOT}}$ might abort after the adversary has seen his outputs of $\mathcal{F}^{\mathcal{D}}_{Corr}$, in which case the simulator needs to come up with random coins for the simulated honest parties which are consistent with the adversary's view. We resolve this by the following technical trick, which ensures that $\mathcal{S}$ needs to invoke $\mathcal{F}^{\mathcal{D}}_{Corr}$ only if the computation of $\Pi^{\Pi_{VOT}}$ was successful: instead of directly computing $\mathcal{F}^{\mathcal{D}}_{Corr}$, we use $\Pi^{\Pi_{VOT}}$ to realize the functionality $\langle \mathcal{F}^{\mathcal{D}}_{Corr} \rangle$ which computes an authenticated (by means of i.t. signatures) n-out-of-n secret sharing of the output of $\mathcal{F}^{\mathcal{D}}_{Corr}$. This sharing is then reconstructed by having every party announce its share. The authenticity of the output sharing ensures that either the reconstruction will succeed or a party that did not announce a properly signed share will be caught, in which case the protocol identifies this party.

Theorem 5. *Assuming Π_{OT}, Π_{VOT}, and $\Pi^{\Pi_{VOT}}$ as in Lemma 2, the protocol $\Pi^{\mathcal{D}}_{Corr}$ securely realizes $\mathcal{F}^{\mathcal{D}}_{Corr}$ with identifiable abort in the CRS model while making black-box use of Π_{OT} and ideal calls to the commitment functionality $\hat{\mathcal{F}}_{COM}$.*

By combining Theorems 4 and 5 with the universal composition theorem, and instantiating $\Pi^{\Pi_{VOT}}$ with the IPS protocol [32] we obtain the following corollary.

Corollary 3. *There exists a protocol which UC realizes any given functionality with identifiable abort, while making black-box use of a protocol for UC realizing $\mathcal{F}_{OT}$ and a protocol for UC realizing $\hat{\mathcal{F}}_{COM}$ in the CRS model.*

The Stand-alone Model. The proof of Theorem 5 does not use the equivocality of the commitments. Therefore, assuming an adaptive 12OT protocol and extractable commitments, it can be carried over to the stand-alone setting. Such extractable commitments can be constructed by making a black-box use of a one-way function [38], which in turns can be obtained via a black-box use of OT [27]. Thus, we get the following result for the stand-alone model (see full version for proof).

Lemma 3 (Stand-alone). *There exists a protocol which securely realizes any given functionality with identifiable abort in the plain model making black-box use of an adaptively secure OT protocol in the plain model.*

References

1. Beimel, A., Lindell, Y., Omri, E., Orlov, I.: 1/p-secure multiparty computation without honest majority and the best of both worlds. In: Rogaway, P. (ed.) CRYPTO 2011. LNCS, vol. 6841, pp. 277–296. Springer, Heidelberg (2011)
2. Ben-Or, M., Goldwasser, S., Wigderson, A.: Completeness theorems for non- cryptographic fault-tolerant distributed computations. In: 20th ACM STOC, pp. 1–10. ACM Press (1988)
3. Bendlin, R., Damgård, I., Orlandi, C., Zakarias, S.: Semi-homomorphic encryption and multiparty computation. In: Paterson, K.G. (ed.) EUROCRYPT 2011. LNCS, vol. 6632, pp. 169–188. Springer, Heidelberg (2011)
4. Canetti, R.: Security and composition of multiparty cryptographic protocols. Journal of Cryptology 13(1), 143–202 (2000)
5. Canetti, R.: Universally composable security: A new paradigm for cryptographic protocols. In: 42nd FOCS, pp. 136–145. IEEE Computer Society Press (2001)
6. Canetti, R., Fischlin, M.: Universally composable commitments. In: Kilian, J. (ed.) CRYPTO 2001. LNCS, vol. 2139, pp. 19–40. Springer, Heidelberg (2001)
7. Canetti, R., Lindell, Y., Ostrovsky, R., Sahai, A.: Universally composable two-party and multi-party secure computation. In: 34th ACM STOC, pp. 494–503. ACM Press (2002)
8. Canetti, R., Rabin, T.: Universal composition with joint state. In: Boneh, D. (ed.) CRYPTO 2003. LNCS, vol. 2729, pp. 265–281. Springer, Heidelberg (2003)
9. Chaum, D., Crépeau, C., Damgård, I.: Multiparty unconditionally secure protocols. In: 20th ACM STOC, pp. 11–19. ACM Press (1988)
10. Choi, S.G., Katz, J., Wee, H., Zhou, H.-S.: Efficient, adaptively secure, and composable oblivious transfer with a single, global CRS. In: Kurosawa, K., Hanaoka, G. (eds.) PKC 2013. LNCS, vol. 7778, pp. 73–88. Springer, Heidelberg (2013)
11. Choi, S.G., Dachman-Soled, D., Malkin, T., Wee, H.: Simple, Black-Box Constructions of Adaptively Secure Protocols. In: Reingold, O. (ed.) TCC 2009. LNCS, vol. 5444, pp. 387–402. Springer, Heidelberg (2009)
12. Cleve, R.: Limits on the Security of Coin Flips when Half the Processors Are Faulty. In: 18th STOC, pp. 364–369 (1986)
13. Cramer, R., Damgård, I.B., Nielsen, J.B.: Multiparty computation from threshold homomorphic encryption. In: Pfitzmann, B. (ed.) EUROCRYPT 2001. LNCS, vol. 2045, pp. 280–299. Springer, Heidelberg (2001)
14. Damgård, I., Keller, M., Larraia, E., Miles, C., Smart, N.P.: Implementing AES via an actively/Covertly secure dishonest-majority MPC protocol. In: Visconti, I., De Prisco, R. (eds.) SCN 2012. LNCS, vol. 7485, pp. 241–263. Springer, Heidelberg (2012)

15. Damgård, I., Keller, M., Larraia, E., Pastro, V., Scholl, P., Smart, N.P.: Practical covertly secure MPC for dishonest majority - or: Breaking the SPDZ limits. In: Crampton, J., Jajodia, S., Mayes, K. (eds.) ESORICS 2013. LNCS, vol. 8134, pp. 1–18. Springer, Heidelberg (2013)
16. Damgård, I., Pastro, V., Smart, N., Zakarias, S.: Multiparty computation from somewhat homomorphic encryption. In: Safavi-Naini, R., Canetti, R. (eds.) CRYPTO 2012. LNCS, vol. 7417, pp. 643–662. Springer, Heidelberg (2012)
17. Frederiksen, T.K., Jakobsen, T.P., Nielsen, J.B., Nordholt, P.S., Orlandi, C.: Mini-LEGO: Efficient secure two-party computation from general assumptions. In: Johansson, T., Nguyen, P.Q. (eds.) EUROCRYPT 2013. LNCS, vol. 7881, pp. 537–556. Springer, Heidelberg (2013)
18. Garay, J.A., MacKenzie, P.D., Prabhakaran, M., Yang, K.: Resource fairness and composability of cryptographic protocols. In: Halevi, S., Rabin, T. (eds.) TCC 2006. LNCS, vol. 3876, pp. 404–428. Springer, Heidelberg (2006)
19. Goldreich, O.: Foundations of Cryptography: Basic Tools, vol. 1. Cambridge University Press, Cambridge (2001)
20. Goldreich, O.: Foundations of Cryptography: Basic Applications, vol. 2. Cambridge University Press, Cambridge (2004)
21. Goldreich, O., Micali, S., Wigderson, A.: How to play any mental game, or a completeness theorem for protocols with honest majority. In: Aho, A. (ed.) 19th ACM STOC, pp. 218–229. ACM Press (1987)
22. Gordon, S.D., Katz, J.: Partial fairness in secure two-party computation. In: Gilbert, H. (ed.) EUROCRYPT 2010. LNCS, vol. 6110, pp. 157–176. Springer, Heidelberg (2010)
23. Goyal, V., Lee, C.-K., Ostrovsky, R., Visconti, I.: Constructing non-malleable commitments: A black-box approach. In: 53rd FOCS, pp. 51–60. IEEE Computer Society (2012)
24. Haitner, I., Ishai, Y., Kushilevitz, E., Lindell, Y., Petrank, E.: Black-Box Constructions of Protocols for Secure Computation. SIAM J. Comput. 40(2), 225–266 (2011)
25. Hirt, M., Maurer, U.M.: Player simulation and general adversary structures in perfect multiparty computation. Journal of Cryptology 13(1), 31–60 (2000)
26. Hirt, M., Maurer, U.M., Zikas, V.: MPC vs. SFE: Unconditional and computational security. In: Pieprzyk, J. (ed.) ASIACRYPT 2008. LNCS, vol. 5350, pp. 1–18. Springer, Heidelberg (2008)
27. Impagliazzo, R., Luby, M.: One-way functions are essential for complexity-based cryptography. In: 30th FOCS, pp. 230–235. IEEE Computer Society Press (1989)
28. Ishai, Y., Kilian, J., Nissim, K., Petrank, E.: Extending oblivious transfers efficiently. In: Boneh, D. (ed.) CRYPTO 2003. LNCS, vol. 2729, pp. 145–161. Springer, Heidelberg (2003)
29. Ishai, Y., Kushilevitz, E., Lindell, Y., Petrank, E.: On combining privacy with guaranteed output delivery in secure multiparty computation. In: Dwork, C. (ed.) CRYPTO 2006. LNCS, vol. 4117, pp. 483–500. Springer, Heidelberg (2006)
30. Ishai, Y., Kushilevitz, E., Ostrovsky, R., Sahai, A.: Zero-knowledge from secure multiparty computation. In: Johnson, D.S., Feige, U. (eds.) 39th ACM STOC, pp. 21–30. ACM Press (2007)
31. Ishai, Y., Ostrovsky, R., Seyalioglu, H.: Identifying cheaters without an honest majority. In: Cramer, R. (ed.) TCC 2012. LNCS, vol. 7194, pp. 21–38. Springer, Heidelberg (2012)
32. Ishai, Y., Prabhakaran, M., Sahai, A.: Founding cryptography on oblivious transfer – efficiently. In: Wagner, D. (ed.) CRYPTO 2008. LNCS, vol. 5157, pp. 572–591. Springer, Heidelberg (2008)

33. Katz, J., Maurer, U., Tackmann, B., Zikas, V.: Universally composable synchronous computation. In: Sahai, A. (ed.) TCC 2013. LNCS, vol. 7785, pp. 477–498. Springer, Heidelberg (2013)
34. Keller, M., Scholl, P., Smart, N.P.: An architecture for practical actively secure MPC with dishonest majority. In: Sadeghi, A.-R., Gligor, V.D., Yung, M. (eds.) 20th ACM CCS, pp. 549–560. ACM Press (2013)
35. Kushilevitz, E., Lindell, Y., Rabin, T.: Information-theoretically secure protocols and security under composition. In: Kleinberg, J.M. (ed.) 38th ACM STOC, pp. 109–118. ACM Press (2006)
36. Nielsen, J.B., Nordholt, P.S., Orlandi, C., Burra, S.S.: A new approach to practical active-secure two-party computation. In: Safavi-Naini, R., Canetti, R. (eds.) CRYPTO 2012. LNCS, vol. 7417, pp. 681–700. Springer, Heidelberg (2012)
37. Ong, S.J., Parkes, D.C., Rosen, A., Vadhan, S.: Fairness with an honest minority and a rational majority. In: Reingold, O. (ed.) TCC 2009. LNCS, vol. 5444, pp. 36–53. Springer, Heidelberg (2009)
38. Pass, R., Wee, H.: Black-box constructions of two-party protocols from one-way functions. In: Reingold, O. (ed.) TCC 2009. LNCS, vol. 5444, pp. 403–418. Springer, Heidelberg (2009)
39. Peikert, C., Vaikuntanathan, V., Waters, B.: A framework for efficient and composable oblivious transfer. In: Wagner, D. (ed.) CRYPTO 2008. LNCS, vol. 5157, pp. 554–571. Springer, Heidelberg (2008)
40. Pfitzmann, B., Waidner, M.: Unconditional byzantine agreement for any number of faulty processors. In: Finkel, A., Jantzen, M. (eds.) STACS 1992. LNCS, vol. 577, pp. 337–350. Springer, Heidelberg (1992)
41. Rabin, M.O.: How to exchange secrets with oblivious transfer. Technical Report TR-81, Aiken Computation Lab, Harvard University (1981), http://eprint.iacr.org/2005/187
42. Rabin, T., Ben-Or, M.: Veri able secret sharing and multiparty protocols with honest majority. In: 21st ACM STOC, pp. 73–85. ACM Press (1989)
43. Seito, T., Aikawa, T., Shikata, J., Matsumoto, T.: Information-theoretically secure key-insulated multireceiver authentication codes. In: Bernstein, D.J., Lange, T. (eds.) AFRICACRYPT 2010. LNCS, vol. 6055, pp. 148–165. Springer, Heidelberg (2010)
44. Seyalioglu, H.: Reducing Trust When Trust is Essential. PhD thesis, UCLA (2012)
45. Shikata, J., Hanaoka, G., Zheng, Y., Imai, H.: Security notions for unconditionally secure signature schemes. In: Knudsen, L.R. (ed.) EUROCRYPT 2002. LNCS, vol. 2332, pp. 434–449. Springer, Heidelberg (2002)
46. Swanson, C., Stinson, D.R.: Unconditionally secure signature schemes revisited. In: Fehr, S. (ed.) ICITS 2011. LNCS, vol. 6673, pp. 100–116. Springer, Heidelberg (2011)
47. Yao, A.C.: Protocols for secure computations. In: 23rd FOCS, pp. 160–164. IEEE Computer Society Press (1982)
48. Zikas, V., Hauser, S., Maurer, U.: Realistic failures in secure multi-party computation. In: Reingold, O. (ed.) TCC 2009. LNCS, vol. 5444, pp. 274–293. Springer, Heidelberg (2009)

Non-Interactive Secure Multiparty Computation[*]

Amos Beimel[1], Ariel Gabizon[2], Yuval Ishai[2], Eyal Kushilevitz[2],
Sigurd Meldgaard[3], and Anat Paskin-Cherniavsky[4]

[1] Dept. of Computer Science, Ben Gurion University, Beer Sheva, Israel
amos.beimel@gmail.com
[2] Computer Science Department, Technion, Haifa, Israel
{arielga,yuvali,eyalk}@cs.technion.ac.il
[3] Google Aarhus, Denmark
stm@cs.au.dk
[4] Computer Science Department, UCLA, Los Angeles, CA, USA
anps83@gmail.com

Abstract. We introduce and study the notion of *non-interactive secure multiparty computation* (NIMPC). An NIMPC protocol for a function $f(x_1, \ldots, x_n)$ is specified by a joint probability distribution $R = (R_1, \ldots, R_n)$ and local encoding functions $\mathrm{Enc}_i(x_i, r_i)$, $1 \leq i \leq n$. Given correlated randomness $(r_1, \ldots, r_n) \in_R R$, each party P_i, using its input x_i and its randomness r_i, computes the message $m_i = \mathrm{Enc}_i(x_i, r_i)$. The messages $m_1, \ldots, m_n$ can be used to decode $f(x_1, \ldots, x_n)$. For a set $T \subseteq [n]$, the protocol is said to be *T-robust* if revealing the messages $(\mathrm{Enc}_i(x_i, r_i))_{i \notin T}$ together with the randomness $(r_i)_{i \in T}$ gives the same information about $(x_i)_{i \notin T}$ as an oracle access to the function f restricted to these input values. Namely, a coalition T can learn no more than the restriction of f fixing the inputs of uncorrupted parties, which, in this non-interactive setting, one cannot hope to hide. For $0 \leq t \leq n$, the protocol is *t-robust* if it is T-robust for every T of size at most t and it is *fully robust* if it is n-robust. A 0-robust NIMPC protocol for f coincides with a protocol in the private simultaneous messages model of Feige et al. (STOC 1994).

In the setting of *computational* (indistinguishability-based) security, fully robust NIMPC is implied by multi-input functional encryption, a notion that was recently introduced by Goldwasser et al. (Eurocrypt 2014) and realized using indistinguishability obfuscation. We consider NIMPC in the *information-theoretic* setting and obtain unconditional positive results for some special cases of interest:

[*] Research by the first three authors and the fifth author received funding from the European Union's Tenth Framework Programme (FP10/2010-2016) under grant agreement no. 259426 ERC-CaC. The first author was also supported by the Frankel center for computer science. Research by the second author received funding from the European Union's Seventh Framework Programme (FP7/2007-2013) under grant agreement no. 257575. The third and fourth authors were supported by ISF grant 1361/10 and BSF grant 2012378.

J.A. Garay and R. Gennaro (Eds.): CRYPTO 2014, Part II, LNCS 8617, pp. 387–404, 2014.
© International Association for Cryptologic Research 2014

- **Group products.** For every (possibly non-abelian) finite group G, the iterated group product function $f(x_1, \ldots, x_n) = x_1 x_2 \ldots x_n$ admits an efficient, fully robust NIMPC protocol.
- **Small functions.** Every function f admits a fully robust NIMPC protocol whose complexity is polynomial in the size of the input domain (i.e., exponential in the total bit-length of the inputs).
- **Symmetric functions.** Every symmetric function $f : X^n \to Y$, where X is an input domain of constant size, admits a t-robust NIMPC protocol of complexity $n^{O(t)}$. For the case where f is a w-out-of-n threshold function, we get a fully robust protocol of complexity $n^{O(w)}$.

On the negative side, we show that natural attempts to realize NIMPC using private simultaneous messages protocols and garbling schemes from the literature fail to achieve even 1-robustness.

Keywords: secure multiparty computation, obfuscation, private simultaneous messages protocols, randomized encoding of functions, garbling schemes, multi-input functional encryption.

1   Introduction

We introduce and study the notion of *non-interactive secure multiparty computation* (NIMPC). This notion can be viewed as a common generalization of several previous notions from the literature, including obfuscation, private simultaneous messages protocols, and garbling schemes. It can also be viewed as a simpler and weaker variant of the recently introduced notion of multi-input functional encryption. Before we define the new notion and discuss its relations with these previous notions, we start with a motivating example.

Consider the following non-interactive scenario for secure multiparty computation. Suppose that each of n "honest but curious" parties holds an input $x_i \in \{0, 1\}$, and the parties wish to conduct a vote by computing the majority value of their inputs. Moreover, the parties want to minimize interaction by each *independently* sending only a *single* message to each other party.[1] It is clear that in this scenario, without any setup, no meaningful notion of security can be achieved: each party can efficiently extract the input x_i from the message of the corresponding party by just simulating incoming messages from all other parties on inputs x_j such that $\sum_{j \neq i} x_j = \lfloor n/2 \rfloor$.

The question we ask is whether it is possible to get better security by allowing a *correlated randomness* setup. That is, the parties get correlated random strings $(r_1, \ldots, r_n)$ that are drawn from some predetermined distribution. Such a setup is motivated by the possibility of securely realizing it during an offline preprocessing phase, which takes place before the inputs are known (see, e.g. [24], for further motivation). The above attack fails in this model, since a party can no

---

[1] Alternative motivating scenarios include each party broadcasting a single message, posting it on a public bulletin board such as a Facebook account, or sending a single message to an external referee who should learn the output.

longer simulate messages coming from the other parties without knowing their randomness. On the other hand, it is still impossible to prevent the following generic attack: any set of parties T can simulate the messages that originate from parties in T on any given inputs. This allows the parties of T to learn the output on any set of inputs that is consistent with the other parties' inputs. In the case of computing majority, this effectively means that the parties in T must learn the sum of the other inputs whenever it is in the interval $[\lfloor n/2 \rfloor - |T|, \lfloor n/2 + 1 \rfloor]$. When T is small, this would still leave other parties with a good level of security. Hence, our goal is to obtain protocols that realize this "best possible" security while completely avoiding interaction.

The above discussion motivates the following notion of *non-interactive secure multiparty computation* (NIMPC). An NIMPC protocol for a function $f(x_1, \ldots, x_n)$ is defined by a joint probability distribution $R = (R_1, \ldots, R_n)$ and local encoding functions $\mathrm{Enc}_i(x_i, r_i)$, where $1 \leq i \leq n$. For a set $T \subseteq [n]$, the protocol is said to be T-*robust* (with respect to f) if revealing the messages $(\mathrm{Enc}_i(x_i, r_i))_{i \notin T}$ together with the randomness $(r_i)_{i \in T}$, where $(r_1, \ldots, r_n)$ is sampled from R, gives the same information about $(x_i)_{i \notin T}$ as an oracle access to the function f restricted to these input values. For $0 \leq t \leq n$, the protocol is said to be t-*robust* if it is T-robust for every T of size *at most* t, and it is said to be *fully robust* if it is n-robust.

Recent work on multi-input functional encryption [13] implies that the existence of a fully robust NIMPC protocol for general functions, with indistinguishability based security, is equivalent to indistinguishability obfuscation (assuming the existence of one-way functions). Combined with the recent breakthrough on the latter problem [11], this gives candidate NIMPC protocols for arbitrary polynomial-time computable functions. (See Section 1.2 for discussion of these and other related works.) The above positive result leaves much to be desired in terms of the underlying intractability assumptions and the potential for being efficient enough for practical use. Motivated by these limitations, we consider the goal of realizing NIMPC protocols with *information-theoretic security* for special cases of interest.

1.1   Our Results

We obtain the following unconditional positive results on NIMPC.

GROUP PRODUCTS. For every (possibly non-abelian) finite group G, the iterated group product function $f_G(x_1, \ldots, x_n) = x_1 x_2 \ldots x_n$ admits an efficient, fully robust NIMPC protocol. The construction makes a simple use of Kilian's randomization technique for iterated group products [26]. While the security analysis in the case of abelian groups is straightforward (see Example 6), the analysis for the general case turns out to be more involved and is deferred to the full version of this paper. We note that this result *cannot* be combined with Barrington's Theorem [4] to yield NIMPC for NC^1. For this, one would need to assign multiple group elements to each party and enforce nontrivial restrictions on the choice of these elements. In fact, efficient information-theoretic NIMPC for NC^1 is impossible, even with indistinguishability-based security, unless the polynomial-time hierarchy collapses [15] (see Section 1.2).

SMALL FUNCTIONS. We show that *every* function f admits a fully robust NIMPC protocol whose complexity is polynomial in the size of the input domain (i.e., exponential in the total bit-length of the inputs). This result can provide a lightweight solution for functions defined over an input domain of a feasible size. This result is described in Section 3. The technique used for obtaining this result also yields *efficient* protocols for computing OR of n bits and, more generally, w-out-of-n threshold functions where either w or $n - w$ are constant.

SYMMETRIC FUNCTIONS. Finally, we show that every symmetric function $h : X^n \to Y$, where X is an input domain of constant size, admits a t-robust NIMPC of complexity $n^{O(t)}$. Thus, we get a polynomial-time protocol for any constant t. More generally, our solution applies to any branching program over an abelian group G, that is, a function $h : X_1 \times \cdots \times X_n \to Y$ of the form $h(x_1, \ldots, x_n) = f(\sum_{i=1}^n x_i)$ for an arbitrary function $f : G \to Y$ (the complexity in this case is $|G|^{O(t)}$). Useful special cases include the above voting example, its generalization to multi-candidate voting (where the output is a partially ordered list such as "$A > B = C > D$"), as well as natural bidding mechanisms. We note that while this construction is only t-robust, larger adversarial sets T can only learn the sum $\sum_{i \notin T} x_i$ (e.g., the sum of all honest votes in the majority voting example) as opposed to all the inputs of honest parties. This construction is more technically involved than the previous constructions. A high level overview of a special case of the construction is given in Section 4, and a formal treatment of the general case appears in the full version of this paper. In the full version we also describe a more efficient variant of the construction for the case $t = 1$.

Inadequacy of Existing Techniques. On the negative side, in the full version we show that natural attempts to realize NIMPC using PSM protocols or garbling schemes from the literature fail to achieve even 1-robustness. This holds even for simple function classes such as symmetric functions.

Applications. Our main motivating application is for scenarios involving secure computations without interaction, such as the one described above. While in the motivating discussion we assumed the parties to be honest-but-curious, offering protection against malicious parties in the above model is in some sense easier than in the standard MPC model. Indeed, malicious parties pose no additional risk to the *privacy* of the honest parties because of the non-interactive nature of the protocol. Moreover, a reasonable level of *correctness* against malicious parties can be achieved via the use of pairwise authentication (e.g., in the case of binary inputs, the correlated randomness setup may give each party MAC-signatures on each of its two possible messages with respect to the verification key of each other party). In the case where multiple parties receive an output, adversarial parties can use their rushing capabilities to make their inputs depend on the information learned on other inputs, unless some simultaneous broadcast mechanism is employed. For many natural functions (such as the majority function) this type of rushing capability in the ideal model is typically quite harmless, especially when T is small. Moreover, this issue does not arise at all in the case where only one party (such as an external server) receives an output.

The goal of eliminating simultaneous interaction in secure MPC protocols was put forward by Halevi, Lindell, and Pinkas (HLP) [20,17]. In contrast to the HLP model, which requires the parties to sequentially interact with a central server, our protocols are completely non-interactive and may be applied with or without a central server. While HLP assume a standard PKI and settle for computational security, we allow general correlated randomness which, in turn, also allows for information-theoretic security.

The NIMPC primitive can also be motivated by the goal of obtaining garbling schemes [30,5] or randomized encodings of functions [22,1] that are robust to leakage of secret randomness. Indeed, in Yao's garbled circuit construction, the secrecy of the input completely breaks down if a pair of input keys is revealed. In the full version of this paper, we show that this is also the case for other garbling schemes and randomized encoding techniques from the literature. The use of t-robust NIMPC can give implementations of garbled circuits and related primitives that are resilient to up to t fully compromised pairs of input keys.

While we did not attempt to optimize the concrete efficiency of our constructions, they seem to be reasonably practical for some natural application scenarios. To give a rough idea of practical feasibility, consider a setting of non-interactive MPC where there are n clients, each holding a single input bit, who send messages to a central server that computes the output. For $n = 20$, our fully robust solution for small functions requires each client to send roughly 6MB of data and store a comparable amount of correlated randomness. In the case of computing a symmetric function, such as the majority function from the above motivating example, one can use an optimized protocol, which appears in the full version of this paper, to get a 1-robust solution with the same message size for $n \approx 1400$ clients (offering full protection against the server and single client and partial protection against larger collusions).

In contrast to the above, solutions that rely on general obfuscation techniques are currently quite far from being efficient enough for practical use. We leave open the question of obtaining broader or stronger positive results for NIMPC, either in the information-theoretic setting or in the computational setting without resorting to general-purpose obfuscation techniques.

1.2 Related Work

In the following, we discuss connections between NIMPC and several related notions from the literature.

Relation with Obfuscation. As was recently observed in the related context of multi-input functional encryption (see below), NIMPC generalizes the notion of obfuscation. The goal of obfuscation is to provide an efficient randomized mapping that converts a circuit (or "program") from a given class into a functionally equivalent circuit that hides all information about the original circuit except its input-output relation. An obfuscation for a given circuit class $\mathcal{C}$ reduces to a fully robust NIMPC for a *universal function* $U_{\mathcal{C}}$ for $\mathcal{C}$. Concretely, $U_{\mathcal{C}}$ takes two types of inputs: input bits specifying a circuit $C \in \mathcal{C}$, and input bits specifying an input to this circuit. An NIMPC protocol for U_C, in which each bit is assigned to a different party, gives rise to the following obfuscation scheme. The obfuscation of a circuit C consists of the message of each party holding a bit of C, together with the randomness of the parties holding the input bits for C. By extending the notion of NIMPC to apply to a class of functions (more accurately, function representations), as we do in the technical sections, it provides a more direct generalization of obfuscation that supports an independent local restriction of each input bit.

In contrast to obfuscation, NIMPC is meaningful and nontrivial to realize even when applied to a single function f (rather than a class of circuits), and even when applied to efficiently learnable functions (in particular, finite functions). Indeed, the requirement of hiding the inputs of uncorrupted parties is hard to satisfy even in such cases.

The relation with obfuscation implies limitations on the type of results on NIMPC one can hope to achieve, as it rules out fully robust protocols with simulation-based security for sufficiently expressive circuit classes [3]. Moreover, it follows from the results of [15] that some functions in NC^1 (in fact, even some families of CNF formulas) do not admit an efficient and fully robust information-theoretic NIMPC protocol, even under an indistinguishability-based definition, unless the polynomial-time hierarchy collapses. However, these negative results on obfuscation do not rule out general solutions with indistinguishability-based security or with a small robustness threshold t, nor do they rule out fully robust solutions with simulation-based security for simple but useful function classes.

Multi-input Functional Encryption. NIMPC can be viewed as a simplified and restricted form of *multi-input functional encryption*, a generalization of functional encryption [29,18,28,6] that was very recently studied in [13] Multi-input functional encryption is stronger than NIMPC in several ways, the most important of which is that it requires the correlated randomness to be *reusable* for polynomially many function evaluations. It was shown in [13] that multi-input functional encryption for general circuits can be obtained from indistinguishability obfuscation and a one-way function. Combined with the recent breakthrough on obfuscation [11], this gives plausible candidates for indistinguishability-based multi-input functional encryption, and hence also fully robust NIMPC, for general circuits. This general positive result can only achieve computational security under strong assumptions. In contrast, by only requiring a one-time use of the correlated randomness, the notion of NIMPC becomes meaningful even in the information-theoretic setting considered in this work.

Private Simultaneous Messages Protocols. A 0-robust NIMPC protocol for f coincides with a protocol for f in the private simultaneous messages (PSM) model of Feige, Kilian, and Naor [10,21]. In this model for non-interactive secure computation, the n parties share a *common* source of randomness that is unknown to an external referee, and they wish to communicate $f(x_1, \ldots, x_n)$ to the referee by sending simultaneous messages depending on their inputs and common randomness. From the messages it received, the referee should be able to recover the correct output but learn no additional information about the inputs. (PSM protocols in which each party has a single input bit are also referred to as *decomposable randomized encodings* [25] or *projective garbling schemes* [5].) While standard PSM protocols are inherently insecure when the referee colludes with even a single party, allowing general correlated randomness (rather than common randomness) gets around this limitation. A natural approach for obtaining NIMPC protocols from PSM protocols is to let the correlated randomness of each party include only the valid messages on its possible inputs. In the full version of this paper, we show that applying this methodology to different PSM protocols and garbling schemes from the literature typically fails to offer even 1-robustness. We also show a case where this methodology does work – using Kilian's PSM protocol for computing the iterated group product [26] yields a fully robust protocol.

Bounded-Collusion Functional Encryption. In the related context of (single-input) functional encryption, Gorbunov et al. [16] have shown how to achieve security against bounded collusions by combining MPC protocols and randomized encoding techniques. Similarly, bounded-collusion identity-based encryption is easier to construct than full-fledged identify-based encryption [9,14]. We do not know how to apply similar techniques for realizing t-robust NIMPC. The difference is likely to be inherent: while the positive results in [16,9,14] apply even for collusion bounds t that are bigger than the security parameter, a similar general result for NIMPC would suffice to imply general (indistinguishability) obfuscation.

2   Preliminaries

Notation 1. *For a set $\mathcal{X} = \mathcal{X}_1 \times \cdots \times \mathcal{X}_n$ and $T \subseteq [n]$ we denote $\mathcal{X}_T \triangleq \prod_{i \in T} \mathcal{X}_i$. For $x \in \mathcal{X}$, we denote by x_T the restriction of x to $\mathcal{X}_T$, and for a function $h : \mathcal{X} \to \Omega$, a subset $T \subseteq [n]$, and $x_{\overline{T}} \in \mathcal{X}_{\overline{T}}$, we denote by $h|_{\overline{T}, x_{\overline{T}}} : \mathcal{X}_T \to \Omega$ the function h where the inputs in $\mathcal{X}_{\overline{T}}$ are fixed to $x_{\overline{T}}$.*

An NIMPC protocol for a family of functions $\mathcal{H}$ is defined by three algorithms: (1) a randomness generation algorithm Gen, which given a description of a function $h \in \mathcal{H}$ generates n correlated random inputs $r_1, \ldots, r_n$, (2) a local encoding function Enc_i $(1 \leq i \leq n)$, which takes an input x_i and a random input r_i and outputs a message, and (3) a decoding algorithm Dec that reconstructs $h(x_1, \ldots, x_n)$ from the n messages. Formally:

Definition 2 (NIMPC: Syntax and Correctness). *Let $\mathcal{X}_1, \ldots, \mathcal{X}_n, \mathcal{R}_1, \ldots, \mathcal{R}_n, \mathcal{M}_1, \ldots, \mathcal{M}_n$ and Ω be finite domains. Let $\mathcal{X} \triangleq \mathcal{X}_1 \times \cdots \times \mathcal{X}_n$ and let $\mathcal{H}$ be a family of functions $h : \mathcal{X} \to \Omega$. A non-interactive secure multiparty computation (NIMPC) protocol for $\mathcal{H}$ is a triplet $\Pi = (\mathrm{Gen}, \mathrm{Enc}, \mathrm{Dec})$ where*

- *Gen $: \mathcal{H} \to \mathcal{R}_1 \times \cdots \times \mathcal{R}_n$ is a randomized function,*
- *Enc is an n-tuple of deterministic functions $(\mathrm{Enc}_1, \ldots, \mathrm{Enc}_n)$, where $\mathrm{Enc}_i : \mathcal{X}_i \times \mathcal{R}_i \to \mathcal{M}_i$,*

– Dec : $\mathcal{M}_1 \times \cdots \times \mathcal{M}_n \to \Omega$ is a deterministic function satisfying the following correctness requirement: for any $x = (x_1, \ldots, x_n) \in \mathcal{X}$ and $h \in \mathcal{H}$,

$$\Pr[r = (r_1, \ldots, r_n) \leftarrow \text{Gen}(h) : \text{Dec}(\text{Enc}(x, r)) = h(x)] = 1,$$

where $\text{Enc}(x, r) \triangleq (\text{Enc}_1(x_1, r_1), \ldots, \text{Enc}_n(x_n, r_n))$.

The communication complexity of Π is the maximum of $\log |\mathcal{R}_1|, \ldots, \log |\mathcal{R}_n|$, $\log |\mathcal{M}_1|, \ldots, \log |\mathcal{M}_n|$.

We next define the notion of t-robustness for NIMPC, which informally states that every t parties can only learn the information they should. Note that in our setting, a coalition T of size t can compute many outputs from the messages of $\overline{T}$, namely, they can repeatedly encode any inputs for the coalition T and decode h with the new encoded inputs and the original encoded inputs of $\overline{T}$. In other words, they have oracle access to $h|_{\overline{T}, x_{\overline{T}}}$ (as defined in Notation 1). Robustness requires that they learn no other information.

Definition 3 (NIMPC: Robustness). For a subset $T \subseteq [n]$, we say that an NIMPC protocol Π for $\mathcal{H}$ is T-robust if there exists a randomized function Sim_T (a "simulator") such that, for every $h \in \mathcal{H}$ and $x_{\overline{T}} \in \mathcal{X}_{\overline{T}}$, we have $\text{Sim}_T(h|_{\overline{T}, x_{\overline{T}}}) \equiv (M_{\overline{T}}, R_T)$, where R and M are the joint randomness and messages defined by $R \leftarrow \text{Gen}(h)$ and $M_i \leftarrow \text{Enc}_i(x_i, R_i)$.

For an integer $0 \le t \le n$, we say that Π is t-robust if it is T-robust for every $T \subseteq [n]$ of size $|T| \le t$. We say that Π is fully robust (or simply refer to Π as an NIMPC for $\mathcal{H}$) if Π is n-robust. Finally, given a concrete function $h : \mathcal{X} \to \Omega$, we say that Π is a (t-robust) NIMPC protocol for h if it is a (t-robust) NIMPC for $\mathcal{H} = \{h\}$.

As the same simulator Sim_T is used for every $h \in \mathcal{H}$ and the simulator has only access to $h|_{\overline{T}, x_{\overline{T}}}$, NIMPC hides both h and the inputs of $\overline{T}$ (to the extent possible).

Remark 4. An NIMPC protocol Π is 0-robust if it is $\emptyset$-robust. In this case, the only requirement is that the messages $(M_1, \ldots, M_n)$ reveal $h(x)$ and nothing else. A 0-robust NIMPC for h corresponds to a private simultaneous messages (PSM) protocol in the model of [10,21]. Note that in a 0-robust NIMPC one can assume, without loss of generality, that the n outputs of Gen are identical. In contrast, it is easy to see that in a 1-robust NIMPC of a nontrivial h more general correlations are required.

While the above definitions treat functions h as finite objects and do not refer to computational complexity, our constructions are computationally efficient in the sense that the total computational complexity is polynomial in the communication complexity. Furthermore, with the exception of the protocol from Lemma 9, the same holds for the efficiency of the simulator Sim_T (viewing the latter as an algorithm having oracle access to $h|_{\overline{T}, x_{\overline{T}}}$). When taking computational complexity into account, the function Gen should be allowed to depend not only on h itself but also on its specific representation (such as a branching program computing h).

Remark 5. (**Statistical and computational variants.**) In this work, we consider NIMPC protocols with perfect security, as captured by Definition 3. However, one could easily adapt the above definitions to capture statistical security and computational security. In the statistical case, we let Gen receive a security parameter κ as an additional input, and require that the two distributions in Definition 3 be $(2^{-\kappa})$-close in statistical distance, rather than identical. In the computational case, we have two main variants corresponding to the two main notions of obfuscation from the literature. In both cases, we require that the two distributions in Definition 3 be computationally indistinguishable. The difference is in the power of the simulator. If the simulator is unbounded, we get an indistinguishability-based NIMPC for which a general construction is implied by indistinguishability obfuscation [11,13]. If the simulator is restricted to probabilistic polynomial time, we get the stronger "virtual black-box" variant to which the impossibility results from [3] apply and one can only hope to get general positive results in a generic model [7,2] or using tamper-proof hardware [19]. We note, however, that the latter impossibility results only apply to function classes that are rich enough to implement pseudo-random functions. In particular, they do not apply to efficiently learnable classes for which obfuscation is trivial. NIMPC is meaningful and nontrivial even in the latter case.

As a simple example, we present an NIMPC protocol for summation in an abelian group.

Example 6. Let G be an abelian group, and define $h : G^n \rightarrow G$ by $h(x_1, \ldots, x_n) = x_1 + \cdots + x_n$ (where the sum is in G). We next define a fully robust NIMPC for h. Algorithm Gen chooses $n-1$ random elements $r_1, \ldots, r_{n-1}$ in G, where each element is chosen independently with uniform distribution, and computes $r_n = -\sum_{i=1}^{n-1} r_i$. The output of Gen is $(r_1, \ldots, r_n)$. Algorithm Enc computes $\mathrm{Enc}_i(x_i, r_i) = x_i + r_i \triangleq m_i$. Algorithm Dec simply sums the n outputs of Enc, that is, computes $\sum_{i=1}^{n} m_i$.

As $\sum_{i=1}^{n} m_i = \sum_{i=1}^{n} x_i + \sum_{i=1}^{n} r_i$, and $\sum_{i=1}^{n} r_i = 0$, correctness follows. We next show that this construction is fully robust. Fix a set $T \subseteq [n]$ and define the simulator Sim_T for T. On inputs x_T, it queries $h|_{\overline{T}, x_{\overline{T}}}(0^{|T|})$ and gets sum $= \sum_{i \in \overline{T}} x_i$. The simulator then chooses $n - 1$ random elements $\rho_1, \ldots, \rho_{n-1}$ in G, each element is chosen independently with uniform distribution, and computes $\rho_n = \mathrm{sum} - \sum_{i=1}^{n-1} \rho_i$. The output of the simulator is $((\rho_i)_{i \in \overline{T}}, (\rho_i)_{i \in T})$.

The following easily verifiable claim states that for functions outputting more than one bit, we can compute each output bit separately. Thus, from now on we will mainly focus on boolean functions.

Claim 7. *Let $\mathcal{X} \triangleq \mathcal{X}_1 \times \cdots \times \mathcal{X}_n$, where $\mathcal{X}_1, \ldots, \mathcal{X}_n$ are some finite domains. Fix an integer $m > 1$. Suppose $\mathcal{H}$ is a family of boolean functions $h : \mathcal{X} \rightarrow \{0, 1\}$ admitting an NIMPC protocol with communication complexity S. Then, the family of functions $\mathcal{H}^m = \{h : \mathcal{X} \rightarrow \{0, 1\}^m \mid h = h_1 \circ \ldots \circ h_m, \ h_i \in \mathcal{H}\}$ admits an NIMPC protocol with communication complexity $S \cdot m$.*

2.1   NIMPC with an Output Server

While an NIMPC protocol Π as defined above can be viewed as an abstract primitive, in the following it will be convenient to describe our constructions in the language of protocols. Such a protocol involves n players $P_1, \ldots, P_n$, each holding an input $x_i \in \mathcal{X}_i$, and an external "output server," a player P_0 with no input. The protocol may have an additional input, a function $h \in \mathcal{H}$. We will let $\mathrm{P}(\Pi)$ denote a protocol that proceeds as follows.

Protocol $\mathrm{P}(\Pi)(h)$

- **Offline preprocessing:** Each player P_i, $1 \le i \le n$, receives the random input $R_i \triangleq \mathrm{Gen}(h)_i \in \mathcal{R}_i$.
- **Online messages:** On input R_i, each player P_i, $1 \le i \le n$, sends the message $M_i \triangleq \mathrm{Enc}_i(x_i, R_i) \in \mathcal{M}_i$ to P_0.
- **Output:** P_0 computes and outputs $\mathrm{Dec}(M_1, \ldots, M_n)$.

We informally note the relevant properties of protocol $\mathrm{P}(\Pi)$:

- For any $h \in \mathcal{H}$ and $x \in \mathcal{X}$, the output server P_0 outputs, with probability 1, the value $h(x_1, \ldots, x_n)$.
- Fix $T \subseteq [n]$. Then, Π is T-robust if in $\mathrm{P}(\Pi)$ the set of players $\{P_i\}_{i \in T} \cup \{P_0\}$ can simulate their view of the protocol (i.e., the random inputs $\{R_i\}_{i \in T}$ and the messages $\{M_i\}_{i \in \overline{T}}$) given oracle access to the function h restricted by the other inputs (i.e., $h|_{\overline{T}, x_{\overline{T}}}$).
- Π is 0-robust if and only if in $\mathrm{P}(\Pi)$ the output server P_0 learns nothing but $h(x_1, \ldots, x_n)$.

In Appendix A we give a more general treatment of non-interactive MPC, including security definitions and extensions to the case where multiple parties may have different outputs and to the case of security against malicious parties.

3   An Inefficient NIMPC for Arbitrary Functions

The main purpose of this section is to present an NIMPC protocol for the set of all functions (though with exponential communication complexity). It will be useful to first present such a protocol for *indicator* functions. For reasons to be clarified later on, it will be convenient to include the zero-function.

Definition 8. *Let $\mathcal{X}$ be a finite domain. For n-tuple $a = (a_1, \ldots, a_n) \in \mathcal{X}$, let $h_a : \mathcal{X} \to \{0,1\}$ be the function defined by $h_a(a) = 1$, and $h_a(x) = 0$ for all $a \ne x \in \mathcal{X}$. Let $h_0 : \mathcal{X} \to \{0,1\}$ be the function that is identically zero on $\mathcal{X}$. Let $\mathcal{H}_{\mathrm{ind}} \triangleq \{h_a\}_{a \in \mathcal{X}} \cup \{h_0\}$ be the set of all indicator functions together with h_0.*

Note that every function $h : \mathcal{X} \leftarrow \{0,1\}$ can be expressed as the sum of indicator functions, namely, $h = \sum_{a \in \mathcal{X}, h(a)=1} h_a$.

Lemma 9. *Fix finite domains $\mathcal{X}_1, \ldots, \mathcal{X}_n$ such that $|\mathcal{X}_i| \le d$ for all $1 \le i \le n$ and let $\mathcal{X} \triangleq \mathcal{X}_1 \times \cdots \times \mathcal{X}_n$. Then, there is an NIMPC protocol Π_{ind} for $\mathcal{H}_{\mathrm{ind}}$ with communication complexity at most $d^2 \cdot n$.*

Proof. For $i \in [n]$, denote $|\mathcal{X}_i| = d_i$. Let $s = \sum_{i=1}^{n} d_i$. We describe a non-interactive protocol, in the output-server model. Fix a function $h \in \mathcal{H}$ that we want to compute. The protocol $P(\Pi_{\text{ind}})(h)$ is as follows.

Preprocessing: If $h = h_0$, then choose s linearly independent random vectors $\{m_{i,b}\}_{i \in [n], b \in \mathcal{X}_i}$ in $\mathbb{F}_2^s$. If $h = h_a$ for some $a = (a_1, \ldots, a_n) \in \mathcal{X}$, choose s random vectors $\{m_{i,b}\}_{i \in [n], b \in \mathcal{X}_i}$ in $\mathbb{F}_2^s$ under the constraint that $\sum_{i=1}^{n} m_{i,a_i} = 0$, and that there are no other linear relations between them (that is, choose all the vectors $m_{i,b}$, except m_{n,a_n}, as random linear independent vectors and set $m_{n,a_n} = -\sum_{i=1}^{n-1} m_{i,a_i}$). For $i \in [n]$, we send the vectors $\{m_{i,b}\}_{b \in \mathcal{X}_i}$ to P_i as the correlated randomness.

Sending messages: For $i \in [n]$, player P_i (on input x_i) sends to P_0 the message $M_i \triangleq m_{i,x_i}$.

Computing $h(x_1, \ldots, x_n)$: P_0 outputs 1 if $\sum_{i=1}^{n} M_i = \mathbf{0}$ and 0 otherwise.

For the correctness, note that $\sum_{i=1}^{n} M_i = \sum_{i=1}^{n} m_{i,x_i}$. If $h = h_a$, for $a \in \mathcal{X}$, this sum equals 0 if and only if $x = a$. If $h = h_0$, this sum is never zero, as all vectors were chosen to be linearly independent in this case.

To prove robustness, fix a subset $T \subsetneq [n]$ and $x_{\overline{T}} \in \mathcal{X}_{\overline{T}}$. The messages $M_{\overline{T}}$ of $\overline{T}$ consist of the vectors $\{m_{i,x_i}\}_{i \in \overline{T}}$. The randomness R_T consists of the vectors $\{m_{i,b}\}_{i \in T, b \in \mathcal{X}_i}$. If $h|_{\overline{T}, x_{\overline{T}}} \equiv 0$, then these vectors are uniformly distributed in $\mathbb{F}_2^s$ under the constraint that they are linearly independent. If $h|_{\overline{T}, x_{\overline{T}}}(x_T) = 1$, for some $x_T \in \mathcal{X}_T$, then $\sum_{i=1}^{n} m_{i,x_i} = 0$ and there are no other linear relations between them. Formally, to prove robustness, we describe a simulator Sim_T: the simulator queries $h|_{\overline{T}, x_{\overline{T}}}$ on all possible inputs in $\mathcal{X}_T$. If all answers are zero, the simulator generates random independent vectors. Otherwise, there is an $x_T \in \mathcal{X}_T$ such that $h|_{\overline{T}, x_{\overline{T}}}(x_T) = 1$, and the simulator outputs random vectors under the constrains described above, that is, all vectors are independent with the exception that $\sum_{i=1}^{n} m_{i,x_i} = 0$.

As for communication complexity, each party P_i receives $d_i \leq d$ binary vectors of length $s \leq dn$ in the preprocessing stage and sends one of them as a message. Hence, at most $d^2 n$ bits. □

We next present an NIMPC for all boolean functions with domain $\mathcal{X} = \mathcal{X}_1 \times \cdots \times \mathcal{X}_n$. The idea is to express any $h : \mathcal{X} \to \{0,1\}$ as a sum of indicator functions, that is, $h = \sum_{a \in \mathcal{X}, h(a)=1} h_a$, and construct an NIMPC for h by using the NIMPC protocols for each h_a. A naive implementation of this idea has two problems. First, it will disclose information on how many 1's the function h has. To overcome this problem, we define $h'_a = h_a$ if $h(a) = 1$ and $h'_a = h_0$ otherwise (this was the motivation of including h_0 in $\mathcal{H}_{\text{ind}}$). Thus, $h = \sum_{a \in \mathcal{X}} h'_a$. The second problem is that if, for example, $h(x) = 1$ and a coalition learns that $h'_a(x) = 1$, then the coalition learns that $x = a$. To overcome this problem, in the preprocessing stage, we permute the domain $\mathcal{X}$.

Theorem 10. *Fix finite domains $\mathcal{X}_1, \ldots, \mathcal{X}_n$ such that $|\mathcal{X}_i| \leq d$ for all $1 \leq i \leq n$ and let $\mathcal{X} \triangleq \mathcal{X}_1 \times \cdots \times \mathcal{X}_n$. Let $\mathcal{H}$ be the set of all functions $h : \mathcal{X} \to \{0,1\}^m$. There exists an NIMPC protocol Π for $\mathcal{H}$ with communication complexity $|\mathcal{X}| \cdot m \cdot d^2 \cdot n$.*

Proof. Let $\Pi_{\text{ind}} = (\text{Gen}, \text{Enc}, \text{Dec})$ be the NIMPC for $\mathcal{H}_{\text{ind}}$, described in Lemma 9. Fix $h \in \mathcal{H}$. Assume for simplicity that $m = 1$ (see Claim 7). Protocol $\text{P}(\Pi)(h)$ is as follows.

Preprocessing:

– Let $I \subseteq \mathcal{X}$ be the set of ones of h (i.e., $I = h^{-1}(1)$). For each $a \in I$, let $r^a = (r_1^a, \ldots, r_n^a) \leftarrow \text{Gen}(h_a)$. For $a \in \mathcal{X} \setminus I$, let $r^a \leftarrow \text{Gen}(h_0)$.
– Choose a random permutation π of $\mathcal{X}$ and define a matrix R, where $R_{i,b} \triangleq r_i^{\pi(b)}$ for $i \in [n]$ and $b \in \mathcal{X}$. Send to P_i the random strings $(R_{i,b})_{b \in \mathcal{X}}$ (that is, the ith row of R).

Sending Messages: Define a matrix M, where $M_{i,b} \triangleq \text{Enc}_i(x_i, R_{i,b})$ for every $i \in [n]$ and $b \in \mathcal{X}$. Each P_i sends to P_0 the message $M_i \triangleq (M_{i,b})_{b \in \mathcal{X}}$.

Computing h: Server P_0 outputs 1 if for some $b \in \mathcal{X}$, $\text{Dec}(M_{1,b}, \ldots, M_{n,b}) = 1$. Otherwise, it outputs 0.

Correctness: Fix $x = (x_1, \ldots, x_n) \in \mathcal{X}$. The server returns 1 if and only if $\text{Dec}(M_{1,b}, \ldots, M_{n,b}) = 1$ for some $b \in \mathcal{X}$, namely, if and only if $\text{Dec}(\text{Enc}_1(x_1, R_{1,b}), \ldots, \text{Enc}_n(x_n, R_{n,b})) = 1$. This happens if and only if $\text{Dec}(\text{Enc}_1(x_1, R_1^a), \ldots, \text{Enc}_n(x_n, R_n^a)) = 1$ for $a = \pi(b)$. By the correctness of Π_{ind} and the protocol description, the above happens if and only if $h_a(x_1, \ldots, x_n) = 1$ for some $a \in I$, that is, if and only if $h(x_1, \ldots, x_n) = 1$. Communication Complexity is obtained by applying Π_{ind} for $|\mathcal{X}|$ times.

Robustness: Fix $T \subseteq [n]$ and $x_{\overline{T}} \in \mathcal{X}_{\overline{T}}$. We wish to simulate the distribution $(M_{\overline{T}}, R_T)$ given $h|_{\overline{T}, x_{\overline{T}}}$. We can think of this distribution as being composed of rows, where each row b is of the form $(M_{\overline{T}}^a, r_T^a)$ for $a = \pi(b)$ for some $b \in \mathcal{X}$, where the permutation π is random.

Observation 11. *If $a_{\overline{T}} = x_{\overline{T}}$ and $h(a) = 1$ then this row was generated for the function h_a, and if $a_{\overline{T}} = x_{\overline{T}}$ and $h(a) = 0$ then this row was generated for h_0. Finally, if $a_{\overline{T}} \neq x_{\overline{T}}$, then this row is distributed as if it was generated for h_0.*

We next construct a simulator Sim_T for the protocol $\text{P}(\Pi)$ on function h. Simulator Sim_T uses the simulator $\text{Sim}_T^{\Pi_{\text{ind}}}$ – the simulator for set T from protocol $\text{P}(\Pi_{\text{ind}})$ of Lemma 9. The simulator Sim_T first queries $h|_{\overline{T}, x_{\overline{T}}}(x_T)$ for every $x_T \in \mathcal{X}_T$. Let $I' \subseteq \mathcal{X}_T$ be the set of ones of $h|_{\overline{T}, x_{\overline{T}}}$. For every $x_T \in I'$, the simulator Sim_T computes $S_{x_T} = \text{Sim}_T^{\Pi_{\text{ind}}}(h_{x_T})$ (where $h_{x_T} : \mathcal{X}_T \to \{0, 1\}$ is such that $h_{x_T}(x) = 1$ if and only if $x = x_T$). Finally, Sim_T samples $\text{Sim}_T^{\Pi_{\text{ind}}}(h_0)$ for $|\mathcal{X}| - |I'|$ times (where $h_0 : \mathcal{X}_T \to \{0, 1\}$ such that $h_0(x) = 0$ for every $x \in \mathcal{X}_T$). Altogether, it obtains $|\mathcal{X}|$ outputs of the simulator $\text{Sim}_T^{\Pi_{\text{ind}}}$. It randomly permutes the order of these outputs, and returns the permuted outputs. The T-robustness of Sim_T follows from the T-robustness of $\text{Sim}_T^{\Pi_{\text{ind}}}$ and Observation 11. □

Remark 12. In the above proof, instead of looking at the set of all functions, we could have looked at the set of functions that are OR's of a fixed subset

$\mathcal{H}' \subset \mathcal{H}_{\text{ind}}$ of indicator functions. For this set of functions, we would get an NIMPC with communication complexity $|\mathcal{H}'| \cdot m \cdot \text{poly}(d, n)$ (rather than the $|\mathcal{X}| \cdot m \cdot \text{poly}(d, n)$ communication complexity above). We could also look at a particular function of this form. Take, for example, $\mathcal{X} = \{0, 1\}^n$ and $\mathcal{H}'$ to be the set of indicator functions of vectors of weight w. Then, we get an NIMPC for the w-out-of-n threshold function with communication complexity $n^{O(w)}$.

4   A t-Robust NIMPC for Abelian Programs

In this section, we present an NIMPC protocol for symmetric functions. In fact, this result is a corollary of a more general result on NIMPC for *abelian group programs*. We next define abelian programs and symmetric functions and formally state our results. The proofs of these results appear in the full version of this paper.

Definition 13. *Let G be an abelian group, $S_1, \ldots, S_n$ be subsets of G, and $\mathcal{H}^G_{S_1,\ldots,S_n}$ be the set of functions $h : S_1 \times \cdots \times S_n \to \{0, 1\}$ of the form $h(x_1, \ldots, x_n) = f(\sum_{i=1}^n x_i)$, for some $f : G \to \{0, 1\}$.*

Definition 14. *A function $h : [d]^n \to \{0, 1\}$ is symmetric if for every $(x_1, \ldots, x_n) \in [d]^n$ and every permutation $\pi : [n] \to [n]$ the following equality holds $h(x_1, \ldots, x_n) = h(x_{\pi(1)}, \ldots, x_{\pi(n)})$.*

The main positive result in this section is an efficient t-robust NIMPC for $\mathcal{H}^G_{S_1,\ldots,S_n}$ whenever G is abelian of poly(n)-size and t is constant.

Theorem 15. *Let t be a positive integer and G an abelian group of size m. Let $S_1, \ldots, S_n$ be subsets of G. Let $d \triangleq \max_{i \in [n]} |S_i|$. Then, there is a t-robust NIMPC protocol for $\mathcal{H}^G_{S_1,\ldots,S_n}$ with communication complexity $O(d^{t+2} \cdot n^{t+2} \cdot m^3)$.*

Corollary 16. *Let d, t and n be positive integers. Let $\mathcal{H}$ be the set of symmetric functions $h : [d]^n \to \{0, 1\}$. There is a t-robust NIMPC protocol for $\mathcal{H}$ with communication complexity $O(d^{t+2} \cdot n^{t+3d-1})$. In particular, for the case of a boolean $h : \{0, 1\}^n \to \{0, 1\}$, the communication complexity is $O(2^t \cdot n^{t+5})$.*

In the rest of this section, we give a high level overview of the construction, focusing for simplicity on the case $t = 1$.

4.1   Group Extension

Recall that a boolean function $h : \{0, 1\}^n \to \{0, 1\}$ is symmetric if and only if there exists a function $f : \{0, \ldots, n\} \to \{0, 1\}$ such that $h(x_1, \ldots, x_n) = f(\sum_{i=1}^n x_i)$. We start by considering a relaxation of the problem where the players are allowed to choose their inputs from a larger domain, which is a group: namely, instead of having an input $x_i \in \{0, 1\}$, we allow each player P_i to have an input $x_i \in \{0, \ldots, n\}$, which can be thought of as an element of the group $G \triangleq \mathbb{Z}_{n+1}$. Given a boolean symmetric function $h : \{0, 1\}^n \to \{0, 1\}$, where $h(x_1, \ldots, x_n) = f(\sum_{i=1}^{n} x_i)$, we extend h in the natural way to a function $h : G^n \to \{0,1\}$, that is, $h(x_1, \ldots, x_n) = f(\sum_{i=1}^{n} x_i)$, where the sum is of elements of G. The first step of our construction is a fully robust NIMPC protocol for the set $\mathcal{H}$ of all functions h as above, namely the group extensions of all symmetric functions. Note that here it is crucial to hide both the function h and the inputs x_i to the extent possible.

To obtain the NIMPC protocol for $\mathcal{H}$, we would like to use the PSM protocol from [21] which provides an efficient solution for symmetric functions. This protocol is defined using a *branching program* representation of h. While this protocol is secure when only P_0 is corrupted, it fails miserably when even a single other party is corrupted. Luckily, there is a simple characterization of the information available to an adversary corrupting P_0 and a set T of other parties P_i: these players learn no more than the graph of the branching program restricted to the inputs $x_{\bar{T}}$. That is, the adversary can learn the labels of all edges it owns (e.g., that such an edge is labeled by the literal $\bar{x}_i$ or the constant $\mathbf{1}$), as well as the *values* of edges it does not own (e.g., that such an edge evaluates to 1) but not their labels. If we apply the protocol to a standard branching program for h, this information will typically reveal to the adversary both the function h and the inputs $x_{\bar{T}}$.

The key idea for realizing $\mathcal{H}$ is to randomize the branching program before applying the protocol from [21]. That is, we start with a standard layered branching program for the symmetric function h, and then (during preprocessing) we randomize it by applying a random cyclic shift to the nodes in each layer. The protocol from [21] is applied to the randomized branching program. With this randomization in place, revealing the branching program restricted by $x_{\bar{T}}$ leaks nothing about $(h, x_{\bar{T}})$ except what must be learned.

4.2   Limiting the Inputs of One Player

The previous subsection gives an NIMPC for the class of (extended) symmetric functions h, with the caveat that the players may use any input in $G = \mathbb{Z}_{n+1}$, rather than just $\{0,1\}$. Let us call this protocol $\Pi_0(h)$.[2]

As mentioned, we need to limit the parties to inputs from $\{0,1\}$. Note that for NIMPC this is relevant also in the honest-but-curious model since the robustness requirement for the extended function allows an adversary, controlling a set T, to compute $h|_{\bar{T}, x_{\bar{T}}}$ on the domain $G^{|T|}$, while for the original function we only allow the adversary to compute $h|_{\bar{T}, x_{\bar{T}}}$ on the domain $\{0,1\}^{|T|}$. In this section, as an intermediate step, we construct a protocol where a specific player, say P_1, is limited to inputs in $\{0,1\}$. The other players, $P_2, \ldots, P_n$, can still choose any inputs in G. Let h_0 and h_1 denote the function h where the first input is fixed to 0 and 1, respectively, that is, $h_i(X_2, \ldots, X_n) \triangleq h(i, X_2, \ldots, X_n)$, for $i \in \{0,1\}$. Consider the following protocol: $P_2, \ldots, P_n$ run the protocols $\Pi_0(h_0)$ and $\Pi_0(h_1)$. At the end of this protocol, the coalition $\{P_0, P_1\}$ – seeing the

---

[2] An important point is that only the preprocessing stage of protocol Π_0 actually depends on h, but we ignore these subtleties here.

messages of $\Pi_0(h_0)$ and $\Pi_0(h_1)$ – knows *exactly* what it is supposed to know: the values $h(0, x_2, \ldots, x_n)$ and $h(1, x_2, \ldots, x_n)$. However, there are two evident problems.

1. On one hand, P_0 alone knows "too much": the same two values $h(0, x_2, \ldots, x_n)$ and $h(1, x_2, \ldots, x_n)$.
2. On the other hand, P_0 does not know which of these two values is the correct one, i.e., $h(x_1, \ldots, x_n)$.

A possible "solution" to the second problem is for P_1 to send its input x_1 to P_0. This is, of course, insecure. Instead, we run $\Pi_0(h_0)$ and $\Pi_0(h_1)$ in a random order, known only to P_1 and given to it in the preprocessing stage (note that $P_2, \ldots, P_n$ need not know which of the two protocols is running to participate). Party P_1 will then send a message stating which one corresponds to its input.

A solution to the first problem is as follows: The (symmetric) functions h_0 and h_1 (which can be though of as $(n + 1)$-bit strings representing their truth tables) are "masked" by $((n + 1)$-bit) random functions α_0 and α_1 (where $\alpha_b : G \to \{0, 1\}$). Let us call these masked versions g_0 and g_1. Specifically, $g_j(X_2, \ldots, X_n) \triangleq h_j(X_2, \ldots, X_n) \oplus \alpha_j(\sum_{i=2}^{n} X_i)$, for $j \in \{0, 1\}$. In the preprocessing stage, we give the masking functions α_0 and α_1 to P_1. Now $P_0, P_2, \ldots, P_n$ run $\Pi_0(g_0)$ and $\Pi_0(g_1)$ (in a random order). Then, P_1 sends to P_0 only the masking α_i corresponding to its input. In terms of security, the problem has been solved: the protocol not corresponding to P_1's input, i.e. $\Pi_0(g_{1-x_1})$, does not reveal any information to P_0, as g_{1-x_1} is a masked version of h, where the mask has not been revealed. However, can P_0 now compute $h(x_1, \ldots, x_n)$? From seeing the messages of $\Pi_0(g_{x_1})$, it knows $g_{x_1}(x_2, \ldots, x_n) = h(x_1, \ldots, x_n) \oplus \alpha_{x_1}(\sum_{i=2}^{n} x_i)$. It also knows α_{x_1}, which was sent by P_1. So now, to "unmask" $h(x_1, \ldots, x_n)$ using α_{x_1} it needs the value $\sum_{i=2}^{n} x_i$, which is more information than we want to give it. Further randomization techniques are needed to solve this problem, and combine the solutions to the two problems above.

4.3   A Secret Sharing Composition

The previous section described a protocol where, for a certain fixed $j \in [n]$, the coalition $\{P_0, P_j\}$ does not learn "too much" – specifically, it could evaluate the function h only on inputs in $\{0, 1\}$ (while a coalition of P_0 with one of the other players is still not restricted to inputs in $\{0, 1\}$). Call this protocol Π_1. Note that h is of the form $h(X_1, \ldots, X_n) = f(\sum_{i=1}^{n} X_i)$ for a function $f : G \to \{0, 1\}$. It is easy to see that Π_1 can work for any function $h'(X_1, \ldots, X_n)$ of this form. We now bootstrap the protocol Π_1 to create one in which *all* players can evaluate h only on inputs in $\{0, 1\}$. For this, we use an additive secret sharing of f. Namely, we choose n random functions $f_1, \ldots, f_n : G \to \{0, 1\}$, such that $\sum_{i=1}^{n} f_i = f$, where the sum is a xor of $|G|$-bit vectors. For $1 \le i \le n$, define $h_i(X_1, \ldots, X_n) \triangleq f_i(\sum_{j=1}^{n} X_j)$. Note that for any $x_1, \ldots, x_n \in G$, we have $h(x_1, \ldots, x_n) = \sum_{i=1}^{n} h_i(x_1, \ldots, x_n)$. For $1 \le i \le n$, we run Π_1 on the function h_i with P_i chosen to be the player that can only use inputs in $\{0, 1\}$. After these protocols are run we have that, on the one hand, P_0 knows $h_1(x_1, \ldots, x_n), \ldots,$ $h_n(x_1, \ldots, x_n)$ and can compute $h(x_1, \ldots, x_n) = \sum_{i=1}^{n} h_i(x_1, \ldots, x_n)$. On the other hand, for any $i \in [n]$ and $a \in G \setminus \{0, 1\}$, parties P_0 and P_i have no information on $h_i(x_1, \ldots, x_{i-1}, a, x_{i+1}, \ldots, x_n)$ and hence no information on $h(x_1, \ldots, x_{i-1}, a, x_{i+1}, \ldots, x_n)$.

References

1. Applebaum, B., Ishai, Y., Kushilevitz, E.: Cryptography in NC^0. In: Proc. FOCS 2004, pp. 166–175 (2004)
2. Barak, B., Garg, S., Kalai, Y.T., Paneth, O., Sahai, A.: Protecting Obfuscation against Algebraic Attacks. In: Nguyen, P.Q., Oswald, E. (eds.) EUROCRYPT 2014. LNCS, vol. 8441, pp. 221–238. Springer, Heidelberg (2014)
3. Barak, B., Goldreich, O., Impagliazzo, R., Rudich, S., Sahai, A., Vadhan, S., Yang, K.: On the (Im)possibility of Obfuscating Programs. In: Kilian, J. (ed.) CRYPTO 2001. LNCS, vol. 2139, pp. 1–18. Springer, Heidelberg (2001)
4. Barrington, D.M.: Bounded-Width Polynomial-Size Branching Programs Recognize Exactly Those Languages in NC^1. In: Proc. STOC 1986, pp. 1–5 (1986)
5. Bellare, M., Hoang, V.T., Rogaway, P.: Foundations of garbled circuits. In: Proc. ACM CCS 2012, pp. 784–796 (2012)
6. Boneh, D., Sahai, A., Waters, B.: Functional Encryption: Definitions and Challenges. In: Ishai, Y. (ed.) TCC 2011. LNCS, vol. 6597, pp. 253–273. Springer, Heidelberg (2011)
7. Brakerski, Z., Rothblum, G.N.: Virtual Black-Box Obfuscation for All Circuits via Generic Graded Encoding. In: Lindell, Y. (ed.) TCC 2014. LNCS, vol. 8349, pp. 1–25. Springer, Heidelberg (2014)
8. Canetti, R.: Security and composition of multiparty cryptographic protocols. Journal of Cryptology 13(1), 143–202 (2000)
9. Dodis, Y., Katz, J., Xu, S., Yung, M.: Key-Insulated Public Key Cryptosystems. In: Knudsen, L.R. (ed.) EUROCRYPT 2002. LNCS, vol. 2332, pp. 65–82. Springer, Heidelberg (2002)
10. Feige, U., Kilian, J., Naor, M.: A Minimal Model for Secure Computation. In: Proc. STOC 1994, pp. 554–563 (1994)
11. Garg, S., Gentry, C., Halevi, S., Raykova, M., Sahai, A., Waters, B.: Candidate Indistinguishability Obfuscation and Functional Encryption for All Circuits. In: Proc. FOCS 2013, pp. 40–49 (2013)
12. Goldreich, O.: Foundations of Cryptography, vol. 2. Cambridge University Press (2004)
13. Goldwasser, S., et al.: Multi-input Functional Encryption. In: Nguyen, P.Q., Oswald, E. (eds.) EUROCRYPT 2014. LNCS, vol. 8441, pp. 578–602. Springer, Heidelberg (2014)
14. Goldwasser, S., Lewko, A.B., Wilson, D.A.: Bounded-Collusion IBE from Key Homomorphism. In: Cramer, R. (ed.) TCC 2012. LNCS, vol. 7194, pp. 564–581. Springer, Heidelberg (2012)
15. Goldwasser, S., Rothblum, G.N.: On Best-Possible Obfuscation. In: Vadhan, S.P. (ed.) TCC 2007. LNCS, vol. 4392, pp. 194–213. Springer, Heidelberg (2007)
16. Gorbunov, S., Vaikuntanathan, V., Wee, H.: Functional Encryption with Bounded Collusions via Multi-party Computation. In: Safavi-Naini, R., Canetti, R. (eds.) CRYPTO 2012. LNCS, vol. 7417, pp. 162–179. Springer, Heidelberg (2012)

17. Gordon, S.D., Malkin, T., Rosulek, M., Wee, H.: Multi-party Computation of Polynomials and Branching Programs without Simultaneous Interaction. In: Johansson, T., Nguyen, P.Q. (eds.) EUROCRYPT 2013. LNCS, vol. 7881, pp. 575–591. Springer, Heidelberg (2013)
18. Goyal, V., Pandey, O., Sahai, A., Waters, B.: Attribute-based encryption for fine-grained access control of encrypted data. In: Proc. ACM CCS 2006, pp. 89–98 (2006)
19. Goyal, V., Ishai, Y., Sahai, A., Venkatesan, R., Wadia, A.: Founding Cryptography on Tamper-Proof Hardware Tokens. In: Micciancio, D. (ed.) TCC 2010. LNCS, vol. 5978, pp. 308–326. Springer, Heidelberg (2010)
20. Halevi, S., Lindell, Y., Pinkas, B.: Secure Computation on the Web: Computing without Simultaneous Interaction. In: Rogaway, P. (ed.) CRYPTO 2011. LNCS, vol. 6841, pp. 132–150. Springer, Heidelberg (2011)
21. Ishai, Y., Kushilevitz, E.: Private simultaneous Messages Protocols with Applications. In: ISTCS 1997, pp. 174–184 (1997)
22. Ishai, Y., Kushilevitz, E.: Randomizing Polynomials: A New Representation with Applications to Round-Efficient Secure Computation. In: FOCS 2000, pp. 294–304 (2000)
23. Ishai, Y., Kushilevitz, E.: Perfect Constant-Round Secure Computation via Perfect Randomizing Polynomials. In: Widmayer, P., Triguero, F., Morales, R., Hennessy, M., Eidenbenz, S., Conejo, R. (eds.) ICALP 2002. LNCS, vol. 2380, pp. 244–256. Springer, Heidelberg (2002)
24. Ishai, Y., Kushilevitz, E., Meldgaard, S., Orlandi, C., Paskin-Cherniavsky, A.: On the Power of Correlated Randomness in Secure Computation. In: Sahai, A. (ed.) TCC 2013. LNCS, vol. 7785, pp. 600–620. Springer, Heidelberg (2013)
25. Ishai, Y., Kushilevitz, E., Ostrovsky, R., Sahai, A.: Cryptography with constant computational overhead. In: Proc. STOC 2008, pp. 433–442 (2008)
26. Kilian, J.: Founding Cryptography on Oblivious Transfer. In: Proc. STOC 1988, pp. 20–31 (1988)
27. Naor, M., Pinkas, B., Sumner, R.: Privacy Preserving Auctions and Mechanism Design. In: Proc. ACM Conference on Electronic Commerce 1999, pp. 129–139 (1999)
28. O'Neill, A.: Definitional Issues in Functional Encryption. IACR Cryptology ePrint Archive 2010: 556
29. Sahai, A., Waters, B.: Fuzzy Identity-Based Encryption. In: Cramer, R. (ed.) EUROCRYPT 2005. LNCS, vol. 3494, pp. 457–473. Springer, Heidelberg (2005)
30. Yao, A.C.C.: How to Generate and Exchange Secrets. In: Proc. 27th FOCS 1986, pp. 162–167 (1986)

A   General Non-Interactive MPC

In this section we extend the treatment of NIMPC to functionalities that may deliver different outputs to different parties as well as to the case of security against malicious parties.

We consider protocols involving n parties, $P_1, \ldots, P_n$, with a correlated randomness setup. That is, we assume an offline preprocessing phase that provides each party P_i with a random input r_i. (This preprocessing can be implemented either using a trusted dealer, by an interactive offline protocol involving the parties themselves, or by an interactive MPC protocol involving a smaller number of specialized servers.) In the online phase, each party P_i, on input (x_i, r_i), may send a single message $m_{i,j}$ to each party P_j. (There is no need to assume secure or authenticated channels, as these can be easily implemented using a correlated randomness setup.)

Let f be a deterministic functionality mapping inputs $(x_1, \ldots, x_n)$ to outputs $(y_1, \ldots, y_n)$. We define security of an NIMPC protocol for such f using the standard "real vs. ideal" paradigm (cf. [8,12]), except that the ideal model is relaxed to capture the best achievable security in the non-interactive setting.

Concretely, for NIMPC in the semi-honest security model we relax the standard ideal model for evaluating f by first requiring all parties to send their inputs to the functionality f, then having f deliver the outputs to the honest parties, and finally allowing the adversary to make repeated oracle queries to f with the same fixed honest inputs. (Similar relaxations of the ideal model were previously considered in other contexts, such as fairness and concurrent or resettable security.) In the malicious security model, one should further relax the ideal model in order to additionally take into account the adversary's capability of rushing[3] (namely, correlating its messages with the messages obtained from honest parties). In the relaxed ideal model, first the honest parties send their inputs to f, then the adversary can repeatedly make oracle calls as above, and finally the adversary can decide on the actual inputs to f that determine the outputs of honest parties.

Given a t-robust NIMPC protocol (according to Definition 2) for each of the n outputs of f, a t-secure protocol for f can be obtained in a straightforward way. In the honest-but-curious model, it suffices to run n independent instances of the protocol described in Section 2.1, where in the i-th instance P_i acts both as a standard party and as the external server P_0. In the malicious model, the correlated randomness setup uses an unconditional one-time MAC to authenticate each of the possible messages sent from P_i to P_j. This is feasible when the input domain of each party is small. In the general case, we can make use of an NIMPC protocol for a functionality f' with a bigger number of parties which is identical to f except for taking a single input bit from each party. Such a functionality f' can be securely realized by a protocol Π' as described above, and then f can be realized by a protocol Π in which each party emulates the corresponding parties in Π'.

---

[3] If some mechanism is available for ensuring that the adversary's messages are independent of the honest parties' messages, this relaxation is not needed.

Feasibility and Infeasibility
of Secure Computation with Malicious PUFs

Dana Dachman-Soled[1], Nils Fleischhacker[2], Jonathan Katz[1],
Anna Lysyanskaya[3], and Dominique Schröder[2]

[1] University of Maryland, College Park, MD, USA
danadach@ece.umd.edu, jkatz@cs.umd.edu
[2] Saarland University, Saarbrücken, Germany
{fleischhacker,schroeder}@cs.uni-saarland.de
[3] Brown University, Providence, RI, USA
anna_lysyanskaya@brown.edu

Abstract. A recent line of work has explored the use of *physically uncloneable functions (PUFs)* for secure computation, with the goals of
(1) achieving universal composability without (additional) setup, and/or
(2) obtaining unconditional security (i.e., avoiding complexity-theoretic
assumptions). Initial work assumed that all PUFs, even those created by
an attacker, are honestly generated. Subsequently, researchers have investigated models in which an adversary can create *malicious* PUFs with
arbitrary behavior. Researchers have considered both malicious PUFs
that might be stateful, as well as malicious PUFs that can have arbitrary behavior but are guaranteed to be stateless.

We settle the main open questions regarding secure computation in
the malicious-PUF model:

– We prove that unconditionally secure oblivious transfer is impossible, even in the stand-alone setting, if the adversary can construct
(malicious) *stateful* PUFs.
– We show that universally composable two-party computation is possible if the attacker is limited to creating (malicious) *stateless* PUFs.
Our protocols are simple and efficient, and do not require any cryptographic assumptions.

1 Introduction

A *physically uncloneable function* (PUF) [19,20,17,1,15] is a physical object generated via a process that is intended to create "unique" objects with "random"
(or at least random-looking) behavior. PUFs can be probed and their response
measured, and a PUF thus defines a function. (We ignore here the possibility of
slight variability in the response, which can be corrected using standard techniques.) At an abstract level, this function has two important properties: it is
random, and it *cannot be copied* even by the entity who created the PUF.

Since their introduction, several cryptographic applications of PUFs have been
suggested, in particular in the area of secure computation. PUFs are especially

J.A. Garay and R. Gennaro (Eds.): CRYPTO 2014, Part II, LNCS 8617, pp. 405–420, 2014.
© International Association for Cryptologic Research 2014

interesting in this setting because they can potentially be used (1) to obtain *universally composable* (UC) protocols [6] without additional setup, thus bypassing known impossibility results that hold for universal composition in the "plain" model [7,8], and (2) to construct protocols with *unconditional* security, i.e., without relying on any cryptographic assumptions.

Initial results in this setting [21,22] showed constructions of oblivious transfer with stand-alone security based on PUFs. Brzuska et al. [5] later formalized PUFs within the UC framework, and showed UC constructions of bit commitment, key agreement, and oblivious transfer (and hence secure computation of arbitrary functionalities) with *unconditional* security. The basic feasibility questions related to PUFs thus seemed to have been resolved.

Ostrovsky et al. [18], however, observe that the previous results implicitly assume that all PUFs, including those created by the attacker, are honestly generated. They point out, correctly, that this may not be a reasonable assumption: nothing forces the attacker to use the recommended process for manufacturing PUFs and it is not clear, in general, how to "test" whether a PUF sent by some party was generated correctly or not. (Assuming a trusted entity who creates the PUFs is not a panacea, as one of the goals of using PUFs is to avoid reliance on trusted parties.) Addressing this limitation, Ostrovsky et al. define a model in which an attacker can create *malicious* PUFs having arbitrary, adversary-specified behavior. The previous protocols can be easily attacked in this new adversarial setting, but Ostrovsky et al. show that it is possible to construct universally composable protocols for secure computation in the malicious-PUF model under additional, number-theoretic assumptions. They explicitly leave open the question of whether unconditional security is possible in the malicious-PUF model. Recently, Damgård and Scafuro [9] have made partial progress on this question by presenting a commitment scheme with unconditional security in the malicious-PUF model.

Stateful vs. Stateless (malicious) PUFs. Honestly generated PUFs are stateless; that is, the output of an honestly generated PUF is independent of its computation history. Ostrovsky et al. note that maliciously generated PUFs might be stateful or stateless. Allowing the adversary to create stateful PUFs is obviously more general. (The positive results mentioned earlier remain secure even against an attacker who can create malicious, stateful PUFs.) Nevertheless, the assumption that the adversary is limited to producing stateless PUFs is meaningful; indeed, depending on the physical technology used to implement the PUFs, incorporating dynamic state in the PUF may simply be infeasible.

1.1 Our Results

Spurred by the work of Ostrovsky et al. and Damgård and Scafuro, we reconsider the possibility of unconditionally secure computation based on malicious PUFs and resolve the main open questions in this setting. Specifically, we show:

1. Unconditionally secure oblivious transfer (and thus unconditionally secure computation of general functions) is impossible when the attacker can create

malicious *stateful* PUFs. Our result holds even with regard to stand-alone security, and even for indistinguishability-based (as opposed to simulation-based) security notions.

2. If the attacker is limited to creating malicious, but *stateless*, PUFs, then universally composable oblivious transfer (OT) and two-party computation of general functionalities are possible. Our oblivious-transfer protocol is efficient and requires each party to create only a single PUF for polynomially many OT executions. The protocol is also conceptually simple, which we view as positive in light of the heavy machinery used in [18].

1.2 Other Related Work

Hardware tokens have also been proposed as a physical assumption on which to base secure computation [14]. PUFs are incomparable to hardware tokens since they are more powerful in one respect and less powerful in another. PUFs have the property that a party cannot query an honestly generated PUF when it is out of that party's possession, whereas in the token model parties place known functionality in the token and can simulate the behavior of the token at any point. On the other hand, tokens can implement arbitrary code, whereas honestly generated PUFs just provide a random function. In any case, known results (such as the fact that UC oblivious transfer is impossible with stateless tokens [11]) do not directly translate from one model to the other.

Impossibility results for (malicious) PUFs are also not implied by impossibility results in the random-oracle model (e.g., [3]). A random oracle can be queried by any party at any time, whereas (as noted above) an honestly generated PUF can only be queried by the party who currently holds it. Indeed, we show that oblivious transfer *is* possible when malicious PUFs are assumed to be stateless; in contrast, oblivious transfer is impossible in the random-oracle model [12].

Ostrovsky et al. [18] consider a second malicious model where the attacker can *query* honestly generated PUFs in a non-prescribed manner. They show that secure computation is impossible if both this and maliciously generated PUFs are allowed. We do not consider the possibility of malicious queries in this work.

In other work, van Dijk and Rührmair [23] show impossibility results in a malicious-PUF model very different from the one considered in [18,9] and here. It is not clear to us how their model corresponds to attacks that could feasibly be carried out in the real world.

2   Physically Uncloneable Functions

A physically uncloneable function (PUF) is a physical device with "random" behavior introduced through uncontrollable manufacturing variations during their fabrication. When a PUF is queried with a stimulus (i.e., a challenge), it produces a physical output (the response). The output of a PUF can be noisy; i.e., querying the PUF twice with the same challenge may yield distinct, but close, responses. Moreover, the response need not be uniform; it may instead only have high min-entropy. Prior work has shown that, by using fuzzy extractors, one can eliminate the noisiness of a PUF and make its output effectively uniform. For simplicity, we assume this in the definition that follows.

Formally, a PUF family is defined by two algorithms S and E. The index-sampling algorithm S, which corresponds to the PUF-fabrication process, takes as input the security parameter 1^λ and returns as output an index id. The evaluation algorithm E takes as input an index id and a challenge[1] c, and generates as output the corresponding response r.

We do not require that S or E can be evaluated efficiently. In fact, these are meant to represent *physical* processes that generate a physical object and measure this object's behavior under various conditions. The index id is simply a formal placeholder that refers to a well-defined physical object; it does not in itself represent any meaningful information about how this object works.

Following [5], we define the two main security properties of PUFs: *unpredictability* and *uncloneability*. As noted earlier, for simplicity we consider only a strong form of unpredictability where the output of the PUF is uniform. Intuitively, uncloneability means that only one party can evaluate a PUF at a time. This is formally modeled using an ideal functionality, $\mathcal{F}_{\mathsf{PUF}}$, that enforces this. Details of this ideal functionality are given in the full version of this work.

Finally, we also allow for the possibility of a maliciously generated PUF whose behavior does not necessarily correspond to (S, E) as described above. We consider two possibilities here: The first possibility is a *malicious-but-stateless PUF* that may use an $\mathsf{E}_{\mathtt{mal}}$ procedure of the adversary's choice in place of the honest algorithm E. Whenever a party in possession of this PUF evaluates it, it receives $\mathsf{E}_{\mathtt{mal}}(c)$ instead of $\mathsf{E}_{\mathtt{id}}(c)$. (As noted in prior work, care must be taken to ensure that the adversary cannot use $\mathsf{E}_{\mathtt{mal}}$ to perform arbitrary exponential-time computation; formally, we restrict $\mathsf{E}_{\mathtt{mal}}$ to be a polynomial-time algorithm with oracle access to E.) The second possibility is a *malicious-and-stateful PUF* that may use a *stateful* $\mathsf{E}_{\mathtt{mal}}$ procedure of the adversary's choice in place of E. Again, $\mathsf{E}_{\mathtt{mal}}$ is limited to polynomial-time computation with oracle access to E.

To simplify notation throughout the rest of the paper, we write $\mathsf{PUF} \leftarrow \mathsf{S}(1^\lambda)$ to denote the fabrication of a PUF, and then write $r := \mathsf{PUF}(c)$.

3   Impossibility Result for Malicious, Stateful PUFs

We prove that any PUF-based oblivious-transfer (OT) protocol is insecure when the attacker has the ability to generate malicious, *stateful* PUFs. Formally:

Theorem 1. *Let Π be a PUF-based OT-protocol where the sender $\mathcal{S}$ and receiver $\mathcal{R}$ each make at most $m = \mathrm{poly}(\lambda)$ PUF queries. Then at least one of the following holds:*

---

[1] We assume the challenge space is just a set strings of a certain length. For some classes of PUFs, this is naturally satisfied (see [17]). For others, this can be achieved using appropriate encoding.

1. *There is an unbounded adversary S^* that uses malicious, stateful PUFs and makes only $\mathrm{poly}(\lambda)$ queries to honestly generated PUFs, and computes the choice bit of R (when R's input is uniform) with probability $1/2 + 1/\mathrm{poly}(\lambda)$.*
2. *There is an unbounded adversary R^* that uses malicious, stateful PUFs and makes only $\mathrm{poly}(\lambda)$ queries to honestly generated PUFs, and correctly guess both secrets of S (when S's inputs are uniform) with probability at least $2/3$.*

3.1   Overview

The starting point for our impossibility result is the impossibility of constructing oblivious transfer in the random-oracle model. The fact that OT is impossible in the random-oracle model follows from the fact that key agreement is impossible in the random-oracle model [12,3,2], and the observation that OT implies key agreement. However, a direct proof ruling out OT in the random-oracle model is also possible, and we sketch such a proof here.

Consider an OT protocol in the random-oracle model between a sender S and receiver R, where S's two input bits are uniform and R's selection bit is uniform. We show that either S or R can attack the protocol. Consider the case where both parties run the protocol honestly, and then at the end of the protocol they each run a variant of the Eve algorithm from [3,2] to obtain a set Q of queries/answers to/from the random oracle. This set Q contains all "intersection queries" between S and R, which are queries made by both parties to the random oracle. However, note that the setting here is different from the key-agreement setting in which a third party (the eavesdropper) runs the Eve algorithm. In fact, in our setting, finding intersection queries is trivial for S and R: all intersection queries are, by definition, already contained in the view of either of the parties. Thus, the point of running the Eve algorithm is for both parties to reconstruct the *same* set of queries Q that contains all intersection queries. As in [3,2], conditioned on the transcript of the protocol and this set Q, the views of S and R are independent. The property of the Eve algorithm we use is that with high probability over random coins of the protocol and the choice of random oracle, the distribution over R's view conditioned on S's view and Q is statistically close to the distribution over R's view conditioned on only the transcript and Q.

To use the above to obtain an attack, we first consider the distribution over R's view conditioned on S's view and Q. We argue that with roughly $1/2$ probability over this distribution, R's view must be consistent with selection bit 0, and with $1/2$ probability is must be consistent with selection bit 1. (If not, then S can compromise R's security by guessing that R's selection bit is the one which is more lilkely.) Next, we consider the distribution over R's view conditioned on only the transcript and Q. Note that R can sample from this distribution, since R knows the transcript and can compute the same set Q. Since this distribution is statistically close to the distribution over R's view conditioned on S's view and

Eve queries, we have that $\mathcal{R}$ can with high probability sample a view consistent with selection bit 0 and $\mathcal{S}$'s view *and* a view consistent with selection bit 1 and $\mathcal{S}$'s view. But correctness of the protocol then implies that $\mathcal{R}$ can with high probability discover both of $\mathcal{S}$'s input bits.

From Random Oracles to PUFs. The problem with extending the above to the PUF model is that, unlike a random oracle, a PUF can only be queried by the party who currently holds the PUF. This mean that the above attack, as described, will not work. In fact, this property is what allows us to *construct* an OT protocol in the case where malicious PUFs are assumed to be stateless! To overcome this difficulty, we will need to use the fact that malicious parties can create *stateful* PUFs.

To illustrate the main ideas, consider a protocol in which four PUFs are used. $\mathsf{PUF}_\mathcal{S}$ and $\mathsf{PUF}'_\mathcal{S}$ are created by $\mathcal{S}$, with $\mathsf{PUF}_\mathcal{S}$ held by $\mathcal{S}$ at the end of the protocol and $\mathsf{PUF}'_\mathcal{S}$ held by $\mathcal{R}$ at the end of the protocol. Similarly, $\mathsf{PUF}_\mathcal{R}, \mathsf{PUF}'_\mathcal{R}$ are created by $\mathcal{R}$, with $\mathsf{PUF}_\mathcal{R}$ held by $\mathcal{R}$ at the end of the protocol and $\mathsf{PUF}'_\mathcal{R}$ held by $\mathcal{S}$ at the end of the protocol. We now want to provide a way for both parties to be able to obtain a set of queries/answers Q for all the PUFs that contains the following "intersection queries":

1. Any query that both parties made to $\mathsf{PUF}'_\mathcal{S}$ or $\mathsf{PUF}'_\mathcal{R}$ (as in [3,2]).
2. All queries that $\mathcal{R}$ made to $\mathsf{PUF}_\mathcal{S}$.
3. All queries that $\mathcal{S}$ made to $\mathsf{PUF}_\mathcal{R}$.

The first of these can be achieved by having $\mathcal{S}$ (resp., $\mathcal{R}$) construct $\mathsf{PUF}'_\mathcal{S}$ (resp., $PUF'_\mathcal{R}$) with known code, such that $\mathcal{S}$ (resp., $\mathcal{R}$) can effectively query $\mathsf{PUF}'_\mathcal{S}$ (resp., $PUF'_\mathcal{R}$) at any time. Formally, we have each party embed a randomly chosen t-wise independent function in the PUF they create, where t is large enough so that the behavior of the PUF is indistinguishable from a random function as far as execution of the protocol (and the attack) is concerned. At the end of the protocol, both parties can then run the Eve algorithm with access to $\mathsf{PUF}'_\mathcal{S}$: $\mathcal{R}$ has access because it holds $\mathsf{PUF}'_\mathcal{S}$, and $\mathcal{S}$ has access because it knows the code in $\mathsf{PUF}'_\mathcal{S}$. An analogous statement holds for $\mathsf{PUF}'_\mathcal{R}$.

To handle the second set of queries, above, we rely on the ability of $\mathcal{S}$ to create stateful PUFs. Specifically, we have $\mathcal{S}$ create $\mathsf{PUF}_\mathcal{S}$ in such a way that it records (in an undetectable fashion) all the queries that $\mathcal{R}$ makes to $\mathsf{PUF}_\mathcal{S}$, in such a way that $\mathcal{S}$ can later recover these queries once $\mathsf{PUF}_\mathcal{S}$ is back in its possession. (This is easy to do by hardcoding in the PUF a secret challenge, chosen in advance by $\mathcal{S}$, to which the PUF responds with the set of all queries made to the PUF.) So, at the end of the protocol, it is trivial for $\mathcal{S}$ to learn all the queries that $\mathcal{R}$ made to $\mathsf{PUF}_\mathcal{S}$. Of course, $\mathcal{R}$ knows exactly the set of queries it made to $\mathsf{PUF}_\mathcal{S}$ throughout the course of the protocol. Queries that $\mathcal{S}$ makes to $\mathsf{PUF}_\mathcal{R}$ are handled in a similar fashion.

To complete the proof, we then show that the set of intersection queries as defined above is enough for the analysis from [3,2] to go through.[2]

3.2   Proof Details

Oblivious Transfer. Oblivious transfer (OT) is a protocol between a sender S with input bits (s_0, s_1) and a receiver R with input bit b. Informally, the receiver wishes to retrieve s_b from S in such a way that (1) S does not "learn" anything about R's choice and (2) R learns nothing about s_{1-b}.

We note that our impossibility holds even for protocols that do not enjoy perfect correctness, i.e., it holds for protocols where correctness holds (over choice of inputs, randomness, and PUFs) with probability $1 - 1/\text{poly}(\lambda)$.

Protocols Based on PUFs. We consider a candidate PUF-based OT protocol Π with ℓ rounds that has 2ℓ *passes* and where in each *pass* a party sends a message. We assume w.l.o.g. that S sends the first message of the protocol and R sends the final message. Let $z = z(\lambda)$ be the total number of PUFs used in protocol Π with security parameter λ. We model the set of all PUFs $\{\text{PUF}_1, \ldots, \text{PUF}_z\}$ utilized by Π as a single random oracle. W.l.o.g. we assume that each query q to a PUF has the form $q = (j, q')$, where j denotes the identity of the PUF being queried, and q' denotes the actual query to this PUF. Note that responses to unique queries $q = (j, q')$ are independent and uniform. We further assume w.l.o.g. that a party can only send a PUF back and forth along with some message m_i of the protocol Π. In particular, we denote by S_{back}^i the set of indices $j \in [z]$ such that PUF_j is sent by S (resp. R) to R (resp. S) immediately after message m_i of Π is sent, and PUF_j was created by R (resp. S). We define S_{PUF}^i to be the set of indices j such that either:

- PUF_j is held by S immediately after message m_i is sent and PUF_j was created by R.
- PUF_j is held by R immediately after message m_i is sent and PUF_j was created by S.

Augmented Transcripts. A full (augmented) transcript of protocol $\Pi = \langle S, R \rangle$ is denoted by $\widetilde{M}$. The "augmented transcript" consists of the transcript $M = m_1, \ldots, m_{2\ell}$ of protocol Π with a set ψ^i appended after each message m_i. If message m_i is sent by S (resp. R), then ψ^i contains all queries made by S (resp. R) up to this point in the protocol to all PUF_j, $j \in S_{\text{back}}^i$. Specifically,

---

[2] In order for our proof to go through, it is crucial to find intersection queries immediately after each message is sent, as opposed to waiting until the end of the protocol. This is necessary in order to ensure the *independence* of the views of S and R. Therefore, we define a variant of the Eve algorithm which, after each protocol message is sent, makes queries to a particular set of PUFs, determined by the sets of PUFs currently held by each party. For example, if immediately after message i is sent S holds $\{\text{PUF}_S, \text{PUF}'_R\}$ and R holds $\{\text{PUF}_R, \text{PUF}'_S\}$, then our Eve variant will make queries only to PUF'_R and PUF'_S.

$M = \{m_1, \ldots, m_{2\ell}\}$ and $\widetilde{M} = \{m_1 || \psi^1, \ldots, m_{2\ell} || \psi^{2\ell}\}$. Note that $\widetilde{M}$ can be computed by both a malicious $\mathcal{S}$ and a malicious $\mathcal{R}$ participating in Π. Intuitively, this is because both malicious $\mathcal{S}$ and $\mathcal{R}$ can program each of their PUFs to record all queries made to it. The following claim formalizes the fact that malicious, stateful PUFs can be used to extract sets of queries made by the opposite party:

Claim 2. *Consider a PUF-based ℓ-round OT protocol, Π. By participating in an execution of Π while using maliciously constructed PUFs, we have that, for all odd $i \in [2\ell]$, both a malicious $\mathcal{S}$ and a malicious $\mathcal{R}$ can find the set of queries ψ^i made by $\mathcal{S}$ up to this point in the protocol to all PUF_j, $j \in S^i_{\mathrm{back}}$. The same claim holds for even $i \in [2\ell]$, with the roles of $\mathcal{S}, \mathcal{R}$ reversed.*

Proof. The malicious $\mathcal{R}$ will create stateful PUFs which record all queries made to them and such that this record can later be retrieved by the creator of the PUF. Since at the end of the i-th pass, for odd $i \in [2\ell]$, $\mathcal{R}$ holds PUF_j, $j \in S^i_{\mathrm{back}}$, we have that the malicious $\mathcal{R}$ can recover the ordered set of queries made to that PUF, and can therefore deduce the set of queries made by $\mathcal{S}$ to that PUF thus far. On the other hand, $\mathcal{S}$ knows the queries it made itself to PUF_j, $j \in S^i_{\mathrm{back}}$. An analogous argument holds for even $i \in [2\ell]$.

We also define the set Ψ^i which is the union of the ψ^j sets for $j \leq i$. Specifically, $\Psi^i = \psi^1 \cup \cdots \cup \psi^i$.

Queries and Views. By $V^i_{\mathcal{S}}$ (resp. $V^i_{\mathcal{R}}$) we denote the view of $\mathcal{S}$ (resp. $\mathcal{R}$) until the end of round i. This includes $\mathcal{S}$'s (resp. $\mathcal{R}$'s) randomness $r_{\mathcal{S}}$ (resp. $r_{\mathcal{R}}$), exchanged messages M^i as well as oracle query-answer pairs known to $\mathcal{S}$ (resp. $\mathcal{R}$) so far. We use $\mathcal{Q}(\cdot)$ as an operator that extracts the set of queries from a set of query-answer pairs or views.

Executions and Distributions. A (full) *execution* of $\mathcal{S}, \mathcal{R}, \mathsf{Eve}$ in protocol Π can be described by a tuple $(r_{\mathcal{S}}, r_{\mathcal{R}}, H)$ where H is a random function. We denote by $\mathcal{E}$ the distribution over (full) executions that is obtained by running the algorithms for $\mathcal{S}, \mathcal{R}, \mathsf{Eve}$ with uniformly chosen random tapes and a sampled oracle H. For a sequence of i (augmented) messages $\widetilde{M}^i = [\tilde{m}_1, \ldots, \tilde{m}_i]$ and a set of query-answer pairs P, by $\mathcal{V}(\widetilde{M}^i, \mathsf{P})$ we denote the joint distribution over the views $(V^i_{\mathcal{S}}, V^i_{\mathcal{R}})$ of $\mathcal{S}$ and $\mathcal{R}$ in their (partial) execution of Π up to the point in the system in which the i-th message is sent (by $\mathcal{S}$ or $\mathcal{R}$) conditioned on: The transcript of messages in the first i passes equals $\widetilde{M}^i$ and $H(j, q') = a$ for all $((j, q'), a) \in \mathsf{P}$ made to H (recall that a query (j, q') to H corresponds to a query q' made to PUF_j). For $(\widetilde{M}^i, \mathsf{P})$ such that $\Pr_{\mathcal{E}}(\widetilde{M}^i, \mathsf{P}) > 0$, the distribution $\mathcal{V}(\widetilde{M}^i, \mathsf{P})$ can be sampled by first sampling $(r_{\mathcal{S}}, r_{\mathcal{R}}, H)$ uniformly at random conditioned on being consistent with $(\widetilde{M}^i, \mathsf{P})$ and then deriving $\mathcal{S}$ and $\mathcal{R}$ views $V^i_{\mathcal{S}}, V^i_{\mathcal{R}}$ from the sampled $(r_{\mathcal{S}}, r_{\mathcal{R}}, H)$.

For $(\widetilde{M}^i, \mathsf{P})$ such that $\Pr_{\mathcal{E}}(\widetilde{M}^i, \mathsf{P}) > 0$, the event $\mathsf{Good}(\widetilde{M}^i, \mathsf{P})$ is defined over the distribution $\mathcal{V}(\widetilde{M}^i, \mathsf{P})$ and holds if and only if $\mathcal{Q}(V^i_{\mathcal{S}}) \cap \mathcal{Q}(V^i_{\mathcal{R}}) \subseteq \mathsf{P}^+$, where $\mathsf{P}^+ = \mathsf{P} \cup \Psi^i$. For $\Pr_{\mathcal{E}}(\widetilde{M}^i, \mathsf{P}) > 0$ we define the distribution $\mathcal{GV}(\widetilde{M}^i, \mathsf{P})$ to be the distribution $\mathcal{V}(\widetilde{M}^i, \mathsf{P})$ conditioned on $\mathsf{Good}(\widetilde{M}^i, \mathsf{P})$.

For complete transcripts $\widetilde{\mathsf{M}}$, the distributions $\mathcal{V}(\widetilde{\mathsf{M}}, \mathsf{P})$ and $\mathsf{Good}(\widetilde{\mathsf{M}}, \mathsf{P})$ are defined similarly.

Transforming the Protocol. We begin by transforming the OT protocol Π into one that has the following properties:

Semi-normal form: We define a semi-normal form for OT protocols, following [3,2]. A protocol is in semi-normal form if it fulfills the following two properties: (1) $\mathcal{S}$ and $\mathcal{R}$ ask at most one query in each protocol round, and (2) the receiver of the last message uses this message to compute its output and it does not query the oracle. We start by converting our OT protocol Π into its semi-normal version. Note that any attack on the semi-normal version of Π can be translated into an attack on the original Π that makes the *same* number of queries [3,2]. Thus, in the following we present our attacks and analysis w.r.t. the semi-normal version of Π.

Using t-wise independent functions: Instead of creating honestly generated PUFs, a malicious $\mathcal{R}$ (resp. $\mathcal{S}$) will create stateful PUFs which behave as t-wise independent hash functions. We define the distribution $\mathcal{V}^t(\widetilde{\mathsf{M}}, \mathsf{P})$ exactly like $\mathcal{V}(\widetilde{\mathsf{M}}, \mathsf{P})$ except some subset of PUFs are instantiated with t-wise independent hash functions for some $t = \mathrm{poly}(m/\varepsilon)$, instead of with random oracles. Since we choose t such that the malicious sender and honest receiver (resp., malicious receiver and honest sender) make a total of at most t queries to all PUFs then: For every setting of random variables $(\widetilde{\mathsf{M}}^i, \mathsf{P}^i)$, the distributions $\mathcal{V}(\widetilde{\mathsf{M}}^i, \mathsf{P}^i)$ and $\mathcal{V}^t(\widetilde{\mathsf{M}}^i, \mathsf{P}^i)$ are identical. Thus, from now on, even when $\mathcal{R}$ or $\mathcal{S}$ are malicious (and create t-wise independent PUFs), we consider only the distribution $\mathcal{V}(\widetilde{\mathsf{M}}, \mathsf{P})$.

Random inputs: In the last step we change the protocol such that both sender and receiver choose their input(s) uniformly at random. Thus, in the following, we consider execution of $\Pi = \langle \mathcal{S}(1^\lambda), \mathcal{R}(1^\lambda) \rangle$ where the parties use their random tapes to choose their inputs.

The Eve Algorithm. Recall that we have converted the protocol Π into semi-normal form. We now present the attacking algorithm, Eve, which will be run by both the malicious sender $(\mathcal{S}')$ and malicious receiver $(\mathcal{R}')$ defined later:

Construction 1. *Let $\varepsilon < 1/100$ be an input parameter. After each message m_i is sent, Eve generates the augmented transcript $\widetilde{\mathsf{M}}^i$ (note that by Claim 2, $\widetilde{\mathsf{M}}^i$ can always be reconstructed by Eve, since Eve is launched by either the malicious $\mathcal{S}$ or $\mathcal{R}$). Given $\widetilde{\mathsf{M}}^i$, Eve attacks the ℓ-round two-party protocol $\Pi = \langle \mathcal{S}, \mathcal{R} \rangle$ as follows. During the attack Eve updates a set P of oracle query-answer pairs as follows: Suppose $\mathcal{S}$ (alternatively $\mathcal{R}$) sends the i-th message in $\widetilde{\mathsf{M}}^i$ which is equal to $\widetilde{m}_i = m_i \| \psi^i$. For $i \in [2\ell]$, Eve does the following: As long as the total number of queries made by Eve is less than $t - 2m$ and there is a query $q = (j, q') \notin \mathsf{P}^+$, where $\mathsf{P}^+ := \Psi^i \cup \mathsf{P}$, such that one of the following holds:*

$$\Pr_{(V_{\mathcal{S}}^i, V_{\mathcal{R}}^i) \leftarrow \mathcal{GV}(\tilde{M}^i, P)}[q' \in \mathcal{Q}(V_{\mathcal{S}}^i) \wedge j \in S_{\text{back}}^i] \geq \frac{\varepsilon^2}{100m}$$

$$or \quad \Pr_{(V_{\mathcal{S}}^i, V_{\mathcal{R}}^i) \leftarrow \mathcal{GV}(\tilde{M}^i, P)}[q' \in \mathcal{Q}(V_{\mathcal{R}}^i) \wedge j \in S_{\text{back}}^i] \geq \frac{\varepsilon^2}{100m}.$$

Eve *queries the lexicographically first such* $q = (j, q')$ *to* H, *adds* $(q, H(q))$ *to* P.

Properties of the Eve Algorithm. We summarize some properties of the Eve algorithm that can be verified by inspection:

Symmetry of Eve: Both $\mathcal{S}$ and $\mathcal{R}$ can run the Eve algorithm, making the same set of queries P to the PUFs. In particular, at the point where message m_i is sent, a party requires only the augmented transcript $\tilde{M}^i$ and oracle access to PUF_j for $j \in S_{\text{PUF}}^i$. Note that for each $j \in S_{\text{PUF}}^i$ a party either holds PUF_j (and so can query it directly) or created PUF_j dishonestly and thus knows the code of PUF_j (and so can simulate responses to queries to PUF_j).

Determinism of Eve: The Eve algorithm is deterministic and so for a fixed transcript $\tilde{M}$ and a fixed set of PUFs, both parties will recover the same set of queries when running Eve.

Number of queries: The number of queries made by Eve is at most $t - 2m$. Thus, since $\mathcal{S}$ and $\mathcal{R}$ each make at most m number of PUF queries, the total number of queries made by $\mathcal{S}$, $\mathcal{R}$ and Eve is at most $(t - 2m) + 2m = t$.

Breaking Oblivious Transfer. Recall that we assume that the honest $\mathcal{S}$ chooses its inputs (s_0, s_1) at random and that the honest $\mathcal{R}$ chooses its input bit at random. Thus, we may consider an execution of OT protocol $\Pi = \langle \mathcal{S}(1^\lambda), \mathcal{R}(1^\lambda) \rangle$ where the parties use their random tapes to choose their inputs.

We now state an alternative version of Theorem 1:

Theorem 3. *Let* $\Pi = \langle \mathcal{S}(1^\lambda), \mathcal{R}(1^\lambda) \rangle$ *be a PUF-based OT-protocol in which the sender and receiver each ask at most* m *queries total to the set of* $z = \text{poly}(\lambda)$ *PUFs,* $\{\text{PUF}_1, \ldots, \text{PUF}_z\}$. *Then, at least one of the following must hold:*

1. *There exists an adversarial* $\mathcal{S}$ *that uses malicious, stateful PUFs to compute the choice bit of* $\mathcal{R}$ *with advantage* $1/\text{poly}(\lambda)$ *and makes* $\text{poly}(\lambda)$ *queries to the PUFs.*
2. *For* $\varepsilon < 1/100$, *there exists an adversarial* $\mathcal{R}$ *that uses malicious, stateful PUFs to correctly guess both secrets of* $\mathcal{S}$ *with probability* $1 - O(\sqrt{\varepsilon})$ *and makes* $\text{poly}(\lambda)$ *queries to the PUFs.*

By choosing constant ε sufficiently small, we may obtain the parameters of Theorem 1.

Proof. We begin with some notation. For a view $V_{\mathcal{R}}$ (resp. $V_{\mathcal{S}}$), we denote by $\text{In}(V_{\mathcal{R}})$ (resp. $\text{In}(V_{\mathcal{S}})$) the input of the corresponding party implicitly contained in its view. We denote by $\text{Out}(V_{\mathcal{R}})$ the output of $\mathcal{R}$ implicitly contained in its view. For a distribution $\mathcal{D}$ and random variables $X_1, \ldots, X_n$, we denote by $\mathcal{D}(X_1, \ldots, X_n)$ the distribution $\mathcal{D}$ conditioned on $X_1, \ldots, X_n$.

Let $p(\cdot)$ be some sufficiently large polynomial. We consider two cases.

Case 1: With probability $1/p(\lambda)$ over $(\widetilde{\mathsf{M}}, \mathsf{P}, \mathsf{V}_{\mathcal{S}})$ generated by a run of $\tilde{\Pi}$ we have that either

$$\Pr_{\mathcal{V}_{\mathsf{V}_{\mathcal{R}}}(\widetilde{\mathsf{M}}, \mathsf{P}, \mathsf{V}_{\mathcal{S}})}[\mathrm{In}(\mathsf{V}_{\mathcal{R}}) = 0 \wedge \mathrm{Out}(\mathsf{V}_{\mathcal{R}}) = s_0] \leq 1/2 - \varepsilon$$

$$\text{or} \quad \Pr_{\mathcal{V}_{\mathsf{V}_{\mathcal{R}}}(\widetilde{\mathsf{M}}, \mathsf{P}, \mathsf{V}_{\mathcal{S}})}[\mathrm{In}(\mathsf{V}_{\mathcal{R}}) = 1 \wedge \mathrm{Out}(\mathsf{V}_{\mathcal{R}}) = s_1] \leq 1/2 - \varepsilon$$

holds, where $(s_0, s_1) = \mathrm{In}(\mathsf{V}_{\mathcal{S}})$.

Case 2: With probability $1 - 1/p(\lambda)$ over $(\widetilde{\mathsf{M}}, \mathsf{P}, \mathsf{V}_{\mathcal{S}})$ generated by a run of $\tilde{\Pi}$ we have that both

$$\Pr_{\mathcal{V}_{\mathsf{V}_{\mathcal{R}}}(\widetilde{\mathsf{M}}, \mathsf{P}, \mathsf{V}_{\mathcal{S}})}[\mathrm{In}(\mathsf{V}_{\mathcal{R}}) = 0 \wedge \mathrm{Out}(\mathsf{V}_{\mathcal{R}}) = s_0] \geq 1/2 - \varepsilon$$

$$\text{and} \quad \Pr_{\mathcal{V}_{\mathsf{V}_{\mathcal{R}}}(\widetilde{\mathsf{M}}, \mathsf{P}, \mathsf{V}_{\mathcal{S}})}[\mathrm{In}(\mathsf{V}_{\mathcal{R}}) = 1 \wedge \mathrm{Out}(\mathsf{V}_{\mathcal{R}}) = s_1] \geq 1/2 - \varepsilon$$

hold, where $(s_0, s_1) = \mathrm{In}(\mathsf{V}_{\mathcal{S}})$.

Clearly, for any PUF-based OT protocol, either Case 1 or Case 2 must hold. We show that if Case 1 holds then a malicious sender may attack receiver privacy making $\mathrm{poly}(m/\varepsilon)$ queries and succeeding with advantage $\varepsilon/4p(\lambda)$, and if Case 2 holds then a malicious receiver may attack sender privacy making $\mathrm{poly}(m/\varepsilon)$ queries and succeeding with probability $1 - O(\sqrt{\varepsilon})$. This is sufficient to prove the theorem.

We next present the attacks on Receiver and Sender privacy. We defer the analysis of the attacks to the full version.

Sender's attack (denoted $\mathcal{S}'$) on receiver privacy:

1. Participate in protocol Π where the PUFs constructed by $\mathcal{S}$ are instantiated with t-wise independent hash functions and maliciously constructed to record $\mathcal{R}$ queries.
2. Convert the resulting transcript M to the augmented transcript $\widetilde{\mathsf{M}}$.
3. Run the Eve algorithm on augmented transcript $\widetilde{\mathsf{M}}$ to generate the set P.
4. Compute the probabilities

$$P_0 = \Pr_{\mathcal{V}_{\mathsf{V}_{\mathcal{R}}}(\widetilde{\mathsf{M}}, \mathsf{P}, \mathsf{V}_{\mathcal{S}})}[\mathrm{In}(\mathsf{V}_{\mathcal{R}}) = 0 \wedge \mathrm{Out}(\mathsf{V}_{\mathcal{R}}) = s_0]$$

$$\text{and} \quad P_1 = \Pr_{\mathcal{V}_{\mathsf{V}_{\mathcal{R}}}(\widetilde{\mathsf{M}}, \mathsf{P}, \mathsf{V}_{\mathcal{S}})}[\mathrm{In}(\mathsf{V}_{\mathcal{R}}) = 1 \wedge \mathrm{Out}(\mathsf{V}_{\mathcal{R}}) = s_1].$$

5. If $P_0 \geq 1/2 + \varepsilon/2$, output 0, if $P_1 \geq 1/2 + \varepsilon/2$, output 1. Otherwise, output 0 or 1 with probability $1/2$.

Receiver's attack (denoted $\mathcal{R}'$) on sender privacy:

1. Participate in protocol Π where the PUFs constructed by $\mathcal{R}$ are instantiated with t-wise independent hash functions and are maliciously constructed to record $\mathcal{S}$ queries.
2. Convert the resulting transcript M to the augmented transcript $\widetilde{\mathsf{M}}$.
3. Run the Eve algorithm on augmented transcript $\widetilde{\mathsf{M}}$ to generate the set P.
4. Compute the probabilities

$$P_0 = \Pr_{\mathcal{V}_{\mathsf{V}_{\mathcal{R}}}(\widetilde{\mathsf{M}},\mathsf{P})}[\mathrm{In}(\mathsf{V}_{\mathcal{R}}) = 0] \quad \text{and} \quad P_1 = \Pr_{\mathcal{V}_{\mathsf{V}_{\mathcal{R}}}(\widetilde{\mathsf{M}},\mathsf{P})}[\mathrm{In}(\mathsf{V}_{\mathcal{R}}) = 1].$$

5. If $P_0 = 0$ or $P_1 = 0$ then abort.
6. Otherwise, draw two views $\mathsf{V}_R(0)$ and $\mathsf{V}_R(1)$ from $\mathcal{V}_{\mathsf{V}_{\mathcal{R}}}(\widetilde{\mathsf{M}}, \mathsf{P}, \mathrm{In}(\mathsf{V}_{\mathcal{R}}) = 0)$ and $\mathcal{V}_{\mathsf{V}_{\mathcal{R}}}(\widetilde{\mathsf{M}}, \mathsf{P}, \mathrm{In}(\mathsf{V}_{\mathcal{R}}) = 1)$, respectively.
7. Output $s_0' = \mathrm{Out}(\mathsf{V}_R(0)), s_1' = \mathrm{Out}(\mathsf{V}_R(1))$.

4   Feasibility Result for Malicious, Stateless PUFs

We show that universally composable two-party computation is possible if the adversary is limited to creating *stateless* malicious PUFs. The core of our result is a construction of an unconditionally secure, universally composable, oblivious-transfer protocol in this model; we describe the protocol here, and defer its proof of security to the full version. In Section 4.2 we briefly discuss how the oblivious-transfer protocol can be used to obtain the claimed result.

4.1   Universally Composable Oblivious Transfer

Our OT protocol adapts the protocol of Brzuska et al. [5], which is secure against attackers limited to honestly generated PUFs. Roughly, we replace the single PUF—generated by one of the parties—in their protocol with a "combined PUF" generated by both parties. Specifically, this "combined PUF" PUF is constructed by having each party generate their own PUFs $\mathsf{PUF}_\mathcal{S}$, $\mathsf{PUF}_\mathcal{R}$, and then defining $\mathsf{PUF}(c) \overset{\mathrm{def}}{=} \mathsf{PUF}_\mathcal{S}(c) \oplus \mathsf{PUF}_\mathcal{R}(c)$. Intuitively, as long as one of the parties generates their PUF honestly, the combined PUF is still unpredictable (the output is random) and uncloneable (without physical access to *both* $\mathsf{PUF}_\mathcal{S}$ and $\mathsf{PUF}_\mathcal{R}$, it is impossible to evaluate PUF).

In our description of the protocol in Figure 1, we have the parties exchange PUFs once, after which they can subsequently execute any pre-determined number N of oblivious-transfer executions. We remark that it is necessary to prevent a malicious $\mathcal{R}$ from substituting a PUF of its own for $\mathsf{PUF}_\mathcal{S}$; this can be done by having $\mathcal{S}$ probe a random point before sending $\mathsf{PUF}_\mathcal{S}$ and then checking it again later. We omit this check from the figure.

Theorem 4. *The protocol in Figure 1 securely realizes $\mathcal{F}_{\mathsf{OT}}$ in the $(\mathcal{F}_{\mathsf{PUF}}, \mathcal{F}_{\mathsf{auth}})$-hybrid model, where malicious parties are limited to generated stateless PUFs (with arbitrary behavior).*

Sender $\mathcal{S}$	session sid	Receiver $\mathcal{R}$
$\mathsf{PUF}_\mathcal{S} \leftarrow \mathsf{S}$		$\mathsf{PUF}_\mathcal{R} \leftarrow \mathsf{S}$
	$\xrightarrow{\quad \mathsf{PUF}_\mathcal{S} \quad}$	
		$i = 1, \ldots, N:$
		$c_i \leftarrow \{0,1\}^\lambda$
		$r_i := \mathsf{PUF}(i\|c_i)$
		store $(c_1, r_1), \ldots, (c_N, r_N)$
	$\xleftarrow{\quad \mathsf{PUF}_\mathcal{S}, \mathsf{PUF}_\mathcal{R} \quad}$	

For $i = 1, \ldots, N$ do:

Input: $s_0, s_1 \in \{0,1\}^\lambda$		Input: $b \in \{0,1\}$
$x_0, x_1 \leftarrow \{0,1\}^\lambda$	$\xrightarrow{\quad x_0, x_1 \quad}$	
		$v := c_i \oplus x_b$
	$\xleftarrow{\quad v \quad}$	
$\hat{r}_0 := \mathsf{PUF}\,(i\|(v \oplus x_0))$		
$\hat{r}_1 := \mathsf{PUF}\,(i\|(v \oplus x_1))$		
$S_0 := s_0 \oplus \hat{r}_0$		
$S_1 := s_1 \oplus \hat{r}_1$		
	$\xrightarrow{\quad S_0, S_1 \quad}$	
		Output: $s_b := S_b \oplus r_i$

Fig. 1. Oblivious transfer protocol. We define $\mathsf{PUF}(c) \overset{\mathrm{def}}{=} \mathsf{PUF}_\mathcal{S}(c) \oplus \mathsf{PUF}_\mathcal{R}(c)$.

Security holds against an unbounded cheating $\mathcal{S}$, and an unbounded cheating $\mathcal{R}$ limited to making polynomially-many queries to $\mathsf{PUF}_\mathcal{S}$. We provide a proof of Theorem 4 in the full version.

4.2 From UC Oblivious Transfer to UC Two-Party Computation

We observe that our UC oblivious-transfer protocol can be used to obtain UC two-party computation of any functionality. The main idea is to first construct a semi-honest secure two-party computation protocol using Yao's garbled-circuit protocol, and to then apply the compiler of Ishai, Prabhakaran, and Sahai [13].

Semi-honest Secure Two-Party Computation. Lindell and Pinkas presented a proof for Yao's two-party secure-computation protocol [16]. They show how to instantiate the garbling part of the protocol with a private-key encryption scheme having certain properties. In addition, the authors show that any pseudorandom function is sufficient to instantiate such a private-key encryption scheme. Our main observation is that we can replace the pseudorandom function with a PUF.[3] This has already been observed before by Brzuska et al. [5] in a

---

[3] Note also that if the circuit generator is malicious, then he cannot violate the circuit evaluator's privacy by generating a malicious PUF.

different context. With this observation, we can apply the result of [16] to obtain a protocol for semi-honest secure two-party computation based on PUFs only (and no computational assumptions).

Theorem 5. *Let f be any functionality. Then there is a (constant-round) protocol that securely computes f for semi-honest adversaries in the $(\mathcal{F}_{\mathsf{PUF}}, \mathcal{F}_{\mathsf{OT}})$-hybrid model.*

We omit the proof since it follows easily from prior work.

Universally Composable Two-Party Computation. In the next step we apply the IPS compiler [13], a black-box compiler that takes

- An "outer" MPC protocol Π with security against a constant fraction of malicious parties.
- An "inner" two-party protocol ρ, in the $\mathcal{F}_{\mathsf{OT}}$-hybrid model, where the security of ρ only needs to hold against semi-honest parties.

and transforms them into a two-party protocol $\Phi_{\Pi,\rho}$ which is secure in the $\mathcal{F}_{\mathsf{OT}}$-hybrid model against malicious corruptions.

In our setting, we must be careful to give information-theoretic instantiations of the "outer" and "inner" protocols so that our final protocol $\Phi_{\Pi,\rho}$ will be unconditionally secure in the $\mathcal{F}_{\mathsf{OT}}$-hybrid model. Fortunately, we may instantiate the "outer" protocol, Π, with the seminal BGW protocol [4] and may instantiate the "inner" protocol, ρ, with the protocol from the previous section. Alternatively, the "inner" protocol can be instantiated with the semi-honest version of the two-party GMW protocol [10] in the $\mathcal{F}_{\mathsf{OT}}$-hybrid model.

Let ψ denote the OT-protocol described in Figure 1 and let $\Phi_{\Pi,\rho}^{\psi}(f)$ denote the IPS-compiled protocol which makes subroutine calls to ψ instead of $\mathcal{F}_{\mathsf{OT}}$ and computes the functionality f. Using Theorems 4 and 5, along with the UC composition theorem, we obtain the following result:

Theorem 6. *For any functionality f, protocol $\Phi_{\Pi,\rho}^{\psi}(f)$ securely computes f in the $(\mathcal{F}_{\mathsf{PUF}}, \mathcal{F}_{\mathsf{auth}})$-hybrid model.*

Acknowledgments. Work of Nils Fleischhacker and Dominique Schröder done in part while visiting the University of Maryland. Their work was supported by the German Federal Ministry of Education and Research (BMBF) through funding for the Center for IT-Security, Privacy, and Accountability (CISPA; see www.cispa-security.org). The visit of Nils Fleischhacker was supported by the Saarbrücken Graduate School of Computer Science funded by the German National Excellence Initiative. Dominique Schröder is also supported by an Intel Early Career Faculty Honor Program Award. The work of Jonathan Katz, as well the visit of Dominique Schröder, was supported in part by NSF award #1223623. Anna Lysyanskaya is supported by NSF awards #0964379 and #1012060.

References

1. Armknecht, F., Maes, R., Sadeghi, A.R., Standaert, F.X., Wachsmann, C.: A formalization of the security features of physical functions. In: IEEE Symposium on Security and Privacy, pp. 397–412. IEEE Computer Society Press (2011)
2. Barak, B., Mahmoody, M.: Merkle's key agreement protocol is optimal: An $o(n^2)$ attack on any key agreement from a random oracle (2013) (manuscript), http://www.cs.virginia.edu/~mohammad/files/papers/MerkleFull.pdf
3. Barak, B., Mahmoody-Ghidary, M.: Merkle puzzles are optimal—An $o(n^2)$-query attack on any key exchange from a random oracle. In: Halevi, S. (ed.) CRYPTO 2009. LNCS, vol. 5677, pp. 374–390. Springer, Heidelberg (2009)
4. Ben-Or, M., Goldwasser, S., Wigderson, A.: Completeness theorems for noncryptographic fault-tolerant distributed computations. In: 20th Annual ACM Symposium on Theory of Computing, pp. 1–10. ACM Press (1988)
5. Brzuska, C., Fischlin, M., Schröder, H., Katzenbeisser, S.: Physically uncloneable functions in the universal composition framework. In: Rogaway, P. (ed.) CRYPTO 2011. LNCS, vol. 6841, pp. 51–70. Springer, Heidelberg (2011)
6. Canetti, R.: Universally composable security: A new paradigm for cryptographic protocols. In: 42nd Annual Symposium on Foundations of Computer Science, pp. 136–145. IEEE Computer Society Press (2001)
7. Canetti, R., Fischlin, M.: Universally composable commitments. In: Kilian, J. (ed.) CRYPTO 2001. LNCS, vol. 2139, pp. 19–40. Springer, Heidelberg (2001)
8. Canetti, R., Kushilevitz, E., Lindell, Y.: On the limitations of universally composable two-party computation without set-up assumptions. Journal of Cryptology 19(2), 135–167 (2006)
9. Damgård, I., Scafuro, A.: Unconditionally secure and universally composable commitments from physical assumptions. In: Sako, K., Sarkar, P. (eds.) ASIACRYPT 2013, Part II. LNCS, vol. 8270, pp. 100–119. Springer, Heidelberg (2013), http://eprint.iacr.org/2013/108
10. Goldreich, O., Micali, S., Wigderson, A.: How to play any mental game, or a completeness theorem for protocols with honest majority. In: Aho, A. (ed.) 19th Annual ACM Symposium on Theory of Computing, pp. 218–229. ACM Press (1987)
11. Goyal, V., Ishai, Y., Mahmoody, M., Sahai, A.: Interactive locking, zero-knowledge pCPs, and unconditional cryptography. In: Rabin, T. (ed.) CRYPTO 2010. LNCS, vol. 6223, pp. 173–190. Springer, Heidelberg (2010)
12. Impagliazzo, R., Rudich, S.: Limits on the provable consequences of one-way permutations. In: 21st Annual ACM Symposium on Theory of Computing, pp. 44–61. ACM Press (1989)
13. Ishai, Y., Prabhakaran, M., Sahai, A.: Founding cryptography on oblivious transfer – efficiently. In: Wagner, D. (ed.) CRYPTO 2008. LNCS, vol. 5157, pp. 572–591. Springer, Heidelberg (2008)
14. Katz, J.: Universally composable multi-party computation using tamper-proof hardware. In: Naor, M. (ed.) EUROCRYPT 2007. LNCS, vol. 4515, pp. 115–128. Springer, Heidelberg (2007)
15. Katzenbeisser, S., Kocabaş, Ü., Rožić, V., Sadeghi, A.-R., Verbauwhede, I., Wachsmann, C.: PUFs: Myth, fact or busted? A security evaluation of physically uncloneable functions (PUFs) cast in silicon. In: Prouff, E., Schaumont, P. (eds.) CHES 2012. LNCS, vol. 7428, pp. 283–301. Springer, Heidelberg (2012)
16. Lindell, Y., Pinkas, B.: A proof of security of Yao's protocol for two-party computation. Journal of Cryptology 22(2), 161–188 (2009)

17. Maes, R., Verbauwhede, I.: Physically unclonable functions: A study on the state of the art and future research directions. In: Towards Hardware-Intrinsic Security, pp. 3–37. Springer (2010)
18. Ostrovsky, R., Scafuro, A., Visconti, I., Wadia, A.: Universally composable secure computation with (Malicious) physically uncloneable functions. In: Johansson, T., Nguyen, P.Q. (eds.) EUROCRYPT 2013. LNCS, vol. 7881, pp. 702–718. Springer, Heidelberg (2013)
19. Pappu, R.S.: Physical One-Way Functions. Phd thesis, Massachusetts Institute of Technology (2001)
20. Pappu, R.S., Recht, B., Taylor, J., Gershenfeld, N.: Physical one-way functions. Science 297, 2026–2030 (2002)
21. Rührmair, U.: Oblivious transfer based on physical unclonable functions. In: Acquisti, A., Smith, S.W., Sadeghi, A.-R. (eds.) TRUST 2010. LNCS, vol. 6101, pp. 430–440. Springer, Heidelberg (2010)
22. Rührmair, U., Katzenbeisser, S., Busch, H.: Strong PUFs: Models, constructions, and security proofs. In: Towards Hardware-Intrinsic Security, pp. 79–96. Springer (2010)
23. van Dijk, M., Rührmair, U.: PUFs in security protocols: attack models and security evaluations. In: IEEE Symposium on Security and Privacy, pp. 286–300. IEEE Computer Society Press (2013)

How to Use Bitcoin to Design Fair Protocols

Iddo Bentov and Ranjit Kumaresan

Department of Computer Science, Technion, Haifa, Israel
{idddo,ranjit}@cs.technion.ac.il

Abstract. We study a model of fairness in secure computation in which
an adversarial party that aborts on receiving output is forced to pay a
mutually predefined monetary penalty. We then show how the Bitcoin
network can be used to achieve the above notion of fairness in the two-
party as well as the multiparty setting (with a dishonest majority). In
particular, we propose new ideal functionalities and protocols for fair
secure computation and fair lottery in this model.

One of our main contributions is the definition of an ideal primitive,
which we call $\mathcal{F}_{CR}^{\star}$ (CR stands for "claim-or-refund"), that formalizes
and abstracts the exact properties we require from the Bitcoin network
to achieve our goals. Naturally, this abstraction allows us to design fair
protocols in a hybrid model in which parties have access to the $\mathcal{F}_{CR}^{\star}$
functionality, and is otherwise independent of the Bitcoin ecosystem. We
also show an efficient realization of $\mathcal{F}_{CR}^{\star}$ that requires only two Bitcoin
transactions to be made on the network.

Our constructions also enjoy high efficiency. In a multiparty setting,
our protocols only require a constant number of calls to $\mathcal{F}_{CR}^{\star}$ per party on
top of a standard multiparty secure computation protocol. Our fair multiparty lottery protocol improves over previous solutions which required
a quadratic number of Bitcoin transactions.

Keywords: Fair exchange, Secure computation, Bitcoin.

1 Introduction

Secure computation enables a set of mutually distrusting parties to carry out
a distributed computation without compromising on privacy of inputs or correctness of the end result. Indeed, secure computation is widely applicable to variety of everyday tasks ranging from electronic auctions to privacy-preserving data
mining. Showing feasibility [50,30,12,19] of this seemingly impossible-to-achieve
notion has been one of the most striking contributions of modern cryptography.
However, definitions of secure computation [29] do vary across models, in part
owing to general impossibility results for fair coin-tossing [22]. In settings where
the majority of the participating parties are dishonest (including the two party
setting), a protocol for secure computation protocols is not required to guarantee
important properties such as guaranteed output delivery or fairness.[1] Addressing

---

[1] Fairness guarantees that if one party receives output then all parties receive output.
  Guaranteed output delivery ensures that an adversary cannot prevent the honest parties from computing the function.

J.A. Garay and R. Gennaro (Eds.): CRYPTO 2014, Part II, LNCS 8617, pp. 421–439, 2014.
© International Association for Cryptologic Research 2014

this deficiency is critical if secure computation is to be widely adopted in practice, especially given the current interest in practical secure computation. Needless to say, it is not very appealing for an honest party to invest time and money to carry out a secure computation protocol until the very end, only to find out that its adversarial partner has aborted the protocol after learning the output.

Fair exchange of digital commodities is a well-motivated and well-studied problem. Loosely speaking, in the problem of fair exchange, there are two (or more) parties that wish to exchange digital commodities (e.g., signed contracts) in a fair manner, i.e., either both parties complete the exchange, or none do. A moment's thought reveals that fair exchange is indeed a special subcase of fair secure computation. Unfortunately, as is the case with fair secure computation, it is known that fair exchange in the standard model cannot be achieved [14,22]. However, solutions for fair exchange were investigated and proposed in a variety of weaker models, most notably in the optimistic model mentioned below. Typically such solutions require cryptosystems with some tailor-made properties, and employ tools of generic secure computation only sparingly (see [15,40]) in part owing to the assumed inefficiency of secure computation protocols. Recent years, however, have witnessed a tremendous momentum shift in practical secure computation (see [36,43] and references therein). Given the zeitgeist, it may seem that solving the problem of fair exchange as a subcase of fair secure computation is perhaps the right approach to take.[2] Unfortunately as described earlier, fair secure computation is impossible.

Workarounds. Indeed, several workarounds have been proposed in the literature to counter adversaries that may decide to abort possibly depending on the outcome of the protocol. The most prominent lines of work include gradual release mechanisms, optimistic models, and partially fair secure computation. Gradual release mechanisms ensure that at any stage of the protocol, the adversary has not learned much more about the output than honest parties. Optimistic models allow parties to pay a subscription fee to a trusted server that can be contacted to restore fairness whenever fairness is breached. Partially fair secure computation provides a solution for secure computation where fairness may be breached but only with some parameterizable (inverse polynomial) probability. In all of the above solutions, one of two things hold: either (1) parties have to run a secure computation protocol that could potentially be much more expensive (especially in the number of rounds) than a standard secure computation protocol, or (2) an external party must be trusted to not collude with the adversary. Further, when an adversary aborts, the honest parties have to expend *extra effort* to restore fairness, e.g., the trusted server in the optimistic model needs to contacted each time fairness is breached. In summary, in all these works,

---

[2] A similar parallel may be drawn to the practicality of secure computation itself. Special purpose protocols for secure computation were exclusively in vogue until very recently. However, a number of recent works have shown that generic secure computation can be much more practical [44,35].

(1) the honest party has to expend extra effort, and (2) the adversary essentially gets away with cheating.[3]

Ideally, rather than asking an honest party to invest additional time and money whenever fairness is (expected to be) breached by the adversary, one would expect "fair" mechanisms to compensate an honest party in such situations. Indeed, this point-of-view was taken by several works [42,41,10]. These works ensure that an honest party would be monetarily compensated whenever a dishonest party aborts. In practice, such mechanisms would be effective if the compensation amount is rightly defined. Note that in contrast to the optimistic model, here the honest party is not guaranteed to get output, but still these works provide a reasonable and practical notion of fairness. Perhaps the main drawback of such works is their dependance on e-cash systems (which unfortunately are not widely adopted yet) or central bank systems which need to be completely trusted.

Bitcoin [47] is a peer-to-peer network that uses the power of cryptography to emulate (among other things) a trusted bank. Its claim to fame is that it is the first practical decentralized digital currency system (which also provides some level of anonymity for its users). A wide variety of electronic transactions take place on the Bitcoin network. As an illustrative example, consider the case of (multiparty) lotteries which are typically conducted by gambling websites (e.g., SatoshiDice). Note that such a lottery requires the participants to trust the gambling website to properly conduct the lottery which may be unreasonable in some cases (and further necessitates paying a house edge). One might wonder if secure computation would provide a natural solution for multiparty lotteries over Bitcoin. Unfortunately, our understanding of Bitcoin is diminished by a lack of abstraction of what the Bitcoin network provides. Consequently there exist relatively very few works that provide any *constructive* uses of Bitcoin [21,2,6].

Our Contributions. Conceptually, our work provides the *missing piece* that simultaneously allows (1) designing protocols of fair secure computation that rely on Bitcoin (and not a trusted central bank), and (2) designing protocols for fair lottery on Bitcoin that use secure computation (and not a trusted gambling website). Our model of fairness is essentially the same as in [2,42,41,1] in that we wish to monetarily penalize an adversary that aborts the protocol after learning the output. We distinguish ourselves from most prior work by providing a *formal treatment*, namely specifying formal security models and definitions, and *proving* security of our constructions. In addition, we extensively consider the *multiparty* setting, and construct protocols that are both more efficient as well as provably secure (in our new model). Our clear abstraction of the functionality that we require from Bitcoin network enables us to not only design modular protocols, but also allow easy adaptations of our solutions to settings other than the

---

[3] This is especially true in today's world where cheap digital pseudonyms [23] are available.

Bitcoin network (e.g., Litecoin, PayPal, or a central trusted bank).[4] Our main contributions include providing formal definitions and efficient realizations for:

- **Claim-or-refund functionality** $\mathcal{F}_{CR}^\star$. A simple yet powerful two-party primitive that accepts deposits from a "sender" and conditionally transfers the deposit to a "receiver." If the receiver defaults, then the deposit is returned to the sender after a prespecified time. In the full version of our paper [13], we describe a Bitcoin protocol for realizing this functionality that requires parties to make only two transactions on the Bitcoin network. We note that variants of $\mathcal{F}_{CR}^\star$ have been constructed and used in [45,7,6].

- **Secure computation with penalties** $\mathcal{F}_f^\star$. In a n-party setting, a protocol for secure computation with penalties guarantees that if an adversary aborts after learning the output but before delivering output to honest parties, then *each* honest party is compensated by a prespecified amount. We show how to construct such a protocol in the $(\mathcal{F}_{OT}, \mathcal{F}_{CR}^\star)$-hybrid model that requires only $O(n)$ rounds[5] and $O(n)$ calls to $\mathcal{F}_{CR}^\star$.

- **Secure lottery with penalties** $\mathcal{F}_{lot}^\star$. In a multiparty setting, a protocol for secure lottery with penalties guarantees that if an adversary aborts after learning the outcome of the lottery but before revealing the outcome to honest parties, then *each* honest party is compensated by a prespecified amount equal to the lottery prize. We show how to construct such a protocol in the $(\mathcal{F}_{OT}, \mathcal{F}_{CR}^\star)$-hybrid model that requires only $O(n)$ rounds and $O(n)$ calls to $\mathcal{F}_{CR}^\star$.

Potential Impact. We hope that our work will encourage researchers to undertake similar attempts at formalizing other important properties of the Bitcoin network, and perhaps even develop a fully rigorous framework for secure computations that involve financial transactions. Also, we design our protocols in a hybrid model, thus enabling us to take advantage of advances in practical secure computation. One reason to do this was because we are somewhat optimistic that our protocols will have a practical impact on the way electronic transactions are conducted over the internet and the Bitcoin network.

Related Work. Most related to our work are the works of Back and Bentov [6] and Andrychowicz *et al.* [2,1]. Indeed, our work is heavily inspired by [6,2] who, to the best of our knowledge, were the first to propose fair two-party (resp. multiparty) lottery protocols over the Bitcoin network. We point out that the n-party lottery protocols of [2] require quadratic number of transactions to be made on the Bitcoin network. In contrast our protocols require only a linear number of Bitcoin transactions. (See full version for a more detailed comparison with [2].) In a followup work [1] that is concurrent to and independent of ours, the authors of [2] propose solutions for fair *two-party* secure computation over the Bitcoin network. In contrast, in this work, we propose *formal security models* for fair computations, and construct fair secure computation and lottery in

---

[4] Indeed, we can readily adapt our constructions to obtain the first multiparty solutions enjoying "legally enforceable" fairness [42].

[5] Contrast this with the gradual release mechanism which require security parameter number of rounds even when $n = 2$.

the *multiparty* setting. As far as fair two-party secure computation is concerned, although the goal of [1] and ours is the same, the means to achieve the goal are significantly different. Specifically, the protocols of [2,1] directly works by building particular Bitcoin transactions (i.e., with no formal definitions of relevant functionalities). In the following, we provide a summary of other related works.

- *Fairness in standard secure computation.* Fair two party coin tossing was shown to be impossible in [22]. Completely fair secure computation for restricted classes of functions was shown in [32,3], while partially fair secure computation for all functions were constructed in [34,9]. Complete primitives for fairness were extensively studied in [33].

- *Gradual release mechanisms.* Starting from early works [8,31], gradual release mechanism have been employed to solve the problem of fair exchange in several settings [14,24,28]. A good survey of this area can be found in [49]. A formal treatment of gradual release mechanisms can also be found in [27].

- *Optimistic model.* There has been a huge body of work starting from [5,4,11] that deals with optimistic models for fair exchange (e.g., [41,46,25]). Optimistic models for secure computation was considered in [15]. [41] consider a model similar to ours where receiving payment in the event of breach of fairness is also considered fair.

- *Legally enforceable fairness.* Chen, Kudla, and Paterson [20] designed protocols for fair exchange of signatures in a model where signatures are validated only in a court-of-law. Following this, Lindell [42] showed how to construct legally enforceable fairness in the two party secure computation where parties have access to a trusted bank (or a court of law).

2   Models and Definitions

Before we begin, we note that our formalization is heavily inspired by prior formalizations in settings similar to ours [42,27]. Let n denote the number of parties and t (resp. h) denote the number of corrupted (resp. honest) parties. We consider settings where $t < n$.[6] In our setting we are interested in dealing with non-standard commodities which we call "coins," that cannot be directly incorporated in standard definitions of secure computation.

Coins. In this paper, we define *coins* as *atomic entities that are fungible and cannot be duplicated.* In particular, we assume coins have the following properties: (1) the owner of a coin is simply the party that possesses it, and further it is guaranteed that *no other party can possess that coin simultaneously*, and (2) coins can be freely transferred from a sender to a receiver (i.e, the sender is no longer the owner of the item while the receiver becomes the new owner of the item), and further, the validity of a received coin can be immediately checked and confirmed. Note we assume that *each coin is perfectly indistinguishable from*

---

[6] Note that even when $t < n/2$, it is not clear how to design a "fair" lottery simply because standard models do not deal with coins.

one another. Further we assume that each party has its own *wallet* and *safe.*[7] All its coins are distributed between its wallet and its safe.

Our definition of coin is intended to capture *physical/cryptographic curren-cies* contained in (individual) physical/cryptographic wallets. As such the above description of a coin does not capture digital cheques or financial contracts (i.e., those that need external parties such as banks or a court-of-law to validate them). However, we chose this definition to keep things simple, and more tech-nically speaking, such a formalization would enable us to consider concurrent composition of protocols that deal with coins (in contrast with the formalization in [42]).

Notation. We use $\mathsf{coins}(x)$ to denote an item whose value is described by $x \in \mathbb{N}$. Suppose a party possesses $\mathsf{coins}(x_1)$ and receives $\mathsf{coins}(x_2)$ from another party, then we say it now possesses $\mathsf{coins}(x_1 + x_2)$. Suppose a party possesses $\mathsf{coins}(x_1)$ and sends $\mathsf{coins}(x_2)$ to another party, then we say it now possesses $\mathsf{coins}(x_1 - x_2)$.

Model. We will prove security of our protocols using the simulation paradigm. To keep things simple:

— Our protocols are designed in a hybrid model where parties have access to two *types* of ideal functionalities which we describe below. In the relevant hybrid model, our protocols will have *straightline* simulators, and thus we can hope for achieving standalone as well as universally composable (UC) security. We chose to provide UC-style definitions [17] of our ideal function-alities.

    — The first type of ideal functionalities are standard ideal functionalities used in secure computation literature These functionalities only provide security with agreement on abort [29]. In particular, they do not provide the notion of fairness that we are interested in.

    — The second type of ideal functionalities are *special* ideal functionalities that deal with coins. These are the ideal functionalities that we will be interested in realizing. Note that only special ideal functionalities deal with coins.

  Special ideal functionalities are denoted by $\mathcal{F}^\star_{\mathrm{xxx}}$ (i.e., with superscript $\star$) to distinguish them notationally from standard ideal functionalities. We will be interested in *secure realization* of these functionalities.

— We work in the standard model of secure computation where parties are assumed to be connected with pairwise secure channels over a synchronous network (i.e., the computation proceeds in "rounds"). See [27,38] on how to make the relevant modifications about synchrony assumptions in the UC-framework [17].

— Our special ideal functionality $\mathcal{F}^\star_{\mathrm{CR}}$ that idealizes Bitcoin transactions, is assumed to be aware of the round structure of the protocol. This choice is inspired by similar assumptions about the "wrapped functionalities" consid-ered in [27].

---

[7] The distinction between wallet and safe will become clear in the description of the ideal/real processes.

On the choice of UC-style definitions. In practice, we expect parties to run variety of electronic transactions concurrently. A natural requirement for proving security would be to consider *universally composable* (UC) security which would in turn also enable modular design of protocols. Perhaps, the main drawback in considering UC security is the fact that to UC realize most (standard) functionalities one typically needs to assume the existence of a trusted setup [18]. To avoid this, one may design concurrently secure protocols based only on pure complexity-theoretic assumptions. Despite this, we chose to work in a UC-like framework (which we describe below) because we believe it enables simpler and cleaner abstraction and description of our ideal functionalities and our protocols. Also we argue that the trusted setup in UC is typically a one-time setup (as opposed to say the optimistic model where trusted help needs to be online).[8] Further, the standalone variant of our protocols require no such setup.

Preliminaries. A function $\mu(\cdot)$ is negligible in λ if for every positive polynomial $p(\cdot)$ and all sufficiently large λ's it holds that $\mu(\lambda) < 1/p(\lambda)$. A probability ensemble $X = \{X(a, \lambda)\}_{a \in \{0,1\}^*, n \in \mathbb{N}}$ is an infinite sequence of random variables indexed by a and $\lambda \in \mathbb{N}$. Two distribution ensembles $X = \{X(a, \lambda)\}_{\lambda \in \mathbb{N}}$ and $Y = \{Y(a, \lambda)\}_{\lambda \in \mathbb{N}}$ are said to be computationally indistinguishable, denoted $X \overset{c}{\equiv} Y$ if for every non-uniform polynomial-time algorithm D there exists a negligible function $\mu(\cdot)$ such that for every $a \in \{0, 1\}^*$,

$$|\Pr[D(X(a, \lambda)) = 1] - \Pr[D(Y(a, \lambda)) = 1]| \leq \mu(\lambda).$$

All parties are assumed to run in time polynomial in the security parameter λ. We follow standard definitions of secure computation [29]. Our main modification is now each party has its own wallet and safe, and further, the view of $\mathcal{Z}$ contains the distribution of coins. We provide a succinct description of our model, which we call "security computation with coins" (SCC), highlighting the differences from standard secure computation. Before that we describe the distinction between wallets and safes.

Wallets vs. safes. Recall that in standard models each party is modeled as an interactive Turing machine. For our purposes, we need to augment the model by providing each party with its own wallet and safe. We allow each party's wallet to be arbitrarily modified by the distinguisher $\mathcal{Z}$ (aka environment). However, parties' safes are out of $\mathcal{Z}$'s control. This is meant to reflect honest behavior in situations where the party has no coins left to participate in a protocol. We require honest parties to simply not participate in such situations. In other words, in order to participate in a protocol, an honest party first locks the required number of coins (specified by the protocol) in its safe. During the course of a protocol, the honest party may gain coins (e.g., by receiving a penalty), or may lose coins (e.g., in a lottery). These gains and losses affect the content of the safes and not the wallets. Finally, at the end of the protocol, the honest party releases the coins associated with that protocol (including new gains) into the wallet.

---

[8] Also note, in practice, one may obtain heuristic UC security in the programmable random oracle model.

Note on the other hand, we give the real/ideal adversary complete control over a corrupt party's wallet *and* safe.

Secure Computation with Coins (SCC security). We now describe the ideal/real processes for SCC. The order of activations is the same as in UC, and in particular, $\mathcal{Z}$ is activated first. In each activation of $\mathcal{Z}$, in addition to choosing (both honest and corrupt) parties' inputs (as in standard UC), $\mathcal{Z}$ also initializes each party's wallet with some number of coins and may activate the hybrid (resp. ideal) adversary $\mathcal{A}$ (resp. $\mathcal{S}$). In every subsequent activation, $\mathcal{Z}$ may read and/or modify (i.e., add coins to or retrieve coins from)[9] the contents of the wallet (but *not* the safe) of each honest party. Further, $\mathcal{Z}$ may also read each honest party's local output tapes, and may write information on its input tape. In the hybrid (resp. ideal) process, the adversary $\mathcal{A}$ (resp. $\mathcal{S}$) has complete access to all tapes, wallets, and safes of a corrupt party. Note that, as in UC, the environment $\mathcal{Z}$ will be an interactive distinguisher.

Let $\text{IDEAL}_{f,\mathcal{S},\mathcal{Z}}(\lambda, z)$ denote the output of environment $\mathcal{Z}$ initialized with input z after interacting in the ideal process with ideal process adversary $\mathcal{S}$ and (standard or special) ideal functionality $\mathcal{G}_f$ on security parameter λ. Recall that our protocols will be run in a hybrid model where parties will have access to a (standard or special) ideal functionality $\mathcal{G}_g$. We denote the output of $\mathcal{Z}$ after interacting in an execution of π in such a model with $\mathcal{A}$ by $\text{HYBRID}^g_{\pi,\mathcal{A},\mathcal{Z}}(\lambda, z)$, where z denotes $\mathcal{Z}$'s input. We are now ready to define what it means for a protocol to SCC realize a functionality.

Definition 1. *Let $n \in \mathbb{N}$. Let π be a probabilistic polynomial-time n-party protocol and let $\mathcal{G}_f$ be a probabilistic polynomial-time n-party (standard or special) ideal functionality. We say that π SCC realizes $\mathcal{G}_f$ with abort in the $\mathcal{G}_g$-hybrid model (where $\mathcal{G}_g$ is a standard or a special ideal functionality) if for every non-uniform probabilistic polynomial-time adversary $\mathcal{A}$ attacking π there exists a non-uniform probabilistic polynomial-time adversary $\mathcal{S}$ for the ideal model such that for every non-uniform probabilistic polynomial-time adversary $\mathcal{Z}$,*

$$\{\text{IDEAL}_{f,\mathcal{S},\mathcal{Z}}(\lambda, z)\}_{\lambda \in \mathbb{N}, z \in \{0,1\}^*} \overset{c}{\equiv} \{\text{HYBRID}^g_{\pi,\mathcal{A},\mathcal{Z}}(\lambda, z)\}_{\lambda \in \mathbb{N}, z \in \{0,1\}^*}.$$

$\diamondsuit$

We have not proven a composition theorem for our definition (although we believe our model should in principle allow composition analogous to the UC composition theorem [17]). For the results in this paper, we only need to *assume* that the Bitcoin protocol realizing $\mathcal{F}^\star_{\text{CR}}$ is concurrently composable. Other than this, we require only standard sequential composition [16]. We stress that our protocols enjoy straightline simulation (both in the way coins and cryptographic primitives are handled), and thus they may be adaptable to a concurrent setting. Finally, we note that we consider only static corruptions.

Next, we define the security notion we wish to realize for fair secure computation and for fair lottery.

---

[9] I.e., we implicitly give $\mathcal{Z}$ the power to create new coins.

Definition 2. *Let* π *be a protocol and* f *be a multiparty functionality. We say that* π securely computes f with penalties *if* π SCC realizes *the functionality* $\mathcal{F}_f^\star$ *according to Definition 1.*

Definition 3. *Let* π *be a protocol. We say that* π *is a* secure lottery with penalties *if* π SCC realizes *the functionality* $\mathcal{F}_{\text{lot}}^\star$ *according to Definition 1.*

2.1   Special Ideal Functionalities

Ideal Functionality $\mathcal{F}_{\text{CR}}^\star$. This is our main special ideal functionality and will serve as a building block for securely realizing more complex special functionalities. (See Figure 1 for a formal description.) At a very basic level, $\mathcal{F}_{\text{CR}}^\star$ allows a sender P_s to *conditionally* send coins(x) to a receiver P_r. The condition is formalized as the revelation of a satisfying assignment (i.e., witness) for a sender-specified circuit $\phi_{s,r}$ (i.e., relation). Further, there is a "time" bound, formalized as a round number τ, within which P_r has to act in order to claim the coins. An important property that we wish to stress is that the satisfying witness is made *public* by $\mathcal{F}_{\text{CR}}^\star$.

The importance of the above functionality is a highly efficient realization via *Bitcoin* that requires only two transactions to be made on the network. See full version [13] for more details. In the Bitcoin realizations of the ideal functionalities, sending a message with coins(x) corresponds to broadcasting a transaction to the Bitcoin network, and waiting according to some time parameter until there is enough confidence that the transaction will not be reversed.

$\mathcal{F}_{\text{CR}}^\star$ with session identifier sid, running with parties $P_1, \ldots, P_n$, a parameter 1^λ, and an ideal adversary $\mathcal{S}$ proceeds as follows:

- *Deposit phase.* Upon receiving the tuple (deposit, $sid, ssid, s, r, \phi_{s,r}, \tau, $ coins(x)) from P_s, record the message (deposit, $sid, ssid, s, r, \phi_{s,r}, \tau, x$) and send it to all parties. Ignore any future deposit messages with the same $ssid$ from P_s to P_r.
- *Claim phase.* In round τ, upon receiving (claim, $sid, ssid, s, r, \phi_{s,r}, \tau, x, w$) from P_r, check if (1) a tuple (deposit, $sid, ssid, s, r, \phi_{s,r}, \tau, x$) was recorded, and (2) if $\phi_{s,r}(w) = 1$. If both checks pass, send (claim, $sid, ssid, s, r, \phi_{s,r}, \tau, x, w$) to all parties, send (claim, $sid, ssid, s, r, \phi_{s,r}, \tau, $ coins(x)) to P_r, and delete the record (deposit, $sid, ssid, s, r, \phi_{s,r}, \tau, x$).
- *Refund phase:* In round $\tau + 1$, if the record (deposit, $sid, ssid, s, r, \phi_{s,r}, \tau, x$) was not deleted, then send (refund, $sid, ssid, s, r, \phi_{s,r}, \tau, $ coins(x)) to P_s, and delete the record (deposit, $sid, ssid, s, r, \phi_{s,r}, \tau, x$).

Fig. 1. The special ideal functionality $\mathcal{F}_{\text{CR}}^\star$

Secure Computation with Penalties. Loosely speaking, our notion of fair secure computation guarantees:

$\mathcal{F}_f^\star$ with session identifier *sid* running with parties $P_1, \ldots, P_n$, a parameter 1^λ, and an ideal adversary $\mathcal{S}$ that corrupts parties $\{P_s\}_{s \in C}$ proceeds as follows: Let $H = [n] \setminus C$ and $h = |H|$. Let d be a parameter representing the safety deposit, and let q denote the penalty amount.

- *Input phase:* Wait to receive a message (input, $sid, ssid, r, y_r,$ coins(d)) from P_r for all $r \in H$. Then wait to receive a message (input, $sid, ssid, \{y_s\}_{s \in C},$ $H',$ coins$(h'q)$) from $\mathcal{S}$ where $h' = |H'|$.
- *Output phase:*
  - Send (return, $sid, ssid,$ coins(d)) to each P_r for $r \in H$.
  - Compute $(z_1, \ldots, z_n) \leftarrow f(y_1, \ldots, y_n)$.
    - If $h' = 0$, then send message (output, $sid, ssid, z_r$) to P_r for $r \in [n]$, and terminate.
    - If $0 < h' < h$, then send (extra, $sid, ssid,$ coins(q)) to P_r for each $r \in H'$, and terminate.
    - If $h' = h$, then send message (output, $sid, ssid, \{z_s\}_{s \in C}$) to $\mathcal{S}$.
  - If $\mathcal{S}$ returns (continue, $sid, ssid, H''$), then send (output, $sid, ssid, z_r$) to P_r for all $r \in H$, and send (payback, $sid, ssid,$ coins$((h - h'')q)$) to $\mathcal{S}$ where $h'' = |H''|$, and send (extrapay, $sid, ssid,$ coins(q)) to P_r for each $r \in H''$.
  - Else if $\mathcal{S}$ returns (abort, $sid, ssid$), send (penalty, $sid, ssid,$ coins(q)) to P_r for all $r \in H$.

Fig. 2. The special ideal functionality $\mathcal{F}_f^\star$ for secure computation with penalties

- An honest party never has to pay any penalty.
- If a party aborts after learning the output and does not deliver output to honest parties, then *every* honest party is compensated.

These guarantees are exactly captured in our description of the ideal functionality $\mathcal{F}_f^\star$ for secure computation with penalties in Figure 2. We elaborate more on the definition of the ideal functionality $\mathcal{F}_f^\star$ below.

Ideal Functionality $\mathcal{F}_f^\star$. In the first phase, the functionality $\mathcal{F}_f^\star$ receives inputs for f from all parties. In addition, $\mathcal{F}_f^\star$ allows the ideal world adverary $\mathcal{S}$ to deposit some coins which may be used to compensate honest parties if $\mathcal{S}$ aborts after receiving the outputs. Note that an honest party makes a fixed deposit coins(d) in the input phase.[10,11] Then, in the output phase, $\mathcal{F}_f^\star$ returns the deposit made by honest parties back to them. If insufficient number of coins are deposited, then $\mathcal{S}$ does not obtain the output, yet may potentially pay penalty to some subset H' of the honest parties. If $\mathcal{S}$ deposited sufficient number of coins, then

---

[10] Ideally, we wouldn't want an honest party to deposit any coins, but we impose this requirement for technical reasons.

[11] To keep the definitions simple (here and in the following), we omitted details involving obvious checks that will be performed to ensure parties provide correct inputs to the ideal functionality, including (1) checks that the provided coins are valid, and (2) deposit amounts are consistent across all parties. If checks fail, then the ideal functionality simply informs all parties and terminates the session.

it gets a chance to look at the output and then decide to continue delivering output to all parties (and further pay an additional "penalty" to some subset H''), or just abort, in which case *all* honest parties are compensated using the penalty deposited by S.

$\mathcal{F}_{\text{lot}}^*$ with session identifier sid running with parties $P_1, \ldots, P_n$, a parameter 1^λ, and an ideal adversary S that corrupts parties $\{P_s\}_{s \in C}$ proceeds as follows: Let $H = [n] \setminus C$ and $h = |H|$ and $t = |C|$. Let d be a parameter representing the safety deposit, and let q be the value of the lottery prize (note: q is also the penalty amount). We assume $d \geq q/n$.

- *Input phase:* Wait to receive a message $(\text{input}, sid, ssid, r, \text{coins}(d))$ from P_r for all $r \in H$. Then wait to receive a message $(\text{input}, sid, ssid, \{y_s\}_{s \in C}, H', \text{coins}(h'q + (tq/n)))$ from S where $h' = |H'|$.
- *Output phase:* Choose $r^* \leftarrow_R \{1, \ldots, n\}$.
  - If $h' = 0$, then send message $(\text{output}, sid, ssid, r^*)$ to P_r for $r \in [n]$, and message $(\text{return}, sid, ssid, \text{coins}(d - q/n))$ to each P_r for $r \in H$. and message $(\text{prize}, sid, ssid, \text{coins}(q))$ to P_{r^*}, and terminate.
  - If $0 < h' < h$, then send $(\text{extra}, sid, ssid, \text{coins}(q))$ to P_r for each $r \in H'$, and message $(\text{return}, sid, ssid, \text{coins}(d))$ to each P_r for $r \in H$, and send $(\text{sendback}, sid, ssid, \text{coins}(tq/n))$ to S, and terminate.
  - If $h' = h$, then send message $(\text{output}, sid, ssid, r^*)$ to S.
  - If S returns $(\text{continue}, sid, ssid, \widetilde{H}', H'')$, then send message $(\text{output}, sid, ssid, r^*)$ to P_r for $r \in [n]$, and message $(\text{return}, sid, ssid, \text{coins}(d - q/n))$ to each P_r for $r \in H$, and message $(\text{prize}, sid, ssid, \text{coins}(q))$ to P_{r^*}, and message $(\text{extrapay}_1, sid, ssid, \text{coins}(q))$ to P_r for $r \in \widetilde{H}'$, and message $(\text{extrapay}_2, sid, ssid, \text{coins}(q/n))$ to P_r for $r \in H''$, and message $(\text{payback}, sid, ssid, \text{coins}((h - \widetilde{h}')q - h''q/n))$ to S where $\widetilde{h}' = |\widetilde{H}'|$ and $h'' = |H''|$, and terminate.
  - Else if S returns $(\text{abort}, sid, ssid)$, send messages $(\text{return}, sid, ssid, \text{coins}(d))$ and $(\text{penalty}, sid, ssid, \text{coins}(q))$ to P_r for all $r \in H$, and messages $(\text{sendback}, sid, ssid, \text{coins}(tq/n))$ to S, and terminate.

Fig. 3. The ideal functionality $\mathcal{F}_{\text{lot}}^*$ for secure lottery with penalties

Secure Lottery with Penalties. Loosely speaking, our notion of fair lottery guarantees the following:

- An honest party never has to pay any penalty.
- The lottery winner has to be chosen uniformly at random.
- If a party aborts *after* learning whether or not it won the lottery without disclosing this information to honest parties, then every honest party is compensated.

These guarantees are exactly captured in our description of the ideal functionality $\mathcal{F}^{\star}_{\text{lot}}$ for secure lottery with penalties in Figure 3. We elaborate more on the definition of the ideal functionality $\mathcal{F}^{\star}_{\text{lot}}$ below.

Ideal Functionality $\mathcal{F}^{\star}_{\text{lot}}$. The high level idea behind the design of $\mathcal{F}^{\star}_{\text{lot}}$ is the same as that for $\mathcal{F}^{\star}_f$. The main distinction is that now the functionality has to ensure that the lottery is conducted properly, in the sense that all parties pay their fair share of the lottery prize (i.e., $\text{coins}(q/n)$). Thus we require that each honest party makes a fixed lottery deposit $\text{coins}(d)$ with $d \geq q/n$. Then, in the second phase, as was the case with $\mathcal{F}^{\star}_f$, the ideal functionality $\mathcal{F}^{\star}_{\text{lot}}$ allows $\mathcal{S}$ to learn the outcome of the lottery only if it made a sufficient penalty deposit (i.e., $\text{coins}(hq + (tq/n))$). As before, if $\mathcal{S}$ decides to abort, then *all* honest parties are compensated using the penalty deposited by $\mathcal{S}$ in addition to getting their lottery deposits back. (I.e., effectively, *every* honest party wins the lottery!)

Remarks. At first glance, it may appear that the sets H', H'' (resp. $H', \widetilde{H}', H''$) in the definition of $\mathcal{F}^{\star}_f$ (resp. $\mathcal{F}^{\star}_{\text{lot}}$) are somewhat unnatural. We stress that we require specification of these sets in the ideal functionalities in order to ensure that we can prove that our protocols securely realize these functionalities. We also stress that it is plausible that a different security definition (cf. Definitions 2, 3) or a different protocol construction may satisfy more "natural" formulations of $\mathcal{F}^{\star}_f$ and $\mathcal{F}^{\star}_{\text{lot}}$. We leave this for future work.

3   Secure Multiparty Computation with Penalties

We design protocols for secure computation with penalties in a hybrid model with (1) a standard ideal functionality realizing an *augmented* version of the unfair underlying function we are interested in computing, and (2) the special ideal functionality $\mathcal{F}^{\star}_{\text{CR}}$ that will enable us to provide fairness. In the following, we assume, without loss of generality, that f delivers the *same* output to all parties. For a function f, the corresponding augmented function $\hat{f}$ performs secret sharing of the output of f using a variant of *non-malleable secret sharing scheme* that is both publicly verifiable and publicly reconstructible (in short, pubNMSS). Secure computation with penalties is then achieved via carrying out "fair reconstruction" for the pubNMSS scheme.[12]

First, we provide a high level description of the semantics of the pubNMSS scheme. The Share algorithm takes as input a secret u, and generates "tag-token" pairs $\{(\text{Tag}_i, \text{Token}_i)\}_{i \in [n]}$. Finally it outputs to each party P_i the i-th token Token_i and $\text{AllTags} = (\text{Tag}_1, \ldots, \text{Tag}_n)$. Loosely speaking, the properties that we

---

[12] Our strategy is similar to the use of non-malleable secret sharing in [33] to construct complete primitives for fair secure computation in the *standard model*. In addition to working in a different model, the main difference is that here we explicitly require public verification and public reconstruction for the non-malleable secret sharing scheme. This requirement is in part motivated by the final Bitcoin realizations where validity of the shares need to be publicly verifiable (e.g., by miners) in order to successfully complete the transactions.

require from pubNMSS are (1) an adversary corrupting $t < n$ parties does not learn any information about the secret unless all shares held by honest parties are disclosed (i.e., in particular, AllTags does not reveal any further information), and (2) for any $j \in [n]$, the adversary cannot reveal $\mathsf{Token}'_j \neq \mathsf{Token}_j$ such that $(\mathsf{Tag}_j, \mathsf{Token}'_j)$ is a valid tag-token pair. Since Share is evaluated inside a secure protocol, we are guaranteed honest generation of tags and tokens. Given this, a natural candidate for a pubNMSS scheme can be obtained via *commitments* that are binding for *honest* sender (exactly as in [26]) and are equivocal. Instantiating a variant of the Naor commitment scheme [48] as done in [26], we obtain a construction of a pubNMSS scheme using only *one-way functions*. (See full version [13] for more details.) We do not attempt to provide a formal definition of pubNMSS schemes. Rather, our approach here is to sketch a specific construction which essentially satisfies all our requirements outlined above. Given a secret u, we generate tag-token pairs in the following way:

- Perform an n-out-of-n secret sharing of u to obtain $u_1, \ldots, u_n$.
- To generate the i-th "tag-token" pair, apply the sender algorithm for a honest-binding commitment using randomness ω_i to secret share u_i to obtain com_i, and set $\mathsf{Tag}_i = \mathsf{com}_i$ and $\mathsf{Token}_i = (u_i, \omega_i)$.

The reconstruction algorithm Rec takes as inputs $(\mathsf{AllTags}', \{\mathsf{Token}'_i\}_{i \in [n]})$ and proceeds in the natural way. First, it checks if $(\mathsf{Tag}'_i, \mathsf{Token}'_i = (u'_i, \omega'_i))$ is a valid tag-token pair (i.e., if Token'_i is a valid decommitment for Tag'_i) for every $i \in [n]$. Next, if the check passes, then it outputs $u' = \oplus_{\ell \in [n]} u'_\ell$, else it outputs $\bot$.

Next we show how to perform "fair reconstruction" for this scheme.

3.1 Fair Reconstruction

Loosely speaking, our notion of fair reconstruction guarantees the following:

- An honest party never has to pay any penalty.
- If the adversary reconstructs the secret, but an honest party cannot, then the honest party is compensated.

In this section, we show how to design a protocol for fair reconstruction in the $\mathcal{F}^\star_{\mathrm{CR}}$-hybrid model. For lack of space, we refer to the full version for intuition, detailed description, and a proof of security of our protocol.

Notation. As discussed before, we assume that the secret has been shared using pubNMSS, i.e., each party P_i now has AllTags and its own token Token_i. Once a party learns all the tokens, then it can reconstruct the secret. On the other hand, even if one token is not revealed, then the secret is hidden. We use T_i as shorthand to denote Token_i. A sender P_s may use (a set of) tags to specify a $\mathcal{F}^\star_{\mathrm{CR}}$ transaction with the guarantee that (except with negligible probability) its deposit can be claimed by a receiver P_r only if it produces the corresponding (set of) tokens. (More precisely, this is captured via the relation $\phi_{s,r}$ specified by P_s). In the following, we use $P_1 \xrightarrow[q,\tau]{T} P_2$ to represent a $\mathcal{F}^\star_{\mathrm{CR}}$ deposit transaction made by P_1 with $\mathsf{coins}(q)$ which can be claimed by P_2 in round τ only if it produces token T, and if P_2 does not claim the transaction, then P_1 gets $\mathsf{coins}(q)$

refunded back after round τ. We use $\tau_1, \ldots, \tau_n$ to denote round numbers. In order to keep the presentation simple and easy to follow, we avoid specifying the exact round numbers, and instead only specify constraints, e.g., $\tau_1 < \tau_2$.

Multiparty Fair Reconstruction via the "Ladder" Construction. We will ask parties to make deposits in two phases. In the first phase, parties $P_1, \ldots, P_n$ simultaneously make a deposit of $\mathsf{coins}(q)$ to recipient P_n that can be claimed only if tokens $T_1, \ldots, T_n$ are produced by P_n. We call these deposits roof deposits. Then, in the second phase, each P_{s+1} makes a deposit of $\mathsf{coins}(s \cdot q)$ to recipient P_s that can be claimed only if tokens $T_1, \ldots, T_s$ are produced by P_s. These deposits are called the ladder deposits. We also force P_{s+1} to make its ladder deposit only if for all $r > s + 1$, party P_r already made its ladder deposit. We present a pictorial description of the deposit phase of the n-party protocol in Figure 4.

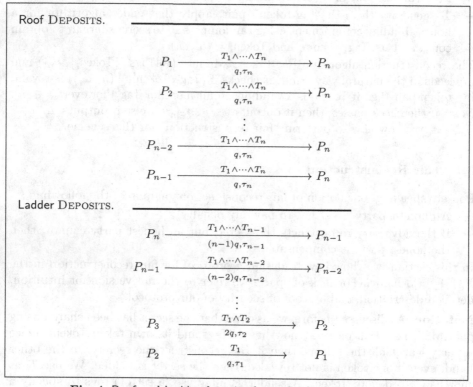

Fig. 4. Roof and Ladder deposit phases for fair reconstruction

We deal with aborts in the deposit phase in the following way. If a corrupt party does not make the roof deposit it is supposed to make, then all parties get their roof deposits refunded following which they terminate the protocol. On the other hand, if a corrupt party P_r fails to make the ladder deposit it is supposed to make, then for all $s < r$, party P_s does not make its ladder deposit at all, while for all $s > r$, party P_s continues to wait until a designated round to see whether its ladder deposit is claimed (and in particular, does not terminate the protocol immediately).

The deposits are then claimed in the reverse direction. Note that the tokens required to claim the i-th ladder deposit consist of tokens possessed by the recipient of the i-th ladder deposit plus the tokens required to claim the $(i+1)$-th ladder deposit (for $i+1 < n$). Therefore, if the $(i+1)$-th ladder deposit is claimed, then the i-th ladder deposit can *always* be claimed. In particular, the above holds even if for some $j > i+1$, (1) the j-th ladder deposit was not claimed by a possibly corrupt party, or (2) the j-th ladder deposit was not even made (which indeed is the reason why we require parties that have made their ladder deposit to wait even if a subsequent ladder deposit was not made). Further, it can be verified that if all parties behave honestly, then across all roof and ladder deposits, the amount deposited equals the amount claimed. See full version for a formal description of the protocol in the $\mathcal{F}^{\star}_{\mathrm{CR}}$-hybrid model. Since $\mathcal{F}_{\mathrm{OT}}$, the ideal functionality for oblivious transfer, is sufficient [37,39] to compute any standard ideal functionality we have the following theorem:

Theorem 1. *Assuming the existence of one-way functions, for every n-party functionality f there exists a protocol that* securely computes f with penalties *in the $(\mathcal{F}_{\mathrm{OT}}, \mathcal{F}^{\star}_{\mathrm{CR}})$-hybrid model. Further, the protocol requires $O(n)$ rounds, a total of $O(n)$ calls to $\mathcal{F}^{\star}_{\mathrm{CR}}$, and each party deposits $O(n)$ times the penalty amount.*

Somewhat surprisingly, minor modifications to the above protocol leads us to a construction for secure lotteries with penalties.

4   Secure Lottery with Penalties

Recall that our notion of fair lottery guarantees the following:
- An honest party never has to pay any penalty.
- The lottery winner has to be chosen uniformly at random.
- If a party aborts after learning whether or not it won the lottery without disclosing this information to honest parties, then every honest party is compensated.

For a formal specification of the ideal functionality see Figure 3. Our protocol proceeds in a similar way to our protocol for secure computation with penalties. Specifically, the parties first engage in a standard secure computation protocol that computes the identity of the lottery winner (i.e., by uniformly selecting an integer from $[n]$), and secret shares this result using pubNMSS (scheme described in Section 3). Now parties need to reconstruct this secret in a fair manner. Note that a malicious party may abort upon learning the outcome of the lottery (say, on learning that it did not win). This is where the fair reconstruction helps, in the sense that parties that did not learn the outcome of the protocol (i.e., the identity of the lottery winner) now receive a penalty payment equal to the lottery prize. However, this alone is not sufficient. One needs to ensure that the lottery winner actually receives the lottery prize too.

Fortunately, by making a minor modification to the "ladder" protocol, we are able to ensure that the lottery winner receives its lottery prize when the reconstruction is completed. Specifically, our modifed ladder protocol now has 3 phases: ridge, roof, and ladder phases. The ladder phase is identical to the ladder phase in the fair reconstruction protocol. We now describe at a high level how this modification works.

First recall that if parties follow the protocol, then at the end of the ladder claims, P_n has lost $(n - 1)q$ coins and every other party has gained q coins (assuming it can get its roof deposits refunded). That is, effectively party P_n has "paid" $(n - 1)q$ coins to learn the outcome of the lottery. Now suppose our roof deposit phase was made w.r.t relations ϕ_{rf}^j by party P_j such that it pays q coins to P_n only if P_j did not win the lottery.[13] Then, at the end of this phase, it is guaranteed that the lottery winner P_j, if $j \neq n$, has won q coins, and (only) P_n has completely paid for the lottery prize. Further even when $j = n$ (i.e., P_n won the lottery) then at the end of the roof deposit phase, party P_n has only "evened out" and in particular has not won the lottery prize. Effectively, P_n has paid the lottery prize to the lottery winner.

Of course, such a situation is highly unsatisfactory. We remedy the situation by introducing "ridge" deposits made by each party P_j except P_n where P_j promises to pay its lottery share q/n to P_n as long as P_n reveals all the tokens. This simple fix allows us to prove the following theorem:

Theorem 2. *Assuming the existence of one-way functions, there exists a n-party protocol for* secure lottery with penalties *in the $(\mathcal{F}_{OT}, \mathcal{F}_{CR}^\star)$-hybrid model. Further, the protocol requires $O(n)$ rounds, a total of $O(n)$ calls to $\mathcal{F}_{CR}^\star$, and each party is required to deposit $O(n)$ times the penalty amount.*

Acknowledgments. We would like to thank Yuval Ishai for many useful discussions. The first author thanks Eli Ben-Sasson for his encouragement and support. We would also like to thank the anonymous referees of Crypto 2014 for their valuable comments.

References

1. Andrychowicz, M., Dziembowski, S., Malinowski, D., Mazurek, L.: Fair two-party computations via the bitcoin deposits, ePrint 2013/837 (2013)
2. Andrychowicz, M., Dziembowski, S., Malinowski, D., Mazurek, L.: Secure multi-party computations on bitcoin. In: IEEE Security and Privacy (2014)
3. Asharov, G.: Towards characterizing complete fairness in secure two-party computation. In: Lindell, Y. (ed.) TCC 2014. LNCS, vol. 8349, pp. 291–316. Springer, Heidelberg (2014)

---

[13] Formally, for $s \in [n]$, define $\phi_{lad}^s(T_1, \ldots, T_s) = \phi(\mathsf{Tag}_1, T_1) \wedge \cdots \wedge \phi(\mathsf{Tag}_s, T_s)$. For all $s \in [n - 1]$, define $\phi_{rf}^s(T_1, \ldots, T_n) = \phi_{lad}^s(T_1, \ldots, T_n) \wedge (\mathsf{Ext}(\mathsf{Tag}_1, T_1) + \cdots + \mathsf{Ext}(\mathsf{Tag}_n, T_n) \neq s \bmod n)$, where Ext extracts the exact share (i.e., the input for the commitment) from token T.

4. Asokan, N., Shoup, V., Waidner, M.: Optimistic protocols for fair exchange. In: ACM CCS, pp. 7–17 (1997)
5. Asokan, N., Shoup, V., Waidner, M.: Optimistic Fair Exchange of Digital Signatures. In: Nyberg, K. (ed.) EUROCRYPT 1998. LNCS, vol. 1403, pp. 591–606. Springer, Heidelberg (1998)
6. Back, A., Bentov, I.: Note on fair coin toss via bitcoin (2013), http://arxiv.org/abs/1402.3698
7. Barber, S., Boyen, X., Shi, E., Uzun, E.: Bitter to better — how to make bitcoin a better currency. In: Keromytis, A.D. (ed.) FC 2012. LNCS, vol. 7397, pp. 399–414. Springer, Heidelberg (2012)
8. Beaver, D., Goldwasser, S.: Multiparty computation with faulty majority. In: IEEE FOCS, pp. 468–473 (1989)
9. Beimel, A., Lindell, Y., Omri, E., Orlov, I.: $1/p$-Secure Multiparty Computation without Honest Majority and the Best of Both Worlds. In: Rogaway, P. (ed.) CRYPTO 2011. LNCS, vol. 6841, pp. 277–296. Springer, Heidelberg (2011)
10. Belenkiy, M., Chase, M., Erway, C., Jannotti, J., Kupcu, A., Lysyanskaya, A., Rachlin, E.: Making p2p accountable without losing privacy. In: Proc. of WPES (2007)
11. Ben-Or, M., Goldreich, O., Micali, S., Rivest, R.: A fair protocol for signing contracts (extended abstract). In: Brauer, W. (ed.) ICALP. LNCS, vol. 194, pp. 43–52. Springer, Heidelberg (1985)
12. Ben-Or, M., Goldwasser, S., Wigderson, A.: Completeness theorems for noncryptographic fault-tolerant distributed computations. In: ACM STOC (1988)
13. Bentov, I., Kumaresan, R.: How to use bitcoin to design fair protocols, ePrint 2014/129 (2014)
14. Boneh, D., Naor, M.: Timed Commitments. In: Bellare, M. (ed.) CRYPTO 2000. LNCS, vol. 1880, pp. 236–254. Springer, Heidelberg (2000)
15. Cachin, C., Camenisch, J.L.: Optimistic Fair Secure Computation. In: Bellare, M. (ed.) CRYPTO 2000. LNCS, vol. 1880, pp. 93–111. Springer, Heidelberg (2000)
16. Canetti, R.: Security and composition of multiparty cryptographic protocols. Journal of Cryptology 13(1), 143–202 (2000)
17. Canetti, R.: Universally composable security: A new paradigm for cryptographic protocols. In: IEEE FOCS, pp. 136–145 (2001)
18. Canetti, R., Kushilevitz, E., Lindell, Y.: On the limitations of universally composable two-party computation without set-up assumptions. In: Biham, E. (ed.) EUROCRYPT 2003. LNCS, vol. 2656, pp. 68–86. Springer, Heidelberg (2003)
19. Chaum, D., Crépeau, C., Damgård, I.: Multiparty unconditionally secure protocols. In: ACM STOC, pp. 11–19 (1988)
20. Chen, L., Kudla, C., Paterson, K.G.: Concurrent Signatures. In: Cachin, C., Camenisch, J.L. (eds.) EUROCRYPT 2004. LNCS, vol. 3027, pp. 287–305. Springer, Heidelberg (2004)
21. Clark, J., Essex, A.: CommitCoin: Carbon Dating Commitments with Bitcoin. In: Keromytis, A.D. (ed.) FC 2012. LNCS, vol. 7397, pp. 390–398. Springer, Heidelberg (2012)
22. Cleve, R.: Limits on the security of coin flips when half the processors are faulty (extended abstract). In: STOC, pp. 364–369 (1986)
23. Friedman, E., Resnick, P.: The social cost of cheap pseudonyms. Journal of Economics and Management Strategy, 173–199 (2000)
24. Garay, J., Jakobsson, M.: Timed release of standard digital signatures. In: Blaze, M. (ed.) FC 2002. LNCS, vol. 2357, pp. 168–182. Springer, Heidelberg (2003)

25. Garay, J.A., Jakobsson, M., MacKenzie, P.D.: Abuse-Free Optimistic Contract Signing. In: Wiener, M. (ed.) CRYPTO 1999. LNCS, vol. 1666, pp. 449–466. Springer, Heidelberg (1999)
26. Garay, J., Katz, J., Kumaresan, R., Zhou, H.-S.: Adaptively secure broadcast, revisited. In: ACM PODC, pp. 179–186 (2011)
27. Garay, J., MacKenzie, P., Prabhakaran, M., Yang, K.: Resource fairness and composability of cryptographic protocols. In: TCC, pp. 404–428 (2006)
28. Garay, J.A., Pomerance, C.: Timed fair exchange of standard signatures. In: Wright, R.N. (ed.) FC 2003. LNCS, vol. 2742, pp. 190–207. Springer, Heidelberg (2003)
29. Goldreich, O.: Foundations of cryptography: Basic Applications, vol. 2 (2004)
30. Goldreich, O., Micali, S., Wigderson, A.: How to play any mental game, or a completeness theorem for protocols with honest majority. In: ACM STOC, pp. 218–229 (1987)
31. Goldwasser, S., Levin, L.A.: Fair Computation of General Functions in Presence of Immoral Majority. In: Menezes, A., Vanstone, S.A. (eds.) CRYPTO 1990. LNCS, vol. 537, pp. 77–93. Springer, Heidelberg (1991)
32. Gordon, S., Hazay, C., Katz, J., Lindell, Y.: Complete fairness in secure two-party computation. In: ACM STOC, pp. 413–422 (2008)
33. Gordon, D., Ishai, Y., Moran, T., Ostrovsky, R., Sahai, A.: On Complete Primitives for Fairness. In: Micciancio, D. (ed.) TCC 2010. LNCS, vol. 5978, pp. 91–108. Springer, Heidelberg (2010)
34. Gordon, S.D., Katz, J.: Partial Fairness in Secure Two-Party Computation. In: Gilbert, H. (ed.) EUROCRYPT 2010. LNCS, vol. 6110, pp. 157–176. Springer, Heidelberg (2010)
35. Huang, Y., Katz, J., Evans, D.: Private set intersection: Are garbled circuits better than custom protocols? In: NDSS (2012)
36. Huang, Y., Katz, J., Kolesnikov, V., Kumaresan, R., Malozemoff, A.J.: Amortizing Garbled Circuits. In: Garay, J.A., Gennaro, R. (eds.) CRYPTO 2014, Part II. LNCS, vol. 8617, pp. 458–475. Springer, Heidelberg (2014)
37. Ishai, Y., Prabhakaran, M., Sahai, A.: Founding Cryptography on Oblivious Transfer – Efficiently. In: Wagner, D. (ed.) CRYPTO 2008. LNCS, vol. 5157, pp. 572–591. Springer, Heidelberg (2008)
38. Katz, J., Maurer, U., Tackmann, B., Zikas, V.: Universally Composable Synchronous Computation. In: Sahai, A. (ed.) TCC 2013. LNCS, vol. 7785, pp. 477–498. Springer, Heidelberg (2013)
39. Ishai, Y., Prabhakaran, M., Sahai, A.: Founding cryptography on oblivious transfer – efficiently. In: Wagner, D. (ed.) CRYPTO 2008. LNCS, vol. 5157, pp. 572–591. Springer, Heidelberg (2008)
40. Küpçü, A., Lysyanskaya, A.: Optimistic Fair Exchange with Multiple Arbiters. In: Gritzalis, D., Preneel, B., Theoharidou, M. (eds.) ESORICS 2010. LNCS, vol. 6345, pp. 488–507. Springer, Heidelberg (2010)
41. Küpçü, A., Lysyanskaya, A.: Usable Optimistic Fair Exchange. In: Pieprzyk, J. (ed.) CT-RSA 2010. LNCS, vol. 5985, pp. 252–267. Springer, Heidelberg (2010)
42. Lindell, A.Y.: Legally-enforceable fairness in secure two-party computation. In: Malkin, T. (ed.) CT-RSA 2008. LNCS, vol. 4964, pp. 121–137. Springer, Heidelberg (2008)
43. Lindell, Y., Riva, B.: Cut-and-Choose Yao-Based Secure Computation in the On-line/Offline and Batch Settings. In: Garay, J.A., Gennaro, R. (eds.) CRYPTO 2014, Part II. LNCS, vol. 8617, pp. 476–494. Springer, Heidelberg (2014)

44. Malkhi, D., Nisan, N., Pinkas, B., Sella, Y.: Fairplay: a secure two-party computation system. In: USENIX, p. 20 (2004)
45. Maxwell, G.: Zero knowledge contingent payment (2011), https://en.bitcoin.it/wiki/Zero_Knowledge_Contingent_Payment
46. Micali, S.: Simple and fast optimistic protocols for fair electronic exchange. In: ACM PODC, pp. 12–19 (2003)
47. Nakamoto, S.: Bitcoin: A peer-to-peer electronic cash system (2008), http://bitcoin.org/bitcoin.pdf
48. Naor, M.: Bit Commitment Using Pseudo-randomness. In: Brassard, G. (ed.) CRYPTO 1989. LNCS, vol. 435, pp. 128–136. Springer, Heidelberg (1990)
49. Pinkas, B.: Fair secure two-party computation. In: Biham, E. (ed.) EUROCRYPT 2003. LNCS, vol. 2656, pp. 87–105. Springer, Heidelberg (2003)
50. Yao, A.C.-C.: How to generate and exchange secrets. In: 27th Annual Symposium on Foundations of Computer Science (FOCS), pp. 162–167. IEEE (1986)

FleXOR: Flexible Garbling for XOR Gates That Beats Free-XOR

Vladimir Kolesnikov[1,*], Payman Mohassel[2], and Mike Rosulek[3,**]

[1] Bell Labs, Murray Hill, NJ, USA
kolesnikov@research.bell-labs.com
[2] University of Calgary, Calgary, AB, Canada
pmohasse@cpsc.ucalgary.ca
[3] Oregon State University, Corvallis, OR, USA
rosulekm@eecs.oregonstate.edu

Abstract. Most implementations of Yao's garbled circuit approach for 2-party secure computation use the *free-XOR* optimization of Kolesnikov & Schneider (ICALP 2008). We introduce an alternative technique called *flexible-XOR* (fleXOR) that generalizes free-XOR and offers several advantages. First, fleXOR can be instantiated under a weaker hardness assumption on the underlying cipher/hash function (related-key security only, compared to related-key and circular security required for free-XOR) while maintaining most of the performance improvements that free-XOR offers. Alternatively, even though XOR gates are not always "free" in our approach, we show that the other (non-XOR) gates can be optimized more heavily than what is possible when using free-XOR. For many circuits of cryptographic interest, this can yield a significantly (over 30%) smaller garbled circuit than any other known techniques (including free-XOR) or their combinations.

1 Introduction

This work proposes efficiency improvements of two-party Secure Function Evaluation (SFE). SFE allows two parties to evaluate any function on their respective inputs x and y, while maintaining privacy of both x and y. SFE of some useful functions today is borderline practical, and first uses of secure computation begin to crop up in industry. The main obstacle in SFE's wider adoption is the cost. Indeed, SFE of most of today's functions of interest is either completely out of

---

* Supported in part by the Intelligence Advanced Research Project Activity (IARPA) via Department of Interior National Business Center (DoI/NBC) contract Number D11PC20194. The U.S. Government is authorized to reproduce and distribute reprints for Governmental purposes notwithstanding any copyright annotation thereon. Disclaimer: The views and conclusions contained herein are those of the authors and should not be interpreted as necessarily representing the official policies or endorsements, either expressed or implied, of IARPA, DoI/NBC, or the U.S. Government.

** Supported by NSF award CCF-1149647.

J.A. Garay and R. Gennaro (Eds.): CRYPTO 2014, Part II, LNCS 8617, pp. 440–457, 2014.
© International Association for Cryptologic Research 2014

reach of practicality, or carries costs sufficient to deter would-be adopters, who instead choose stronger trust models, entice users to give up their privacy with incentives, or use similar crypto-workarounds. We believe that truly practical efficiency is required for SFE to see use in real-life applications.

Our results and motivation. We improve both the required assumptions and efficiency, albeit not both simultaneously, of a commonly used SFE tool, Garbled Circuit (GC).

On the practical side, our construction results in savings of GC size of over 30% (in garbled circuits typically analyzed in the literature) as compared to the state-of-the-art GC variant using the free-XOR technique of Kolesnikov & Schneider [15]. For a fundamental protocol, which has been studied and optimized for over three decades, this is a significant improvement. We emphasize that the fleXOR approach is more general than the specific instantiations we show, and we expect better optimizations to be discovered later on. At the same time, we prove that computing optimal instantiations (i.e. those minimizing the GC size) is NP-complete.

On the theoretical side, we aim to remove the Random Oracle (RO) requirement of the free-XOR technique without sacrificing efficiency. We weaken the RO assumption to that of correlation-robustness (CR) while retaining most of the performance improvements associated with free-XOR (only $10 - 20\%$ loss for analyzed circuits).[1] This choice is natural, motivated by several pragmatic considerations:

(1) Perhaps most importantly, today an efficient GC protocol will almost certainly use the OT extension of Ishai et al. [11]. Indeed, the orders of magnitude efficiency improvement brought by the IKNP OT extension transformed the field of secure computation. The OT extension, as well as its follow-up constructions, requires CR hash functions. Thus, our choice allows to avoid the introduction of any additional assumptions in most cases.

(2) Another important factor is the degree of analysis of the candidate implementations of the employed function. Cryptanalysts study at length related-key attacks for real-world block ciphers/primitives, but, to our knowledge, key circularity attacks are less researched.

Further, the question of understanding and reducing/eliminating the RO assumption associated with free-XOR is motivated by recent work. Choi et al. [5] shows that circular-correlation robustness is a sufficient condition for free-XOR. It also presents a black-box separation which demonstrates that CR is strictly weaker than circular-correlation robustness (which, in turn, is weaker than RO). Choi et al. [5] explicitly leave open the question: "is there a garbled-circuit variant where certain gates can be evaluated for free, without relying on assumptions beyond CPA-secure encryption?" Addressing this question, Applebaum [1] showed that free-XOR can be realized in the standard model under the learning

---

[1] In fact, there is no penalty at all for *formulas* (circuits with fan-out 1). That is, our approach matches the performance of free-XOR on formulas, but under the weaker correlation-robustness assumption.

parity with noise (LPN) assumption. While novel at the fundamental level, the efficiency of the protocol of [1] is far from practical.

Our work raises and addresses related questions: *Can the efficiency improvement of free XOR be extended? Can it be achieved under weaker assumptions?*

Our metric: computation vs communication. In this work we focus on measuring performance by the size of the GC, a very clean and expressive metric. Since the associated computations are fast, we believe that in many (but not all, of course) practical scenarios communication complexity of our constructions will correlate with their total execution time. Indeed, in this work, we use aggressive (2-row) garbled row reduction (GRR2) due to Pinkas et al. [19], which involves computing polynomial interpolation. While more expensive than the standard PRF or hash function garbling, GRR2 nevertheless is a very efficient technique as evidenced by the performance experiments in [19]. GRR2 approach (denoted PRF-SS in the performance tables in [19]) is about 1x-3x times slower than the fastest experiment. However, note that a very fast 1Gbps network and a slow 2-core computer was used in [19]. Today, 1Gbps channel is still state-of-the-art, but computational power of a typical machine grew by factor of 4-6, mainly due to increased number of cores. Thus, we expect that today, the bottleneck of the [19] experiments would be in the network traffic, and not in the CPU load. This is even more likely to be so in the future, as historical hardware trends indicate faster advances in computational power than in network speeds.

At the same time, of course, specific use cases may dictate an extremely low-power CPU with an available fast network, which would imply different cost structure of our protocols. However, as argued above, today and in the expected future, communication performance is a good metric for our protocols.

1.1   Overview of Our Approach

In a garbled circuit, each wire receives a pair (A, B) of (bitstring) labels which conceptually encode TRUE and FALSE. Let us call $A \oplus B$ the **offset** of the wire. The idea behind the free XOR technique is to ensure that all wires have the same (secret) offset. Then the garbled gate can be evaluated by simply XOR-ing the wire labels.

FleXOR. With the idea of "wire offsets" in mind, consider the case where an XOR gate's input wires do not have the same wire offset. Intuitively, the free-XOR approach can be applied if we "translate" the incoming wire labels to bring them to the desired output offset. Namely, let the two input wires have wire labels $(A, A \oplus \Delta_1)$ and $(B, B \oplus \Delta_2)$, and suppose we would like the output wire labels to have offset Δ_3. We then select random "translated" wire values $\widetilde{A}, \widetilde{B}$. Let E be gate encryption function. Then we can garble this XOR gate with the following ciphertexts:

$$E_A(\widetilde{A}); \quad E_{A \oplus \Delta_1}(\widetilde{A} \oplus \Delta_3); \quad E_B(\widetilde{B}); \quad E_{B \oplus \Delta_2}(\widetilde{B} \oplus \Delta_3);$$

Now, the first two ciphertexts allow the evaluator to translate wire labels $(A, A \oplus \Delta_1)$ with offset Δ_1 into new ones $(\widetilde{A}, \widetilde{A} \oplus \Delta_3)$ of the desired offset Δ_3. Similarly the last two ciphertexts permit $(B, B \oplus \Delta_2) \rightsquigarrow (\widetilde{B}, \widetilde{B} \oplus \Delta_3)$. Now, these "translated" wire labels share the same offset Δ_3 and so the output labels $(\widetilde{A} \oplus \widetilde{B}, \widetilde{A} \oplus \widetilde{B} \oplus \Delta_3)$ can be obtained simply by XORing the "translated" labels.

So far we did not save anything: this method requires 4 ciphertexts to garble an XOR gate. However, we can reduce this cost with two simple observations:

- If we can arrange the wire label assignments so that $\Delta_1 = \Delta_3$, then the first two ciphertexts are not needed at all (the labels on this wire already have the correct offset). If $\Delta_2 = \Delta_3$, then the second two ciphertexts are not needed. Indeed, $\Delta_1 = \Delta_2 = \Delta_3$, corresponds to the free-XOR case.
- Next, we can apply a standard garbled row-reduction technique (GRR) of [19]. The idea is that ciphertexts 1 & 3 above can always be set to the string 0^λ, implicitly setting $\widetilde{A} = D_A(0^\lambda)$ and $\widetilde{B} = D_B(0^\lambda)$, where D is the gate decryption function. Hence, ciphertexts 1 & 3 never need to be sent.

As a result, we obtain a method to garble XOR gates that requires 0, 1, or at most 2 ciphertexts total, depending on how many of $\{\Delta_1, \Delta_2, \Delta_3\}$ are unique.[2] We call this method *flexible-XOR*, or **fleXOR** for short.

FleXOR application. We show how the fleXOR tool can be used to achieve the two goals motivating this work.

Consider grouping circuit wires into *equivalence classes*, where wires in the same equivalence class have the same offset. Since the arrangement of equivalence classes affects the cost of garbling each XOR gate, we are interested in assignments that minimize the total cost for all XOR gates.

If minimizing cost of XOR gates was the *only* constraint, then we could simply place all wires into a single equivalence class, and our construction in fact collapses to standard free-XOR. However, we consider additional constraints in class assignment, which result in the following improvements over the state-of-the-art GC (with free-XOR + GRR):

- **Performance improvement.** Recall, *row reduction* [19] is a technique for "compressing" a standard garbled gate from a size of 4 ciphertexts down to either 3 or 2. Free-XOR is compatible with the milder 3-ciphertext row reduction (which we call GRR3), but not with the more aggressive 2-ciphertext variant (GRR2). The problem is that gates garbled under GRR2 will have output wire labels with an unpredictable offset — it is not possible to force them to use the global wire offset Δ used by free-XOR. In contrast, our

---

[2] Our high-level description does not indicate how to garble an XOR gate using just one ciphertext in the case that $\Delta_1 = \Delta_2 \neq \Delta_3$. This is indeed possible using similar techniques (perform free XOR on the input wires, since they share a common offset, and then, with one ciphertext, adjust the result to Δ_3). However, our wire-ordering heuristics never produce XOR gates with this property, hence we do not consider this case throughout the writeup.

fleXOR generalization does not force any specific wires to share the same offset hence there is no inherent incompatibility with using GRR2. Nevertheless it is necessary to put some constraints on the class assignment (a "safety" property that we define). We propose a heuristic algorithm for obtaining a safe assignment, and use it to obtain significant reduction in the GC size, in the experiments we run.

- **Weakened assumptions.** In the free-XOR world, the non-XOR gates are garbled by encrypting plaintexts $X, X \oplus \Delta$ using combinations of keys $Y, Y \oplus \Delta$. The appearance of a secret value Δ as both a key and plaintext requires a circularity assumption on the gate-level cipher [5]. With an appropriate constraint (i.e. monotonicity property) on wire equivalence classes, we can ensure that wire labels from the class indexed i are used as keys to encrypt wire labels only from a class indexed $j > i$. Under this additional constraint, our construction can be instantiated under a significantly weaker (related-key only, not circular) hardness assumption than free-XOR. At the same time, our experiments show only mild performance loss as compared to state-of-the-art algorithms needing circularity assumption.

Recall that fleXOR easily collapses to free-XOR when grouping all wires in the same class. We view this as an important feature of our scheme. In terms of size of garbled circuits, free-XOR performs better in some settings while the new fleXOR method performs better in others. By adopting and implementing fleXOR, one can always have available both options, and seamlessly choose the best method via appropriate choice of wire equivalence classes.

1.2   Organization of Paper

After discussing related work (Section 1.3) and preliminaries (Section 2), we set up the required technical details. In Section 3, we formalize the notion of gate cipher and show that it can be instantiated with RO and correlation-robust (CR) functions. In Section 4, we explicitly write our circuit garbling scheme in the recent "garbling schemes" convention [3], and provide a proof of security with a concrete reduction to the security of the underlying gate cipher. In Section 5 we explicitly integrate garbled row reductions from [19] into the garbling protocols and prove security via concrete reductions.

Once this set up is in place, in Section 6 we present two algorithms for assigning wire classes. One, achieving what we call monotone ordering, allows us to avoid circularity in key applications. The second, more performance-oriented, achieving what we call safe ordering, allows our garbling protocols to generate GC up to and over 30% smaller than currently best known.

In Section 7, we provide detailed performance comparison of both of our heuristic algorithms.

1.3   Related work

Garbled circuit is a general and an extremely efficient technique of secure computation, requiring only one round of interaction in the semi-honest model. Due to this generality and practicality, GC and related protocols have been receiving a lot of attention in the literature.

The basic GC is so simple and minimal that it has proven hard to improve. Most of the GC research considers its application to solving problems at hand, such as set intersection, auction design, etc. A much smaller number of papers deal with technical improvements to GC-based two-party SFE, such as OT extension [11,14] or cut-and-choose improvements for malicious case [10,16,17].

Our work belongs to a third category, aiming to improve and understand the garbling scheme itself. Since the original paper of Yao over 30 years ago, only a few works fit into this category. Beaver et al. [2] introduced the point-and-permute idea, which allows the evaluator to decrypt just a single ciphertext in the garbled gate. Naor et al. [18] introduced 3-row garbled row reduction optimization. Kolesnikov and Schneider [15] introduced the popular free-XOR technique allowing XOR gates to be evaluated without cost. Pinkas et al. [19] introduced 2-row GRR and observed that GRR3 is compatible with free-XOR. Choi et al. and Appelbaum helped clarify the underlying assumptions for free-XOR, now seen as a natural part of GC. Choi et al. [5] weakened the free-XOR assumption, by defining a sufficient gate cipher property, circular security. Applebaum [1] showed how to implement free-XOR in the standard model (using the LPN assumption, and hence not competitive with today's standard GC).

In related but incomparable work, Kolesnikov and Kumaresan [13] obtained approximately 3x factor performance improvement over state-of-the-art GC by evaluating slices of information-theoretic GC of Kolesnikov [12]. Their protocol has linear number of rounds and is not secure against malicious evaluator. We also mention, but do not discuss in detail multi-party SFE such as [9,8,6].

Bellare et al. [3] introduced the *garbling schemes* abstraction, which we use here.

2 Preliminaries

2.1 Code-Based Games

We use the convention of code-based games [4]: A game $\mathcal{G}$ starts by executing the Initialize procedure. Then the adversary $\mathcal{A}$ is invoked and allowed to query the procedures that comprise the game. When the adversary halts, the Finalize procedure is called with the output of the adversary. The output of the Finalize procedure is taken to be the outcome of the game, whose random variable we denote by $\mathcal{G}^{\mathcal{A}}(\lambda)$, where λ is the global security parameter.

2.2 Garbling Schemes

Bellare, Hoang, and Rogaway [3] introduce the notion of a garbling scheme as a cryptographic primitive. We refer the reader to their work for a complete treatment and give a brief summary here.[3] A garbling scheme consists of the

---

[3] Their definitions apply to any kind of garbling, but we specify the notation for *circuit* garbling.

following algorithms: Garble takes a circuit f as input and outputs (F, e, d) where F is a garbled circuit, e is encoding information, and d is decoding information. Encode takes an input x and encoding information e and outputs a garbled input X. Eval takes a garbled circuit F and garbled input X and outputs a garbled output Y. Finally, Decode takes a garbled output Y and decoding information d and outputs a plain circuit-output (or an error $\perp$).

Our work uses the prv.sim (privacy), obv.sim (obliviousness), and aut (authenticity) security definitions from [3], which we state below. In the prv.sim and obv.sim games, the Initialize procedure chooses $\beta \leftarrow \{0, 1\}$, and the Finalize($\beta'$) procedure returns $\beta \stackrel{?}{=} \beta'$. In all three games, the adversary can make a single call to the Garble procedure, which is defined below. Additionally, the function Φ denotes the information about the circuit that is allowed to be leaked by the garbling scheme; the function S is a simulator, and G denotes a garbling scheme.

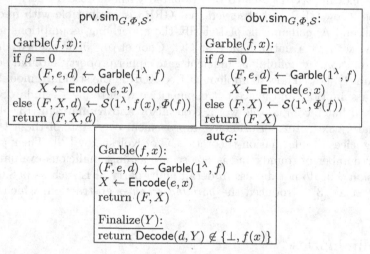

We then define the advantage of the adversary in the three security games:

$$\mathsf{Adv}^{\mathsf{prv.sim}}_{G,\Phi,S}(\mathcal{A}, \lambda) := \left| \Pr[\mathsf{prv.sim}^{\mathcal{A}}_{G,\Phi,S}(\lambda) = 1] - \frac{1}{2} \right|;$$

$$\mathsf{Adv}^{\mathsf{obv.sim}}_{G,\Phi,S}(\mathcal{A}, \lambda) := \left| \Pr[\mathsf{obv.sim}^{\mathcal{A}}_{G,\Phi,S}(\lambda) = 1] - \frac{1}{2} \right|.$$

$$\mathsf{Adv}^{\mathsf{aut}}_{G}(\mathcal{A}, \lambda) := \Pr[\mathsf{aut}^{\mathcal{A}}_{G}(\lambda) = 1];$$

3   Our Gate-Level Cipher Abstraction

Yao's technique conceptually garbles each gate with "boxes locked via two keys." We adopt the approach used by [19] and elsewhere, in which gates are garbled as $H(w_i \| w_j \| T) \oplus w_k$, where w_i, w_j are wire labels on input wires, T is a tweak/nonce, w_k is a wire label of an output wire, and H is a key-derivation function. We now describe more specifically what property is needed of H.

3.1 Definitions

We define two security games formally. They are parameterized by a KDF $H : \{0,1\}^* \to \{0,1\}^{\lambda+1}$. Game $\mathsf{kdf.rk}_{H,n}$ includes the $\boxed{\text{boxed}}$ statement, and $\mathsf{kdf.circ}_{H,n}$ excludes the boxed statement.

Initialize:
$\Delta_1, \ldots, \Delta_n \leftarrow \{0,1\}^\lambda$
$\Delta_0 := \Delta_\infty := 0^\lambda$
$\beta \leftarrow \{0,1\}$

Finalize(β'):
return $\beta' \overset{?}{=} \beta$

$\underline{\text{Fn}(X, Y, a, b, c, T):}$
return $\bot$ if T previously used in any Fn query
  or $\{a, b\} \subseteq \{0, \infty\}$ $\boxed{\text{or } c \leq \max\{a, b\}}$
if $\beta = 0$ then $Z := H(X \oplus \Delta_a, Y \oplus \Delta_b, T) \oplus (\Delta_c \| 0)$
  else $Z \leftarrow \{0,1\}^{\lambda+1}$
return Z

Briefly, the games proceed as follows. The challenger generates n random (secret) wire offsets $\{\Delta_i\}_i$, where n is a parameter of the game. The values $\Delta_0 := \Delta_\infty := 0^\lambda$ are set as a convenience.

The adversary can then make queries of the form $H(X \oplus \Delta_a, Y \oplus \Delta_b, T) \oplus \Delta_c$, provided that at least one of $\{\Delta_a, \Delta_b\}$ is unknown (i.e., $a, b \notin \{0, \infty\}$), and the tweak values T are never reused. The result of this expression should be indistinguishable from random.

In the $\mathsf{kdf.rk}$ variant of the game, there is an additional "monotonicity" restriction, that $c > \max\{a, b\}$, which prevents the adversary from invoking "key cycles" among the secret Δ_i values. It is in this setting that having two values Δ_0 and Δ_∞ is convenient. A query of the form $H(X, Y \oplus \Delta_i, T)$ can be made via $a = 0$, $b = i$, $c = \infty$, so that the monotonicity condition is satisfied ($c = 0$, for example, would break monotonicity).

Definition 1. *Let $H : \{0,1\}^* \to \{0,1\}^{\lambda+1}$ be a KDF, $\mathcal{A}$ be a PPT adversary, and the games $\mathsf{kdf.rk}_{H,n}$, $\mathsf{kdf.circ}_{H,n}$ be defined as above. We then define the advantage of the adversary in these games as:*

$$\mathsf{Adv}^{\mathsf{kdf.rk}}_{H,n}(\mathcal{A}, \lambda) := \left| \Pr[\mathsf{kdf.rk}^{\mathcal{A}}_{H,n}(\lambda) = 1] - \frac{1}{2} \right|;$$

$$\mathsf{Adv}^{\mathsf{kdf.circ}}_{H,n}(\mathcal{A}, \lambda) := \left| \Pr[\mathsf{kdf.circ}^{\mathcal{A}}_{H,n}(\lambda) = 1] - \frac{1}{2} \right|$$

Single-key vs. Dual-key. In our main construction, we garble XOR gates using only one key (wire label) and non-XOR gates using two keys (wire labels). We let H^2 be a synonym for H, and define shorthand:

$$H^1(K, T) \overset{\text{def}}{=} H^2(K, K, T)[1..\lambda]$$

We take only the first λ bits of the output for H^1 because we do not need the 1 extra bit in our construction when using H^1 (the extra bit is used for the permute bit, which is easier to handle for XOR gates).

Since H^1 takes a shorter input than H^2, it is conceivable that H^1 could be implemented more efficiently than H^2 in practice (e.g., invoking a hash function with a smaller input and hence fewer iterations). However, this kind of optimization not the focus of our work.

3.2 Instantiation from a Random Oracle

Lemma 1. *Let* $H : \{0,1\}^* \to \{0,1\}^{\lambda+1}$ *be a random oracle. Then for all* $\mathcal{A}$, *we have* $\mathsf{Adv}_{H,n}^{\mathsf{kdf.circ}}(\mathcal{A}, \lambda) \leq 16n(q_A + q_C)^2/2^\lambda$, *where* q_A, q_C *are the number of queries made to the random oracle (locally) and to the Fn procedure, respectively, by* $\mathcal{A}$ *(and* n *is the parameter of the security game).*

3.3 Instantiation from Correlation-Robustness

The free-XOR approach was formally proven secure in the RO model, and believed secure under some (unspecified) variant of correlation-robustness [11]. Choi et al [5] showed that the most natural variant of correlation-robustness (called *2-correlation-robust*) was in fact insufficient for free-XOR. Below we have translated their definition to the framework of code-based games. We then show that 2-correlation-robustness is sufficient for kdf.rk security.

Definition 2 (adapted from [5]). *Let* $H : \{0,1\}^* \to \{0,1\}^{\lambda+1}$ *be a hash function.*[4] *Define* $\mathsf{Adv}_H^{\mathsf{2corr}}(\mathcal{A}, \lambda) := |\Pr[\mathsf{2corr}_H^{\mathcal{A}}(\lambda) = 1] - \frac{1}{2}|$, *where* $\mathsf{2corr}_H$ *is the game defined as follows:*

<u>Initialize:</u>
$\Delta \leftarrow \{0,1\}^\lambda$
$\beta \leftarrow \{0,1\}$

<u>Finalize(β'):</u>
return $\beta \overset{?}{=} \beta'$

$\mathsf{Fn}(X, Y, T)$:
return $\perp$ if this query previously made
if $\beta = 0$ then $Z_1 := H(X \oplus \Delta, Y \oplus \Delta, T)$
$\qquad\qquad\qquad Z_2 := H(X \oplus \Delta, Y, \qquad T)$
$\qquad\qquad\qquad Z_3 := H(X, \qquad Y \oplus \Delta, T)$
else $Z_1, Z_2, Z_3 \leftarrow \{0,1\}^\lambda$
return Z_1, Z_2, Z_3

Lemma 2. *For all probabilistic polynomial-time* $\mathcal{A}$, *we have* $\mathsf{Adv}_{\Sigma,n}^{\mathsf{kdf.rk}}(\mathcal{A}, \lambda) \leq n \cdot \mathsf{Adv}_R^{\mathsf{2corr}}(\mathcal{A}', \lambda)$, *where* $\mathcal{A}'$ *has comparable runtime to* $\mathcal{A}$.

4 Baseline Construction

We now present our "basic" fleXOR garbling scheme. It requires some auxiliary information about the circuit, defined below:

Definition 3. *A* **wire ordering** *for a boolean circuit* C *is a function* $\mathcal{L}$ *that assigns an integer to each wire in* C. *Without loss of generality, we assume that* $\mathrm{im}(\mathcal{L}) = \{1, \ldots, L\}$ *for some integer* L, *and we denote* $|\mathcal{L}| = L$. *We say that* $\mathcal{L}$ *is* **monotone** *if:*

---

[4] H may be drawn from a family of hash functions, but for simplicity we refer to H as a single function.

Garble$(1^\lambda, C, \mathcal{L})$:
for $\ell = 1$ to $|\mathcal{L}|$: $\Delta_\ell \leftarrow \{0,1\}^\lambda$
for each input bit i corresponding to wire j of C:
  $b_j \leftarrow \{0,1\}$
  $w_j^0 \leftarrow \{0,1\}^\lambda$; $w_j^1 := w_j^0 \oplus \Delta_{\mathcal{L}(j)}$
  for $v \in \{0,1\}$: $e[i,v] := w_j^{v \oplus b_j} \| b_j$
for each gate g in C, in topological order:
  let i, j denote g's input wires
  let k denote g's output wire
  if g is an XOR gate:
    if $\mathcal{L}(i) \neq \mathcal{L}(k)$:
      $\widetilde{w}_i^0 \leftarrow \{0,1\}^\lambda$; $\widetilde{w}_i^1 := \widetilde{w}_i^0 \oplus \Delta_{\mathcal{L}(k)}$
      for $b \in \{0,1\}$: $c_{0,b} := H^1(w_i^b, g\|0\|b) \oplus \widetilde{w}_i^b$
    else for $b \in \{0,1\}$: $\widetilde{w}_i^b := w_i^b$; $c_{0,b} := \bot$
    if $\mathcal{L}(j) \neq \mathcal{L}(k)$:
      $\widetilde{w}_j^0 \leftarrow \{0,1\}^\lambda$; $\widetilde{w}_j^1 := \widetilde{w}_j^0 \oplus \Delta_{\mathcal{L}(k)}$
      for $b \in \{0,1\}$: $c_{1,b} := H^1(w_j^b, g\|1\|b) \oplus \widetilde{w}_j^b$
    else for $b \in \{0,1\}$: $\widetilde{w}_j^b := w_j^b$; $c_{1,b} := \bot$
    $w_k^0 := \widetilde{w}_i^0 \oplus \widetilde{w}_j^0$; $w_k^1 := \widetilde{w}_i^1 \oplus \widetilde{w}_j^1$;
    $b_k := b_i \oplus b_j$
  else g computes logic $G : \{0,1\}^2 \to \{0,1\}$:
    $b_k \leftarrow \{0,1\}$
    $w_k^0 \leftarrow \{0,1\}^\lambda$; $w_k^1 := w_k^0 \oplus \Delta_{\mathcal{L}(k)}$
    for $a, b \in \{0,1\}^2$:
      $v := b_k \oplus G(a \oplus b_i, b \oplus b_j)$
      $c_{a,b} := H^2(w_i^a, w_j^b, g\|a\|b) \oplus w_k^v \| v$
    $F[g] := (c_{00}, c_{01}, c_{10}, c_{11})$
for each output bit i corresponding to wire j of C:
  for $v \in \{0,1\}$: $d[i,v] := H^1(w_j^{v \oplus b_j}, \mathsf{out}\|j\|v)$
return (F, e, d)

Encode(e, x) :
for $i = 1$ to $|x|$: $X[i] := e[i, x_i]$
return X

Eval(F, X) :
for each input wire i in C:
  $w_i^* \| b_i^* \leftarrow X[i]$
for each gate g in C, in topological order:
  let i, j denote g's input wires
  let k denote g's output wire
  parse $F[g]$ as $(c_{00}, c_{01}, c_{10}, c_{11})$
  if g is an XOR gate:
    if $c_{01} = \bot$ then $\widetilde{w}_i^* := w_i^*$
      else $\widetilde{w}_i^* := H^1(w_i^*, g\|0\|b_i^*) \oplus c_{0,b_i^*}$
    if $c_{11} = \bot$ then $\widetilde{w}_j^* := w_j^*$
      else $\widetilde{w}_j^* := H^1(w_j^*, g\|1\|b_j^*) \oplus c_{1,b_j^*}$
    $w_k^* := w_i^* \oplus w_j^*$; $b_k^* := b_i^* \oplus b_j^*$
  else:
    $w_k^* \| b_k^* := H^2(w_i^*, w_j^*, g\|b_i^*\|b_j^*) \oplus c_{b_i^*, b_j^*}$
for each output bit i in C:
  let j be the corresponding wire
  $Y[i] := H^1(w_j^*, \mathsf{out}\|j\|b_j^*)$
return Y

Decode(Y, d) :
for $i = 1$ to $Y.len$:
  if $Y[i] = d[i, 0]$ then $y_i = 0$
  elsif $Y[i] = d[i, 1]$ then $y_i = 1$
  else return $\bot$
return y

Fig. 1. Our baseline garbling scheme

1. *for each XOR gate, with input wires i & j and output wire k: $\mathcal{L}(k) \geq \max\{\mathcal{L}(i), \mathcal{L}(j)\}$, and*
2. *for each non-XOR gate, with input wires i & j and output wire k: $\mathcal{L}(k) > \max\{\mathcal{L}(i), \mathcal{L}(j)\}$.*

We now give the complete description of our garbling scheme. Following [3], the scheme consists of 4 algorithms: Garble, Encode, Eval, Decode. We make one syntactic change, and allow Garble to accept as input auxiliary information $\mathcal{L}$, which is a wire ordering of the given circuit.

The scheme is described formally in Figure 1. It follows the typical Yao approach for garbling a circuit. Briefly, for each wire i, the garbler chooses two wire labels w_i^0, w_i^1 such that $w_i^0 \oplus w_i^1 = \Delta_{\mathcal{L}(i)}$. We use the point-and-permute bit optimization of [18], where a permute bit b_i is chosen so that $w_i^{b_i}$ encodes FALSE on wire i, and $w_i^{1 \oplus b_i}$ encodes TRUE. Non-XOR gates are garbled in the usual way.

XOR gates use the approach described in the introduction. Namely, suppose an XOR gate has input wires i, j and output wire k. If $\mathcal{L}(i) = \mathcal{L}(k)$, then no action is required for wire i in this gate (and no ciphertexts are included in the garbled circuit). Otherwise, we choose "adjusted" wire labels $\widetilde{w}_i^0, \widetilde{w}_i^1$ whose offset is the target value $\Delta_{\mathcal{L}(k)}$ and provide two ciphertexts that allow the evaluator to obtain $\widetilde{w}_i^b$ from w_i^b. The same logic applies for input wire j, and finally a "free XOR" is performed on these adjusted wire labels.

Theorem 1. *Let $G[H]$ denote our garbling scheme (Figure 1), where H is a KDF. Let Φ denote the side information function that leaks the circuit topology, distinction between XOR vs non-XOR gates (but not distinctions among non-XOR gates), and the wire ordering function $\mathcal{L}$ used. Then, for all probabilistic polynomial-time $\mathcal{A}$, there exists a polynomial-time simulator $\mathcal{S}$ such that:*

$$\mathsf{Adv}^{\mathsf{prv.sim}}_{G[H],\Phi,\mathcal{S}}(\mathcal{A},\lambda) \leq \mathsf{Adv}^{\mathsf{kdf.circ}}_{H,|\mathcal{L}|}(\mathcal{A}',\lambda)$$

*where $\mathcal{A}'$ has runtime essentially the same as $\mathcal{A}$. Furthermore, when the wire ordering function $\mathcal{L}$ is **monotone**, we have:*

$$\mathsf{Adv}^{\mathsf{prv.sim}}_{G[H],\Phi,\mathcal{S}}(\mathcal{A},\lambda) \leq \mathsf{Adv}^{\mathsf{kdf.rk}}_{H,|\mathcal{L}|}(\mathcal{A}',\lambda)$$

5   Incorporating Row Reductions

Row-reduction optimizations were introduced by Naor et al. [18] and later formalized and extended by Pinkas et al. [19]. They describe two flavors of row reduction, which we discuss and adapt to our fleXOR technique.

5.1   Optimization 1: Mild Row Reduction

In the first variant of row reduction, Naor et al. describe how to reduce standard 4-ciphertext garbled gates to 3 ciphertexts. Conceptually, this is done by fixing one of the ciphertexts to be the all-zeroes string. The idea is that if, say, c_{00} is known to always consist of all zeroes, then it does not actually need to be included in the garbled output.

For example, when garbling a non-XOR gate we see that ciphertext c_{00} will be zero if the appropriate output wire label (concatenated with its permute bit) is chosen to be $H^2(w_i^0, w_j^0, g\|00)$, which is the value that would be used to mask that wire label.

Hence, instead of choosing wire labels and permute bits uniformly, we choose one wire label to be an output of the KDF H and set the other label so that the two labels have the desired offset. We can use this idea with our XOR gates as well, following the ideas described in the introduction. Recall that to garble an XOR gate, we choose random "adjusted" wire labels for each input wire (whose offset requires adjusting). Instead of choosing these adjusted wire labels uniformly, we choose them to be the appropriate output of the KDF.

The formal description of this optimization is given in the full version. When garbling XOR gates, the ciphertexts c_{00}, c_{10} are always empty (implicitly set to all zeroes). Hence, XOR gates require 0, 1, or 2 ciphertexts. For non-XOR gates, the ciphertext c_{00} is always empty (implicitly set to all zeroes), so these gates require 3 ciphertexts.

That this optimization requires no additional properties of the wire ordering, and it achieves essentially identical security to our baseline construction:

Theorem 2. *Let $G^1[H]$ denote our "optimization #1" garbling scheme described above. Let Φ be as in Theorem 1. Then, for all probabilistic polynomial-time $\mathcal{A}$, there exists a polynomial-time simulator $\mathcal{S}$ such that:*

$$\mathsf{Adv}^{\mathsf{prv.sim}}_{G^1[H],\Phi,\mathcal{S}}(\mathcal{A},\lambda) \leq \mathsf{Adv}^{\mathsf{kdf.circ}}_{H,|\mathcal{L}|}(\mathcal{A}',\lambda)$$

where $\mathcal{A}'$ has runtime essentially the same as $\mathcal{A}$. Furthermore, when the wire ordering function $\mathcal{L}$ is **monotone**, *we have:*

$$\mathsf{Adv}^{\mathsf{prv.sim}}_{G^1[H],\Phi,\mathcal{S}}(\mathcal{A},\lambda) \leq \mathsf{Adv}^{\mathsf{kdf.rk}}_{H,|\mathcal{L}|}(\mathcal{A}',\lambda)$$

5.2   Optimization 2: Aggressive Row Reduction

The second variant of row reduction reduces each garbled gate from 4 to 2 ciphertexts. Here we consider applying this optimization to the non-XOR gates in our scheme. This optimization has the effect of setting both output wire labels (and hence, their offset) implicitly. Superficially, this seems at odds with our approach, in which we always choose wire labels to have some desired offset.

However, suppose that g is a non-XOR gate with output wire k. If we process this gate before any other wire i with $\mathcal{L}(i) = \mathcal{L}(k)$, then we can indeed set the offset $\Delta_{\mathcal{L}(k)}$ implicitly based on the result of the row-reduction applied to this gate. If we process the gates in a topological order, one can capture this property by requiring that $\mathcal{L}(k) > \mathcal{L}(j)$ for every wire j that influences k (i.e. j has to be processed before k). We will also require that no other non-XOR gate in the circuit has output wire k' with $\mathcal{L}(k) = \mathcal{L}(k')$, though XOR gates can safely have this property.

The necessary properties on the wire ordering are summarized in the following definition:

Definition 4. *We say that $\mathcal{L}$ is* **safe** *if:*

1. *for each non-XOR gate g with output wire k, and each wire j that influences[5] g, we have $\mathcal{L}(k) > \mathcal{L}(j)$.*
2. *for each value ℓ, there is at most one non-XOR gate whose output wire k satisfies $\mathcal{L}(k) = \ell$.*

Note that a wire ordering may be any combination of safe/non-safe, monotone/ non-monotone.

We say that a topological ordering of gates in a circuit C is **safety-respecting of $\mathcal{L}$** *if for every non-XOR gate g with output wire k, g appears earlier in the ordering than any other gate g' with output wire k' satisfying $\mathcal{L}(k) = \mathcal{L}(k')$.*

Assuming that $\mathcal{L}$ is safe, we can garble all non-XOR gates using only two ciphertexts, plus 4 additional bits. XOR gates still require 0, 1, or 2 ciphertexts as in the previous section.

---

[5] A wire j influences a wire k if there is a directed path in the circuit that contains wire j before wire k.

Our approach for row-reduction is the same as [19], but we give a short overview here in the interest of completeness. For simplicity, we assume that all non-XOR gates compute boolean-AND logic. Briefly, for each (a, b), we compute $V_{ab} = H^2(w_i^a, w_j^b, g\|a\|b)$. Hence, only one V_{ab} value is accessible to the evaluator. If the evaluator obtains V_{ab} with $(a, b) = (\bar{b}_i, \bar{b}_j)$, then the evaluator has TRUE on both input wires and hence this V_{ab} should allow the evaluator to obtain the "TRUE" output wire label $w_k^{1\oplus b_k}$. All other V_{ab} values should allow the evaluator to obtain the "FALSE" label $w_k^{b_k}$.

To make this work, let P be the degree-2 polynomial that passes through the 3 points of the form $(2a + b, V_{ab})$, for the (a, b) pairs which are supposed to yield $w_k^{b_k}$. Then let Q be the degree-2 polynomial that passes through the points $(4, P(4))$, $(5, P(5))$, and the point $(2a + b, V_{ab})$ for the "other" pair (a, b). The idea is that we can give the evaluator the values $P(4)$ and $P(5)$. When combined with his unique V_{ab} value, he can interpolate to obtain either the polynomial P or Q, depending on the output logic of the gate. Hence, we can set the two wire labels to be points on P and Q respectively, say, $P(-1)$ and $Q(-1)$.

The formal description of this optimization is given in the full version. We must also account for the permute bits, which require 4 extra bits. Overall, each AND-gate requires $2\lambda + 4$ bits, while XOR-gates still require 0, λ, or 2λ bits. We require the garbling procedure to process gates in a *safety-respecting* topological order, which ensures that Δ_ℓ gets set (while garbling an AND-gate) before it is used when later garbling an XOR gate.

Theorem 3. *Let $G^2[H]$ denote our "optimization #2" garbling scheme described above. Let Φ be as in Theorem 1. Then, for all probabilistic polynomial-time $\mathcal{A}$, there exists a polynomial-time simulator $\mathcal{S}$ such that:*

$$\mathsf{Adv}_{G^2[H], \Phi, \mathcal{S}}^{\mathsf{prv.sim}}(\mathcal{A}, \lambda) \le (n + 1) \cdot \mathsf{Adv}_{H, |\mathcal{L}|}^{\mathsf{kdf.circ}}(\mathcal{A}', \lambda)$$

*where $\mathcal{A}'$ has runtime essentially the same as $\mathcal{A}$. Furthermore, when the wire ordering function $\mathcal{L}$ is **monotone**, we have:*

$$\mathsf{Adv}_{G^2[H], \Phi, \mathcal{S}}^{\mathsf{prv.sim}}(\mathcal{A}, \lambda) \le (n + 1) \cdot \mathsf{Adv}_{H, |\mathcal{L}|}^{\mathsf{kdf.rk}}(\mathcal{A}', \lambda)$$

5.3 GRR2-Salvaging

In general, it is not possible to combine fleXOR garbling with aggressive row reduction if the wire ordering is non-safe. Nevertheless, we observe that it is possible to garble *one* non-XOR gate in each $\mathcal{L}$-equivalence class using aggressive row reduction. Roughly speaking, for each value ℓ, we identify the topologically first non-XOR gates g whose output wire i satisfies $\mathcal{L}(i) = \ell$. We ensure that g is processed before any other such gates, garble it with GRR2, and use the result to implicitly set Δ_ℓ. The remaining gates in g's equivalence class can then be garbled using GRR3.

This approach slightly generalizes our construction in the previous section. It provides a modest reduction in size, which we discuss in Section 7.

6   Optimizing the Choice of Wire Orderings

We have identified two types of wire orderings for use with our fleXOR construction: *monotone* and *safe* ordering. In this section, we consider the problem of optimizing the choice of wire ordering: i.e., a safe/monotone wire ordering that minimizes the size of the fleXOR-garbled circuit. In particular, we need only consider the total size of garbled XOR gates. An XOR gate with input wires i and j and output wire k, requires two ciphertexts if $\mathcal{L}(i) \neq \mathcal{L}(k)$ and $\mathcal{L}(j) \neq \mathcal{L}(k)$, requires one ciphertext if only one of the inequalities holds, and is "free" (no ciphertexts) if $\mathcal{L}(i) = \mathcal{L}(j) = \mathcal{L}(k)$.

6.1   Monotone Orderings

We start by showing that the problem of finding an *optimal monotone ordering* of a circuit is NP-complete. In particular, we prove the following theorem in the full version, via a simple reduction to 3SAT.

Theorem 4. *The following problem is NP-complete: Given a circuit C and integer N, determine whether there is a monotone wire ordering of C for which garbling the XOR gates using the fleXOR scheme requires at most N ciphertexts.*

It is, however, easy to find at least *some* monotone wire ordering, using an elementary linear-time algorithm. First, assign each input wire i to $\mathcal{L}(i) = 1$. Then process the gates in topological order and assign to each output wire the minimum $\mathcal{L}$ allowed by the monotonicity condition. We mention this simple approach only because it can be computed on the fly at basically no expense, in the same pass that garbles the circuit. This may be important in memory-critical applications where circuits are processed via streaming.

In Figure 2, we propose a better heuristic for monotone orderings, inspired by the following observation. Note that it is only the non-XOR gates which necessarily increase the wire ordering number between input and output wires of a gate. Define the **non-XOR-depth** of a wire i in a circuit C as the maximum number of non-XOR gates among all directed paths from i to an output wire. The non-XOR-depth of every gate in a circuit can be computed via a simple dynamic programming approach. Then, we define a wire-ordering function $\mathcal{L}$ so that $\mathcal{L}(i) + \text{non-XOR-depth}(i)$ is constant for all wires i. Hence, wires closer to the outputs receive higher wire-ordering. This heuristic is in fact optimal, and results in *all* XOR gates free, when the circuit has fan-out 1 (i.e., the circuit encodes a *formula*). It is also not hard to prove that it minimizes the size of the range of the wire-ordering function hence (intuitively) increasing the likelihood of the input and output wires of an XOR gate being in the same class.

We further refine this heuristic by revisiting each XOR gate one more time, in topological order, and reducing the order of each output wire to maximum of orders of its input wires (if this is not already the case). If done in topological order, this does not affect the monotonicity of the ordering.

Proposition 5. *The algorithm of Figure 2 computes a monotone wire ordering in linear time.*

```
for every wire i:
    compute non-XOR-depth[i]
set Λ = 1 + num wires in circuit
for each wire i:
    set L[i] := Λ − non-XOR-depth[i]
for each XOR gate g in topo. order:
    denote g's inputs wires by i, j
    denote g's output wire by k
    if L[k] > max{L[i], L[j]}
        set L[k] := max{L[i], L[j]}
```

```
for each input wire i:
    set L[i] := 1
set count := 2
for each gate g, in topo. order:
    denote g's output wire by k
    if g is an XOR gate:
        set L[k] := 1
    else:
        set L[k] := count
        count := count + 1
```

Fig. 2. Monotone wire ordering heuristic   **Fig. 3.** Safe wire ordering heuristic

We implemented both heuristic algorithms for monotone orderings, and tested them on a wide range of circuits. In general, our second heuristic algorithm outperforms the elementary one by 20-40% (in terms of average cost per XOR gate).

6.2   Safe Orderings

The constraints for safe wire ordering are fairly strict, making it challenging to devise good heuristic algorithms that minimize the number ciphertexts needed to garble XOR gates. Nevertheless, we introduce a simple and intuitive algorithm that performs well in practice as demonstrated in our analysis in the following section.

Since the output wires of non-XOR gates must have distinct $\mathcal{L}$-values in a safe ordering, our idea is to assign such wires values incrementally, and in topological order, starting from 2. Then, for each XOR gate, we let the $\mathcal{L}$-value of its output wire be 1 (see Figure 3). The resulting ordering will always satisfy the definition of a safe ordering. In particular, if wire i influences a non-XOR gate with output wire j, then $\mathcal{L}(i) < \mathcal{L}(j)$, either by the topological constraint (when wire i emanates from a non-XOR gate), or because $\mathcal{L}(i) = 1 < \mathcal{L}(j)$ (when i emanates from an XOR gate).

Proposition 6. *The algorithm of Figure 3 computes a safe wire ordering in linear time.*

6.3   Other Constraints for Wire Orderings

Here we considered safe and monotone orderings separately, but we note that it is possible (and interesting) to consider their combination i.e. optimization problems for orderings that are both safe and monotone. We leave open the problem of designing good heuristics for this problem.

As mentioned earlier, using a trivial wire ordering (all wires assigned the same index) causes fleXOR construction to collapse to free-XOR.

Most 2PC protocols based on garbled circuits require only what is provided by the "garbling schemes" abstraction of [3] which we use here. The fleXOR

circuit	GRR2	free-XOR	fleXOR monotone	safe	best
DES	2.0 (2.0)	2.79 (0.0)	2.84 (0.93)	**1.89** (0.38)	1.89
AES	2.0 (2.0)	**0.64** (0.0)	0.76 (0.15)	0.72 (0.37)	0.64
SHA-1	2.0 (2.0)	1.82 (0.0)	2.02 (0.75)	**1.39** (0.45)	1.39
SHA-256	2.0 (2.0)	2.05 (0.0)	2.26 (0.76)	**1.56** (0.60)	1.56
Hamming distance	2.0 (2.0)	**0.50** (0.0)	0.67 (0.20)	**0.50** (0.20)	0.50
minimum in set	2.0 (2.0)	**0.87** (0.0)	1.01 (0.41)	**0.87** (0.41)	0.87
32 × 32 fast mult	2.0 (2.0)	**0.90** (0.0)	1.15 (0.36)	0.94 (0.49)	0.90
1024-bit millionaires	2.0 (2.0)	**1.00** (0.0)	1.08 (0.25)	**1.00** (0.50)	1.00

Fig. 4. Comparison of standard garbling (with GRR2 row reduction), free-XOR, and fleXOR instantiations. The main number in each cell shows average number of ciphertexts per gate; the number in the parentheses shows average number of ciphertexts per XOR gate only.

construction is thus automatically compatible with these protocols. However, some protocols [20,16] "break the abstraction boundary" of garbling schemes and include optimizations that take advantage of specific properties of free-XOR. In particular, they only require that either the input wires or output wires all share a common offset (sometimes across several garbled circuits); they do not require anything of the internal wires. It is easy to include such a constraint on input/output wires in a fleXOR wire ordering, allowing fleXOR to be compatible with these protocols as well.

7   Performance Comparison

In this section we empirically evaluate the performance of our fleXOR approach against free-XOR and standard (GRR2) garbling. We obtained several circuits of interest [21,7] and evaluated the performance of our garbling schemes on them. As outlined in the introduction, our primary metric is the size (number of ciphertexts) needed to garble a circuit. The results are summarized in Figure 4.

Eliminating the circularity assumption. As discussed earlier, fleXOR avoids the strong circular-security assumption of free-XOR, when instantiated with a monotone wire ordering. Weakening the assumption does come at a cost, since not all XOR gates are free as a result. Comparing the 2nd and 3rd colums in Figure 4 illustrates the cost savings of circularity. In general, we show that the circularity assumption can be eliminated with a typical increase in garbled circuit size of around 10% (and never more than 20% in our analysis).

We used the heuristic method of Figure 2 for finding good monotone wire orderings (it performed better than the elementary method, on all circuits we tried). The numbers for free-XOR and for fleXOR+monotone both reflect mild (GRR3) row reduction for the non-XOR gates, except that we apply GRR2-salvaging (Section 5.3) for fleXOR. The gain from GRR2-salvaging varies considerably, but is sometimes noticeable. For example, the numbers in Figure 4 reflect a savings from GRR2-salvaging of 3976 ciphertexts for SHA256, but only 40 for the AES circuit.

Beating (and matching) free-XOR efficiency. As discussed earlier, fleXOR is compatible with aggressive (GRR2) row reduction when it is instantiated with a safe wire ordering. We used the heuristic of Figure 3 to compute good safe orderings for all circuits. The last column of Figure 4 shows the size of the resulting garbled circuits. We point out that the fleXOR-garbled circuit was larger than the free-XOR garbled circuit in only two cases: For the AES circuit (which contained a significantly higher proportion of XOR gates than any other circuit we obtained), the fleXOR garbling was 12% larger than free-XOR; for the fast multiplication circuit, fleXOR was 5% larger. Our best performance was from the DES circuit, whose fleXOR-garbled circuit was 32% smaller than free-XOR.

Again we emphasize that any implementation of fleXOR matches the performance of free-XOR when assigning all wires the same index in the wire ordering. Hence, any implementation of fleXOR would easily be able to be provide whichever of the two wire orderings — safe fleXOR or free-XOR — was preferable, on a per-circuit basis, to realize the column labeled "best" in Figure 4.

(Sub)Optimality. Finally, we emphasize that we did not attempt to find **optimal** orderings for any circuit (which is NP-hard in general), only "good enough" wire orderings found by our simple heuristics. Hence, fleXOR has potential to produce garbled circuits even smaller than the ones reflected in our empirical results here. It is also possible that the circuits themselves could be optimized for fleXOR, similar to how some circuits are currently optimized for free-XOR (i.e., to minimize the number of non-XOR gates).

References

1. Applebaum, B.: Garbling XOR gates "For free" in the standard model. In: Sahai, A. (ed.) TCC 2013. LNCS, vol. 7785, pp. 162–181. Springer, Heidelberg (2013)
2. Beaver, D., Micali, S., Rogaway, P.: The round complexity of secure protocols (extended abstract). In: 22nd ACM STOC, pp. 503–513. ACM Press (1990)
3. Bellare, M., Hoang, V.T., Rogaway, P.: Foundations of garbled circuits. In: Yu, T., Danezis, G., Gligor, V.D. (eds.) ACM CCS 2012, pp. 784–796. ACM Press (2012)
4. Bellare, M., Rogaway, P.: The security of triple encryption and a framework for code-based game-playing proofs. In: Vaudenay, S. (ed.) EUROCRYPT 2006. LNCS, vol. 4004, pp. 409–426. Springer, Heidelberg (2006)
5. Choi, S.G., Katz, J., Kumaresan, R., Zhou, H.-S.: On the security of the "Free-XOR" technique. In: Cramer, R. (ed.) TCC 2012. LNCS, vol. 7194, pp. 39–53. Springer, Heidelberg (2012)
6. Damgård, I., Ishai, Y., Krøigaard, M., Nielsen, J.B., Smith, A.: Scalable multi-party computation with nearly optimal work and resilience. In: Wagner, D. (ed.) CRYPTO 2008. LNCS, vol. 5157, pp. 241–261. Springer, Heidelberg (2008)
7. Henecka, W., Schneider, T.: Memory efficient secure function evaluation, https://code.google.com/p/me-sfe/

8. Hirt, M., Maurer, U., Lucas, C.: A dynamic tradeoff between active and passive corruptions in secure multi-party computation. In: Canetti, R., Garay, J.A. (eds.) CRYPTO 2013, Part II. LNCS, vol. 8043, pp. 203–219. Springer, Heidelberg (2013)

9. Hirt, M., Tschudi, D.: Efficient general-adversary multi-party computation. In: Sako, K., Sarkar, P. (eds.) ASIACRYPT 2013, Part II. LNCS, vol. 8270, pp. 181–200. Springer, Heidelberg (2013)

10. Huang, Y., Katz, J., Evans, D.: Efficient secure two-party computation using symmetric cut-and-choose. In: Canetti, R., Garay, J.A. (eds.) CRYPTO 2013, Part II. LNCS, vol. 8043, pp. 18–35. Springer, Heidelberg (2013)

11. Ishai, Y., Kilian, J., Nissim, K., Petrank, E.: Extending oblivious transfers efficiently. In: Boneh, D. (ed.) CRYPTO 2003. LNCS, vol. 2729, pp. 145–161. Springer, Heidelberg (2003)

12. Kolesnikov, V.: Gate evaluation secret sharing and secure one-round two-party computation. In: Roy, B. (ed.) ASIACRYPT 2005. LNCS, vol. 3788, pp. 136–155. Springer, Heidelberg (2005)

13. Kolesnikov, V., Kumaresan, R.: Improved secure two-party computation via information-theoretic garbled circuits. In: Visconti, I., De Prisco, R. (eds.) SCN 2012. LNCS, vol. 7485, pp. 205–221. Springer, Heidelberg (2012)

14. Kolesnikov, V., Kumaresan, R.: Improved OT extension for transferring short secrets. In: Canetti, R., Garay, J.A. (eds.) CRYPTO 2013, Part II. LNCS, vol. 8043, pp. 54–70. Springer, Heidelberg (2013)

15. Kolesnikov, V., Schneider, T.: Improved garbled circuit: Free XOR gates and applications. In: Aceto, L., Damgård, I., Goldberg, L.A., Halldórsson, M.M., Ingólfsdóttir, A., Walukiewicz, I. (eds.) ICALP 2008, Part II. LNCS, vol. 5126, pp. 486–498. Springer, Heidelberg (2008)

16. Lindell, Y.: Fast cut-and-choose based protocols for malicious and covert adversaries. In: Canetti, R., Garay, J.A. (eds.) CRYPTO 2013, Part II. LNCS, vol. 8043, pp. 1–17. Springer, Heidelberg (2013)

17. Mohassel, P., Riva, B.: Garbled circuits checking garbled circuits: More efficient and secure two-party computation. In: Canetti, R., Garay, J.A. (eds.) CRYPTO 2013, Part II. LNCS, vol. 8043, pp. 36–53. Springer, Heidelberg (2013)

18. Naor, M., Pinkas, B., Sumner, R.: Privacy preserving auctions and mechanism design. In: Proceedings of the 1st ACM Conference on Electronic Commerce, EC 1999, pp. 129–139. ACM, New York (1999)

19. Pinkas, B., Schneider, T., Smart, N.P., Williams, S.C.: Secure two-party computation is practical. In: Matsui, M. (ed.) ASIACRYPT 2009. LNCS, vol. 5912, pp. 250–267. Springer, Heidelberg (2009)

20. Shelat, A., Shen, C.-H.: Two-output secure computation with malicious adversaries. In: Paterson, K.G. (ed.) EUROCRYPT 2011. LNCS, vol. 6632, pp. 386–405. Springer, Heidelberg (2011)

21. Tillich, S., Smart, N.: Circuits of basic functions suitable for MPC and FHE, http://www.cs.bris.ac.uk/Research/CryptographySecurity/MPC/

Amortizing Garbled Circuits

Yan Huang[1], Jonathan Katz[1], Vladimir Kolesnikov[2],
Ranjit Kumaresan[3], and Alex J. Malozemoff[1]

[1] University of Maryland, College Park, MD, USA
{yhuang,jkatz,amaloz}@cs.umd.edu
[2] Bell Labs, Murray Hill, NJ, USA
kolesnikov@research.bell-labs.com
[3] Technion, Haifa, Israel
ranjit@cs.technion.ac.il

Abstract. We consider secure two-party computation in a *multiple-execution* setting, where two parties wish to securely evaluate the same circuit multiple times. We design efficient garbled-circuit-based two-party protocols secure against *malicious* adversaries. Recent works by Lindell (Crypto 2013) and Huang-Katz-Evans (Crypto 2013) have obtained optimal complexity for cut-and-choose performed over garbled circuits in the single execution setting. We show that it is possible to obtain much lower *amortized* overhead for cut-and-choose in the multiple-execution setting.

Our efficiency improvements result from a novel way to combine a recent technique of Lindell (Crypto 2013) with LEGO-based cut-and-choose techniques (TCC 2009, Eurocrypt 2013). In concrete terms, for 40-bit statistical security we obtain a 2× improvement (per execution) in communication and computation for as few as 7 executions, and require only 8 garbled circuits (i.e., a 5× improvement) per execution for as low as 3500 executions. Our results suggest the exciting possibility that secure two-party computation in the malicious setting can be less than an order of magnitude more expensive than in the semi-honest setting.

1   Introduction

Two-party secure computation (2PC) is a rapidly developing area of cryptography. While the basic approach for semi-honest security, *garbled circuits* (GC) [27], is extensively studied and is largely settled, security against malicious players has seen recent significant improvements. The classical technique for lifting the GC approach to work in the malicious setting is *cut-and-choose* (C&C), formalized and proven secure by Lindell and Pinkas [15]. Until recently, this approach required significant overhead: to guarantee probability of cheating $< 2^{-s}$, approximately $3s$ garbled circuits needed to be generated and sent. However, in Crypto 2013 two works reduced the number of garbled circuits required in cut-and-choose to $s + O(\log s)$ [9] and to s [14].

Our Contribution. We further significantly reduce the replication factor for C&C-based protocols in the *multiple execution* setting, where the same function

J.A. Garay and R. Gennaro (Eds.): CRYPTO 2014, Part II, LNCS 8617, pp. 458–475, 2014.
© International Association for Cryptologic Research 2014

(possibly with different inputs) is evaluated multiple times either in parallel or sequentially. To achieve this, we combine in a novel way the "fast C&C" technique of Lindell [14] with the "LEGO C&C" technique [6,22].

Our Setting and Motivation. We consider the *multiple execution* setting, where two parties compute the same function on possibly different inputs either in parallel or sequentially. Here we argue that multiple evaluations of the same function is indeed a natural and frequently-occurring important scenario.

Today, 2PC is only beginning to enter practical deployment. However, we can reasonably speculate on likely future use cases. In the commercial setting, 2PC is natural in both business-to-business and business-to-customer interactions. For example, a bank customer could perform financial transactions (e.g., payments or transfers), a cell phone customer could perform private location-based queries, two businesses or government agencies might query their joint databases of customers, etc. In all of these scenarios, many of the securely evaluated functions are the same, only differing on their inputs. In fact, we conjecture that single-execution functions may be *less likely* to be used in commercial settings. This is because, as a rule-of-thumb of security, externally-accessible interfaces need to be clean and standardized. Allowing a small number of predetermined customer actions allows for more manageable overall security.

Additionally, many complex protocols from the research literature include multiple executions of the same function evaluated on different inputs. For example, Gordon et al. [8] propose sublinear 2PC based on oblivious RAM (ORAM). In their protocol, each ORAM step is executed by evaluating the same function using 2PC. Another frequently used subroutine is oblivious PRF, used, e.g., in the previously mentioned sublinear 2PC work [8] as well as in private database searches [4,12]. A recent such work [23] traverses the database search tree by evaluating the same match function at each tree node. Finally, any two universal circuits (of the same size) are implementing the same function.

1.1 Preliminaries

Let s denote the statistical security parameter; namely, an adversary can succeed in cheating with probability up to 2^{-s}. Let n denote the computational security parameter. We let t denote the total number of times the parties wish to evaluate a given circuit, and let $\rho = \rho(s, t)$ represent the number of circuits, per evaluation, that need to be generated to achieve an error probability of 2^{-s}. Before discussing our specific technical contribution, we recall the main ideas of our building blocks.

Fast Cut-and-Choose Using Cheating Punishment [14]. Cut-and-choose (C&C) protocols for GCs work by letting circuit constructor P_1 generate and send a number of GCs to the evaluator P_2, who then chooses a subset of circuits to open and check for correctness. If the checks pass, the remaining circuits are evaluated as in Yao's protocol [27], and the final output is obtained by taking majority over the individual outputs. In concrete terms, prior works [15,25] required at least 125 circuits to be sent by P_1 to guarantee security 2^{-40}. Lindell's improved technique [14] achieves 2^{-s} security while requiring P_1 to send only s circuits (i.e., 40 circuits for 2^{-40} security).

Lindell's protocol (which we call the "fast C&C" protocol) has two phases. In the first phase, P_1 with input x and P_2 with input y run a modified C&C which ensures that P_2 obtains a proof of cheating ϕ if it receives two inconsistent output values in any two evaluation circuits. Now, if all evaluation circuits produce the same output z, P_2 locally stores z as its output. Both parties *always* continue to the second *cheating-punishment* phase. In it, P_1 and P_2 securely evaluate a *smaller* circuit C', which takes as inputs P_1's input x and P_2's proof ϕ. (P_2 inputs random values if he does not have ϕ.) P_1 proves in zero-knowledge the consistency of its input x between the two phases. C' outputs x to P_2 if ϕ is a valid proof of cheating; otherwise P_2 receives nothing. The efficiency improvement is due to the fact that cheating is *punished* if there is any inconsistency in outputs.

LEGO Cut-and-Choose [6,22]. These works take a different approach by implementing a two-stage C&C at the *gate* level. The evaluation circuit is then constructed from the unopened garbled gates. In the first stage, P_1 sends multiple garbled gates and P_2 performs a standard C&C with replication factor $\rho(s) = O(s/\log|C|)$. P_2 aborts if any opened gate is garbled incorrectly. In the next stage, P_2 partitions the $\rho(s)|C|$ garbled gates into *buckets* such that each bucket contains $O(\rho(s))$ garbled gates. This two-stage C&C ensures that, except with probability 2^{-s}, each bucket contains a *majority* of correctly constructed garbled gates.

To connect gates with one another, Nielsen and Orlandi [22] use homomorphic Pedersen commitments. The resulting computational efficiency is relatively poor as they perform several expensive public-key operations *per gate*. This is addressed in the miniLEGO work [6], where the authors (among other things) construct homomorphic commitments from oblivious transfer (OT), whose cost can be amortized by OT extension [10]. However, the overall efficiency of this construction is still lacking in concrete terms due to large constants inside the big-O notation. In particular, the communication efficiency is adversely affected by the use of asymptotically constant-rate codes that are concretely inefficient.

1.2   Overview of Our Approach

Our main idea for the multiple execution setting is to run two-stage LEGO C&C at the *circuit* level, and then use fast C&C in the second stage (thereby requiring only a single correctly constructed circuit from each bucket). In particular, now the size of C' used in each execution depends only on the input and output lengths of C, and is no longer proportional to $|C|$. In this section, we focus only on the cut-and-choose aspect of the protocol; namely, on preventing P_1's cheating by submitting incorrect garbled circuits. More detailed protocol descriptions for both the parallel and sequential settings can be found in Section 2 and Section 3.

In the first-stage cut-and-choose, P_1 constructs and sends to P_2 a total of ρt GCs. Next, P_2 requests that P_1 open a random $\rho t/2$-sized subset of the garbled circuits. If P_2 discovers that any opened garbled circuit is incorrectly constructed,

Table 1. The number of garbled circuits required *per execution* in order to guarantee a security loss of $< 2^{-40}$. For comparison, the last two columns show the number of circuits required by the fast C&C protocol [14] in the parallel and sequential settings. Note that when using the fast C&C protocol for sequential executions we need to increase the replication factor from s to $s + \log t$.

# of Executions	Replication parallel/sequential	Replication for Fast C&C parallel	sequential
2	32	40	41
4	24	40	42
7	20	40	42
20	16	40	44
100	12	40	46
3500	8	40	51

it aborts. Otherwise, P_2 proceeds to the second stage cut-and-choose, where it randomly assigns unopened circuits to t buckets such that each bucket contains $\rho/2$ circuits. Now, as in the fast C&C protocol [14], each of the t evaluations are executed in two phases. In the first phase of the kth execution, party P_2 evaluates the $\rho/2$ evaluation circuits contained in the kth bucket. The circuits are designed such that if P_2 obtains different outputs from evaluating circuits in the kth bucket, then it obtains a proof of cheating ϕ_k. Next, both parties continue to the cheating-punishment phase, where P_1 and P_2 securely evaluate a smaller circuit that outputs P_1's input x_k if P_2 provides a valid proof ϕ_k.

Clearly, P_1 succeeds in cheating only if (1) it constructed $m \geq \rho/2$ bad circuits, (2) none of these m bad circuits were caught in the first cut-and-choose stage (i.e., $m \leq \rho t/2$), and (3) in the second stage, there exists a bucket that contains all bad circuits. It is easy to see that the probability with which m bad circuits escape detection in the first stage cut-and-choose is $\binom{\rho t - m}{\rho t/2} / \binom{\rho t}{\rho t/2}$. Conditioned on this event happening, the probability that a particular bucket contains all bad circuits is $\binom{m}{\rho/2} / \binom{\rho t/2}{\rho/2}$. Applying the union bound, we conclude that the probability that P_1 succeeds in cheating is bounded by

$$t \binom{\rho t - m}{\rho t/2} \binom{m}{\rho/2} / \binom{\rho t}{\rho t/2} \binom{\rho t/2}{\rho/2}.$$

For any given t and s, the smallest ρ, hinging on the maximal probability of P_1's successful attack, can be determined by enumerating over all possible values of m (i.e., $\{\rho/2, \rho/2 + 1, \ldots, \rho t/2\}$).

As an example, for $t = 20$ in a parallel execution setting with $s = 40$, using our protocol the circuit generator needs to construct $16 \cdot t = 320$ garbled circuits, whereas using a naïve application of Lindell's protocol [14] requires $40 \cdot t = 800$ garbled circuits.

Parallel vs. Sequential Executions. As will be evident, it is important to distinguish between the settings where multiple evaluations are carried out in parallel (e.g., when all inputs are available at the start of the protocol) and where these evaluations are carried out sequentially (e.g., when not all inputs are available as they, for example, depend on the outputs of previous executions). Below, we provide an overview of the main challenges of each setting, and an outline of our solutions.

Parallel executions. Under the DDH assumption, we apply our C&C technique in the parallel execution setting by modifying Lindell's protocol [14] as follows. We construct a generalized *cut-and-choose oblivious transfer* (C&C OT) functionality that supports *multi-stage* cut-and-choose. We call this functionality $\mathcal{F}_{\text{mcot}}$. Asymptotically, we can realize $\mathcal{F}_{\text{mcot}}$ using general secure computation, since the circuit for $\mathcal{F}_{\text{mcot}}$ depends only on the length of P_2's input and is otherwise independent of the circuit. However, such a realization is extremely inefficient in practice (the size of the circuit for realizing $\mathcal{F}_{\text{mcot}}$ needs to accept inputs of length at least $n\rho t\ell$, where n is the computational security parameter and ℓ is the input length). Instead, we show an efficient realization that is only a factor $\rho t^2/s$ less efficient (per execution) than the modified C&C OT realization of Lindell [14]. We elaborate more on this, and other important details, in Section 2.

Sequential executions. To prevent a malicious evaluator from choosing its inputs based on the garbled circuit, GC-based 2PC protocols perform OT *before* the constructor sends its GCs to the evaluator (i.e., before the cut-and-choose phase). This forces the parties, and in particular the evaluator, to "commit" to their inputs before performing the cut-and-choose. This, however, does not work in the sequential setting, where the parties may not know all their inputs at the beginning of the protocol. Standard solutions used in previous works [1,7,20] include assuming the garbled-circuit construction is adaptively secure or using adaptively-secure garbling [3] explicitly, assuming the programmable random-oracle model. Another issue is that since now we perform OTs for each execution separately, we can no longer use C&C OT or its variants; instead we rely on the "XOR-tree" approach of Lindell and Pinkas [15] to avoid selective failure attacks. We elaborate more on this, and other details, in Section 3.

Our solution for the sequential setting readily carries over to the parallel setting. In particular, adapting our protocol from the sequential to the parallel setting may address situations where the cost incurred by the use of $\mathcal{F}_{\text{mcot}}$ outweighs the cost of using both the XOR-tree approach and adaptively secure garbled circuits.

1.3   Related Work

Lindell and Pinkas [15] gave the first[1] rigorous 2PC protocol based on cut-and-choose. For $s = 40$, their protocol required at least $17s = 680$ garbled circuits. Subsequent work by the same authors [16] reduced the number of circuits to 128. This was later improved by shelat and Shen [25] to 125 using a more precise analysis of the C&C approach. In Crypto 2013, two works [9,14] proposed (among other things) dramatic improvements to the number of garbled circuits that need

---

[1] C&C mechanisms were previously employed in works by Pinkas [24] and Malkhi et al. [18] but these approaches were later shown to be flawed [13,19].

to be sent. In more detail, for achieving statistical security 2^{-s}, Huang et al.'s protocol [9] requires $2s + O(\log s)$ circuits, where each party generates half of them, and Lindell's protocol [14] requires exactly s circuits.

While all of the above works perform cut-and-choose over circuits, applying cut-and-choose at the gate-level has also been considered [5,6,21,22]. As discussed above, this approach naturally extends to the multiple execution setting, and furthermore is not inherently limited to considering settings where the same function is evaluated multiple times. Nielsen et al. [21] indeed show concrete efficiency improvements using gate-level cut-and-choose techniques. However, the number of rounds grows linearly with the depth of the evaluated circuit.

Finally, in independent and concurrent work, Lindell and Riva [17] also investigate the multiple execution setting, and obtain performance improvements similar to ours. An interesting difference between our works is that while we always let the evaluator pick half the circuits to check, they show that varying the number of check circuits can lead to an additional performance improvement.

2   The Parallel Execution Setting

Consider a setting where two parties wish to securely evaluate the same function multiple times in parallel. Let f denote the function of interest, and let t denote the number of times the parties wish to evaluate f. Let P_1's (resp., P_2's) input in the kth execution be x_k (resp., y_k), and let $x = (x_1, \ldots, x_t)$ and $y = (y_1, \ldots, y_t)$. We define $f^{(t)}(x, y) = (f(x_1, y_1), \ldots, f(x_t, y_t))$.

We adapt Lindell's protocol [14] to support our cut-and-choose technique in the parallel execution setting. The main difficulty is the design and construction of a generalization of cut-and-choose oblivious transfer [16] which we use to avoid the "selective failure attack" where a malicious P_1 constructs invalid keys for P_2's input wires to try and deduce P_2's inputs based on if P_2 aborts execution or not. We discuss this more in Section 2.1. We note that the naïve idea of using the XOR-tree approach [15] in our setting does not appear to work without using adaptively secure garbled circuits. Specifically, it is no longer clear how P_1, without any knowledge of which circuits will end up as evaluation circuits, can batch P_2's input keys together in a way that lets P_2 learn different sets of input keys corresponding to different evaluation circuits and yet within each evaluation bucket guaranteeing that P_2 can learn only input keys corresponding to the same set of inputs.

We give details of our protocol construction for the parallel executions setting in Section 2.2.

2.1   Generalizing Cut-and-Choose Oblivious Transfer

Cut-and-choose oblivious transfer (C&C OT) [16] is an extension of standard one-out-of-two oblivious transfer (OT). The sender inputs L pairs of strings, and the receiver inputs L selection bits to select one string out of each pair of sender strings. The receiver also inputs a set J of size $L/2$ that consists of indices

Inputs:

- P_1 inputs ℓ vectors $\boldsymbol{x}_i$, each containing s pairs of values $x_0^{i,j}, x_1^{i,j} \in \{0,1\}^{n \times n}$, $i \in [\ell]$, $j \in [s]$. In addition, P_1 inputs s "check values" $\chi_1, \ldots, \chi_s \in (\{0,1\}^n)^s$.
- P_2 inputs $\sigma_1, \ldots, \sigma_\ell \in \{0,1\}$ and a set of indices $J \subseteq [s]$.

Outputs: P_1 receives no output. P_2 receives the following:

- For every $i \in [\ell]$ and $j \in J$, P_2 receives $(x_0^{i,j}, x_1^{i,j})$.
- For every $i \in [\ell]$, P_2 receives $\langle x_{\sigma_i}^{i,1}, \ldots, x_{\sigma_i}^{i,s} \rangle$.
- For every $k \notin J$, P_2 receives χ_k.

In other words, P_2 receives $\{\chi_j\}_{j \in [s] \setminus J}$ and $\{\{x_{\sigma_i}^{i,j}\}_{j \in [s] \setminus J}, \{(x_0^{i,j}, x_1^{i,j})\}_{j \in J}\}_{i \in [\ell]}$.

Fig. 1. Modified batch single-choice cut-and-choose OT functionality $\mathcal{F}_{\text{ccot}}$ [14]

where it wants *both* the sender's inputs to be revealed. Note that for indices not contained in J, only those sender inputs that correspond to the receiver's selection bits are revealed. In applications to secure computation, and in particular when transferring input keys corresponding to a particular input wire across all evaluation circuits, one needs *single-choice* cut-and-choose oblivious transfer, where the receiver is restricted to inputting the *same* selection bit in all the $L/2$ instances where it receives exactly one out of two sender strings. Furthermore, when transferring input keys for multiple input wires, it is crucial that the subset J input by the receiver is the same across each instance of single-choice C&C OT executed for all input wires. This variant, called *batch single-choice* C&C OT, can be realized from the decisional Diffie-Hellman problem [16].

Lindell [14] presented a variant of batch single-choice C&C OT [16] in order to address settings where the check set J input by the receiver may be of arbitrary size. We denote this variant by $\mathcal{F}_{\text{ccot}}$; see Figure 1 for the formal description. In this variant, in addition to obtaining one of the two sender inputs for pairs whose indices are not in J, the receiver also obtains a "check value" for each index not in J. These check values are used to confirm whether or not a circuit is an evaluation circuit.

For our purposes, we introduce a new variant of $\mathcal{F}_{\text{ccot}}$, which we call batch single-choice *multi-stage* C&C OT. We denote this primitive by $\mathcal{F}_{\text{mcot}}$ and present its formal description in Figure 2. At a high level, our variant differs from $\mathcal{F}_{\text{ccot}}$ in that receiver P_2 can now input multiple sets $J_1, \ldots, J_t$ (where J is now implicitly defined as $[\rho t] \setminus \cup_{k \in [t]} J_k$) and make independent selections for each of $J_1, \ldots, J_t$. Unlike in Lindell's scheme [14], we only need to consider sets $J_1, \ldots, J_t$ whose sizes are pre-specified in order to provide the desired security guarantees. However, as in the $\mathcal{F}_{\text{ccot}}$ functionality, $\mathcal{F}_{\text{mcot}}$ (1) does not require sets $J_1, \ldots, J_t$ to be of a particular size, and (2) delivers "check values" for indices contained in each of $J_1, \ldots, J_t$. These check values are used to confirm whether a circuit is an evaluation circuit in the kth bucket for some $k \in [t]$.

Inputs:

- P_1 inputs ℓ vectors $\boldsymbol{x}_i$, each containing ρt^2 pairs $x_0^{i,j}, x_1^{i,j} \in \{0,1\}^n$. In addition, P_1 inputs ρt^2 "check values" $\chi_1^1, \ldots, \chi_{\rho t}^1; \ldots; \chi_1^t, \ldots, \chi_{\rho t}^t \in \{0,1\}^n$.
- P_2 inputs $\boldsymbol{\sigma}_1 = (\sigma_{1,1}, \ldots, \sigma_{1,\ell}), \ldots, \boldsymbol{\sigma}_t = (\sigma_{t,1}, \ldots, \sigma_{t,\ell}) \in \{0,1\}^\ell$ and sets $J_1, \ldots, J_t$ that are pairwise non-intersecting subsets of $[\rho t]$.

Outputs: Party P_1 receives no output. Party P_2 receives the following:

- For every $k \in [t]$ and for every $j \in J_k$, party P_2 receives χ_j^k.
- Let $J = [\rho t] \setminus \cup_{k \in [t]} J_k$. For every $i \in [\ell]$ and $j \in [\rho t]$:
  - If $j \in J$, then P_2 receives $(x_0^{i,j}, x_1^{i,j})$.
  - Otherwise, if there exists a (unique) $k \in [t]$ such that $j \in J_k$, then P_2 receives $x_{\sigma_{k,i}}^{i,j}$.

In other words, P_2 receives sets $\{\chi_j^1\}_{j \in J_1}, \ldots, \{\chi_j^t\}_{j \in J_t}$ and $\{\{x_{\sigma_{1,i}}^{i,j}\}_{j \in J_1}, \ldots, \{x_{\sigma_{t,i}}^{i,j}\}_{j \in J_t}, \{(x_0^{i,j}, x_1^{i,j})\}_{j \in J}\}_{i \in [\ell]}$.

Fig. 2. Batch single-choice multi-stage cut-and-choose OT functionality $\mathcal{F}_{\mathrm{mcot}}$

Designing the $\mathcal{F}_{\mathrm{mcot}}$ Functionality. As in $\mathcal{F}_{\mathrm{ccot}}$, the sender P_1 inputs ℓ vectors $\boldsymbol{x}_1, \ldots, \boldsymbol{x}_\ell$ each of length ρt, where each element in the vector is a pair of values (corresponding to the 0-key and the 1-key of a given garbled wire). In addition, P_1 inputs ρt^2 "check values". Receiver P_2 inputs t vectors $\boldsymbol{\sigma}_1, \ldots, \boldsymbol{\sigma}_t$ each of length ℓ and pairwise non-intersecting sets $J_1, \ldots, J_t$. Upon receiving these inputs from P_1 and P_2, the functionality computes $J = [\rho t] \setminus \cup_{k \in [t]} I_k$, and delivers, for each $j \in J$, the jth element (i.e., both values in the jth pair) in each of the ℓ vectors. Next, for every $k \in [t]$ and for each $j \in J_k$, the functionality delivers to P_2 the $\sigma_{k,i}$ value in the jth pair of vector $\boldsymbol{x}_i$ for every $i \in [\ell]$ along with the check value χ_j^k.

Realizing $\mathcal{F}_{\mathrm{mcot}}$ in the $\mathcal{F}_{\mathrm{ccot}}$-hybrid model. We now proceed to construct a protocol for $\mathcal{F}_{\mathrm{mcot}}$. Our goal is to provide an information-theoretic reduction from $\mathcal{F}_{\mathrm{mcot}}$ to $\mathcal{F}_{\mathrm{ccot}}$. We first consider a naïve approach which serves as a warm-up to our final construction and provides intuition behind our definition of $\mathcal{F}_{\mathrm{mcot}}$.

The naïve approach. We propose the following natural approach to realizing $\mathcal{F}_{\mathrm{mcot}}$ from $\mathcal{F}_{\mathrm{ccot}}$: P_1 first performs a t-out-of-t additive secret sharing of all input keys corresponding to P_2's inputs. In addition, P_1 chooses ρt^2 check values. Next, P_1 and P_2 interact with the $\mathcal{F}_{\mathrm{ccot}}$ functionality t times in parallel. In the kth interaction, P_1 provides the kth additive share of its input plus ρt check values $\chi_1^k, \ldots, \chi_{\rho t}^k$ (i.e., a check value for each circuit that could potentially be an evaluation circuit in the kth execution), while P_2 provides its inputs for the kth execution along with a set $[\rho t] \setminus J_k$, where J_k indicates the indices of the evaluation circuits to be used in the kth execution. Let $J = [\rho t] \setminus \cup_{k \in [t]} J_k$. At the end of the interaction, P_2 obtains (1) all t additive shares of input keys, and therefore all input keys, for circuits GC_j with $j \in J$, and (2) all t additive shares of input keys that *correspond to its actual input* in the kth execution, and therefore its input keys, along with check values for circuits GC_j with $j \notin J$.

Note, in particular, that for the check circuits, P_2 does not obtain the check values, and for the evaluation circuits, P_2 does not obtain both input keys. Thus, the above protocol seems to successfully fulfill our requirements from the $\mathcal{F}_{\text{mcot}}$ functionality. However, note that there is no mechanism in place to enforce that P_2 supplies non-intersecting sets $J_1, \ldots, J_k$. In the following we show that this prevents the above protocol from realizing $\mathcal{F}_{\text{mcot}}$.

Suppose $t = 2$. A malicious P_2 may input overlapping sets J_1, J_2 to $\mathcal{F}_{\text{ccot}}$. The consequence of this is that P_2 now possesses check values χ_j^1 and χ_j^2 for $j \in J_1 \cap J_2$. Clearly, the functionality $\mathcal{F}_{\text{mcot}}$ does not allow this. On the other hand, recall that the input keys are all additively shared, and as a result P_2 does not possess input keys corresponding to its input in circuit GC_j unless its input in both executions are identical. At the surface, there does not seem to be any attack due to this malicious strategy. Sure, P_2 can now equivocate on assigning GC_j to either the first evaluation bucket or the second evaluation. However, as observed earlier, it either has no corresponding keys, or it is going to evaluate both circuits on the same input, say y (in which case it seems immaterial whether j is revealed as part of J_1 or J_2). Unfortunately, we show that the above strategy for malicious P_2 is not simulatable. In particular, at the end of the interaction with $\mathcal{F}_{\text{ccot}}$, the simulator successfully extracts P_2's input in the first and second execution, but is now unable to decide on how to fake the garbled circuit GC_j. On the one hand, if $j \in J_1$, then the fake garbled circuit has to output $z_1 = f(x_1, y)$. On the other hand, if $j \in J_2$, then the fake garbled circuit has to output $z_2 = f(x_2, y)$. Therefore, the simulator has to choose on how to fake GC_j in the dark. Note that a simulation strategy for this specific case that decides to fake GC_j to output z_1 with probability $1/2$, and to output z_2 with probability $1/2$, does indeed succeed with probability $1/2$. However, this strategy does not extend well to the case when t is large.

The discussion above motivates our definition of $\mathcal{F}_{\text{mcot}}$; in particular, it reinforces why $\mathcal{F}_{\text{mcot}}$ must deliver at most one check value per circuit. In the following, we explain how to modify the naïve construction to enforce this.

Our approach. The high level idea behind our protocol is to let P_1 perform independent additive sharings of both the input values as well as the check values. Then P_1 and P_2 query the $\mathcal{F}_{\text{ccot}}$ functionality t times to transfer the values as required by $\mathcal{F}_{\text{mcot}}$. We detail this below, explaining it in the context of our secure computation protocol.

Let $(x_0^{i,j}, x_1^{i,j})$ be the input keys corresponding to P_2's ith input wire in GC_j. First, P_1 performs a t-out-of-t additive secret sharing of all input values corresponding to P_2's inputs; i.e., for each $i \in [\ell], j \in [\rho t]$, P_1 secret shares $x_0^{i,j}$ (resp., $x_1^{i,j}$) into $\{x_0^{i,j,k}\}_{k \in [t]}$ (resp., $\{x_1^{i,j,k}\}_{k \in [t]}$). P_1 then chooses ρt^2 check values $\{\chi_1^k, \ldots, \chi_{\rho t}^k\}_{k \in [t]}$. It then performs a $(2\ell(t-1)+1)$-out-of-$(2\ell(t-1)+1)$ additive sharing of each value χ_j^k to obtain shares denoted $\widetilde{\chi}_j^k$, $\{\chi_{0,k}^{i,j,k'}, \chi_{1,k}^{i,j,k'}\}_{k' \in [t] \setminus \{k\}, i \in [\ell]}$. Then, instead of creating inputs to $\mathcal{F}_{\text{ccot}}$ using $x_c^{i,j,k}$

shares alone, P_1 instead creates a "share block" $X_c^{i,j,k} = (x_c^{i,j,k}, \chi_{c,1}^{i,j,k}, \ldots, \chi_{c,t}^{i,j,k})$. That is, a share block $X_c^{i,j,k}$ contains, in addition to a share of the input key, a share of all check values corresponding to circuit GC_j.

Next, P_1 and P_2 run t instances of $\mathcal{F}_{\text{ccot}}$ in parallel. In the kth interaction, in addition to the ρt check value shares $\widetilde{\chi}_1^k, \ldots, \widetilde{\chi}_{\rho t}^k$, P_1 provides its kth share block while P_2 provides its inputs for the kth execution along with a set $[\rho t] \setminus J_k$, where J_k indicates the indices of the evaluation circuits to be used in the kth execution. Let $J = [\rho t] \setminus \cup_{k \in [t]} J_k$. At the end of the interaction, P_2 obtains (1) all t share blocks of input keys, and therefore all input keys, for circuits GC_j with $j \in J$, and (2) all t share blocks of input keys that *correspond to its actual input* in the kth execution, and therefore its input keys, along with a check value $\widetilde{\chi}_j^k$ for circuits GC_j with $j \in J_k$.

Note, in particular, that for each check circuit GC_j, P_2 does not obtain the check value χ_j^k for any k, because it always misses the check value share $\widetilde{\chi}_j^k$. For each evaluation circuit GC_j with $j \in J_k$, P_2 does not obtain both input keys, and more importantly can obtain at most one check value (which is χ_j^k). This is because share blocks contain shares of input keys as well as shares of check values. For an evaluation circuit, party P_2 always misses a share block, and consequently shares of all values $\chi_j^{k'}$ with $k' \neq k$. Furthermore, if P_2 wants to ensure it receives χ_j^k, then it should never input $J_{k''}$ such that $k'' \neq k$ and yet $j \in J_{k''}$. This is because for $j \in J_{k''}$, P_2 is guaranteed to miss a share block that contains an additive share of χ_j^k. Note that the above observations suffice to deal with a malicious P_2 that inputs overlapping sets since in this case P_2 fails to obtain any check values corresponding to indices in the intersection.

The formal description of the protocol in the $\mathcal{F}_{\text{ccot}}$-hybrid model can be found in Figure 3. We prove the following in the full version.

Theorem 1. *There exists a protocol perfectly realizing $\mathcal{F}_{\text{mcot}}$ in the $\mathcal{F}_{\text{ccot}}$-hybrid model.*

2.2   Using $\mathcal{F}_{\text{mcot}}$ in the Parallel Execution Setting

The input vectors x_i, for $i \in [\ell]$, contain the key pairs associated with the ith input wire for P_2 in each of the ρt circuits. The vector σ_k corresponds to the inputs used by P_2 in the kth execution. An honest P_2 chooses sets $J_1, \ldots, J_t$ such that they are pairwise non-intersecting and each set is of size exactly $\rho/2$. The main observation is that, for a given execution $k \in [t]$, P_2 obtains check values χ_j^k from $\mathcal{F}_{\text{mcot}}$ only for $j \in J_k$. Therefore, once the parties complete the interaction with $\mathcal{F}_{\text{mcot}}$ and P_1 sends all the garbled circuits, we let P_1 determine the evaluation circuits in each bucket based on whether P_2 sends the corresponding check values. At this point, P_1 checks that each bucket of evaluation circuits is well-defined and that these buckets are of equal size, i.e., $\rho/2$. If not, P_1 aborts. To overcome technical difficulties, we also require P_2 to provide "check values" for the check circuits as well. A check value for check circuit GC_j, denoted χ_j, may simply be the set of all input keys (i.e., both the 0-key and the 1-key) on all wires in circuit GC_j.

Inputs:

- P_1 inputs ℓ vectors of pairs $\boldsymbol{x}_i = \langle(x_0^{i,1}, x_1^{i,1}), \ldots, (x_0^{i,\rho t}, x_1^{i,\rho t})\rangle$ for $i \in [\ell]$. In addition, P_1 inputs ρt^2 "check values" $(\chi_1^1, \ldots, \chi_{\rho t}^1), \ldots, (\chi_1^t, \ldots, \chi_{\rho t}^t)$. All values are in $\{0, 1\}^n$.
- P_2 inputs $\boldsymbol{\sigma}_1 = (\sigma_{1,1}, \ldots, \sigma_{1,\ell}), \ldots, \boldsymbol{\sigma}_t = (\sigma_{t,1}, \ldots, \sigma_{t,\ell}) \in \{0, 1\}^\ell$ and sets $J_1, \ldots, J_t$.

Protocol:

- For all $i \in [\ell]$, P_1 performs a t-out-of-t additive secret sharing of $\boldsymbol{x}_i$ to obtain shares $\boldsymbol{x}_{i,1}, \ldots, \boldsymbol{x}_{i,t}$. For $k \in [t]$, let $\boldsymbol{x}_{i,k} = \langle(x_0^{i,1,k}, x_1^{i,1,k}), \ldots, (x_0^{i,\rho t,k}, x_1^{i,\rho t,k})\rangle$. Let $X_0^{i,j,k} = (x_0^{i,j,k}, \chi_{0,1}^{i,j,k}, \ldots, \chi_{0,t}^{i,j,k})$ and $X_1^{i,j,k} = (x_1^{i,j,k}, \chi_{1,1}^{i,j,k}, \ldots, \chi_{1,t}^{i,j,k})$, where $\chi_{0,1}^{i,j,k}, \ldots, \chi_{0,t}^{i,j,k}$ and $\chi_{1,1}^{i,j,k}, \ldots, \chi_{1,t}^{i,j,k}$ are random independent values in $\{0, 1\}^n$. Let $\boldsymbol{X}_{i,k} = \langle(X_0^{i,1,k}, X_1^{i,1,k}), \ldots, (X_0^{i,\rho t,k}, X_1^{i,\rho t,k})\rangle$.
- For all $k \in [t]$ and $j \in [\rho t]$, set $\tilde{\chi}_j^k = \chi_j^k \oplus \bigoplus_{k' \in [t] \setminus \{k\}, i \in [\ell]} (\chi_{0,k}^{i,j,k'} \oplus \chi_{1,k}^{i,j,k'})$.
- P_1 and P_2 run t instances of $\mathcal{F}_{\mathrm{ccot}}$ in parallel as follows. In the kth instance:
  - P_1 inputs ℓ vectors of pairs $\boldsymbol{X}_{i,k}$ of length ρt for $i \in [\ell]$ and ρt "check values" $\tilde{\chi}_1^k, \ldots, \tilde{\chi}_{\rho t}^k$. P_2 inputs $\sigma_{k,1}, \ldots, \sigma_{k,\ell} \in \{0, 1\}$ and the set $[\rho t] \setminus J_k$.
  - P_2 receives $\{\tilde{\chi}_j^k\}_{j \in J_k}$ and $\{\{X_{\sigma_{k,i}}^{i,j,k}\}_{j \in J_k} \cup \{(X_0^{i,j,k}, X_1^{i,j,k})\}_{j \in [\rho t] \setminus J_k}\}_{i \in [\ell]}$.
- For all $k \in [t]$ and $j \in J_k$, P_2 reconstructs $\chi_j^k = \tilde{\chi}_j^k \oplus \bigoplus_{k' \in [t] \setminus \{k\}, i \in [\ell]} (\chi_{0,k}^{i,j,k'} \oplus \chi_{1,k}^{i,j,k'})$.
- Let $J = [\rho t] \setminus \cup_{k \in [t]} J_k$. For all $i \in [\ell]$ and $j \in [\rho t]$, P_2 does the following:
  - If $j \in J$: set $x_0^{i,j} = \bigoplus_{k \in [t]} x_0^{i,j,k}$, and $x_1^{i,j} = \bigoplus_{k \in [t]} x_1^{i,j,k}$.
  - If there exists (unique) $k \in [t]$ such that $j \in J_k$: set $x_{\sigma_{k,i}}^{i,j} = \bigoplus_{k \in [t]} x_{\sigma_{k,i}}^{i,j,k}$.
- P_2 outputs sets $\{\chi_j^1\}_{j \in J_1}, \ldots, \{\chi_j^t\}_{j \in J_t}$ and $\{\{(x_0^{i,j}, x_1^{i,j})\}_{j \in J}, \{x_{\sigma_{1,i}}^{i,j}\}_{j \in J_1}, \ldots, \{x_{\sigma_{t,i}}^{i,j}\}_{j \in J_t}\}_{i \in [\ell]}$.

Fig. 3. Realizing $\mathcal{F}_{\mathrm{mcot}}$ in the $\mathcal{F}_{\mathrm{ccot}}$-hybrid model

Applying the Cheating-Punishment Technique. Inspired by Lindell's protocol [14], we use the knowledge of two different garbled values for a single output wire as a "proof" that P_2 received inconsistent outputs in a given execution. P_2 can use this proof to obtain P_1's input in a cheating-punishment phase. This cheating-punishment phase is implemented via a secure computation protocol, and thus it is important that the second phase functionality has a small circuit. We employ several optimizations proposed by Lindell [14] to keep the size of this circuit small. One important difference in our setting is that, unlike in Lindell's protocol [14], we cannot have, for a given output wire w, the same output keys b_w^0, b_w^1 across all garbled circuits. This is because in our setting garbled circuits are assigned to different evaluation buckets, and the circuits in each bucket can be evaluated with different input values, and thus can produce different outputs. Thus (even an honest) P_2 could potentially learn, say, output key b_w^0 in one execution and output key b_w^1 in another. We address this by simply removing the requirement that the set of output keys across different garbled circuits are the same. Thus, the circuit for the cheating-punishment phase for the kth execution must now take as input from P_1 *all* of the output keys in *all* of the evaluation circuits in the kth bucket, and from P_2 a pair of output keys that serve as proof of cheating. Somewhat surprisingly, we show that the size of the circuit (measured as the number of non-XOR gates) for the cheating-punishment phase is essentially the same as the circuit in Lindell's protocol [14].[2]

Another detail we wish to point out is that in our protocol we need to run separate cheating-punishment phases for each execution. This is a restriction imposed by the way in which P_1 proves consistency of its inputs [14,16]. However, we can run all of the t cheating-punishment phases *in parallel*. For this reason we use the universally composable variant of Lindell and Pinkas's protocol [16] (which is essentially obtained by replacing oblivious transfers and zero-knowledge subprotocols with their universally composable variants) to implement each cheating-punishment phase.

Other Details. We now describe other important details of our protocol.

- *Input consistency across multiple executions.* It is important to guarantee that P_1 provides consistent inputs across all circuits in the kth execution. Fortunately, existing mechanisms [14,16] for ensuring input consistency in the single execution setting can be readily extended to the multiple execution setting as well.
- *Encoded translation tables for garbled circuits.* As in Lindell's protocol [14], we modify the output translation tables used in the garbled circuits. Specifically, for keys k_i^0, k_i^1 on output wire i, we create an *encoded* output table $[h(k_i^0), h(k_i^1)]$, where h is some one-way function. We require that the output keys (or more precisely, the output of h applied to the output keys) corresponding to 0 and 1 are distinct. This encoding gives us the following two properties: (1) P_2 after evaluating a garbled circuit can use the encoded translation tables to determine whether the output is 0 or 1, and (2) the encoded translation table does not reveal the other output key (since this is equivalent to inverting the one-way function) to P_2.
- *Optimizing the cheating-punishment circuit.* We can apply similar techniques as shown by Lindell [14] to optimize the size of the cheating-punishment circuit to contain only ℓ non-XOR gates. We leave the details to the full version.

Formal Description. We proceed to the formal description of our protocol.

Inputs: P_1 has input $x = (x_1, \ldots, x_t)$, where $x_k \in \{0,1\}^\ell$, and P_2 has input $y = (y_1, \ldots, y_t)$, where $y_k \in \{0,1\}^\ell$.

---

[2] Of course, the cost of realizing our cheating-punishment phase is more than the corresponding cost in Lindell's protocol [14], mainly due to P_1's input being larger (but only by a factor of $\rho/2$).

Auxiliary Inputs: A statistical security parameter s, a computational security parameter n, the description of a circuit C where $C(x, y) = f(x, y)$, the number of evaluations t of the function f, and $(\mathbb{G}, q, g)$ where $\mathbb{G}$ is a cyclic group with generator g and prime order q, where q is of length n. Let $\mathsf{Ext} : \mathbb{G} \to \{0, 1\}^n$ be a function mapping group elements to bitstrings. In the following, $\rho = \rho(s, t)$ is the replication factor defined as being the smallest $u \in \mathbb{N}$ such that for all $m \in \{u/2, \ldots, ut/2\}$ it holds that $t \cdot \binom{ut-m}{ut/2} \binom{m}{u/2} / \binom{ut/2}{ut/2} \binom{ut/2}{u/2} \leq 2^{-s}$. If no such u exists or if $\rho \geq s$, then parties abort this protocol, and instead run the fast C&C protocol [14] for the function $f^{(t)}$.

Outputs: P_2 receives $f^{(t)}(x, y)$ and P_1 receives no output. Let ℓ' denote the length of the output of $f(x, y)$.

Protocol:

1. **Input key choice and circuit preparation:**
   - P_1 chooses random values $a_1^0, a_1^1, \ldots, a_\ell^0, a_\ell^1 \in_R \mathbb{Z}_q$, $r_1, \ldots, r_{\rho t} \in_R \mathbb{Z}_q$ and $(b_{1,1}^0, b_{1,1}^1, \ldots, b_{1,\ell'}^0, b_{1,\ell'}^1), \ldots, (b_{\rho t,1}^0, b_{\rho t,1}^1, \ldots, b_{\rho t,\ell'}^0, b_{\rho t,\ell'}^1) \in_R \{0,1\}^{n\ell'}$ such that for every $c_1, c_2 \in \{0, 1\}, j_1, j_2 \in [\rho t], i_1, i_2 \in [\ell']$ it holds that $b_{j_1,i_1}^{c_1} = b_{j_2,i_2}^{c_2}$ iff $i_1 = i_2$ and $j_1 = j_2$ and $c_1 = c_2$.
   - Let $w_1, \ldots, w_\ell$ denote the input wires corresponding to P_1's input, let $w_{i,j}$ denote the ith input wire in the jth garbled circuit, and let $k_{i,j}^b$ denote the key associated with bit b on wire $w_{i,j}$. P_1 sets $k_{i,j}^b$ as follows:
     $$k_{i,j}^0 = \mathsf{Ext}(g^{a_i^0 \cdot r_j}) \quad \text{and} \quad k_{i,j}^1 = \mathsf{Ext}(g^{a_i^1 \cdot r_j}).$$
   - Let $w_1', \ldots, w_{\ell'}'$ denote the output wires. The keys for wire w_i' in the jth garbled circuit are set to $b_{j,i}^0$ and $b_{j,i}^1$.
   - P_1 constructs ρt independent garblings, $GC_1, \ldots, GC_{\rho t}$, of circuit C, using random keys except for wires $w_1, \ldots, w_\ell$ and $w_1', \ldots, w_m'$, where the keys are set as above.

2. **Oblivious transfers:** P_1 and P_2 run $\mathcal{F}_{\mathrm{mcot}}$ as follows:
   - For $i \in [\ell]$, let $\mathbf{z}_i$ denote a vector containing the ρt pairs of keys associated with P_2's ith input bit in all the garbled circuits. P_1 inputs $\mathbf{z}_1, \ldots, \mathbf{z}_\ell$, as well as random values $\chi_1^1, \ldots, \chi_{\rho t}^1; \ldots; \chi_1^t, \ldots, \chi_{\rho t}^t$.
   - P_2 inputs random sets $J_1, \ldots, J_t$ which are pairwise non-intersecting subsets of $[\rho t]$ such that for all $k \in [t]$ it holds that $|J_k| = \rho/2$. Let $J = [\rho t] \setminus \cup_{k \in [t]} J_k$. P_2 also inputs bits $(\sigma_{1,1}, \ldots, \sigma_{1,\ell}), \ldots, (\sigma_{t,1}, \ldots, \sigma_{t,\ell}) \in \{0, 1\}^\ell$, where $\sigma_{k,i} = y_{k,i}$ for every $i \in [\ell]$ and $k \in [t]$.
   - For $j \in J$, P_2 receives both input keys associated with its input wires in garbled circuit GC_j, and for each $k \in [t]$ and $j \in J_k$, P_2 receives the keys associated with its input y_k on its input wires in garbled circuit GC_j. Also, for every $k \in [t]$ and $j \in J_k$, P_2 receives χ_j^k.

3. **Send circuits and commitments:** P_1 sends P_2 the garbled circuits $GC_1, \ldots, GC_{\rho t}$, the "seed" for the randomness extractor Ext, the following commitment to the garbled values associated with P_1's input wires:
   $$\{(i, 0, g^{a_i^0}), (i, 1, g^{a_i^1})\}_{i \in [\ell]} \quad \text{and} \quad \{(j, g^{r_j})\}_{j=1}^{\rho t}$$
   and the encoded output translation tables:
   $$\{[(h(b_{j,1}^0), h(b_{j,1}^1)), \ldots, (h(b_{j,\ell'}^0), h(b_{j,\ell'}^1))]\}_{j \in [\rho t]}.$$
   If $h(b_{j,i}^0) = h(b_{j,i}^1)$ for any $1 \leq i \leq \ell', 1 \leq j \leq \rho t$, then P_2 aborts.

4. **Send cut-and-choose challenge:** P_2 sends P_1 the sets $J, J_1, \ldots, J_t$ along with values $\{\chi_j^1\}_{j \in J_1}, \ldots, \{\chi_j^t\}_{j \in J_t}$, and all the keys associated with its input wires in all circuits GC_j for $j \in J$. If the values received by P_1 are (1) incorrect, or (2) the sets $J_1, \ldots, J_t$ are not pairwise non-intersecting, or (3) the input keys associated with P_2's input wires in circuits GC_j are revealed incorrectly, or (4) there exists some $k \in [t]$ such that $|J_k| \neq \rho/2$, then it outputs $\perp$ and aborts. Circuits GC_j for $j \in J$ are called *check circuits* and circuits GC_j for $j \in J_k$ are called *evaluation circuits* in the kth bucket.

5. **Send garbled input values in the evaluation circuits:** For each $k \in [t]$: P_1 sends the input keys associated with input x_k for the evaluation circuits in the kth bucket: For each $j \in J_k$ and every wire $i \in [\ell]$, P_1 sends the value $k'_{i,j} = g^{a_i^{x_{k,i}} \cdot r_j}$ and P_2 sets $k_{i,j} = \mathsf{Ext}(k'_{i,j})$.

6. **Circuit evaluation:** For each $k \in [t]$, P_2 does the following:
   - For each $j \in J_k$ and every wire $i \in [\ell']$, P_2 computes $b'_{j,i}$ by evaluating GC_j. If P_2 receives exactly *one* valid output value per output wire, then let z_k denote this output. In this case, it chooses random values $b_0^k, b_1^k \in_R \{0,1\}^n$. If P_2 receives *two* valid outputs on any output wire then it sets $b_0^k = b'_{j_1,i}$ and $b_1^k = b'_{j_2,i}$, where $j_1, j_2 \in J_k$ denote the conflicting circuit indices. If P_2 receives *no* valid output values on any output wire, then P_2 aborts.

7. **Run secure computation to detect cheating:** For each $k \in [t]$, P_1 and P_2 do the following *in parallel*:
   P_1 defines a circuit with the values $\{b_{j,1}^0, b_{j,1}^1, \ldots, b_{j,\ell'}^0, b_{j,\ell'}^1\}_{j \in J_k}$ hardcoded. The circuit computes the following function:
   - P_1 inputs $x_k \in \{0,1\}^\ell$ and has no output.
   - P_2 inputs a pair of values b_0^k, b_1^k.
   - If there exists values $i \in [\ell']$ and $j_1, j_2 \in J_k$ such that $b_0^k = b_{j_1,i}^0$ and $b_1^k = b_{j_2,i}^1$, then P_2's output is x_k; otherwise it receives no output.
   P_1 and P_2 run the UC-secure protocol of Lindell and Pinkas [16] on this circuit (except for the proof of P_1's input values), as follows:
   - P_1 inputs x_k; P_2 inputs b_0^k and b_1^k as computed in Step 6.
   - The garbled circuits constructed by P_1 use the same a_i^0, a_i^1 values as were chosen in Step 1, and the parties use $3(s + \log t)$ copies of the circuit for the cut-and-choose.
   If this computation results in an abort, then both parties halt.

8. **Check circuits for computing** $f^{(t)}(x, y)$**:**
   - For $j \in J$, P_1 sends r_j to P_2, and P_2 checks that these values are consistent with the pairs $\{(j, g^{r_j})\}_{j \in J}$ received in Step 3. If not, P_2 aborts.
   - For every $j \in J$, P_2 uses the $g^{a_i^0}, g^{a_i^1}$ values received in Step 3 and the r_j values received above to compute the keys for P_1's input wires as $k_{i,j}^0 = \mathsf{Ext}(g^{a_i^0 \cdot r_j}), k_{i,j}^1 = \mathsf{Ext}(g^{a_i^1 \cdot r_j})$. In addition, P_2 uses the keys obtained from $\mathcal{F}_{\mathrm{mcot}}$ in Step 2 for its own input wires. P_2 verifies that GC_j is a correct garbling of C. If there exists a circuit for which this does not hold, then P_2 aborts.

9. **Verify consistency of** P_1**'s input:** For each $k \in [t]$: Let $\widehat{J}_k$ be the set of check circuits used in the 2PC computation in Step 7 for the kth bucket, let $\widehat{r}_{j,k}$ be the value used in that computation, and let $\widehat{k}_{i,j}$ be the analogous value of $k'_{i,j}$ in Step 5 received by P_2 in the computation in Step 7. For each $k \in [t]$, P_1 and P_2 do the following *in parallel*:

- For every input wire $i \in [\ell']$, P_1 proves a zero-knowledge proof-of-knowledge that there exist some $\sigma_{k,i} \in \{0,1\}$ such that for every $j \in J_k$ and every $j' \notin \hat{J}_k$, it holds that $k'_{i,j} = g^{a_i^{\sigma_{k,i}} \cdot r_j}$ and $\hat{k}_{i,j} = g^{a_i^{\sigma_{k,i}} \cdot \hat{r}_{j',k}}$. If any of the t proofs fail, then P_2 aborts.

10. **Output evaluation:** For each $k \in [t]$, P_2 does the following:
    - If P_2 received no inconsistent outputs in Step 6, then it uses the encoded translation tables to decode the outputs it received, and sets z_k to that value. If P_2 received inconsistent output, then let x_k be the output that P_2 received from the circuit in Step 7. Let $z_k = f(x_k, y_k)$ be the output in this case.

    P_2 outputs $z = (z_1, \ldots, z_t)$ and terminates.

We prove the following theorem in the full version.

Theorem 2. *Let s (resp., n) be the statistical (resp., computational) security parameter. If the decisional Diffie-Hellman assumption holds in $(\mathbb{G}, g, q)$, h is a one-way function, and the underlying circuit garbling procedure is secure, then for all $t = \mathsf{poly}(n)$, the protocol described above securely computes $f^{(t)}$ in the presence of a malicious adversary with error at most $2^{-s} + \mu(n)$ for some negligible function $\mu(\cdot)$.*

3   The Sequential Execution Setting

We now consider the setting where the parties securely evaluate the same function f multiple times sequentially. Let t denote the number of times the parties wish to evaluate f. Let P_1's (resp., P_2's) input in the kth execution be denoted by x_k (resp., y_k). Let $f^{[t]}$ denote the reactive functionality that computes f a total of t times sequentially.

The main difference between this setting and the parallel setting discussed in Section 2 is that in the sequential setting the parties may not know their inputs to all executions at the start of the protocol. In particular, inputs may depend on outputs from previous executions. Thus, the parallel execution protocol does not immediately carry over to the sequential setting. To see why, observe for instance that $\mathcal{F}_{\mathrm{mcot}}$ requires P_2 to submit all of its inputs at once[3]. This is not possible since in the sequential setting we cannot assume that P_2 has all its inputs at the beginning of the protocol. Instead, we take a different route; namely, we use the "XOR-tree" approach [15,26] to protect against the so-called "selective failure attack" [13,19,25]. (In the parallel execution setting, this attack was implicitly avoided due to the use of $\mathcal{F}_{\mathrm{mcot}}$.) In this approach, the circuit C to be evaluated is first modified into an equivalent circuit C_{XT} (to include an "XOR-tree" for

---

[3] Standard oblivious transfer precomputation/"correction" techniques [2] still apply to $\mathcal{F}_{\mathrm{mcot}}$ as well; however, it is not clear how to "correct" $\mathcal{F}_{\mathrm{mcot}}$ correlations in a way suitable for the sequential setting.

P_2's inputs). Then, P_1 sends commitments to input keys corresponding to P_2's input wires in C_{XT}. The corresponding decommitments are revealed to P_2 via a standard one-out-of-two oblivious transfer. In order to prevent P_2 from using different inputs across evaluation circuits within the same bucket, P_1 batches together the decommitments corresponding to a particular input wire across all evaluation circuits in a given bucket. Note that herein lies an opportunity for a malicious P_1 to force P_2 to abort the protocol depending on its input. (This can be done for instance by sending incorrect decommitments for say only the 0-key on a particular wire.) However, the modified circuit C_{XT} is such that the success of any such selective OT attack is statistically independent of P_2's actual input value. Therefore, if an honest P_2 receives an invalid decommitment and is unable to decrypt the evaluation circuit, then it simply aborts knowing that its privacy is not compromised. Finally, we note that since we use one-out-of-two oblivious transfer (as opposed to $\mathcal{F}_{\mathrm{mcot}}$), we can leverage oblivious transfer extension techniques [10,11,21] to obtain better efficiency.

We stress that the oblivious transfer step happens *after* P_1 sends all the GCs to P_2. This is because P_2's inputs to all t executions are not available at the beginning of the protocol. Further, P_2's inputs may depend on previous outputs, which can be obtained only by decrypting evaluation circuits, i.e., after the evaluation bucket for the current execution is fully determined. Note that our cut-and-choose technique guarantees that there is at least one good evaluation circuit in every bucket under the assumption that P_1 has already committed to all its (good and bad) garbled circuits before the check sets and the evaluation sets are determined. Unfortunately, the above ordering of the oblivious transfer step and the garbled circuit sending step now allows a malicious P_2 to choose its input as a function of the garbled circuits it receives. To counter this, we need to use *adaptively secure garbling schemes* [3] instead of standard garbled circuits; adaptively secure garbling schemes can be constructed efficiently in the programmable random oracle model [3]. Note that we do not need the use of adaptively secure garbling schemes for implementing the cheating-punishment phase. Indeed, all the inputs for that subprotocol are known before the phase begins, and therefore, the oblivious transfer step can be carried out before P_1 sends its garbled circuits for that phase.

Due to lack of space, we leave both the formal description and the proof of the following theorem to the full version.

Theorem 3. *Let s (resp., n) be the statistical (resp., computational) security parameter. If the decisional Diffie-Hellman assumption holds in $(\mathbb{G}, g, q)$, h is a one-way function, and the circuit is garbled using an adaptively secure garbling scheme, then for all polynomial values of t, the protocol described above securely computes $f^{[t]}$ in the presence of a malicious adversary with error at most $2^{-s} + \mu(n)$ for some negligible function $\mu(\cdot)$.*

Acknowledgments. Work of Yan Huang and Jonathan Katz supported in part by NSF award #1111599. Work of Vladimir Kolesnikov supported in part by the Intelligence Advanced Research Project Activity (IARPA) via Department of Interior National Business Center (DoI/NBC) contract Number D11PC20194. Work of Ranjit Kumaresan supported by funding from the European Community's Seventh Framework Programme (FP7/2007–2013) under grant agreement number 259426. Work of Alex J. Malozemoff conducted with Government support through the National Defense Science and Engineering Graduate (NDSEG) Fellowship, 32 CFG 168a, awarded by DoD, Air Force Office of Scientific Research. The U.S. Government is authorized to reproduce and distribute reprints for Governmental purposes notwithstanding any copyright annotation thereon. Disclaimer: The views and conclusions contained herein are those of the authors and should not be interpreted as necessarily representing the official policies or endorsements, either expressed or implied, of IARPA, DoI/NBC, or the U.S. Government.

References

1. Applebaum, B., Ishai, Y., Kushilevitz, E., Waters, B.: Encoding functions with constant online rate or how to compress garbled circuits keys. In: Canetti, R., Garay, J.A. (eds.) CRYPTO 2013, Part II. LNCS, vol. 8043, pp. 166–184. Springer, Heidelberg (2013)

2. Beaver, D.: Precomputing oblivious transfer. In: Coppersmith, D. (ed.) CRYPTO 1995. LNCS, vol. 963, pp. 97–109. Springer, Heidelberg (1995)

3. Bellare, M., Hoang, V.T., Rogaway, P.: Adaptively secure garbling with applications to one-time programs and secure outsourcing. In: Wang, X., Sako, K. (eds.) ASIACRYPT 2012. LNCS, vol. 7658, pp. 134–153. Springer, Heidelberg (2012)

4. Cash, D., Jarecki, S., Jutla, C., Krawczyk, H., Roşu, M.-C., Steiner, M.: Highly-scalable searchable symmetric encryption with support for boolean queries. In: Canetti, R., Garay, J.A. (eds.) CRYPTO 2013, Part I. LNCS, vol. 8042, pp. 353–373. Springer, Heidelberg (2013)

5. Damgård, I., Orlandi, C.: Multiparty computation for dishonest majority: From passive to active security at low cost. In: Rabin, T. (ed.) CRYPTO 2010. LNCS, vol. 6223, pp. 558–576. Springer, Heidelberg (2010)

6. Frederiksen, T.K., Jakobsen, T.P., Nielsen, J.B., Nordholt, P.S., Orlandi, C.: Mini-LEGO: Efficient secure two-party computation from general assumptions. In: Johansson, T., Nguyen, P.Q. (eds.) EUROCRYPT 2013. LNCS, vol. 7881, pp. 537–556. Springer, Heidelberg (2013)

7. Gennaro, R., Gentry, C., Parno, B.: Non-interactive verifiable computing: Outsourcing computation to untrusted workers. In: Rabin, T. (ed.) CRYPTO 2010. LNCS, vol. 6223, pp. 465–482. Springer, Heidelberg (2010)

8. Gordon, S.D., Katz, J., Kolesnikov, V., Krell, F., Malkin, T., Raykova, M., Vahlis, Y.: Secure two-party computation in sublinear (amortized) time. In: Yu, T., Danezis, G., Gligor, V.D. (eds.) ACM CCS 2012, pp. 513–524. ACM Press (2012)

9. Huang, Y., Katz, J., Evans, D.: Efficient secure two-party computation using symmetric cut-and-choose. In: Canetti, R., Garay, J.A. (eds.) CRYPTO 2013, Part II. LNCS, vol. 8043, pp. 18–35. Springer, Heidelberg (2013)

10. Ishai, Y., Kilian, J., Nissim, K., Petrank, E.: Extending oblivious transfers efficiently. In: Boneh, D. (ed.) CRYPTO 2003. LNCS, vol. 2729, pp. 145–161. Springer, Heidelberg (2003)

11. Ishai, Y., Prabhakaran, M., Sahai, A.: Founding cryptography on oblivious transfer – efficiently. In: Wagner, D. (ed.) CRYPTO 2008. LNCS, vol. 5157, pp. 572–591. Springer, Heidelberg (2008)
12. Jarecki, S., Jutla, C.S., Krawczyk, H., Rosu, M.C., Steiner, M.: Outsourced symmetric private information retrieval. In: Sadeghi, A.R., Gligor, V.D., Yung, M. (eds.) ACM CCS 2013, pp. 875–888. ACM Press (2013)
13. Kiraz, M., Schoenmakers, B.: A protocol issue for the malicious case of Yao's garbled-circuit construction. In: 27th Symposium on Information Theory in the Benelux, pp. 283–290 (2006)
14. Lindell, Y.: Fast cut-and-choose based protocols for malicious and covert adversaries. In: Canetti, R., Garay, J.A. (eds.) CRYPTO 2013, Part II. LNCS, vol. 8043, pp. 1–17. Springer, Heidelberg (2013)
15. Lindell, Y., Pinkas, B.: An efficient protocol for secure two-party computation in the presence of malicious adversaries. In: Naor, M. (ed.) EUROCRYPT 2007. LNCS, vol. 4515, pp. 52–78. Springer, Heidelberg (2007)
16. Lindell, Y., Pinkas, B.: Secure two-party computation via cut-and-choose oblivious transfer. In: Ishai, Y. (ed.) TCC 2011. LNCS, vol. 6597, pp. 329–346. Springer, Heidelberg (2011)
17. Lindell, Y., Riva, B.: Cut-and-choose secure computation in the online/offline and batch settings. In: Garay, J.A., Gennaro, R. (eds.) CRYPTO 2014, Part II. LNCS, vol. 8617, pp. 476–494. Springer, Heidelberg (2014)
18. Malkhi, D., Nisan, N., Pinkas, B., Sella, Y.: Fairplay — a secure two-party computation system. In: Blaze, M. (ed.) 13th USENIX Security Symposium. USENIX Association (August 2004)
19. Mohassel, P., Franklin, M.: Efficiency tradeoffs for malicious two-party computation. In: Yung, M., Dodis, Y., Kiayias, A., Malkin, T. (eds.) PKC 2006. LNCS, vol. 3958, pp. 458–473. Springer, Heidelberg (2006)
20. Mohassel, P., Riva, B.: Garbled circuits checking garbled circuits: More efficient and secure two-party computation. In: Canetti, R., Garay, J.A. (eds.) CRYPTO 2013, Part II. LNCS, vol. 8043, pp. 36–53. Springer, Heidelberg (2013)
21. Nielsen, J.B., Nordholt, P.S., Orlandi, C., Burra, S.S.: A new approach to practical active-secure two-party computation. In: Safavi-Naini, R., Canetti, R. (eds.) CRYPTO 2012. LNCS, vol. 7417, pp. 681–700. Springer, Heidelberg (2012)
22. Nielsen, J.B., Orlandi, C.: LEGO for two-party secure computation. In: Reingold, O. (ed.) TCC 2009. LNCS, vol. 5444, pp. 368–386. Springer, Heidelberg (2009)
23. Pappas, V., Vo, B., Krell, F., Choi, S.G., Kolesnikov, V., Bellovin, S., Keromytis, A., Malkin, T.: Blind seer: A scalable private DBMS. In: 2014 IEEE Symposium on Security and Privacy. IEEE Computer Society Press (May 2014)
24. Pinkas, B.: Fair secure two-party computation. In: Biham, E. (ed.) EUROCRYPT 2003. LNCS, vol. 2656, pp. 87–105. Springer, Heidelberg (2003)
25. shelat, A., Shen, C.-h.: Two-Output Secure Computation with Malicious Adversaries. In: Paterson, K.G. (ed.) EUROCRYPT 2011. LNCS, vol. 6632, pp. 386–405. Springer, Heidelberg (2011)
26. Woodruff, D.P.: Revisiting the efficiency of malicious two-party computation. In: Naor, M. (ed.) EUROCRYPT 2007. LNCS, vol. 4515, pp. 79–96. Springer, Heidelberg (2007)
27. Yao, A.C.C.: How to generate and exchange secrets (extended abstract). In: 27th FOCS, pp. 162–167. IEEE Computer Society Press (October 1986)

Cut-and-Choose Yao-Based Secure Computation in the Online/Offline and Batch Settings*

Yehuda Lindell and Ben Riva

Dept. of Computer Science,
Bar-Ilan University, Israel
lindell@biu.ac.il, benr.mail@gmail.com

Abstract. Protocols for secure two-party computation enable a pair of mistrusting parties to compute a joint function of their private inputs without revealing anything but the output. One of the fundamental techniques for obtaining secure computation is that of Yao's garbled circuits. In the setting of malicious adversaries, where the corrupted party can follow any arbitrary (polynomial-time) strategy in an attempt to breach security, the cut-and-choose technique is used to ensure that the garbled circuit is constructed correctly. The cost of this technique is the construction and transmission of multiple circuits; specifically, s garbled circuits are used in order to obtain a maximum cheating probability of 2^{-s}.

In this paper, we show how to reduce the amortized cost of cut-and-choose based secure two-party computation in the batch and online/offline settings to $\mathcal{O}\left(\frac{s}{\log N}\right)$ garbled circuits when N secure computations are run. Although $\mathcal{O}(\frac{s}{\log N})$ may seem to be a mild efficiency improvement asymptotically, it is a *dramatic improvement* for concrete parameters since s is a statistical security parameter and so is typically small. Specifically, instead of 40 circuits to obtain an error of 2^{-40}, when running 2^{10} executions we need only 7.06 circuits on average per secure computation, and when running 2^{20} executions this is reduces to an average of just 4.08. In addition, in the online/offline setting, the online phase per secure computation consists of evaluating only 6 garbled circuits for 2^{10} executions and 4 garbled circuits for 2^{20} executions (plus some small additional overhead). In practice, when using fast implementations (like the JustGarble framework of Bellare et al.), the resulting protocol is remarkably fast.

We present a number of variants of our protocols with different assumptions and efficiency levels. Our basic protocols rely on the DDH assumption alone, while our most efficient variants are proven secure in the random-oracle model. Interestingly, the variant in the random-oracle model of our protocol for the online/offline setting has online communication that is independent of the size of the circuit in use. None of the previous protocols in the online/offline setting achieves this property, which is very significant since communication is usually a dominant cost in practice.

* This work was funded by the European Research Council under the European Union's Seventh Framework Programme (FP/2007-2013) / ERC Grant Agreement n. 239868 (LAST), and under the European Union's Seventh Framework Program (FP7/2007-2013) under grant agreement n. 609611 (PRACTICE). A full version of this work appears in the *Cryptology ePrint Archive*, 2014.

J.A. Garay and R. Gennaro (Eds.): CRYPTO 2014, Part II, LNCS 8617, pp. 476–494, 2014.
© International Association for Cryptologic Research 2014

1   Introduction

1.1   Background

In the setting of secure two-party computation, a pair of parties with private inputs wish to compute a joint function of their inputs. The computation should maintain privacy (meaning that the legitimate output but nothing else is revealed), correctness (meaning that the output is correctly computed), and more. These properties should be maintained even if one of the parties is corrupted. The feasibility of secure computation was demonstrated in the 1980s, where it was shown that any probabilistic polynomial-time functionality can be securely computed [Yao86, GMW87].

The two main adversary models that have been considered in the literature are *semi-honest* and *malicious*. A semi-honest adversary follows the protocol specification but attempts to learn more than allowed by inspecting the transcript. In contrast, a malicious adversary can follow any arbitrary (probabilistic polynomial-time) strategy in an attempt to break the security guarantees of the protocol. On the one hand, the security guarantees in the semi-honest case are rather weak, but there exist extraordinarily efficient protocols [HEKM11, BHR12b, ALSZ13]. On the other hand, the security guarantees in the malicious case are very strong, but they come at a significant computational cost.

The goal of constructing efficient secure two-party (2PC) computation protocols in the presence of malicious adversaries has been an active area of research in the recent years. [JS07, NO09] construct 2PC protocols with a small number of exponentiations per gate of the circuit, which is quite inefficient in practice. [IPS08, IKO+11] construct 2PC protocols based on the MPC-in-the-head approach which (asymptotically) requires only a small number of symmetric-key operations per gate of the circuit, though no implementation has been presented yet to clarify the concrete complexity of this approach in practice. [NNOB12, FJN+13] construct 2PC protocols in the random-oracle model with (amortized) $\mathcal{O}(s/\log(|C|))$ symmetric-key operations per gate of the circuit, where s is a security parameter and $C(\cdot)$ is a boolean circuit that computes the function of interest. [DPSZ12, DKL+143] construct secure multi-party computation protocols with security against *all-but-one* corrupted parties, and thus, could be used in the two-party setting as well. These protocols use somewhat homomorphic encryption. The protocols of [NNOB12, DPSZ12, DKL+143] all require a number of rounds of communication that is in the order of the depth of the circuit being computed.[1] Thus, their performance is limited in the case of deep circuits, and when parties are geographically far and so communication latency is significant.

A different approach that has received a lot of attention is based on applying the *cut-and-choose* technique to Yao's garbled-circuit protocol. In this technique, one of the parties prepares many garbled circuits, and the other asks to open a random subset of them in order to verify that they are correct; the parties

---

[1] The protocol of [FJN+13] *is* constant round. However, its concrete efficiency has not been established.

then evaluate the remaining, unchecked circuits. This forces the party generating the garbled circuits to make most of them correct, or it will be caught cheating (solving perhaps the biggest problem in applying Yao's protocol to the malicious setting, which is that an incorrect garbled circuit that computes the wrong function cannot be distinguished from a correct garbled circuit). [MF06, LP07, LP11, SS11, Lin13, MR13, SS13] present different 2PC protocols based on this approach, and several implementations have been presented to study the concrete efficiency of it in practice (e.g.[PSSW09, SS11, KSS12, SS13]). *In this work we focus on the cut-and-choose approach.*

Is It Possible to Go Below s Garbled Circuits with 2^{-s} Error? Until the recent work of [Lin13], protocols that use the cut-and-choose technique required approximately $3s$ garbled circuits to obtain a bound of 2^{-s} on the cheating probability by the adversary. Recently, [Lin13] showed that by executing another light 2PC, the number of garbled circuits can be reduced to s, which seems optimal given that 2^{-s} is the probability that a "cut" is as bad as possible (meaning that all the checked circuits are good and all the unchecked circuits are bad). The number of garbled circuits affects both computation time and communication. In most applications, when $|C|$ is large, sending s garbled circuits becomes the dominant overhead. (For example, [HMSG13] showed a prototype for garbling a circuit on GPUs, which generates more than 30 million gates per second. The communication size of this number of gates is about 15GB, and transferring 15GB of data most likely takes much more than a second.) Thus, further reducing the number of circuits is an important goal. *This goal is the focus of this paper.*

2PC with Offline and Online Stages. In the online/offline setting, the parties try to push as much work as possible to an offline stage in which they do not know their inputs. Later, in the online stage, when they have their inputs, they use the results of the offline stage to run a very efficient online phase, possibly with much lower latency than their standard counterparts.

The protocols of [NNOB12, DPSZ12, DKL+143] are especially well suited to the online/offline setting, and have extremely efficient online stages.[2] However, these protocols require many rounds of interaction in the online stage (i.e., $\mathcal{O}(\text{depth}(C))$ rounds). They therefore become considerably slower for deep circuits and over high-latency networks.

Previous cut-and-choose based protocols work only in the regular setting, in which both parties run the protocol from beginning to its end. Note that cut-and-choose based 2PC protocols are constant-round, which is another reason for trying to apply them in the online/offline setting.

---

[2] In fact, the protocols of [NNOB12, DPSZ12, DKL+143] allow the parties to choose the function also in the online stage. In this work we assume that the function is known in the offline stage, and it is only the inputs that are obtained later.

1.2   Our Contributions

As we have mentioned, the goal of this paper is to reduce the number of circuits in cut-and-choose on Yao's garbled circuits. We achieve this goal in the multiple-execution setting, where a pair of parties run many executions of the protocol. As we will see, this enables the parties to *amortize* the cost of the check-circuits over many executions.

Amortizing Checks over Multiple Executions. In the single-execution setting, party P_1 constructs s circuits and party P_2 asks to open a random subset of them. If P_1 makes some of them incorrect and some correct, then it can always succeed in cheating if P_2 opens all of the good circuits and the remaining are all bad. Since this bad event can happen with probability 2^{-s}, this approach to cut-and-choose seems to have a limitation of s circuits for 2^{-s} error. However, consider now the case that the parties wish to run N executions. One possibility is to simply prepare $N \cdot s$ circuits and work as in the single execution case. Alternatively, P_1 can prepare $c \cdot N$ circuits (for some constant c); then P_1 can ask to open a subset of the circuits; finally, P_2 randomly assigns the remaining circuits to N small buckets of size B (where one bucket is used for every execution). The protocol that we use, which is based on [Lin13], has the property that P_1 can cheat only if there is a bucket in which *all* of the circuits are bad. The probability of this happening when not too many bad circuits are constructed by P_1 is very small, but if P_1 does construct many bad circuits then it will be caught even if a relatively small subset of circuits is checked.

This idea is very powerful and it enables us to obtain an extraordinary speedup over the single-execution case. Asymptotically, only $\mathcal{O}(\frac{s}{\log N})$ garbled circuits are needed per execution (on average). Concretely, if the parties wish to run $N = 1024$ executions and maintain an error of 2^{-40}, then it suffices to construct 7229 circuits, check 15% of them, and randomly map the remaining into buckets of size 6. The number of circuits per execution is thus reduced from 40 to 7.06, which is a considerable improvement. As the number of executions grows, the improvement is more significant. Specifically, for $N = 1,048,576$ and an error of 2^{-40}, it suffices to construct 4,279,903 circuits, check 2% of them, and randomly map the remaining into buckets of size 4. The number of circuits per execution is thus reduced to just 4.08, which is almost a tenfold improvement! Finally, we note that improvements are obtained even for small numbers of N; e.g., for $N = 10$ the number of circuits per execution is reduced to 20, which is half the cost.

The Batch Setting – Parallel Executions. In this setting, the parties run N executions in parallel. Formally, they compute the functionality $F(\boldsymbol{x}, \boldsymbol{y}) = (f(x_1, y_1), \ldots, f(x_N, y_N))$ where $\boldsymbol{x} = (x_1, \ldots, x_N)$ and $\boldsymbol{y} = (y_1, \ldots, y_N)$. We start with the protocol of [Lin13] and apply our amortized checking technique in order to use only $\mathcal{O}\left(\frac{s}{\log N}\right)$ garbled circuits per execution. However, the protocol of [Lin13] does not work in a setting where the circuits are constructed without knowing which circuits will be placed together in a single bucket. In Section 2.2 we describe the problems that arise and how we overcome them.

The Online/Offline Setting. Next, we turn to the online/offline setting, with the aim of constructing an efficient 2PC protocol with a constant-round online stage and low latency. In order to achieve this, we show how to adapt the protocol of [Lin13] to the online/offline setting, and then use the amortized checking technique described above to significantly reduce the number of circuits needed. There are many issues that arise when trying to run cut-and-choose based protocols in the online/offline setting, mainly due to the fact that many of the techniques used to prevent cheating when cut-and-choose is used assume that the parties inputs are fixed even before the cut-and-choose takes place. In Section 2.3 we present a high-level description of our protocol, and our solutions to the problems that arise in this setting with cut-and-choose.

Our protocol achieves very high efficiency. First, the overall time (offline and online) is much lower than running a separate execution for every computation. Thus, we do not obtain a very fast online time at the expense of a very slow offline time. Rather, the overall protocol is highly efficient, and most of the work can be carried out in the offline phase. Second, our online phase requires very little communication, the evaluation of a small number of circuits, and little overhead. Concretely, when 1,000 executions are prepared in the offline phase, then the online phase requires evaluating only 5 circuits; in modern implementations like [BHR12b] and [HMSG13], this is extremely fast (with more executions, this is even further reduced).

Our basic protocol for the online/offline setting is the first (efficient) 2PC protocol in that setting with a constant-round online phase and security in the standard model (with security under the DDH assumption). In the full version, we show how to further reduce the complexity of the online stage, including a method for significantly reducing the communication of the online stage to be independent of $|C|$, in the random-oracle model. We stress that the most efficient protocols of [NNOB12, DPSZ12, DKL+143], which also work in the random-oracle model, require at least $\mathcal{O}(|C|)$ communication in the online stage, and at least depth(C) rounds.

Concurrent Work. In independent concurrent work, [HKK+13] show how to amortize the number of garbled circuits for multiple-executions of secure computation in a similar fashion to ours. However, here, we additionally focus on reducing the overhead of the cheating-recovery step (e.g. by amortizing its number of garbled circuits as well, and by moving most of its cost to the offline stage) and on minimizing the number of exponentiations in the online stage. We note that in the cut-and-choose of [HKK+13], P_2 always checks half of the circuits. In contrast, we show that better results can be obtained using different parameters; we believe that our analysis can be used in their protocol in a straightforward way.

1.3 Organization

Due to the lack of space in this abstract, we provide only an outline and high-level description of our techniques. A full description of our protocols, proofs of security, and a full combinatorial analysis of the number of circuits needed appears in the full version.

2 High Level Description of Our Techniques

We describe the main ideas behind our protocols. For simplicity, we focus here on specific parameters, though in Section 3 and in the full version we give a more general analysis of the possible parameters.

We begin by describing how cut-and-choose on Yao's protocol can be made more efficient (with low amortized cost) in batch settings where many computations take place. Then, we show how to achieve security in the online/offline setting where parties' inputs are fixed in the online phase. The low amortized cost for the batch setting is relevant both to the online/offline setting and to a setting where many computations take place in parallel.

2.1 Amortized Cut-and-Choose in Multiple Executions

We now describe how the number of circuits in cut-and-choose can be dramatically reduced in the case that many secure computation executions are run between two parties (either in parallel or in an online/offline setting). Assume that P_1 and P_2 would like to execute N protocols with maximum error probability of 2^{-s}, where s is a statistical security parameter. The naive approach of running the protocol of [Lin13] N times would require them to use a total number of garbled circuits of $N \cdot s$. As discussed earlier, our main goal in this paper is to reduce the number of garbled circuits by amortizing the overhead when many invocations of 2PC are executed.[3] The ideas described here will be used in both the batch protocol (Section 2.2) and the online/offline protocol (Section 2.3).

Recall that in cut-and-choose based two-party computation, P_1 prepares s garbled circuits, P_2 asks P_1 to open a random subset of them which are then checked by P_2, and then P_2 evaluates the remaining circuits. The main idea behind our technique is to run the cut-and-choose on *many* circuits, and then *randomly combine* the remaining ones into N sets (or "buckets"), where each set will be used for a single evaluation. The intuition behind this idea is as follows. The cheating recovery method of [Lin13] (described below in Section 2.2)

---

[3] We remark that it is possible to increase the number of check circuits and reduce the number of evaluated circuits in an online/offline version of the protocol of [Lin13], in order to improve the online time. For example, in order to maintain error of 2^{-40}, one can construct 80 circuits overall, and can check 70 and evaluate only 10. This will reduce the online time from approximately 20 to 10 (since in [Lin13] approximately half the circuits are evaluated). However, as we can see from this example, the total number of circuits grows very fast, rendering this approach ineffective.

ensures that security is preserved unless all evaluation circuits in a single set are incorrect. Now, by checking many circuits together and randomly combining them, the probability that one set will have all incorrect circuits (but yet no incorrect circuits were checked) is very small.

In more detail, in our technique P_1 prepares $2N \cdot B$ garbled circuits and sends them to P_2, where B is a parameter we define later. For each circuit, P_2 chooses with probability $1/2$ whether to check it or to use it later for evaluation. (This means that on average, P_2 checks $N \cdot B$ circuits. In our actual protocol we make sure that *exactly* $N \cdot B$ circuits remain. In addition, as we discuss below, we will typically not check half of the circuits and lower probabilities give better results.) Then, P_2 chooses a random mapping function $\pi : [N \cdot B] \to [N]$ that maps each of the remaining circuits in a "bucket" of B circuits, which will later be used as the evaluation-circuits of a single two-party protocol execution. Clearly, a malicious P_1 could prepare a small number of incorrect garbled circuits (say $\mathcal{O}(\beta)$), and not be caught in the checks with good probability (here $\beta < s$ and so $2^{-\beta}$ probability is too high). However, since π is chosen at random by P_2, we show that unless there are *many* incorrect circuits, the probability that any one of the buckets contains only incorrectly constructed garbled circuits is smaller than 2^{-s}. We prove that when $B \geq \frac{s}{1+\log N} + 1$, the probability that any bucket contains B incorrect circuits (and so all are incorrect) is at most 2^{-s}. Thus, the total number of circuits is $2N \cdot B = \frac{2Ns}{1+\log N} + 2N$. When $\log N > \frac{2s}{s-2} - 1$ we have that $2N \cdot B < N \cdot s$ and so a concrete improvement is obtained from just using [Lin13] even for just a few executions. Asymptotically, the number of circuits per execution is $\mathcal{O}(\frac{s}{\log N})$, which shows that when N gets larger, the amortized number of circuits becomes small. When plugging in concrete numbers that are relevant in practice, the improvement is striking. For example, consider $s = 40$ and $N = 512$ executions (observe that $\log N = 9$ and $\frac{2s}{s-2} - 1 = 1.10$ and so the condition is fulfilled). Now, for these parameters we have $B = \lceil \frac{s}{1+\log N} + 1 \rceil = 5$, and so only 512×10 garbled circuits are needed overall, with just 5 circuits evaluated in each execution. This is better by a factor of 4 compared to the Ns option. When many executions are run, even better numbers are obtained. For example, with $N = 524288$ we obtain that only 524288×6 circuits are needed overall (better by a factor of $6\frac{2}{3}$ than the naive option).

We remark that the probability of checking or evaluating a circuit greatly influences the number of circuits. Above, we have assumed that this probability is $\frac{1}{2}$. In Section 3 we analyse the above parameters in the general case. As we will see, better parameters are typically achieved with lower probabilities of checking a circuit. In addition, when working in the online/offline setting, this flexibility actually provides a tradeoff between the number of circuits in the online and in the offline phases. This is due to the fact that checking more circuits in the offline stage reduces the number of circuits to be evaluated in the online stage but increases the number of circuits checked in the offline phase.

In the protocol of [Lin13] secure computation is also used for the cheating recovery mechanism (described below in Section 2.2). This mechanism works as long as a *majority* of the circuits in a bucket are good. In the multiple-execution setting, we use a similar method for bucketizing these circuits, while guaranteeing that a majority of the circuits in any bucket be good (rather than just ensuring at least one good circuit). Using this method we significantly reduce the number of circuits needed for the cheating recovery. E.g., for $N = 1024$ protocol executions we need only buckets of size $B = 12$, and a total number of circuits of 24576 (i.e,, 24 circuits per execution). The protocol of [Lin13] requires about 125 circuits per execution, and thus we obtain an improvement of a factor of 5 in this part of the protocol (for these parameters).

More Concrete Examples. In Section 3 we provide a full analysis of the cheating probability for different choices of parameters. We describe some concrete examples here with $s = 40$, in order to provide more of an understanding of the efficiency gains obtained; in the full version of this paper, we show the cost for many different choice of parameters. When considering 2^{10} and 2^{20} executions, the best choices and the resulting cost is summarized in the following table (the bucket size is the number of circuits evaluated in the online phase):

Table 1. Best parameters for $s = 40$ (p is the probability that a circuit is *not* checked)

Number of executions N	p	Bucket size(B)	Overall number of circuits ($\lceil B \cdot N/p \rceil$)	Average # circuits per execution
2^{10}	0.1	4	40,960	40.00
2^{10}	0.65	5	7,877	7.69
2^{10}	0.85	6	7,229	7.06
2^{20}	0.65	3	4,839,582	4.62
2^{20}	0.98	4	4,279,903	4.08

Observe that in the case of $p = 0.1$, the average number of circuits is the same as in a single execution. However, it has the lowest online time. In contrast, at the price of just a single additional circuit in the online time, the offline time is reduced by a factor of over 5. In general, the bigger p is, the smaller the total number of balls is (up to a certain limit). However, the number of balls in each bucket grows proportionally with p. This means that using p it is possible to obtain a tradeoff between online and offline time. Specifically, a higher p means less circuits overall but more circuits in the online stage (where each bucket is evaluated), thereby reducing the offline time at the expense of increasing the online time. Conversely, a lower p means more circuits in the offline stage and smaller bucket and so less computation in the online stage.

We remark that improvements are not only obtained for large values of N. In the case of $N = 32$, with $p = 0.75$ we obtain buckets of size 10 (so 10 evaluations in the online phase) and an average of 13.34 circuits overall per execution. This is a considerable improvement over 40 circuits as required in [Lin13]. Of course, as N becomes smaller, the improvement is less significant. Nevertheless, for $N = 10$, with $p = 0.55$ we obtain an average of 20 circuits per execution, which is half the cost of [Lin13]. Going to the other extreme, with a huge number of executions the amortized cost becomes very small. Taking $N = 2^{30}$ (which isn't practical today but may be in the future), we can take $p = 0.99$ and obtain buckets of size 3 and an overall overage of just 3.03 circuits per execution. In the full version of the paper we also present graphs of the dependence of B and the total number of circuits in p, and how the average number of balls per bucket decreases as the number of buckets grows.

Regarding the number of circuits required for the cheating-recovery mechanism, for $N = 2^{10}$ we get that $B = 12$, and that the total number of circuits is $12 \times 1024 \times 2 = 24576$ (i.e., 24 circuits per execution). For $N = 2^{20}$ we get that $B = 6$, and that the total number of circuits is $6 \times 1048576 \times 2 = 12,582,912$ (i.e., 12 circuits per execution). This is in contrast to 125 circuits, as required in [Lin13].

2.2 Batch Two-Party Computation

The protocol of [Lin13] requires s garbled circuits per 2PC execution for achieving soundness of 2^{-s}. Here we would like to reduce this overhead when multiple executions of 2PC are executed in a batch setting; i.e., run in parallel. In this section, we assume that the reader is familiar with the protocol of [Lin13].

If we try to use the protocol of [Lin13] as-is in the batch setting, and take advantage of the ideas presented in Section 2.1, two problematic issues arise. We now describe these issues and how we solve them.

First, in the cut-and-choose oblivious transfer of [Lin13], the receiver uses only one input to all OTs, whereas in the batch setting, P_2 should be able to input many different inputs, and they have to be consistent in each bucket. This consistency of P_2's input is enforced by having P_2 prove in zero knowledge that its OT queries are for the same input in all circuits. In order to enable P_2 to use separate inputs in each bucket, we modify the protocol as follows. First, P_2 privately selects which circuits to use and how to bucket them before the OTs are executed. Then, the parties run the cut-and-choose OT, where P_2 inputs its j-th input in the circuits that it chose to be in the j-th bucket. However, P_2 does *not* prove consistency of its input at this point (since the buckets are not yet known to P_1), but rather postpones this proof until after it sends the cut and random mapping to buckets to P_1. After the mapping to buckets has been given to P_1, it is possible for P_2 to separately prove in zero knowledge for every bucket that its OT queries in the j-th bucket are for the same input. Observe also that since this proof is given before P_2 can evaluate any circuit, no information can be gained if P_2 tries to cheat.

A second issue that arises when trying to use the protocol of [Lin13] in the batch setting is what P_2 does in the case that it gets different outputs in some of the evaluated circuits. We call this mechanism of [Lin13] cheating recovery since it enables P_2 to obtain correct output when P_1 has tried to cheat. In order for this mechanism to work, [Lin13] uses the same output labels in *all* circuits, and in case P_2 gets different labels for the same wire (meaning different outputs), the two labels allow it to recover P_1's input. Unfortunately, this technique cannot work in the batch setting, since there, naturally, P_2 would get different outputs from different buckets, and thus will always learn two labels of some output wire. This would enable a cheating P_2 to learn P_1's input.

Our solution to this problem is as follows. For simplicity, assume that there is only one output wire, and assume that D is a special constant that is revealed to P_2 in the case that it receives different output values on the wire in different circuits (we later describe how this "magic" happens). Recall that in [Lin13], a second, lighter, two-party computation is executed with a boolean circuit C', where P_1 inputs (x, D) (with x being the value used in computing the actual circuit), P_2 inputs d, and $C'(x, D, d) = x$ if $d = D$, and 0 otherwise. Thus, if P_2 obtained D due to receiving different outputs in different circuits, then in the second two-party computation it inputs $d = D$ and learns x, thereby enabling it to locally compute the correct output $f(x, y)$. Otherwise, if learns nothing about x; in addition, P_1 does not know if P_2 learned x or not.

Instead of using the same output labels in all garbled circuits, P_1 uses random ones (as in the standard Yao's circuit). After P_2 announces the "cut" in the offline stage and the mapping to the buckets, P_1 opens the checked circuits and P_2 verifies them as described before. Then in the online stage the parties follow the next steps. For every bucket (separately), P_1 chooses a random D. Concretely, consider the j-th bucket; then P_1 chooses random values D_j and R_j. Denote the garbled circuits in the j-th bucket by $gc_1, gc_2, \ldots gc_B$. Furthermore, denote the output-wire labels of circuit gc_i by W_i^0, W_i^1. P_1 sends the encryptions $\{\ \mathsf{Enc}_{W_i^0}(R_j),\ \mathsf{Enc}_{W_i^1}(R_j \oplus D_j)\ \}_{i=1,\ldots,B}$. P_1 also sends P_2 the hash $\mathsf{Hash}(D_j)$. The purpose of these encryptions and hash is that in case P_2 learns two output labels that correspond to different outputs, P_2 can learn both R_j and $R_j \oplus D_j$ and can use it to recover D_j. It then verifies that it has the right D_j using $\mathsf{Hash}(D_j)$. (In the case of many output wires, each output wire in a bucket encrypts in the above way using a different R_j. Thus, D_j can be obtained from any *pair* of output wire labels in the j-th bucket.)

After P_2 evaluates the circuits of C, it learns a set of labels $W' = \{W_1', \ldots, W_B'\}$. P_2 uses the values of W' to decrypt the corresponding $c_i^0 = \mathsf{Enc}_{W_i^0}(R_j)$ or $c_i^1 = \mathsf{Enc}_{W_i^1}(R_j \oplus D_j)$. In case P_2 learns both W_i^0 and W_i^1, it can recover $d_j = \mathsf{Dec}_{W_i^0}(c_i^0) \oplus \mathsf{Dec}_{W_i^1}(c_i^1)$ (which should equal $D_j = R_j \oplus (R_j \oplus D_j)$). In case P_2 gets many "potential" D's (which can happen if P_1 does not construct the values honestly), it can identify the correct one using the value $\mathsf{Hash}(D_j)$. Next, the parties execute the 2PC protocol with the circuit $C'(x, D, d)$, and P_2 learns x in case it learned the correct D_j earlier. Finally, P_2 verifies that P_1 constructed all of the values for the cheating recovery correctly. This check is carried out after the 2PC protocol for C' has concluded, since at this point revealing D_j to P_2 can cause no damage. For this check, P_1 reveals all of the pairs W_i^0, W_i^1, allowing P_2 to check that the encryptions $\{\mathsf{Enc}_{W_i^0}(R_j), \mathsf{Enc}_{W_i^1}(R_j \oplus D_j)\}_{i=1,\ldots,B}$ and $\mathsf{Hash}(D_j)$ are consistent. Since P_1 can cheat regarding the output labels W_i^0, W_i^1, we require that when it sends a garbled circuit (before the cut is revealed), it also sends commitments on all the output wire labels of that circuit. These commitments are checked if the circuit is chosen to be checked in the cut-and-choose. Thus, any good circuit has the property that the output labels encrypt R_j and $R_j \oplus D_j$.

Unfortunately, the above does not suffice to ensure that P_2 learns D_j in the case that there are two different outputs. This is due to the fact that it is only guaranteed that *one* circuit in the bucket is good. Now, if P_2 receives two different outputs in two different circuits, then the second circuit may *not* be good and so P_2 may obtain the correct R_j from the good circuit but some value $S_j \neq R_j \oplus D_j$ from the other.

Nevertheless, in the case that P_2 received different outputs, but did not obtain D_j that is consistent with the hashed value $\mathsf{Hash}(D_j)$ sent by P_1, party P_2 simply outputs the output of the garbled circuit for which the output labels it received from the evaluation are all consistent with the output labels that were decommitted. To see why this suffices, observe that P_2 receives two different outputs, and one of them is from a good circuit. Denote the two circuits from which P_2 receives different outputs by gc_1, gc_2, and denote by gc_1 the circuit that was correctly garbled. Then, there are two possibilities: (1) P_2 obtained the correct D_j, and thus recovers x using the second 2PC (and can output the correct $f(x, y)$ by just computing the function f with P_1's input x); (2) P_2 did not recover the correct D_j, meaning that the output labels it received do not decrypt R_j and $R_j \oplus D_j$. However, since gc_1 is correct, including the commitments on its output labels, and since $\mathsf{Enc}_{W_i^0}(R_j)$ and $\mathsf{Enc}_{W_i^1}(R_j \oplus D_j)$ are checked, gc_1 gives P_2 the correct value (either R_j or $R_j \oplus D_j$, depending on the output bit in question). Now, if the output label that P_2 received from gc_2 also decrypts its corresponding R_j or $R_j \oplus D_j$, then P_2 should have learnt the correct D_j. This means that the label that P_2 received in gc_2 does *not* match the label that P_1 revealed from the decommitment on gc_2's output labels. Thus, P_2 knows that gc_1 is the correct circuit and not gc_2, and can take the output of the computation to be the output of gc_1. (Note that by what we have explained, if P_2 does not obtain D_j and the checks on the commitments and encryptions passed, then there is only *one circuit* in which the output labels obtained by P_2 are consistent with the commitments. Thus, there is no ambiguity regarding the output.)

Although the above issues are the main parts of the cheating-recovery process of our protocols, there are other small steps that are needed in order to make sure that the protocol is secure. For example, P_2 should verify that P_1 inputs the correct D to C'. Also, efficiency-wise, recall that $3s$ garbled circuits of C' are used in the protocol of [Lin13]; here, we amortize their cut-and-choose as well, as described above. These issues are dealt with in the detailed description of the protocol in the full version.

2.3   Two-Party Computation with Online/Offline Stages

Protocols for secure computation in the presence of malicious adversaries via cut-and-choose on garbled circuits employ a number of methods to prevent cheating. First, many circuits are sent and a fraction checked, in order to ensure that some of the garbled circuits are correct (this is the basic cut-and-choose). Second, since many circuits are evaluated in the evaluation phase, it is necessary to force

P_1 and P_2 to use the same input in every circuit in an evaluation. Third, so-called selective OT attacks must be thwarted (where a cheating P_1 provides correct circuits but partially incorrect values in the oblivious transfer phase where P_2 receives keys to decrypt the circuits, based on its input). Finally, the cheating recovery technique described in Section 2.2 is used to enable P_2 to complete the computation correctly in case some of the evaluation circuits are correct and some are incorrect. In all existing protocols, some (if not all) of the aforementioned checks utilize the fact that the parties' inputs are given and fixed before the checks are carried out (in fact, in [Lin13] even the basic cut-and-choose on circuits is intertwined with the selective OT attack prevention and so requires the inputs to already be fixed). Thus, these protocols do not work in the online/offline setting.

In this section, we describe how to deploy these methods in an online/offline setting where the checks are carried out in the offline setting, and the online setting should be very fast.[4] Ideally, the online setting should have no exponentiations, and should involve some minimal communication (that is independent of the circuit size) and the evaluation of the circuits in the bucket only. Our protocol achieves this goal, with some small additional work in the online stage. We note that in the standard model we do require some exponentiations in the online phase, but just *two per circuit* which in practice is insignificant. In addition, P_1 needs to transmit B garbled circuits to P_2 for evaluation in the online phase, where B is the bucket size (in practice, a small constant of between 4 and 6). We also present a variant of our protocol in the random oracle model that requires no exponentiations whatsoever in the online phase, and has very little communication; in particular, the communication is independent of the circuit size. The use of a random oracle is due to problems that arise when adaptively-secure garbled circuits [BHR12a] are needed. This issue is discussed separately in Section 2.4.

Ensuring Correctness of the Garbled Circuit. Intuitively, the aim of the cut-and-choose process is to verify that the garbled circuits are correct. Thus, it is possible to run this process (send all circuits and then open and check a fraction of them) in an offline stage even before the parties have inputs. Then, in the online stage, when the parties have inputs and would like to compute the output of the computation as fast as possible, they only need to evaluate the remaining "evaluation" circuits, which results in a much lower latency.

Enforcing P_1's Input Consistency. We start with the approach taken in [MF06, LP11, SS11]. Let wire j be an input-wire of P_1. In a standard garbling process, two random strings are chosen as the labels of wire j. However, here, the two labels are chosen to be commitments to the actual value they represent, e.g., the label that corresponds to the bit 0 is actually a commitment to 0 (more exactly, the label is the output of a hash function, which is also a randomness

---

[4] Our aim here is to reduce the work of the online stage as much as possible, in order to achieve very fast computation in the online stage. Tradeoffs between the offline and online stages are of course possible, and we leave this for future work.

extractor, on the appropriate commitment). In addition, the commitments used have the property that one can prove equality of multiple committed messages with high efficiency, without revealing the actual messages.

This solution can be used in the online/offline setting in a straightforward way. Namely, when a circuit is checked, these commitments are checked as well. In contrast, when a set of circuits is used for evaluation, P_1 sends the commitments that correspond to its input, along with a proof that they are all commitments to the same bit 0 or 1. However, the disadvantage of this method is that it requires a few exponentiations per bit of P_1's input, and we would like to move all exponentiations possible to the offline stage. In order to achieve this, instead of directly computing $f(x, y)$, we modify the garbled circuit to compute the function $f'(x^{(1)}, x^{(2)}, y) = f(x^{(1)} \oplus x^{(2)}, y)$, where $x^{(1)}$ and $x^{(2)}$ are P_1's inputs and are chosen randomly by P_1 under the constraint that $x^{(1)} \oplus x^{(2)} = x$. In the garbling process, the garbled labels of the wires of $x^{(1)}$ are constructed using the commitment method of [MF06, LP11, SS11], while the labels of the wires of $x^{(2)}$ are standard (i.e., random strings). In addition, for each wire of $x^{(2)}$, P_2 sends commitments on the two input-wire labels (i.e., if the labels are W^0, W^1, P_1 sends $\mathsf{Com}(0\|W^0), \mathsf{Com}(1\|W^1)$). Now, in the offline stage, when a circuit is checked, P_2 verifies that all of the above was followed correctly. Furthermore, in the circuits that are to be evaluated, P_1 chooses a *random* $x^{(1)}$ and sends the commitments that correspond to $x^{(1)}$ along with the proof of message equality. This proves to P_2 that P_1's input $x^{(1)}$ is the same in all evaluated circuits (of course, at least in the properly constructed circuits). All this is carried out in the offline phase.

In the online stage, when P_1 knows x, it sends P_2 the actual value of $x^{(2)} = x^{(1)} \oplus x$, along with the decommitments of the labels that correspond to $x^{(2)}$ (the decommitments prove that the same $x^{(2)}$ is sent in all circuits). We stress that $x^{(2)}$ is sent in the clear, and is the same for all evaluated circuits (this reveals nothing about x since $x^{(1)}$ is random and not revealed). As a result, the same $x^{(1)}$ and $x^{(2)}$ is used in all circuits (the consistency of $x^{(1)}$ is enforced in the offline phase, and the consistency of $x^{(2)}$ is immediate since it is sent in the clear) and so the same x is used in all evaluated circuits. Note that no exponentiations are needed in the online stage, and only a small number of decommitments and decryptions are computed.

In summary, online/offline consistency of P_1's input is obtained by randomly splitting P_1's input into a secret part $x^{(1)}$ (which is dealt with in the offline stage), and a public part $x^{(2)}$ which can be revealed in the online stage. Since $x^{(2)}$ can be chosen to equal $x \oplus x^{(1)}$ in the online phase, after x is known, the correct result is obtained and consistency is preserved at very little online cost.

Protecting against Selective-OT Attacks. We use a variant of the cut-and-choose oblivious transfer protocols of [LP11, Lin13], and modify it to work in the online/offline setting. The modification is similar to the method used for P_1's input; i.e., instead of computing the function $f'(x^{(1)}, x^{(2)}, y) = f(x^{(1)} \oplus x^{(2)}, y)$ as above, the parties compute $f''(x^{(1)}, x^{(2)}, y^{(1)}, y^{(2)}) = f(x^{(1)} \oplus x^{(2)}, y^{(1)} \oplus y^{(2)})$, where P_2 uses a random value for $y^{(1)}$ in the offline stage, and later uses $y^{(2)} =$

$y^{(1)} \oplus y$ once it knows its input y in the online stage. The cut-and-choose oblivious transfer protocol is used for protecting against selective OT attacks on the OTs that are used for P_2 to learn the garbled labels of $y^{(1)}$. In contrast, the labels of $y^{(2)}$ are obtained by having P_2 send $y^{(2)}$ in the clear and having P_1 send the associated garbled labels (these labels are committed in the offline phase and thus the labels are sent to P_2 as decommitments, which prevents P_1 from changing them). As before, all exponentiations are carried out in the offline stage alone.

Cheating Recovery. The protocol of [Lin13] uses a cheating recovery process for allowing P_2 to learn x in case P_2 obtains different outputs from the evaluated circuits. This method allows for only s circuits to be used in order to obtain 2^{-s} cheating probability, since an adversary can only cheat if *all* checked circuits are correct and *all* evaluated circuits are incorrect. However, the protocol of [Lin13] requires the parties to run the cheating recovery process *before* the checked circuits are opened, which obviously is unsatisfactory in the online/offline setting since now P_2 does all the expensive checking in the online stage again.

Our solution for this problem is the same solution as described above for the batch setting; see Section 2.2. Namely, assume that D is a special constant that is revealed to P_2 in the case that it receives different output values on the wire in different circuits, and for simplicity assume that there is only one output wire. We would like to securely compute the boolean circuit $C'(x^{(1)}, D, d)$, where $(x^{(1)}, D)$ are P_1's input, d is P_2's input, and $C'(x^{(1)}, D, d) = x^{(1)}$ if $d = D$, and 0 otherwise. We note that only P_2 receives output (since the method requires that P_1 not know if P_2 learned D or not). Recall that $x^{(1)}$ is the secret part of P_1's input, and so if $x^{(1)}$ is obtained by P_2 then it can compute $x = x^{(1)} \oplus x^{(2)}$ and obtain P_1's real input. Everything else in this solution is identical to the solution described in Section 2.2; the use of $x^{(1)}$ instead of x enables us to check the circuits used in the cheating-recovery mechanism in the offline phase.

There are several other subtle issues to take care of regarding the secure computation of C'. First, we require P_1 to use the same $x^{(1)}$ in C and C'. This is solved by using commitments for the input-wire labels for $x^{(1)}$ as described above. Second, we need to protect the OTs for P_2 to learn the labels of d from selective-OT attacks. This is solved using the variant of cut-and-choose OT we use for the OTs for C. Third, in order to push all the expensive exponentiations to the offline stage, we split the parties inputs in the cheating-recovery circuit C' into random inputs in the offline stage and public inputs in the online stage as we did with the inputs of C. Note that the above issues are only part of the cheating-recovery process of our protocols, and additional steps are needed in order to make sure that the protocol secure.

2.4 On Adaptively Secure Garbled Circuits in the Online/Offline Setting

The standard security notion of garbled circuits considers a *static* adversary who chooses its input before seeing the garbled circuit. While this notion suffices for standard 2PC protocols (e.g., [LP07, LP11, SS11] where the oblivious transfers that determine P_2's input can be run before the garbled circuits are sent), it causes a problem in the online/offline setting. This is due to the fact that we would like to send all the garbled circuits in the offline stage in order to reduce the online stage communication. However, this means that the circuits are sent before the parties (and in particular the adversary) have chosen their inputs.

Recently, [BHR12a, AIKW13] introduced an *adaptive* variant of garbled circuits, in which the adversary is allowed to choose its input *after* seeing the garbled circuit. Indeed, adaptively secure garbling scheme would allow us to send all the garbled circuits in the offline stage before the parties have chosen their inputs. However, the only known efficient constructions of adaptively secure garbled circuit are in the random-oracle model [BHR12a, AIKW13].[5]

We do not try to present new solutions to the adaptively-secure garbled-circuit problem in this work. Rather, we present two options based on current constructions. Our first solution is in the standard model and works by having P_1 send only the checked garbled circuits in the offline stage. In contrast, the evaluation garbled circuits are sent in the online stage. These latter circuits are committed (using a trapdoor commitment) in the offline stage, and this enables the simulator to actually construct the garbled circuit after the input is given, solving the adaptive problem. The drawback of this solution is that significant communication is needed in the online stage, incurring considerable cost. Our second solution is to use the random-oracle construction of [PSSW09, BHR12a]. In this case, *all* of the garbled circuits are sent in the offline stage, and the communication of the online stage depends only on the number of inputs and outputs of the circuits (and the security parameters). Thus, we obtain a clear tradeoff between the security model and efficiency. We believe that any future construction of efficient adaptively secure garbled circuits in the standard model may be plugged into the second construction in order to maintain its low communication and remove the random-oracle.

3   Combinatorics of Multiple Cut-and-Choose: Balls and Buckets

In this section we deal with *balls* and *buckets*. A ball can be either normal or cracked. Similarly to cut-and-choose, we describe a game in which party P_1 prepares a bunch of balls, P_2 checks a subset of them and aborts if some of them are cracked, and otherwise randomly places the remaining ones in buckets. Our goal is to bound the probabilities that (a) one of the buckets consists of only cracked balls (i.e., a fully-cracked bucket), and (b) there is a bucket in which the majority of the balls are cracked (i.e., a majority-cracked bucket). We follow

---

[5] [BHR12a] also present a construction in the standard model which requires the online stage communication to be the same size as the garbled circuit, but this does not help us to reduce the online communication. In addition, [BHK13] presents a construction in the standard model based on *UCE-hash* functions. However, the only known proven construction of UCE-hash is in the ROM.

the analysis of [Nor13, Theorem 4.4] and [Nor13, Theorem 6.2], while handling different and slightly more general parameters.

3.1 The Fully-Cracked Bucket Game

Let Game 1 be the following game. P_2 chooses three parameters p, N and B, and sets $M = \left\lceil \frac{NB}{p} \right\rceil$ and $m = NB$. A potentially adversarial P_1 (who we will denote by $\mathcal{A}$) prepares M balls and sends them to P_2. Then, party P_2 chooses at random a subset of the balls of size $M - m$; these balls are checked by P_2 and if one of them is cracked then P_2 aborts. Index the balls that are not checked by $1, \ldots, m$. P_2 chooses a random mapping function $\pi : [m] \to [N]$ that places the unchecked balls in buckets of size B. We define that $\mathsf{Game}_1(\mathcal{A}, N, B, p) = 1$ if and only if P_2 does not abort and there is a fully cracked bucket (note that $M = \left\lceil \frac{NB}{p} \right\rceil$ and $m = NB$ and so are not separate parameters in the game). The proof of the following theorem can be found in the full version of this paper:

Theorem 1. *Let s be a statistical security parameter, and let $B, N \in \mathbb{N}$ and $p \in (0, 1)$ be as above. If*

$$B \geq \frac{s + \log N - \log p}{\log(N - Np) - \frac{\log p}{1-p}}, \tag{1}$$

then for every adversary $\mathcal{A}$ it holds that $\Pr\left[\mathsf{Game}_1(\mathcal{A}, N, B, p) = 1\right] < 2^{-s}$.

We remark that in the proof of Theorem 1 we show that the probability that the adversary wins in the game is at most

$$\frac{\binom{M-t}{m-t}}{\binom{M}{m}} \cdot N \cdot \binom{t}{B}\binom{m}{B}^{-1} \tag{2}$$

and then proceed to show that this is less than 2^{-s} as long as Eq. (1) holds, for general parameters. However, for concrete sets of parameters we can compute slightly tighter bounds or more optimized parameters. For example, Theorem 1 states that for $s = 40$, $N = 1024$ and $p = 0.7$, B should be 6. However, by analytic calculation, for this set of parameters we actually have that the maximal cheating probability is at most $2^{-51.07}$. If we take $B = 5$ we have that the maximal cheating probability is at most $2^{-40.85}$. This means that instead of using $\frac{1024 \times 6}{0.7} = 8778$ balls, we can use only $\frac{1024 \times 5}{0.7} = 7315$ balls for the same p and N! This "gap" is significant even for smaller values of N. For parameters $s = 40$, $N = 32$ and $p = 0.75$, Theorem 1 requires B to be 10. The maximum of Eq. (2) for these parameters is at most 2^{-44}, which, again, is much smaller than the 2^{-40} bound given by Theorem 1. In fact, if we take $N = 32$, $p = 0.8$ and $B = 10$, we get that the maximum of Eq. (2) is at most $2^{-40.1}$, without increasing B as required if we had used Theorem 1 with $p = 0.8$. This reduces the expected number of balls per bucket from 13.34 (for $p = 0.75$) to only 12.5 (for $p = 0.8$).

We leave further optimizations and analysis of the above bounds for future work, and recommend computing analytically the exact bounds based on the above analysis whenever performance is critical. More examples of the parameters obtained and discussion on recommended values appears in the full version of the paper.

3.2 The Majority-Cracked Bucket Game

Let Game 2 be the same game as Game 1, but where $\mathcal{A}$ wins if P_2 is left with a bucket that consists of at least $\frac{B}{2}$ cracked balls. Define that $\mathsf{Game}_2(\mathcal{A}, N, B, p) = 1$ if and only if P_2 does not abort the game and there is a majority-cracked bucket. (Recall that $\mathsf{Game}_1(\mathcal{A}, N, B, p) = 1$ only if *all* of the balls in some bucket are cracked.)

In the full version of this paper, we prove the following theorem:

Theorem 2. *Let s be a security parameter, and let B, $N \in \mathbb{N}$ $p \in (0, 1)$ be as above. If*

$$B \geq \frac{2s + 2\log N - \log(-1.25 \log p) - 1}{\log N + \log(-1.25 \log p) - 2} ,$$

then for every adversary $\mathcal{A}$ it holds that $\Pr\left[\,\mathsf{Game}_2(\mathcal{A}, N, B, p) = 1\,\right] < 2^{-s}$.

In the full version, we discuss in depth what parameters this yields with $s = 40$. Briefly, we can see that the effect of p on B and the total number of balls is similar to those dependences in Game 1, although the concrete numbers are different. For $N = 1024$ and $p = 0.7$, only 20 garbled circuits are needed on average per execution (as opposed to 125 in the cut-and-choose of [Lin13]) and only 14 circuits are used in the online stage. For larger values of N, these numbers decrease significantly, e.g. for $N = 1048576$ and $p = 0.9$ only 8.89 circuits are needed on average per execution, where only 8 are used in the online stage. In addition, we obtain a significant improvement over the cut-and-choose of [Lin13] also for small values of N, e.g., for $N = 32$ and $p = 0.6$, only 51.69 circuits are needed on average per execution (which is less than half than needed in [Lin13]).

References

[AIKW13] Applebaum, B., Ishai, Y., Kushilevitz, E., Waters, B.: Encoding functions with constant online rate or how to compress garbled circuits keys. In: Canetti, R., Garay, J.A. (eds.) CRYPTO 2013, Part II. LNCS, vol. 8043, pp. 166–184. Springer, Heidelberg (2013)

[ALSZ13] Asharov, G., Lindell, Y., Schneider, T., Zohner, M.: More efficient oblivious transfer and extensions for faster secure computation. In: CCS, pp. 535–548. ACM (2013)

[BHK13] Bellare, M., Hoang, V.T., Keelveedhi, S.: Instantiating random oracles via uCEs. In: Canetti, R., Garay, J.A. (eds.) CRYPTO 2013, Part II. LNCS, vol. 8043, pp. 398–415. Springer, Heidelberg (2013)

[BHR12a] Bellare, M., Hoang, V.T., Rogaway, P.: Adaptively secure garbling with applications to one-time programs and secure outsourcing. In: Wang, X., Sako, K. (eds.) ASIACRYPT 2012. LNCS, vol. 7658, pp. 134–153. Springer, Heidelberg (2012)

[BHR12b] Bellare, M., Hoang, V.T., Rogaway, P.: Foundations of garbled circuits. In: CCS, pp. 784–796. ACM (2012)

[DKL+143] Damgård, I., Keller, M., Larraia, E., Pastro, V., Scholl, P., Smart, N.P.: Practical covertly secure MPC for dishonest majority – or: Breaking the SPDZ limits. In: Crampton, J., Jajodia, S., Mayes, K. (eds.) ESORICS 2013. LNCS, vol. 8134, pp. 1–18. Springer, Heidelberg (2013)

[DPSZ12] Damgård, I., Pastro, V., Smart, N., Zakarias, S.: Multiparty computation from somewhat homomorphic encryption. In: Safavi-Naini, R., Canetti, R. (eds.) CRYPTO 2012. LNCS, vol. 7417, pp. 643–662. Springer, Heidelberg (2012)

[FJN+13] Frederiksen, T.K., Jakobsen, T.P., Nielsen, J.B., Nordholt, P.S., Orlandi, C.: MiniLEGO: Efficient secure two-party computation from general assumptions. In: Johansson, T., Nguyen, P.Q. (eds.) EUROCRYPT 2013. LNCS, vol. 7881, pp. 537–556. Springer, Heidelberg (2013)

[GMW87] Goldreich, O., Micali, S., Wigderson, A.: How to play any mental game. In: STOC, pp. 218–229. ACM (1987)

[HEKM11] Huang, Y., Evans, D., Katz, J., Malka, L.: Faster secure two-party computation using garbled circuits. In: USENIX Security Symposium (2011)

[HKK+13] Huang, Y., Katz, J., Kolesnikov, V., Kumaresan, R., Malozemoff, A.J.: Amortizing garbled circuits. In: CRYPTO, Springer, Heidelberg (2014)

[HMSG13] Husted, N., Myers, S., Shelat, A., Grubbs, P.: Gpu and cpu parallelization of honest-but-curious secure two-party computation. In: Proceedings of the 29th Annual Computer Security Applications Conference, pp. 169–178. ACM (2013)

[IKO+11] Ishai, Y., Kushilevitz, E., Ostrovsky, R., Prabhakaran, M., Sahai, A.: Efficient non-interactive secure computation. In: Paterson, K.G. (ed.) EUROCRYPT 2011. LNCS, vol. 6632, pp. 406–425. Springer, Heidelberg (2011)

[IPS08] Ishai, Y., Prabhakaran, M., Sahai, A.: Founding cryptography on oblivious transfer – efficiently. In: Wagner, D. (ed.) CRYPTO 2008. LNCS, vol. 5157, pp. 572–591. Springer, Heidelberg (2008)

[JS07] Jarecki, S., Shmatikov, V.: Efficient two-party secure computation on committed inputs. In: Naor, M. (ed.) EUROCRYPT 2007. LNCS, vol. 4515, pp. 97–114. Springer, Heidelberg (2007)

[KSS12] Kreuter, B., Shelat, A., Shen, C.-H.: Billion-gate secure computation with malicious adversaries. In: USENIX Security, p. 14 (2012)

[Lin13] Lindell, Y.: Fast cut-and-choose based protocols for malicious and covert adversaries. In: Canetti, R., Garay, J.A. (eds.) CRYPTO 2013, Part II. LNCS, vol. 8043, pp. 1–17. Springer, Heidelberg (2013)

[LP07] Lindell, Y., Pinkas, B.: An efficient protocol for secure two-party computation in the presence of malicious adversaries. In: Naor, M. (ed.) EUROCRYPT 2007. LNCS, vol. 4515, pp. 52–78. Springer, Heidelberg (2007)

[LP11] Lindell, Y., Pinkas, B.: Secure two-party computation via cut-and-choose oblivious transfer. In: Ishai, Y. (ed.) TCC 2011. LNCS, vol. 6597, pp. 329–346. Springer, Heidelberg (2011)

[MF06] Mohassel, P., Franklin, M.K.: Efficiency tradeoffs for malicious two-party computation. In: Yung, M., Dodis, Y., Kiayias, A., Malkin, T. (eds.) PKC 2006. LNCS, vol. 3958, pp. 458–473. Springer, Heidelberg (2006)

[MR13]   Mohassel, P., Riva, B.: Garbled circuits checking garbled circuits: More efficient and secure two-party computation. In: Canetti, R., Garay, J.A. (eds.) CRYPTO 2013, Part II. LNCS, vol. 8043, pp. 36–53. Springer, Heidelberg (2013)

[NNOB12]   Nielsen, J.B., Nordholt, P.S., Orlandi, C., Burra, S.S.: A new approach to practical active-secure two-party computation. In: Safavi-Naini, R., Canetti, R. (eds.) CRYPTO 2012. LNCS, vol. 7417, pp. 681–700. Springer, Heidelberg (2012)

[NO09]   Nielsen, J.B., Orlandi, C.: LEGO for two-party secure computation. In: Reingold, O. (ed.) TCC 2009. LNCS, vol. 5444, pp. 368–386. Springer, Heidelberg (2009)

[Nor13]   Nordholt, P.S.: New Approaches to Practical Secure Two-Party Computation. Institut for Datalogi, Aarhus Universitet (2013)

[PSSW09]   Pinkas, B., Schneider, T., Smart, N.P., Williams, S.C.: Secure two-party computation is practical. In: Matsui, M. (ed.) ASIACRYPT 2009. LNCS, vol. 5912, pp. 250–267. Springer, Heidelberg (2009)

[SS11]   shelat, A., Shen, C.-h.: Two-output secure computation with malicious adversaries. In: Paterson, K.G. (ed.) EUROCRYPT 2011. LNCS, vol. 6632, pp. 386–405. Springer, Heidelberg (2011)

[SS13]   Shelat, A., Shen, C.-H.: Fast two-party secure computation with minimal assumptions. In: CCS, pp. 523–534. ACM (2013)

[Yao86]   Yao, A.C.-C.: How to generate and exchange secrets. In: SFCS, pp. 162–167. IEEE Computer Society (1986)

Dishonest Majority Multi-Party Computation for Binary Circuits

Enrique Larraia, Emmanuela Orsini, and Nigel P. Smart

Dept. Computer Science, University of Bristol, UK
{Enrique.LarraiadeVega,Emmanuela.Orsini}@bristol.ac.uk,
nigel@cs.bris.ac.uk

Abstract. We extend the Tiny-OT two party protocol of Nielsen et al (CRYPTO 2012) to the case of n parties in the dishonest majority setting. This is done by presenting a novel way of transferring pairwise authentications into global authentications. As a by product we obtain a more efficient manner of producing globally authenticated shares, in the random oracle model, which in turn leads to a more efficient two party protocol than that of Nielsen et al.

1 Introduction

In recent years actively secure MPC has moved from a theoretical subject into one which is becoming more practical. In the variants of multi-party computation which are based on secret sharing the major performance improvement has come from the technique of authenticating the shared data and/or the shares themselves using information theoretic message authentication codes (MACs). This idea has been used in a number of works: In the case of two-party MPC for binary circuits in [14], for n-party dishonest majority MPC for arithmetic circuits over a "largish" finite field [4,7], and for n-party dishonest majority MPC over binary circuits [8]. All of these protocols are in the pre-processing model, in which the parties first engage in a function and input independent offline phase. The offline phase produces various pieces of data, often Beaver style [3] "multiplication triples", which are then consumed in the online phase when the function is determined and evaluated.

In the case of the protocol of [14], called Tiny-OT in what follows, the authors use the technique of applying information theoretic MACs to the oblivious transfer (OT) based GMW protocol [10] in the two party setting. In this protocol the offline phase consists of producing a set of pre-processed random OTs which have been authenticated. The offline phase is then executed efficiently using a variant of the OT extension protocol of [12]. For a detailed discussion on OT extension see [2,12,14]. In this work we shall take OT extension as a given sub-procedure.

One can think of the Tiny-OT protocol as applying the authentication technique of [4] to the two party, binary circuit case, with a pre-processing which is based on OT as opposed to semi-homomorphic encryption. For two party protocols over binary circuits practical experiments show that Tiny-OT far outperforms other protocols, such as those based on Yao's garbled circuit technique.

J.A. Garay and R. Gennaro (Eds.): CRYPTO 2014, Part II, LNCS 8617, pp. 495–512, 2014.
© International Association for Cryptologic Research 2014

This is because of the performance of the offline phase of the Tiny-OT protocol. Thus a natural question is to ask, whether one can extend the Tiny-OT protocol to the n-party setting for binary circuits.

Results and Techniques. In this paper we mainly address ourselves to the above question, i.e. how can we generalize the two-party protocol from [14] to the n-party setting?

We first describe what are the key technical difficulties we need to overcome. The Tiny-OT protocol at its heart has a method for authenticating random bits via pairwise MACs, which itself is based on an efficient protocol for OT-extension. In [14] this protocol is called aBit. Our aim is to use this efficient two-party process as a black-box. Unfortunately, if we extend this procedure naively to the three party case, we would obtain (for example) that parties P_1 and P_2 could execute the protocol so that P_1 obtains a random bit and a MAC, whilst P_2 obtains a key for the MAC used to authenticate the random bit. However, party P_3 obtains no authentication on the random bit obtained by P_1, nor does it obtain any information as to the MAC or the key.

To overcome this difficulty, we present a protocol in which we fix an unknown global random key and where each party holds a share of this key. Then by executing the pairwise aBit protocol, we are able to obtain a secret shared value, as well as a shared MAC, by all n-parties. This resulting MAC is identical to the MAC used in the SPDZ protocol from [6]. This allows us to obtain authenticated random shares, and in addition to permit parties to enter their inputs into the MPC protocol.

The online phase will then follow similarly to [6], if we can realize a protocol to produce "multiplication triples". In [14] one can obtain such triples by utilizing a complex method to produce authenticated random OTs and authenticated random ANDs (called aOTs and aANDs)[1]. We notice that our method for obtaining authenticated bits also enables us to obtain a form of authenticated OTs in a relatively trivial manner, and such authenticated OTs can be used directly to implement a multiplication gate in the online phase.

Our contribution is twofold. First, we generalize the two-party Tiny-OT protocol to the n-party setting, using a novel technique for authentication of secret shared bits, and completely new offline and online phases. Thus we are able to dispense with the protocols to generate aOTs and aANDs from [14], obtaining a simple and efficient online protocol. Second, and as a by product, we obtain a more efficient protocol than the original Tiny-OT protocol, in the two party setting when one measures efficiency in terms of the number of aBit's needed per multiplication gate. The security of our protocols are proven in the standard universal composability (UC) framework [5] against a malicious adversary and static corruption of parties. The definitional properties of an MPC protocol are implicit in this framework: output indistinguishability of the ideal and the

---

[1] In fact the paper [14] does not produce such multiplication triples, but they follow immediately from the presentation in the paper and would result in a more efficient online phase than that described in [14].

real process gives *correctness*, and the fact that any information gathered by a real adversary is obtainable by an ideal adversary gives *privacy*. Although not explicitly stated, we work with the random oracle model, as we need to implement commitments to check the correctness of the MACs, more precisely, we work with programmable random oracles. See the Appendix of [6] for details.

Related Work. For the case of n party protocols, where $n > 2$, there are three main techniques using such MACs. In [4] each share of a given secret is authenticated by pairwise MACs, i.e. if party P_i holds a share a_i, then it will also hold a MAC $M_{i,j}$ for every $j \neq i$, and party P_j will hold a key $K_{i,j}$. Then, when the value a_i is made public, party P_i also reveals the $n - 1$ MAC values, that are then checked by other parties using their private keys $K_{i,j}$. Note that each pair of parties holds a separate key/MAC for each share value. In [7] the authors obtain a more efficient online protocol by replacing the MACs from [4] with global MACs which authenticate the shared values a, as opposed to the shares themselves. The authentication is also done with respect to a fixed global MAC key (and not pairwise and data dependent). This method was improved in [6], where it is shown how to verify these global MACs without revealing the secret global key. In [8] the authors adapt the technique from [7] for the case of small finite fields, in a way which allows one to authenticate multiple field elements at the same time, without requiring multiple MACs. This is performed using a novel application of ideas from coding theory, and results in a reduced overhead for the online phase.

Future Directions. We end this introduction by describing two possible extensions to our work. Firstly, each bit in our protocol is authenticated by an element in a finite field $\mathbb{F}_{2^\kappa}$. Whilst such values are never transmitted in our online phase due to our MACCheck protocol, they do provide an overhead in the computation. In [8] the authors show how to reduce this overhead using coding theory techniques. It would be interesting to see how such techniques could be applied to our protocol, and what advantage if any they would bring.

Secondly, our protocol requires $n \cdot (n - 1)/2$ executions of the aBit protocol from [14]. Each pairwise invocation requires the execution of an OT-extension protocol, and hence we require $O(n^2)$ such OT-channels. In [11], in the context of traditional MPC protocols, the authors present techniques and situations in which the number of OT-channels can be reduced to $O(n)$. It would be interesting to see how such techniques could be applied in practice to the protocol described in this paper.

2 Notation

In this section we settle the notation used throughout the paper. We use κ to denote the security parameter. We let $\mathsf{negl}(\kappa)$ denote some unspecified function $f(\kappa)$, such that $f = o(\kappa^{-c})$ for every fixed constant c, saying that such a function is *negligible* in κ. We say that a probability is *overwhelming* in κ if it is $1 - \mathsf{negl}(\kappa)$.

We consider the sets $\{0, 1\}$ and $\mathbb{F}_2^\kappa$ endowed with the structure of the fields $\mathbb{F}_2$ and $\mathbb{F}_{2^\kappa}$, respectively. Let $\mathbb{F} = \mathbb{F}_{2^\kappa}$, we will denote elements in $\mathbb{F}$ with greek letters and elements in $\mathbb{F}_2$ with roman letters.

We will additively secret share bits and elements in $\mathbb{F}$, among a set of parties $\mathcal{P} = \{P_1, \ldots, P_n\}$, and sometimes abuse notation identifying subsets $\mathcal{I} \subseteq \{1, \ldots, n\}$ with the subset of parties indexed by $i \in \mathcal{I}$. We write $\langle a \rangle^{\mathcal{I}}$ if a is shared amongst the set $\mathcal{I} = \{i_1, \ldots, i_t\}$ with party P_{i_j} holding a value a_{i_j}, such that $\sum_{i_j \in \mathcal{I}} a_{i_j} = a$. Also, if an element $x \in \mathbb{F}_2$ (resp. $\beta \in \mathbb{F}$) is additively shared among *all* parties we write $\langle x \rangle$ (resp. $\langle \beta \rangle$). We adopt the convention that if $a \in \mathbb{F}_2$ (resp. $\beta \in \mathbb{F}$) then the shares $a_i \in \mathbb{F}_2$ (resp. $\beta_i \in \mathbb{F}$).

(Linear) arithmetic on the $\langle \cdot \rangle^{\mathcal{I}}$ sharings can be performed as follows. Given two sharings $\langle x \rangle^{\mathcal{I}_x} = \{x_{i_j}\}_{i_j \in \mathcal{I}_x}$ and $\langle y \rangle^{\mathcal{I}_y} = \{y_{i_j}\}_{i_j \in \mathcal{I}_y}$ we can compute the following linear operations

$$a \cdot \langle x \rangle^{\mathcal{I}_x} = \{a \cdot x_{i_j}\}_{i_j \in \mathcal{I}_x},$$
$$a + \langle x \rangle^{\mathcal{I}_x} = \{a + x_{i_1}\} \cup \{x_{i_j}\}_{i_j \in \mathcal{I}_x \setminus \{i_1\}},$$
$$\langle x \rangle^{\mathcal{I}_x} + \langle y \rangle^{\mathcal{I}_y} = \langle x + y \rangle^{\mathcal{I}_x \cup \mathcal{I}_y}$$
$$= \{x_{i_j}\}_{i_j \in \mathcal{I}_x \setminus \mathcal{I}_y} \cup \{y_{i_j}\}_{i_j \in \mathcal{I}_y \setminus \mathcal{I}_x} \cup \{x_{i_j} + y_{i_j}\}_{i_j \in \mathcal{I}_x \cap \mathcal{I}_y}.$$

Our protocols will make use of pseudo-random functions, which we will denote by $\mathsf{PRF}_s^{X,t}(\cdot)$ where for a key s and input $m \in \{0, 1\}^*$ the pseudo-random function is defined by $\mathsf{PRF}_s^{X,t}(m) \in X^t$, where X is some set and t is a non-negative integer.

Authentication of Secret Shared Values. As described in the introduction the literature gives two ways to authenticate a secret globally held by a system of parties, one is to authenticate the shares of each party, as in [4], the other is to authenticate the secret itself, as in [7]. In addition we can also have authentication in a pairwise manner, as in [4,14], or in a global manner, as in [7]. Both combinations of these variants can be applied, but each implies important practical differences, e.g., the total amount of data each party needs to store and how checking of the MACs is performed. In this work we will use a combination of different techniques, indeed the main technical trick is a method to pass from the technique used in [14] to the technique used in [7].

Our main technique for authentication of secret shared bits is applied by placing an *information theoretic tag* (MAC) on the shared bit x. The authenticating key is a random line in $\mathbb{F}$, and the MAC on x is its corresponding line point, thus, the linear equation $\mu_\delta(x) = \nu_\delta(x) + x \cdot \delta$ holds, for some $\mu_\delta(x), \nu_\delta(x), \delta \in \mathbb{F}$. We will use these lines in various operations[2], for various values of δ. In particular, there will be a special value of δ, which we denote by α and assume to be $\langle \alpha \rangle^{\mathcal{P}}$ shared, which represents the *global* key for our online MPC protocol. This will be the same key for every bit that needs to be authenticated. It will turn

---

[2] For example, we will also use lines to generate OT-tuples, i.e. quadruples of authenticated bits which satisfy the algebraic equation for a random OT.

out that for the key α we always have $\nu_\alpha(x) = 0$. By abuse of notation we will sometimes refer to a general δ also as a *global* key, and then the corresponding $\nu_\delta(x)$, is called the *local* key.

Distinguishing between parties, say $\mathcal{I}$, that can reconstruct bits (together with the line point), and those parties, say $\mathcal{J}$, that can reconstruct the line gives a natural generalization of both ways to authenticate, and it also allows to move easily from one to another. We write $[x]_{\delta,\mathcal{J}}^{\mathcal{I}}$ if there exist $\mu_\delta(x), \nu_\delta(x) \in \mathbb{F}$ such that:

$$\mu_\delta(x) = \nu_\delta(x) + x \cdot \delta,$$

where we have that x and $\mu_\delta(x)$ are $\langle \cdot \rangle^{\mathcal{I}}$ shared, and $\nu_\delta(x)$ and δ are $\langle \cdot \rangle^{\mathcal{J}}$ shared, i.e. there are values x_i, μ_i, and ν_j, δ_j, such that

$$x = \sum_{i \in \mathcal{I}} x_i, \qquad \mu_\delta(x) = \sum_{i \in \mathcal{I}} \mu_i, \qquad \nu_\delta(x) = \sum_{j \in \mathcal{J}} \nu_j, \qquad \delta = \sum_{j \in \mathcal{J}} \delta_j.$$

Notice that $\mu_\delta(x)$ and $\nu_\delta(x)$ depend on δ and x: we can fix δ and so obtain *key-consistent* representations of bits, or we can fix x and obtain different *key-dependant* representations for the same bit x. To ease the reading, we drop the sub-index $\mathcal{J}$ if $\mathcal{J} = \mathcal{P}$, and, also, the dependence on δ and x when it is clear from the context. We note that in the case of $\mathcal{I}_x = \mathcal{J}_x$ then we can assume $\nu_j = 0$.

When we take the fixed global key α and we have $\mathcal{I}_x = \mathcal{J}_x = \mathcal{P}$, we simplify notation and write $[\![x]\!] = [x]_{\alpha,\mathcal{P}}^{\mathcal{P}}$. By our comment above we can, in this situation, set $\nu_j = 0$ [3], this means that a $[\![x]\!]$ sharing is given by two sharings $(\langle x \rangle^{\mathcal{P}}, \langle \mu \rangle^{\mathcal{P}})$. Notice that the $[\![\cdot]\!]$-representation of a bit x implies that x is *authenticated* with the global key α and that it is $\langle \cdot \rangle$-shared, i.e. its value is actually unknown to the parties.

This notation does not quite align with the previous secret sharing schemes used in the literature, but it is useful for our purposes. For example, with this notation the MAC scheme of [4] is one where each data element x is shared via $[x_i]_{\alpha_j,j}^i$ sharings. Thus the data is shared via a $\langle x \rangle$ sharing and the authentication is performed via $[x_i]_{\alpha_j,j}^i$ sharings, i.e. we are using two sharing schemes simultaneously. In [7] the data is shared via our $[\![x]\!]$ notation, except that the MAC key value ν is set equal to $\nu = \nu'/\alpha$, where ν' being a *public value*, as opposed to a shared value. Our $[\![x]\!]$ sharing is however identical to that used in [6], bar the differences in the underlying finite fields.

Looking ahead we say that a bit $[\![x]\!]$ is *partially opened* if $\langle x \rangle$ is opened, i.e. the parties reveal the shares of x, but not the shares of the MAC value $\mu_\alpha(x)$.

Arithmetic on $[\![x]\!]$ Shared Values. Given two representations $[x]_{\delta,\mathcal{J}_x}^{\mathcal{I}_x} = (\langle x \rangle^{\mathcal{I}_x}, \langle \mu_\delta(x) \rangle^{\mathcal{I}_x}, \langle \nu_\delta(x) \rangle^{\mathcal{J}_x})$ and $[y]_{\delta,\mathcal{J}_y}^{\mathcal{I}_y} = (\langle y \rangle^{\mathcal{I}_y}, \langle \mu_\delta(y) \rangle^{\mathcal{I}_y}, \langle \nu_\delta(y) \rangle^{\mathcal{J}_y})$, under same the δ, the parties can locally compute $[x + y]_{\delta,\mathcal{J}_x \cup \mathcal{J}_y}^{\mathcal{I}_x \cup \mathcal{I}_y}$ as $(\langle x \rangle^{\mathcal{I}_x} + \langle y \rangle^{\mathcal{I}_y}, \langle \mu_\delta(x) \rangle^{\mathcal{I}_x} + \langle \mu_\delta(y) \rangle^{\mathcal{I}_y}, \langle \nu_\delta(x) \rangle^{\mathcal{J}_x} + \langle \nu_\delta(y) \rangle^{\mathcal{J}_y})$ using the arithmetic on $\langle \cdot \rangle^{\mathcal{I}}$ sharings above.

---

[3] Otherwise one can subtract ν_j from μ_j, before setting ν_j to zero.

Let $[\![x]\!] = (\langle x \rangle, \langle \mu(x) \rangle)$ and $[\![y]\!] = (\langle y \rangle, \langle \mu(y) \rangle)$ be two different authenticated bits. Since our sharings are linear, as well as the MACs, it is easy to see that the parties can locally perform linear operations:

$$[\![x]\!] + [\![y]\!] = (\langle x \rangle + \langle y \rangle, \langle \mu(x) \rangle + \langle \mu(y) \rangle) = [\![x + y]\!]$$
$$a \cdot [\![x]\!] = (a \cdot \langle x \rangle, a \cdot \langle \mu(x) \rangle) = [\![a \cdot x]\!],$$
$$a + [\![x]\!] = (a + \langle x \rangle, \langle \mu(a + x) \rangle) = [\![a + x]\!].$$

where $\langle \mu(a + x) \rangle$ is the sharing obtained by each party $i \in \mathcal{P}$ holding the value $\alpha_i \cdot a + \mu_i(x)$.

This means that the only remaining question to enable MPC on $[\![\cdot]\!]$-shared values is how to perform multiplication and how to generate the $[\![\cdot]\!]$-shared values in the first place. Note, that a party P_i that wishes to enter a value into the MPC computation is wanting to obtain a $[x]^i_{\alpha, \mathcal{P}}$ sharing of its input value x, and that this is a $[\![x]\!]$-representation if we set $x_i = x$ and $x_j = 0$ for $j \neq i$.

3   MPC Protocol for Binary Circuit

We start presenting a high level view of the protocols that allow us to perform multi-party computation for binary circuits. We assume synchronous communication and authentic point-to-point channels. Our protocol is in the pre-processing model in which we allow a function (and input) independent pre-processing, or offline, phase which produces correlated randomness. This enables a lightweight online phase, that does not need public-key machinery.

In the following sections we will describe a protocol, Π_{Online}, implementing the actual function evaluation in the $(\mathcal{F}_{\mathsf{Comm}}, \mathcal{F}_{\mathsf{Prep}})$-hybrid model; a protocol, Π_{Prep}, implementing the offline phase in the $(\mathcal{F}_{\mathsf{Comm}}, \mathcal{F}_{\mathsf{Bootstrap}})$-hybrid model; and a novel way to authenticate bits to more than two parties, which takes as starting point the aBit command of [14], and which we model with the $\mathcal{F}_{\mathsf{Bootstrap}}$ functionality.

The online phase implements the standard functionality $\mathcal{F}_{\mathsf{Online}}$

It is based on the $[\![\cdot]\!]$-representation of bits described in Section 2, and it is very similar to the online phase of other MPC protocols [6,7,8,14]. We compute a function represented as a

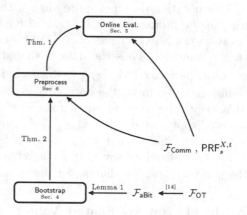

Fig. 1. Overview of Protocols Enabling MPC

binary circuit, where private inputs are additively shared among the parties, and correctness is guaranteed by using additive secret sharings of linear MACs with

---

Functionality $\mathcal{F}_{\mathsf{Online}}$

Initialize: On input ($init$) the functionality activates and waits for an input from the environment. Then it does the following: if it receives Abort, it waits for the environment to input a set of corrupted parties, outputs it to the parties, and aborts; otherwise it continues.

Input: On input ($input, P_i, varid, x$) from P_i and ($input, P_i, varid, ?$) from all other parties, with $varid$ a fresh identifier, the functionality stores ($varid, x$).

Add: On command ($add, varid_1, varid_2, varid_3$) from all parties (if $varid_1, varid_2$ are present in memory and $varid_3$ is not), the functionality retrieves ($varid_1, x$), ($varid_2, y$) and stores ($varid_3, x + y$).

Multiply: On input ($multiply, varid_1, varid_2, varid_3$) from all parties (if $varid_1, varid_2$ are present in memory and $varid_3$ is not), the functionality retrieves ($varid_1, x$), ($varid_2, y$) and stores ($varid_3, x \cdot y$).

Output: On input ($output, varid$) from all honest parties (if $varid$ is present in memory), the functionality retrieves ($varid, y$) and outputs it to the environment. The functionality waits for an input from the environment. If this input is Deliver then y is output to all players. Otherwise it outputs $\varnothing$ is output to all players.

Fig. 2. Secure Function Evaluation

Protocol Π_{MACCheck}

Usage: The parties have a set of $[\![a_i]\!]$, sharings and public bits b_i, for $i = 1, \ldots, t$, and they wish to check that $a_i = b_i$, i.e. they want to check whether the public values are consistent with the shared MACs held by the parties.

As input the system has sharings $(\langle \alpha \rangle, \{b_i, \langle a_i \rangle, \langle \mu(a_i) \rangle\}_{i=1}^{t})$. If the MAC values are correct then we have that $\mu(a_i) = b_i \cdot \alpha$, for all i.

MACCheck($\{b_1, \ldots, b_t\}$):

1. Every party P_i samples a seed s_i and asks $\mathcal{F}_{\mathsf{Comm}}$ to broadcast $\tau_i = \mathsf{Comm}(s_i)$.
2. Every party P_i calls $\mathcal{F}_{\mathsf{Comm}}$ with $\mathsf{Open}(\tau_i)$ and all parties obtain s_j for all j.
3. Set $s = s_1 + \cdots + s_n$.
4. Parties sample a random vector $\boldsymbol{\chi} = \mathsf{PRF}_s^{\mathbb{F},t}(0) \in \mathbb{F}^t$; note all parties obtain the same vector as they have agreed on the seed s.
5. Each party computes the public value $b = \sum_{i=1}^{t} \chi_i \cdot b_i \in \mathbb{F}$.
6. The parties locally compute the sharings $\langle \mu(a) \rangle = \chi_1 \cdot \langle \mu(a_1) \rangle + \cdots + \chi_t \cdot \langle \mu(a_t) \rangle$ and $\langle \sigma \rangle = \langle \mu(a) \rangle - b \cdot \langle \alpha \rangle$.
7. Party i asks $\mathcal{F}_{\mathsf{Comm}}$ to broadcast his share $\tau_i' = \mathsf{Comm}(\sigma_i)$.
8. Every party calls $\mathcal{F}_{\mathsf{Comm}}$ with $\mathsf{Open}(\tau_i')$, and all parties obtain σ_j for all j.
9. If $\sigma_1 + \cdots + \sigma_n \neq 0$, the parties output $\varnothing$ and abort, otherwise they accept all b_i as valid authenticated bits.

Protocol 3. Method to Check MACs on Partially Opened Values

global secret key α. For simplicity we assume one single input for each party and one public output. The online protocol, presented in Section 5, uses the linearity of the $[\![\cdot]\!]$-sharings to perform additions and scalar multiplications locally. For general multiplications we need utilize data produced during the offline phase, in particular the output of the GaOT (Global authenticated OT) command of Section 6. Refer to Figure 4 for a complete description of the functionality for preprocessing data. The aforementioned command GaOT builds upon $\Pi_{\mathsf{Bootstrap}}$ protocol, described in Section 4, to generate random authenticated OTs and, as we noted above, we skip the less efficient procedures of [14].

The Functionality $\mathcal{F}_{\mathsf{Prep}}$

Let A be the set of indices of corrupt parties.

Initialize: On input (Init) from honest parties, the functionality samples random α_i for each $i \notin A$. It waits for the environment to input corrupt shares $\{\alpha_j\}_{j \in A}$ If any $j \in A$ outputs abort, then the functionality aborts and returns the set of $j \in A$ which returned abort. Otherwise the functionality sets $\alpha = \alpha_1 + \cdots + \alpha_n$, and outputs α_k to honest P_k.

Share: On input (i, x, Share) from party P_i, and (i, Share) from all other parties. The functionality produces an authentication $[\![x]\!] = (\langle x \rangle, \langle \mu \rangle)$. It sets $x_j = 0$ if $j \neq i$. Also, the MAC might be shifted by a value Δ_H, i.e. $\mu = x \cdot \alpha + \Delta_H$, where Δ_H is an $\mathbb{F}_2$-linear combination of $\{\alpha_k\}_{k \notin A}$ not known to the environment. It proceeds as follows:
  - Set $\mu = x \cdot \alpha$. If $i \in A$, the environment specifies x.
  - Wait for the environment to specify MAC shares $\{\mu_j\}_{j \in A}$, and generate $\langle \mu \rangle$ where the portion of honest shares is consistent with the adversarial shares, but otherwise random.
  - Set $x_k = 0$ if $k \neq i$, $k \notin A$. If the environment inputs shift-P_k set $\mu_k = \mu_k + \alpha_k$.
  - Output $\{x_k, \mu_k\}$ to honest P_k

GaOT: On input (GaOT) from the parties, the functionality waits for the environment to input "Abort" or "Continue". If it is told to abort, it outputs the special symbol $\varnothing$ to all parties. Otherwise it samples three random bits e, x_0, x_1, and sets $z = x_e$. Then, for every bit $y \in \{e, z, x_0, x_1\}$ the functionality produces an authentication $[\![y]\!] = (\langle y \rangle, \langle \mu(y) \rangle)$, but let the environment to specify shares for corrupt P_j. It proceeds as follows:
  - Set $\mu(y) = y \cdot \alpha$.
  - Wait for the environment to input bit shares $\{y_j\}_{j \in A}$, and MAC shares $\{\mu_j\}_{j \in A}$, and creates sharings $\langle y \rangle$, $\langle \mu \rangle$ where the portion of honest shares is consistent with adversarial shares.
  - Output $\{y_k, \mu_k\}$ to honest P_k.

Fig. 4. Ideal Preprocessing

Notice that, as in [6,7,8,14], during the online computation of the circuit we do not know if we are working with the correct values, since we do not check the MACs of partially opened values during the computation. This check is postponed to the end of the protocol, where we call the MACCheck procedure as in [6] (see Protocol 3). Note this procedure enables the checking of multiple sets of values partially opened during the computation without revealing the global secret key α, thus our MPC protocol can implement reactive functionalities.

The MAC checking protocol is called in both the offline and the online phases, it requires access to an ideal functionality for commitments $\mathcal{F}_{\mathsf{Comm}}$ in the random oracle model (see full version), and it is not intended to implement any functionality. Also, note that the algebraic correctness of the output of the GaOT command in the offline phase is checked in the offline phase and not in the online phase.

4 From Tiny-OT aBit's to $[\![\cdot]\!]$-Sharings

At the heart of our MPC protocol is a method to translate from the two party aBits produced by the offline phase of the Tiny-OT protocol in [14], to the $[\![\cdot]\!]$-sharings under some global shared key α from Section 2. We note that the protocol to produce aBit's is the only sub-protocol from [14] which we use in this paper, and thus the more complex protocols in [14] for producing aOT's and

---

The Functionality $\mathcal{F}_{\mathsf{aBit}}$

Authenticated Bit(P_i, P_j): This functionality selects a random $\delta_j \in \mathbb{F}$ and a random bit r, and returns a sharing $[r]^i_{\delta_j, j}$.

- On input (aBit, i, j) from honest P_i and P_j, the functionality samples a random δ_j and a random sharing $[r]^i_{\delta_j, j} = (r, \mu_i, \nu_j)$, such that $\mu_i = \nu_j + r \cdot \delta_j$. It then outputs $\{r, \mu_i\}$ to P_i and $\{\delta_j, \nu_j\}$ to P_j.
- If P_i is corrupted, the functionality waits for the environment to input the pair $\{r, \mu_i\}$ and it sets $\nu_j = \mu_i + r \cdot \delta_j$ for some randomly chosen δ_j, and $\{\delta_j, \nu_j\}$ is returned to party P_j.
- If P_j is corrupted, the functionality waits for the environment to input the pair $\{\delta_j, \nu_j\}$, r is selected at random and μ_i is set to be $\nu_j - r \cdot \delta_j$. The pair $\{r, \mu_i\}$ is returned to party P_i.

Fig. 5. Two-party Bit Authentication [14]

aAND's we discard. We first deal with the underlying two party sub-protocols, and then we use these to define our multi-party protocols.

4.1 Two-Party [·]-Representations

Thus throughout we assume access to an ideal functionality $\mathcal{F}_{\mathsf{aBit}}$, given in Figure 5, that produces a substantially unbounded number of (oblivious) authenticated *random* bits for two parties, under some *randomly* chosen key δ_j known by one of the parties. This functionality can be implemented assuming a functionality $\mathcal{F}_{\mathsf{OT}}$ and using OT-extension techniques as in [14]. For ease of exposition we present the functionality as returning single bits for single requests. In practice the functionality is implemented via OT-extension and so one is able to obtain *many* aBits on each invocation of the functionality, for a given value of δ_j. Adapting our protocols to deal with multiple aBit production for a single random fixed δ_j chosen by the functionality is left to the reader[4].

Using the protocol $\Pi_{\mathsf{2\text{-}Share}}$, described in Protocol 6, we can obtain a "two-party" representation $[r]^i_{\delta_j, j}$ of a random bit known to P_i, under the key *chosen* by P_j. This extension is needed because we need to adapt the aBit command to the multi-party case. For example, if two parties, P_i and P_j, run the command (aBit, i, j), they obtain a random $[r]^i_{\delta'_j, j}$, with respect to δ'_j; when P_j calls (aBit, k, j) with a different party $P_k, k \neq j$, then they obtain a random $[s]^k_{\tilde{\delta}_j, j}$, with a different $\tilde{\delta}_j$. Thus allowing the parties to select their own values of δ_j means that we can obtain key-consistent [·]-representations, in which each party P_j uses the same fixed δ_j. The security of the protocol $\Pi_{\mathsf{2\text{-}Share}}$ follows from the security of the original aBit in [14]: intuitively the changes required to obtain a

---

[4] Note, that in this situation we (say) produce $1,000,000$ aBits per invocation with a fixed random value of δ_j, then on the next invocation we obtain another $1,000,000$ aBits but with a new random δ_j value. This is not explicit in the ideal functionality description of aBit presented in [14], but is implied by their protocol.

The Subprotocol $\Pi_{\text{2-Share}}$

2Share$(i, j; \delta_j)$: On input (2-Share, i, j, δ_j), where P_j has $\delta_j \in \mathbb{F}$ as input, this command produces a $[r]^i_{\delta_j, j}$ sharing of a random bit r.

1. P_i and P_j call $\mathcal{F}_{\text{aBit}}$ on input (aBit, i, j): The box samples a random δ'_j and then produces

$$[r]^i_{\delta'_j, j} = (r, \mu'_i, \nu_j),$$

such that $\mu'_i = \nu_j + r \cdot \delta'_j$, and outputs $\{r, \mu'_i\}$ to P_i and $\{\delta'_j, \nu_j\}$ to P_j.
2. P_j computes $\sigma_j = \delta_j + \delta'_j$ and sends σ_j to party P_i.
3. P_i sets $\mu_i = \mu'_i + r \cdot \sigma_j = \nu_j + r \cdot \delta_j$.

Protocol 6. Switching to Fixed δ-shares

The Functionality $\mathcal{F}_{\text{Bootstrap}}$

Let A be the indices of corrupt parties.

Initialize: On input (Init) from honest parties, the functionality activates and waits for the environment to input a set of shares $\{\delta_j\}_{j \in A}$. It samples random $\delta \in \mathbb{F}$ and prepares sharing $\langle \delta \rangle$, where the portions of honest shares are consistent with the adversarial shares, but otherwise random. If any $j \in A$ outputs abort, then the functionality aborts and returns the set of $j \in A$ which returned abort, otherwise it continues.

Share: On input (i, x, Share) from party P_i, and (i, Share) from all other parties. The functionality produces a representation $[x]^i_\delta = (\langle x \rangle^i, \langle \mu \rangle^i, \langle \nu \rangle^{\mathcal{P}})$, except that ν might be shifted by a value Δ_H, i.e. $\mu = x \cdot \delta + \nu + \Delta_H$, where Δ_H is an $\mathbb{F}_2$-linear combination of $\{\delta_k\}_{k \notin A}$, which is not known to the environment. It proceeds as follows:
   - It samples random $\mu \in \mathbb{F}$. If $i \in A$ waits for the environment to input $\{\mu, x\}$.
   - The functionality sets $\nu = x \cdot \delta + \mu$.
   - The functionality waits for the environment to input shares $\{\nu_j\}_{j \in A}$, and prepares sharing $\langle \nu \rangle^{\mathcal{P}}$ consistent with the adversarial shares. The portion of honest shares are otherwise random.
   - If the environment inputs shift-P_k, the functionality sets $\nu_k = \nu_k + \delta_k$, $k \notin A$.
   - It outputs $\{\nu_k, \delta_k\}$ to honest P_k.

Fig. 7. Ideal Generation of $[\cdot]^i_{\delta, \mathcal{P}}$-representations

consistent $[\cdot]$-representation do not compromise security, because δ_j is one-time-padded with the random δ'_j produced by $\mathcal{F}_{\text{aBit}}$. See [13] for details. Notice that the command 2-Share takes δ_j as the input of P_j. In particular the value δ_j may not be used to authenticate bits. Thus we could use the protocol $\Pi_{\text{2-Share}}$ to obtain a sharing of the *scalar product* $r \cdot \delta_j$, where P_i obtains the random bit r, and the other party decides what field element $\delta_j \in \mathbb{F}$ gets multiplied in. Then party P_i obtains the result μ_i masked by a one-time pad value ν_j known only to P_j. This application of the subprotocol $\Pi_{\text{2-Share}}$ is going to be crucial in our method to obtain authenticated OT's in our pre-processing phase. As a consequence we *do not always see* δ_j as an authentication key.

4.2 Multiparty $[\cdot]$-Representation

Here we show how to generalize the $\Pi_{\text{2-Share}}$ protocol in order to obtain an n-party representation $[x]^i_\delta$ of a bit x chosen by P_i. This is what the functionality

---

The Protocol $\Pi_{\text{Bootstrap}}$

Initialize: Each party P_i samples a random δ_i. Define $\delta = \delta_1 + \cdots + \delta_n$.
Share: On input (i, x, Share) from P_i and (i, Share) from all other parties, do:
1. For each $j \neq i$, call $\Pi_{\text{2-Share}}$ with $(2\text{-Share}, i, j, \delta_j)$. Party P_i obtains $\{r_{i,j}, \mu_{i,j}\}_{j \neq i}$ whilst party P_j obtains $\nu_{i,j}$, such that $\mu_{i,j} = \nu_{i,j} + r_{i,j} \cdot \delta_j$.
2. Party P_i samples ϵ at random and sets $\mu_i = \epsilon + \sum_{j \neq i} \mu_{i,j}$ and $\nu_i = \epsilon + x \cdot \delta_i$.
3. Party P_i sends $d_j = x + r_{i,j}$ to party P_j for all $j \neq i$.
4. For $j \neq i$, P_j sets $\nu_j = \nu_{i,j} + d_j \cdot \delta_j$.
5. Output $\{\mu_i, \nu_i\}$ to P_i and $\{\nu_j\}$ to party P_j, for $j \neq i$. The system now has $[x]_\delta^i$.

Protocol 8. Transforming Two-party Representations onto $[\cdot]_{\delta, \mathcal{P}}^i$-representations

$\mathcal{F}_{\text{Bootstrap}}$ models in Figure 7. It bootstraps from a two party authentication to a multi-party authentication of the shared bit. As before for $\Pi_{\text{2-Share}}$, we can see the outputs of $\mathcal{F}_{\text{Bootstrap}}$ as the shares of scalar products $x \cdot \delta$, where one party P_i chooses the scalar (bit) x, but now the field element δ is unknown and additively shared among all the parties. An interesting feature of this functionality is that the adversary can only influence *honest* outputs in a small way, that we model with the shift-P_k flag. Additionally, we can not prevent corrupt parties from outputting what they wish, this is reflected on the fact that the functionality leaves their outputs undefined. The main difference between this functionality and the equivalent in the SPDZ protocol [7], is that in [7] the functionality takes as input an offset known to the adversary who adjusts his shares to obtain an invalid MAC value by this linear amount. We do not model this in our functionality, instead we allow the adversary to choose his shares arbitrarily (which obtains the same effect). However, in our protocol the adversary can also introduce an unknown (to the adversary) error into the MAC values. In particular the adversary can decide whether to shift honest shares, but he cannot choose the shifting, namely, an element on the $\mathbb{F}_2$-span of secrets δ_k of honest parties P_k. Later, we manage to determine whether there are any errors (both adversarially known and unknown ones) using an *information-theoretic* MACCheck procedure that we borrow from [6]. See full version for details.

The protocol $\Pi_{\text{Bootstrap}}$, described in Protocol 8, realizes the ideal functionality $\mathcal{F}_{\text{Bootstrap}}$ in a hybrid model in which we are given access to $\mathcal{F}_{\text{aBit}}$. It permits to obtain $[x]_\delta^i$ and it is implemented by sending to each $P_j, j \neq i$, a mask of x using the random bits given by **2-Share**$(i, j; \delta_j)$ as paddings, and then allowing P_j to adjust his share to the right value. In total the protocol needs to execute $n - 1$ aBit per scalar product.

Lemma 1. *In the $\mathcal{F}_{\text{aBit}}$-hybrid model, the protocol $\Pi_{\text{Bootstrap}}$ implements $\mathcal{F}_{\text{Bootstrap}}$ with perfect security against any static adversary corrupting up to $n - 1$ parties.*

Proof. See full version.

5   The Online Phase

In this section we present the protocol Π_{Online}, described in Protocol 9, which implements the online functionality in the $(\mathcal{F}_{\text{Comm}}, \mathcal{F}_{\text{Prep}})$-hybrid model. The

---

Protocol Π_{Online}

Initialize: The parties call Init on the $\mathcal{F}_{\text{Prep}}$ functionality to get the shares α_i of the global MAC key α. If $\mathcal{F}_{\text{Prep}}$ aborts outputting a set of corrupted parties, then the protocol returns this subset of A. Otherwise the operations specified below are performed according to the circuit.

Input: To share his input bit x, P_i calls $\mathcal{F}_{\text{Prep}}$ with input (i, x, Share) and party P_j for $i \neq j$ calls $\mathcal{F}_{\text{Prep}}$ with input (i, Share). The parties obtain $[\![x]\!]$ where the x-share of P_j is set to zero if $j \neq i$.

Add: On input $([\![a]\!], [\![b]\!])$, the parties locally compute $[\![a + b]\!] = [\![a]\!] + [\![b]\!]$.

Multiply: On input $([\![a]\!], [\![b]\!])$, the parties call $\mathcal{F}_{\text{Prep}}$ on input (GaOT), obtaining a random GaOT tuple $\{[\![e]\!], [\![z]\!], [\![x_0]\!], [\![x_1]\!]\}$. The parties then perform:
1. The parties locally compute $[\![f]\!] = [\![b]\!] + [\![e]\!]$ and $[\![g]\!] = [\![x_0]\!] + [\![x_1]\!] + [\![a]\!]$.
2. The shares $[\![f]\!]$ and $[\![g]\!]$ are partially opened.
3. The parties locally compute

$$[\![c]\!] = [\![x_0]\!] + f \cdot [\![a]\!] + g \cdot [\![e]\!] + [\![z]\!].$$

Output: This procedure is entered once the parties have finished the circuit evaluation, but still the final output $[\![y]\!]$ has not been opened.
1. The parties call the protocol Π_{MACCheck} on input of all the partially opened values so far. If it fails, they output $\varnothing$ and abort. $\varnothing$ represents the fact that the corrupted parties remain undetected in this case.
2. The parties partially open $[\![y]\!]$ and call Π_{MACCheck} on input y to verify its MAC. If the check fails, they output $\varnothing$ and abort, otherwise they accept y as a valid output.

Protocol 9. Secure Function Evaluation in the $\mathcal{F}_{\text{Comm}}, \mathcal{F}_{\text{Prep}}$-hybrid Model

basic idea behind our online phase is to use the set of GaOTs output in the offline phase to evaluate each multiplication gate. To see how this is done, consider that we want to multiply two authenticated bits $[\![a]\!], [\![b]\!]$. The parties take a GaOT tuple $\{[\![e]\!], [\![z]\!], [\![x_0]\!], [\![x_1]\!]\}$ off the pre-computed list. Recall we have for such tuples $z = x_e$. It is then relatively straightforward to compute authenticated shares of $[\![c]\!]$, where $c = a \cdot b$, as follows: First, the parties partially open $[\![f]\!] = [\![b]\!] + [\![e]\!]$ and $[\![g]\!] = [\![x_0]\!] + [\![x_1]\!] + [\![a]\!]$, and then set $[\![c]\!] = [\![x_0]\!] + f \cdot [\![a]\!] + g \cdot [\![e]\!] + [\![z]\!]$. To see why this is correct, note that since, $x_e + x_0 + e \cdot (x_0 + x_1) = 0$, we have $c = x_0 + (b + e) \cdot a + (x_0 + x_1 + a) \cdot e + z = a \cdot b$.

Theorem 1. *In the $(\mathcal{F}_{\text{Comm}}, \mathcal{F}_{\text{Prep}})$-hybrid model, the protocol Π_{Online} securely implements $\mathcal{F}_{\text{Online}}$ against any static adversary corrupting up to $n - 1$ parties, assuming protocol* MACCheck *utilizes a secure pseudo-random function* $\text{PRF}_s^{\mathbb{F}, t}(\cdot)$.

Proof. See full version.

6 The Offline Phase

Here we present our offline protocol Π_{Prep} (Protocol 10). The key part of this protocol is the GaOT command. In [14] the authors give a two-party protocol to enable one party, say A, to obtain two authenticated bits e, z, and the other party, say B, to obtain two authenticated secret bits x_0, x_1, such that $z = x_e$ and e, x_0 and x_1 are chosen at random. We generalize such a procedure to many parties and we obtain sharings $[\![e]\!]$, $[\![z]\!]$, $[\![x_0]\!]$, $[\![x_1]\!]$, subject to $z = x_e$. Notice that the values e, z, x_0, x_1 are not known so they can be used in the online phase to implement multiplication gates.

The idea behind the GaOT command it is to exploit the relation between "affine functions" and "selector functions", in which a bit e selects one of two elements (χ_0, χ_1) in $\mathbb{F}$. This connection was already noted in [1] on the context of garbling arithmetic circuits via randomized encodings. Thus, on one hand we have authentications, that are essentially evaluations of affine functions, and on the other we have OT quadruples, that can be seen as selectors. Seeing both as the same object means that a way to authenticate bits also gives us a way to generate OTs, and the other way around. The procedure is broken into three steps, **Share OT**, **Authenticate OT** and **Sacrifice OT**. We examine these three stages in turn. To produce bit quadruples (e, z, x_0, x_1), such that $z = x_e$, the parties will use a (secret) affine line in $\mathbb{F}$ parametrized by (ϑ, η). Note that with our functionality $\mathcal{F}_{\mathsf{Bootstrap}}$ we get $[e_i]_\eta^i$, where e_i is known to P_i, and an additive sharing $\langle \eta \rangle$ is held by the system. We denote this concrete execution of the functionality as $\mathcal{F}_{\mathsf{Bootstrap}}(\eta)$, since we shall use fresh copies of $\mathcal{F}_{\mathsf{Bootstrap}}$ to generate more OT quadruples and also for authentication purposes. Note, that η is not an input to the functionality but a shared random value produced when initialising the functionality. Now, performing n independent queries of Share command on this copy $\mathcal{F}_{\mathsf{Bootstrap}}(\eta)$, the parties can generate

$$[e]_\eta^P = [e_1]_\eta^1 + \cdots + [e_n]_\eta^n. \tag{1}$$

Thus, the system obtains two (secret) elements $\langle e \rangle$, $\langle \zeta \rangle$, such that $\zeta = \vartheta + e \cdot \eta$, for line $(\langle \vartheta \rangle, \langle \eta \rangle)$. Define $\chi_0 = \vartheta$ and $\chi_1 = \vartheta + \eta$, so it holds $\zeta = \chi_e$. The quadruple (e, z, x_0, x_1) is then given by the least significant bits of the corresponding field elements $(e, \zeta, \chi_0, \chi_1)$. This conclude the **Share OT** step.

To add MACs to each bit of the quadruple that the parties just generated, the protocol uses the $\mathcal{F}_{\mathsf{Bootstrap}}(\alpha)$ instance to obtain a sharing $\langle \alpha \rangle$ of the global key. Each party can now authenticate his shares of (e, z, x_0, x_1) querying Share command and obtaining $[\![e]\!]$, $[\![z]\!]$, $[\![x_0]\!]$, $[\![x_1]\!]$. We emphasize that the same α is used to authenticate all OT quadruples, thus $\mathcal{F}_{\mathsf{Bootstrap}}(\alpha)$ is fixed once and for all.

After the **Authenticate OT** step the parties have sharings $[\![e]\!]$, $[\![z]\!]$, $[\![x_0]\!]$, $[\![x_1]\!]$, which could suffer from two possible errors induced by the corrupted parties: Firstly the algebraic equation $z = x_e$ may not hold, and second the MAC values may be inconsistent. For the latter problem we will check all the partially opened values using the MACCheck procedure at the end of the offline phase. For the former case we use the **Sacrifice OT** step. We use the same methodology as in [4,7,6], i.e. one quadruple is checked by "sacrificing" another quadruple. The idea involving sacrificing can be seen as follows: We associate to each pair of quadruples a polynomial $S(t)$ over the field of secrets ($\mathbb{F}_2$ in our case), which is the zero polynomial only if both quadruples are correct. Thus, proving correctness of quadruples is equivalent to proving that $S(t)$ is the zero polynomial. This is done by securely evaluating $S(t)$ on a random public challenge bit t via a combination of addition gates and two openings (plus one extra opening to check the

---

The Protocol Π_{Prep}

Let A be the set of indices of corrupt parties.

Initialize: On input (Init) from honest parties and adversary, the system runs a copy of $\mathcal{F}_{\text{Bootstrap}}$ which is denoted $\mathcal{F}_{\text{Bootstrap}}(\alpha)$. Then it calls Init on $\mathcal{F}_{\text{Bootstrap}}(\alpha)$. If $\mathcal{F}_{\text{Bootstrap}}(\alpha)$ aborts, outputting a set of corrupted parties, then the protocol returns this subset of A and aborts. Otherwise, the values δ_i returned by $\mathcal{F}_{\text{Bootstrap}}(\alpha)$ are labelled as α_i. Set $\alpha = \alpha_1 + \cdots + \alpha_n$, and output α_i to honest parties P_i.

Share: On input (i, x, Share) from party i and (j, Share) from all parties $j \neq i$. The protocol calls Share command of $\mathcal{F}_{\text{Bootstrap}}(\alpha)$ to obtain $[x]_\alpha^i$, given by $\{\langle \mu \rangle^i, \langle \nu \rangle^{\mathcal{P}}\}$. Then, for $j \neq i$, party P_j sets his share of x to be zero, and $\mu_j(x) = \nu_j$. Party P_i sets $\mu_i(x) = \mu + \nu_i$. Thus, the parties obtain $[\![x]\!]$.

GaOT: On input (GaOT) from all P_i, execute the following sub-procedures:

Share OT. This generates sharings $(\langle e \rangle, \langle z \rangle, \langle x_0 \rangle, \langle x_1 \rangle)$ such that x_0, x_1 and e are random bits. If all parties are honest then it holds $z = x_e$.

1. The system runs a fresh copy of $\mathcal{F}_{\text{Bootstrap}}$ on Init command getting an additive sharing $\langle \eta \rangle$ for some random $\eta \in \mathbb{F}$. Denote this copy as $\mathcal{F}_{\text{Bootstrap}}(\eta)$.
2. Each party samples a random bit e_i. Define $e = e_1 + \cdots + e_n$.
3. For each $i = 1, \ldots, n$, the system calls $\mathcal{F}_{\text{Bootstrap}}(\eta)$ on input (i, e_i, Share) from party P_i and input (i, Share) from any other P_j, to obtain $[e_i]_\eta^i$. That is, (in an honest execution) P_i gets $\zeta_i \in \mathbb{F}$, and the parties gets an additive sharing $\langle \vartheta_i \rangle$ of some unknown $\vartheta_i \in \mathbb{F}$, such that $\zeta_i = \vartheta_i + e_i \cdot \eta$. The parties compute $[e]_\eta^{\mathcal{P}} = [e_1]_\eta^1 + \cdots + [e_n]_\eta^n$.
4. At this point of the protocol, the system holds sharings $\langle e \rangle, \langle \zeta \rangle, \langle \vartheta \rangle, \langle \eta \rangle$, so it can derive $\langle \chi_0 \rangle = \langle \vartheta \rangle$, and $\langle \chi_1 \rangle = \langle \vartheta \rangle + \langle \eta \rangle$. Note that (for an honest execution) $\zeta = \vartheta + e \cdot \eta$, or in other words $\zeta = \chi_e$.
5. Each party P_i sets $z_i, x_{0,i}, x_{1,i}$ to be the least significant bits of $\zeta_i, \chi_{0,i}, \chi_{1,i}$ respectively, so as to obtain sharings $\langle z \rangle, \langle x_0 \rangle$ and $\langle x_1 \rangle$.

Authenticate OT. This step produces authentications on the bits previously computed. For every bit $y \in \{e, z, x_0, x_1\}$ it does the following:

6. Call $\mathcal{F}_{\text{Bootstrap}}(\alpha)$ on input (i, y_i, Share) from P_i and (j, Share) for party P_j to obtain $[y_i]_\alpha^i$.
7. Compute $[\![y]\!]$ by forming $\sum_{i \in \mathcal{P}} [y_i]_\alpha^i$, and then computing $\mu(y) - \nu(y)$.

Sacrifice OT. This step checks that the authenticated OT-quadruples are correct. Let $[\![e]\!], [\![z]\!], [\![x_0]\!], [\![x_1]\!]$ be the quadruple to check, and κ a security parameter:

8. Every party P_i samples a seed s_i and asks $\mathcal{F}_{\text{Comm}}$ to broadcast $\tau_i = \text{Comm}(s_i)$.
9. Every P_i calls $\mathcal{F}_{\text{Comm}}$ with $\text{Open}(\tau_i)$ and all parties obtain s_j for all j. Set $s = s_1 + \cdots + s_n$.
10. Parties sample a random vector $\mathbf{t} = \text{PRF}_s^{\mathbb{F}_2, \kappa}(0) \in \mathbb{F}_2^\kappa$. Note all parties obtain the same vector as they have agreed on the seed s.
11. For $i = 1, \ldots, \kappa$, repeat the following:
    - Take one fresh quadruple $[\![e_i]\!], [\![z_i]\!], [\![x_{0,i}]\!], [\![x_{1,i}]\!]$, and partially open the values $p_i = t_i \cdot ([\![x_0]\!] + [\![x_1]\!]) + [\![x_{0,i}]\!] + [\![x_{1,i}]\!]$ and $q_i \Leftarrow [\![e]\!] + [\![e_i]\!]$.
    - Locally evaluate c_i such that $[\![c_i]\!] = t_i \cdot ([\![z]\!] + [\![x_0]\!]) + [\![z_i]\!] + [\![x_{0,i}]\!] + p_i \cdot [\![e]\!] + q_i \cdot ([\![x_{0,i}]\!] + [\![x_{1,i}]\!])$, and check it partially opens to zero. If it does not, then abort.
12. The parties call Π_{MACCheck} on the values partially opened in step 11.
13. If no abort occurs, output $[\![e]\!], [\![z]\!], [\![x_0]\!], [\![x_1]\!]$ as a valid quadruple.

Protocol 10. Preprocessing: Input Sharing and Creation of OT Quadruples in the $\mathcal{F}_{\text{Bootstrap}}$-hybrid Model

evaluation), and then checking that the result of the evaluation partially opens to zero. In this way we would waste κ quadruples to check one quadruple, to get security of $2^{-\kappa}$; we refer the reader to Section 7 for a more efficient sacrifice procedure.

Theorem 2. *Let κ be the security parameter and $t \in \mathbb{N}$. In the $(\mathcal{F}_{\text{Comm}}, \mathcal{F}_{\text{Bootstrap}})$-hybrid model, the protocol Π_{Prep} securely implements $\mathcal{F}_{\text{Prep}}$ with statistical security* *on κ against any static adversary corrupting up to $n - 1$ parties, assuming the existence of $\mathsf{PRF}_s^{X,m}(\cdot)$ with domain $X = \mathbb{F}$ (resp. $\mathbb{F}_2$) and $m = t$ (resp. κ).*

Proof. See full version.

7   Batching the Sacrifice Step

This technique (an adaptation of a technique to be found originally in [15,6,9]) permits to check a batch of OT quadruples for algebraic correctness using a *smaller* number of "sacrificed" quadruples than the basic version we described in Section 6. Recall, the idea is to check that an authenticated OT-quadruple $\mathsf{GaOT}_i = (\llbracket e_i \rrbracket, \llbracket z_i \rrbracket, \llbracket x_i \rrbracket, \llbracket y_i \rrbracket)$ verifies the "multiplicative" relation $m_i = z_i + x_i + e_i \cdot (x_i + y_i) = 0$.

At a high level, Protocol 11 essentially consists of two different phases. Let $(\mathsf{GaOT}_1, \ldots, \mathsf{GaOT}_N)$ be a set of OT quadruples, in the first phase a fixed portion of these GaOTs are partially opened as in a classical cut-and-choose step. If any of the opened OT quadruples does not satisfy the multiplicative relation the protocol aborts. Otherwise it runs the second phase: the remaining GaOTs are permuted and uniformly distributed into t buckets of size T. Then, for each of the buckets, the protocol selects a BucketHead, i.e. the first (in the lex order) GaOT in the bucket (as in [9]), and uses the remaining GaOTs in the same bucket to check that BucketHead correctly satisfies the multiplicative relation.

We call CheckGaOTs the GaOTs used to check the BucketHead, and we denote them by $\mathsf{CheckGaOT} = (\llbracket \mathfrak{e} \rrbracket, \llbracket \mathfrak{z} \rrbracket, \llbracket \mathfrak{h} \rrbracket, \llbracket \mathfrak{g} \rrbracket)$, with $\mathfrak{z} = \mathfrak{h} + \mathfrak{e}(\mathfrak{h} + \mathfrak{g})$.

If any BucketHead does not pass the test, then we know that some parties are corrupted and the protocol aborts. If all the checks pass then we obtain t algebraically correct BucketHeads, i.e. t OT quadruples, with overwhelming probability.

Theorem 3. *For $T \geq \frac{\kappa + \log_2(t)}{\log_2(t)}$ the previous protocol provide t correct GaOTs with error probability $2^{-\kappa}$.*

Proof. See full version.

We can replace the **Sacrifice OT** step in Π_{Prep} with the above Bucket-Cut-and-Choose Protocol and, for an appropriate choice of the parameters, Theorem 2 (and relative proof) still holds.

Notice, how the value h has little effect on the final probability (we suppressed the effect in the statement of the Theorem since it is so low). This means we can take $h = 1$ to obtain the most efficient protocol, which means the amount of cut-and-choose performed is relatively low.

To measure the efficiency of this protocol we can consider the ratio $r = \frac{(T+h)\cdot t}{t} = T + h$: it measures the number of GaOTs that we need to produce one actively secure OT quadruple. Setting $h = 1$ and an error probability of 2^{-40}, we obtain Table 1 for different values of $t = 2^{10}, 2^{14}, 2^{20}$.

Bucket Cut-and-Choose Protocol

Input : Let $N = (T + h) \cdot t$ be the number of input GaOTs and T the size of the buckets, with $T \geq 2$. We let $1 \leq h \leq T$ denote an additional parameter controlling how much cut-and-choose we perform.

Phase-I *Cut-And-Choose* :
1. Every P_i samples a seed s_i and asks $\mathcal{F}_{\mathsf{Comm}}$ to broadcast $\tau_i = \mathsf{Comm}(s_i)$.
2. Every party P_i calls $\mathcal{F}_{\mathsf{Comm}}$ with $\mathsf{Open}(\tau_i)$ and all parties obtain s_j for all j. Set $s = s_1 + \cdots + s_n$.
3. Using a $\mathsf{PRF}_{s}^{\mathbb{F}_2, N}$, parties sample a random vector $\mathbf{v} \in \mathbb{F}_2^N$, such that the number of its non-zero entries is $h \cdot t$ (i.e. the Hamming weight of $\mathbf{v}$ is $h \cdot t$).
4. Let $\mathcal{J}$ be the set of indices j such that $v_j \neq 0$, and, $\forall j \in \mathcal{J}$, the parties partially open GaOT_j and check that it satisfies the algebraic relation $z_j + x_j = e_j \cdot (x_j + y_j)$. If there exists an algebraically incorrect GaOT_j quadruple, then the protocol aborts.

Phase-II *Bucket-Sacrifice* :
5. Permute the unopened GaOTs according to a random permutation π on $T \cdot t$ indices, again using a PRF_s. Then renumber the permuted unopened GaOT_j, such that $j = 1, \ldots, T \cdot t$, and, for $i = 1, \ldots, t$, create the ith bucket as $\{\mathsf{GaOT}_j\}_{j=iT-T+1}^{iT}$.
6. Parties compute a $\mathsf{BucketHead}(i)$ for each $i = 1, \ldots, t$, i.e. return the first (in the lex order) element in the ith bucket.
7. For $i = 1, \ldots, t$, parties check that $\mathsf{BucketHead}(i) = \mathsf{GaOT}_i = (\llbracket e_i \rrbracket, \llbracket z_i \rrbracket, \llbracket x_i \rrbracket, \llbracket y_i \rrbracket)$ is correct using the other GaOTs in the bucket: For $j = iT - T + 2, \ldots, iT$ do
   - Set $\mathsf{CheckGaOT}_j = \mathsf{GaOT}_j = (\llbracket \mathfrak{e}_j \rrbracket, \llbracket \mathfrak{z}_j \rrbracket, \llbracket \mathfrak{h}_j \rrbracket, \llbracket \mathfrak{g}_j \rrbracket)$.
   - Parties open $\langle e_i + \mathfrak{e}_j \rangle$ and $\langle x_i + y_i + \mathfrak{h}_j + \mathfrak{g}_j \rangle$.
   - Parties locally compute

   $$\llbracket c_{i,j} \rrbracket = \llbracket z_i + x_i \rrbracket + \llbracket \mathfrak{z}_j + \mathfrak{h}_j \rrbracket + (e_i + \mathfrak{e}_j)\llbracket \mathfrak{h}_j + \mathfrak{g}_j \rrbracket + (x_i + y_i + \mathfrak{h}_j + \mathfrak{g}_j)\llbracket e_i \rrbracket,$$

   and check it partially opens to zero.
   - If all checks go through output GaOT_i as valid quadruples; otherwise abort.
8. The parties execute the protocol Π_{MACCheck} to check all partially opened values.

Protocol 11. Bucket Cut-and-Choose Protocol

Table 1. Number of GaOTs we need to check t quadruples

r	$T = r - h$	t	$\frac{40 + \log_2(t)}{\log_2(t)}$
4	3	2^{20}	3
5	4	2^{14}	3.85
6	5	2^{10}	5

8 Efficiency Analysis

Here we briefly examine the cost of a multiplication in terms of the number of aBits required in the case of two parties. We use the Bucket-Cut-and-Choose Protocol described in Section 7. We notice that each GaOT requires us to consume ten aBits; we need to execute the **Share OT** step to determine e, z, x_0, x_1 (which requires one aBit consumption per player, i.e. two in total when $n = 2$); in addition each of these four bits needs to be authenticated in **Authenticate OT** in Protocol 10 (which again requires one aBit consumption per player, i.e. eight in total when $n = 2$). Since we need one checked GaOT to perform a secure multiplication, and we sacrifice $r - 1$ GaOT to obtain a checked one; this means we require $r \cdot 10$ aBits per secure multiplication in the two party case. Depending on the parameters we use for our sacrifice step in Appendix 7, this equates to

40, 50 or 60 aBits per secure multiplication, setting $t = 2^{20}, 2^{14}, 2^{10}$, respectively, and an error probability of 2^{-40}.

We now compare this to the number of aBits needed in the Tiny-OT protocol [14]. In this protocol each secure multiplication requires two aBits, two aANDs and two aOTs. Assuming a bucket size T in the protocols to generate aANDs and aOTs; each aAND (resp. aOT) requires four LaANDs (resp LaOTs). Each LaAND requires four aBits and each LaOT requires three aBits. Thus the total number of aBits per secure multiplication is $2 \cdot (1 + T \cdot 4 + T \cdot 3) = 14 \cdot T + 2$. To achieve the same error probability of 2^{-40}, with same values of $t = 2^{20}, 2^{14}, 2^{10}$, they need 44, 58 and 72 aBits, respectively. We see therefore that we can make our protocol (in the two party case) more efficient than the Tiny-OT protocol, when we measure efficiency in terms of the number of aBits consumed.

Acknowledgements. This work has been supported in part by ERC Advanced Grant ERC-2010-AdG-267188-CRIPTO, by EPSRC via grant EP/I03126X and by research sponsored by Defense Advanced Research Projects Agency (DARPA) and the Air Force Research Laboratory (AFRL) under agreement number FA8750-11-2-0079[5].

References

1. Applebaum, B., Ishai, Y., Kushilevitz, E.: How to garble arithmetic circuits. In: Ostrovsky, R. (ed.) FOCS, pp. 120–129. IEEE (2011)
2. Asharov, G., Lindell, Y., Schneider, T., Zohner, M.: More efficient oblivious transfer and extensions for faster secure computation. In: ACM Conference on Computer and Communications Security, pp. 535–548. ACM (2013)
3. Beaver, D.: Efficient multiparty protocols using circuit randomization. In: Feigenbaum, J. (ed.) CRYPTO 1991. LNCS, vol. 576, pp. 420–432. Springer, Heidelberg (1992)
4. Bendlin, R., Damgård, I., Orlandi, C., Zakarias, S.: Semi-homomorphic encryption and multiparty computation. In: Paterson, K.G. (ed.) EUROCRYPT 2011. LNCS, vol. 6632, pp. 169–188. Springer, Heidelberg (2011)
5. Canetti, R.: Universally composable security: A new paradigm for cryptographic protocols. In: FOCS, pp. 136–145. IEEE Computer Society (2001)
6. Damgård, I., Keller, M., Larraia, E., Pastro, V., Scholl, P., Smart, N.P.: Practical covertly secure MPC for dishonest majority – or: Breaking the SPDZ limits. In: Crampton, J., Jajodia, S., Mayes, K. (eds.) ESORICS 2013. LNCS, vol. 8134, pp. 1–18. Springer, Heidelberg (2013)
7. Damgård, I., Pastro, V., Smart, N.P., Zakarias, S.: Multiparty computation from somewhat homomorphic encryption. In: Safavi-Naini, R., Canetti, R. (eds.) [16], pp. 643–662 (2012)

---

[5] The US Government is authorized to reproduce and distribute reprints for Government purposes notwithstanding any copyright notation thereon. The views and conclusions contained herein are those of the authors and should not be interpreted as necessarily representing the official policies or endorsements, either expressed or implied, of Defense Advanced Research Projects Agency (DARPA) or the U.S. Government.

8. Damgård, I., Zakarias, S.: Constant-overhead secure computation of boolean circuits using preprocessing. In: Sahai, A. (ed.) TCC 2013. LNCS, vol. 7785, pp. 621–641. Springer, Heidelberg (2013)
9. Frederiksen, T.K., Jakobsen, T.P., Nielsen, J.B., Nordholt, P.S., Orlandi, C.: Minilego: Efficient secure two-party computation from general assumptions. In: Johansson, T., Nguyen, P.Q. (eds.) EUROCRYPT 2013. LNCS, vol. 7881, pp. 537–556. Springer, Heidelberg (2013)
10. Goldreich, O., Micali, S., Wigderson, A.: How to play any mental game or a completeness theorem for protocols with honest majority. In: Aho, A.V. (ed.) STOC, pp. 218–229. ACM (1987)
11. Harnik, D., Ishai, Y., Kushilevitz, E.: How many oblivious transfers are needed for secure multiparty computation? In: Menezes, A. (ed.) CRYPTO 2007. LNCS, vol. 4622, pp. 284–302. Springer, Heidelberg (2007)
12. Ishai, Y., Kilian, J., Nissim, K., Petrank, E.: Extending oblivious transfers efficiently. In: Boneh, D. (ed.) CRYPTO 2003. LNCS, vol. 2729, pp. 145–161. Springer, Heidelberg (2003)
13. Larraia, E., Orsini, E., Smart, N.P.: Dishonest majority multi-party computation for binary circuits. Cryptology ePrint Archive, Report 2014/101
14. Nielsen, J.B., Nordholt, P.S., Orlandi, C., Burra, S.S.: A new approach to practical active-secure two-party computation. In: Safavi-Naini, R., Canetti, R. (eds.) [16], pp. 681–700 (2012)
15. Nielsen, J.B., Orlandi, C.: LEGO for two-party secure computation. In: Reingold, O. (ed.) TCC 2009. LNCS, vol. 5444, pp. 368–386. Springer, Heidelberg (2009)
16. Safavi-Naini, R., Canetti, R. (eds.): CRYPTO 2012. LNCS, vol. 7417. Springer, Heidelberg (2012)

Efficient Three-Party Computation
from Cut-and-Choose

Seung Geol Choi[1,*], Jonathan Katz[2], Alex J. Malozemoff[2],
and Vassilis Zikas[3,**]

[1] Computer Science Department, United States Naval Academy, Annapolis, MD, USA
choi@usna.edu
[2] Computer Science Department, University of Maryland, College Park, MD, USA
{jkatz,amaloz}@cs.umd.edu
[3] Computer Science Department, ETH Zurich, Switzerland
vzikas@inf.ethz.ch

Abstract. With relatively few exceptions, the literature on efficient
(practical) secure computation has focused on secure two-party computation (2PC). It is, in general, unclear whether the techniques used
to construct practical 2PC protocols—in particular, the *cut-and-choose*
approach—can be adapted to the multi-party setting.

In this work we explore the possibility of using cut-and-choose for
practical secure *three-party* computation. The three-party case has been
studied in prior work in the semi-honest setting, and is motivated by
the observation that real-world deployments of multi-party computation
are likely to involve few parties. We propose a constant-round protocol
for three-party computation tolerating any number of malicious parties,
whose computational cost is only a small constant worse than that of
state-of-the-art two-party protocols.

1   Introduction

The past few years have seen a tremendous amount of attention devoted to making secure computation truly practical (e.g., [19, 23, 24]). With only a few exceptions [13, 14, 23], however, this work has tended to focus on secure *two-party*
computation (2PC). In the semi-honest setting, a series of papers [3, 17–19]
showed that Yao's garbled circuit technique [38] can yield very efficient protocols for the computation of boolean circuits. In the malicious setting, Lindell and
Pinkas [27] initiated use of the *cut-and-choose* technique, also based on Yao's
garbled circuits, for constructing efficient, constant-round protocols. This technique was developed further in several subsequent works [20, 24, 25, 28, 29, 31,
33, 34, 36, 37], and yields the fastest known protocols for (malicious) secure
two-party computation (2PC) of boolean circuits.

2PC protocols with malicious security can also be based on the GMW protocol [15] (e.g., the TinyOT protocol [32]). Although this approach yields protocols

---

* Portions of this work were done while at the University of Maryland.
** Portions of this work were done at the University of Maryland and UCLA.

J.A. Garay and R. Gennaro (Eds.): CRYPTO 2014, Part II, LNCS 8617, pp. 513–530, 2014.
© International Association for Cryptologic Research 2014

with round complexity linear in the (multiplicative) depth of the circuit, it offers the advantage that much of the computation can be pushed to an *offline*, pre-processing phase that is executed before the parties receive their inputs. The subsequent *online* computation is very fast and uses mainly information-theoretic techniques.

In the setting of *multi*-party computation (MPC) with security against an arbitrary number of corruptions, the situation is somewhat different. While there has been much recent work on optimizing MPC for *semi-honest* adversaries [3, 5–8, 10], less work has focused on security against malicious corruptions. The work of Ishai, Prabhakaran, and Sahai [22] gives protocols with good *asymptotic* efficiency; however, despite some promising optimizations [26], it has not yet produced practical instantiations. The SPDZ protocol [4, 12–14, 23], which handles arithmetic circuits, has extremely fast online running time at the cost of a very slow offline phase. However, unlike protocols based on garbled circuits, SPDZ runs for a linear (instead of constant) number of (online) rounds, and in each such round every party needs to utilize a broadcast channel. To our knowledge, SPDZ's implementation experiments [12–14] were run on a local-area network where physical broadcast is available, and thus the delay due to accounting for round-timeouts and/or running a multi-party broadcasting protocol when operating in a wide-area network environment has not been taken into account. This delay may be non-trivial depending on circumstances: Schneider and Zohner [35] have shown that as the latency between machines increases, the cost of each round becomes more and more significant.

Finally, the work of Goyal, Mohassel, and Smith [16] uses the cut-and-choose technique to construct a multi-party protocol secure in the *covert* setting.

Multi-party Computation for a Small Number of Parties. Research on secure computation has traditionally been divided into two classes: work focusing on two-party computation, and work focusing on multi-party computation for an arbitrary number of parties.[1] Yet, in practice, it seems that the most likely scenarios for secure MPC would involve a small number of parties [5]. In general, as the number of parties increases, the cost of communication amongst the parties increases as well. In a wide-area network setting, this may have a huge impact on the running time of the protocol.

In addition, the three-party setting is interesting in its own right. For example, suppose the government would like to run some privacy preserving computation on a company's dataset, such as flight manifests. Now, suppose the public does not trust that these parties are not colluding. Thus, we could add a third party, trusted by the public, into the computation to enforce that the two main parties are not simply sharing all their information.

Our Contributions. We construct the first practical, constant-round protocol for secure three-party computation of boolean circuits. Our protocol uses player-simulation techniques in order to compile existing (cut-and-choose-based) 2PC

---

[1] Here we are interested in protocols tolerating an arbitrary number of corruptions. One could further distinguish work on MPC that assumes an honest majority.

protocols into three-party protocols. We instantiate our compiler with state-of-the-art 2PC constructions and show that the addition of a third party comes at the cost of roughly a factor eight overhead over the underlying 2PC protocol in terms of computation, and a factor sixteen overhead in terms of communication. This running time appears to be superior to the state-of-the-art MPC protocols in terms of *start-to-finish* running time. Of course, computing the exact overhead requires implementations of both our protocol and the underlying 2PC protocol and is a subject of future research. As a further optimization point, our protocol makes *only three calls overall* to a broadcast channel (one with each party as sender), as opposed to existing practical MPC solutions (for more than two parties) which use broadcast for communicating all protocol messages. This may be important in certain wide-area network settings where communication (and broadcast specifically) is very expensive. The most efficient instantiation of our protocol requires the random oracle model.

Overview of Our Protocol. Denote the three parties by P_1, P_2, and P_3. The high-level idea of our construction is to execute a two-party protocol $\widehat{\pi}$, where one of the two parties (say $\widehat{P}_1$) is emulated by P_1 and P_2 via a two-party protocol π, and the other party is played by P_3.

Naïvely applying the above idea yields an inefficient construction even when state-of-the-art 2PC protocols are used for π and $\widehat{\pi}$. Assume, for example, that the most efficient 2PC protocol is used for both π and $\widehat{\pi}$, where π simply computes the circuit of $\widehat{P}_1$ among P_1 and P_2. The security of the resulting construction follows trivially from the composition theorem. However, unless the size of the circuit is very small, this approach results in a huge blowup on the overall runtime; in particular, if t is the time π needs to compute the circuit of $\widehat{P}_1$ and $\widehat{t}$ is the time that $\widehat{\pi}$ needs to compute the three-party circuit, then the runtime of the above naïve construction is $t \cdot \widehat{t}$, yielding at least a quadratic blowup.

Emulating the Sender vs. Emulating the Receiver. In most cut-and-choose-based 2PC protocols, the parties have distinct roles: one is the *sender*, or circuit generator, and the other is the *receiver*, or circuit verifier. One might be tempted to think that, because the role of the verifier in the protocol is more "passive" (in the sense that the computation is less complicated), the most natural approach would be to emulate the verifier among P_1 and P_2 (and have P_3 locally do the heavier work doing circuit generation and opening over broadcast). This seemingly direct approach fails as one needs a mechanism for P_1 and P_2 to include their inputs into the garbled circuits. Clearly, doing so by having P_1 first receive his input-keys via OT (as in the original Yao-based constructions) and then handing them to P_2 yields an insecure protocol; indeed, an adversary corrupting P_2 and P_3 can then trivially learn P_1's inputs.

Instead, in this work we have P_1 and P_2 emulate the sender, and we have P_3 play the role of the receiver. More precisely, we adapt the distributed circuit-garbling technique [1, 11] to the two-party setting, allowing P_1 and P_2 to compute a sharing of a garbled circuit which they then reconstruct for P_3. By appropriate optimizations, we ensure that distributed garbling requires P_1 and P_2 to compute and communicate roughly as much as the sender in an execution of the Yao protocol (plus some OT calls per gate); P_3 needs to do nothing during the circuit garbling. Most interestingly, our construction features a mechanism which allows P_3 to receive the keys corresponding to his input bits for evaluating the garbled circuit by only one invocation of OT per input-bit with each of P_1 and P_2.

Our distributed garbling scheme is secure against malicious adversaries, which ensures that an adversary corrupting only one of the parties P_1 or P_2 cannot produce a maliciously constructed garbled circuit. In order to protect against an adversary who corrupts both P_1 *and* P_2, we rely on the cut-and-choose technique. We give concrete instantiations (in the random oracle model) of our protocol using a combination of two 2PC protocols by Lindell and Pinkas [27, 28]. In the full version [9], we present a construction based on the more recent protocol by Lindell [25] which drastically reduces the number of circuit garblings required for cut-and-choose.

Interestingly, the cut-and-choose technique does not only protect against corrupting both P_1 and P_2, but allows a considerable efficiency improvement. More precisely, it allows us to avoid using costly authenticated shares (towards P_3) for the computed (shared) garbled circuit. Instead, our distributed garbling scheme outputs, even in the malicious setting, a plain two-out-of-two sum sharing of the garbled circuit.

2   Preliminaries

We let k denote the computational security parameter and let s denote the statistical security parameter. We use $x \overset{\$}{\leftarrow} S$ to denote choosing a value x uniformly at random from the set S, and use $\|$ to denote concatenation.

Circuit Notation. We follow the circuit notation of Bellare, Hoang, and Rogaway [2]. A circuit C is defined by parameters $(\mathsf{n}, \mathsf{m}, \mathsf{q}, L, R, G)$, where n is the number of input wires, m is the number of output wires, and q is the number of gates, where each gate is indexed by its output wire. Thus, the total number of wires in the circuit is $\mathsf{n} + \mathsf{q}$. The numbering of wires starts with the inputs and ends with the outputs; i.e., we have inputs $\{1, \ldots, \mathsf{n}\}$ and outputs $\{\mathsf{n}+\mathsf{q}-\mathsf{m}+1, \ldots, \mathsf{n}+\mathsf{q}\}$. The function L (resp., R) takes as input a gate index and returns the left (resp., right) input wire to the gate. We require $L(\gamma) < R(\gamma) < \gamma$ for any gate index γ. The function G encodes the functionality of a given gate, e.g., $G_\gamma(0, 1) = 0$ if the gate with index γ is an AND gate. Because we consider circuits with inputs from multiple parties, let $\{\mathsf{n}_{i-1}+1, \ldots, \mathsf{n}_i\}$ denote the input wires "controlled" by party P_i, with $\mathsf{n}_0 = 0$.

We denote *input gates* as those gates with one or more input wires, *inner gates* as those gates with no input or output wires, and *output gates* as those gates with an output wire.

Secret Sharing. Our constructions use two-out-of-two secret sharing. In the semi-honest setting, we use a standard (linear) sharing of strings: the secret $x \in \{0, 1\}^*$ is split into two random *summands* x_1 and x_2 such that $x_1 \oplus x_2 = x$, with P_i holding the summand x_i. We denote the *sharing* of x by $[x] = ([x]^{(1)}, [x]^{(2)})$, where we refer to each $[x]^{(i)} = x_i$ as P_i's *share* of x. This sharing is linear: If $[x]$ and $[y]$ are sharings of x and y respectively, then $[x] \oplus [y]$ is a sharing of $x \oplus y$; that is, $[x \oplus y] = [x] \oplus [y]$ and thus P_i can locally compute his share as $[x \oplus y]^{(i)} = [x]^{(i)} \oplus [y]^{(i)}$. It is straight-forward to verify that the above secret-sharing is *private* provided that the summands x_1 and x_2 are uniformly chosen (restricted only on $x_1 \oplus x_2 = x$); i.e., any single share $[x]^{(i)}$ contains no information about the secret x. Reconstructing a sharing $[x]$ is done by having each party announce his share $[x]^{(i)}$ and taking x to be the exclusive-or of the announced shares.

Our protocols use shares of two types of secrets: k-bit strings $x \in \{0,1\}^k$ and bits $b \in \{0,1\}$. For clarity in the presentation, we use the bracket notation introduced above for sharings of $x \in \{0,1\}^k$, and use the notation $\langle \cdot \rangle$ for sharings of bits; i.e., if $b \in \{0,1\}$ then a sharing of b is denoted as $\langle b \rangle = (\langle b \rangle^{(1)}, \langle b \rangle^{(2)})$.

In the malicious setting we need the sharings of bits to be *authenticated*; i.e., in addition to his summand b_i, each party P_i holds an authentication tag t_i for a message authentication code (MAC), with another party P_j holding the corresponding verification key k_j. More precisely, in a sharing $\langle b \rangle = (\langle b \rangle^{(1)}, \langle b \rangle^{(2)})$ of b, each party's share is now a tuple $\langle b \rangle^{(i)} := (b_i, t_i, k_j)$, where $b_1 \oplus b_2 = b$, and t_i is a valid MAC on b_i with key k_j. This ensures that the adversary cannot make the reconstruction output any value other than the secret b. In particular, to reconstruct some sharing $\langle b \rangle = (\langle b \rangle^{(1)}, \langle b \rangle^{(2)})$, each party P_i first announces his summand b_i and the corresponding authentication tag t_i; subsequently, each party P_i checks that the other party P_j announced a validly authenticated summand matching his own verification key and if this is not the case he rejects. The inability of an adversarial P_i to announce a summand other than b_i follows from the unforgeability of the MAC, as P_i does not know the key k_j matching his authentication tag.

We also assume this authentication is linear in the following sense: Given $\langle b \rangle$ and $\langle b' \rangle$, the parties can compute $\langle b \rangle \oplus \langle b' \rangle$ *locally*. Namely, $\langle b \rangle \oplus \langle b' \rangle = (\langle b \oplus b' \rangle^{(1)}, \langle b \oplus b' \rangle^{(2)})$, where $\langle b \oplus b' \rangle^{(i)} = (b_i \oplus b'_i, t_i \oplus t'_i, k_j \oplus k'_j)$ is a valid authentication. We can construct such authenticated sharings using the TinyOT protocol [32]; see the full version [9] for details.

3 Two-Party Distributed Garbling Scheme

In this section we describe our construction of a two-party distributed garbling scheme. Our protocol combines the standard Yao garbling circuit technique with the distributed garbling ideas from Damgård and Ishai [11]. The main idea is the following: The players jointly compute a garbled circuit, where the gates are garbled by use of a distributed encryption scheme which takes, for each encryption, one key from each party.

We describe our construction in several steps. In Section 3.1 we give a description of our garbling scheme; i.e., the code of the sender in our version of Yao's protocol. This section gives the reader familiarity with our notation and is used as a reference in the distributed protocol. Next, in Section 3.2 we describe an efficient (semi-honest) protocol that allows parties P_1 and P_2 to securely emulate

---

Auxiliary Inputs: Security parameter k, circuit $(\mathsf{n}, \mathsf{m}, \mathsf{q}, L, R, G) \leftarrow C$.

1. **Generate masks:**
   - For $w \in \{1, \ldots, \mathsf{n} + \mathsf{q} - \mathsf{m}\}$: set $\lambda_w \xleftarrow{\$} \{0,1\}$.
   - For $w \in \{\mathsf{n} + \mathsf{q} - \mathsf{m} + 1, \ldots, \mathsf{n} + \mathsf{q}\}$: set $\lambda_w \leftarrow 0$.
2. **Generate sub-keys:**
   - For $w \in \{1, \ldots, \mathsf{n} + \mathsf{q}\}$ and $b \in \{0,1\}$: set $s^1_{w,b}, s^2_{w,b} \xleftarrow{\$} \{0,1\}^k$.
3. **Construct garbled circuit:**
   - For $\gamma \in \{\mathsf{n} + 1, \ldots, \mathsf{n} + \mathsf{q}\}$: Let $\alpha \leftarrow L(\gamma)$ and $\beta \leftarrow R(\gamma)$ be the index of the left and right input wires, respectively, of the gate indexed by γ. Letting $K_{w,b} = (s^1_{w,b}, s^2_{w,b})$, for $i, j \in \{0,1\}^2$, compute the following:

$$P[\gamma, i, j] \leftarrow \mathsf{Enc}_{K_{\alpha,i}, K_{\beta,j}} \left(K_{\gamma, G_\gamma(\lambda_\alpha \oplus i, \lambda_\beta \oplus j) \oplus \lambda_\gamma} \| G_\gamma(\lambda_\alpha, \lambda_\beta) \oplus \lambda_\gamma \right)$$

4. **Output circuit:**
   - Set $GC \leftarrow (\mathsf{n}, \mathsf{m}, \mathsf{q}, L, R, P)$, and output:

$$\left(GC, \{ (s^1_{w, b \oplus \lambda_w}, s^2_{w, b \oplus \lambda_w}, b \oplus \lambda_w) : w \in \{1, \ldots, \mathsf{n}\}, b \in \{0,1\} \} \right).$$

Fig. 1. Circuit garbling scheme

the circuit-garbling procedure from Section 3.1. Finally, in Section 3.3, we show how to make the garbling procedure maliciously secure.

3.1 Single-Party Garbling Scheme

Our garbling scheme is a slight variant of the Damgård and Ishai protocol [11] adapted to two parties. This should be regarded as an initial step towards our ultimate goal of a *distributed* garbling scheme. Here, we describe the high-level construction; see Figure 1 for the detailed protocol.

We associate two random keys $K_{w,0}, K_{w,1}$ with each wire w in the circuit; key $K_{w,0}$ corresponds to the value '0' and $K_{w,1}$ corresponds to the value '1'. Each key $K_{w,b}$ consists of two sub-keys $s^1_{w,b}$ and $s^2_{w,b}$; that is, $K_{w,b} = (s^1_{w,b}, s^2_{w,b})$. In addition, for each wire w we choose a random mask bit λ_w. Each key has an associated tag, derived from the mask bit, which acts as a blinding of the true value the key represents.

Now, consider gate G_γ in the circuit with input wires α and β. The garbled gate of G_γ consists of an array of four encryptions: for each $(b_\alpha, b_\beta) \in \{0,1\} \times \{0,1\}$, the row (b_α, b_β) consists of an encryption of $K_{\gamma, G_\gamma(b_\alpha \oplus \lambda_\alpha, b_\beta \oplus \lambda_\beta) \oplus \lambda_\gamma}$ and its corresponding tag $G_\gamma(b_\alpha \oplus \lambda_\alpha, b_\beta \oplus \lambda_\beta) \oplus \lambda_\gamma$ under keys K_{α, b_α} and K_{β, b_β}. Let P denote a table that stores all the garbled gates; in particular, the entry $P[\gamma, b_\alpha, b_\beta]$ contains an encryption corresponding to row (b_α, b_β) of the garbled gate for G_γ.

Evaluation proceeds as follows. Let α and β be input wires connected to gate G with index γ. The evaluator is given $(K_{\alpha, b_\alpha \oplus \lambda_\alpha}, b_\alpha \oplus \lambda_\alpha)$ and $(K_{\beta, b_\beta \oplus \lambda_\beta}, b_\beta \oplus \lambda_\beta)$, along with P. He takes the row $P[\gamma, b_\alpha \oplus \lambda_\alpha, b_\beta \oplus \lambda_\beta]$ and decrypts it using the keys $K_{\alpha,b_\alpha \oplus \lambda_\alpha}$ and $K_{\beta,b_\beta \oplus \lambda_\beta}$, resulting in $(K_{\gamma,G(b_\alpha,b_\beta) \oplus \lambda_\gamma}, G(b_\alpha, b_\beta) \oplus \lambda_\gamma)$. It is straightforward to verify that by continuing this evaluation, the output of each gate will be revealed masked by its corresponding mask. By picking masks of the output wires to be '0' we ensure that the evaluator receives the (unmasked) output of the circuit.

3.2 Distributing the Garbling Scheme between Two Parties

We now show how to emulate the above garbling scheme between two parties in the *semi-honest* setting. We assume the parties have access to the following two-party ideal functionalities:

- *Gate computation* $\mathcal{F}_{\text{gate}}^G(\langle a \rangle, \langle b \rangle)$: The functionality takes as input sharings $\langle a \rangle$ and $\langle b \rangle$ of bits a and b, respectively, and is parameterized by a binary gate G; it outputs a sharing $\langle G(a,b) \rangle$ of the output of G on input (a,b).
- *One-out-of-two oblivious secret sharing* $\mathcal{F}_{\text{oshare}}^i(\langle b \rangle, m_0, m_1)$: The functionality takes as input a sharing $\langle b \rangle$ of a bit b (i.e., each party inputs his share), along with two messages m_0, m_1 from P_i, and outputs a random two-out-of-two sharing $[m_b]$ of m_b.
- *Constant bit sharing* $\mathcal{F}_{\text{const}}^b()$: The functionality is parameterized by a bit $b \in \{0,1\}$, and outputs a random sharing $\langle b \rangle$ of b.
- *Random bit sharing* $\mathcal{F}_{\text{rand}}()$: The functionality chooses a random bit $r \xleftarrow{\$} \{0,1\}$ and computes and outputs a random sharing $\langle r \rangle$ of r.
- *Bit secret sharing* $\mathcal{F}_{\text{ss}}^i(b)$: The functionality takes input bit $b \in \{0,1\}$ from P_i and outputs a random two-out-of-two sharing $\langle b \rangle$ of b.

Each of these can be instantiated efficiently in the semi-honest setting; see the full version [9] for details.

Distributed Encryption Scheme. We utilize Damgård and Ishai's distributed encryption scheme [11]. Suppose the message and the key for the encryption scheme are distributed as follows:

- The message m is secret-shared; i.e., P_1 holds $[m]^{(1)}$ and P_2 holds $[m]^{(2)}$.
- The encryption key $K = (s^1, s^2)$ is distributed such that P_1 holds s^1 and P_2 holds s^2.

The encryption of the secret-shared message m with tweak T under key $K = (s^1, s^2)$ is:

$$\mathsf{Enc}_K^T(m) = (\mathsf{Enc}_{s^1,T}^1(m), \mathsf{Enc}_{s^2,T}^2(m)) = \left([m]^{(1)} \oplus F_{s^1}^1(T), [m]^{(2)} \oplus F_{s^2}^1(T) \right),$$

where F_k^1 is a PRF keyed by key k. To decrypt a ciphertext $c := \mathsf{Enc}_K^T(m)$, each party P_i sends his sub-key s^i to the decrypter, who uses them to recover the shares of m and reconstruct m.

Double encryption is defined analogously. For keys $K_\alpha = (s_\alpha^1, s_\alpha^2)$ and $K_\beta = (s_\beta^1, s_\beta^2)$, where P_i holds (s_α^i, s_β^i), encryption with tweak T works as follows:

$$\mathsf{Enc}_{K_\alpha,K_\beta}^T(m) = \left([m]^{(1)} \oplus F_{s_\alpha^1}^1(T) \oplus F_{s_\beta^1}^2(T), [m]^{(2)} \oplus F_{s_\alpha^2}^1(T) \oplus F_{s_\beta^2}^2(T) \right).$$

Auxiliary Inputs: Security parameter k, circuit $(\mathsf{n},\mathsf{m},\mathsf{q},L,R,G) \leftarrow C$.

P_1 and P_2 compute $\langle 1 \rangle \leftarrow \mathcal{F}_{\mathsf{const}}^1$, which they use throughout.

1. **Generate mask bits:**
   - For $w \in \{1,\dots,\mathsf{n}_1\}$: P_1 sets $\lambda_w \stackrel{\$}{\leftarrow} \{0,1\}$ and $\langle \lambda_w \rangle \leftarrow \mathcal{F}_{\mathsf{ss}}^1(\lambda_w)$.
   - For $w \in \{\mathsf{n}_1+1,\dots,\mathsf{n}\}$: P_2 sets $\lambda_w \stackrel{\$}{\leftarrow} \{0,1\}$ and $\langle \lambda_w \rangle \leftarrow \mathcal{F}_{\mathsf{ss}}^2(\lambda_w)$.
   - For $w \in \{\mathsf{n}+1,\dots,\mathsf{n}+\mathsf{q}-\mathsf{m}\}$: set $\langle \lambda_w \rangle \leftarrow \mathcal{F}_{\mathsf{rand}}$.
   - For $w \in \{\mathsf{n}+\mathsf{q}-\mathsf{m}+1,\dots,\mathsf{n}+\mathsf{q}\}$: set $\langle \lambda_w \rangle \leftarrow \mathcal{F}_{\mathsf{const}}^0$.
2. **Generate sub-keys:**
   - For $w \in \{1,\dots,\mathsf{n}+\mathsf{q}\}$ and $b \in \{0,1\}$: P_i sets $s_{w,b}^i \stackrel{\$}{\leftarrow} \{0,1\}^k$.
3. **Construct garbled circuit:**
   - For $\gamma \in \{\mathsf{n}+1,\dots,\mathsf{n}+\mathsf{q}\}$: Let $\alpha \leftarrow L(\gamma)$ and $\beta \leftarrow R(\gamma)$ be the indices of the left and right input wires, respectively, of the gate indexed by γ. For $i,j \in \{0,1\}^2$, compute the following selector bits:

   $$\langle \sigma_{\gamma,i,j} \rangle \leftarrow \mathcal{F}_{\mathsf{gate}}^{G_\gamma}(\langle \lambda_\alpha \rangle \oplus \langle i \rangle, \langle \lambda_\beta \rangle \oplus \langle j \rangle) \oplus \langle \lambda_\gamma \rangle.$$

   Next, for $i,j \in \{0,1\}^2$, compute sharings of the appropriate sub-keys to use for each row:

   $$\begin{bmatrix} \hat{s}_{\gamma,i,j}^1 \end{bmatrix} \leftarrow \mathcal{F}_{\mathsf{oshare}}^1(\langle \sigma_{\gamma,i,j} \rangle, s_{\gamma,0}^1, s_{\gamma,1}^1),$$
   $$\begin{bmatrix} \hat{s}_{\gamma,i,j}^2 \end{bmatrix} \leftarrow \mathcal{F}_{\mathsf{oshare}}^2(\langle \sigma_{\gamma,i,j} \rangle, s_{\gamma,0}^2, s_{\gamma,1}^2).$$

   Finally, for $i,j \in \{0,1\}^2$, compute the distributed encryptions of the (permuted) sub-keys and selector bits. That is, letting $K_{w,b} = (s_{w,b}^1, s_{w,b}^2)$, compute:

   $$(P^1[\gamma,i,j], P^2[\gamma,i,j]) \leftarrow \mathsf{Enc}_{K_{\alpha,i},K_{\beta,j}}^{\gamma \| i \| j}\left(\begin{bmatrix} \hat{s}_{\gamma,i,j}^1 \end{bmatrix} \| \begin{bmatrix} \hat{s}_{\gamma,i,j}^2 \end{bmatrix} \| \langle \sigma_{\gamma,i,j} \rangle \right).$$

4. **Output circuit:**
   - Let $C^i \leftarrow (\mathsf{n},\mathsf{m},\mathsf{q},L,R,P^i)$, $S^i \leftarrow \left\{ (s_{w,0}^i, s_{w,1}^i) : w \in \{1,\dots,\mathsf{n}\} \right\}$.
   - P_1 outputs $\left(C^1, S^1, \left\{ (\langle b_w \rangle^{(1)}, \langle \lambda_w \rangle^{(1)}, b_w, \lambda_w) : w \in \{1,\dots,\mathsf{n}_1\} \right\} \right)$.
   - P_2 outputs $\left(C^2, S^2, \left\{ (\langle b_w \rangle^{(2)}, \langle \lambda_w \rangle^{(2)}, b_w, \lambda_w) : w \in \{\mathsf{n}_1+1,\dots,\mathsf{n}\} \right\} \right)$.

Fig. 2. Two-party distributed circuit-garbling protocol $\Pi_{\mathbf{GC}}(P_1,P_2)$. For semi-honest security, use standard secret sharing; for malicious security use authenticated secret sharing.

Distributed Garbling Scheme. We now give a high-level description of our two-party distributed garbling scheme $\Pi_{\mathbf{GC}}(P_1,P_2)$; see Figure 2 for details. As before, for each wire w in the circuit we associate keys $K_{w,0} = (s_{w,0}^1, s_{w,0}^2)$ and $K_{w,1} = (s_{w,1}^1, s_{w,1}^2)$ corresponding to bits '0' and '1', respectively. However, in the distributed setting, each sub-key is only known to one of the two parties; i.e., P_i only knows $(s_{w,0}^i, s_{w,1}^i)$. Each wire is also associated with a mask bit λ_w which is secret shared between the two parties such that no party knows λ_w.

Consider gate G_γ in the circuit with input wires indexed by α and β. As in the non-distributed case, we construct an array containing four rows corresponding to a random permutation of the four possible outcomes of gate G_γ applied to bits b_α and b_β. However, in the distributed case neither party should know what is being encrypted. Recall that in the non-distributed setting, the circuit generator can easily compute $G_\gamma(\lambda_\alpha \oplus b_\alpha, \lambda_\beta \oplus b_\beta)$ to construct the array. However, in the distributed setting, neither party knows (and should *not* know) λ_α or λ_β. Thus, the parties utilize the $\mathcal{F}_{\mathbf{gate}}$ functionality, which takes as input the shares $\langle\lambda_\alpha\rangle \oplus \langle b_\alpha\rangle$ and $\langle\lambda_\beta\rangle \oplus \langle b_\beta\rangle$, and computes a sharing of $G_\gamma(\lambda_\alpha \oplus b_\alpha, \lambda_\beta \oplus b_\beta)$. Let $\langle\sigma_{\gamma,b_\alpha,b_\beta}\rangle = \mathcal{F}_{\mathbf{gate}}^G(\langle b_\alpha\rangle \oplus \langle\lambda_\alpha\rangle, \langle b_\beta\rangle \oplus \langle\lambda_\beta\rangle) \oplus \langle\lambda_\gamma\rangle$. The value $\sigma_{\gamma,b_\alpha,b_\beta}$ denotes which key to encrypt; that is, in row (b_α, b_β) we encrypt key $K_{\gamma,\sigma_{\gamma,b_\alpha,b_\beta}}$. However, we must still enforce that neither party knows what key $K_{\gamma,\sigma_{\gamma,b_\alpha,b_\beta}}$ represents. We handle this by utilizing another functionality, $\mathcal{F}_{\mathbf{oshare}}$. For each of the four $\sigma_{\gamma,b_\alpha,b_\beta}$ values, and for each party P_i, the parties compute $\mathcal{F}_{\mathbf{oshare}}^i(\langle\sigma_{\gamma,b_\alpha,b_\beta}\rangle, s_{\gamma,0}^i, s_{\gamma,1}^i)$. This produces a share of the appropriate sub-key for party P_i, with the crucial fact that P_i does not know which of his sub-keys was shared. The results of $\mathcal{F}_{\mathbf{oshare}}$ are used as the shares to be encrypted.

Note that we can use this two-party distributed garbling scheme as a building block for a somewhat efficient semi-honest two-party secure computation protocol. See the full version [9] for the detailed construction. We do not claim that this scheme is superior to existing 2PC protocols; however, it serves as an important building-block to our end goal of an efficient 3PC protocol.

Also note that this distributed garbling scheme can scale to more than two parties, given access to multi-party variants of the necessary functionalities. Thus, we can also achieve (semi-honest) *multi*-party secure computation using this approach; we leave the development of efficient instantiations of these functionalities as future work.

3.3 Achieving Malicious Security

The semi-honest distributed garbling scheme described in Section 3.2 can be directly adapted to work against a malicious adversary by modifying the hybrid functionalities to work in an authenticated manner; namely, we use authenticated sharings in place of standard secret sharings:

- $\mathcal{F}_{\mathbf{const}}^1()$ and $\mathcal{F}_{\mathbf{rand}}()$: The output share is authenticated.
- $\mathcal{F}_{\mathbf{gate}}^G(\langle a\rangle, \langle b\rangle)$: The inputs and outputs are all authenticated sharings.
- $\mathcal{F}_{\mathbf{oshare}}^i(\langle b\rangle, m_0, m_1)$: The selection bit b is an authenticated sharing.
- $\mathcal{F}_{\mathbf{ss}}^i(b)$: The output is an authenticated sharing of b.

Observe that we only authenticate sharings of bits and *not* sharings of the sub-keys $s_{w,b}^i$. This complicates the proof, as the sharing does not provide means of protecting against a malicious party sending inconsistent key-shares, but yields a more efficient construction; see the full version [9] for details.

We also need a notion of *encrypting* authenticated shares. Recall that for an authenticated share $\langle b\rangle = (\langle b\rangle^{(1)}, \langle b\rangle^{(2)})$, we have $\langle b\rangle^{(i)} = (b_i, t_i, k_j)$, where party P_i holds b_i and t_i, and party P_j holds k_j. Thus, letting $K = (s^1, s^2)$, we define

$$\mathsf{Enc}_K^T(\langle b\rangle) = (\mathsf{Enc}_{s^1, T}^1(b_1\|t_1\|k_1), \mathsf{Enc}_{s^2, T}^2(b_2\|t_2\|k_2)).$$

On decryption, each party's ciphertext is decrypted and the authenticity of b_1 and b_2 are verified using the (encrypted) tags and keys. Thus, when evaluating a garbled circuit, the party checks the authenticity of the share from the decrypted row of each garbled gate; if the check fails, the party aborts.

Again, we can convert this garbling scheme into a (now *maliciously*-secure) 2PC scheme; see the full version [9] for the details. Likewise, we could also construct an MPC variant with efficient *multi-party* instantiations of the underlying functionalities which we leave as future work.

4   Three-Party Computation from Cut-and-Choose

As mentioned above, we can directly adapt the distributed garbling scheme to work over multiple parties, and thus construct a 3PC scheme; however, in this case the underlying functionalities need to support multiple parties rather than just two parties and are thus unlikely to be more efficient in practice. Thus, in this section we show how to utilize the maliciously secure two-party distributed garbling scheme from Section 3 to construct a maliciously secure *three*-party secure computation protocol, using almost entirely two-party constructs (the only three-party functionality needed is that of coin-tossing).

We first cover preliminary notions, such as the ideal functionalities we need, in Section 4.1. Then, in Section 4.2 we show how to adapt a combination of two existing cut-and-choose protocols [27, 28] to the three-party setting. In the full version [9] we use this "generic" protocol to show how to adapt Lindell's protocol [25] (the current state-of-the-art garbled-circuit-based protocol at the time of writing) to the three-party setting. The cost of each of these three-party protocols is roughly *eight times* the computational cost of the underlying two-party protocol they are based on, and roughly *sixteen times* the communication cost (plus the cost of a small number of OTs per gate, which can be efficiently amortized using OT extension [21, 32]), and thus we show that we can achieve efficient secure three-party computation at only a small factor of the cost of the most efficient Yao-based two-party protocol.

4.1   Preliminaries

Ideal Functionalities. In addition to the ideal functionalities used in the two-party distributed garbling scheme, we need the following additional (maliciously secure) functionalities:

- *Three-party coin-flipping* $\mathcal{F}_{\mathbf{cf}}()$: The functionality outputs a random bit-string $\rho \stackrel{\$}{\leftarrow} \{0,1\}^s$ to each party.
- *One-out-of-two oblivious transfer* $\mathcal{F}_{\mathbf{ot}}^{i,j}(b, m_0, m_1)$: The functionality takes as input a choice bit b from party P_i and messages m_0, m_1 from P_j, and outputs m_b to party P_i.
- *ZKPoK of extended Diffie-Hellman tuple* $\mathcal{F}_{\mathbf{zkpok}}^{i,j}(a, (g, h_0, h_1, \{u_i, v_i\}_i))$: The functionality takes as input a from party P_i, and tuple $(g, h_0, h_1, \{u_i, v_i\}_i)$

from party P_j, and outputs 1 to party P_j if either all tuples in $\{(g, h_0, u_i, v_i)\}_i$ are Diffie-Hellman tuples with $h_0 = g^a$ or all tuples in $\{(g, h_1, u_i, v_i)\}_i$ are Diffie-Hellman tuples with $h_1 = g^a$, and 0 otherwise.

These can all be efficiently instantiated in a standard fashion; see the full version [9] for the details.

Distributed Garbled Circuits for Three Parties. Note that the garbling protocol $\Pi_{\mathbf{GC}}$ in Figure 2 only garbles a circuit containing inputs from two parties. We can easily adapt this to support input from a third (external) party as follows. Let $\Pi'_{\mathbf{GC}}(P_1, P_2)$ be the same as $\Pi_{\mathbf{GC}}(P_1, P_2)$ except for the following modifications:

- All operations over P_2's input now operate over wires $w \in \{n_1 + 1, \ldots, n_2\}$.
- In Step 1, we add the following for generating shares for P_3's input wires: For $w \in \{n_2 + 1, \ldots, n\}$: generate $\langle \lambda_w \rangle \leftarrow \mathcal{F}_{\mathbf{rand}}$.
- In Step 4, party P_i outputs $\{\langle \lambda_w \rangle^{(i)} : w \in \{n_2 + 1, \ldots, n\}\}$ in addition to his normal outputs.

4.2   Achieving Malicious Security for Three Parties

Note that our two-party distributed garbling scheme has the property that if at most one of the two parties is corrupt, the garbling of circuit C either correctly evaluates C on P_1's and P_2's inputs, or causes the evaluator to abort. That is, a malicious party cannot "alter" the garbling to evaluate some circuit other than C. Now, if both P_1 and P_2 are corrupt, they can of course garble an arbitrary circuit. This suggests the following approach to three-party computation: If either P_1 or P_2 are honest, we need only construct a single garbled circuit, which is sent to P_3 to be evaluated. To cover the case where both P_1 and P_2 are corrupt, we use cut-and-choose to prevent P_3 from evaluating a maliciously constructed circuit. In what follows, we utilize existing cut-and-choose protocols from the literature [27, 28] and "plug in" our distributed garbling scheme as necessary. Thus, security mostly follows from the security proofs of the underlying cut-and-choose protocols. In the full version [9] we show how we can use this protocol in an adaptation of Lindell's protocol [25] to the three-party setting.

The basic intuition for security is as follows. Cut-and-choose is used to prevent P_3 from evaluating maliciously constructed circuits when both P_1 and P_2 are malicious. For the case where either P_1 or P_2 is honest, $\Pi'_{\mathbf{GC}}(P_1, P_2)$ assures us that the garbled circuit constructed between P_1 and P_2 is either correctly constructed or causes P_3 to abort (independent of any party's input).

Protocol Description. We assume the reader is familiar with the cut-and-choose technique; here we briefly discuss the main technical challenges that result from a naïve application of cut-and-choose and how we address them.

- *Input inconsistency.* The use of cut-and-choose produces multiple garbled circuits to be evaluated by P_3. The idea with this attack is that a given party (either P_1 or P_2 in the three-party case) can give inconsistent sub-keys in each of these circuits such that P_3 ends up evaluating different inputs for P_1/P_2 instead of consistent inputs across all garbled circuits. This is a well-known attack, and there are multiple solutions in the two-party setting. Here, we use the Diffie-Hellman pseudorandom synthesizer trick [30, 28] and adapt it in a straightforward manner to the three-party setting.

– *Selective failure.* This attack arises when the parties execute OT to send the sub-keys for P_3's input. Note that if the sender in the OT inputs one valid label and one invalid label, he can learn a bit of P_3's input by learning whether the garbled-circuit evaluation fails or not. We circumvent this problem by directly applying the "XOR-tree" approach [27].

We now give a high-level description of our protocol.

1. The parties first replace the input circuit C^0 with a circuit C, where the only difference is each of P_3's input wires is replaced by an XOR of s new input wires, preventing either party P_1 or P_2 from launching a selective failure attack on P_3's input choices.
2. P_1 and P_2 generate the required commitments needed for input consistency, as is done in the protocol of Lindell and Pinkas [28].
3. P_1 and P_2 construct s garbled circuits using $\Pi'_{\mathbf{GC}}$ and the input sub-keys generated as in the protocol of Lindell and Pinkas [28].
4. P_1 and P_2 compute authenticated sharings (between each other; P_3 is not involved here) of their input bits.
5. P_1 and P_2 both run (separately) an OT protocol with P_3 for each of P_3's input wires, where P_1/P_2 input their sub-keys and P_3 chooses based on his input. (Note that any cheating by P_1/P_2 here will be caught with high-probability by the cut-and-choose step below.) Thus, P_3 now has keys for each of his input bits.
6. P_1 and P_2 send the (distributed) garbled circuits, along with the input consistency commitments, to P_3.
7. All three parties run a coin-tossing protocol to determine which circuits for P_3 to open and which to evaluate.
8. For the evaluation circuits, P_1 and P_2 send the sub-keys and selector bits for their inputs to P_3. Note that we need to be careful in this step, as we need to enforce that, for example, P_1 uses the same input as was shared in Step 2 above. This is accomplished as follows. Recall that P_1 and P_2 have sharings of each other's inputs and mask bits, all of which are authenticated. Thus, P_1 can send the (authenticated) share of her masked input to P_2, who can verify its authenticity, and thus reconstruct the masked input bit using his own share (and likewise for P_2). This allows an honest P_2 to send the correct sub-key (correct in the sense that it corresponds to P_1's input shared in Step 2) to P_3, even with a malicious P_1.
9. For the check circuits, P_1 and P_2 send the required information for P_3 to decrypt the check circuits and verify correctness. If any of these check circuits are incorrectly constructed, P_3 aborts; otherwise, he has high confidence that the majority of the evaluation circuits are correctly constructed.

10. For the evaluation circuits, P_3 checks for input consistency against the sub-keys sent by P_1 and P_2 in Step 8 using a zero-knowledge proof-of-knowledge protocol [28], aborting on any inconsistency.
11. Finally, P_3 evaluates the evaluation circuits, outputting the majority over the circuits' output.

See below for the full protocol description.

Protocol $\Pi_{\mathbf{3PC}}^m(P_1, P_2, P_3)$

Auxiliary Inputs: Security parameter k, statistical security parameter s, circuit C^0, cyclic group $\mathbb{G}$ with (prime) order q and generator g, and randomness extractor H.

Inputs: For $w \in \{1, \ldots, \mathsf{n}_1\}$, P_1 has inputs b_w; for $w \in \{\mathsf{n}_1 + 1, \ldots, \mathsf{n}_2\}$, P_1 has inputs b_w; for $w \in \{\mathsf{n}_2 + 1, \ldots, \mathsf{n}\}$, P_3 has inputs b_w.

1. Each party replaces C^0 with a circuit C where each of P_3's input wires is replaced by an exclusive-or of s new input wires. We let $(\mathsf{n}, \mathsf{m}, \mathsf{q}, L, R, G) \leftarrow C$, and denote P_3's new inputs by $\hat{b}_w$.

2. For $w \in \{1, \ldots, \mathsf{n}_1\}$: P_1 sets $a_{w,0}^1, a_{w,1}^1 \xleftarrow{\$} \mathbb{Z}_q$ and constructs $\left\{ (w, 0, g^{a_{w,0}^1}), (w, 1, g^{a_{w,1}^1}) \right\}$.

   For $w \in \{\mathsf{n}_1 + 1, \ldots, \mathsf{n}_2\}$: P_2 sets $a_{w,0}^2, a_{w,1}^2 \xleftarrow{\$} \mathbb{Z}_q$ and constructs $\left\{ (w, 0, g^{a_{w,0}^2}), (w, 1, g^{a_{w,1}^2}) \right\}$.

   For $j \in \{1, \ldots, s\}$: P_i, for $i \in \{1, 2\}$, sets $r_j^i \xleftarrow{\$} \mathbb{Z}_q$ and constructs $\left\{ (j, g^{r_j^i}) \right\}$.

   For $j \in \{1, \ldots, s\}$: P_1 and P_2 run up to Step 2 ("Generate sub-keys") of $\Pi_{\mathbf{GC}}^3(P_1, P_2)$, where the parties do the following in the jth iteration:
   - For $w \in \{1, \ldots, \mathsf{n}_1\}$: P_1 sets $s_{w, b \oplus \lambda_{w,j}, j}^1 \leftarrow H(g^{a_{w,b}^1 \cdot r_j^1})$ for $b \in \{0, 1\}$.
   - For $w \in \{\mathsf{n}_1 + 1, \ldots, \mathsf{n}_2\}$: P_2 sets $s_{w, b \oplus \lambda_{w,j}, j}^2 \leftarrow H(g^{a_{w,b}^2 \cdot r_j^2})$ for $b \in \{0, 1\}$.
   - All other sub-keys are generated in the normal fashion.

3. For $j \in \{1, \ldots, s\}$: P_1 and P_2 continue their executions of $\Pi_{\mathbf{GC}}^3(P_1, P_2)$, producing garbled circuit GC_j.

4. For $w \in \{1, \ldots, \mathsf{n}_1\}$: P_1 and P_2 compute $\langle b_w \rangle \leftarrow \mathcal{F}_{\mathbf{ss}}^1(b_w)$.
   For $w \in \{\mathsf{n}_1 + 1, \ldots, \mathsf{n}_2\}$: P_1 and P_2 compute $\langle b_w \rangle \leftarrow \mathcal{F}_{\mathbf{ss}}^2(b_w)$.

5. For $j \in \{1, \ldots, s\}$ and $w \in \{\mathsf{n}_2 + 1, \ldots, \mathsf{n}\}$: P_1 and P_2 exchange $\langle \lambda_{w,j} \rangle$ with each other, reconstructing $\lambda_{w,j}$ locally. Both P_1 and P_2 send $\lambda_{w,j}$ to P_3.
   For $w \in \{\mathsf{n}_2 + 1, \ldots, \mathsf{n}\}$: P_i, for $i \in \{1, 2\}$, and P_3 run $\mathcal{F}_{\mathbf{ot}}$, with P_i as the sender inputting $\left(\left\{ s_{w, \lambda_{w,j}, j}^i \right\}_{j \in \{1, \ldots, s\}}, \left\{ s_{w, \lambda_{w,j} \oplus 1, j}^i \right\}_{j \in \{1, \ldots, s\}} \right)$ and P_3 as the receiver inputting $\hat{b}_w$.

6. P_i, for $i \in \{1, 2\}$, sends the sets from Step 2, along with $\{GC_j^i\}_{i=1}^s$, to P_3.

7. The parties set $\rho \leftarrow \mathcal{F}_{\mathbf{cf}}$. Let $\mathcal{CC} = \{i : \rho_i = 1\}$, and $\mathcal{EC} = \{1, \ldots, s\} \setminus \mathcal{CC}$.

8. For $j \in \mathcal{EC}$:
   - For $w \in \{1, \ldots, \mathsf{n}_1\}$: P_1 sends $\langle b_w \rangle^{(1)} \oplus \langle \lambda_{w,j} \rangle^{(1)}$ to P_2, who reconstructs $b_w \oplus \lambda_{w,j}$ locally. P_1 sends $(s_{w, b_w \oplus \lambda_{w,j}, j}^1, b_w \oplus \lambda_{w,j})$ to P_3, and P_2 sends $(s_{w, b_w \oplus \lambda_{w,j}, j}^2, b_w \oplus \lambda_{w,j})$ to P_3.

- For $w \in \{n_1+1, \ldots, n\}$: P_2 sends $\langle b_w \rangle^{(2)} \oplus \langle \lambda_{w,j} \rangle^{(2)}$ to P_1, who reconstructs $b_w \oplus \lambda_{w,j}$ locally. P_1 sends $(s^1_{w, b_w \oplus \lambda_{w,j}, j}, b_w \oplus \lambda_{w,j})$ to P_3, and P_2 sends $(s^2_{w, b_w \oplus \lambda_{w,j}, j}, b_w \oplus \lambda_{w,j})$ to P_3.

9. For $j \in \mathcal{CC}$:
   - P_i, for $i \in \{1, 2\}$, does the following:
     - Sends r^i_j to P_3, and P_3 checks that these values are consistent with the pairs $\left\{ (j, g^{r^i_j}) \right\}$ sent before.
     - For $w \in \{1, \ldots, n\}$: Sends sub-keys $s^i_{w,0,j}$ and $s^i_{w,1,j}$, mask bit share $\lambda^{(i)}_{w,j}$, and the keys to the authenticated bits to P_3.
   - Given the above information, P_3 reconstructs all input labels and verifies they match with those labels sent previously. Also, using said labels, P_3 verifies that the garbled circuit is correctly constructed.

10. For $j \in \mathcal{EC}$:
    - For $w \in \{1, \ldots, n_1\}$: P_1 sends $g^{a^1_{w,b_w} \cdot r^1_j}$ to P_3, who sets $s^1_{w, b_w \oplus \lambda_{w,j}, j} \leftarrow H(g^{a^1_{w,b_w} \cdot r^1_j})$.
    - For $w \in \{n_1+1, \ldots, n_2\}$: P_2 sends $g^{a^2_{w,b_w} \cdot r^2_j}$ to P_3, who sets $s^2_{w, b_w \oplus \lambda_{w,j}, j} \leftarrow H(g^{a^2_{w,b_w} \cdot r^2_j})$.

    For $w \in \{1, \ldots, n_1\}$: P_1 and P_3 run $\mathcal{F}_{\mathsf{zkpok}}$, with P_1 acting as the prover inputting a^1_{w,b_w} and P_3 acting as the verifier inputting $\left(g, g^{a^1_{w,0}}, g^{a^1_{w,1}}, \left\{ (g^{r^1_j}, g^{a^1_{w,b_w} \cdot r^1_j}) \right\}_{j \in \mathcal{EC}} \right)$.

    For $w \in \{n_1 + 1, \ldots, n_2\}$: P_2 and P_3 run $\mathcal{F}_{\mathsf{zkpok}}$, with P_2 acting as the prover inputting a^2_{w,b_w} and P_3 acting as the verifier inputting $\left(g, g^{a^2_{w,0}}, g^{a^2_{w,1}}, \left\{ (g^{r^2_j}, g^{a^2_{w,b_w} \cdot r^2_j}) \right\}_{j \in \mathcal{EC}} \right)$.

11. For $j \in \mathcal{EC}$:
    - P_3 evaluates GC_j using $\left\{ (s^1_{w, b_w \oplus \lambda_{w,j}, j}, s^2_{w, b_w \oplus \lambda_{w,j}, j}, b_w \oplus \lambda_{w,j}) \right\}_{w \in \{1, \ldots, n\}}$ as inputs.

    P_3 outputs the majority output over the evaluated circuits.

In the full version [9] we prove the following.

Theorem 1. *Let C be an arbitrary polynomial-size circuit and let $\mathbb{G}$ be a cyclic group with prime order. Given access to ideal functionalities $\mathcal{F}_{\mathsf{const}}$, $\mathcal{F}_{\mathsf{gate}}$, $\mathcal{F}_{\mathsf{oshare}}$, $\mathcal{F}_{\mathsf{ot}}$, $\mathcal{F}_{\mathsf{rand}}$, and $\mathcal{F}_{\mathsf{ss}}$, and assuming that the decisional Diffie-Hellman problem is hard in $\mathbb{G}$, then $\Pi^m_{\mathbf{3PC}}(P_1, P_2, P_3)$ securely computes the circuit C in the presence of an adversary corrupting an arbitrary number of parties.*

4.3 Efficiency

We now argue why our 3PC protocol is roughly eight times as expensive in terms of computation as the underlying 2PC protocol we utilize, and roughly sixteen times as expensive in terms of communication. Both protocols are very similar to the underlying 2PC protocol they are based on; the major changes in terms of computational cost are that (1) the cost of encrypting a single row increases due to the use of the distributed encryption scheme, and (2) P_3 needs to do twice the work (due to communicating with *both* P_1 and P_2) as compared to the evaluator in the underlying 2PC protocol. Indeed, it takes about eight PRF calls (where one PRF call equals outputting k bits) to encrypt a single row of the garbled circuit, and thus the cost and size of a garbled circuit increases by a factor of eight. The cost for P_1 and P_2 to distributively garble a circuit is a small number of OTs per gate, and this can be amortized using OT extension techniques [21].

In terms of communication cost, both P_1 and P_2 need to send their half of the distributed garbled circuit to P_3, and the communication cost of actually constructing a distributed garbled circuit is roughly the cost of a standard garbled circuit. Since each garbled circuit is eight times larger than in the underlying 2PC protocol, we find that the overall communication size increases my approximately sixteen.

Comparison with SPDZ. We compare our three-party protocol with the SPDZ protocol [4, 12–14, 23], an efficient protocol over arithmetic circuits that works for n parties and arbitrary corruptions, and uses the preprocessing paradigm. SPDZ represents the state-of-the-art in terms of efficiency in the multi-party setting. Here we focus on the differences between both SPDZ and our protocol, and discuss their strengths and weaknesses. Due to the different characteristics of each protocol (e.g., arithmetic versus boolean, linear versus constant round, etc.), these protocols are somewhat "incomparable". However, we hope to give a general idea of the efficiency trade-offs of both protocols.

There are several key differences between the SPDZ protocol and our own. For one, SPDZ works over arithmetic circuits, whereas our protocol works over boolean circuits. In terms of communication, the SPDZ protocol requires rounds linear in the depth of the circuit, whereas our protocol is constant-round. While it is difficult to compare the impact of this without an implementation and experiments, it seems intuitive that as the latency between machines increases, the cost of each additional communication round increases as well; this intuition has been backed up by experiments in the semi-honest setting [35]. And while SPDZ works in the standard model, the most efficient instantiation of our protocol requires the random oracle model.

Finally, we consider the start-to-finish execution time (i.e., including the cost of preprocessing) for running an AES circuit. The preprocessing in our protocol is basically that found in the TinyOT protocol [32], and, using the numbers presented there, is fairly efficient (around 1 minute [32, Figure 21]). Efficiency comes from the fact that the preprocessing is only between two parties, namely, the circuit generators. The on-line running time is conjectured to be around that of maliciously secure two-party protocols using cut-and-choose. The SPDZ protocol, on the other hand, has a very efficient (information-theoretic) online phase but a much costlier offline phase (around 17 minutes for three parties [12, Table 2]). In addition, it has a one-time setup phase which is very costly: the parties need to execute an MPC protocol for a circuit which generates a key pair with the secret key secret-shared among the parties. Executing this on its own would likely eclipse the running time of our protocol.[2] Thus, given preprocessing, it seems likely that SPDZ would out-perform our protocol; however, in the setting of executing the protocol from start to finish, we conjecture that our protocol would be more efficient.

Acknowledgments. Work of Seung Geol Choi supported in part by the Office of Naval Research under Grant Number N0001414WX20588. Work of Jonathan Katz supported in part by NSF awards #0964541 and #1111599. Work of Alex J. Malozemoff supported by a National Defense Science and Engineering Graduate (NDSEG) Fellowship, 32 CFG 168a, awarded by DoD, Air Force Office of Scientific Research. Work of Vassilis Zikas supported in part by NSF awards #09165174, #1065276, #1118126, and #1136174, US-Israel BSF grant #2008411, OKAWA Foundation Research Award, IBM Faculty Research Award, Xerox Faculty Research Award, B. John Garrick Foundation Award, Teradata Research Award, Lockheed-Martin Corporation Research Award, the Defense Advanced Research Projects Agency through the U.S. Office of Naval Research under Contract N00014-11-1-0392, and Swiss National Science Foundation (SNF) Ambizione grant PZ00P2_142549. The views expressed are those of the authors and do not reflect the official policy or position of the Department of Defense or the U.S. Government.

References

1. Beaver, D., Micali, S., Rogaway, P.: The round complexity of secure protocols. In: 22nd ACM STOC, pp. 503–513. ACM Press (1990)
2. Bellare, M., Hoang, V.T., Rogaway, P.: Foundations of garbled circuits. In: Yu, T., Danezis, G., Gligor, V.D. (eds.) ACM CCS 2012, pp. 784–796. ACM Press (2012)
3. Ben-David, A., Nisan, N., Pinkas, B.: FairplayMP: a system for secure multi-party computation. In: Ning, P., Syverson, P.F., Jha, S. (eds.) ACM CCS 2008, pp. 257–266. ACM Press (2008)
4. Bendlin, R., Damgård, I., Orlandi, C., Zakarias, S.: Semi-homomorphic encryption and multiparty computation. In: Paterson, K.G. (ed.) EUROCRYPT 2011. LNCS, vol. 6632, pp. 169–188. Springer, Heidelberg (2011)
5. Bogdanov, D., Laur, S., Willemson, J.: Sharemind: A framework for fast privacy-preserving computations. In: Jajodia, S., Lopez, J. (eds.) ESORICS 2008. LNCS, vol. 5283, pp. 192–206. Springer, Heidelberg (2008)
6. Bogetoft, P., et al.: Secure multiparty computation goes live. In: Dingledine, R., Golle, P. (eds.) FC 2009. LNCS, vol. 5628, pp. 325–343. Springer, Heidelberg (2009)
7. Burkhart, M., Strasser, M., Many, D., Dimitropoulos, X.: SEPIA: Privacy-preserving aggregation of multi-domain network events and statistics. In: Goldberg, I. (ed.) 19th USENIX Security Symposium. USENIX Association, Washington (2010)

---

[2] We note that Damgård et al. [13] present an efficient protocol for this one-time setup phase in the weaker *covert* security model.

8. Choi, S.G., Hwang, K.-W., Katz, J., Malkin, T., Rubenstein, D.: Secure multi-party computation of boolean circuits with applications to privacy in on-line marketplaces. In: Dunkelman, O. (ed.) CT-RSA 2012. LNCS, vol. 7178, pp. 416–432. Springer, Heidelberg (2012)

9. Choi, S.G., Katz, J., Malozemoff, A.J., Zikas, V.: Efficient three-party computation from cut-and-choose. Cryptology ePrint Archive, Report 2014/128 (2014), https://eprint.iacr.org/

10. Damgård, I., Geisler, M., Krøigaard, M., Nielsen, J.B.: Asynchronous multiparty computation: Theory and implementation. In: Jarecki, S., Tsudik, G. (eds.) PKC 2009. LNCS, vol. 5443, pp. 160–179. Springer, Heidelberg (2009)

11. Damgård, I., Ishai, Y.: Constant-round multiparty computation using a black-box pseudorandom generator. In: Shoup, V. (ed.) CRYPTO 2005. LNCS, vol. 3621, pp. 378–394. Springer, Heidelberg (2005)

12. Damgård, I., Keller, M., Larraia, E., Miles, C., Smart, N.P.: Implementing AES via an actively/Covertly secure dishonest-majority MPC protocol. In: Visconti, I., De Prisco, R. (eds.) SCN 2012. LNCS, vol. 7485, pp. 241–263. Springer, Heidelberg (2012)

13. Damgård, I., Keller, M., Larraia, E., Pastro, V., Scholl, P., Smart, N.P.: Practical Covertly Secure MPC for Dishonest Majority – Or: Breaking the SPDZ Limits. In: Crampton, J., Jajodia, S., Mayes, K. (eds.) ESORICS 2013. LNCS, vol. 8134, pp. 1–18. Springer, Heidelberg (2013)

14. Damgård, I., Pastro, V., Smart, N.P., Zakarias, S.: Multiparty computation from somewhat homomorphic encryption. In: Safavi-Naini, R., Canetti, R. (eds.) CRYPTO 2012. LNCS, vol. 7417, pp. 643–662. Springer, Heidelberg (2012)

15. Goldreich, O., Micali, S., Wigderson, A.: How to play any mental game or A completeness theorem for protocols with honest majority. In: Aho, A. (ed.) 19th ACM STOC, pp. 218–229. ACM Press (May 1987)

16. Goyal, V., Mohassel, P., Smith, A.: Efficient two party and multi party computation against covert adversaries. In: Smart, N.P. (ed.) EUROCRYPT 2008. LNCS, vol. 4965, pp. 289–306. Springer, Heidelberg (2008)

17. Henecka, W., Kögl, S., Sadeghi, A.R., Schneider, T., Wehrenberg, I.: TASTY: Tool for automating secure two-party computations. In: Al-Shaer, E., Keromytis, A.D., Shmatikov, V. (eds.) ACM CCS 2010, pp. 451–462. ACM Press (2010)

18. Huang, Y., Evans, D., Katz, J.: Private set intersection: Are garbled circuits better than custom protocols? In: NDSS 2012, The Internet Society (February 2012)

19. Huang, Y., Evans, D., Katz, J., Malka, L.: Faster secure two-party computation using garbled circuits. In: Wagner, D. (ed.) 20th USENIX Security Symposium. USENIX Association, San Francisco (2011)

20. Huang, Y., Katz, J., Evans, D.: Efficient secure two-party computation using symmetric cut-and-choose. In: Canetti, R., Garay, J.A. (eds.) CRYPTO 2013, Part II. LNCS, vol. 8043, pp. 18–35. Springer, Heidelberg (2013)

21. Ishai, Y., Kilian, J., Nissim, K., Petrank, E.: Extending oblivious transfers efficiently. In: Boneh, D. (ed.) CRYPTO 2003. LNCS, vol. 2729, pp. 145–161. Springer, Heidelberg (2003)

22. Ishai, Y., Prabhakaran, M., Sahai, A.: Founding cryptography on oblivious transfer – efficiently. In: Wagner, D. (ed.) CRYPTO 2008. LNCS, vol. 5157, pp. 572–591. Springer, Heidelberg (2008)

23. Keller, M., Scholl, P., Smart, N.P.: An architecture for practical actively secure MPC with dishonest majority. In: Sadeghi, A.R., Gligor, V.D., Yung, M. (eds.) ACM CCS 2013, pp. 549–560. ACM Press (November 2013)

24. Kreuter, B., Shelat, A., Shen, C.H.: Towards billion-gate secure computation with malicious adversaries. In: Kohno, T. (ed.) 21st USENIX Security Symposium. USENIX Association, Bellevue (2012)

25. Lindell, Y.: Fast cut-and-choose based protocols for malicious and covert adversaries. In: Canetti, R., Garay, J.A. (eds.) CRYPTO 2013, Part II. LNCS, vol. 8043, pp. 1–17. Springer, Heidelberg (2013)

26. Lindell, Y., Oxman, E., Pinkas, B.: The IPS compiler: Optimizations, variants and concrete efficiency. In: Rogaway, P. (ed.) CRYPTO 2011. LNCS, vol. 6841, pp. 259–276. Springer, Heidelberg (2011)

27. Lindell, Y., Pinkas, B.: An efficient protocol for secure two-party computation in the presence of malicious adversaries. In: Naor, M. (ed.) EUROCRYPT 2007. LNCS, vol. 4515, pp. 52–78. Springer, Heidelberg (2007)

28. Lindell, Y., Pinkas, B.: Secure two-party computation via cut-and-choose oblivious transfer. In: Ishai, Y. (ed.) TCC 2011. LNCS, vol. 6597, pp. 329–346. Springer, Heidelberg (2011)

29. Lindell, Y., Pinkas, B., Smart, N.P.: Implementing two-party computation efficiently with security against malicious adversaries. In: Ostrovsky, R., De Prisco, R., Visconti, I. (eds.) SCN 2008. LNCS, vol. 5229, pp. 2–20. Springer, Heidelberg (2008)

30. Mohassel, P., Franklin, M.: Efficiency tradeoffs for malicious two-party computation. In: Yung, M., Dodis, Y., Kiayias, A., Malkin, T. (eds.) PKC 2006. LNCS, vol. 3958, pp. 458–473. Springer, Heidelberg (2006)

31. Mohassel, P., Riva, B.: Garbled circuits checking garbled circuits: More efficient and secure two-party computation. In: Canetti, R., Garay, J.A. (eds.) CRYPTO 2013, Part II. LNCS, vol. 8043, pp. 36–53. Springer, Heidelberg (2013)

32. Nielsen, J.B., Nordholt, P.S., Orlandi, C., Burra, S.S.: A new approach to practical active-secure two-party computation. In: Safavi-Naini, R., Canetti, R. (eds.) CRYPTO 2012. LNCS, vol. 7417, pp. 681–700. Springer, Heidelberg (2012)

33. Nielsen, J.B., Orlandi, C.: LEGO for two-party secure computation. In: Reingold, O. (ed.) TCC 2009. LNCS, vol. 5444, pp. 368–386. Springer, Heidelberg (2009)

34. Pinkas, B., Schneider, T., Smart, N.P., Williams, S.C.: Secure two-party computation is practical. In: Matsui, M. (ed.) ASIACRYPT 2009. LNCS, vol. 5912, pp. 250–267. Springer, Heidelberg (2009)

35. Schneider, T., Zohner, M.: GMW vs. Yao? Efficient secure two-party computation with low depth circuits. In: Sadeghi, A.-R. (ed.) FC 2013. LNCS, vol. 7859, pp. 275–292. Springer, Heidelberg (2013)

36. Shelat, A., Shen, C.H.: Two-output secure computation with malicious adversaries. In: Paterson, K.G. (ed.) EUROCRYPT 2011. LNCS, vol. 6632, pp. 386–405. Springer, Heidelberg (2011)

37. Shelat, A., Shen, C.H.: Fast two-party secure computation with minimal assumptions. In: Sadeghi, A.R., Gligor, V.D., Yung, M. (eds.) ACM CCS 2013, pp. 523–534. ACM Press (November 2013)

38. Yao, A.C.C.: How to generate and exchange secrets (extended abstract). In: 27th FOCS, pp. 162–167. IEEE Computer Society Press (October 1986)

Author Index